OXFORD MATHEMATICAL MONOGRAPHS

OXFORD MATHEMATICAL MONOGRAPHS

A. Belleni-Moranti: *Applied semigroups and evolution equations*
A. M. Arthurs: *Complementary variational principles* 2nd edition
M. Rosenblum and J. Rovnyak: *Hardy classes and operator theory*
J. W. P. Hirschfeld: *Finite projective spaces of three dimensions*
A. Pressley and G. Segal: *Loop groups*
D. E. Edmunds and W. D. Evans: *Spectral theory and differential operators*
Wang Jianhua: *The theory of games*
S. Omatu and J. H. Seinfeld: *Distributed parameter systems: theory and applications*
J. Hilgert, K. H. Hofmann, and J. D. Lawson: *Lie groups, convex cones, and semigroups*
S. Dineen: *The Schwarz lemma*
S. K. Donaldson and P. B. Kronheimer: *The geometry of four-manifolds*
D. W. Robinson: *Elliptic operators and Lie groups*
A. G. Werschulz: *The computational complexity of differential and integral equations*
L. Evens: *Cohomology of groups*
G. Effinger and D. R. Hayes: *Additive number theory of polynomials*
J. W. P. Hirschfeld and J. A. Thas: *General Galois geometries*
P. N. Hoffman and J. F. Humphreys: *Projective representations of the symmetric groups*
I. Györi and G. Ladas: *The oscillation theory of delay differential equations*
J. Heinonen, T. Kilpelainen, and O. Martio: *Non-linear potential theory*
B. Amberg, S. Franciosi, and F. de Giovanni: *Products of groups*
M. E. Gurtin: *Thermomechanics of evolving phase boundaries in the plane*
I. Ionescu and M. Sofonea: *Functional and numerical methods in viscoplasticity*
N. Woodhouse: *Geometric quantization* 2nd edition
U. Grenander: *General pattern theory*
J. Faraut and A. Koranyi: *Analysis on symmetric cones*
I. G. Macdonald: *Symmetric functions and Hall polynomials* 2nd edition
B. L. R. Shawyer and B. B. Watson: *Borel's methods of summability*
M. Holschneider: *Wavelets: an analysis tool*
Jacques Thévenaz: *G-algebras and modular representation theory*
Hans-Joachim Baues: *Homotopy type and homology*
P. D. D'Eath: *Black holes: gravitational interactions*
R. Lowen: *Approach spaces: the missing link in the topology–uniformity–metric triad*
Nguyen Dinh Cong: *Topological dynamics of random dynamical systems*
J. W. P. Hirschfeld: *Projective geometries over finite fields* 2nd edition
K. Matsuzaki and M. Taniguchi: *Hyperbolic manifolds and Kleinian groups*
David E. Evans and Yasuyuki Kawahigashi: *Quantum symmetries on operator algebras*
Norbert Klingen: *Arithmetical similarities: prime decomposition and finite group theory*
Isabelle Catto, Claude Le Bris, and Pierre-Louis Lions: *The mathematical theory of thermodynamic limits: Thomas–Fermi type models*
D. McDuff and D. Salamon: *Introduction to symplectic topology* 2nd edition
William M. Goldman: *Complex hyperbolic geometry*
Charles J. Colbourn and Alexander Rosa: *Triple systems*

Triple Systems

CHARLES J. COLBOURN

Department of Computer Science
University of Vermont

and

ALEXANDER ROSA

Department of Mathematics and Statistics
McMaster University

CLARENDON PRESS • OXFORD

This book has been printed digitally and produced in a standard specification in order to ensure its continuing availability

OXFORD
UNIVERSITY PRESS

Great Clarendon Street, Oxford OX2 6DP

Oxford University Press is a department of the University of Oxford.
It furthers the University's objective of excellence in research, scholarship, and education by publishing worldwide in

Oxford New York

Auckland Cape Town Dar es Salaam Hong Kong Karachi
Kuala Lumpur Madrid Melbourne Mexico City Nairobi
New Delhi Shanghai Taipei Toronto

With offices in

Argentina Austria Brazil Chile Czech Republic France Greece
Guatemala Hungary Italy Japan South Korea Poland Portugal
Singapore Switzerland Thailand Turkey Ukraine Vietnam

Oxford is a registered trade mark of Oxford University Press in the UK and in certain other countries

Published in the United States
by Oxford University Press Inc., New York

Reprinted 2007

ISBN 978-0-19-853576-8

To our wives and children:

Karen Colbourn
Susie Colbourn
Sarah Colbourn

Nadia Rosa
Daniela Rosa

PREFACE

A hundred and fifty years ago, combinatorics was a disjointed collection of problems arising from research at the infancy of abstract algebra, finite geometry, and topology. Triple systems were among the first objects systematically studied, arising in the geometry of Plücker, Steiner, Hesse, Klein, and others; the algebra of Cayley and Sylvester; the group theory of Jordan and Netto; and in many ways most importantly in the celebrated Kirkman schoolgirl problem that fascinated mathematicians, amateur and professional, for a hundred and twenty years. From these varied beginnings, the study of triple systems has grown into a major part of the study of combinatorial designs. Yet, at the same time, perhaps because triple systems are among the simplest combinatorial designs, their study has a remarkably diverse nature.

Triple systems are a natural generalization of graphs, and indeed much of their study has a decidedly graph-theoretic flavour. Connections with geometry, algebra, group theory, finite fields, and cyclotomy provide other perspectives. Applications to coding theory, cryptography, and computer science have suggested new problems on triple systems. Triple systems often provide the prototype for deep results in combinatorial design theory; indeed design theory is permeated by problems that were first understood for triple systems, and then generalized.

This rich set of connections has made the study of triple systems a large, but often disjointed, one. This book is an attempt to report on the current knowledge about triple systems, to collect together common themes, and to provide an accurate portrait of the incredible variety of problems and results. We have tried to convey more than merely statements of results, to ensure that representative samples of major styles of proof techniques are provided. We have chosen to include some technical and lengthy proofs in order to provide a truer picture of the diversity and complexity of research on triple systems. This has the unintended consequence of placing emphasis on specific topics, while other no less important ones receive an abbreviated treatment. The lengthy bibliography stands as evidence that much more could be said. Nevertheless, one of our goals was to provide a relatively complete treatment, and so we have included many pointers into this large body of literature when a fuller treatment was not possible.

The book is intended to be read primarily by those with a basic knowledge of combinatorial design theory. Some experience in graph theory is particularly useful. The book can be profitably read in conjunction with a text in combinatorial design theory; although self-contained in this area, generalizations to designs with larger block size are omitted for the most part. We expect the book to be of primary value to researchers, graduate students, and those in allied fields needing a guide to this large and varied topic.

Chapter 0 provides an historical background, while Chapters 1, 2, and 3 detail the central material on constructions and existence that is used throughout the book. Every chapter thereafter supposes that the reader is familiar with material in the first three chapters, and makes occasional references as well to results from other chapters. Chapters 4–8 describe topics which, in our view, are principally concerned with triple systems themselves: isomorphism, enumeration, subsystems, and automorphisms. Chapters 9–23 treat a number of challenging problems on triple systems in some detail. Often these problems are motivated by questions in related disciplines, and often the techniques borrow heavily from other disciplines. Chapters 24 and 25 provide a guide to two related classes of triple systems in which the triples contain ordered pairs. A comprehensive bibliography on triple systems is provided. Not all references in the bibliography are cited in the text, but are included nonetheless in order to permit the diligent reader to explore further.

Perhaps the largest surprise for the reader in opening a book such as this is that a subject like triple systems has been studied so extensively, by such a variety of techniques, and in such a variety of settings. We hope that the juxtaposition of classical and modern techniques; of algebraic, geometric, graph-theoretic, and computational ideas; and of pure, applied, and some recreational mathematics will promote a continued appreciation for triple systems.

Before setting you free to explore, and enjoy, the world of triple systems, we convey some heartfelt thanks. The main thanks, now and always, go to our families. Their love and support, and forgiveness for hours spent poring over papers and reflected in a screen, are in fact what makes such a book possible. It will not be possible to express adequately our appreciation to all those who made suggestions, corrections, and the like. We therefore content ourselves with singling out a few people whose contributions, in a variety of ways, were crucial. Thanks to Jeff Dinitz, Jean Doyen, Peter Gibbons, Mike Grannell, Terry Griggs, Zhike Jiang, Don Kreher, Alan Ling, Rudi Mathon, Kevin Phelps, Doug Stinson, Anne Street, and Zhu Lie. Jeff, Zhike, and Alan deserve special thanks for commenting on drafts of chapters.

Burlington, Vermont Charles J. Colbourn
Hamilton, Ontario Alexander Rosa

August 1998

CONTENTS

AN HISTORICAL INTRODUCTION

0.1 Beginnings

We shall likely never know who first noticed that given seven elements, seven sets of three elements each can be found so that every pair of two elements appears in exactly one of the seven chosen sets; or who first noticed that writing down the rows, columns, forward diagonals, and back diagonals of a 3×3 matrix of nine distinct symbols yields twelve sets of three elements chosen from nine, again so that every pair of two elements appears in exactly one of the twelve sets.

Plücker (1835) encountered the latter set of twelve sets in a study of algebraic curves, but went on to generalize this type of set system. In a footnote, he remarks:

Nicht jede Anzahl von Elementen ist von der Art, dass sie sich so zu drei gruppiren lassen, dass in den verschiedenen Gruppen alle Combinationen zweier Elemente vorkommen und jede derselben nur ein einziges Mal. Wenn m die Anzahl solcher Elemente und n irgend eine ganze Zahl, Null nicht ausgeschlossen, bedeutet, so überzeugt man sich leicht, dass m von der Form $6n+3$ sein muss. Die Anzahl der verschiedenen Gruppen beträgt alsdann ein Drittel der Anzahl der Combinationen von m Elementen zu zwei, mithin $\frac{1}{3}m\frac{1}{2}(m-1)$, und jedes Element kommt in $\frac{1}{2}(m-1)$ verschiedenen Gruppen vor.

In this passage, Plücker observes that, given m elements, a set of 3-element subsets in which every pair of two elements appears in exactly one of the chosen sets will contain $\frac{1}{3}m\frac{1}{2}(m-1)$ sets. He arrives from this at the incorrect conclusion that m must be an odd multiple of 3, which he says is 'easy to see' ('so überzeugt man sich leicht'). Plücker (1839) subsequently corrected this statement to observe that $m \equiv 1, 3 \pmod 6$ is the necessary condition; he also posed the more general question of the existence of $(t+1)$-element subsets of an m-set that contain every t-set exactly once.

Subsequent writers typically do not refer to Plücker when ascribing the source of these questions, but rather to Woolhouse (1844) or to Steiner (1853); we return to this in the next section. It is of interest to modern readers that triple systems find their origins in studies of cubic curves, rather than in recreational problems as is often thought. Plücker's discovery of the nine point configuration formed an advance in the basic theory of projective geometry, and especially of plane algebraic curves. The remarkable connection results from the fact that every nonsingular cubic has nine points of inflection, and these define twelve lines each containing three of them, every pair defining one line. As de Vries (1984) notes, this theorem of Plücker has been repeated often in the literature, but in recent times has rarely been attributed to Plücker (or anyone else). We do not plan to take the detour into algebraic geometry needed to explain Plücker's result, but instead refer the reader to Semple and Roth (1949) and Semple and Kneebone

(1959) for general background, and to Walker (1950) and Seidenberg (1968) for more detailed expositions of the results. Plücker's work formed an important advance in the development of algebraic geometry; an informative treatment of both Plücker and his work appears in Kolmogorov and Yushkevich (1981).

The discovery of the structure on the nine inflection points on nonsingular cubics was soon repeated, both by Hesse (1844) and Steiner (1846). Hesse (1847, 1848) develops substantial theory related to the nine point configuration in studying the intersections of cubic curves. Plücker (1847) asserted his priority in making the discovery, apparently without effect. Aronhold (1850) in a survey article credits Hesse, and the dissenting voice of Salmon (1850) asserts that the discovery is implicit in earlier work of Chasles.

At the same time, a more remarkable geometric object had attracted the attention of geometers, the double tangents of fourth order curves. Steiner (1846a) examined these. But again, Plücker (1839) had explored the 28 double tangents and constructed a configuration of 819 4-subsets of 28 points with the property that every 3-subset appears in exactly one of the 4-subsets. Geometric, algebraic, and tactical problems all arose from these beginnings. Cayley (1849) and Salmon (1849) showed that there are 27 lines on the cubic surface, and Geiser (1869) later established the connection between the fourth order curves and cubic surfaces. Jordan (1870), Klein (1879), and Dickson (1916) report on the role these geometric objects played in the development of group theory, and Henderson (1911) gives an interesting geometric treatment. Clebsch (1876) devotes much effort to the interplay of the geometry and the algebra. In one sense, then, these are more properly the beginnings of projective geometry and the theory of finite groups. However, the combinatorial (tactical) problems that arose made the time ripe for general questions concerning arrangements of finite numbers of elements into sets exhibiting some 'balance' property.

0.2 Kirkman's "On a problem in combinations"

In the *Lady's and Gentleman's Diary* of 1844, W.S.B. Woolhouse (who elsewhere published on the moons of Jupiter, comets, eclipses, gradients on railways, and the deposit of submarine cables (Royal Society of London, 1870)) stated the prize question for that year:

> Determine the number of combinations that can be made out of n symbols, p symbols in each; with this limitation, that no combination of q symbols which may appear in any one of them may appear in any other.

In modern language, given an n-set V, we are to determine the largest number of subsets of V each of size p so that no subset of q elements of V appears in more than one of the p-subsets. When every q-subset appears in *exactly one* of the p-subsets, such a configuration is known as a *Steiner system*, and denoted $S(q,p,n)$. The question of Plücker (1839) concerns the $S(t,t+1,m)$ systems.

Steiner (1853) later asked a series of questions for $q = 2$, 3, 4, and so on. His first is:

Welche Zahl, N, von Elementen hat die Eigenschaft, dass sich die Elementen so zu dreien ordnen lassen, dass je zwei in einer, aber nur in einer Verbindung vorkommen?

Although later, Steiner's name has come to be firmly associated with Steiner systems. Ironically, a more careful examination of Steiner's series of questions reveals that he was, in fact, asking about designs different from those named for him, and different from those about which Woolhouse (1844) asked. While Steiner's first question quoted above indeed asks about $S(2,3,N)$ designs, for $i \geq 2$ his ith question supposes that sets of blocks (of size $j+2$) have been chosen for each $1 \leq j < i$, and asks for a set of blocks each of size $i+2$ so that no block of size less than $i+2$ appears in one of the chosen blocks, but every $(i+1)$-set not containing a block of size at most $i+1$ appears in exactly one of the chosen blocks of size $i+2$. As Hanani (1984) observes, when Steiner's first two questions are answered, the solution is equivalent to an $S(3,4,N+1)$. However, thereafter the correspondence fails: answering Steiner's first three questions is *not*, in general, equivalent to finding an $S(4,5,N+2)$. The difference is critical, since Bussey (1909, 1914) and Hanani (1984) showed that when $N = 2^n - 1$, the first $n-1$ Steiner questions admit a solution; in contrast, no $S(q,p,n)$ is known for $q \geq 6$. Fortunately, for triple systems, Steiner's question does coincide with Plücker's and Woolhouse's, so this variation in definitions is just an entertaining diversion here.

Plücker, Woolhouse, and Steiner all surely underestimated the difficulty of the problems that they posed, as they remain unsolved to this day. In fact, not a single example of a Steiner system having $q \geq 6$ is known at the present time. Despite this, the mathematical progeny of their simply stated questions in combinatorial design theory is enormous.

In 1847, a rector of a small parish in rural England, the Reverend Thomas Penyngton Kirkman, published a landmark paper in the *Cambridge and Dublin Mathematical Journal* entitled "On a problem in combinations". Kirkman was a remarkable man and a remarkable mathematician; indeed, Macfarlane (1916) names him one of the ten most influential British mathematicians of the nineteenth century, although he worked in relative isolation. Mills (1980) paints a picture of an energetic but frustrated man who worked ceaselessly to advance both mathematics and theology. Biggs (1981) gives an appraisal of his mathematical work, and shows us Kirkman as a mathematical visionary whose work was too often overlooked or forgotten.

Kirkman had a knack for argument. Among his ecclesiastical arguments, Gridgeman (1973) reports that

> [Kirkman's] satiric paraphrase of Spencer's definition of evolution is a notable example of a Kirkmannerism. Spencer, he wrote, was really defining the concept as 'a change from a nohowish untalkaboutable all-likeness, to a somehowish and in-general-talkaboutable not-all-likeness, by continuous somethingelseifications and sticktogetherations.'

Kirkman (1847) settled the existence question for Steiner triple systems, which are $S(2,3,v)$ systems, thereby settling Woolhouse's question for all n when

$p = 3$ and $q = 2$ completely. However, for many years, Kirkman was to receive little credit for this result; well into the current century, Reiss (1859), writing in response to the question of Steiner (1853) in *Crelle's Journal*, was mostly taken to be the first to settle existence of Steiner triple systems. This is a pattern with Kirkman's work. For example, Kirkman (1850) found projective and affine planes of prime orders years before the standard reference, Fano (1892), but to this day receives little of the recognition due him.

The coincidence between Kirkman's proof and Reiss's is remarkable. In fact, Kirkman (1887) stated his feelings on the matter, asking 'how did the *Cambridge and Dublin Mathematical Journal*, Vol. II, p. 191, contrive to steal so much from a later paper in *Crelle's Journal*, Vol. LVI, p. 326, on exactly the same problem in combinations?' Louise D. Cummings (1918) wrote:

> A comparison of the Reiss paper with that of Kirkman reveals a remarkable identity of method in the work of these two men.

She goes on, somewhat charitably, to say:

> This curious duplication of Reiss of the Kirkman work gives rise to the question as to whether Reiss had a knowledge of the Kirkman paper, or whether the problem was of such a type that two investigators would naturally have followed the same line of argument.

There are now many different existence proofs for Steiner triple systems; while one of the constructions used by Kirkman is a standard textbook exercise in combinatorial design theory, the second Kirkman construction is apparently little known. In fact, in over a thousand published works on triple systems to date, Kirkman's second construction is used apparently in only two papers: Kirkman's original work, and that of Reiss. Anderson (1990) revived interest in the Kirkman constructions.

How did Kirkman settle the existence of Steiner triple systems? We give a complete development of Kirkman's proof, which remains interesting for its ingenuity and simplicity. Let V be a finite nonempty set of v elements, and let $\mathcal{B}$ be a set of 3-subsets (*triples*) of V so that every 2-subset (*pair*) of V occurs in exactly one triple of $\mathcal{B}$. Evidently there are $\binom{v}{2}$ pairs, and every triple contains three of them. Thus $\binom{v}{2} \equiv 0 \pmod 3$ and hence $v \equiv 0, 1 \pmod 3$. Considering all pairs containing a fixed element $x \in V$, there must be $(v-1)/2$ triples containing x. Hence $v \equiv 1 \pmod 2$, and thus $v \equiv 1, 3 \pmod 6$. Kirkman proves that the condition that $v \equiv 1, 3 \pmod 6$ is not only necessary, but is sufficient. His proof employs two recursive constructions. The first constructs an $S(2, 3, 2v + 1)$ from an $S(2, 3, v)$; the second constructs an $S(2, 3, 2v - 5)$ from an $S(2, 3, v)$.

Together with the trivial $S(2, 3, 1)$ $(\{x\}, \varnothing)$, these two constructions produce an $S(2, 3, v)$ whenever $v \equiv 1, 3 \pmod 6$. For writing $v = 2n + 1$, if n is odd then $n \equiv 1, 3 \pmod 6$ and the first construction applies; and if n is even then $n + 3 \equiv 1, 3 \pmod 6$ and the second construction applies.

Kirkman enforces an additional property of the Steiner triple systems used. Let $(V, \mathcal{B})$ be an $S(2, 3, n)$, and consider two elements x and y. There is a unique element z defined by $\{x, y, z\} \in \mathcal{B}$. Let $\widehat{V} = V \setminus \{x, y, z\}$. If there is a permutation

$\pi : \mathbb{Z}_{n-3} \mapsto \widehat{V}$ so that

$$\{\{x, \pi(2i), \pi(2i+1)\}, \{y, \pi(2i+1), \pi(2i+2)\} : 0 \le i < (n-3)/2\} \subseteq \mathcal{B},$$

performing arithmetic modulo $n-3$, then the $S(2,3,v)$ is *crofty* with respect to x and y, and the element ordering $(\pi(0), \pi(1), \ldots, \pi(n-4))$ is a *crofty ordering.*

In addition to crofty $S(2,3,v)$, Kirkman's recursions employ a second ingredient. Let X be a set of $2n$ elements. A partition of the set $\binom{X}{2}$ of all unordered pairs of elements into $2n-1$ classes called *1-factors*, $F_0, \ldots, F_{2n-2}$, is a *1-factorization* of order $2n$ if for every $x \in X$, and each F_i, there is exactly one pair in F_i that contains x. Let $X = \{\sigma_0, \ldots, \sigma_{2n-2}\} \cup \{\alpha\}$. Define

$$F_j = \{\{\sigma_{i+j}, \sigma_{2n-1-i+j}\} : 1 \le i < n\} \cup \{\{\sigma_j, \alpha\}\},$$

subscripts computed modulo $2n-1$. Now $(F_0, \ldots, F_{2n-2})$ is a 1-factorization of order $2n$; it is the well-known 1-factorization GK_{2n} (see Mendelsohn and Rosa, 1985). Reordering the 1-factors does not change the 1-factorization. So define $H_j = F_{(j+1)n}$, subscripts modulo $2n-1$, for $j \in \mathbb{Z}_{2n-1}$. $(H_0, \ldots, H_{2n-2})$ is the *Kirkman 1-factorization* of order $2n$. For our purposes, it is important that $\{\sigma_0, \sigma_1\} \in H_0$,

$$\{\{\sigma_0, \sigma_{i+1}\}, \{\sigma_1, \sigma_i\}\} \subseteq H_i \text{ for } 2 \le i < 2n-2,$$

$$\{\{\sigma_0, \sigma_2\}, \{\sigma_1, \alpha\}\} \subseteq H_1,$$

and $\{\{\sigma_0, \alpha\}, \{\sigma_1, \sigma_{2n-2}\}\} \subseteq H_{2n-2}$.

Kirkman's first construction is:

Theorem 0.1 *If an $S(2,3,2n-1)$ exists, then a crofty $S(2,3,4n-1)$ exists.*

Proof Let $(H_0, \ldots, H_{2n-2})$ be the Kirkman 1-factorization of order $2n$. Let $(V, \mathcal{B})$ be an $S(2,3,2n-1)$, with $V = \{v_0, \ldots, v_{2n-2}\}$. Define

$$\mathcal{D} = \{\{v_i, a, b\} : \{a, b\} \in H_i, 0 \le i \le 2n-2\}.$$

Then $(V \cup \{\sigma_0, \ldots, \sigma_{2n-2}\} \cup \{\alpha\}, \mathcal{B} \cup \mathcal{D})$ is an $S(2,3,4n-1)$. In addition, $\{\sigma_0, \sigma_1, v_0\}$ is in $\mathcal{D}$, and the $S(2,3,4n-1)$ is crofty with respect to σ_0 and σ_1; $(v_1, \sigma_2, v_2, \sigma_3, v_3, \sigma_4, \ldots, \sigma_{2n-2}, v_{2n-2}, \alpha)$ gives the crofty ordering. □

Without the condition on croftiness, this construction is very well known. Kirkman's second construction, however, is not.

Theorem 0.2 *If a crofty $S(2,3,2n+1)$ exists, and $n \ge 1$, then there exists a crofty $S(2,3,4n-3)$.*

Proof Let $(H_0, \ldots, H_{2n-2})$ be the Kirkman 1-factorization of order $2n$. Let $(V, \mathcal{B})$ be an $S(2,3,2n+1)$ with $V = \{x, y, z\} \cup \{v_i : 0 \le i \le 2n-3\}$, which is crofty with respect to x and y, having crofty ordering $(v_0, v_1, \ldots, v_{2n-4}, v_{2n-3})$. Form

$$\begin{aligned}
\widehat{\mathcal{B}} &= \{\{a,b,c\} : \{a,b,c\} \in \mathcal{B}, \{a,b,c\} \cap \{x,y\} = \varnothing\} \\
\mathcal{C} &= \{\{v_i, a, b\} : \{a,b\} \in H_{i+1}, \{a,b\} \cap \{\sigma_0, \sigma_1\} = \varnothing, \\
&\qquad 0 \le i \le 2n-3\} \\
\mathcal{D} &= \{\{z, a, b\} : \{a,b\} \in H_0, \{a,b\} \neq \{\sigma_0, \sigma_1\}\} \\
\mathcal{E} &= \{\{v_i, v_{i+1}, \sigma_{i+2}\} : 0 \le i < 2n-3\} \cup \{\{\alpha, v_0, v_{2n-3}\}\}
\end{aligned}$$

Define $W = \{v_0, \ldots, v_{2n-3}\} \cup \{z\} \cup \{\sigma_2, \sigma_3, \ldots, \sigma_{2n-2}\} \cup \{\alpha\}$. Then $(W, \widehat{\mathcal{B}} \cup \mathcal{C} \cup \mathcal{D} \cup \mathcal{E})$ is an $S(2,3,4n-3)$. Moreover, $\{\sigma_2, \sigma_3, v_3\} \in \mathcal{C}$, and the $S(2,3,4n-3)$ is crofty with respect to σ_2 and σ_3;

$$(v_0, v_1, v_2, \alpha, v_4, \sigma_4, v_5, \sigma_5, \ldots, v_{2n-3}, \sigma_{2n-3}, z, \sigma_{2n-2})$$

is the crofty ordering. □

Kirkman's construction of Steiner triple systems is all the more remarkable in that it constructs a system of every admissible order starting only with the trivial system on one element.

0.3 Kirkman's "On the puzzle of the fifteen young ladies"

Kirkman (1850) posed a simple problem that was destined to change combinatorial design theory:

Fifteen young ladies in a school walk out three abreast for seven days in succession: it is required to arrange them daily, so that no two shall walk twice abreast.

Without the requirement of arranging the triples in days, the configuration is a Steiner triple system of order 15, and hence was known to Kirkman.

Cayley (1850, 1863) published the first solution to this *Kirkman schoolgirl problem*:

	a	b	c	d	e	f	g
abc				35	17	82	64
ade		62	84			15	37
afg		13	57	86	42		
bdf	47		16		38		25
bge	58		23	14		67	
cdg	12	78			56	34	
cef	36	45		27			18

The girls are $\{a, b, c, d, e, f, g, 0, 1, 2, 3, 4, 5, 6, 7, 8\}$; each row of the array gives the 'parade' for one day, by taking the triple in the first column together with the triples formed by adjoining the column index to the pair, when one appears.

Kirkman (1850b) also gives a solution, repeated in Kirkman (1851) accompanied by a solution attributed to four others. Apparently, the Kirkman schoolgirl problem excited great interest among amateur mathematicians. Although a number of professional mathematicians studied the problem and its generalizations as well, the recreational nature of much of the early work probably explains, more than any other cause, the oft-held view that such problems are parlour tricks.

Remarkably, Kirkman's priority in stating his celebrated problem was disputed by Sylvester (1861), who mused that

> I may also take occasion to observe that, in connexion with my researches in combinatorial aggregration, long before the publication of my unfinished paper in the *Magazine*, I had fallen upon the question of forming a heptatic aggregate of triadic synthemes comprising all the duads to the base 15, which has become so well known, and fluttered so many a gentle bosom, under the title of the fifteen school-girl's problem; and it is not improbable that the question, under its existing form, may have originated through channels which can no longer be traced in the oral communications made by myself to my fellow-undergraduates at the University of Cambridge long years before its first appearance, which I believe was in the *Ladies' Diary* for some year which my memory is unable to furnish.

Predictably, Kirkman (1862) reacted:

> No man can doubt, after reading [Professor Sylvester's] words, that he was in possession of the number 15 when he was an Undergraduate in Cambridge. But the difficulty of tracing the origin of the puzzle, from my own brains to the fountain named at that University, is considerably enhanced by the fact that, when I proposed the question in 1849, I had never had the pleasure of seeing either Cambridge or Professor Sylvester.

Woolhouse (1862) defended Kirkman's priority, and Sylvester's musings were for the most part of no consequence. The ingenuity with which amateurs and professionals alike set upon the Kirkman problem is the real story, however. Anstice (1852, 1853) developed a fundamental understanding of designs acted upon by cyclic groups, which was pursued by Peirce (1860) and Kirkman (1862). Anderson (1995) provides a modern treatment of Anstice's work. We must appreciate that these researches were being undertaken at the dawn of group theory. The mathematical consequences of research ancillary to the Kirkman problem prompted Peirce (1860) to write about the Kirkman schoolgirl problem:

> In closing my researches upon this puzzle, which has interested me by the curious and novel analysis into which it has led me, I cannot regard it as quite useless.

Woolhouse (1862, 1863) collected and systematized all known solutions to the schoolgirl problem, and found seven nonisomorphic ones. Carpmael (1881) subsequently reported solutions that he believed different from any of those catalogued by Woolhouse. Finally, Cole (1922) established that there are exactly seven nonisomorphic solutions.

Naturally, the story does not end there. Problems on 21 schoolgirls (Bray, 1883) and the like recur in the recreational literature. Lucas (1883) includes the Kirkman schoolgirl problem in his widely read book on mathematical recreations.

Kirkman (1850a) asked for what numbers of girls the analogous problem admits a solution. Peirce (1860) posed the general question in this way:

A given number, f, of girls are required to walk for a given number, h, of days in a given number, g, or ranks, of which each rank consists of a given number, k, of girls; subject to the condition that each girl is to walk once, and only once, in the same rank with every other girl.

Kirkman (1850a) reports a construction of a Reverend James Mease establishing that solutions for 3^m girls, $m \geq 1$, can always be found. Using this guide, he determines solutions for orders of the form $5 \cdot 3^m$ with $m \geq 1$. A large literature on specialized constructions evolved from these beginnings; see Eckenstein (1912) or Ray-Chaudhuri and Wilson (1973). Ahrens (1901) served as an influential source of material on the Kirkman problem.

Ultimately, the Kirkman schoolgirl problem for triples was solved by Lu Jiaxi, a self-taught mathematician in Inner Mongolia, in two papers submitted for publication in 1961–65. It appears that the editors did not understand the importance of Lu's contributions; Kang (1995) refers to their 'careless, hasty dealing' with the papers. Indeed, the loss was compounded by the onset of the ten years of extreme leftist trend of thought in China, known to many as the Cultural Revolution. Many journals ceased publication altogether, and Lu was sent to be reformed through labour. By the time that Lu could resume submitting papers, the first published solution of the Kirkman schoolgirl problem, due to Ray-Chaudhuri and Wilson (1973), had appeared. Lu's prior solution came to light only upon his untimely death in 1983. Fortunately, his paper (Lu, 1965) is now available in his collected works. Undaunted, Lu went on to make remarkable advances in the large set problem for Steiner triple systems.

Ray-Chaudhuri and Wilson's complete solution firmly restored the Kirkman problem as a mathematical problem of importance, rather than a collection of recreational puzzles; in the process, much of the theory they developed has had wide-reaching consequences.

0.4 Netto, Moore, and Heffter

Jordan (1870), in his pioneering studies in the nascent field of group theory, calculated the automorphism group of the projective triple system of order 15, giving an explicit construction of it. He also examined the groups acting on the inflection points of the cubic curve, and on the 27 points on the cubic surface. Max Noether (1879), father of the better known Emma Noether, undertook a study of quartic curves, and encountered both the Steiner triple systems of order 7 and of order 9. He also examined the Steiner quadruple system $S(3,4,8)$, calculated the group of $168 = 7{\cdot}6{\cdot}4$ automorphisms of the $S(2,3,7)$, and observed that there are precisely 30 distinct systems. Apparently unaware of the earlier solutions by Kirkman and Reiss, Netto (1881, 1893) asked for which orders such a triple system exists. In 1893, he gave three new constructions. He established a direct product construction. In addition, he gave direct constructions whenever the order of the system is a prime, and when it is three times a prime. He reports on orders less than 100 which his techniques are able to construct, states that 25 and 85 are the only values missed in this range, and observes: 'Die

Construction für $n = 25$ ist leicht zu geben'. Moore (1893) completed Netto's existence proof for Steiner triple systems, by developing the notion of subsystems and the singular direct product construction. Indeed, using these techniques, he established that for every admissible order which is at least 13, there are at least two nonisomorphic solutions. Moore (1896) treats his work on triple systems in a more general setting.

During the same period, Heffter (1891) undertook a study of triangulations of orientable surfaces. He considers the duals of such triangulations, and asks that every two faces share exactly one edge. In this way, the triangles of his triangulations form a set of triples in which every pair occurs exactly twice, i.e. twofold triple systems. Heffter (1891) gives an elegant construction for twofold triple systems when the order is of the form $12x + 7$ as follows. Let $\mathcal{T}$ be the set of triples $\{\{2i + 1, 2i + 2, 12x + 4 - 4i\} : 0 \leq i < 4x + 2\}$. Then form a set of triples

$$\{\{j, a + j, b + a + j\} : \{a, b, c\} \in \mathcal{T}, j \in \mathbb{Z}_{12x+7}\}$$

with arithmetic modulo $12x + 7$. He observes that the differences of the elements in the triples formed are (plus or minus) those in the corresponding triple of $\mathcal{T}$. Since every difference occurs (as plus or minus) twice in $\mathcal{T}$, every pair occurs twice in the triples so formed. It seems that when Heffter soon thereafter saw the work of Netto, he realized that their different investigations had brought them to a similar problem. Heffter (1896) observes that the construction of Netto (1893) gives such cyclic systems (i.e. those generated over a cyclic group). He saw that Netto's requirement that the order be prime could be relaxed. To do so, he stated:

Aus den Zahlen $1, 2, \ldots, 3m$ sind m Gruppen von je dreien zu bilden, so dass
1) die Summe der 3 Zahlen stets $\equiv 0$ mod. $6m + 1$ ist,
2) alle $3m$ Zahlen voneinander verschieden sind,
3) nicht 2 der $3m$ Zahlen sich zu $6m + 1$ ergänzen.

This is *Heffter's first difference problem*, and it gives a simple structural characterization of systems of orders congruent to 1 modulo 6 which admit the cyclic automorphism. Heffter (1897) stated his second difference problem, to treat cases in the 3 modulo 6 class. He settled all orders less than 100 using solutions to his difference problems. Much later, the Heffter difference problems were solved completely by Peltesohn (1938). Heffter's original direction, the study of triangulations, evolved into a difficult and important investigation. Emch (1929, 1930a), Ringel (1961, 1965), and others studied the relationships between twofold triple systems and triangulations. This work culminated in the Map Colour Theorem; see Ringel (1975).

The work of Netto, Moore, and Heffter was done without knowledge of the earlier heritage from Plücker, Kirkman, Cayley, and others. Netto (1901) partially repairs this oversight by referring to Reiss's work. The omission perhaps had the benefit of furnishing a larger repertoire of constructions than would otherwise have been found. Of course, it also resulted in Moore being credited as the first to settle the existence question by later authors, compounding the confusion. It is also of interest to see that, although Kirkman (1853) had already remarked on a set of 21 triples on 7 points containing each pair three times (and indeed that this system did not contain a Steiner triple system of order 7 within it), there was little interest in triple systems in which pairs occur in more than one triple prior to Heffter (1891). The geometric ancestry of much of the research is demonstrated in the almost singular interest in having pairs each 'define' one triple, and to this day one finds a strong emphasis on this case in the literature.

0.5 Cole, Cummings, and White

With more and more constructions being found, systematic classifications of the designs produced became of critical importance. It was apparently well known that the Steiner triple systems of order 7 and 9 are unique up to isomorphism. Zulauf (1897) showed that the known systems of order 13 fell into two isomorphism classes. De Pasquale (1899) and Brunel (1902) showed that only two isomorphism classes are possible. Cole (1913) developed techniques for establishing this enumeration in an elementary way. Cummings (1914) began the effort to classify the Steiner triple systems of order 15. She found that two such systems can have the same group and yet be nonisomorphic. She further classified all systems on 15 elements having a subsystem on 7 elements (which she refers to as the Reiss systems, being unaware until later of Kirkman's prior work; see Cummings, 1918).

At the same time, White (1913) introduced a structural isomorphism invariant for Steiner triple systems, and reports on how to distinguish the two systems of order 13. Each of Cole, Cummings, and White was poised to determine the number of systems of order 15, and in a remarkable memoir (White, Cole, and Cummings, 1919), they succeeded in determining that there are precisely 80 nonisomorphic solutions.

White (1915a) had concluded that there are in excess of 37,000,000 Steiner triple systems of order 31, and hence White, Cole, and Cummings (1919) closed their memoir with a review of their four primary methods for distinguishing systems, and remarked:

> In the present state of the theory the most desirable forward step would be a demonstration that some one of these methods is (or is not) a sufficient means of proving congruency for triad systems of any number of elements above 15.

Essentially, they thereby indicated the importance of simple computational tools for the analysis of triple systems, a theme which has often been returned to subsequently.

The complete classification of systems on 15 elements stands as an important milestone, particularly because for no larger order has the number of Steiner triple systems been calculated exactly. Stinson and Ferch (1985) establish that more than 2,000,000 nonisomorphic systems exist for the next possible order, 19.

Fisher (1940), unaware of the catalogue for order 15, repeated the generation but missed one system and hence reported only 79. Hall and Swift (1955) repeated the calculation, aware of both of the earlier catalogues. Theirs is one of the first cases in which digital computers were used to catalogue combinatorial designs. They explain:

> The computations were performed on SWAC. This is a binary machine using 36 digits plus a sign in a word. The high speed memory consists of 256 words and the medium speed memory was a drum holding 4096 words. These sizes limited the choice of routines and the method of storage.

Hall and Swift confirmed the generation by White, Cole, and Cummings (1919).

0.6 Fisher, Yates, and Bose

The study of combinatorial designs in the first part of this century remained for the most part one of specific problems motivated by geometry, group theory, and algebraic curves. Advances were made particularly in the study of automorphisms of triple systems. Peltesohn (1939) solved Heffter's difference problems, Skolem (1927) developed simpler techniques to construct triple systems, and Bays (1931) developed an extensive theory of multiplier automorphisms of cyclic triple systems.

However, the most influential development during this period arose from the pioneering work of Fisher, Yates, and their colleagues in the design of statistical experiments. Fisher (1925) laid out mathematical principles of experimental design, but Yates (1936) was the first to state clearly the need for block designs:

> Most biological workers are probably by now familiar with the methods of experimental design known as randomized blocks and the Latin square. These were originally developed by Prof. R. A. Fisher, when Chief Statistician at Rothamsted Experimental Station, for use in agricultural field trials. The same principles of design are of general utility, wherever the basic material is at all variable, for comparing the effects of different treatments, or the growth of different strains of an organism, or other similar problems.

Yates went on to define balanced incomplete block designs, but provided no references to the by then substantial literature on block designs. Fisher and Yates (1938) produced a large table of designs for use in experimental situations.

The remarkable confluence of a rapidly developing application area, and a century of research in geometry and algebra, was effected in a paper by Professor R.C. Bose, then in Calcutta. Bose (1939) is an exceptional contribution, in that it presented many previously disjointed methods and applications as a coherent theory (albeit one in its infancy). At once, it demonstrated to statisticians the importance of the mathematical research, and established combinatorial design theory as a mathematical subject in its own right.

Bose (1939) builds on the results of Heffter, Moore, and Netto, and on the earlier work of Reiss, the Reverend Kirkman having been forgotten once again. Applications in experimental design made triple systems in which pairs are each covered λ times for some fixed number $\lambda > 1$ of interest, and Bose extends the work of Emch (1929, 1930a) in this area.

Bose concentrated on direct constructions of designs, and gave a detailed exposition of a method of 'pure and mixed differences'. He examined the designs that arise from projective and affine geometries. He established a new method for the direct construction of Steiner triple systems. For the first time, the existence problem for triple systems with $\lambda = 2$ was systematically studied.

Sixty years on, the paper of Bose forms a watershed. Before then, while combinatorial design theory arose with some frequency in other researches, it was an area without a basic unifying theme, and without a substantial application. After the paper of Bose, the themes of the area that recur today were clarified, and the importance of constructing designs went far beyond either recreational interest, or as a secondary tactical problem in a larger algebraic or geometric study.

After Bose's seminal work, design theory and the study of triple systems have grown at an accelerating pace. Applications in error correction provided a second influx of research after the Second World War, and continued refinements of connections with experimental design, finite geometry, and group theory advanced the field. It is perhaps best left to later writers to identify the importance of the advances made. However, the importance of the solution of the Kirkman schoolgirl problem in Ray-Chaudhuri and Wilson (1971) cannot be overlooked, both for the techniques introduced and for its closing a 120-year-old chapter, and opening many more.

1

DESIGN-THEORETIC FUNDAMENTALS

1.1 Designs

Combinatorial designs appear in a wide variety of disguises, and with a number of different restrictions. Generically, a combinatorial design consists of a ground set of elements and a collection of subsets of these elements. What distinguishes most combinatorial designs from an arbitrary collection of subsets is the requirement for *balance* or *regularity* in some regard. For example, equality in the sizes of the subsets, or the number of subsets in which each t-subset of elements occurs, or the sizes of the intersection of pairs of subsets, may be required. In our investigation, the central restriction is that the design is *t-balanced*: every unordered set of t elements is required to appear in the same number λ of subsets in the collection. We are particularly concerned with designs that are 2-balanced.

1.1.1 *Balanced incomplete block designs*

A *balanced incomplete block design*, or simply *block design*, is a pair $(V, \mathcal{B})$. V is a set of *elements*, usually taken to be of finite cardinality. The number of elements of a block design is the *order* of the design, and is denoted by v. $\mathcal{B}$ is a collection of subsets of V called *blocks*; the number of blocks is usually denoted by b. Each block in $\mathcal{B}$ has the same cardinality, termed the *block size* and denoted by k. Finally, each unordered pair of elements occurs as a subset of the same number λ of blocks in $\mathcal{B}$; λ is the *index* of the design. Since every element occurs with each of the remaining $v-1$ elements in λ blocks, and in any single block it appears with $k-1$ other elements, it follows that every element occurs in the same number $\lambda(v-1)/(k-1)$ of blocks. This is the *replication number* of the design and is denoted r.

Block designs are typically specified by giving these five parameters: order or number of elements (v), number of blocks (b), replication number (r), block size (k), and index (λ). Hence a block design is often called a (v, b, r, k, λ)*-design*. These five parameters are not independent; as we have seen,

$$r(k-1) = \lambda(v-1),$$

obtained by counting pairs involving a fixed element in two different ways. Similarly, we can count all unordered pairs in two different ways. There are $\binom{v}{2}$ unordered pairs, and each occurs λ times. On the other hand, each block contains $\binom{k}{2}$ unordered pairs, and there are b blocks. Thus

$$\lambda\binom{v}{2} = b\binom{k}{2}.$$

Given v, k, and λ, the number of blocks and the replication number can be calculated; hence a (v, b, r, k, λ)-design is often called a (v, k, λ)-*design* or a $B[k, \lambda; v]$.

Generalizations and restrictions of block designs have been very widely studied, and we see many of these in subsequent chapters. Nevertheless, our primary goal is to study one particular restriction. When the block size is three, a block design is a *triple system*. The four remaining parameters of a triple system are determined by giving the order and the index of the system; hence we employ the notation $\mathrm{TS}(v, \lambda)$ for triple systems. When $\lambda = 1$, we often use the notation $\mathrm{STS}(v)$ and call the system a *Steiner triple system* of order v. Similarly, when $\lambda = 2$, we use the notation $\mathrm{TTS}(v)$ and call the system a *twofold triple system* of order v.

1.1.2 *Pairwise balanced designs*

We employ a number of generalizations of block designs. A more general type of design is obtained by relaxing the requirement for constant block size. Let K be a set of positive integers, each at least two. A *pairwise balanced design* with block sizes K, order v, and index λ, or simply (v, K, λ)-*design*, is a pair $(V, \mathcal{B})$ with $|V| = v$. $\mathcal{B}$ is a collection of subsets of V, in which each subset $B \in \mathcal{B}$ has $|B| \in K$. Every pair of elements occurs in precisely λ blocks as before. When the index of a pairwise balanced design is not specified, it is taken to be one.

An *incomplete pairwise balanced design* of order v, index λ, block sizes K, and with a *hole* of size w, or simply (v, w, K, λ)-IPBD is a v set V together with a collection $\mathcal{B}$ of blocks. There is a subset W of V of size w with the property that every unordered pair of elements from V, not both from w, occurs in precisely λ blocks in $\mathcal{B}$. Unordered pairs of elements in W occur in no block of $\mathcal{B}$. A $(v, w, \{3\}, \lambda)$-IPBD is an *incomplete triple system*, and is denoted $\mathrm{ITS}(v, w; \lambda)$. When $\lambda = 1$, the notation $\mathrm{ITS}(v, w)$ is used.

1.1.3 *Group divisible designs*

Group divisible designs are a further relaxation permitting a spanning set of holes. Let V be a set of v elements. Let $\mathcal{G} = \{G_1, \ldots, G_s\}$ be a partition of V; each class of the partition is (by an unfortunate choice of terminology) called a *group*. Let $\mathcal{B}$ be a set of subsets of V with the property that every unordered pair of elements occurs either within one of the groups, or in precisely λ blocks of $\mathcal{B}$, but not in both. If the block sizes all occur in the set K, then $(V, \mathcal{G}, \mathcal{B})$ is a *group divisible design* (*GDD*) of order V, index λ, and block sizes K. The *type* of the group divisible design is the multiset of group sizes in $\mathcal{G}$. Often types are written in exponential notation. Type $g_1^{u_1} \cdots g_s^{u_s}$ indicates that there are u_i groups of size g_i for $1 \le i \le s$.

1.1.4 *Transversal designs and orthogonal arrays*

A *transversal design* is a restricted type of group divisible design. Specifically, a transversal design of *order* n, block size k, and index one, or simply $\mathrm{TD}(k, n)$, has kn elements partitioned into k groups of size n. Every block has size k, and thus meets every group in a single element. Thus it is a GDD with $K = \{k\}$ of type n^k.

An *orthogonal array* OA(k, s) is a $k \times s^2$ array with entries from a set of s elements, having the property that in any two rows, each ordered pair of symbols occurs exactly once. An OA(k, n) exists if and only if a TD(k, n) exists.

To see this, let A be an OA(k, n) on the n symbols in X. On $V = X \times \{1, \ldots, k\}$ (a set of size kn), form a set $\mathcal{B}$ of k-sets as follows. For $1 \leq j \leq n^2$, include $\{(a_{i,j}, i) : 1 \leq i \leq k\}$ in $\mathcal{B}$. Then let $\mathcal{G}$ be the partition of V whose classes are $\{X \times \{i\} : 1 \leq i \leq k\}$. Then $(V, \mathcal{G}, \mathcal{B})$ is a TD(k, n). This process can be reversed to recover an OA(k, n) from a TD(k, n).

1.2 Latin squares and quasigroups

Consider a transversal design TD(k, n) with $k \geq 3$. Choose any three distinct groups, say $\{x_1, \ldots, x_n\}$, $\{y_1, \ldots, y_n\}$, and $\{z_1, \ldots, z_n\}$. Deleting all occurrences of the points in the remaining $k - 3$ groups leaves a TD$(3, n)$. A convenient device for examining a TD$(3, n)$ is to form an $n \times n$ array, placing z_k in position (i, j) whenever $\{x_i, y_j, z_k\}$ is a block of the transversal design. The result is a widely-studied combinatorial structure, a *latin square*.

A latin square of *side* n is an $n \times n$ array $L = (\ell_{ij})$. Each entry ℓ_{ij} contains a single symbol from an n-set $\mathcal{S}$ of symbols. Each symbol occurs in each row precisely once, and in each column precisely once. For example, a latin square of order 5 is depicted here, using symbol set $\mathcal{S} = \{a, b, c, d, e\}$.

a	b	c	d	e
e	a	b	c	d
d	e	a	b	c
c	d	e	a	b
b	c	d	e	a

Independently permuting the rows, columns, and symbols transforms a latin square into a (possibly distinct) latin square. Similarly, renaming symbols, and using different sets of size n to index the rows and columns of a latin square, transforms the latin square into a (possibly distinct) latin square. Thus we often choose to employ symbol set $\mathbb{Z}_n$, and to use the same set to index rows and columns.

A *quasigroup* $(Q, \otimes)$ of order n is a set Q of size n equipped with a binary operation $\otimes$. For given $a, b \in Q$, the equations $a \otimes b = x$, $a \otimes y = b$, and $z \otimes a = b$ have unique solutions. Now suppose that $(Q, \otimes)$ is a quasigroup, and let $Q = \{q_1, \ldots, q_n\}$. Define an $n \times n$ array $L(Q) = (\ell_{ij})$, setting $\ell_{ij} = q_i \otimes q_j$. Evidently $L(Q)$ is a latin square on symbol set Q. Conversely, given a latin square $L = (\ell_{ij})$ on symbol set $\mathcal{S} = \{s_1, \ldots, s_n\}$ with rows and columns indexed by $\mathcal{S}$, define a binary operation $\oplus$ by setting $s_i \oplus s_j = \ell_{ij}$. Then $(S, \oplus)$ is a quasigroup.

Latin squares exist whenever the side is a nonnegative integer:

Lemma 1.1 *For all $n \geq 0$, there is a latin square of side n.*

Proof Define a quasigroup $(\mathbb{Z}_n, \oplus)$ by setting $i \oplus j = (i + j) \pmod n$ for $i, j \in \mathbb{Z}_n$. $\square$

When a quasigroup $(Q, \otimes)$ satisfies $a \otimes a = a$ for all $a \in Q$, it is *idempotent*. The corresponding latin square has each symbol occurring exactly once on the main diagonal. In general, a *transversal* in a latin square of side n is a subset of n cells in the square, so that one is from each row, one is from each column, and each cell contains a different symbol. The diagonal of an idempotent latin square is a transversal.

When a latin square L of side n contains a transversal T, the square can be *bordered* to form a latin square L' of side $n+1$. To do this, we first choose a new symbol σ not appearing in L. We place all entries of L not in the transversal T in the square L'. For each entry of T, we proceed as follows. If the entry is from row i, column j, and contains symbol k, we place symbol k in entries $(n+1, j)$ and $(i, n+1)$ of L', and place σ in entry (i, j) of L'. Finally, we place symbol σ in entry $(n+1, n+1)$. Then L' is a latin square of side $n+1$. We illustrate the bordering technique here. First we show a latin square of side 5, with the elements of a transversal shown in bold. Then we border using the transversal:

a	b	c	d	e
e	a	**b**	c	d
d	e	a	b	**c**
c	**d**	e	a	b
b	c	d	**e**	a

σ	b	c	d	e	a
e	a	σ	c	d	b
d	e	a	b	σ	c
c	σ	e	a	b	d
b	c	d	σ	a	e
a	d	b	e	c	σ

Lemma 1.2 *For every $n \geq 0$, $n \neq 2$, there exists an idempotent quasigroup of order n.*

Proof When $n = 2z+1$ is odd, the binary operation $i \odot j = (z+1)(i+j) \pmod{2z+1}$ defines an idempotent quasigroup. When $n = 2z+2 \geq 4$ is even, first form the idempotent quasigroup of order $2z+1$ as above. Now the entries in positions $(i, i+1)$ for $0 \leq i < 2z$, together with the entry $(2z, 1)$, form a transversal in the corresponding square. Use this transversal to border the square, using new symbol $2z+1$, to obtain an idempotent latin square. □

When a quasigroup $(Q, \otimes)$ satisfies $a \otimes b = b \otimes a$ for all $a, b \in Q$, the quasigroup is *commutative*, and the corresponding latin square is *symmetric*.

Lemma 1.3 *A commutative quasigroup of side n exists for all $n \geq 0$. A commutative idempotent quasigroup of side n exists if and only if $n = 0$ or n is odd.*

Proof Lemma 1.1 produces a commutative quasigroup for every order. Lemma 1.2 constructs a commutative idempotent quasigroup for all odd orders. Thus we need only treat commutative idempotent quasigroups of even order. Consider the corresponding latin square of even side. Suppose that the side is greater than zero, and consider a fixed symbol. It occurs once on the diagonal, and an even number of times off the diagonal as a consequence of symmetry. Thus each symbol must appear an odd number of times. Hence the side must be odd except when it is zero. □

When the side is even, a symmetric latin square must have each symbol occurring an even number of times on the main diagonal. A latin square is *half-idempotent* if every symbol occurs on the main diagonal either zero or two times.

Lemma 1.4 *There exists a half-idempotent symmetric latin square for every even side.*

Proof Lemma 1.1 constructs such a square. □

A *subsquare* of a latin square is a square subarray that is itself a latin square. A latin square is an N_2-*latin square* if it contains no subsquare of order 2. Kotzig, Lindner, and Rosa (1975), Kotzig and Turgeon (1975), and McLeish (1975) establish:

Theorem 1.5 *An N_2-latin square of order n exists for all $n \geq 3$ and $n \neq 4$.*

We typically blur the distinctions among quasigroups, latin squares, and transversal designs with block size three.

There is an extensive literature on embedding partial latin squares and rectangles into latin squares. A fundamental result in this direction is from Ryser (1951):

Theorem 1.6 *Let L be a partial latin square of order n in which cell (i, j) is filled if and only if $i \leq r$ and $j \leq s$. Then L can be completed to a latin square of order n if and only if $N(x) \geq r + s - n$ for $i = 1, 2, \ldots, n$, where $N(x)$ denotes the number of elements in L that are equal to x.*

1.3 Factorizations and graph decompositions

It is often convenient to examine designs in a graph-theoretic setting. We develop some basic observations here concerning graphs and multigraphs of which we have need later. A *multigraph* is a set of *vertices*, and a collection of unordered pairs of vertices called *edges*. It is a *graph* if no pair of vertices occurs more than once as an edge.

The multigraph (W, F) is a *submultigraph* of the multigraph (V, E) when $W \subseteq V$ and $F \subseteq E$ (as multisets); it is *spanning* if $W = V$, and it is *induced* if $F = \{e \in E : e \subseteq W\}$.

A vertex is *incident* with an edge when the vertex occurs in the edge. Two vertices are *adjacent* when there is an edge with which both are incident. The *degree* of a vertex is the number of edges with which it is incident (occasionally, however, if the multigraph has loops, we employ the convention that a loop contributes twice to the degree, once for each end of the edge). A multigraph is *regular* of degree r if every vertex has degree r.

An r-regular spanning submultigraph of a multigraph G is an r-*factor* of G. When $G = (V, E)$ admits a partition $E = E_1 \cup \cdots \cup E_\ell$ in which (V, E_i) is an r-factor of G, the set $\{E_1, \ldots, E_\ell\}$ is an r-*factorization* of G.

1.3.1 *Edge-colourings*

An *edge-colouring* of a multigraph $G = (V, E)$ is a mapping from E to a set C of colours. It is *proper* if no vertex is incident with two or more edges of the same colour. In a proper edge-colouring, each colour class of edges forms a subgraph with maximum degree at most one; thus a colour class is a partial 1-factor, also termed a *matching*. A 1-factor is often called a *perfect matching*.

We recall some fundamental results concerning matchings in graphs, referring to Lovász and Plummer (1986) for a more detailed exposition. König (1916) proved:

Theorem 1.7 *In a bipartite multigraph $G = (V, E)$, the maximum size of a matching equals the minimum size of a set X of vertices for which $e \cap X \neq \varnothing$ for all $e \in E$.*

Frobenius (1917) proved:

Theorem 1.8 *A bipartite multigraph $G = (V, E)$ with bipartition $A \cup B$, $|A| = |B|$, has a perfect matching if and only if for every $X \subseteq A$, the number of neighbours of vertices in X is at least $|X|$.*

P. Hall (1935) proved:

Theorem 1.9 *A bipartite multigraph $G = (V, E)$ with bipartition $A \cup B$ has a matching in which every vertex of A is covered if and only if for every $X \subseteq A$, the number of neighbours of vertices in X is at least $|X|$.*

These theorems are all equivalent: each implies the other two. For our purposes, however, their importance lies in two consequences proved by Steinitz (1894) but typically attributed to König (1916):

Corollary 1.10 *A regular bipartite multigraph has a 1-factor.*

Corollary 1.11 *A regular bipartite multigraph has a 1-factorization.*

Turning to graphs in general, Tutte (1947) proved the central result:

Theorem 1.12 *A multigraph $G = (V, E)$ has a perfect matching if and only if for every set $S \subseteq V$, the number of connected components of $G \setminus S$ that have an odd number of vertices is at most $|S|$.*

Unlike the bipartite case, regular graphs need not have 1-factors; and even when they have 1-factors, they need not have 1-factorizations. This has led to a substantial effort to establish the existence of 1-factorizations for various classes of graphs.

Vizing (1964) proved a fundamental theorem:

Theorem 1.13 *A simple graph with maximum degree Δ has a proper edge-colouring with $\Delta + 1$ colours.*

For multigraphs, by vertex-colouring the line graph using the theorem of Brooks (1941), we obtain:

Theorem 1.14 *A multigraph with maximum degree* Δ *has a proper edge-colouring with* $\lfloor 3\Delta/2 \rfloor$ *colours.*

Folkman and Fulkerson (1969) prove:

Theorem 1.15 *If* G *is an* m*-edge graph that admits a proper edge-colouring with* c *colours, and* $c' \geq c$*, then* G *admits a proper edge-colouring in* c' *colours in which the sizes of the edge colour classes differ by at most one.*

1.3.2 *1-factorizations*

A *circulant* graph of order n and *spectrum* $d_1, \ldots, d_s$ is a graph with vertex set $\{0, \ldots, n-1\}$, and edge set

$$\{\{j, (j + d_i) \pmod n\} : 0 \leq j < n, 1 \leq i \leq s\}.$$

Such a graph is also termed a *cyclic graph* with differences $d_1, \ldots, d_s$. A difference d is n*-good* if n is even and $n/(\gcd(d,n))$ is even. Evidently the cyclic graph with a single difference d is a 2-factor unless $2d = n$; in fact, this cyclic graph has a 1-factorization precisely when d is n-good. Stern and Lenz (1980) and Hartman (1980) proved a much stronger statement:

Theorem 1.16 *A cyclic graph (of even order) with differences* $d_1, \ldots, d_s$ *has a 1-factorization if and only if at least one of the differences* $d_1, \ldots, d_s$ *is* n*-good.*

We obtain Theorem 1.16 here as a corollary of one of the theorems due to Stong (1985). If Γ is a group, and $S \subseteq \Gamma \setminus \{1\}$, the *Cayley graph* of S in Γ, $Cay(S : \Gamma)$, has vertex set Γ and edge set

$$\{\{g, gs\} : g \in \Gamma, \text{ and } s \in S \text{ or } s^{-1} \in S\}.$$

A cyclic graph is a Cayley graph of the cyclic group. Stong (1985) establishes for many classes of groups that any connected Cayley graph has a 1-factorization. We treat only one case, when the group Γ is abelian.

Two simple technical lemmas are needed, using standard notions of graph products (see Bondy and Murty (1976), for example).

Lemma 1.17 *The product of a regular graph* G *and* K_2 *has a 1-factorization.*

Proof Let $G = (V, E)$ be a regular graph. We produce a 1-factorization for the product of G and K_2 on vertex set $V \times \{0, 1\}$. By Vizing's theorem (1.13), G has a proper edge-colouring in $r + 1$ colours, where r is the degree of G. Use this edge-colouring to colour identically the edges of the copies of G on $V \times \{0\}$ and $V \times \{1\}$. Now we colour the remaining 1-factor of edges. Assign the edge $\{(v, 0), (v, 1)\}$ the unused colour at v in the edge-colouring of G. This extends the $r + 1$-edge-colouring of G to a proper $r + 1$-edge-colouring of the product of G and K_2, an $r + 1$-regular graph. □

Lemma 1.18 *Let* $S_1, S_2 \subseteq \Gamma \setminus \{1\}$ *be arbitrary subsets of nonidentity group elements. If* $S_2 \subseteq S_1$*, all elements of* $S_1 \backslash S_2$ *have even order in* Γ*, and* $Cay(S_2 : \Gamma)$

has a 1-factorization, then $Cay(S_1 : \Gamma)$ has a 1-factorization. In particular, any Cayley graph of a group Γ of order 2^k admits a 1-factorization.

Proof For each $s \in S_1 \setminus S_2$, the edges generated by s form a 1-factor if s is of order two, and they form a 2-factor in which every cycle has even length if the order of s is even and greater than two. □

Theorem 1.19 *If Γ is a group, and $S \subset \Gamma$ generates an abelian subgroup of Γ of even order, then $Cay(S : \Gamma)$ has a 1-factorization.*

Proof If S generates a proper subgroup $\langle S \rangle$ of Γ, $Cay(S : \Gamma)$ consists of isomorphic copies of $Cay(S : \langle S \rangle)$. Thus we need only treat the case when Γ is abelian, and S generates Γ. Since Γ has even order in this case, it is either a 2-group or is a semidirect product of a group H with a 2-group G. Let $s \in S \setminus H$, and $S' = S \cap H$. By Lemma 1.18, we need only establish that $Cay(S' \cup \{s\} : \langle S' \cup \{s\} \rangle)$ has a 1-factorization. Now $\langle S' \cup \{s\} \rangle$ is a semidirect product of $\langle S' \rangle$ with a 2-group G'. If s has order two, then $Cay(S' \cup \{s\} : \langle S' \cup \{s\} \rangle)$ is the product of $Cay(S' : \langle S' \rangle)$ and K_2; hence by Lemma 1.17, it has a 1-factorization. So suppose that s has even order greater than two, and let g be the image of s under the natural homomorphism from the semidirect product to G'. By Lemma 1.18, $Cay(\{g\} : G')$ is 1-factorizable. Choose one 1-factor from the factorization and extend it to $Cay(S' \cup \{s\} : \langle S' \cup \{s\} \rangle)$ by including every edge generated by s whose projection lies in the 1-factor. The remaining edges form disjoint copies of the product of $Cay(S' : \langle S' \rangle)$ and K_2; hence by Lemma 1.17, it has a 1-factorization. □

We need a simple recursive construction for 1-factorizations. Given a 1-factorization $F_1, \ldots, F_{n-1}$ on a set X, we can *double* the 1-factorization to produce a 1-factorization on $X \times \{0, 1\}$:

Lemma 1.20 *If $F_1, \ldots, F_{n-1}$ is a 1-factorization on X, then a 1-factorization on $X \times \{0, 1\}$ exists with 1-factors:*

1. *$\{\{(x, a), (y, a + b)\} : \{x, y\} \in F_j$ and $a \in \{0, 1\}\}$ for $1 \leq j < n$ and $b \in \{0, 1\}$, with arithmetic on the second coordinate modulo 2, and*
2. *$\{\{(x, 0), (x, 1)\} : x \in X\}$.*

The verification that the result is a 1-factorization is straightforward.

1.3.3 *Regular factors*

Tutte (1952) developed a general theory of factors in graphs, in which every vertex v has a specified degree $f(v)$ that is required in the factor. Tutte proved a comprehensive theorem, the 'f-factor theorem', for this case. We give the statement here just for r-factors:

Theorem 1.21 *A multigraph $G = (V, E)$ has an r-factor if and only if for every two disjoint subsets X and Y of V, the number of connected components C of $G \setminus (X \cup Y)$ for which $rV(C) + \alpha$ is odd does not exceed $r|X| + (|V| - r)|Y| - \beta$,* *where α is the number of edges connecting vertices of C with those of Y, and β is the number of edges connecting vertices of x with those of Y.*

Theorem 1.21 imposes no restriction when r is even.

1.3.4 *2-factorizations*

Theorem 1.21 establishes a fundamental theorem of Petersen (1891); we give a direct proof:

Theorem 1.22 *Every regular multigraph G of even degree has a 2-factorization.*

Proof We may treat the case when G is connected. Form an eulerian circuit C of G, and orient the edges of C to form a directed eulerian circuit $\widehat{C}$. Now form a new graph H with vertex set $\{v_I, v_O : v \in V\}$, and arcs $\{\{x_O, y_I\} : (x, y) \in \widehat{C}\}$. H is a bipartite graph, and hence by Corollary 1.11 admits a 1-factorization $F_1, \ldots, F_d$. Partition the edges of G into classes $H_1, \ldots, H_d$ by replacing each occurrence of v_I and v_O in $F_1, \ldots, F_d$ by v. Then $H_1, \ldots, H_d$ is a 2-factorization of G. □

1.3.5 *Absence of factors*

Regular graphs on an even number of vertices need not have a 1-factor. In fact, Wallis (1981) proved:

Theorem 1.23 *The smallest simple d-regular graph having no connected component with an odd number of vertices, and having no 1-factor, has v vertices, where*

$$\begin{aligned} v &= 3d + 7 && \text{if } d \text{ is odd, } d \geq 3; \\ v &= 3d + 4 && \text{if } d \text{ is even, } d \geq 6; \text{ and} \\ v &= 22 && \text{if } d = 4. \end{aligned}$$

Hoffman, Rodger, and Rosa (1993) determined the smallest order for d-regular graphs that admit *no* regular factor:

Theorem 1.24 *A d-regular simple graph on v vertices having no regular factor exists if and only if $d \geq 3$ is odd, v is even, and $v \geq (d+1)^2$.*

1.3.6 *Room squares*

A *Room square* of *side* n on a set S of $n+1$ symbols is an $n \times n$ array F for which:

1. each cell of F is either empty or contains an unordered pair of symbols from S;
2. every symbol occurs exactly once in each row and in each column of F; and
3. every unordered pair of symbols occurs in exactly one cell of F.

The rows of a Room square give a 1-factorization of K_{n+1} on vertex set S. The columns give a second 1-factorization. Evidently, the two factorizations are *orthogonal*, since each factor of one meets each factor of the other in at most one edge.

The existence problem for Room squares was settled by Mullin and Wallis (1975); see Dinitz and Stinson (1992) for an extensive survey.

Theorem 1.25 *A Room square of side n exists if and only if n is odd and $n \notin \{3,5\}$.*

2

EXISTENCE: DIRECT METHODS

Constructions for designs fall into two main classes. *Direct* constructions produce a design without requiring a (smaller) design of the same type in the construction, while *recursive* constructions produce larger designs from smaller ones of the same type. In this chapter, we examine direct constructions for triple systems. Then in Chapter 3, we turn to recursive methods.

Some examples of small triple systems are shown in Table 2.1.

TS(3,1)	TS(5,3)	TS(6,2)	TS(7,1)	TS(9,1)
012	012	012	013	012
	013	023	124	345
	014	034	235	678
TS(4,2)	023	045	346	036
012	024	015	045	147
013	034	135	156	258
023	123	124	026	048
123	124	235		156
	134	134		237
	234	245		057
				138
				246

Table 2.1 *Some small triple systems*

We frequently choose to view a triple system TS(v, λ) as an edge-partition of λK_v into triangles, and use the terms triple and triangle, the terms pair and edge, and the terms point, vertex, and element synonymously. In this graph context, some necessary conditions for the existence of a TS(v, λ) are immediate. Since the degree of λK_v is $\lambda(v-1)$, and every triangle accounts for degree 0 or 2 at a vertex, we require $\lambda(v-1) \equiv 0 \pmod 2$. Since λK_v has $\lambda\binom{v}{2}$ edges, and every triangle accounts for three of these, we require $\lambda v(v-1) \equiv 0 \pmod 6$. Finally, since λK_2 contains no triangle whatsoever, we must have $v \neq 2$.

This gives the basic necessary conditions:

Lemma 2.1 *A TS(v, λ) exists only if* $\lambda \equiv 0 \pmod{\gcd(v-2, 6)}$ *and* $v \neq 2$, *or equivalently,*

$$v \equiv 1, 3 \pmod 6 \text{ if } \lambda \equiv 1, 5 \pmod 6$$

$$v \equiv 0, 1 \pmod{3} \text{ } \mathit{if}\, \lambda \equiv 2, 4 \pmod{6}$$
$$v \equiv 1 \pmod{2} \text{ } \mathit{if}\, \lambda \equiv 3 \pmod{6}$$
$$v \neq 2 \text{ } \mathit{if}\, \lambda \equiv 0 \pmod{6}$$

2.1 Constructions from latin squares

2.1.1 *Equivalence with Steiner quasigroups*

Any Steiner triple system may be provided with a structure of a *Steiner quasigroup* by setting $z = x \circ y$ when $\{x, y, z\}$ is a block of the design or when $x = y = z$. A *Steiner quasigroup* is a commutative quasigroup whose binary operation $\circ$ satisfies $x \circ x = x$ and $(x \circ y) \circ y = x$. For the STS(7) in Table 2.1, the Steiner quasigroup is shown on the right. Given a Steiner quasigroup we recover a Steiner triple system by letting the blocks be the sets $\{x, y, z\}$ when $z = x \circ y$ and x and y are two distinct points.

$\circ$	0	1	2	3	4	5	6
0	0	3	6	1	5	4	2
1	3	1	4	0	2	6	5
2	6	4	2	5	1	3	0
3	1	0	5	3	6	2	4
4	5	2	1	6	4	0	3
5	4	6	3	2	0	5	1
6	2	5	0	4	3	1	6

Theorem 2.2 *The correspondence $(S, \mathcal{L}) \longrightarrow (S, \circ)$ between Steiner triple systems and Steiner quasigroups is one-to-one up to isomorphism.*

Given an STS(v) $(V, \mathcal{B})$, we can also add a new element ∞ to the set V and define a binary operation $\oplus$ on $V \cup \{\infty\}$ by setting $\infty \oplus x = x \oplus \infty = x$ for all $x \in V$, and $x \oplus y = z$ whenever $\{x, y, z\} \in \mathcal{B}$. Then $(V \cup \{\infty\}, \oplus)$ is a *totally symmetric* loop, i.e. a loop satisfying $x \oplus y = y \oplus x$, $\infty \oplus x = x$, $x \oplus x = \infty$, and $x \oplus (x \oplus y) = y$. Here ∞ is the loop identity. Conversely, given a totally symmetric loop we can recover an STS in a unique manner (see, for example, Bruck, 1963). A totally symmetric loop is often called a *Steiner loop*.

Equivalently, given a Steiner quasigroup $(V, \circ)$, a *Steiner near-1-factorization* can be formed on vertex set V, defining near 1-factors to be

$$\{\{\{x, y\} : x, y \in V, x \neq y, x \circ y = z\} : z \in V\}.$$

The addition of an infinite point enables us to extend this near-1-factorization to a *Steiner 1-factorization* on $V \cup \{\infty\}$, in which the 1-factors are

$$\{\{\{\infty, z\}\} \cup \{\{x, y\} : x, y \in V, x \neq y, x \circ y = z\} : z \in V\}.$$

Equivalently, from a Steiner quasigroup of order v, we can form a unipotent (constant diagonal) quasigroup of order $v + 1$ by adding a new column and a new row whose entries are the elements of V in order, and then placing the new symbol ∞ on the diagonal. The process for the STS(7) is illustrated in Table 2.2.

$\oplus$	0	1	2	3	4	5	6	∞
0	∞	3	6	1	5	4	2	0
1	3	∞	4	0	2	6	5	1
2	6	4	∞	5	1	3	0	2
3	1	0	5	∞	6	2	4	3
4	5	2	1	6	∞	0	3	4
5	4	6	3	2	0	∞	1	5
6	2	5	0	4	3	1	∞	6
∞	0	1	2	3	4	5	6	∞

Table 2.2 *The unipotent quasigroup from the STS(7)*

2.1.2 *Bose's construction and variants*

Bose (1939) devised one of the most widely used direct constructions for triple systems; it has been extensively adapted to various problems. We present three basic forms of it here.

Commutative $3v$ *construction*
Let $Q_i = (V, \otimes_i)$, $i = 0, 1, 2$, be commutative idempotent quasigroups of order v. On $V \times \{0,1,2\}$, form a collection $\mathcal{B}$ of triples

$$\{\{x_i, y_i, (x \otimes_i y)_{i+1}\} : x, y \in V, x < y, i \in \mathbb{Z}_3\}.$$

(Here and henceforth we typically use subscript notation x_i to denote the ordered pair (x, i).) Then $(V \times \{0,1,2\}, \mathcal{B})$ is a $\{3\}$-GDD of type 3^v.

Evidently in $\mathcal{B}$ every pair $\{x_i, y_i\}$ appears in exactly one triple. For a pair $\{x_i, y_j\}$, we may suppose without loss of generality that $j = i+1 \pmod 3$, for if not then we interchange x_i and y_j. Now $x \otimes_i z = y$ has a unique solution for z, since Q_i is a quasigroup. Now suppose that $x \neq y$; then since Q_i is idempotent, $z \notin \{x, y\}$. For $x \neq y$, then, we find $\{x_i, y_{i+1}\}$ in the triple $\{x_i, z_i, y_{i+1}\}$. The pairs $\{\{x_i, x_{i+1}\} : x \in V, i \in \mathbb{Z}_3\}$ appear in no triple of $\mathcal{B}$; these form v groups $\{\{x_0, x_1, x_2\} : x \in V\}$.

Lemma 2.3 *For all* $v \equiv 3 \pmod 6$, *there exists an STS*(v).

Proof Write $m = v/3$. By Lemma 1.3, there exists a commutative idempotent quasigroup of order m. Apply the commutative $3v$ construction to produce a $\{3\}$-GDD of type 3^m, and replace every group by a triple on the three elements of the group. □

Noncommutative $3v$ *construction*
Let $Q = (V, \oplus)$ be an idempotent quasigroup of order v. Define a set $\mathcal{B}$ of triples on $V \times \{0,1,2\}$ as

$$\{\{x_i, y_i, (x \oplus y)_{i+1}\} : x, y \in V, x \neq y, i \in \mathbb{Z}_3\}.$$

Then $(V \times \{0,1,2\}, \mathcal{B})$ is a $\{3\}$-GDD of index two and type 3^v.

As before the groups are $\{\{x_0, x_1, x_2\} : x \in V\}$. Each pair $\{x_i, y_i\}$ is evidently in exactly two triples. Considering a pair $\{x_i, y_{i+1}\}$ with $x \neq y$, let us suppose that $x \oplus z = y$, and that $w \oplus x = y$. Then $\{x_i, y_{i+1}\}$ appears in the triples $\{x_i, z_i, y_{i+1}\}$ and $\{w_i, x_i, y_{i+1}\}$.

Lemma 2.4 *For all $m \geq 3$, there exists a TTS$(3m)$ and a TTS$(3m+1)$.*

Proof There exists an idempotent quasigroup for all $m \geq 3$ (Theorem 1.2). Apply the noncommutative $3v$ construction to this quasigroup to obtain a $\{3\}$-GDD of index 2 and type 3^m. To obtain a TTS$(3m)$, take every group as a triple twice. To obtain instead a TTS$(3m+1)$, add a new element ∞, and the triples $\{\{\infty, x_i, x_j\} : 0 \leq i < j < 3\}$. Finally, include each group as a triple once. □

Commutative $3v+1$ construction
Let $Q = (V, \odot)$ be a commutative half-idempotent quasigroup of (even) order v. On $(V \times \{0,1,2\}) \cup \{\infty\}$, form triples $\mathcal{B}$ to be

1. $\{\{x_i, y_i, (x \odot y)_{i+1}\} : x, y \in V, x < y, i \in \mathbb{Z}_3\}$;
2. $\{\{\infty, x_i, (x \odot x)_{i+1}\} : x \in V, x \odot x \neq x, i \in \mathbb{Z}_3\}$; and
3. $\{\{x_0, x_1, x_2\} : x \in V, x \odot x = x\}$.

Then $((V \times \{0,1,2\}) \cup \{\infty\}, \mathcal{B})$ is an STS$(3v+1)$.

Lemma 2.5 *For all $m \geq 1$, there exists an STS$(6m+1)$.*

Proof By Theorem 1.4 there exists a commutative half-idempotent quasigroup of order $2m$ for all $m \geq 1$. Apply the commutative $3v+1$ construction to this quasigroup. □

The Lindner $3v+2$ construction
Let $Q = (V, \oplus)$ be an idempotent commutative quasigroup of order v, and let $V = \{1, \ldots, v\}$. Let α be the permutation of $V \setminus \{1\}$ mapping i to $i+1$ for $2 \leq i < v$, and mapping v to 2. Since v is necessarily odd, α is of even length. Define a set of triples $\mathcal{B}$ on $V \times \{0,1,2\}$ by

$$\mathcal{B} = \{\{x_i, y_i, (x \oplus y)_{i+1} : x, y \in V, x \neq y, i \in \{0,1\}\} \;\cup$$

$$\{\{x_i, y_i, ((x \oplus y)\alpha)_{i+1}\} : x, y \in V, i = 2, x \neq y\}.$$

Let further

$$\mathcal{C} = \{\{\infty_1, (2t)_0, (2t)_1\}, \{\infty_1, (2t+1)_1, (2t+1)_2\}, \{\infty_1, (2t)_2, (2t+1)_0\},$$

$$\{\infty_2, (2t)_1, (2t)_2\}, \{\infty_2, (2t+1)_0, (2t+1)_1\} : 1 \leq t \leq (v-1)/2\} \;\cup$$
$$\{\{\infty_2, (2t+1)_2, (2t+2)_0\} : 1 \leq t \leq (v-3)/2\} \cup \{\{\infty_2, v_2, 2_0\}\},$$

where ∞_1 and ∞_2 are two new elements.

Lemma 2.6 *For all $m \geq 1$, there exists an ITS$(6m+5, 5)$ and a TS$(6m+5, 3)$.*

Proof Let $v = 2m+1$. The union of $\mathcal{B}$ and $\mathcal{C}$ given above is an ITS$(6m+5, 5)$. Now include every triple of $\mathcal{B} \cup \mathcal{C}$ three times. Add to this the triples of a TS$(5,3)$ on $\{\infty_1, \infty_2, 1_0, 1_1, 1_2\}$. The result is a TS$(6m+5, 3)$. □

The general $3v+2$ construction
Let $Q = (V, \otimes)$ be an idempotent quasigroup of order v. Define a set $\mathcal{B}$ of triples on $V \times \{0,1,2\}$ as in the noncommutative $3v$ construction, to obtain a $\{3\}$-GDD of type 3^v.

Lemma 2.7 *For all $m \geq 3$, there exists a TS$(3m+2, 6)$.*

Proof Let $a \in V$ be a fixed element, and let ∞_1 and ∞_2 be two new elements. To obtain the TS$(3m+2, 6)$, include the triples of $\mathcal{B}$ (with $v = m$) three times each, $\{\{\infty_\ell, x_i, x_j\} : 0 \leq i < j < 3, \ell \in \{1,2\}, x \in V \setminus \{a\}\}$ three times each, and the triples $\{\{\infty_1, \infty_2, a_i\} : i \in \{0,1,2\}\}$ and $\{a_1, a_2, a_3\}$ twice each. □

2.1.3 *Idempotent latin squares*

Let $Q = (V, \otimes)$ be a commutative quasigroup of (odd) order v. A *perpendicular array* of *strength* three is a $\binom{v+1}{2} \times 3$ array L, whose rows are the ordered triples $(x, y, x \otimes y)$ for $x, y \in V$, $x \leq y$. When Q is idempotent, the perpendicular array contains v rows of the form (x,x,x). Deleting all such rows, and dropping the ordering on the rows, we obtain a collection of triples. Evidently, every pair of elements of V occurs exactly three times in this collection of triples, arising from the rows $(x, y, x \otimes y)$, (x, w, y) when $x \otimes w = w \otimes x = y$, and (z, y, x) when $z \otimes y = y \otimes z = x$. Thus using Lemma 1.3, we have:

Lemma 2.8 *Every commutative idempotent quasigroup of order v underlies a TS$(v,3)$. Hence a TS$(v,3)$ exists for all odd v.*

Instead consider an idempotent quasigroup $(V, \oplus)$ of order v, and define the orthogonal array OA$(3, v)$ associated with the quasigroup. Again striking out the v rows of the form (x,x,x), and treating the rows as unordered, we have a triple system. In this case, however, every pair occurs precisely six times. Thus using Lemma 1.2, we have:

Lemma 2.9 *Every idempotent quasigroup of order v underlies a TS$(v,6)$. Hence a TS$(v,6)$ exists for all v except for $v = 2$.*

2.1.4 *A complete existence proof for triple systems*

These constructions underlie a proof of the sufficiency of the basic necessary conditions.

Theorem 2.10 *A TS(v, λ) exists if and only if $\lambda \equiv 0 \pmod{\gcd(v-2, 6)}$ and $v \neq 2$.*

Proof Necessity is from Lemma 2.1. For sufficiency, write $\lambda_{\min} = \gcd(v-2, 6)$, and $\ell = \lambda/\lambda_{\min}$. We first form a TS$(v, \lambda_{\min})$:

$v \pmod 6$	Construction
0,4	Lemma 2.4 if $v \geq 10$
	Table 2.1 for $v \in \{4, 6\}$
1	Lemma 2.5
2	Lemma 2.9
3	Lemma 2.3
5	Lemma 2.8

Replicate the triples of this $\mathrm{TS}(v, \lambda_{\min})$ ℓ times each to obtain a $\mathrm{TS}(v, \lambda)$. □

We could legitimately treat the existence problem as solved, and ask whether further constructions are of much interest. However, each construction lends itself to the construction of triple systems with additional properties.

2.2 Primes and prime powers

Finite fields have been used extensively in the construction of designs. We examine their application in the direct construction of triple systems. Netto (1893) established the basic construction.

Theorem 2.11 *Let p be a prime, $n \geq 1$, and $p^n \equiv 1 \pmod 6$. Let $\mathbb{F}_{p^n}$ be a finite field on a set X of size $p^n = 6t+1$ with 0 as its zero element, and ω a primitive root of unity. Then*

$$\{\{\omega^i + j, \omega^{2t+i} + j, \omega^{4t+i} + j\} : 0 \leq i < t, j \in X\}$$

(with computations in $\mathbb{F}_{p^n}$) is the set of blocks of an STS(p^n) on X.

Proof Consider two distinct elements $x, y \in \mathbb{F}_{p^n}$. Let $d = x - y$ (arithmetic in $\mathbb{F}_{p^n}$). Now since $d \neq 0$ and $\omega^{2t} - 1 \neq 0$, d can be uniquely written in the form $(\omega^{2t} - 1)\omega^{2jt}\omega^i$ for $j \in \{0,1,2\}$ and $0 \leq i < 2t$. Since $\omega^{3t} = -1$, if $i \geq t$, we may write

$$-d = y - x = (\omega^{2t} - 1)\omega^{2(j+1)t}\omega^{i-t}.$$

Thus we suppose without loss of generality that $d = x - y$ and $i < t$. Then $\{x, y\}$ appears in the triple $\{\omega^i, \omega^{2t+i}, \omega^{4t+i}\} + (x - \omega^{2(j+1)t+i})$. Consequently, every pair of distinct elements in $\mathbb{F}_{p^n}$ appears in at least one of the triples defined. Since the total number of pairs in the triples defined is precisely $\binom{p^n}{2}$, every pair occurs in exactly one triple. □

A variant of this construction is the one that is more commonly associated with Netto (perhaps inaccurately; Dembowski (1968) appears to have attached Netto's name to these systems):

Theorem 2.12 *Let p be a prime, $n \geq 1$, and $p^n \equiv 7 \pmod{12}$. Let $\mathbb{F}_{p^n}$ be a finite field on a set X of size $p^n = 6t+1 = 12s+7$ with 0 as its zero element and ω a primitive root of unity. Then*

$$\{\{\omega^{2i} + j, \omega^{2t+2i} + j, \omega^{4t+2i} + j\} : 0 \leq i < t, j \in X\}$$

forms the blocks of an STS(p^n) on X.

The triple systems from Theorem 2.12 are the *Netto triple systems.*

Netto (1893) also gave a construction for orders of the form $3p$ when $p \equiv 5 \pmod 6$ is prime; in this case, $\mathbb{F}_p$ can be taken to be the integers modulo p, $\mathbb{Z}_p$. This was generalized by Heffter (1897):

Theorem 2.13 *Let $p = 2t + 1$ be an odd prime. Let ω be a primitive root of unity in $\mathbb{Z}_p$ satisfying $\omega \equiv 1 \pmod 3$. Then*

$$\{\{\omega^i + j, \omega^{2t+i} + j, \omega^{4t+i} + j\} : 0 \leq i < t, j \in \mathbb{Z}_{3p}\}$$

(with computations modulo $3p$) is the set of blocks of $\{3\}$-GDD of type 3^p with groups $\{\{j, j+p, j+2p\} : j \in \mathbb{Z}_p\}$.

Filling groups (see §3.4) yields an STS($3p$).

2.3 Projective and affine spaces

In this section, we use two specific finite fields, $\mathbb{F}_2$ and $\mathbb{F}_3$, to define some infinite classes of triple systems.

Let W_n be an $(n + 1)$-dimensional vector space over $\mathbb{F}_2$. A *punctured* subspace of W_n is obtained from a subspace by removing the zero vector. The *n-dimensional projective space* PG$(n, 2)$ is the set of all punctured subspaces of W_n, ordered by inclusion. Punctured subspaces of dimension one are *points*, and those of dimension two are *lines*.

Theorem 2.14 *The points and lines of PG$(n, 2)$ form the elements and triples of an STS$(2^{n+1} - 1)$.*

Proof Let $\oplus$ be the operation of vector addition on W_n. For any two nonzero $(n+1)$-vectors x and y, we observe that $\{0, x, y, x \oplus y\}$ is a subspace of dimension two. Thus every pair of nonzero vectors defines a single punctured subspace containing three vectors. Taking nonzero vectors as elements, and punctured subspaces of dimension two as triples, gives an STS$(2^{n+1} - 1)$. □

Often the STS$(2^{n+1} - 1)$ produced in Theorem 2.14 is itself denoted by PG$(n, 2)$ and called a *projective space.*

In the same manner, when V_n is an n-dimensional vector space over $\mathbb{F}_3$, the *n-dimensional affine space* AG$(n, 3)$ is the set of all subspaces of V_n *and their cosets*, ordered by set inclusion. Points are vectors, and lines are translates of 1-dimensional subspaces.

Theorem 2.15 *The points and lines of AG$(n, 3)$ form the elements and triples of an STS(3^n).*

Proof For two vectors $\mathbf{x}$ and $\mathbf{y}$, a line is formed by $\{\mathbf{x}, \mathbf{y}, \mathbf{x}+\mathbf{y}\}$, where addition is performed modulo 3 componentwise. This line is a translate of the subspace $\{0, \mathbf{y} - \mathbf{x}, \mathbf{y}\}$. Evidently every pair of distinct points defines a unique line in this way. □

Again, we often denote the STS(3^n) as AG$(n, 3)$, and call it an *affine space* or *affine triple system.*

A *boolean 1-factorization* is a 1-factorization on 2^n vertices, obtained by taking vertices to be binary vectors of length n, and a 1-factor is obtained for each nonzero vector by including the edges for which the modulo 2 sum of the end-vertices is the specified vector. This is just the affine space AG$(n, 2)$.

The boolean 1-factorization on 2^n vertices is obtained by applying Lemma 1.20 to the boolean 1-factorization on 2^{n-1} vertices. More importantly, we find:

Lemma 2.16 *A 1-factorization is the Steiner 1-factorization of a projective triple system if and only if it is boolean.*

Proof The equivalence between the Steiner 1-factorization from PG$(n, 2)$ and the boolean 1-factorization AG$(n, 2)$ follows from the definitions using binary vectors of length n. □

2.4 Peltesohn's construction

We describe here a particularly simple direct construction. A *difference triple* modulo v is a triple in which either two numbers sum to the third or all three numbers sum to v. Heffter (1897) observes that a partition of the set $\{1, \dots, 3m\}$ into a collection HDP$_1(m)$ of m difference triples enables us to form an STS$(6m+1)$; forming such a partition is *Heffter's first difference problem.* Let us see how to form the triple system.

Let $\{\{a_i, b_i, c_i\} : 1 \leq i \leq m\}$ be an HDP$_1(m)$. Then form a collection $\mathcal{S}$ of m triples $\{\{0, a_i, a_i + b_i\} : 1 \leq i \leq m\}$. In $\mathcal{S}$, computing every *difference* between elements within a triple, we find every nonzero number in $\mathbb{Z}_{6m+1}$ exactly once. Thus if we take

$$\mathcal{B} = \{\{j, a_i + j, a_i + b_i + j\} : 1 \leq i \leq m, j \in \mathbb{Z}_{6m+1}\},$$

we obtain the triples of an STS$(6m + 1)$. We see this idea many times, so it warrants some explanation here. Consider any pair $\{x, y\}$ with $x, y \in \mathbb{Z}_{6m+1}$, $x \neq y$. Compute $d = y - x$ modulo $6m + 1$; without loss of generality, we may assume that $1 \leq d \leq 3m$, for if this does not happen, we may interchange x and y. Now d appears as a difference precisely once in the HDP$_1(m)$ used. If $d = a_i$, $\{x, y\}$ appears in the triple $\{x, a_i + x = y, a_i + b_i + x\}$. If $d = b_i$, it appears in the triple $\{-a_i, x, b_i + x = y\}$. Finally, if $d = c_i$, then if $a_i + b_i = c_i$, $\{x, y\}$ appears in the triple $\{x, a_i + x, a_i + b_i + x = y\}$; and if $a_i + b_i + c_i = 6m + 1$, it appears in $\{x - a_i - b_i = y, x - b_i, x\}$. We examine less tedious ways to establish this when we examine automorphisms of triple systems.

Heffter (1897) also observes that if we can partition $\{1, \dots, 3m+1\} \setminus \{2m+1\}$ into m difference triples modulo $6m + 3$, we obtain an STS$(6m + 3)$ by the same transformation as above, and in addition including blocks $\{j, 2m+1+j, 4m+2+j\}$ for $0 \leq j \leq 2m$. This partitioning problem is *Heffter's second difference problem,* and a solution is denoted by HDP$_2(m)$.

$m = 3s,\ s \geq 2$:

$(3r+1, 4s-r+1, 4s+2r+2)$	$r = 0, \ldots, s-1$
$(3r+2, 8s-r, 8s+2r+2)$	$r = 0, \ldots, s-1$
$(3r+3, 6s-2r-1, 6s+r+2)$	$r = 0, \ldots, s-2$
$(3s, 3s+1, 6s+1)$	

$m = 3$:

(1,5,6), (2,8,9), (3,4,7)

$m = 3s+1,\ s \geq 1$:

$(3r+1, 8s-r+3, 8s+2r+4)$	$r = 0, \ldots, s-1$
$(3r+2, 6s-2r+1, 6s+r+3)$	$r = 0, \ldots, s-1$
$(3r+3, 4s-r+1, 4s+2r+4)$	$r = 0, \ldots, s-1$
$(3s+1, 4s+2, 7s+3)$	

$m = 1$:

(1,2,3)

$m = 3s+2,\ s \geq 1$:

$(3r+2, 6s-2r+3, 6s+r+5)$	$r = 0, \ldots, s-1$
$(3r+3, 8s-r+5, 8s+2r+8)$	$r = 0, \ldots, s-1$
$(3r+1, 4s-r+3, 4s+2r+4)$	$r = 0, \ldots, s$
$(3s+2, 7s+5, 8s+6)$	

$m = 2$:

(1,3,4), (2,5,6)

Table 2.3 *Solutions to Heffter's first difference problem*

We now prove two remarkable theorems due to Peltesohn (1939), using simplified proofs from M. Colbourn and Mathon (1980).

Theorem 2.17 *For all $m \geq 1$, there exists an HDP$_1(m)$ and an STS$(6m+1)$.*

Proof Table 2.3 shows the difference triples used in each of three cases, $m \equiv 0, 1, 2 \pmod 3$. □

Theorem 2.18 *For all $m \geq 2$, there exists an HDP$_2(m)$ and an STS$(6m+3)$. There is no HDP$_2(1)$.*

Proof The impossibility for $m = 1$ is straightforward: there is no difference triple modulo 9 in $\{1, 2, 4\}$. For the remaining cases, Table 2.4 gives the solutions explicitly. □

These constructions are by no means the only ones of their type that produce solutions to the Heffter difference problems. Peltesohn's constructions have the remarkable appeal that they construct a Steiner triple system of every order

$m = 3s,\ s \geq 1$:

$(3r+1, 8s-r+1, 8s+2r+2)$	$r = 0, \ldots, s-1$
$(3r+2, 4s-r, 4s+2r+2)$	$r = 0, \ldots, s-1$
$(3r+3, 6s-2r-1, 6s+2r+2)$	$r = 0, \ldots, s-1$

$m = 3s+1,\ s \geq 4$:

$(3r+1, 4s-r+3, 4s+2r+4)$	$r = 0, \ldots, s$
$(3r+2, 8s-r+2, 8s+2r+4)$	$r = 2, \ldots, s-2$
$(3r+3, 6s-2r+1, 6s+r+4)$	$r = 1, \ldots, s-2$

$(2, 8s+3, 8s+5)$, $(3, 8s+1, 8s+4)$, $(5, 8s+2, 8s+7)$,
$(3s-1, 3s+2, 6s+1)$, $(3s, 7s+3, 8s+6)$

$m = 4$:

(1,12,13), (2,5,7), (3,8,11), (4,6,10)

$m = 7$:

(1,11,12), (2,17,19), (3,20,22), (4,10,14),
(5,8,13), (6,18,21), (7,9,16)

$m = 10$:

(1,15,16), (2,27,29), (3,25,28), (4,14,18), (5,26,31)
(6,17,23), (7,13,20), (8,11,19), (9,24,30), (10,12,22)

$m = 3s+2,\ s \geq 1$:

$(3r+1, 4s-r+3, 4s+2r+4)$	$r = 0, \ldots, s$
$(3r+2, 8s-r+6, 8s+2r+8)$	$r = 0, \ldots, s$
$(3r+3, 6s-2r+3, 6s+r+6)$	$r = 0, \ldots, s-1$

$m = 2$:

(1,3,4), (2,6,7)

Table 2.4 *Solutions to Heffter's second difference problem*

except for $v = 9$ using a simple direct pattern. As we see next, the pattern can be simpler yet.

2.5 Constructions from integer sequences

Skolem (1957, 1958) studied a restricted version of Heffter's first difference problem: when does a partition of the set $\{1, \ldots, 2n\}$ into ordered pairs $\{(a_r, b_r) : r = 1, \ldots, n \text{ and } b_r - a_r = r\}$ exist? Such a partition is a *pure Skolem sequence* of order n. If such a sequence exists, then $\{(r, n+a_r, n+b_r) : r = 1, \ldots, n\}$ is an $\mathrm{HDP}_1(n)$.

Skolem (1957) proved:

$n = 4s$:

$(4s + r - 1, 8s - r + 1)$	$r = 1, \ldots, 2s$
$(r, 4s - r - 1)$	$r = 1, \ldots, s - 2$
$(s + r + 1, 3s - r)$	$r = 1, \ldots, s - 2$
$(s - 1, 3s)$, $(s, s + 1)$, $(2s, 4s - 1)$, $(2s + 1, 6s)$	

$n = 4s + 1$:

$(4s + r + 1, 8s - r + 3)$	$r = 1, \ldots, 2s$
$(r, 4s - r + 1)$	$r = 1, \ldots, s$
$(s + r + 2, 3s - r + 1)$	$r = 1, \ldots, s - 2$
$(s + 1, s + 2)$, $(2s + 1, 6s + 2)$, $(2s + 2, 4s + 1)$	

Table 2.5 *Pure Skolem sequences*

$n = 4s + 2$:

$(r, 4s - r + 2)$	$r = 1, \ldots, 2s$
$(4s + r + 3, 8s - r + 4)$	$r = 1, \ldots, s - 1$
$(5s + r + 2, 7s - r + 3)$	$r = 1, \ldots, s - 1$
$(2s + 1, 6s + 2)$, $(4s + 2, 6s + 3)$, $(4s + 3, 8s + 5)$, $(7s + 3, 7s + 4)$	

$n = 4s - 1$:

$(4s + r, 8s - r - 2)$	$r = 1, \ldots, 2s - 2$
$(r, 4s - r - 1)$	$r = 1, \ldots, s - 2$
$(s + r + 1, 3s - r)$	$r = 1, \ldots, s - 2$
$(s - 1, 3s)$, $(s, s + 1)$, $(2s, 4s - 1)$, $(2s + 1, 6s - 1)$, $(4s, 8s - 1)$	

Table 2.6 *Hooked Skolem sequences*

Theorem 2.19 *A pure Skolem sequence of order n exists if and only if $n \equiv 0, 1$* (mod 4).

Proof For necessity, observe that when $b_r - a_r$ is even, b_r and a_r are of the same parity; and when $b_r - a_r$ is odd, they have opposite parity. Consequently, the number of even integers less than or equal to n must itself be even; hence $n \equiv 0, 1 \pmod 4$.

For sufficiency, Table 2.5 presents constructions. □

Skolem (1958) also observed that such a partition of $\{1, \ldots, 2n + 1\} \setminus \{2n\}$ would give a solution to $\mathrm{HDP}_1(n)$. Such a partition is a *hooked Skolem sequence.* O'Keefe (1961) proved:

Theorem 2.20 *There exists a hooked Skolem sequence of order n if and only if* $n \equiv 2, 3 \pmod 4$.

Proof For sufficiency, Table 2.6 gives constructions. □

$n = 4s$:

$(r, 4s - r + 1)$	$r = 1, \ldots, s - 1$
$(s + r - 1, 3s - r)$	$r = 1, \ldots, s - 1$
$(4s + r + 1, 8s - r + 1)$	$r = 1, \ldots, s - 1$
$(5s + r + 1, 7s - r + 1)$	$r = 1, \ldots, s - 1$
$(2s - 1, 2s)$, $(3s, 5s + 1)$, $(3s + 1, 7s + 1)$, $(6s + 1, 8s + 1)$	

$n = 4s - 1$:

$(r, 4s - r)$	$r = 1, \ldots, 2s - 1$
$(4s + r + 1, 8s - r)$	$r = 1, \ldots, s - 2$
$(5s + r, 7s - r - 1)$	$r = 1, \ldots, s - 2$
$(2s, 6s - 1)$, $(5s, 7s + 1)$, $(4s + 1, 6s)$, $(7s - 1, 7s)$	

Table 2.7 *Split Skolem sequences*

Pure and hooked Skolem sequences thus yield a complete solution to Heffter's first difference problem, and thus a complete existence proof for Steiner triple systems of orders congruent to 1 modulo 6. Analogous constructions for Heffter's second difference problem were obtained by Rosa (1966). A *split Skolem sequence* of order n is a partition of $\{1, \ldots, 2n+1\} \setminus \{n+1\}$ into ordered pairs $\{(a_r, b_r) : r = 1, \ldots, n$ and $b_r - a_r = r\}$. Given such a split Skolem sequence $\{\{r, n+a_r, n+b_r\} : r = 1, \ldots, n\}$ is an $\mathrm{HDP}_2(n)$.

Rosa (1966) proved:

Theorem 2.21 *A split Skolem sequence of order n exists if and only if $n \equiv 0, 3 \pmod 4$.*

Proof For sufficiency, Table 2.7 gives constructions. □

For the remaining case, a *split hooked Skolem sequence* is a similar partition of $\{1, \ldots, 2n + 2\} \setminus \{n + 1, 2n + 1\}$; again this leads to an $\mathrm{HDP}_2(n)$.

Theorem 2.22 *A split hooked Skolem sequence of order n exists if and only if $n \equiv 1, 2 \pmod 4$ and $n \neq 1$.*

Proof For sufficiency, Table 2.8 gives constructions. □

This completes an alternate proof of the existence of solutions to Heffter's second difference problem for all $n > 1$, and hence for the existence of STS$(6n + 3)$s.

2.6 The Schreiber–Wilson construction

Here we describe a construction from Schreiber (1973) and Wilson (1974a). It applies when, for every prime divisor p of $n = v - 2$, the order of $-2 \pmod p$ is singly even. To form an STS(v), write $n = v - 2$ and choose an abelian group Γ of order n with binary operation $\oplus$ and identity 0. Since $n \equiv 1, 5 \pmod 6$, Γ has no elements of order two or three. Now consider all triples of group elements

$n = 4s+1,\ s \geq 2$:

$(r, 4s-r+2)$	$r = 1, \ldots, 2s$
$(5s+r, 7s-r+3)$	$r = 1, \ldots, s$
$(4s+r+2, 8s-r+3)$	$r = 1, \ldots, s-2$
$(2s+1, 6s+2)$, $(6s+1, 8s+4)$, $(7s+3, 7s+4)$	

$n = 5$:

(3,4), (8,10), (9,12), (1,5), (2,7)

$n = 4s+2,\ s \geq 1$:

$(r, 4s-r+3)$	$r = 1, \ldots, 2s$
$(4s+r+4, 8s-r+4)$	$r = 1, \ldots, s-1$
$(5s+r+3, 7s-r+3)$	$r = 1, \ldots, s-2$
$(2s+1, 6s+3)$, $(2s+2, 6s+2)$, $(4s+4, 6s+4)$, $(7s+3, 7s+4)$, $(8s+4, 8s+6)$	

$n = 2$:

(1,2), (4,6)

Table 2.8 *Hooked split Skolem sequences*

$\{a, b, c\}$ with $a \oplus b \oplus c = 0$. Let $\mathcal{D}$ be the set of such triples in which all three group elements are distinct, and let $\mathcal{T}$ be those in which there are two distinct elements. Finally, let $\mathcal{S}$ be those containing a single element; in Γ, $a \oplus a \oplus a = 0$ has the unique solution $a = 0$, since Γ has no elements of order three. Thus $\mathcal{S}$ contains the single triple $\{0, 0, 0\}$.

To form a triple system $(\Gamma \cup \{\alpha, \beta\}, \mathcal{B})$:

1. Include all triples of $\mathcal{D}$ in $\mathcal{B}$.
2. First replace every unordered pair $\{a, b\}$ by the ordered pair (a, b) if $a \oplus a \oplus b = 0$, and by (b, a) otherwise. Partition the pairs of $\mathcal{T}$ into orbits under the mapping $\phi : z \mapsto z'$ where $z \oplus z' \oplus z' = 0$ (if $(a, b) \in \mathcal{T}$, $a = \phi(b)$). Now consider an orbit of pairs of $\mathcal{T}$ under ϕ. Under the stated restriction that for every prime divisor p of n, the order of $-2 \pmod{p}$ is singly even, each orbit of elements from $\mathcal{T}$ is of even length. Suppose that the orbit has length $2k$, and that the pair (a, b) is in the orbit, then add to $\mathcal{B}$ the triples $\{\{\alpha, \phi^{2i}(a), \phi^{2i+1}(a)\}, \{\beta, \phi^{2i+1}(a), \phi^{2i+2}(a)\}\}$ for $0 \leq i < k$.
3. Include the triple $\{\alpha, \beta, 0\}$ in $\mathcal{B}$.

2.7 Computational methods

Often we require a triple system or $\{3\}$-GDD with additional properties, but no simple direct or recursive construction for the design is known. In such cases, we often resort to computational methods. We outline the two main techniques employed.

2.7.1 *Backtracking*

Backtracking is an exhaustive method that can be used in principle to generate all solutions to a problem. Let us consider as a generic problem the task of partitioning the edges of a multigraph into triangles. Let $G = (V, E)$ be an m-edge, n-vertex multigraph. Evidently, if $m \not\equiv 0 \pmod 3$ or G has an odd degree vertex, then G has no edge-partition into triangles, and we can report a negative answer.

Otherwise, let us suppose that the vertices of G are totally ordered: $v_1 \prec v_2 \prec \cdots \prec v_n$. This vertex ordering induces an ordering on sets of vertices; we use reverse lexicographic order, namely for $\{v_{i_1}, \dots, v_{i_k}\}$ $(i_\ell < i_{\ell+1})$ precedes $\{v_{j_1}, \dots, v_{j_k}\}$ $(j_\ell < j_{\ell+1})$ whenever there exists an x, $1 \leq x \leq k$, for which $v_{i_\ell} = v_{j_\ell}$ for all $x < \ell \leq k$ and $v_{i_x} \prec v_{j_x}$.

The algorithm then proceeds as follows. At any moment, some triangles have been chosen; initially this set is empty. We *advance* by finding the first edge (under the ordering) that does not yet appear in a triangle; suppose it is $\{v_i, v_j\}$. If no such edge exists, we have found a solution, so we report it and then backtrack. If there is an unused edge, we form a set of *candidates* for this edge by finding all vertices v_k for which there is an unused edge $\{v_i, v_k\}$ and an unused edge $\{v_j, v_k\}$. When the previously chosen triple is $\{v_i, v_j, v_\ell\}$, all candidates v_k with $k < \ell$ are immediately removed from the candidate set.

We then *extend*: if there is no remaining candidate for the current pair, we backtrack. Otherwise, we choose the first remaining candidate v_k and remove it from the list of candidates. We add the triple $\{v_i, v_j, v_k\}$ and mark its edges as used. Then we advance again.

We *backtrack* as follows. If no triangles are currently chosen, we report the end of the search and stop. Otherwise we remove the last chosen triangle $\{v_i, v_j, v_k\}$ $(i < j < k)$ from the solution, marking its edges unused. We set the current pair to $\{v_i, v_j\}$ and extend. This has the effect of trying the next candidate for the first remaining pair.

This method, if run to completion, produces all edge-partitions of G into triangles. Moreover, it produces each edge-partition exactly once. To see this, observe that for any set of triangles there is a unique way to order each triangle $\{v_i, v_j, v_k\}$ so that $i < j < k$. Sorting the resulting triangles so that the first two positions are in reverse lexicographic order, and when equality in the first two positions occurs, the third is in nondecreasing order, gives a unique representative of the edge-partition. Only this unique representative is generated. This is a special case of the *orderly method* due to Read (1977).

The drawback of backtracking as a direct construction method is that it typically is computationally intensive. Unless some additional structure can be assumed, the number of partial edge-partitions examined is often prohibitive.

Mathon (1991, 1996) and Gibbons (1996) describe the method in more detail. Gibbons (1976) and Denny (1998) describe implementations of constructive techniques using backtracking. Moura (1999) develops a branch-and-cut technique for exhaustive generation.

2.7.2 *Hillclimbing*

Often we want to find a single solution, rather than to exhaust all solutions. In this case, Stinson (1985) has devised a remarkably effective method. Again we explain it in terms of edge-partition into triangles.

Let $\mathcal{B}$ be a collection of triangles that have been already chosen, and suppose that the edges in triangles of $\mathcal{B}$ have been marked as used. In the *hillclimbing* method, certain operations are performed repeatedly:

1. Choose an unused edge $\{v_i, v_j\}$; if none can be chosen, an edge-partition has been found, so stop.
2. Choose a second unused edge $\{v_i, v_k\}$; if none can be chosen, report failure and stop.
3. Consider the edges (if any) between v_j and v_k.
   - (a) If there is an unused edge $\{v_j, v_k\}$, add triangle $\{v_i, v_j, v_k\}$ to $\mathcal{B}$, and mark the three edges as used.
   - (b) Otherwise, if there is a used edge $\{v_j, v_k\}$, choose a triangle $\{v_j, v_k, v_\ell\}$ in $\mathcal{B}$, remove it and mark its three edges unused, add the triangle $\{v_i, v_j, v_k\}$ to $\mathcal{B}$, and mark its three edges used.
   - (c) Otherwise, if there is no edge $\{v_j, v_k\}$ at all, do nothing.

For every pair chosen in step (1), we find that if no edge can be chosen in step (2), then vertex v_i has odd degree. For in the set $\mathcal{B}$, the number of edges used at v_i must be even, and thus if only one unused edge remains, the total number of edges incident at v_i must be odd.

Now in step (3), three outcomes are possible. Either a new triangle is added (3a), an old triangle is replaced by a new one (3b), or no change occurs. Thus as the algorithm proceeds, the number of triangles chosen *never decreases*, and hence the term hillclimbing. In the algorithm described, no method is yet specified for making the choices in steps (1), (2), and (3b). In each case, the choice is made *randomly*.

Stinson (1985) observes that the method can fail to terminate if implemented as above. We give a simple example of this. Suppose that in partitioning K_{15}, we arrive at a configuration where the 25 triples of a TD(3, 5) have been chosen. Six further triples can be added, two inside each group of size five. However, at that point, step (3b) will be repeatedly executed! No triples of the TD(3, 5) are ever removed, and there is no way to obtain a set of more than 31 triples containing these 25 on the 15 points. We leave the verification of this to the reader.

Thus Stinson (1985) suggests choosing two *iteration thresholds*, ℓ and s. We attempt to construct the edge-partition up to ℓ times, each time starting from the empty set of triangles. For each attempt, we maintain a count of the number of times step (1) is executed *without* an increase in the number of triples. If this count ever exceeds s, we abandon the current attempt in failure.

In this manner, the algorithm may terminate in failure although an edge-partition into triangles exists. However, in a wide variety of applications this appears not to occur provided the graph to be partitioned is in some sense 'dense'. For example, Colbourn and Mathon (1987) report on a method of this type for edge-partitioning graphs whose complements are triangle-free.

Other 'local optimization' techniques for generating designs, such as simulated annealing and genetic algorithms, are discussed in Gibbons (1996).

3

EXISTENCE: RECURSIVE METHODS

3.1 Direct and indirect products

Generalized direct product
Let $(V, \mathcal{B})$ be a TS(v, λ), and let $(W, \mathcal{D})$ be a TS(w, μ). Let $\mathcal{D} = \{D_1, \ldots D_b\}$, and form b quasigroups on V, $Q_1 = (V, \oplus_1), \ldots, Q_b = (V, \oplus_b)$.

We form a TS$(vw, \lambda\mu)$ on $V \times W$, with blocks:

1. for each $\{x, y, z\} \in \mathcal{B}$ and each $i \in W$, let $\{x_i, y_i, w_i\}$ be a triple μ times;
2. for each $D_r = \{i, j, k\} \in \mathcal{D}$, $i < j < k$, and all elements $x, y \in V$ (not necessarily distinct), let $\{x_i, y_j, (x \oplus_r y)_k\}$ be a triple λ times.

Direct product
Suppose that $\lambda = 1$. In the generalized direct product, take $Q_1 = \cdots = Q_b$ to be the Steiner quasigroup of $(V, \mathcal{B})$.

Often, when referring to the direct product, we mean the generalized direct product. An important special case occurs when $|W| = 3$, and $\mu = 1$. This is a *tripling construction* that forms a TS$(3v, \lambda)$ from a TS(v, λ) and an arbitrary quasigroup of order v (or a latin square of order v, or a TD$(3, v)$).

Singular direct product
Let $(W, \mathcal{D})$ be a TS(w, μ) with triples $\mathcal{D} = \{D_1, \ldots, D_b\}$. Let $(V, \mathcal{B})$ be a TS(v, λ), containing a sub-TS(v', λ) $(V', \mathcal{B}')$, and let $\overline{V} = V \setminus V'$ and $\overline{\mathcal{B}} = \mathcal{B} \setminus \mathcal{B}'$. For each $D_r \in \mathcal{D}$, define a quasigroup $Q_r = (\overline{V}, \otimes_r)$ on $\overline{V}$. We define a TS$(w(v - v') + v', \mu\lambda)$ on $(W \times \overline{V}) \cup V'$ with the triples:

1. for each $\{a, b, c\} \in \mathcal{B}'$, include $\{a, b, c\}$ μ times;
2. for $w \in W$, $a \in V'$, $b, c \in \overline{V}$ and $\{a, b, c\} \in \mathcal{B}$, take $\{a, b_w, c_w\}$ as a triple μ times;
3. for $w \in W$, $a, b, c \in \overline{V}$ and $\{a, b, c\} \in \mathcal{B}$, take $\{a_w, b_w, c_w\}$ as a triple μ times;
4. for $\{i, j, k\} = D_r \in \mathcal{D}$, $i < j < k$, and $a, b \in \overline{V}$, take $\{a_i, b_j, (a \otimes_r b)_k\}$ as a triple λ times.

When the subsystem $(V', \mathcal{B}')$ is empty, this is the generalized direct product. An important special case also arises when $|W| = 3$, $|V'| = 1$, and $\mu = 1$: using a TS(v, λ) and an arbitrary quasigroup of order $v - 1$, we obtain a TS$(3v - 2, \lambda)$.

Indirect product
One potential drawback of the generalized direct product is its requirement for a TS(v, λ). This can be overcome to a large degree by the use of an 'indirect product'.

Let $(V, \mathcal{B})$ be an incomplete triple system ITS$(v, w; \lambda)$ with a hole on $W \subset V$, $|W| = w$. Let $(X, \mathcal{D})$ be a TS(x, μ), and suppose there is a TS$(wx, \lambda\mu)$. Let $\mathcal{D} = \{D_1, \ldots, D_b\}$, and define quasigroups $Q_i = (V, \oplus_i)$ for $1 \le i \le b$ so that each has a subquasigroup on W. Then a TS$(vx, \lambda\mu)$ on $V \times X$ has triples

1. for $\{a, b, c\} \in \mathcal{B}$ and $i \in X$, $\{a_i, b_i, c_i\}$ is a triple μ times;
2. for $a, b \in V$, $|\{a, b\} \cap W| \le 1$, $B_r = \{i, j, k\} \in \mathcal{D}$, $i < j < k$, $\{a_i, b_j, (a \oplus_r b)_k\}$ is a triple λ times; and
3. include all triples of a TS$(wx, \lambda\mu)$ on $W \times X$.

Taking $x = 3$ and using an ITS(11, 5), a quasigroup of order 11 having a subquasigroup of order 5, and an STS(15), for example, we obtain an STS(33).

Singular indirect product

In a similar way, the singular indirect product generalizes both the indirect product and the singular direct product.

Let $(V, \mathcal{B})$ be an incomplete triple system ITS$(v, w; \lambda)$ with a hole on $W \subset V$, $|W| = w$. Let $0 \le h \le w < v$. Let $H \subseteq W$, $|H| = h$, and define $\overline{V} = V \setminus H$. Let $(X, \mathcal{D})$ be a TS(x, μ), and suppose there is a TS$((w - h)x + h, \lambda\mu)$. Let $\mathcal{D} = \{D_1, \ldots, D_b\}$, and suppose there exist quasigroups $Q_i = (\overline{V}, \oplus_i)$ for $1 \le i \le b$ so that each has a subquasigroup on $W \setminus H$. Then a TS$((v - h)w + h, \lambda\mu)$ on $V \times X$ has triples

1. for $\{a, b, c\} \in \mathcal{B}$, $\{a, b, c\} \cap H = \varnothing$ and $i \in W$, $\{a_i, b_i, c_i\}$ is a triple μ times;
2. for $a \in H$, $b, c \in \overline{V}$, $\{a, b, c\} \in \mathcal{B}$ and $i \in W$, $\{a, b_i, c_i\}$ is a triple μ times;
3. for $a, b \in \overline{V}$, $|\{a, b\} \cap W| \le 1$, $B_r = \{i, j, k\} \in \mathcal{D}$, $i < j < k$, $\{a_i, b_j, (a \oplus_r b)_k\}$ is a triple λ times; and
4. include all triples of a TS$((w - h)x + h, \lambda\mu)$ on $((W \setminus H) \times X) \cup H$.

For example, taking an ITS(11, 5), a quasigroup of order 10 with a subquasigroup of order four, and an STS(13), we choose $v = 11$, $w = 5$, $h = 1$ and $x = 3$ to obtain an STS(31).

3.2 Doubling constructions

Let $(W, \mathcal{D})$ be a TS(w, λ), and $(V, \{\{0, 1, 2\}\})$ be an STS(3) having the subsystem $(\{0\}, \varnothing)$. Apply the singular direct product to produce a TS$(2w+1, \lambda)$. Theorem 0.1 is a special case of this.

In order to generalize, let us consider the structure of the resulting TS$(2w+1)$. In the singular direct product, we must choose a quasigroup $Q = (\{1, 2\}, \odot)$; without loss of generality, we may take $1 \odot 1 = 2 \odot 2 = 1$ and $1 \odot 2 = 2 \odot 1 = 2$. Then it is easy to check that the TS$(2w + 1, \lambda)$ on $\{0\} \cup (\{1, 2\} \times W)$ contains a copy of the original TS(w, λ) on the elements $\{1\} \times W$.

So let us suppose that a TS$(2v+1, \lambda)$ on V, $|V| = 2v+1$, contains the triples of a TS(v, λ) on $W \subset V$, $|W| = v$; such a contained triple system is a subsystem (see Chapter 6). Let us classify the triples of the larger system into four types:

call a triple *type* i if it contains exactly i elements from W, and let n_i be the number of triples of type i. Now there are $\lambda\binom{v}{2}$ pairs on W, and hence

$$3n_3 + n_2 = \lambda\binom{v}{2}. \tag{3.1}$$

By the same token,

$$3n_0 + n_1 = \lambda\binom{v+1}{2}. \tag{3.2}$$

In addition, since there are $\lambda v(v+1)$ pairs with one element in W and the other in $V \setminus W$, we have

$$2n_1 + 2n_2 = \lambda v(v+1). \tag{3.3}$$

These three equations in four unknowns yield

$$n_2 = \lambda\binom{v}{2} - 3n_3, n_1 = 3n_3 + \lambda v, n_0 = \lambda\binom{v}{2}/3 - n_3.$$

Now since a subsystem is present on W, $n_3 = \lambda\binom{v}{2}/3$ in our case. Thus $n_2 = 0, n_1 = \lambda\binom{v+1}{2}$, and $n_0 = 0$. Only two types of triples arise: those in the subsystem, and those meeting the subsystem in exactly one point!

Now consider all triples of the latter type that meet the subsystem in a fixed $w \in W$. Consider the pairs F_w obtained by deleting w from each such triple. Since w occurs with every element of $V \setminus W$ exactly λ times, and such occurrences are only in triples of the latter type, F_w is a λ-regular multigraph on $V \setminus W$. Still more is true. Since the $\mathrm{TS}(2v+1, \lambda)$ is pairwise balanced with index λ, every pair of elements from $V \setminus W$ has λ occurrences in $\{F_w : w \in W\}$; in other words, $\{F_w : w \in W\}$ is a λ-factorization of a λK_{v+1} whose vertex set is $V \setminus W$.

With these observations in hand, we can reverse the process to obtain a much simpler view of the $2v+1$ construction. For an element x and a collection $\mathcal{P}$ of pairs, define $x \cdot \mathcal{P}$ to be $\{\{x, y, z\} : \{y, z\} \in \mathcal{P}\}$.

The $2v+1$ *construction*
Let $(V, \mathcal{B})$ be a $\mathrm{TS}(v, \lambda)$, $V = \{1, \ldots, v\}$, and let $\{F_1, \ldots, F_v\}$ be a λ-factorization of λK_{v+1} on vertex set X, $X \cap V = \varnothing$. Then

$$\left(V \cup X, \mathcal{B} \cup \bigcup_{i \in V} i \cdot F_i\right)$$

is a $\mathrm{TS}(2v+1, \lambda)$ having a sub-$\mathrm{TS}(v, \lambda)$.

Now let us consider the structure in general of a $\mathrm{TS}(2v+s, \lambda)$ containing a sub-$\mathrm{TS}(v, \lambda)$. Defining n_i to be the number of triples containing i elements of the subsystem, we find:

$$3n_3 + n_2 = \lambda\binom{v}{2}$$

$$3n_0 + n_1 = \lambda\binom{v+s}{2}$$
$$2n_1 + 2n_2 = \lambda v(v+s)$$

Solving these we obtain

$$n_2 = \lambda\binom{v}{2} - 3n_3, n_1 = \frac{\lambda v(s+1)}{2} + 3n_3,$$
$$n_0 = \lambda\left[\frac{s-1}{2}\frac{v+s}{3} + \binom{v}{2}\right] - n_3$$

Since $n_3 = \lambda\binom{v}{2}$, we have

$$n_2 = 0, n_1 = \frac{\lambda v(v+s)}{3}, n_0 = \frac{\lambda(s-1)(v+s)}{6}.$$

The generic $2v+s$ *construction*

Let $(V, \mathcal{B})$ be a TS(v, λ) with $V = \{1, \ldots, v\}$. Let X be a $(v+s)$-set disjoint from V, and let G be a λK_{v+s} on X. Suppose that G has an edge-partition into a λv-regular multigraph $\widehat{G}$ and a $\lambda(s-1)$-regular subgraph $\widetilde{G}$ so that

1. $\widehat{G}$ has a λ-factorization $F_1, \ldots, F_v$; and
2. $\widetilde{G}$ has an edge-partition $\mathcal{D}$ into triangles.

Then

$$\mathcal{B} \cup \mathcal{D} \cup \bigcup_{w \in V} w \cdot F_w$$

is the set of triples of a TS$(2v+s, \lambda)$.

Any application of this construction requires that we choose a suitable partition into $\widehat{G}$ and $\widetilde{G}$. We give some examples next.

A $2v+3$ *construction*

Suppose that $v \equiv 0 \pmod 3$, $\lambda \geq 1$, and a TS(v, λ) exists. Then choose $X = \{x_0, \ldots, x_{v+2}\}$, and let $\widetilde{G}$ contain edges $\{\{x_i, x_{i+v/3}\} : i \in \mathbb{Z}_{v+3}\}$ λ times each. Now $\widetilde{G}$ has an edge-partition into triangles, namely $\{\{x_i, x_{i+v/3}, x_{i+2v/3}\} : 0 \leq i < v/3\}$ λ times each. The remaining graph $\widehat{G}$ is a λ-fold copy of a circulant graph, that contains the good difference 1. Thus the Stern–Lenz theorem (Theorem 1.16) ensures that $\widehat{G}$ has a λ-factorization. Applying the generic $2v+s$ construction, we obtain a TS$(2v+3, \lambda)$.

A $2v+4$ *construction*

Suppose that $\lambda \equiv 0 \pmod 2$, and a TS(v, λ) exists. Let $X = \{x_0, \ldots, x_{v+3}\}$, and let $\widetilde{G}$ contain all edges

$$\{\{x_i, x_j\} : i - j \equiv d \pmod{v+4}, d \in \{-3, -2, -1, 1, 2, 3\}\},$$

λ times each. Then $\widetilde{G}$ has an edge-partition into triangles: $\{\{x_i, x_{i+1}, x_{i+3}\} : i \in \mathbb{Z}_{v+4}\}$ and $\{\{x_i, x_{i+2}, x_{i+3}\} : i \in \mathbb{Z}_{v+4}\}$ $\lambda/2$ times each. The remaining graph

$\widehat{G}$ is λv-regular, and since λ is even, Petersen's theorem (Theorem 1.22) ensures that it has a λ-factorization. Apply the generic $2v+s$ construction to produce a TS$(2v+4,\lambda)$.

A $2v+7$ construction

There are many possible $2v+7$ constructions; we show one particularly interesting one, from Rosa (1975a). Suppose that a TS(v,λ) exists, and let $X = A \cup B$, $A = \{a_1,\ldots,a_v\}$, $B = \{b_1,\ldots,b_7\}$. Choose $\widetilde{G}$ to be a λ-fold copy of a K_7 on B together with the edges $\{\{a_i,a_j\} : i-j \equiv d \pmod v, d \in \{-3,-2,-1,1,2,3\}\}$. Then $\widetilde{G}$ has an edge partition into triangles, obtained by placing a TS$(7,\lambda)$ on B, and taking the triples $\{\{a_i,a_{i+1},a_{i+3}\} : i \in \mathbb{Z}_v\}$ λ times each. The real issue here is how to λ-factorize the remaining graph $\widehat{G}$. If λ is even, Petersen's theorem again provides the required factorization. So suppose that λ is odd, and hence that v is odd. Write $v = 2m+1$, and form either a Skolem sequence or a hooked Skolem sequence $\mathcal{S} = \{(p_r,q_r) : p_r - q_r = r, r = 1,\ldots,m\}$. Now define $Y = A \setminus \{a_i : i \in \{q_r,p_r\}, 4 \le r \le m\}$. Evidently $|Y| = 7$; let $Y = \{a_{j_i} : 1 \le i \le 7\}$. Form v edge-disjoint 1-factors of $\widehat{G}$, $H_1,\ldots,H_v$, as

$$H_k = \{\{a_{j_i+k-1}, b_i\} : 1 \le i \le 7\} \cup \{\{a_{p_r+k-1}, a_{q_r+k-1}\} : 4 \le r \le m\}.$$

Then $\{\lambda H_i : 1 \le i \le v\}$ is a λ-factorization of $\widehat{G}$. Apply the generic $2v+s$ construction to produce a TS$(2v+7,\lambda)$.

This $2v+7$ construction produces an STS$(2v+7)$ with not only a sub-TS(v,λ), but also with a disjoint sub-TS$(7,\lambda)$.

Of course, there is a wide variety of partitions into $\widehat{G}$ and $\widetilde{G}$ from which to choose in general, and for each the choice of edge-partition or factorization is typically quite flexible. Perhaps this flexibility accounts for the pervasiveness of doubling constructions in research on triple systems.

3.3 The $2v-2$ and $3v-4$ constructions

Here we describe some of the basic recursive constructions that increase the index.

The $2v-2$ construction

This construction produces a TS$(2v-2,2\lambda)$ from a TS(v,λ) and a λ-factorization of λK_{v-1}.

Let $X = \{x_1,\ldots,x_{v-2}\}$, $Y = \{y_1,\ldots,y_{v-2}\}$ and $V = \{\infty_1,\infty_2\} \cup X \cup Y$. Let $(X \cup \{\infty_1,\infty_2\},\mathcal{B}_1)$ and $(Y \cup \{\infty_1,\infty_2\},\mathcal{B}_2)$ be TS(v,λ). Form a λ-factorization $F_1,\ldots,F_{v-2}$ of λK_{v-1} on $Y \cup \{\infty_1\}$, and form a λ-factorization $H_1,\ldots,H_{v-2}$ of λK_{v-1} on $X \cup \{\infty_2\}$.

Then

$$\mathcal{B}_1 \cup \mathcal{B}_2 \cup \bigcup_{i=1}^{v-2} x_i \cdot F_i \cup \bigcup_{i=1}^{v-2} y_i \cdot H_i$$

forms the set of blocks of a TS$(2v-2,2\lambda)$.

The $3v-4$ construction
This construction produces a $\mathrm{TS}(3v-4,3\lambda)$ from a $\mathrm{TS}(v,\lambda)$ and a λ-factorization of λK_{v-1}.

Let $V=\{\infty_1,\infty_2\}\cup(X\times\{0,1,2\})$, $X=\{1,2,\dots,v-2\}$. For $i=0,1,2$ form a $\mathrm{TS}(v,\lambda)$ $(\{\infty_1,\infty_2\}\cup(X\times\{i\}),\mathcal{B}_i)$. Then form six λ-factorizations of λK_{v-1}: $F_{i,1},\dots,F_{i,v-2}$, $i\in\{0,1,2\}$ on $\{\infty_1\}\cup(X\times\{i\})$, and $H_{i,1},\dots,H_{i,v-2}$, $i\in\{0,1,2\}$ on $\{\infty_2\cup(X\times\{i\})$.

Then a $\mathrm{TS}(3v-4,3\lambda)$ exists on element set V with triples

$$\bigcup_{i\in\{0,1,2\}}\left[\mathcal{B}_i\cup\bigcup_{x\in X}x_i\cdot F_{i-1,x}\cup\bigcup_{x\in X}x_i\cdot H_{i+1,x}\right].$$

3.4 PBDs, GDDs, and PBD-closure

In this section, we examine methods to obtain triple systems from other designs. While our objective is simply to produce various systems with block size three, we treat the general case.

Filling in blocks
If there is a $(v,K;\lambda)$-PBD, and for every $k\in K$ there is a $(k,K';\mu)$-PBD, then there exists a $(v,K';\lambda\mu)$-PBD.
Let $(V,\mathcal{B})$ be a $(v,K;\lambda)$-PBD. For each block $B\in\mathcal{B}$, define a $(|B|,K';\mu)$-PBD $(B,\mathcal{D}_B)$. Then $(V,\bigcup_{B\in\mathcal{B}}\mathcal{D}_B)$ is a $(v,K';\lambda\mu)$-PBD.

PBD closure
Let $B(K,\lambda)$ be the set $\{v:\exists(v,K;\lambda)\text{-PBD}\}$. When $\lambda=1$, often one just writes $B(K)$. A set K of integers is *PBD-closed* when $K=B(K,1)$; the *PBD closure* of a set K is the smallest PBD-closed set $K'\supseteq K$ for which $B(K,1)=B(K',1)$. A *basis* of a PBD-closed set K' is a subset $K\subseteq K'$ for which $B(K,1)=B(K',1)$, and for no subset $J\subset K$ does $B(J,1)=B(K',1)$.

Wilson (1972) proved a remarkable theorem:

Theorem 3.1 *Let K be a PBD-closed set. Let $\alpha(K)=\gcd(\{k-1:k\in K\})$ and $\beta(K)=\gcd(\{k(k-1):k\in K\})$. Then there is a constant c depending only on $\alpha(K)$ and $\beta(K)$ for which if $v\geq c$, $v-1\equiv 0 \pmod{\alpha(K)}$ and $v(v-1)\equiv 0 \pmod{\beta(K)}$, then $v\in K$.*

Suppose that one considers a finite set J and its PBD closure K. It is easy to verify that $\alpha(J)=\alpha(K)$ and that $\beta(J)=\beta(K)$. Thus from Theorem 3.1, we obtain:

Theorem 3.2 *For any set J of integers, there is a fixed constant c for which if $v\geq c$, $v-1\equiv 0 \pmod{\alpha(J)}$ and $v(v-1)\equiv 0 \pmod{\beta(J)}$, then v is in the PBD closure of J.*

The application of Theorem 3.2 depends upon giving some combinatorial construction to show that a set is PBD-closed. Using the simple operation of filling in blocks, we see that the set $B(k,\lambda)=\{v:\exists(v,K;\lambda)\text{-PBD}\}$ is trivially PBD-closed.

Let us mention some easy examples. If $J = \{3\}$, $\alpha(J) = 2$ and $\beta(J) = 6$. Theorem 3.2 then asserts that the PBD closure of $\{3\}$ contains all sufficiently large integers congruent to 1 or 3 modulo 6, as we would expect. Theorem 3.2 alone does not determine $B(3,1)$ but rather ensures only asymptotic existence. Turning to $B(\{3\},\lambda)$, using only block size three in the PBD, we cannot produce a PBD of order other than 1 or 3 modulo 6 in which to fill blocks. If λ is even, however, we may take $J = \{3,4\}$, using a TS$(3,\lambda/2)$ or a TS$(4,\lambda/2)$ to fill blocks. In this case, $\alpha(J) = 1$ and $\beta(J) = 6$, whence Theorem 3.2 establishes that a TTS(v) exists for all sufficiently large $v \equiv 0,1 \pmod 3$.

Of course, Wilson's theorems on PBD closure applied to triple systems are, for the most part, machinery that is at once much too powerful for the asymptotics, and much too weak for the smaller cases.

Often it is useful to know closures for a specific set K of block sizes. Gronau, Mullin, and Pietsch (1995) determined exact closures for some sets containing the block size three. Their results are summarized in Table 3.1. Grüttmüller (1999, 1999a) and Rees (1999) address related results.

Since $7,9 \in B(\{3\})$, and $10 \in B(\{3,4\})$, the list given suffices to settle all cases when $\{3\} \subseteq K \subseteq \{3,4,5,6,7,8,9,10\}$. The exceptions listed are all proved to be definite exceptions. Bennett, Gronau, Ling, and Mullin (1996) list related results.

Filling in holes

If there exists a $(v,u,K;\lambda)$-IPBD and a $(u,K';\lambda)$-PBD, then there exists a $(v,K\cup K';\lambda)$-PBD. Simply place the blocks of the PBD on the hole of the IPBD.

Filling in groups

Let $(X,\mathcal{G},\mathcal{B})$ be a $(v,K;\lambda)$-GDD of type $g_1^{t_1},\cdots g_\ell^{t_\ell}$. Suppose that for $1 \le i \le \ell$, there exists a $(u+g_i,u,K_i;\lambda)$-IPBD. Then a $\left(u+\sum_{i=1}^{\ell} g_t t_i, u, K \cup \bigcup_{i=1}^{\ell} K_i;\lambda\right)$-IPBD also exists.

Let $G_1,\ldots,G_s$ be the groups of the GDD, and add u new points U. On $G_i \cup U$, place a copy of a $(u+|G_i|,u,K;\lambda)$-IPBD with the hole on U.

A special case occurs when $U = \varnothing$ and each $(0+g_i,0,K_i;\lambda)$-IPBD has $K_i = \{g_i\}$; in this case, we are simply taking each group to be a block λ times.

Extending parallel classes

Suppose that there is a $(v+u,u,K;\lambda)$-IPBD on $V \cup U$ with its hole on U. Further suppose that there is a collection $\mathcal{R}$ of blocks that contain every point in V λ times, and contain no point in U, for which the block sizes in $\mathcal{R}$ are from $K' \subseteq K$. Then there is a $(v+u+1,u+1,K\cup(K'+1);\lambda)$-IPBD.

Simply add a new point ∞ to each block of $\mathcal{R}$. Then ∞ is added to the hole of the IPBD.

Truncating a PBD

If there is a (v,K)-PBD containing a point in blocks of sizes $K' \subseteq K$, there is a $(v-1,K)$-GDD with group sizes from $K'-1$. To see this, delete every occurrence of the point.

For example, since there exists an STS(v) for all $v \equiv 1,3 \pmod 6$, truncating it yields a $\{3\}$-GDD of type $2^{(v-1)/2}$.

Truncating a GDD
If there is a (v, K)-GDD with group sizes from G, and some point in a group of size $g \in G$ occurs in blocks of sizes from $K' \subset K$, then there is a $(v-1, K \cup (K'-1))$-GDD with group sizes from $G \cup \{g-1\}$.

Simply delete the point from each block and group in which it occurs.

3.5 Fundamental construction of Wilson

Wilson (1972) generalized the notion of products of designs in a substantial way. We treat his generalization in some detail here.

Theorem 3.3 *If there exist*

1. *a GDD $(X, \mathcal{G}, \mathcal{B})$ with groups $\mathcal{G} = \{G_1, \ldots, G_s\}$, blocks $\mathcal{B} = \{B_1, \ldots, B_b\}$, and index λ (called the master design);*
2. *an assignment to each element $x \in X$ of a nonnegative integer weight w_x; and*
3. *for every $B_i \in \mathcal{B}$, an 'ingredient'GDD with block sizes from K_i, group sizes $(w_x : x \in B_i)$, and index μ,*

then there exists a GDD on $\sum_{x \in X} w_x$ points, with block sizes from $\bigcup_{i=1}^{b} K_i$, group sizes $(\sum_{x \in G_j} w_x : 1 \leq j \leq s)$, having index $\lambda\mu$.

Proof Associate with each $x \in X$ a distinct set W_x of w_x points. For each block $B_i \in \mathcal{B}$, form a GDD of index μ with block sizes from K_i having groups $(W_x : x \in B_i)$ (one or more of the sets in $\{W_x\}$ may be empty, in which case the GDD in fact has fewer groups than $|B_i|$). Then take as groups $(\bigcup_{x \in G_j} W_x : j = 1, \ldots, s)$. □

The generalized direct product is an easy special case of this: since a $\{3\}$-GDD of index μ of type m^3 exists for all $m \geq 1$ (this is a TD$(3, m)$ repeated μ times, for example), and a TS(v, λ) is equivalent to a $\{3\}$-GDD of index λ of type 1^v, taking the latter to be the master design and giving each point weight m, Theorem 3.3 gives a $\{3\}$-GDD of index $\lambda\mu$ of type m^v. When a TS$(m, \lambda\mu)$ exists, we can fill in groups to obtain a TS$(vm, \lambda\mu)$.

The singular direct product is also a special case. Simply fill in groups with $(m+u, u, \{3\})$-IPBD, and fill the hole with a TS$(u, \lambda\mu)$.

Theorem 3.3 can be further generalized to *incomplete* GDDs, yielding the indirect products as special cases. We do not need this level of generality for our discussions of triple systems.

3.5.1 *An application to GDDs*

To illustrate the use of the Fundamental Construction, we prove:

Theorem 3.4 *There exists a $\{3\}$-GDD of type g^t if and only if $g(t-1) \equiv 0 \pmod 2$, $g^2t(t-1) \equiv 0 \pmod 3$, and $t \neq 2$.*

Proof Necessity follows from the observation that every element occurs in $g(t-1)$ pairs to be covered, and the total number of pairs is $g^2\binom{t}{2}$. Moreover, if $t = 2$, no triple can be formed at all.

For sufficiency, we always use $\{3\}$-GDDs of type m^3 (equivalently, TD$(3, m)$s) as ingredients in an application of the Fundamental Construction. We treat various cases for the number t of groups. First suppose that $t \equiv 1, 3 \pmod 6$. Then since an STS(t) exists (Theorem 2.10), we have a $\{3\}$-GDD of type 1^t. Use this as the master design, giving every element weight g. Apply the Fundamental Construction using TD$(3, g)$s as ingredients to obtain a $\{3\}$-GDD of type g^t.

When $t \equiv 0, 4 \pmod 6$, a necessary condition is that g is even. Write $g = 2m$. Now an STS$(2t+1)$ exists, and truncating one point from this gives a $\{3\}$-GDD of type 2^t. Use this as the master design and give each point weight m. Apply the Fundamental Construction using TD$(3, m)$ to obtain a $\{3\}$-GDD of type $(2m)^t = g^t$.

When $t \equiv 5 \pmod 6$, it is necessary that $g \equiv 0 \pmod 3$. Write $g = 3m$. By Lemma 2.3, there exists a $\{3\}$-GDD of type 3^t. Use this as the master design, giving every point weight m. Apply the Fundamental Construction, using TD$(3, m)$, to obtain a $\{3\}$-GDD of type $(3m)^t = g^t$.

Finally, when $t \equiv 2 \pmod 6$ $(t > 2)$, a necessary condition is that $g \equiv 0 \pmod 6$. Write $g = 6m$. We must construct a $\{3\}$-GDD of type 6^t, $t \equiv 2 \pmod 6$. We produce such a GDD: let $Q = (T, \otimes)$ be an idempotent quasigroup of order t. On $T \times \{0, 1\}$, we define a binary operation $\oplus$. For $x, y \in T$, $x \neq y$, let $w = x \otimes y$ and $z = y \otimes x$. Then define

$$\begin{aligned} x_0 \oplus y_0 &= y_0 \oplus x_0 = z_0 \\ x_0 \oplus y_1 &= y_1 \oplus x_0 = z_1 \\ x_1 \oplus y_0 &= y_0 \oplus x_1 = w_1 \\ x_1 \oplus y_1 &= y_1 \oplus x_1 = w_0 \end{aligned}$$

This gives a partial commutative quasigroup $(T \times \{0, 1\}, \oplus)$ in which all products are defined except $x_i \oplus x_j$ for $x \in T$, $i, j \in \{0, 1\}$. Apply the commutative $3v$ construction to this *partial* quasigroup; when a product is undefined in the quasigroup, no triple is produced. It is then easy to verify that the result is a $\{3\}$-GDD of type 6^t. Use this as the master design, giving every element weight m. Apply the Fundamental Construction, using TD$(3, m)$, to obtain the $\{3\}$-GDD of type $(6m)^t = g^t$. □

Sizes	Closure	Exceptions
$B(\{3\})$	$1,3 \pmod 6$	–
$B(\{3,4\})$	$0,1 \pmod 3$	6
$B(\{3,5\})$	$1 \pmod 2$	–
$B(\{3,6\})$	$0,1 \pmod 3$	4, 10, 12, 22
$B(\{3,8\})$	$\mathbb{Z}$	4, 5, 6, 10, 11, 12, 14, 16, 17, 18, 20, 23, 26, 28, 29, 30, 34, 35, 36, 38
$B(\{3,10\})$	$0,1 \pmod 3$	4, 6, 12, 16, 18, 22, 24, 34, 36, 42
$B(\{3,4,5\})$	$\mathbb{Z}$	6, 8
$B(\{3,4,6\})$	$0,1 \pmod 3$	–
$B(\{3,4,8\})$	$\mathbb{Z}$	5, 6, 11, 14, 17
$B(\{3,5,6\})$	$\mathbb{Z}$	4, 8, 10, 12, 14, 20, 22
$B(\{3,5,8\})$	$\mathbb{Z}$	4, 6, 10, 12, 14, 16, 18, 20, 26, 28, 30, 34
$B(\{3,5,10\})$	$\mathbb{Z}$	4, 6, 8, 12, 14, 16, 18, 20, 22, 24, 26, 32, 34, 36, 38, 42, 44
$B(\{3,6,8\})$	$\mathbb{Z}$	4, 5, 10, 11, 12, 14, 17, 20, 23
$B(\{3,6,10\})$	$0,1 \pmod 3$	4, 12, 22
$B(\{3,8,10\})$	$\mathbb{Z}$	4, 5, 6, 11, 12, 14, 16, 17, 18, 20, 23, 26, 29, 35, 36
$B(\{3,4,5,6\})$	$\mathbb{Z}$	8
$B(\{3,4,5,8\})$	$\mathbb{Z}$	6
$B(\{3,4,6,8\})$	$\mathbb{Z}$	5, 11, 14, 17
$B(\{3,5,6,8\})$	$\mathbb{Z}$	4, 10, 12, 14, 20
$B(\{3,5,6,10\})$	$\mathbb{Z}$	4, 8, 12, 14, 20, 22
$B(\{3,5,8,10\})$	$\mathbb{Z}$	4, 6, 12, 14, 16, 18, 20, 26
$B(\{3,6,8,10\})$	$\mathbb{Z}$	4, 5, 11, 12, 14, 17, 20, 23
$B(\{3,4,5,6,8\})$	$\mathbb{Z}$	–
$B(\{3,5,6,8,10\})$	$\mathbb{Z}$	4, 12, 14, 20

Table 3.1 *Closure* $B(K)$ *when* $3 \in K$

4

ISOMORPHISM AND INVARIANTS

In this chapter, we introduce the fundamental notion of isomorphism. We examine the complexity of algorithms for deciding whether two triple systems are isomorphic. We then explore properties of triple systems that are invariant up to isomorphism, that can be used with varying degrees of success in distinguishing nonisomorphic triple systems.

Let $(V, \mathcal{B})$ and $(W, \mathcal{D})$ be two triple systems. Let $\phi : V \mapsto W$ be a one-to-one mapping. Under ϕ, every triple $B = \{x, y, z\} \in \mathcal{B}$ maps to a triple $\phi(B) = \{\phi(x), \phi(y), \phi(z)\}$. If ϕ *preserves the triples*, i.e. $\mathcal{D} = \{\phi(B) : B \in \mathcal{B}\}$, then ϕ is an *isomorphism* from $(V, \mathcal{B})$ to $(W, \mathcal{D})$. Evidently, when ϕ is such an isomorphism, ϕ^{-1} is an isomorphism from $(W, \mathcal{D})$ to $(V, \mathcal{B})$. The two designs are *isomorphic*, denoted $(V, \mathcal{B}) \cong (W, \mathcal{D})$, if there is an isomorphism from one to the other, and are *nonisomorphic* otherwise.

In order to discover whether or not two designs are isomorphic, it surely suffices to check every one-to-one mapping to see whether it is an isomorphism. Given two designs with v points and b triples, such a computation would require the examination of $v!$ mappings in the worst case. A major goal of the design and analysis of algorithms generally is to avoid such exponential computation times, and to treat an algorithm as *efficient* if its computation time in the worst case grows no faster than some polynomial in the size of the input. In order to specify a triple system for input to an algorithm, we can list the triples, so that the input has size $3b$. Evidently, the naive method of trying all mappings gives a computation time that grows faster than any polynomial in b. In fact, its computation time grows more quickly than $(\sqrt{b})!$. The inefficiency of the naive approach here has led to a substantial amount of research on isomorphism of triple systems as a computational problem.

4.1 The computational complexity of isomorphism

The *incidence graph* $\text{Inc}(V, \mathcal{B})$ is a bipartite graph with vertex set $\{v_x : x \in V\} \cup \{b_B : B \in \mathcal{B}\}$, and edge set $\{\{v_x, b_B\} : x \in V, B \in \mathcal{B}, x \in B\}$. When the design is a $\text{TS}(v, \lambda)$, its incidence graph has v vertices of degree $\lambda(v-1)/2$ and $\lambda v(v-1)/6$ vertices each of degree three. Now $\lambda(v-1)/2 = 3$ only when $v = 7$ and $\lambda = 1$, $v = 4$ and $\lambda = 2$, or $v = 3$ and $\lambda = 3$. With these few exceptions, for two $\text{TS}(v, \lambda)$s $(V, \mathcal{B})$ and $(W, \mathcal{D})$, *every* isomorphism from $\text{Inc}(V, \mathcal{B})$ to $\text{Inc}(W, \mathcal{D})$ maps the vertices $\{v_x : x \in V\}$ to the vertices $\{v_w : w \in W\}$. This isomorphism between the incidence graphs then induces an isomorphism on the triple systems. Conversely, any isomorphism of the triple systems corresponds to a single isomorphism of their incidence graphs. For the $\text{TS}(3,3)$, $\text{TS}(4,2)$, and

TS(7, 1), the number of isomorphisms of the incidence graph is twice the number of isomorphisms of the triple systems themselves, obtained by interchanging the vertices $\{v_x : x \in V\}$ and $\{b_B : B \in \mathcal{B}\}$. Thus for TS$(v, \lambda)$s $(V, \mathcal{B})$ and $(W, \mathcal{D})$:

$$\mathrm{Inc}(V, \mathcal{B}) \cong \mathrm{Inc}(W, \mathcal{D}) \iff (V, \mathcal{B}) \cong (W, \mathcal{D}). \tag{4.1}$$

Moreover, the size of Inc$(V, \mathcal{B})$ is bounded by a constant times the size of $(V, \mathcal{B})$ itself. Consequently, determining whether two triple systems are isomorphic is polynomial time reducible to deciding whether two simple graphs are isomorphic, the widely studied *graph isomorphism problem.* In one sense, this provides little information, since the complexity of the graph isomorphism problem itself is open. In fact, the graph isomorphism problem is probably the most important problem for which no proof of NP-completeness is currently known, although the problem is easily seen to be in NP. On the other hand, no efficient algorithm for graph isomorphism is known. See Garey and Johnson (1979) for an introduction to NP and to NP-completeness. In an effort to classify isomorphism problems with respect to computational complexity, we call a problem $\mathcal{P}$ *isomorphism easy* if $\mathcal{P}$ can be polynomial time reduced to graph isomorphism, *isomorphism hard* if graph isomorphism can be polynomial time reduced to $\mathcal{P}$, and *isomorphism complete* if $\mathcal{P}$ is polynomial time equivalent to graph isomorphism. Isomorphism completeness is evidence that a problem does not admit an efficient algorithm, but must be treated at the present time as substantially weaker evidence of this than is NP-completeness.

We have seen that isomorphism of triple systems is isomorphism easy. This leaves open a number of possibilities. First, it may be that graph isomorphism has an efficient algorithm, in which case so do all isomorphism easy and isomorphism complete problems. Second, it may be that neither graph isomorphism nor isomorphism of triple systems admit efficient algorithms. Third, it may be that while graph isomorphism has no efficient algorithm, isomorphism of triple systems does. We rule out the third possibility here, proving that isomorphism of triple systems is in fact isomorphism complete using the methods of Colbourn and Colbourn (1981a).

A *k-graph* is a complete graph $K_n = (V, E)$ together with a mapping $E \mapsto K$, where K is a set of k colours. A k-graph is *colour-regular* if the number of edges of a particular colour incident at a particular vertex is independent of the colour and vertex chosen. It is *colour-complementary* if there is a mapping $\phi : V \mapsto V$ and a cyclic permutation $\psi : K \mapsto K$ so that the k-graph (V, E) with colouring $E \mapsto K$ is the same as the k-graph $(\phi(V), \phi(E))$ with colouring $\phi(E) \mapsto \psi(K)$. Finally, a graph is *k-regular colour complementary* (k-rcc) if it is a colour-regular, colour-complementary k-graph. Colbourn and Colbourn (1981a) show that the problems of deciding isomorphism of 2-rcc graphs (*self-complementary graphs*) and 3-rcc graphs are isomorphism complete. This underlies their main construction.

Theorem 4.1 *Deciding isomorphism of triple systems is isomorphism complete.*

Proof As a result of our earlier observations, it remains only to show that there is a transformation Φ from 3-rcc graphs to triple systems for which

1. $\Phi(G) \cong \Phi(H) \iff G \cong H$ for any two 3-rcc graphs G and H; and
2. $\Phi(G)$ can be produced in time bounded by a fixed polynomial in the size of the 3-rcc graph G.

Start with an n-vertex 3-rcc graph $G = (V, E)$ with colouring $\rho : E \mapsto K$, $|K| = 3$. Let E_1, E_2, E_3 be the three colour classes of edges of G; since G is colour-regular, for $i = 1, 2, 3$, (V, E_i) is a regular graph of degree $d = (n-1)/3$, and has $q = n(n-1)/6$ edges. We form a TS$(6d+1, d)$ $\Phi(G)$.

Let $A = \{x_i, y_i, z_i : 1 \leq i \leq d\}$. The element set of $\Phi(G)$ is $V \cup A$, and its collection of triples is:

1. For $1 \leq i \leq d$ and all $\{v, w\} \in E_1$, $\{x_i, v, w\}$ is a triple.
2. For $1 \leq i \leq d$ and all $\{v, w\} \in E_2$, $\{y_i, v, w\}$ is a triple.
3. For $1 \leq i \leq d$ and all $\{v, w\} \in E_3$, $\{z_i, v, w\}$ is a triple.
4. On A, place the triples of a TS$(3d, d)$.

It is straightforward to check that $\Phi(G)$ is a TS$(6d+1, d)$. It is also easy to check that provided that identical TS$(3d, d)$s are used in step (4), $G \cong H \implies \Phi(G) \cong \Phi(H)$. To obtain the converse, we establish that given $\Phi(G)$ we can recover G uniquely up to isomorphism. Given $\Phi(G)$, we must distinguish the elements in A from those in V. To do this, consider the *pair-closure* of a set of elements, that is the smallest superset for which there is no triple containing precisely two elements of the superset (equivalently, the smallest subdesign containing the given set). For any two elements $a, b \in V \cup A$, their pair-closure is $V \cup A$ *unless* $a, b \in A$, in which case the closure is just A or a subset thereof. Thus to find A, we simply find all pairs whose pair-closure is not $V \cup A$, and take the union of all such pairs. This enables us to locate all triples added as part of the TS$(3d, d)$ and delete them. Now choose any element of A and call it x_1. Suppose that $\{x_1, a, b\}$ is a triple. The remaining $\{x_i\}$ are precisely those appearing in a triple with $\{a, b\}$. Choose one of the remaining elements in A and call it y_1, and find the remaining $\{y_i\}$ similarly. What remains in A is now the $\{z_i\}$. Finally, to recover G, colour each pair on V red if it appears with x_1, white if it appears with y_1, and blue if it appears with z_1. This gives a 3-rcc graph that is isomorphic to G. □

A few remarks are in order. $\Phi(G)$ and $\Phi(H)$ may have more isomorphisms than do G and H, depending on the structure and placement of the TS$(3d, d)$s used. Essentially, this proof relies on the $2v + 1$ construction (from §3.2), using a 3-rcc graph to dictate the d-factorization of dK_{3d+1} to use. This makes the structure of the subsystem relatively unimportant, as long as the same subsystem is used in each application of the transformation.

We might have expected that the regularity and pairwise balance of a triple system would enable an algorithm to obtain better performance, but Theorem

4.1 shows that in this sense, triple systems are as unstructured as graphs in general.

An essential feature of the proof here is that the index of the triple system grows linearly as a function of the number of elements. Thus it says nothing about the case when the index is fixed. We examine this restricted case next.

4.2 Isomorphism algorithms for fixed index

A naive method for testing isomorphism of two TS(v, λ)s is to simply list all possible $v!$ mappings, and check for each whether it is an isomorphism. Suppose instead that we define an arbitrary total ordering $\prec$ on all distinct TS(v, λ)s, and given a single TS(v, λ) we try all $v!$ ways of relabelling the elements. This results in a number of distinct TS(vs, λ)s, all isomorphic to the given one. Now define the TS(v, λ) that comes first under the order $\prec$ to be the *canonical form* of the triple system. Evidently, two isomorphic triple systems have identical canonical forms. Thus to test isomorphism, it suffices to find such a canonical form.

Of course, if we must try all $v!$ relabellings to find the canonical form, the method remains inefficient. However, if we can find the canonical form by looking at a small subset of the possible labellings, we may obtain a more efficient method. This is the essential idea in an algorithm of Miller (1978), that we examine next.

4.2.1 *Steiner triple systems: Miller's algorithm*

Fix the index to be one. We define a total order on distinct Steiner triple systems as follows. Assume an arbitrary total order $<$ on the element set. A natural order on pairs is *reverse lexicographic order* $\prec_r$: for $w < x$, $y < z$, $\{w, x\} \prec_r \{y, z\}$ if $x < z$, or if $x = z$ and $w < y$. For two triples of elements, $X = \{x_1, x_2, x_3\}$ and $W = \{w_1, w_2, w_3\}$ with $x_1 < x_2 < x_3$ and $w_1 < w_2 < w_3$, let $X \prec W$ if $\{x_1, x_2\} \prec_r \{w_1, w_2\}$, or $\{x_1, x_2\} = \{w_1, w_2\}$ and $x_3 < w_3$. Then for two distinct Steiner triple systems $\mathcal{B}$ and $\mathcal{D}$, let $\mathcal{B} \prec \mathcal{D}$ if the first triple under $\prec$ in $\mathcal{B} \setminus \mathcal{D}$ precedes the first triple of $\mathcal{D} \setminus \mathcal{B}$.

Our goal is to generate the labellings of the element set that lead to the canonical form under the order $\prec$. A *labelling* is a one–one mapping $\psi : V \mapsto \{1, 2, \ldots, v\}$. A *form* of a TS$(v, \lambda)$ $(V, \mathcal{B})$ is the result of renaming the elements under a labelling ψ. The *canonical form* is then the first form under $\prec$ obtained from any labelling. A labelling is *canonical* if its form is the canonical form.

We search for canonical labellings by starting with no elements labelled, and labelling one element at a time until all elements are labelled. A *partial labelling* is a pair (X, ϕ), where $X \subseteq V$ and $\phi : X \mapsto \{1, \ldots, |X|\}$ is one–one. A partial labelling (X, ψ) is the *s-prefix* of a partial labelling (Y, ϕ) if $|X| = s$, $X \subseteq Y$ and $\psi(x) = \phi(x)$ for all $x \in X$. For a partial labelling (X, ϕ) with $X \neq V$, the *extension* $\text{ext}(X, \phi)$ is the set of $|V \setminus X|$ partial labellings

$$\{(X \cup \{u\}, \phi_u) : u \in V \setminus X\},$$

where $\phi_u(x) = \phi(x)$ for $x \in X$, and $\phi_u(u) = |X| + 1$. For a labelling (V, ϕ), $\text{ext}(V, \phi) = \{(V, \phi)\}$.

Let $\mathcal{S}_0 = \{(\varnothing, \phi)\}$, the initial partial labelling in which no elements are labelled. Define $\mathcal{S}_i = \bigcup_{(X,\phi)\in\mathcal{S}_{i-1}} \text{ext}(X, \phi)$ for $i \geq 1$. Evidently $\mathcal{S}_v$ contains all $v!$ labellings. Can we guarantee *in advance* that some of these labellings are not canonical?

Consider a partial labelling (X, ϕ) in which some triple $\{x, y, z\} \in \mathcal{B}$ has $\{x, y\} \subseteq X$ and $z \in V \setminus X$; let $\{x, y\}$ be the first pair under $\prec_r$ for which this holds. Now consider the extension of (X, ϕ); it contains $(X \cup \{z\}, \phi_z)$. If it also contains a second partial labelling $(X \cup \{u\}, \phi_u)$, we observe that *any* labelling consistent with the partial labelling ϕ_u gives a form that follows (under $\prec$) *every* form from a labelling that is consistent with ϕ_z. In other words, all canonical labellings are generated if the extension is replaced by the single partial labelling $\{(X \cup \{z\}, \phi_z)\}$.

With this in mind, we define a second operation. The *closure* $\text{cl}(X, \phi)$ of a partial labelling (X, ϕ) is a partial labelling (X', ϕ'):

$X' \leftarrow X$; $\phi' \leftarrow \phi$;
while some triple $\{x, y, z\} \in \mathcal{B}$ has $\{x, y\} \subseteq X'$
  and $z \in V \setminus X'$ **do**
  let $\{x, y\}$ be the first triple under $\prec_r$ for
    which $\{x, y, z\} \in \mathcal{B}$, $\{x, y\} \subseteq X'$, and
    $z \in V \setminus X'$
  $X' \leftarrow X \cup \{z\}$
  extend ϕ' to map z to $|X|$
od

When $(X, \phi) = \text{cl}(X, \phi)$, (X, ϕ) is *closed.* When (X, ϕ) is closed, the triples induced on X themselves form a Steiner triple system. Equivalently, closure produces the smallest subdesign containing the given elements. In order to obtain all canonical labellings, we can always replace a partial labelling (X, ϕ) by $\text{cl}(X, \phi)$. Thus if we define $\mathcal{T}_0 = \{(\varnothing, \phi)\}$, and for $i \geq 1$,

$$\mathcal{T}_i = \bigcup_{(X,\phi)\in\mathcal{T}_{i-1}} \{\text{cl}(X', \phi') : (X', \phi') \in \text{ext}(X, \phi)\},$$

then for some $\ell \leq v$, all canonical labellings appear in $\mathcal{T}_\ell$. Now $|\mathcal{T}_i| \leq v|\mathcal{T}_{i-1}|$ for all $i \geq 1$, and so $|\mathcal{T}_\ell| \leq v^\ell$. To bound the number of labellings examined, we can bound ℓ.

Lemma 4.2 *If* $(X, \phi) = cl(X, \phi)$, $X \neq V$, $(X', \phi') \in ext(X, \phi)$ *and* $(\widehat{X}, \widehat{\phi}) = cl(X', \phi')$, *then* $|\widehat{X}| \geq 2|X| + 1$. *Consequently, if* $(X, \phi) \in \mathcal{T}_i$, *then* $|X| \geq \min(v, 2^i - 1)$.

Proof Let $X' \setminus X = \{u\}$. For each $x \in X$, define o_x to be the *unique* element for which $\{u, x, o_x\} \in \mathcal{B}$. Now $o_x \in V \setminus X$, $o_x \neq u$, and $o_x = o_y$ precisely when

$x = y$. So $X \cup \{u\} \cup \{o_x : x \in X\} \subseteq \widehat{X}$, establishing the first inequality. That every partial labelling in $\mathcal{T}_i$ has at least $\min(v, 2^i - 1)$ labelled elements follows by induction. □

Lemma 4.2 ensures that $\mathcal{T}_\ell$ contains all canonical labellings for some $\ell \leq \log_2(v+1)$. Thus we have:

Theorem 4.3 *A canonical form of an STS(v) can be computed in $O(v^{\log v})$ time.*

Proof Simply compute $\mathcal{T}_\ell$ for ℓ the largest integer less than or equal to $\log_2(v+1)$. Then $\mathcal{T}_\ell$ has at most v^ℓ labellings, from which we extract the first form under $\prec$. □

In any practical implementation, a number of improvements are possible. We examine some of these here. From a practical standpoint, our goals are to make termination occur as soon as possible, that is to make ℓ as small as possible, and to make the sets $\mathcal{T}_i$, $i = 0, \ldots, \ell$, as small as possible.

From a partial labelling (X, ϕ), define a *partial form* to be

$$\{\{\phi(x), \phi(y), \phi(z)\} : \{x, y, z\} \in \mathcal{B}, \{x, y, z\} \subseteq X\}.$$

For two distinct partial forms $\mathcal{D}$ and $\mathcal{D}'$, we let $\mathcal{D} \prec \mathcal{D}'$ if $|\mathcal{D}| = |\mathcal{D}'|$ and the first triple in $\mathcal{D} \setminus \mathcal{D}'$ precedes (under $\prec$) the first triple of $\mathcal{D}' \setminus \mathcal{D}$. Now suppose that in forming $\mathcal{T}_i$, some partial labelling yields a partial form $\mathcal{D}'$ that is preceded by the partial form $\mathcal{D}$ of another partial labelling in $\mathcal{T}_i$. No extension of the partial labelling giving $\mathcal{D}'$ can be canonical, as such extensions always follow the extensions of $\mathcal{D}$ under $\prec$. Thus in $\mathcal{T}_i$, we need retain only those partial labellings whose partial forms are minimal under $\prec$. This has the effect of reducing the sizes of the sets $\{\mathcal{T}_i\}$, but appears not to reduce ℓ, the number of such sets.

In $\mathcal{T}_i$, the number of labelled elements in a partial labelling is at least $2^i - 1$, but may in fact be much larger. In many cases, already in $\mathcal{T}_3$ we find some full labellings. We would like to take advantage of the 'extra' labelled elements that closure sometimes introduces. Evidently, it would be preferable to consider partial labellings in which as many elements as possible are labelled. To accomplish this, Colbourn, Magliveras, and Stinson (1992) suggest using a different total order $\lhd$ in place of $\prec$. For two closed partial labellings (X, ϕ) and (Y, ψ) of the same triple system $(V, \mathcal{B})$, let $\mathcal{D}$ and $\mathcal{D}'$ be the corresponding partial forms. Let $\mathcal{D}_s$ be the partial form of the closure of the s-prefix of (X, ϕ), and let $\mathcal{D}'_s$ be the partial form of the closure of the s-prefix of (Y, ψ). We say $\mathcal{D} \lhd \mathcal{D}'$ if there is an integer ℓ, $1 \leq \ell \leq \min(|X|, |Y|)$ for which

1. $\mathcal{D}_s = \mathcal{D}'_s$ for $0 \leq s < \ell$;
2. $|\mathcal{D}_\ell| > |\mathcal{D}'_\ell|$, or $|\mathcal{D}_\ell| = |\mathcal{D}'_\ell|$ and $\mathcal{D}_\ell \prec \mathcal{D}'_\ell$.

Now we proceed as before, but use the order $\lhd$ in place of $\prec$. Using a different order, the canonical forms need not be the same. However, there can be a significant advantage to the proposed order. In forming $\mathcal{T}_i$, we eliminate all partial labellings that were eliminated using $\prec$, and more: using $\lhd$ we also retain only those partial labellings that have as many labelled elements as possible. This typically has the effect of reducing the number of extension steps required. In fact, if *any* three elements do not appear together in a proper subdesign, then all canonical labellings appear already in $\mathcal{T}_3$.

We might hope that these improvements might reduce the running time given in Theorem 4.3. However, the projective triple system of order $2^i - 1$ admits $\prod_{j=0}^{i-1}(2^i - 2^j)$ canonical labellings, so in the worst case, these heuristics make no improvement.

This algorithm can also be implemented in a parallel environment. The difficult step in doing this is to implement the closure operation so that new elements are labelled 'in parallel'. Colbourn, Stinson, and Teirlinck (1992) describe a fast parallel implementation of closure.

4.2.2 *Automorphisms*

An *automorphism* of a triple system is an isomorphism of a triple system onto itself. Composing two automorphisms gives an automorphism, the identity is always an automorphism, and the inverse of an automorphism is again an automorphism. Thus the set of all automorphisms forms a permutation group under composition, often called the *full automorphism group* $\mathrm{Aut}(V, \mathcal{B})$. An *automorphism group* of the design is a subgroup of its full automorphism group.

Automorphisms indicate which relabellings lead to identical forms of a design. How many automorphisms does a Steiner triple system have? A method to compute this is implicit in our discussion of Miller's algorithm. If a Steiner triple system has an automorphism group of order γ, there are γ different labellings that map the system into any fixed form. Thus γ is just the number of canonical labellings that Miller's algorithm finds.

4.2.3 *Fixed index: Babai–Luks algorithm*

Let us first take a somewhat more algebraic view of Miller's technique. Given a Steiner triple system $(V, \mathcal{B})$, let us consider the Steiner quasigroup $(V, \otimes)$ that it underlies, as described in §2.1.1. In essence, Miller's algorithm chooses (in the extension phases) a subset $S \subseteq V$ with the property that each element of V is the product under $\otimes$ of elements of S. Thus S is a generating set for $(V, \otimes)$; it need not be minimal. Miller's algorithm rests on the observation that any automorphism fixing all elements of S must be the identity, and the timing rests on the observation that the smallest generating set has at most $\log(v + 1)$ elements.

How can we extend these ideas to higher index? Babai and Luks (1983) addressed this question. For a $\mathrm{TS}(v, \lambda)$ $(V, \mathcal{B})$ and a set $S \subseteq V$, let $S_0 = S$,

$$S_{i+1} = \bigcup \{x, y, z : \{x, y, z\} \in \mathcal{B}, |\{x, y, z\} \cap S_i| \geq 2\}$$

for $i \geq 1$, and let $\ell = \min(i : S_i = S_{i+1})$. Then the *span* of S is S_ℓ, and S is a *generating set* if its span is V.

The first key observation of Miller's method generalizes:

Lemma 4.4 *A* $TS(v, \lambda)$ *has a generating set of size at most* $\log(v+1)$.

Proof Construct a generating set as follows. Let T_0 be empty. For $i \geq 1$, if $T_{i-1} \neq V$, choose an element $z_i \in V \setminus T_{i-1}$ and set $T_i = \text{span}(T_{i-1} \cup \{z_i\})$. In this way a set $Z = \{z_1, \dots, z_\ell\}$ is determined that is a generating set.

Now the span of any set of elements is a subdesign (that is, a triple system of index λ on the elements of the span). Thus by elementary counting, $|T_{i+1}| \geq 2|T_i| + 1$, and hence $\ell \leq \log(v+1)$. □

For a generating set T, let us examine the automorphisms of the triple system that fix all elements of the generating set. This is the *stabilizer* $\text{Stab}_T(V, \mathcal{B})$. For a Steiner triple system, $\text{Stab}_T(V, \mathcal{B})$ is trivial; however, for index greater than one, the stabilizer may be nontrivial. Nevertheless, Babai and Luks (1983) remark that it has a very simple structure.

Lemma 4.5 *Let* $T \subset V$ *be a generating set for a* $TS(v, \lambda)$ $(V, \mathcal{B})$. *Let* $\Gamma_T = \text{Stab}_T(V, \mathcal{B})$. *Then there exists an element* $x \in V \setminus T$ *so that* $T' = T \cup \{x\}$ *is a generating set and* $\Gamma_{T'} = \text{Stab}_{T'}(V, \mathcal{B})$ *is a subgroup of* Γ_T *of index at most* λ.

Proof Find $s, t \in T$ so that there is a triple $\{s, t, x\} \in \mathcal{B}$ with $x \notin T$; such a pair exists since T is a generating set and $T \neq V$. Now set $T' = T \cup \{x\}$. The index of $\Gamma_{T'}$ in Γ_T is the number of elements of V to which x is mapped under an automorphism of Γ_T. Any such automorphism must map the triple $\{s, t, x\}$ to a triple $\{s, t, y\} \in \mathcal{B}$ and since the triple system has index λ, the number of elements to which x can map under Γ_T is at most λ. But then $\Gamma_{T'}$ has index at most λ in Γ_T. □

Babai and Luks (1983) then establish that a $\text{TS}(v, \lambda)$ with a fixed generating set has an automorphism group whose structure is similar to that of the automorphism group of a connected $(\lambda+1)$-regular graph with one edge fixed. Using this analogy, they apply techniques of Luks (1982) to obtain:

Theorem 4.6 *Isomorphism of* $TS(v, \lambda)$*s with a fixed generating set can be tested in polynomial time when the index* λ *is a fixed constant. Consequently, isomorphism of* $TS(v, \lambda)$*s can be tested in* $v^{\log v + c(\lambda)}$ *time, for fixed* λ, *where* $c(\lambda)$ *is a constant depending only on* λ.

The isomorphism algorithm for triple systems proceeds by finding *all* minimal generating sets; since each has at most $\log(v+1)$ elements, there are at most $v^{\log v}$ of them. For each, a triple system with the generating set fixed is formed.

From a practical standpoint, improvements similar to those used for Miller's algorithm are possible. In particular, we may choose to examine only the smallest generating sets, or only those with some isomorphism invariant property such as the sizes of subsystems generated.

The details of testing isomorphism once a generating set is fixed are elegant, but require a substantial diversion into computational group theory; see Luks (1982) and references therein.

4.3 Isomorphism invariants

In order to certify that two designs are isomorphic, it suffices to exhibit one isomorphism. However, in order to certify that two designs are nonisomorphic, we are faced with a seemingly more complicated task. Of course, all mappings could be listed; then we could show, for each, a triple that is not preserved by the mapping. However, such a certificate of nonisomorphism would have length that is exponential in the number of elements.

Shorter certificates of nonisomorphism are sometimes available. For example, two triple systems on different numbers of elements cannot be isomorphic. Similarly, two triple systems with different numbers of triples cannot be isomorphic. In the same vein, two triple systems having different numbers of *distinct* triples cannot be isomorphic. In each case, there is a property of a triple system (number of elements, triples, and distinct triples in our examples) that is invariant under isomorphisms.

An *isomorphism invariant* $\mathcal{I}$ of a triple system $(V, \mathcal{B})$ is a function that satisfies $(V, \mathcal{B}) \cong (W, \mathcal{D}) \Longrightarrow \mathcal{I}(V, \mathcal{B}) = \mathcal{I}(W, \mathcal{D})$. An isomorphism invariant is *complete* if $(V, \mathcal{B}) \cong (W, \mathcal{D})$ if and only if $\mathcal{I}(V, \mathcal{B}) = \mathcal{I}(W, \mathcal{D})$. If two triple systems have different values of an isomorphism invariant, they are surely nonisomorphic and we have the desired certificate of nonisomorphism. If two systems have the same value of the invariant, we have no information about whether or not they are nonisomorphic, unless the invariant is complete. Isomorphism invariants may be useful in distinguishing nonisomorphic systems, but we typically have no guarantee that the invariant will succeed.

Naturally, we are only interested in invariants that are *efficient*, in that the value of the invariant can be computed in polynomial time in the size of a given triple system. Since no efficient complete invariant is currently known, our goal is to find invariants that are effective, that is, appear to be successful in distinguishing nonisomorphic triple systems. One measure of effectiveness of an invariant, proposed by Petrenjuk and Petrenjuk (1980), is its *sensitivity*. This is the ratio of the number of nonisomorphic systems distinguished by the invariant to the total number of nonisomorphic systems, so that a complete invariant has sensitivity one.

4.3.1 *Trains and compact trains*

White (1913) developed perhaps the first invariant for Steiner triple systems that looks beyond simple structural properties. The *train* of a Steiner triple system $(V, \mathcal{B})$ is a directed graph with vertex set $\binom{V}{3}$, consisting of all 3-subsets of elements. Each vertex has exactly one arc leaving it, determined as follows. Let $(V, \odot)$ be the Steiner quasigroup of $(V, \mathcal{B})$. Then the arc leaving vertex $\{x, y, z\}$ is placed to vertex $\{x \odot y, x \odot z, y \odot z\}$.

Evidently, in the train of a Steiner triple system the arc exiting a vertex corresponding to a block is a loop in the digraph, and this is the only place that loops arise. Thus every vertex corresponding to a block is the sink of a directed tree with all arcs directed towards the sink. The remaining components of the train, if any, are directed cycles with pendant directed trees that are directed in towards the cycle.

The isomorphism type of the train is an invariant of the isomorphism type of the Steiner triple system, but nonisomorphic systems may have isomorphic trains. At first, it appears that we have simply taken a complex isomorphism problem and transformed it into another complex isomorphism problem. However, trains have a very simple structure. Since trees and cycles admit a number of simple linear time coding schemes, the isomorphism type of the train can be simply encoded (see Colbourn, Colbourn, and Rosenbaum, 1982). It appears to be rare for nonisomorphic Steiner triple systems to have isomorphic trains. Nevertheless, the requirement of storing $O(v^3)$ subsets in the computation of a train limits their use. Colbourn, Colbourn, and Rosenbaum (1982) suggest instead retaining information about component sizes and number of sources (vertices of indegree 0), and show that for small Steiner triple systems these *compact trains* are effective isomorphism invariants. Stinson (1983) proposes instead using the indegree sequence of the train; while the outdegrees are all 1, the indegrees show remarkable variation.

4.3.2 *Fragments and cycle structure*

Cole (1913) and Cummings (1914, 1914a) proposed a very different invariant. For a Steiner triple system $(V, \mathcal{B})$, the *neighbourhood* $N(x)$ of an element $x \in V$ is:

$$N(x) = \{\{y, z\} : \{x, y, z\} \in \mathcal{B}\}.$$

Thus the neighbourhood of an element x can be viewed as a 1-regular graph on $V \setminus \{x\}$. Form the *double neighbourhood* $DN(x, y)$ of two elements x and y as the subgraph of $N(x) \cup N(y)$ induced on $V \setminus \{x, y\}$. Now the double neighbourhood of two elements x and y contains one vertex z of degree 0 (the element z completes the block $\{x, y, z\} \in \mathcal{B}$), but all other elements have degree two). Thus the double neighbourhood is a union of simple cycles, each of even length at least four. The isomorphism type of a double neighbourhood can then be encoded directly as a multiset of cycle lengths. The *cycle structure* of a Steiner triple system is the multiset of $\binom{v}{2}$ double neighbourhoods, each encoded as a multiset of cycle lengths. Isomorphic systems lead to identical cycle structures, and not surprisingly, the converse does not hold.

As with trains, the space to store an entire cycle structure is $O(v^3)$. For this reason, Gibbons (1976) examined a simpler invariant in which only the 4-cycles are recorded; a 4-cycle arises from a set of four triples

$$\{\{a, w, x\}, \{a, y, z\}, \{b, w, z\}, \{b, x, y\}\},$$

called a *fragment, quadrilateral,* or *Pasch configuration.* Gibbons (1976) found fragments to be an effective invariant for small orders (up to 15), but for larger orders it appears that many Steiner triple systems have no fragment at all (these quadrilateral-free STSs are discussed in §13.2).

In §12.6, we establish that every possible union of cycles of even length actually arises as a double neighbourhood; thus we expect that cycle structure remains an effective invariant for large orders. Beyond the characterization of double neighbourhoods, however, there is no theoretical basis for this expectation.

4.3.3 *Neighbourhood graphs*

When the index is one, every element in a triple system has an isomorphic neighbourhood, since there is a unique 1-regular multigraph. However, for $\lambda > 1$, the neighbourhood of an element in a $\mathrm{TS}(v, \lambda)$ could in principle be any λ-regular multigraph on $v-1$ vertices. Thus properties of the v neighbourhood graphs can be employed as isomorphism invariants. They have been used by Mathon and Rosa (1977), Morgan (1977), Harms, Colbourn, and Ivanov (1987), and others.

Isomorphism of neighbourhood graphs for index two is straightforward. Simply record the cycle lengths. Thus for index two, neighbourhood graphs admit simpler algorithms than triple systems apparently do. For index at least three, neighbourhood graphs themselves pose difficult isomorphism problems. We can use the algorithm of Babai and Luks (1983), but in practice we resort to *ad hoc* techniques.

In Chapter 12, an extensive partial characterization of neighbourhood graphs of triple systems is developed. The variety of multigraphs that arise supports the belief that neighbourhood graphs are useful isomorphism invariants.

4.3.4 *Intersections and clique analysis*

Many of the isomorphism invariants employed involve finding a graph that represents the triple system in some way. A natural mechanism for this is to form a *block intersection graph.* Let $(V, \mathcal{B})$ be a $\mathrm{TS}(v, \lambda)$. For $S \subset \{0, 1, 2, 3\}$, the *S-intersection graph* of $(V, \mathcal{B})$ has a vertex for every block of $\mathcal{B}$. Two vertices are adjacent if the size of the intersection of the corresponding blocks occurs in S. Two isomorphic triple systems have isomorphic S-intersection graphs for each S. Thus invariants of the intersection graphs yield invariants of the triple system. Gibbons (1976) and Gibbons, Mathon, and Corneil (1977) suggest determining the number of cliques (complete subgraphs) of small sizes, and using a refinement procedure to distinguish vertices based upon the number of small cliques in which they and their neighbours occur.

The 3-intersection graph G_3 of a triple system consists of vertex-disjoint complete subgraphs. Each subgraph corresponds to blocks that are identical as subsets of elements. The 2-intersection graph G_2 has the property that $G_2 \cup 3G_3$ is $3(\lambda - 1)$-regular, since every pair in a fixed block occurs in $\lambda - 1$ other blocks. The 1-intersection graph G_1 has the property that every vertex has degree $3(\lambda(v-1)/2 - \lambda + 1)$, since every element occurs in $\lambda(v-1)/2$ blocks.

When the index is one, intersection graphs exhibit a stronger form of regularity. In this case, G_3 and G_2 are both void, and G_1 is $3(v-1)/2$-regular. Now consider two blocks B_1 and B_2. If they intersect, the corresponding two vertices are adjacent and have $(v-1)/2+2$ common neighbours. If they do not intersect, the corresponding two vertices are nonadjacent and have nine common neighbours. Since the number of common neighbours of two vertices depends only on whether they are adjacent or not, G_1 is a *strongly regular graph.* This uniformity of vertices in intersection graphs indicates the importance of considering more than the neighbourhoods of two vertices in the intersection graphs.

The strongly regular graphs arising from Steiner triple systems can be recognized using a criterion of Neumaier (1979). This criterion plays an important role in an algorithm for isomorphism of strongly regular graphs due to Spielman (1996).

4.4 Work points

1. Is there an efficient algorithm for (a) isomorphism of Steiner triple systems? (b) isomorphism of triple systems of fixed index?
2. Is there a reduction from isomorphism of TS(v, λ)s with a fixed generating set to isomorphism of connected graphs of maximum degree $\lambda+1$ with an edge fixed?
3. Can trains be generalized to be effective isomorphism invariants for triple systems of index greater than one?

5

ENUMERATION

5.1 Small orders

In §0.4 and §0.5, some history of hand computations of catalogues of triple systems is outlined. In this chapter, we give catalogues of triple systems for small orders. Each block is presented vertically.

```
0000011122
1123423433
2345554545
```

Table 5.1 *The unique TS(6, 2) design*

```
0001123
1242534
3654656
```

Table 5.2 *The unique TS(7, 1) design*

```
1: 00000011112222   2: 00000011112222
   11335533443344      11335533443344
   22446655666655      22446656565656
3: 00000011112222   4: 00000011112223
   11334533453344      11234523453344
   22456646565656      23456664565656
```

Table 5.3 *The four nonisomorphic TS(7, 2) designs*

The order of the smallest nontrivial Steiner triple system is 7; a system is given in Table 5.2. This design is unique up to isomorphism, and is the design of points and lines of PG(2,2), the projective plane of order 2. It is cyclic, and its full automorphism group is the simple group of order 168 isomorphic to PGL(3,2) and to PSL(2,7). Hence on any 7-element set there are $(7!)/(7 \cdot 6 \cdot 4) = 30$ distinct STS(7)s.

Nandi (1946) enumerated the nonisomorphic TS(7,2) designs in Table 5.3. Morgan (1977) generated the TS(7,3)s in Table 5.4.

There exists a unique STS(9) up to isomorphism, displayed in Table 5.5. It is isomorphic to the points and lines of AG(2,3), the affine plane of order 3. It has no 9-cycle as an automorphism, but is acted upon by $\mathbb{Z}_3 \times \mathbb{Z}_3$. Its full automorphism group has order 432, and it has a concise presentation as the rows, columns, front diagonals, and back diagonals of the square

$$\begin{pmatrix} 1\,2\,3 \\ 4\,5\,6 \\ 7\,8\,9 \end{pmatrix}.$$

On any 9-element set, there are $(9!)/(9 \cdot 8 \cdot 6) = 840$ distinct STS(9)s. Mathon and Rosa (1977), building on a list of Morgan (1977), generated the TS(9,2)s, shown in Table 5.6.

1:	000000000111111222222	2:	000000000111111222222
	111333555333444333444		111333555333444333444
	222444666555666666555		222444666556566566556
3:	000000000111111222222	4:	000000000111111222222
	111333455333445333444		111333455333445333444
	222445666456566566556		222445666466556556566
5:	000000000111111222222	6:	000000000111111222223
	111333445333445333445		111233455233445333444
	222456566456566456566		223445666546566566565
7:	000000000111111222223	8:	000000000111111222223
	111233455233445333444		111233445233455333444
	223445666645566566556		223455666644566566556
9:	000000000111111222223	10:	000000000111111222233
	111233445233445333454		111223445223345334544
	223456566546566456665		234356566465656456656

Table 5.4 *The 10 nonisomorphic TS(7,3) designs*

000011122236
134534534547
268787676858

Table 5.5 *The unique TS(9,1) design*

The two TS(13,1)s, shown in Table 5.7, were first exhibited by Zulauf (1897). The 80 nonisomorphic STS(15)s are given in Tables 5.8–5.12. $|G|$ is the order of the automorphism group. CI is the chromatic index, the minimum number of colours with which the blocks can be coloured so that no two intersecting blocks receive the same colour; when CI = 7, the design is resolvable. PC is the number of parallel classes. Sub is the number of (7,3,1)-subdesigns, and Pa is the number of Pasch configurations (four triples on six points).

Other complete enumeration results for triple systems have been undertaken. For TS(6,λ)s, Ivanov (1983) and Gronau and Prestin (1982) found four designs

1:	000000001111112222223344	2:	000000001111112222223344
	113355773344663344556655		113355773344663344556655
	224466885577888866777788		224466885578787866787878
3:	000000001111112222223344	4:	000000001111112222223344
	113355673344673344556655		113355773344663344555656
	224467885568787867687878		224466885758786768788877
5:	000000001111112222223344	6:	000000001111112222223344
	113355773344663344555656		113355673344673344555656
	224466885758786867787887		224467885657887868678787
7:	000000001111112222223344	8:	000000001111112222223344
	113355673344673344555656		113355673344673344555656
	224467885658787867687887		224467885658787867688778
9:	000000001111112222223344	10:	000000001111112222223344
	113355673344663344555756		113355773344563344565656
	224467885858776767886878		224466885768786857877887
11:	000000001111112222223344	12:	000000001111112222223344
	113355673344573344565656		113355673344573344565656
	224467885678687856788787		224467885678687856877887
13:	000000001111112222223344	14:	000000001111112222223344
	113355673344573344565656		113355673344573344565656
	224467885867686758788787		224467885867686758877887
15:	000000001111112222223344	16:	000000001111112222223344
	113346673345573344555656		113345673345673344556655
	224557884676887868677887		224567884586787867687878
17:	000000001111112222223344	18:	000000001111112222223344
	113345673345673344555656		113345673345663344555756
	224567884856786778687887		224567884857786778686887
19:	000000001111112222223344	20:	000000001111112222223344
	113345673345663344555657		113345673345563344566755
	224567884857787867686788		224567884876785678877868
21:	000000001111112222223344	22:	000000001111112222223344
	113345673345563344576655		113345673345563344575656
	224567884876785867687788		224567884786876857687887
23:	000000001111112222223344	24:	000000001111112222233333
	113345673345563344565756		112445672445674455644556
	224567884876786758876887		233567883567887868778687
25:	000000001111112222233333	26:	000000001111112222233334
	112445672445674455644556		112445672345673455644565
	233567883568787867878687		233567884567888768778678
27:	000000001111112222233334	28:	000000001111112222233334
	112445672345673455644565		112445672345673455644565
	233567884568787868778687		233567884568788767878678
29:	000000001111112222233334	30:	000000001111112222233334
	112445672345673455644565		112445672345673455644565
	233567884576886878778876		233567884576888678778678
31:	000000001111112222233334	32:	000000001111112222233334
	112445672345673455644565		112445672345673455644565
	233567884586787768868877		233567884586788768767788
33:	000000001111112222233344	34:	000000001111112222233344
	112355672344673345645655		112355672344673345645556
	234467885568787868776878		234467885658786778887867
35:	000000001111112222233344	36:	000000001111112222233344
	112345672345673345645655		112345672345673345645655
	234567885468787876886778		234567885468787886776878

Table 5.6 *The 36 nonisomorphic* $TS(9,2)$ *designs*

1:	`000000111112222233334445556`
	`13579b3469a3467867868a7897`
	`2468ac578bc95acbbacc9bbac9`
2:	`000000111112222233334445556`
	`13579b3469a3467867868a7897`
	`2468ac578bc95abcbcac9babc9`

Table 5.7 *The two nonisomorphic TS*(13, 1) *designs*

when $\lambda = 4$, and six designs when $\lambda = 6$. Ivanov (1983, 1985) found 13 when $\lambda = 8$, 19 when $\lambda = 10$, and 34 when $\lambda = 12$. Denny (1998) produced all 48 TS(6,14)s and all 76 TS(6,16)s.

For TS(7,λ)s, we have already seen 1, 4, and 10 systems for $\lambda = 1, 2, 3$, respectively. Ivanov (1985) found 35, 109, and 417 systems for $\lambda = 4, 5, 6$, respectively; the number of TS(7,4)s was also determined by Gronau and Prestin (1982). Pietsch (1994) corrected the number of TS(7,6)s to 418, and produced all 1508, 5413, and 17785 TS(7,λ)s for $\lambda = 7, 8, 9$, respectively. Mathon and Pietsch (1999) and Denny (1998) produced all 54613, 155118, and 412991 for $\lambda = 10, 11, 12$, respectively. Denny (1998) produced all 1033129, 2449592, 5528257, and 11944627 TS(7,λ)s for $\lambda = 13, 14, 15, 16$, respectively. Mathon and Rosa (1996) report that Spence determined the number of TS(8,6)s to be 3,077,244.

Mathon and Lomas (1992) establish that there are precisely 22,521 nonisomorphic TS(9,3)s. TS(9,3)s without repeated blocks were earlier enumerated by Harms, Colbourn, and Ivanov (1987), and 332 such triple systems exist (this corrects earlier counts of 330 in Ivanov (1983) and 329 in Harnau (1980)). Mathon and Rosa (1996) report that Spence determined the number of TS(9,4)s to be 16,585,031.

Ganter, Gülzow, Mathon, and Rosa (1978) found all 566 TS(10,2)s with repeated triples. Colbourn, Colbourn, Harms, and Rosa (1983) determined that there are 394 TS(10,2)s without repeated triples, for a total of 960 nonisomorphic TS(10,2)s; this concurs with Ivanov (1983).

Explicit formulas for the numbers of TS(v, λ)s are, for the most part, unknown. However, when v is small, some explicit results are known. Engel and Gronau (1988) determined numbers of TS(6,λ)s, and Mathon (1991) proves:

Theorem 5.1 *The number of nonisomorphic triple systems of order* 6 *and index* $2n$ *is* $\sum_{k=0}^{5} a_{k,n \bmod 12} \lfloor n/12 \rfloor^k$, *where* $A = (a_{ij})$ *is a* 6×12 *matrix with rows and columns indexed from* 0, *as follows:*

$$A = \frac{1}{10} \begin{pmatrix} 10 & 10 & 40 & 60 & 130 & 190 & 340 & 480 & 760 & 1050 & 1530 & 2060 \\ 111 & 136 & 321 & 476 & 821 & 1176 & 1771 & 2436 & 3411 & 4536 & 6061 & 7836 \\ 295 & 620 & 1005 & 1480 & 2075 & 2820 & 3745 & 4880 & 6255 & 7900 & 9845 & 12120 \\ 1220 & 1400 & 1700 & 2120 & 2660 & 3320 & 4100 & 5000 & 6020 & 7160 & 8420 & 9800 \\ 360 & 720 & 1080 & 1440 & 1800 & 2160 & 2520 & 2880 & 3240 & 3600 & 3960 & 4320 \\ 864 & 864 & 864 & 864 & 864 & 864 & 864 & 864 & 864 & 864 & 864 & 864 \end{pmatrix}.$$

#	System	$\|G\|$	CI	PC	Sub	Pa
1:	00000001111112222223333444455556666	20160	7	56	15	105
	13579bd3478bc3478bc789a789a789a789a					
	2468ace569ade65a9edbcdecbeddebcedcb					
2:	00000001111112222223333444455556666	192	8	24	7	73
	13579bd3478bc3478bc789a789a789a789a					
	2468ace569ade65a9edbcdecbededcbdebc					
3:	00000001111112222223333444455556666	96	9	8	3	57
	13579bd3478bc3478bc789a789a789a789a					
	2468ace569ade65a9edbcdedebcedcbcbed					
4:	00000001111112222223333444455556666	8	9	8	3	49
	13579bd3478bc3478bc789a789a789a789a					
	2468ace569ade65a9edbcdedbecedcbcebd					
5:	00000001111112222223333444455556666	32	8	16	3	49
	13579bd3478bc3478bc789a789a789a789a					
	2468ace569ade65a9edbceddecbcbdeedbc					
6:	00000001111112222223333444455556666	24	8	12	3	37
	13579bd3478bc3478bc789a789a789a789a					
	2468ace569ade65a9edbceddecbcdbeebdc					
7:	00000001111112222223333444455556666	288	7	32	3	33
	13579bd3478bc3478bc789a789a789a789a					
	2468ace569ade65a9edbdececbdcedbdbce					
8:	00000001111112222223333444455556666	4	9	4	1	37
	13579bd3478bc34789c78ab789a789a789a					
	2468ace569ade65abedc9deedcbdebcbcde					
9:	00000001111112222223333444455556666	2	9	2	1	31
	13579bd3478bc34789c78ab789a789a789a					
	2468ace569ade65abedc9deedbcdecbbcde					
10:	00000001111112222223333444455556666	2	9	6	1	31
	13579bd3478bc34789c78ab789a789a789a					
	2468ace569ade65abedc9deedcbdcbebedc					
11:	00000001111112222223333444455556666	2	9	6	1	23
	13579bd3478bc34789c789a789a78ab789a					
	2468ace569ade65abedcdbeecdbd9cebecd					
12:	00000001111112222223333444455556666	3	9	1	1	32
	13579bd3478bc34789c789a789a789a78ab					
	2468ace569ade65abedcdbeecdbbecdd9ce					
13:	00000001111112222223333444455556666	8	9	4	1	33
	13579bd3478bc34789c789a789a78ab789a					
	2468ace569ade65abedcebdedcbd9cebcde					
14:	00000001111112222223333444455556666	12	9	0	1	37
	13579bd3478bc34789c789a78ab789a789a					
	2468ace569ade65abedcebdd9ceedcbbcde					
15:	00000001111112222223333444455556666	4	8	8	1	25
	13579bd3478bc34789a78ac789a789b789a					
	2468ace569ade65bcdee9bddbecadcecebd					
16:	00000001111112222223333444455556666	168	9	0	1	49
	13579bd3478bc34789a789a78bc789a789a					
	2468ace569ade65bcdeedcba9eddebccbed					

Table 5.8 *The 80 STS*(15)*s: part 1*

#	System	$\|G\|$	CI	PC	Sub	Pa
17:	000000011111122222233334444555556666 13579bd3478bc34789a789a789c78ab789a 2468ace569ade65bcdeedcbabedd9cecebd	24	8	12	1	25
18:	000000011111122222233334444555556666 13579bd3478bc34789a78ac789b789a789a 2468ace569ade65bcede9bdadcedebccbde	4	9	4	1	25
19:	000000011111122222233334444555556666 13579bd3478bc34789a789a78ab789c789a 2468ace569ade65bdceebdcc9deaebddceb	12	7	16	1	17
20:	000000011111122222233334444555556666 13579bd3478bc34789a78ac789b789a789a 2468ace569ade65bdece9bdacdedbcecebd	3	9	1	1	20
21:	000000011111122222233334444555556666 13579bd3478ac34789b789a789a78ab789c 2468ace569bde65acedbdcecedbd9ceeabd	3	9	1	1	20
22:	000000011111122222233334444555556666 13579bd3478ac34789b789a789c789a78ab 2468ace569bde65acedbdceeabdcedbd9ce	3	8	4	1	17
23:	000000011111122222233334444555566667 13579bd3478bc34589c789a58ab789789aa 2468ace569ade67abedcdbed9cebececdbd	1	9	1	0	18
24:	000000011111122222233334444555566667 13579bd3478bc34589c789a589a78978aba 2468ace569ade67abedcdbebecdecdd9ceb	1	9	0	0	19
25:	000000011111122222233334444555566667 13579bd3478bc34589c789a589a78978aba 2468ace569ade67abedcedbecbdbdcd9cee	1	9	1	0	20
26:	000000011111122222233334444555566667 13579bd3478bc3458ac789a589a789789ab 2468ace569ade67b9edcbededbcacddecbe	1	9	0	0	23
27:	000000011111122222233334444555566667 13579bd3478bc34589a78ab589c789789aa 2468ace569ade67bcded9ceaebdcdeebcdb	1	9	3	0	14
28:	000000011111122222233334444555566667 13579bd3478bc34589a78ab589c789789aa 2468ace569ade67bcedd9ceaebdedccbdeb	1	9	2	0	15
29:	000000011111122222233334444555566667 13579bd3478bc34589a789a58bc789789aa 2468ace569ade67bcedebdca9eddeccdbeb	3	9	0	0	19
30:	000000011111122222233334444555566667 13579bd3478bc34589a78ab589c789789aa 2468ace569ade67bdced9ceabedcedecbdb	2	9	3	0	14
31:	000000011111122222233334444555566667 13579bd3478bc34589a789a58bc789789aa 2468ace569ade67bdcedbeca9edcedecbdb	4	9	5	0	18
32:	000000011111122222233334444555566667 13579bd3478bc34589a78ab589c789789aa 2468ace569ade67bdcec9deaebddceebdcb	1	9	2	0	13

Table 5.9 *The 80 STS(15)s: part 2*

#	System	|G|	CI	PC	Sub	Pa
33:	000000011111122222233333444455566667 13579bd3478bc34589a78ab589c789789aa 2468ace569ade67bdecc9deaebdecddbceb	1	9	1	0	12
34:	000000011111122222233333444455566667 13579bd3478bc34589a789a589c78978baa 2468ace569ade67bdecdbceaebdecdc9edb	1	9	1	0	12
35:	000000011111122222233333444455566678 13579bd3478bc34569a689a579a78a789bc 2468ace569ade78bcdedbecedcbc9daebed	3	9	0	0	13
36:	000000011111122222233333444455566678 13579bd3478bc34569a689a579a78a789bc 2468ace569ade78bceddbceecdbd9caebed	4	9	1	0	10
37:	000000011111122222233333444455566678 13579bd3478bc34569a68ab579c78978aa9 2468ace569ade78bced9dceaebdceddbebc	12	9	5	0	6
38:	000000011111122222233333444455566678 13579bd3478bc34569a68ab579a78978ac9 2468ace569ade78bdce9cdecedbadebecdb	1	9	4	0	9
39:	000000011111122222233333444455566678 13579bd3478bc34569a68ab579c78978aa9 2468ace569ade78bdce9dceabedecdcebdb	1	9	1	0	12
40:	000000011111122222233333444455566678 13579bd3478bc34569a689a579a78a789bc 2468ace569ade78bdcecbedecdbd9caebed	1	9	0	0	13
41:	000000011111122222233333444455566678 13579bd3478bc34569a689c57ab78a789a9 2468ace569ade78bdceabed9dceecdcebbd	1	9	1	0	12
42:	000000011111122222233333444455566678 13579bd3478bc34569a68ab579c78978aa9 2468ace569ade78bdec9cdeaebddeccbebd	2	9	5	0	8
43:	000000011111122222233333444455566678 13579bd3478bc34569a68ac579b78978aa9 2468ace569ade78bdec9ebdacdeedcbcedb	6	9	3	0	10
44:	000000011111122222233333444455566678 13579bd3478bc34569a689a579a78a789bc 2468ace569ade78bdeccbdeedcbc9daebed	2	9	1	0	8
45:	000000011111122222233333444455566678 13579bd3478bc34569a68ac579a78978ab9 2468ace569ade78becd9bededbcacdcdbee	1	9	2	0	9
46:	000000011111122222233333444455566678 13579bd3478bc34569a68ab579c78978aa9 2468ace569ade78becd9dceabedceddcbeb	1	9	2	0	7
47:	000000011111122222233333444455566678 13579bd3478bc34569a68ab579c78978aa9 2468ace569ade78bedc9cdeabeddeccdbeb	1	9	1	0	10
48:	000000011111122222233333444455566678 13579bd3478bc34568a69ab578c78979aa9 2468ace569ade79becd8dceabedcdedcbeb	1	9	1	0	8

Table 5.10 *The 80 STS(15)s: part 3*

#	System	$\|G\|$	CI	PC	Sub	Pa
49:	00000001111112222223333444455566679 13579bd3478bc34568a689a578a78978abc 2468ace569ade79becddebccdbeadec9bed	1	9	2	0	7
50:	00000001111112222223333444455566678 13579bd3478bc34568a69ac578a78979ab9 2468ace569ade79becd8bededbcadccdbee	1	8	7	0	6
51:	00000001111112222223333444455566678 13579bd3478bc34568968ab579c79a78aa9 2468ace569ade7abecd9dce8ebddcecbdbe	1	9	2	0	9
52:	00000001111112222223333444455566678 13579bd3478bc345689689a579c79a78aab 2468ace569ade7abecdcdeb8ebddceb9dce	1	9	0	0	9
53:	00000001111112222223333444455566678 13579bd3478bc345689689a579c79a78aab 2468ace569ade7abecdbdec8ebddcec9dbe	1	9	1	0	10
54:	00000001111112222223333444455566679 13579bd3478bc345689689a578c78a78aab 2468ace569ade7abecdbecd9ebdcded9cbe	1	9	2	0	11
55:	00000001111112222223333444455566678 13579bd3478bc34568969ab578c78a79aa9 2468ace569ade7abecd8cde9ebdcdedbcbe	1	9	2	0	9
56:	00000001111112222223333444455566678 13579bd3478bc345689689a579c79a78aab 2468ace569ade7abedcdceb8ebdcdeb9cde	1	9	1	0	8
57:	00000001111112222223333444455566678 13579bd3478bc34568968ab579c79a78aa9 2468ace569ade7abedc9cde8ebdcdedbcbe	1	8	4	0	5
58:	00000001111112222223333444455566678 13579bd3478bc345689689a579c79a78aab 2468ace569ade7abedcbced8ebdcded9cbe	1	9	3	0	8
59:	0000000111111222222333344445556667a 13579bd3478bc345689689a578a789789cb 2468ace569ade7ebacdcbeddb9caecedbde	3	9	0	0	13
60:	0000000111111222222333344445556667a 13579bd3478bc34568968a9578a789789cb 2468ace569ade7ebacddecbcb9dadeebcde	1	9	6	0	7
61:	00000001111112222223333444455556666 13579bd3478ac34789b789c789a78ab789a 2468ace569bde65aecdbaedecdbd9cecdbe	21	7	7	1	14
62:	00000001111112222223333444455566667 13579bd3478ac34589b78ab589a789789ca 2468ace569bde67adcec9edbedcdceeabdb	3	9	0	0	7
63:	00000001111112222223333444455566667 13579bd3478ac34589b78ab589a789789ca 2468ace569bde67aecdd9cebcdecdeeabdb	3	8	6	0	7
64:	00000001111112222223333444455566667 13579bd3478ac34589a789a589b78b789ca 2468ace569bde67cdbeecdbaecdd9ebaedc	3	9	3	0	10

Table 5.11 *The 80 STS*(15)*s: part 4*

#	System	$\|G\|$	CI	PC	Sub	Pa
65:	000000011111122222233333444455566678 13579bd3478ac34569b68ac579a789789ba 2468ace569bde78dacee9bdbedcacecdbde	1	9	2	0	7
66:	000000011111122222233333444455566678 13579bd3478ac34568968ac579a79b789ba 2468ace569bde7bdacee9bd8edcacecdbde	1	9	3	0	6
67:	000000011111122222233333444455566679 13579bd3478ac34568a68ac578b789789ab 2468ace569bde79cbdee9bdaecdbeddacce	1	8	4	0	5
68:	000000011111122222233333444455566678 13579bd3478ac34569b68ac579a789789ba 2468ace569bde78adcee9bdbedccdeacbde	1	9	1	0	6
69:	000000011111122222233333444455566679 13579bd3478ac3456a868ac578b789789ab 2468ace569bde79bcede9bdaecddecbadce	1	8	4	0	5
70:	00000001111112222223333344445556667a 13579bd3478ac34568968ab5789789789cb 2468ace569bde7dbaece9cdceabadebcdde	1	9	2	0	9
71:	00000001111112222223333344445556667a 13579bd3478ac34568968ab5789789789cb 2468ace569bde7cabdee9cddeabbecacdde	1	9	2	0	5
72:	00000001111112222223333344445556667a 13579bd3478ac34568968ab5789789789cb 2468ace569bde7dcbaee9cdaecbbedadcde	1	9	4	0	5
73:	000000011111122222233333444455566679 13579bd3478ac34568a689a578b78c789ab 2468ace569bde7c9bdeecdbae9dbeddacce	4	9	9	0	6
74:	00000001111112222223333344445556667a 13579bd3478ac345698689a578b789789cb 2468ace569bde7abdecedbcce9daedbacde	4	9	3	0	8
75:	00000001111112222223333344445556667a 13579bd3478ac345689689a578b789789cb 2468ace569bde7cdbaeedbcae9dbecacdde	3	8	6	0	7
76:	000000011111122222233333444455566678 13579bd3478ac34569b68ab578978979aac 2468ace569bde7da8cee9cdcbaedebcdbed	5	9	1	0	10
77:	000000011111122222233333444455566678 13579bd3478ac34569a68ab578979b789ac 2468ace569bde7d8cebe9cdacebdcebaded	3	9	1	0	2
78:	000000011111122222233333444455566678 13579bd3478ac34569b689c578b78a79aa9 2468ace569bde7ac8edeabd9cdedebbdcec	4	9	9	0	6
79:	000000011111122222233333444455566678 13579bd3478ac34568b689c57ab79a789a9 2468ace569bde79ecadaebd8dcecdbbdeec	36	8	17	0	6
80:	000000011111122222233333444455566678 13579bd3469ac34578b678a58ab78979c9a 2468ace578bde96aecdbcded9cebecaeddb	60	9	11	0	0

Table 5.12 *The 80 STS(15)s: part 5*

Goulden and Pietsch (1999) give an exponential generating function for the number of TS(6,$2n$)s.

5.2 Asymptotics

Let $NN(n)$ denote the number of nonisomorphic STS(n)s, and let $ND(n)$ denote the number of distinct STS(n)s. Assume throughout this section that $n \equiv 1, 3 \pmod 6$. Kirkman (1847) showed that $NN(n) \geq 1$, and Moore (1893) showed that $NN(n) \geq 2$ for $n \geq 13$. Moore (1895) suggested that $NN(n)$ grows rapidly as a function of n, and this sentiment was repeated by Bays (1923a) and Witt (1938). White (1915a) obtained a large lower bound on $NN(31)$ by considering systems having a subsystem of order 15.

Assmus and Mattson (1966) determined that $NN(n) \to \infty$ as $n \to \infty$ in the special case that n is of the form $2^\alpha - 1$. Doyen and Valette (1971) established that $NN(n) \to \infty$ as $n \to \infty$ for all admissible orders. Aleksejev (1974) obtained the lower bound that

$$NN(n) \geq n^{\frac{n^2}{12} - O(\frac{n^2}{\log(n)})}.$$

Wilson (1974) used similar techniques, but obtained a more precise lower bound, which we discuss in some detail.

First let us consider the (much easier) upper bound. Evidently, $NN(n) \leq ND(n) \leq n!NN(n)$, so we consider upper bounds on $ND(n)$. Doyen and Valette (1971) proved:

Lemma 5.2 $ND(n) \leq (n-2)^{n^2/6}$.

Proof Let $(V, \mathcal{B})$ be an STS(n). Place an arbitrary total order $\prec$ on V, and let $\sqsubset$ be the total order on $\binom{V}{2}$ defined by $\{u, v\} \sqsubset \{x, y\}$ whenever $u \prec v$, $x \prec y$, and either $v \prec y$, or $v = y$ and $u \prec x$. This is the colexicographic order on pairs induced by the total order on elements. For each triple $B_i = \{x, y, z\} \in \mathcal{B}$, call the first pair in B_i in the order $\sqsubset$ the *representative pair*, P_i. Now order the blocks $B_1, \ldots, B_b$ of $\mathcal{B}$ so that $P_i \sqsubset P_{i+1}$ for $1 \leq i < b$. For $1 \leq i \leq b$, let x_i be the element for which $P_i \cup \{x_i\} = B_i$. Form the b-tuple $(x_1, \ldots, x_b)$. Since every distinct STS(n) leads to a distinct b-tuple in this way, $ND(n)$ does not exceed the number of ways to form a b-tuple of elements from V, not using either of the first two elements under $\prec$. □

Lemma 5.2, by mapping into b-tuples, constructs an overestimate that appears to be quite loose. Wilson (1974) improved the bound, by examining the process of constructing a triple system by repeatedly choosing a triple which is consistent (i.e., shares no pair) with those already chosen. To obtain a bound, a method is needed to bound the number of ways in which the next triple can be chosen:

Lemma 5.3 *Let G be a simple n-vertex, m-edge graph. Then the number of triangles in G is bounded by $((\sqrt{2})/3)m^{1.5}$.*

Proof (Sketch) Write $m = \binom{k}{2} + j$ where $0 \le j < k$. We can verify that the number of triangles in an m-edge graph is maximized at $\binom{k}{3} + \binom{j}{2}$ by the graph consisting of a complete subgraph (clique) on k vertices, one more vertex adjacent to j of the clique vertices, and all other vertices isolated. The bound follows by some simple algebra. □

Wilson (1974) draws the relevant conclusion:

Theorem 5.4 $NN(n) \le ND(n) \le \left(\frac{n}{\sqrt{e}}\right)^{n^2/6}$.

Proof Let $\widehat{D}(n)$ be the number of b-tuples of triples on an n-set, so that the unordered set of triples is an STS(n). Then $\widehat{D}(n) = b!ND(n)$. We bound $\widehat{D}(n)$ by considering the number of ways in which we can build up a b-tuple of triples corresponding to a partial triple system, one triple at a time. For $0 \le t < b$, to choose the $(t+1)$st triple, we must avoid the $3t$ edges already used in previously chosen triples and hence we must choose a triangle in a graph with $3(b-t)$ edges. By Lemma 5.3, this can be done in at most $\sqrt{6}(b-t)^{1.5}$ ways. In choosing all t triples sequentially, we obtain

$$\widehat{D}(n) \le \prod_{t=0}^{b-1} \sqrt{6}(b-t)^{1.5} = 6^{b/2}(b!)^{1.5}.$$

Consequently, $ND(n) \le 6^{b/2}(b!)^{0.5}$, or equivalently $[ND(n)]^2 \le 6^b b!$. Using Stirling's formula for $b!$, we obtain $[ND(n)]^2 \le (6b/e)^{b+1}$, and the bound follows upon taking square roots. □

It appears at first that this upper bound remains quite loose, since the greedy selection of triples is done under extremely optimistic assumptions about the pairs that remain. Wilson (1974) conjectures that

$$\lim_{n\to\infty} n^{-1} NN(n)^{6/n^2} = e^{-2},$$

while Theorem 5.4 shows only that the limit is at most $e^{-0.5}$.

Now we turn to the lower bound of Aleksejev (1974) and Wilson (1974). Let $ND_w(n)$ be the number of distinct Steiner triple systems of order n which contain a subsystem of order w on w fixed points. Let $L(s)$ be the number of distinct latin squares of order s.

Theorem 5.5 *For $n \ge 19$, there exist u and s satisfying $(n+14)/3 \ge u \ge n/3 \ge s \ge (n-7)/3$ for which $ND_7(n) \ge L(s)(ND_7(u))^3$.*

Proof Write $n = 3s + w$ with $w \in \{0,1,3,7\}$ and $s + w \equiv 1,3 \pmod 6$. Apply singular direct product using a latin square of order s, and incomplete triple systems of order $s+w$ each having a hole of size w. Fill the final hole with an STS(w). □

Theorem 6.7 establishes that $ND_7(n) \geq 1$ when $n \geq 15$ is admissible. In addition, every nontrivial STS has subsystems of orders 0, 1, and 3 which can be used in the singular direct product. The effective application of Theorem 5.5 requires a good lower bound on the number of latin squares. Using the inequality of Theorem 5.5, Wilson (1974) proves:

Theorem 5.6 *If* $\log L(n) \geq \alpha n^2(\log n - 2)$ *then for* $\delta > 0$ *and* n *sufficiently large,* $\log NN(n) \geq (\alpha v^2)/6\,(\log v - (3\log 3)/2 - 2 - \delta)$.

When Theorem 5.6 was proved initially, Wilson (1974) observed that $\alpha = 0.5$ can be chosen by bounds in Ryser (1969). He further noted from Ryser (1969) that if the van der Waerden Permanent Conjecture ('the permanent of a doubly stochastic $n \times n$ matrix is at least $(n!)/(n^n)$') holds, we could then take $\alpha = 1$. Subsequently, Friedland (1979) made a significant breakthrough towards the van der Waerden Permanent Conjecture, which implies that $L(n) \sim (e^{-2}n)^{n^2}$ for $n \to \infty$ (see van Lint and Wilson, 1992). Indeed, the van der Waerden Permanent Conjecture has been proved by Falikman (1982) and Egorychev (1981). This improved bound on $L(n)$, along with Theorem 5.6, establishes a lower bound on $NN(n)$ which enables us to determine the asymptotic growth rate:

Theorem 5.7 $NN(n) = n^{n^2(\frac{1}{6}+o(1))}$.

5.3 Computational lower bounds

We have seen explicit results for small orders, and asymptotic results indicating an exponential growth rate. The production of larger catalogues of triple systems, and the determination of reasonable lower bounds on numbers of systems which are effective for small orders have also been the subject of much research.

White (1915a) developed a simple technique for establishing some surprisingly large lower bounds. He examined $NN(31)$, and for purposes of illustration we examine the same case here. When an STS(15) has exactly g automorphisms, it can evidently be written in $(15!)/g$ distinct ways. Examining the 80 STS(15)s, we therefore find exactly 60,230,793,823,200 distinct STS(15)s. From this, we find at least the same number of distinct Steiner 1-factorizations of order 16 (here the 15 1-factors are not to be permuted). Thus we find 60230793823200^2 distinct STS(31)s using the $v \to 2v+1$ construction, each having a subsystem of order 15 on the same set of points. Next, White observes that an STS(31) can have at most 31 subsystems of order 15. Now consider an STS(31) having a subsystem of order 15. After selecting a subsystem of order 15 (in one of at most 31 ways), the subsystem on 15 specified elements can be written in at most 15! ways, thereby fixing the 15 elements. This fixes the remaining elements, and thus the number of ways to write the system is bounded by $31 \cdot 15!$. Thus we find that

$$NN(31) \geq \frac{60230793823200^2}{31 \cdot 15!} > 89,490,280,468,632.$$

A more careful analysis can be done. If we select two particular STS(15)s, one having group order d and s sub-STS(7)s, and the other having group order d' and s' sub-STS(7)s, we obtain at least $(15!)/(dd') - ss'7!$ nonisomorphic STS(31)s by the $v \to 2v+1$ construction. Since these each have a unique subsystem of order 15, we obtain the much better bound that $NN(31) \geq 2,778,891,257,165,377$. White (1915a) obtained a much smaller bound as a result of having less complete information about the STS(15)s. While this bound for $NN(31)$ seems extremely large, the upper bound from Lemma 5.2 exceeds 10^{234}, and that from Lemma 5.4 exceeds 10^{169}. Even using the exact values from the proof of Lemma 5.3 reduces the bound only to 1.72×10^{166}. These are upper bounds on the number of distinct STS(31)s, but since 31! is approximately 8.22×10^{33}, the lower and upper bounds remain far apart.

Mathon and Rosa (1985) improve the lower bound for $NN(31)$ by observing that there are at least 36 nonisomorphic automorphism-free 1-factorizations of K_{16} having no sub-1-factorization of K_8, and hence that $NN(31) \geq 80 \cdot 36 \cdot 16! \geq 6 \times 10^{16}$. Indeed, we can do better. Consider the 1-factorization of $K_{8,8}$:

{1,9}, {2,10}, {3,11}, {4,12}, {5,16}, {6,14}, {7,13}, {8,15}
{1,11}, {2,9}, {3,10}, {4,13}, {5,12}, {6,15}, {7,16}, {8,14}
{1,10}, {2,11}, {3,9}, {4,14}, {5,15}, {6,16}, {7,12}, {8,13}
{1,12}, {2,14}, {3,13}, {4,9}, {5,10}, {6,11}, {7,15}, {8,16}
{1,13}, {2,16}, {3,14}, {4,15}, {5,9}, {6,10}, {7,11}, {8,12}
{1,14}, {2,12}, {3,15}, {4,16}, {5,13}, {6,9}, {7,10}, {8,11}
{1,15}, {2,13}, {3,16}, {4,11}, {5,14}, {6,12}, {7,9}, {8,10}
{1,16}, {2,15}, {3,12}, {4,10}, {5,11}, {6,13}, {7,14}, {8,9}

This is automorphism-free and contains no sub-1-factorization of $K_{4,4}$. Hence on each of the symbols $\{1,\ldots,8\}$ and $\{9,\ldots,16\}$, one of the 6240 1-factorizations of K_8 can be placed in order to obtain an automorphism-free 1-factorization of K_{16}. Employing each of these with an STS(15) having no sub-STS(7) (of which there are 57 such STS(15)s) gives then a nonisomorphic STS(31). Hence we find $NN(31) \geq 57 \times 6240^2 \times 16! \geq 4.64 \times 10^{22}$.

Lower bounds arise from a number of constructions in this way. More accurate estimates appear to require some partial enumeration by computer. Stinson and Ferch (1985) employed a hill-climbing technique to find 2,111,276 nonisomorphic STS(19)s; Royle (1989) used a similar technique to generate over 1,000,000 nonisomorphic TS(12,2)s. These illustrate the remarkable 'combinatorial explosion' that occurs from one order to the next. Stinson and Ferch (1985), by examining numbers of STS(19)s generated having distinct isomorphism invariants, and then by examining numbers of STS(19)s generated having nontrivial subsystems, obtain estimates of $NN(19)$ in the range from 0.354×10^9 to 1.309×10^9. McKay (see Mathon and Rosa, 1996) has since established by computational search that $NN(19) \geq 1.1 \times 10^9$, but the true value remains unknown.

Theorem 7.81 establishes that 'most' STSs have no nontrivial automorphisms. For obvious reasons, focus on generating triple systems has more recently been on those systems exhibiting extra properties. For example, Stinson and Seah (1986) generated all STS(19)s having a nontrivial subsystem, and Colbourn, Magliveras, and Stinson (1992) generated all STS(19)s having a nontrivial group of automorphisms.

A permutation can be specified simply by giving the collection of cycle lengths in the permutation. We write this collection in exponential notation: $\ell_1^{n_1} \cdots \ell_s{}^{n_s}$ is the collection containing n_i entries equal to ℓ_i for $1 \leq i \leq s$. We classify nonidentity permutations as follows. A *nonbasic* permutation π is one for which there exists an i so that π^i has a different collection of cycle lengths than π, and π^i is not the identity. The permutation is *basic* otherwise. Our main concern, naturally, is with basic permutations.

We present for completeness some data about the STS(19)s having a nontrivial group of automorphisms. Colbourn, Magliveras, and Stinson (1992) focus on automorphisms of type $1^n a^m$, where $n \in \{0, 1, 3, 7\}$ and $a \geq 2$ is prime. These are the *basic* automorphisms. The numbers of nonisomorphic STS(19)s with each type of basic automorphism are:

19^1	4
$1^1 2^9$	184
$1^1 3^6$	12021
$1^3 2^8$	80591
$1^7 2^6$	80558
$1^7 3^4$	124

Of course, many STS(19)s admit more than one of these automorphisms. Table 5.13 gives the numbers of STS(19)s admitting each subset of the basic automorphisms. In Table 5.14, we give the same information for the nonbasic automorphisms. We use the partitioning of Table 5.13 again in Table 5.14; thus we omit the information about basic automorphisms, since this can be determined by looking in Table 5.13. Although two STS(19)s may have the same group order and class (same subset of the basic automorphisms), they need not have the same subset of nonbasic automorphisms. Hence some classes of Table 5.13 partition further in Table 5.14. In listing the types of the nonbasic automorphisms, we omit the fixed points. Systems whose group is generated by a single basic automorphism are omitted. Any type of automorphism not explicitly mentioned in Table 5.13 or 5.14 does not arise as the type of an automorphism of an STS(19).

5.4 Work points

1. Prove Wilson's conjecture: $\lim_{n\to\infty} n^{-1} NN(n)^{6/n^2} = e^{-2}$.
2. Determine asymptotically the number of TS(n, λ)s as $n \to \infty$ for fixed $\lambda > 1$.
3. Complete the constructive enumeration of the STS(19)s.

Table 5.13 *Basic automorphisms of STS(19)s*

Order	Class	19^1	$1^1 2^9$	$1^1 3^6$	$1^3 2^8$	$1^7 2^6$	$1^7 3^4$	#STSs
432				$\star$	$\star$	$\star$	$\star$	1
171		$\star$		$\star$				1
144				$\star$	$\star$	$\star$		1
108				$\star$	$\star$	$\star$	$\star$	1
96				$\star$	$\star$	$\star$		1
57		$\star$		$\star$				2
54				$\star$		$\star$	$\star$	2
32					$\star$	$\star$		3
24				$\star$	$\star$	$\star$		11
19		$\star$						1
18	a		$\star$	$\star$				1
	b			$\star$	$\star$		$\star$	2
	c			$\star$		$\star$	$\star$	6
	d			$\star$		$\star$		2
16					$\star$	$\star$		13
12	a			$\star$	$\star$	$\star$		8
	b			$\star$	$\star$			7
	c			$\star$		$\star$		12
	d				$\star$	$\star$	$\star$	10
9				$\star$				19
8	a				$\star$	$\star$		84
	b				$\star$			17
6	a		$\star$	$\star$				14
	b			$\star$	$\star$			14
	c			$\star$		$\star$		116
	d				$\star$		$\star$	10
	e					$\star$	$\star$	28
4	a				$\star$	$\star$		839
	b				$\star$			662
	c					$\star$		620
3	a			$\star$				11800
	b						$\star$	64
2	a		$\star$					169
	b				$\star$			78907
	c					$\star$		78800
totals		4	184	12021	80591	80558	124	172248

Table 5.14 *Nonbasic automorphisms of STS(19)s*

Class	9^2	6^3	3^26^2	2^14^4	2^18^2	8^2	4^4	2^26^2	2^24^3	#
432			⋆			⋆	⋆	⋆		1
171	⋆									1
144			⋆	⋆	⋆		⋆			1
108			⋆				⋆			1
96			⋆					⋆		1
57										2
54			⋆							2
32				⋆			⋆			3
24			⋆							11
18a		⋆								1
18b			⋆							2
18c			⋆							6
18d								⋆		2
16						⋆	⋆			1
							⋆			1
				⋆	⋆		⋆			5
				⋆						6
12a			⋆							8
12b										7
12c										12
12d								⋆		10
9	⋆									9
										10
8a										84
8b					⋆		⋆			10
				⋆			⋆			5
						⋆	⋆			2
6a		⋆								14
6b										14
6c			⋆							104
										12
6d								⋆		10
6e										28
4a										839
4b				⋆						498
							⋆			153
										11
4c									⋆	48
										572
totals	10	15	137	518	16	4	183	24	48	

6

SUBSYSTEMS AND HOLES

A *partial triple system* PTS(v, λ) is a set V of v elements and a collection $\mathcal{B}$ of triples, so that each unordered pair of elements occurs in at most λ triples of $\mathcal{B}$. Its *leave* is the multigraph on vertex set V in which the edge $\{x, y\}$ appears $\lambda - s$ times when there are precisely s triples of $\mathcal{B}$ containing $\{x, y\}$.

An *incomplete triple system* of order v, index λ, and with a *hole* of size w, is a PTS(v, λ) $(V, \mathcal{B})$ with the property that for some $W \subseteq V$, $|W| = w$,

1. if $x, y \in W$, no triple of $\mathcal{B}$ contains $\{x, y\}$; and
2. if $x \in V \setminus W$ and $y \in V$, then exactly λ triples of $\mathcal{B}$ contain $\{x, y\}$.

Such a partial system is denoted ITS$(v, w; \lambda)$. When an ITS$(v, w; \lambda)$ $(V, \mathcal{B})$ exists having a hole on $W \subseteq V$, if there is a TS(w, λ) $(W, \mathcal{D})$, the system $(V, \mathcal{B} \cup \mathcal{D})$ is a TS(v, λ). In this system, the TS(w, λ) is a *subsystem.*

One method to produce incomplete triple systems is to remove all triples of a subsystem to form the hole. However, only those incomplete triple systems in which the size of the hole is λ-admissible can be produced in this way. Hence many results originally proved for triple systems with subsystems have been generalized to triple systems with holes.

6.1 One subsystem or hole

If an ITS$(v, w; \lambda)$ is to exist, evidently $w \leq v$; and $w = v$ exactly when the system contains no triples at all. If $w = 0$, the ITS is just a TS(v, λ). Suppose that $0 < w < v$ and that $\lambda > 0$. Let the elements in the hole be $\{h_1, \ldots, h_w\}$ and the elements not in the hole be $\{x_1, \ldots, x_{v-w}\}$. Consider the triples involving the fixed element x_1. Since no pair $\{h_i, h_j\}$ $(i \neq j)$ appears in a triple, there must be λw distinct triples of the form $\{x_1, h_i, x_{j_i}\}$. Since x_1 appears with each of $\{x_2, \ldots, x_{v-w}\}$ in λ triples, it follows that $v - w - 1 \geq w$. Thus we obtain a basic necessary condition for a hole:

Lemma 6.1 *If an ITS$(v, w; \lambda)$ exists, and $\lambda > 0$, then $v \geq 2w + 1$ or $v = w$.*

The size of a hole is also constrained by the requirement that the number of pairs not in the hole be a multiple of three, and that every element occurs in an even number of such pairs. Thus we obtain the further conditions when $w > 0$ that

$$\lambda\left(\binom{v}{2} - \binom{w}{2}\right) \equiv 0 \pmod{3}, \text{and}$$

$$\lambda(v-1) \equiv \lambda(v-w) \equiv 0 \pmod{2}.$$

For example, an ITS$(11, 5)$ can (and does) exist, but no ITS$(13, 5)$ can exist.

Summarizing these in a more convenient form, an ITS$(v, w; \lambda)$ can exist for $0 < w < v$ only if $v \geq 2w + 1$ and w is as in Table 6.1.

	λ (mod 6)			
v (mod 6)	0	1, 5	2, 4	3
0, 4	$w > 0$	no w	$w \equiv 0, 1(3)$	no w
1, 3	$w > 0$	$w \equiv 1, 3(6)$	$w \equiv 0, 1(3)$	$w \equiv 1(2)$
2	$w > 0$	no w	$w \equiv 2(3)$	no w
5	$w > 0$	$w \equiv 5(6)$	$w \equiv 2(3)$	$w \equiv 1(2)$

Table 6.1 *Necessary conditions for ITS$(v, w; \lambda)$s*

Now suppose that an ITS$(v, w; \lambda)$ $(V, \mathcal{B})$ exists with hole W. Let $X = V \setminus W$, and define a multigraph G with vertex set X, and edges $E = \bigcup_{w \in W} E_w$, where

$$E_w = \{\{y, z\} : \{w, y, z\} \in \mathcal{B}\},$$

including an edge as many times as the pair occurs in $\mathcal{B}$ with w. G is a λw-regular multigraph; and G has a λ-factorization given by $\{E_w : w \in W\}$. Moreover, the triples $\{B \in \mathcal{B} : B \cap W = \varnothing\}$ partition the remaining edges of λK_{v-w} into triangles. Thus an ITS$(v, w; \lambda)$ is equivalent to a partition of λK_{v-w} into w λ-factors, and $\lambda v(v-2w-1)/6$ triangles. Let us call such a partition of λK_n having w λ-factors an *$(n, w; \lambda)$-decomposition.*

Lemma 6.2 *There is an ITS$(v, w; \lambda)$ if and only if a $(v-w, w; \lambda)$-decomposition exists.*

Now we develop some constructions for $(n, w; \lambda)$-decompositions.

Lemma 6.3 *If $s \equiv 0, 1 \pmod 4$, $n = 6s + t$ for $t \geq 1$, and $t\lambda$ is even, then an $(n, w; \lambda)$-decomposition exists for all $t - 1 \leq w \leq 6s + t - 1$ for which $w \equiv t - 1$ $(\mathrm{mod}\ 6/\gcd(6, \lambda))$.*

Proof Let (a_i, b_i), $1 \leq i \leq s$ be a pure Skolem sequence (Theorem 2.19). Form a collection of s sets of blocks $\{\mathcal{O}_i : i = 1, \ldots, s\}$ by taking

$$\mathcal{O}_i = \{\{j, s + a_i + j, s + b_i + j\} : 0 \leq j < 6s + t\},$$

arithmetic modulo $6s+t$. Extend this collection by defining $\mathcal{O}_i = \mathcal{O}_k$ when $i \equiv k \pmod s$. Let $\ell = \lambda s - (\lambda(w - (t - 1)))/6$. Now define a collection of triples $\mathcal{B} = \bigcup_{i=1}^{\ell} \mathcal{O}_i$. The pairs occurring in triples of $\mathcal{B}$ form a $(6\lambda s - \lambda(w - t + 1))$-regular multigraph. Hence the pairs that remain form a λw-regular multigraph. When λ is even, this multigraph has a λ-factorization by Petersen's theorem (Theorem 1.22). When λ is odd, a little more effort is involved. Since λ is odd, t is even. The multigraph of remaining pairs is the union of λ cyclic graphs on $6s+t$ vertices. Since edges of difference $3s + (t/2)$ all remain, each of the λ cyclic graphs admits a 1-factorization by Theorem 1.16. Taking unions of the resulting 1-factors, λ at a time, yields the λ-factorization required. □

Using hooked Skolem sequences instead (from Theorem 2.20), we obtain a similar result:

Lemma 6.4 *If $s \equiv 2,3 \pmod 4$, $n = 6s + t$ for $t \geq 3$, and $t\lambda$ is even, then an $(n, w; \lambda)$-decomposition exists for all $t - 1 \leq w \leq 6s + t - 1$ for which $w \equiv t - 1 \pmod{6/\gcd(6, \lambda)}$.*

Using split Skolem sequences, we obtain a similar construction:

Lemma 6.5 *Let $s \geq 0$, $n = 6s + t$ for $t \equiv 0 \pmod 3$, $t \geq 3$. Let $\delta \in \{0, 1, 2\}$, with $\delta\lambda$ even. If $t\lambda$ is even, then an $(n, w; \lambda)$-decomposition exists for all $t - 1 - \delta \leq w \leq 6s + t - 1 - \delta$ for which $w \equiv t - 1 - \delta \pmod{6/\gcd(6, \lambda)}$.*

Proof Let (a_i, b_i), $i = 1, \ldots, s$ be a split (Theorem 2.21) or split hooked (Theorem 2.22) Skolem sequence of order s. In either case, define the set of triples

$$\mathcal{O}_i = \left\{ \left\{ j, j + \frac{6s+t}{3} + a_i, j + \frac{6s+t}{3} + b_i \right\} : 0 \leq j < 6s + t \right\},$$

arithmetic modulo $6s + t$, for $i = 1, \ldots, s$. Let $\ell = \lambda s - (\lambda(w - (t - 3)))/6$. Now define a collection of triples $\mathcal{B} = \bigcup_{i=1}^{\ell} \mathcal{O}_i$. Add to $\mathcal{B}$ $\delta\lambda/2$ copies of each of the triples $\{j, j + (6s + t)/3, j + 2((6s + t)/3)\}$ for $0 \leq j < (6s + t)/3$. The pairs occurring in triples of $\mathcal{B}$ form a $(6\lambda s - \lambda(w - t + 3) + (2 - \delta)\lambda)$-regular multigraph. Hence the pairs that remain form a λw-regular multigraph. Proceed now to λ-factorize the remaining multigraph, using Theorem 1.22 when λ is even. When λ is odd, Theorem 1.16 gives the factorization. □

A more complicated construction is also needed; we prove a preliminary result.

Lemma 6.6 *If the integers $\mathcal{D} = \{1, \ldots, 3s - 1\} \setminus \{2s\}$ can be partitioned into classes $C_1, \ldots, C_m, X$ where*

1. *$C_i = \{a_i, b_i, c_i\}$ with $a_i + b_i \equiv \pm c_i \pmod{6s}$; and*
2. *$a_1 \equiv 1 \pmod 6$, $b_1 \equiv 1 \pmod 3$,*

then there is a $(6s, 6s - 6m + 1; 1)$-decomposition.

Proof Produce $6s - 6m - 3$ 1-factors $F_5, \ldots, F_{6s-6m+1}$ by applying Theorem 1.16 to the cyclic graph with differences $X \cup \{2s\}$. Then for $2 \leq i \leq m$, let

$$\mathcal{T}_i = \{\{j, a_i + j, b_i + j\} : j \in \mathbb{Z}_{6s}\}.$$

Let $\mathcal{T}_{m+1} = \{\{j, 2s + j, 4s + j\} : j \in \mathbb{Z}_{2s}\}$. It remains to treat the pairs whose difference occurs in C_1. Suppose without loss of generality that $b_1 \equiv 1 \pmod 3$. Let

$$\mathcal{T}_1 = \{\{3j, 3j + a_i, 3j + a_i + b_i\} : i \in \mathbb{Z}_{2s}\}.$$

Form a 2-regular graph G on $\mathbb{Z}_{6s}$ with edges

$$\{\{3j, 3j + b_1\}, \{3j + 1, 3j + 1 + b_1\}, \{3j + 2, 3j + 2 + a_1\} : j \in \mathbb{Z}_{2s}\}$$

v	w	C_1	$C_2,\ldots,C_{(v-w+1)/6}$
25	7	$\{1,4,5\}$	$\{3,7,8\}$
31	13	$\{1,7,8\}$	
37	13	$\{1,10,11\}$	$\{2,7,9\}$
43	13	$\{1,13,14\}$	$\{4,5,9\}$, $\{2,6,8\}$
49	13	$\{1,16,17\}$	$\{2,11,13\}$, $\{4,10,14\}$, $\{6,9,15\}$

Table 6.2 *Small cases for subsystems*

and a 2-regular graph H on $\mathbb{Z}_{6s}$ with edges

$$\{\{3j,3j+b_1\},\{3j+1,3j+1+b_1\},\{3j+2,3j+2+a_1\} : j \in \mathbb{Z}_{2s}\}.$$

G and H consist of even length cycles, since $2b_1 + a_1 \equiv 1 \pmod 2$, so they can be 1-factored to give four additional 1-factors $F_1,\ldots,F_4$.

The triples $\cup_{i=1}^{m+1}T_i$ together with the 1-factors $F_1,\ldots,F_{6s-6m+1}$ form the required decomposition. □

These constructions establish the theorem of Doyen and Wilson (1972):

Theorem 6.7 *An ITS*(v,w) *exists whenever* $w \equiv 1,3 \pmod 6$ *or* $w = 0$, $v \equiv 1,3 \pmod 6$, *and* $v \geq 2w+1$.

Proof If $v \not\equiv w \pmod 6$, then $w \equiv v-w-1 \pmod 6$. Apply Lemma 6.3 or 6.4 provided that $w \geq 3$. If $v \equiv w \equiv 3 \pmod 6$, apply Lemma 6.4 with $\delta = 2$. For $v,w \equiv 1 \pmod 6$, we require more work. We need only treat cases with $v < 4w+3$, since $2w+1 \equiv 3 \pmod 6$ and there exists an ITS$(2w+1,w)$. We settle the relevant cases when $w \in \{7,13\}$ by appealing to Lemma 6.6 using the partitions given in Table 6.2, and applying the $3v-2$ construction to produce an ITS$(19,7)$. Thus we may assume that $w \geq 19$. Write $v-w = 6s$ and $w = 6s-6m+1$, noting that $s-m \geq 3$. Form a split or split hooked Skolem sequence (Theorem 2.21 or 2.22) $\{(a_r,b_r) : 1 \leq r \leq m\}$. Let

$$\mathcal{M}_i = \{i, 2s+m+1-a_i, 2s+m+1-b_i\}.$$

Let $C_1 = \{1,3s-2,3s-1\}$, and $C_i = \mathcal{M}_i$ for $2 \leq i \leq m$. Let

$$X = \{1,\ldots,3s-1\} \setminus \left(\{2s\} \cup \bigcup_{i=1}^{m} C_i\right).$$

Then $C_1,\ldots,C_m,X$ gives a partition of $\{1,\ldots,3s-1\}\setminus\{2s\}$, noting that $2s+m < 3s-2$ since $m \leq s-3$. Apply Lemma 6.6 to complete the proof. □

The constructions also settle the case when the hole size in an incomplete STS is not the order of a subsystem. This case was first settled by Mendelsohn and Rosa (1983):

Theorem 6.8 *An ITS*(v,w) *exists whenever* $w \equiv v \equiv 5 \pmod 6$ *and* $v \geq 2w+1$.

Proof Apply Lemma 6.3 or 6.4. □

Stern (1980) generalized the Doyen–Wilson theorem (6.7) to all indices; the more general determination of possible sizes of a hole is *Stern's theorem:*

Theorem 6.9 *An ITS$(v, w; \lambda)$ exists if and only if*

1. $w = 0$ *and* v *is* λ*-admissible;*
2. $v = w$*; or*
3. $0 < w < v$ *and*
   (a) $v \geq 2w + 1$;
   (b) $\lambda\left(\binom{v}{2} - \binom{w}{2}\right) \equiv 0 \pmod 3$; *and*
   (c) $\lambda(v-1) \equiv \lambda(v-w) \equiv 0 \pmod 2$.

Hoffman (1999) provides a further generalization.

Theorem 6.10 *Let* V *be a set of* v *elements, and* $W \subset V$. *Let* $w = |W|$. *A partial triple system on* V*, in which each pair* P *occurs in* 0, λ, *or* μ *triples depending on whether* $P \cap W$ *has* 0, 1, *or* 2 *elements, exists if and only if these conditions are satisfied:*

1. *if* $w < v$ *then* $\lambda w \equiv 0 \pmod 2$;
2. *if* $w > 0$ *then* $\mu(w-1) - \lambda(v-w) \equiv 0 \pmod 2$;
3. $w(\mu(w-1) - \lambda(v-w)) \equiv 0 \pmod 3$; *and*
4. *if* $w > 0$ *then* $\lambda(v-w) \leq \mu(w-1)$, *with equality if* $w = 2$.

Generalizing to cases with two prescribed holes, intersecting or disjoint, appears to be substantially more difficult. We defer this to Chapter 21, to develop the required auxiliary results. Theorems concerning subsystems admit a number of important generalizations in the context of pairwise balanced designs. A $(K \cup \{k^\star\})$-PBD denotes a PBD with at least one block of size k if $k \in K$, and with exactly one block of size k if $k \notin K$. With this notation, Theorem 6.8 concerns $\{3, w^\star\}$-PBDs with $w \equiv 5 \pmod 6$. Hartman and Heinrich (1993) handled most cases when $K = N_{\geq 3}$, the set of all integers at least three, leaving 16 possible exceptions. Heathcote (1993) settled one of these, and Chee, Colbourn, Gallant, and Ling (1997) settled the remainder, to obtain:

Theorem 6.11 *An $(N_{\geq 3} \cup \{k^\star\})$-PBD of order* v *exists if and only if* $v \geq 2k+1$ *except when*

1. $v = 2k + 1$ *and* $k \equiv 0 \pmod 2$;
2. $v = 2k + 2$ *and* $k \not\equiv 4 \pmod 6$, $k > 1$;
3. $v = 2k + 3$ *and* $k \equiv 0 \pmod 2$, $k > 6$;
4. $v \in \{7, 8, 9, 10, 12, 13\}$ *and* $k = 2$;
5. $(v, k) \in \{(11, 4), (17, 6)\}$.

Colbourn, Haddad, and Linek (1996) prove an analogous result for $\mathcal{O}$, the set of all positive odd integers:

Theorem 6.12 *Let v, k be odd positive integers, and $v \geq 2k+1$. Then there exists an $(\mathcal{O} \cup \{k^\star\})$-PBD of order v. In fact, there exists a $\{3,5,k^\star\}$-PBD of order v.*

Another theorem along similar lines is very nearly proved in Hartman and Heinrich (1993), but apparently not stated explicitly:

Theorem 6.13 *A $\{3,4,k^\star\}$-PBD of order v exists if and only if $v \geq 2k+1$ and*

$$v \equiv k \equiv 2 \pmod 3, \text{ or } v \equiv 0,1 \pmod 3 \text{ and } k \equiv 0,1 \pmod 3,$$

except when

1. $v = 2k+1$ *and* $k \equiv 0 \pmod 2$;
2. $v = 2k+2$, $k \equiv 1 \pmod 6$, *and* $k > 1$;
3. $v = 2k+3$, $k \equiv 0 \pmod 6$, *and* $k > 6$; *or*
4. $(v,k) = (8,2)$.

6.2 Group divisible designs

In this section, we examine the more general situation in which a partial triple system has a spanning set of holes. Throughout this section, we assume that $\lambda = 1$.

The fundamental existence question for $\{3\}$-GDDs asks when a $\{3\}$-GDD (of index one) and group-type $g_1^{t_1} \cdots g_s^{t_s}$ exists. This question is far from settled. Even the necessary conditions appear elusive. Colbourn (1993) gives a number of necessary conditions, which we examine first.

Suppose that a $\{3\}$-GDD exists having s groups, with group sizes $\ell_1, \ldots, \ell_s$; by convention, we order the group sizes so that $\ell_1 \geq \ell_2 \geq \cdots \geq \ell_s$. What conditions must the group sizes $\ell_1, \ldots, \ell_s$ satisfy? We establish certain necessary conditions:

Lemma 6.14 *A $\{3\}$-GDD on v elements with group sizes $\ell_1 \geq \ell_2 \geq \cdots \geq \ell_s$ exists only if:*

1. $\binom{v}{2} \equiv \sum_{i=1}^{s} \binom{\ell_i}{2} \pmod 3$;
2. $v \equiv \ell_i \pmod 2$ *for all* $i = 1, \ldots, s$;
3. $\ell_1 \leq \sum_{i=3}^{s} \ell_i$;
4. $v_0(v - v_0) \leq 2\left[\binom{v_0}{2} + \binom{v-v_0}{2} - \sum_{i=1}^{s} \binom{\ell_i}{2}\right]$, *for* $v_0 = \sum_{i=1}^{s} \alpha_i \ell_i$ *and* $\alpha_i \in \{0,1\}$, $i = 1, \ldots, s$;
5. $2\ell_2\ell_3 \geq \ell_1\left[\ell_2 + \ell_3 - \sum_{i=4}^{s} \ell_i\right]$; *and*
6. *if* $\ell_1 = \sum_{i=3}^{s} \ell_i$, $2\ell_3\ell_4 \geq (\ell_1 - \ell_2)\left[\ell_3 + \ell_4 - \sum_{i=5}^{s} \ell_i\right]$.

Proof Condition (1) follows from the fact that each triple uses three pairs, so the number of pairs must be a multiple of three. For condition (2), observe that every element must be in an even number of pairs, and hence $v - \ell_i$ must be even. For condition (3), consider one of the ℓ_2 elements in the second group. It appears in ℓ_1 distinct triples with elements of the first group, and in each such triple the third element is not in the first two groups. For condition (4), consider the partition of the v elements into two sets, one of size v_0 and the other of size $v - v_0$, with each group entirely within one of the sets. Thus the right-hand side of the inequality gives twice the number of pairs inside the two sets that must appear in triples, while the left-hand side gives the number of pairs that contain one element from each set. Since every triple must use at least one edge inside one of the sets, the inequality follows.

For condition (5), we proceed as follows. If we delete all elements of the first group, the result must be a $\{2,3\}$-GDD with group sizes $\ell_2, \ldots, \ell_s$ having ℓ_1 parallel classes of edges. Consider the ℓ_1-regular graph that is the union of these parallel classes. Partition the elements into three classes V_1, V_2, V_3, where V_1 contains the ℓ_2 elements of the second group, V_2 contains the ℓ_3 elements of the third group, and V_3 contains the $\sum_{i=4}^{s} \ell_i$ elements of the remaining groups. Let ε_{ij} be the number of edges with one endpoint in V_i and the other in V_j. Evidently, $\varepsilon_{11} = \varepsilon_{22} = 0$ and $\varepsilon_{ij} = \varepsilon_{ji}$. We obtain a system of equations:

$$\begin{aligned} \varepsilon_{12} + \varepsilon_{13} &= \ell_1 \ell_2 \\ \varepsilon_{12} + \varepsilon_{23} &= \ell_1 \ell_3 \\ \varepsilon_{13} + \varepsilon_{23} + 2\varepsilon_{33} &= \ell_1 \left(\sum_{i=4}^{s} \ell_i \right) \end{aligned}$$

Now $\varepsilon_{12} \le \ell_2 \ell_3$, and thus we obtain $\varepsilon_{23} \ge \ell_3(\ell_1 - \ell_2)$ and $\varepsilon_{13} \ge \ell_2(\ell_1 - \ell_3)$. Thus $2\varepsilon_{33} \le \ell_1(\sum_{i=4}^{s} \ell_i - \ell_2 - \ell_3) + 2\ell_2\ell_3$, or $2(\ell_2\ell_3 - \varepsilon_{33}) \ge \ell_1(\ell_2 + \ell_3 - \sum_{i=4}^{s} \ell_i)$. Since $\varepsilon_{33} \ge 0$, the inequality follows.

Condition (6) is obtained similarly: If $\ell_1 = \sum_{i=3}^{s} \ell_i$, deleting all elements of the first two groups leaves a $\{2,3\}$-GDD with group sizes $\ell_3, \ldots, \ell_s$ in which the pairs form a $(\ell_1 - \ell_2)$-regular graph. □

It is not clear in advance that conditions (5) and (6) are not subsumed by condition (4). However, group sizes (13,7,7,1,1,1,1,1,1) meet conditions (1)–(4) but fail to meet condition (5) (since $2 \cdot 7 \cdot 7 \not\ge 13(7 + 7 - 6)$. In addition, it is not clear in advance that condition (6) is needed given conditions (1)–(5). However, (31,11,11,11,1,1,1,1,1,1,1,1,1) meets conditions (1)–(5), but fails to meet condition (6) (since $2 \cdot 11 \cdot 11 \not\ge (31 - 11) \cdot (11 + 11 - 9)$).

It appears likely that these conditions are not sufficient in general, as further structural conditions such as (5) and (6) can be expected. In fact, conditions (4), (5), and (6) all result from the requirement that one must be able to find ℓ_1 disjoint 1-factors in a complete multipartite graph $K_{\ell_2,\ldots,\ell_s}$, but are not by themselves sufficient to ensure that such a set of 1-factors exists.

The related question of when a complete multipartite graph on an even number of vertices with part sizes $\ell_1, \ldots, \ell_m$ admits a specified number η of edge-disjoint 1-factors appears to be open, but for recent progress, see Hoffman and Rodger (1996); of course, a $\{3\}$-GDD with group sizes $\eta, \ell_1, \ldots, \ell_m$ leads to such a set of 1-factors.

Let us turn to sufficiency. The easiest case arises when all groups have the same size; we have already seen the complete solution for this case, Theorem 3.4.

Colbourn, Hoffman, and Rees (1992) generalize this to permit one different group:

Theorem 6.15 *Let g, t, and u be nonnegative integers. There exists a $\{3\}$-GDD of the type $g^t u^1$ if and only if these conditions are all satisfied:*

1. *if $g > 0$, then $t \geq 3$ or $t = 2$ and $u = g$, or $t = 1$ and $u = 0$ or $t = 0$;*
2. *$u \leq g(t-1)$ or $gt = 0$;*
3. *$g(t-1) + u \equiv 0 \pmod 2$ or $gt = 0$;*
4. *$gt \equiv 0 \pmod 2$ or $u = 0$;*
5. *$\frac{1}{2}g^2 t(t-1) + gtu \equiv 0 \pmod 3$.*

When $g = 1$, this is just the Doyen–Wilson theorem (6.7) when $u \equiv 1, 3 \pmod 6$, or the Mendelsohn–Rosa theorem (6.8), when $u \equiv 5 \pmod 6$. By removing a point in the hole of an ITS(v, w), one obtains Theorem 6.15 with $g = 2$. Rees (1987) settled the case when $g = 3$, and Rosa and Hoffman (1986) settled the case when $g = 4$. The complete proof by Colbourn, Hoffman, and Rees (1992) employs these along with recursions to increase group sizes, and to increase the number of groups. A number of direct constructions are also employed.

Surprisingly, even the extension to two different group sizes, with any number of groups of each size, remains open. In this direction, Colbourn, Cusack, and Kreher (1995) prove:

Theorem 6.16 *Let u, r, and t be positive integers. Then there exists a $\{3\}$-GDD of type $u^r 1^t$ with $t \geq 1$ and $r \geq 1$ if and only if these conditions are satisfied:*

1. *$u \equiv 1 \pmod 2$;*
2. *$r + t \equiv 1 \pmod 2$;*
3. *if $r = 1$, $t \geq u + 1$;*
4. *if $r = 2$, $t \geq u$;*
5. *$\binom{t}{2} + rut + \binom{r}{2}u^2 \equiv 0 \pmod 3$.*

6.3 Simple triple systems

A TS(v, λ) is *simple* when it has no repeated triples. Our existence proofs until this point have permitted repeated triples. A subsystem of order three in a TS(v, λ) is, in fact, a λ-times repeated triple. In a wide variety of situations, the presence or absence of repeated triples is of concern.

When repeated triples are not permitted, we cannot simply take unions of systems with smaller indices; hence the existence questions become substantially more complex. Let us first establish a basic observation.

Lemma 6.17 *A simple TS(v, λ) exists if and only if a simple TS$(v, v-2-\lambda)$ exists.*

Proof Let $(V,\mathcal{B})$ be a simple TS(v,λ). Forming the *complement* $\overline{\mathcal{B}} = \binom{V}{3} \setminus \mathcal{B}$ with respect to the set of *all* triples $\binom{V}{3}$ on V, we produce a simple TS$(v, v-2-\lambda)$, $(V,\overline{\mathcal{B}})$. □

Evidently, $0 \le \lambda \le v-2$ in a simple TS(v,λ). Dehon (1983) was the first to give a complete determination of the existence of simple TS(v,λ)s:

Theorem 6.18 *A simple TS(v,λ) exists if and only if v is λ-admissible and $0 \le \lambda \le v-2$.*

Since that time, a number of ways to establish this result have arisen; for example, it is a consequence of Theorems 15.15, 15.4, and 15.5 on large sets.

6.4 Subsystems of simple systems

Shen (1992a) proved the analogue of Stern's theorem (6.9) in the case that the triple system and its subsystem are both simple. An $(n,w;\lambda)$-decomposition is *simple* if it contains no repeated triples, and the w λ-factors of λK_n each contain no repeated edges. Evidently $\lambda \le n-1$ for factors to be simple, and if $w=0$, then $\lambda \le n-2$ if no repeated triples are present. For a simple subsystem of order w to be adjoined to the decomposition, however, we require the much stronger condition that $\lambda \le w-2$. For this reason, we examine simple systems with a subsystem here, rather than the more general problem of simple systems with a hole.

We follow the development of Shen (1992a) closely. The approach is similar in spirit to the proof of the Doyen–Wilson theorem (6.7) given earlier. Let $D_{n,\lambda}$ be a multiset containing the integers $1,\ldots,\lfloor (n-1)/2 \rfloor$ each λ times, and containing the integer $n/2$ $\lambda/2$ times when n is even (and λ is even).

Lemma 6.19 *If $D_{n,\lambda}$ has a partition into sets $T_1,\ldots,T_m, X$ where*

1. *either $T_i = \{a_i, b_i, c_i\}$ with $a_i+b_i \equiv \pm c_i \pmod n$ and $T_i \ne \{n/3, n/3, n/3\}$; or $T_i = \{n/3\}$;*
2. *a set $\{a,b,c\}$ occurs among the $\{T_i\}$ at most twice if a, b, and c are all distinct; and it occurs at most once if two of $\{a,b,c\}$ are equal;*
3. *a set $\{n/3\}$ occurs at most once among the $\{T_i\}$; and*
4. *the cyclic multigraph with differences specified by X has a λ-factorization in which there are w λ-factors and each is simple,*

then there is a simple $(n,w;\lambda)$-decomposition.

Proof The only verification required is that when T_i and T_j contain the same differences, the triples produced from them can be chosen to contain no repetitions. So suppose that $T_i = T_j = \{a,b,c\}$. Since the three differences are all distinct, the sets of triples $\{\{j, j+a, j+a+b\} : j \in \mathbb{Z}_n\}$ and $\{\{j, j+b, j+a+b\} : j \in \mathbb{Z}_n\}$ (arithmetic $\pmod n$) are disjoint. □

A simpler version of Lemma 6.6 is also needed:

Lemma 6.20 *If* $n \equiv 0 \pmod 3$, $a, b, c \in D_{n,\lambda}$, $a + b \equiv \pm c \pmod n$, $a \equiv b \not\equiv 0 \pmod n$, *and either* $a \not\equiv b \pmod n$ *or* $a \not\equiv c \pmod n$, *then from the differences* $\{a, b, c\}$ *we can form*

1. n *triples; or*
2. $2n/3$ *triples and one 2-factor; or*
3. $n/3$ *triples and two 2-factors; or*
4. *three 2-factors.*

Proof Let $\mathcal{B}_i = \{\{i + 3j, i + 3j + a, i + 3j + a + b\} : 0 \leq j < n/3\}$, arithmetic (mod n). Each of $\mathcal{B}_0$, $\mathcal{B}_1$, and $\mathcal{B}_2$ can be taken either as a set of triples, or the pairs therein can be taken to form a 2-factor. □

In order to illustrate Shen's techniques, we treat the case when $n = v - w$ is odd. Evidently, one of v or w is even; since both are λ-admissible, λ is even. Write $\lambda = 2\widehat{\lambda}$.

Lemma 6.21 *If* $n = v - w \equiv 1 \pmod 2$, v *and* w *are* λ*-admissible, and* $4w + 2 \geq v \geq 2w + 1$, *then there is a simple* $(n, w; \lambda)$*-decomposition.*

Proof We consider cases for n modulo 6. When $n \equiv 1 \pmod 6$, write $n = 6s + 1$, and observe that $2s \leq w \leq 6s$ and $\widehat{\lambda} \leq (w - 2)/2$. Form a set $\mathcal{S}$ of $2s\widehat{\lambda} - 1 - \widehat{\lambda}(\widehat{\lambda} - 1)/2$ difference triples:

$$\begin{array}{ll} (2i, 3s - i, 3s + 1 + i) & 1 \leq i < s \\ (2i - j, 3s - i, 3s + j + 1 - i) & 1 \leq j < \widehat{\lambda},\ \lfloor (j + 2)/2 \rfloor \leq i \leq s \\ (2i - j, s + i, s + j - i) & 1 \leq j < \widehat{\lambda},\ \lfloor (j + 2)/2 \rfloor \leq i \leq s \end{array}$$

(When a difference $\alpha > 3s$ is indicated, choose instead $6s + 1 - \alpha$.) Now every difference appears at most λ times in $\mathcal{S}$; and every triple that contains a repeated difference occurs at most once in $\mathcal{S}$. Moreover, since $2s \leq w \leq 6s$, and $\widehat{\lambda} \leq (w - 2)/2$, there remain $3 + (3\widehat{\lambda}(\widehat{\lambda} - 1))/2 < \widehat{\lambda}w$ occurrences of differences in $D_{n,\lambda}$ not accounted for by $\mathcal{S}$. As w and v are λ-admissible, $\lambda w \equiv 0 \pmod 3$. Form a set $\mathcal{S}'$ by removing $(\widehat{\lambda}w)/3 - 1 - (\widehat{\lambda}(\widehat{\lambda} - 1))/2$ triples from $\mathcal{S}$; then $\mathcal{S}'$ covers all but $\widehat{\lambda}w$ differences from $D_{n,\lambda}$. Since no difference appears more than λ times, and $w > \lambda$, the $\widehat{\lambda}w$ differences can be partitioned into w sets, each containing $\widehat{\lambda}$ distinct differences, and thus give a set of w simple λ-factors. Apply Lemma 6.19 to complete the case $n \equiv 1 \pmod 6$.

When $n \equiv 5 \pmod 6$, we proceed similarly. Let $n = v - w = 6s - 1$, and form a set $\mathcal{S}$ of $2(s - 1)\widehat{\lambda} - \widehat{\lambda}(\widehat{\lambda} - 1)/2$ triples:

$$\begin{array}{ll} (2i - j, 3s - i - 1, 3s + j - i) & 0 \leq j < \widehat{\lambda},\ \lfloor (j + 2)/2 \rfloor \leq i \leq s - 1 \\ (2i - j, s + i, s + j - i) & 1 \leq j < \widehat{\lambda},\ \lfloor (j + 2)/2 \rfloor \leq i \leq s - 1 \end{array}$$

Proceed as before, applying Lemma 6.19.

When $n \equiv 3 \pmod 6$, write $n = 6s + 3$. Form a set $\mathcal{S}$ of $2s\widehat{\lambda} - \widehat{\lambda}(\widehat{\lambda} - 1)/2$ triples:

$$\begin{array}{ll} (2i-j, 3s-i+1, 3s+j-i+2) & 0 \le j < \widehat{\lambda},\ \lfloor (j+2)/2 \rfloor \le i \le s-1 \\ (2i-j, s+i, s+j-i) & 1 \le j < \widehat{\lambda},\ \lfloor (j+2)/2 \rfloor \le i \le s-1 \end{array}$$

There remain $\widehat{\lambda}(3\widehat{\lambda}+1)/2$ uncovered differences; when $\widehat{\lambda}w \equiv \widehat{\lambda}(3\widehat{\lambda}+1)/2 \pmod 3$, we proceed as before. When $\widehat{\lambda} \not\equiv 0 \pmod 3$, there remain cases in which an integral number of difference triples cannot be dismantled into 2-factors. Let $T \in \mathcal{S}$ be the triple $\{1, s, s+1\}$ when $s \equiv 0, 1 \pmod 3$, or the triple $\{2, s, s+2\}$ when $s \equiv 2 \pmod 3$. Form $\mathcal{S}'$ from $\mathcal{S} \setminus \{T\}$ by removing

$$\left\lfloor \frac{\widehat{\lambda}w}{3} - \widehat{\lambda}(3\widehat{\lambda}+1)/2 \right\rfloor$$

triples. $\mathcal{S}'$ covers all but $\widehat{\lambda}w + \beta$ differences of $D_{n,\lambda}$, for $\beta \in \{1, 2\}$. Apply Lemma 6.20 to partition the pairs in the triples in the orbit of T into $3 - \beta$ 2-factors and $(\beta n)/3$ triples. Form the remaining triples using difference triples from $\mathcal{S}'$. The uncovered $\widehat{\lambda}w - 3 + \beta$ differences can be partitioned into 2-factors, and all 2-factors can then be combined to form w simple λ-factors. □

Shen (1992a) employs similar techniques in the case when $n = v - w$ is even. In that case, however, when λ is odd, more caution must be taken to ensure that *w simple* λ-factors can be produced. In summary, Shen (1992a) obtained:

Theorem 6.22 *A simple TS(v, λ) exists having a simple sub-TS(w, λ) if and only if v is λ-admissible, w is λ-admissible, $v \ge 2w+1$, and $\lambda \le w-2$.*

This does not settle existence of simple $(n, w; \lambda)$-decompositions completely; in particular, it does not settle the existence of simple incomplete triple systems.

6.5 Relatives of 3-GDDs

In this section, we examine related types of design. For one of their applications, see §17.4.

6.5.1 *Modified 3-GDDs*

A *modified group divisible design* with block sizes in K (K-MGDD) of *type* (u, v) and *index* λ is a set $U \times V$ with $|U| = u$ and $|V| = v$, partitioned into *first groups* $\{\{u\} \times V : u \in U\}$, and into *second groups* $\{U \times \{v\} : v \in V\}$, and equipped with a collection $\mathcal{B}$ of blocks with sizes from K, so that every pair of elements appears either in a first or second group together, or in exactly λ blocks in $\mathcal{B}$, but not both. Assaf (1990) introduced the notion of K-MGDD and settled existence of 3-MGDDs; we give a short proof here.

Lemma 6.23 *Let $u, v \ge 3$. A 3-MGDD of type (u, v) and index λ exists if and only if $\lambda \equiv 0 \pmod{\gcd(u-2, v-2, 6)}$.*

Proof The necessity of the condition is easily checked, and sufficiency rests only on the existence when $\lambda = \gcd(u-2, v-2, 6)$. Let $|U| = u$, $|V| = v$, and form the MGDD on $U \times V$ as follows. Suppose without loss of generality that

$\gcd(u-2,6) \le \gcd(v-2,6)$, and when equal, that $u \le v$. When $u=3$, take an idempotent quasigroup $(V,\oplus)$ of order v; define the triples of the 3-MGDD, taking $U = \{1,2,3\}$, by $\{\{(1,x),(2,y),(3,x\oplus y)\} : x,y \in V,\ x \ne y\}$. Now if $\gcd(u-2,6)$ divides $\gcd(v-2,6)$, employ a $\mathrm{TS}(u,\gcd(u-2,6))$ $(U,\mathcal{B})$ and let $|V| = v$. For each $B \in \mathcal{B}$, form a 3-MGDD with index 1 on $B \times V$. The result is a 3-MGDD on $U \times V$.

The final case is when $u \equiv 0,4 \pmod 6$ and $v \equiv 5 \pmod 6$. There is a {3,4,6}-PBD $(U,\mathcal{B})$ and a {3,5}-PBD $(V,\mathcal{D})$ (see §3.4). So when $B \in \mathcal{B}$ and $D \in \mathcal{D}$, form a 3-MGDD of index 1 on $B \times D$; among these, we require 3-MGDDs of index 1 and types (4,5) and (6,5) to complete the proof. For $(4,5)$, on $\mathbb{Z}_5 \times \{0,1,2,3\}$, form the triples

$$\{0_0,1_1,2_2\},\ \{0_0,2_1,3_3\},\ \{0_0,1_2,4_3\},\ \{0_1,2_2,3_3\};$$

multiply each first component by -1 to produce four further blocks, and develop the eight blocks mod $(5,-)$ to produce the MGDD. For $(6,5)$, on $(\mathbb{Z}_5 \cup \{\infty\}) \times Z_5$, form the blocks

$$\{\infty_0,0_1,4_3\},\ \{\infty_0,0_2,1_4\},\ \{0_0,1_1,3_2\},\ \{0_0,3_1,2_2\};$$

develop mod (5,5) to produce the MGDD. □

6.5.2 *Uniform holey 3-GDDs*

Wei (1993) introduced a generalization of modified 3-GDDs. A *holey group divisible design* $(V,\mathcal{G},\mathcal{H},\mathcal{B})$ with block size 3, or 3-HGDD, is a set V of elements, together with two partitions $\mathcal{G}$ and $\mathcal{H}$ of V into first and second groups, with the property that for $G, G' \in \mathcal{G}$ and $H \in \mathcal{H}$, we have $|G \cap H| = |G' \cap H|$. The blocks $\mathcal{B}$ are 3-subsets of V containing every pair exactly λ times if the pair does not lie in the same first or second group; blocks contain no pairs from a first or second group. A holey group divisible design is a 3-HGDD of type $(u : g_1^{r_1} \cdots g_s^{r_s})$ when the first groups, $\mathcal{G}$, consist of a partition of V into u groups of size $\sum_{i=1}^{s} g_i r_i$; and the second groups, $\mathcal{H}$, consist of r_i groups of size ug_i for $1 \le i \le s$, each meeting each first group in g_i elements. A 3-HGDD of type $(u : t^v)$ is *uniform*.

Wei (1993) proves a complete existence theorem for uniform 3-HGDDs of index λ:

Theorem 6.24 *A 3-HGDD of index λ and type $(u : t^v)$ exists if and only if $u \ge 3$, $v \ge 3$, $\lambda(v-1)(u-1)t \equiv 0 \pmod 2$, and $\lambda t^2 u(u-1)v(v-1) \equiv 0 \pmod 6$.*

One proof of Theorem 6.24 employs two relatively easy constructions:

Lemma 6.25 *If a 3-MGDD of type (u,v) and index λ exists, and $t \ge 0$, then a 3-HGDD of type $(u : t^v)$ exists.*

Proof **(Sketch)** This is an application of Wilson's fundamental construction. □

Lemma 6.26 *If a 3-GDD of type T and index λ exists, and $u \geq 3$, then a 3-HGDD of type $(u : T)$ and index λ exists.*

Proof (Sketch) Give weight u to the points of the 3-GDD, using 3-MGDDs of type $(3, u)$ as ingredient designs. □

Using Lemma 6.26 with uniform 3-GDDs handles all cases in the proof of Theorem 6.24 except when $\lambda \equiv 1, 5 \pmod 6$, $u \equiv 5 \pmod 6$, $v \equiv 0, 4 \pmod 6$, and $t \equiv 1, 5 \pmod 6$; Lemma 6.25 handles these cases.

6.5.3 *Holey 3-GDDs*

Holey 3-GDDs have also been considered for certain nonuniform types. Heinrich (1982) proved that a latin square of order $gt + hs$, with $g < h$, having $t + s$ disjoint subsquares, of which t have size g and s have size h, exists whenever $t \geq 3$ and $s \geq 3$; whenever $t \geq 3$ and $s = 1$; and whenever $t = 1$, $s \geq 3$, and $h \leq (t-1)g$. In addition, she settles the cases when s or t is 2 when the number of groups is 'large'. Such a latin square leads directly to a holey 3-GDD of type $(3 : g^t h^s)$.

Wang and Yin (1999) examine a generalization:

Theorem 6.27 *A 3-HGDD of type $(u : g^t w^1)$ exists if and only if:*

1. $u \geq 3$, $t \geq 2$,
2. $w = g$ *when* $t = 2$ *and* $0 \leq w \leq g(t-1)$ *when* $t \geq 3$,
3. $(u-1)gt \equiv 0 \pmod 2$,
4. $(u-1)(g(t-1) + w) \equiv 0 \pmod 2$,
5. $u(u-1)(g^2 t(t-1) + 2gtw) \equiv 0 \pmod 6$.

6.5.4 *Incomplete 3-GDDs*

An *incomplete 3-GDD*, or 3-IGDD, of type $(g : h)^u$ and index λ is a set X of gu elements, a partition $\mathcal{G}$ of X into u *groups* of size g, a *hole* $H \subset X$ satisfying $|H \cap G| = h$ for each $G \in \mathcal{G}$, and a collection $\mathcal{B}$ of *blocks*, so that every pair of elements not in a group together, and not both in the hole, lies in exactly λ blocks of $\mathcal{B}$. Pairs of elements in a group together or in the hole together appear in no blocks.

Miao and Zhu (1989) settle the existence of 3-IGDDs of index one completely:

Theorem 6.28 *A 3-IGDD of type $(g : h)^u$ and index one exists if and only if*

$$\begin{array}{llll}
u \equiv 1,3 \pmod 6, & g > 0, & h > 0 & \textit{or} \\
u \equiv 0,4 \pmod 6, & g \equiv 0 \pmod 2, & h \equiv 0 \pmod 2 & \textit{or} \\
u \equiv 5 \pmod 6, & g \equiv 0 \pmod 3, & h \equiv 0 \pmod 3 & \textit{or} \\
u \equiv 5 \pmod 6, & g \equiv 1,2 \pmod 3, & h \equiv 1,2 \pmod 3 & \textit{or} \\
u \equiv 2 \pmod 6, & g \equiv 0 \pmod 6, & h \equiv 0 \pmod 6 & \textit{or} \\
u \equiv 2 \pmod 6, & g \equiv 2,4 \pmod 6, & h \equiv 2,4 \pmod 6; & \textit{and}
\end{array}$$

$g \geq 2h$, $u \geq 3$.

Necessity of the conditions of Theorem 6.28 is by examining triples, both in total and those involving a fixed element. Sufficiency is the more difficult. A main construction is a simple application of Wilson's fundamental construction:

Lemma 6.29 *If a K-GDD of type m^u exists, and for every $k \in K$, a 3-IGDD of type $(v : n)^k$ exists, then a 3-IGDD of type $(mv : mn)^u$ exists.*

To produce needed ingredients, we must address cases with small number u of groups. When $u = 3$, we have:

Lemma 6.30 *A 3-IGDD of type $(g : h)^3$ exists whenever $g \geq 2h$ and $h \geq 0$.*

Proof Ryser's theorem (Theorem 1.6) shows that there is a latin square of side g containing a subsquare of side h under the stated conditions. Form groups consisting of rows, columns, and symbols, respectively. Form triples $\{r, c, s\}$ whenever symbol s occurs in row r, column c, *not* in the subsquare. □

Corollary 6.31 *For $u \equiv 1, 3 \pmod 6$, a 3-IGDD of type $(g : h)^u$ exists whenever $g \geq 2h$ and $h \geq 0$.*

Proof Apply Lemma 6.29 using $m = 1$ and $K = \{3\}$. □

Corollary 6.32 *For $u \equiv 0, 4 \pmod 6$, a 3-IGDD of type $(g : h)^u$ exists whenever $g \geq 2h$, $h \geq 0$, and $g \equiv h \equiv 0 \pmod 2$.*

Proof Apply Lemma 6.29 using $m = 2$ and $K = \{3\}$. Theorem 3.4 gives the 3-GDD of type 2^u, and Lemma 6.30 gives the ingredient IGDDs. □

It remains to treat cases when $u \equiv 2 \pmod 3$, the first of which is $u = 5$. Indeed, Miao and Zhu (1989) observe that:

Lemma 6.33 *If the conditions of Theorem 6.28 are sufficient for $u = 5$, then they are sufficient for all $u \equiv 2 \pmod 3$.*

Proof Form a $\{3,5^\star\}$-PBD of order $2u+1$, and delete a point not on the block of size 5 to form a $\{3,5\}$-GDD of type 2^u. Apply Lemma 6.29 using $m = 2$, using ingredients with three groups from Lemma 6.30, and the hypothesized ingredients with five groups. □

It remains only to treat the cases when $u = 5$. The solution of Miao and Zhu (1989) is a combination of a number of recursive and direct constructions. With subsequent advances on PBDs, it can be simplified.

Lemma 6.34 *There is a 3-IGDD of type $(2x : x)^5$ for all $x \geq 0$.*

Proof When $x = 1$, form a $\{3, 5^\star\}$-PBD on 11 elements. Delete a point not on the block of size 5 to form a $\{3, 5^\star\}$-GDD of type 2^5; remove the block of size 5 to form the hole in a 3-IGDD of type $(2 : 1)^5$. When $x > 1$, give weight x to the points of a 3-IGDD of type $(2 : 1)^5$, using 3-GDDs of type x^3 as ingredient designs. □

Lemma 6.35 *If* $g - h \equiv 3 \pmod 6$ *and* $g - h - 3 \leq h \leq g - h$, *then a 3-IGDD of type* $(g : h)^5$ *exists.*

Proof Write $x = (g - h - 1)/2$. Let $\sigma = h - 2x + 2$, so that $0 \leq \sigma \leq 3$. Let X be a set of $g - h$ elements, $Y_1, \ldots, Y_{x-1}$ be sets of 2 elements each, and W be a set of σ elements, so that all of the chosen sets are disjoint. Let $V = X \cup W \cup \bigcup_{i=1}^{x-1} Y_i$. We form the 3-IGDD on $V \times \{0,1,2,3,4\}$, with the hole on $(W \cup \bigcup_{i=1}^{x-1} Y_i) \times \{0,1,2,3,4\}$ as follows.

There is an STS$(g - h)$ on X whose triples can be partitioned into x parallel classes $P_1, \ldots, P_x$, i.e. a Kirkman triple system. When $B \in P_i$, we place on $(B \cup Y_i) \times \{0,1,2,3,4\}$ a 3-HGDD of type $(5 : 1^3 2^1)$, aligning the hole of 2^5 on $Y_i \times \{0,1,2,3,4\}$ and the other holes in the obvious manner. When $B \in P_x$, we place on $(B \cup W) \times \{0,1,2,3,4\}$ a 3-IGDD of type $(3 + \sigma : \sigma)^5$, placing the hole of size σ on $W \times \{0,1,2,3,4\}$. The small ingredients needed here are all easily produced. □

Lemma 6.36 *A 3-IGDD of type* $(g : h)^5$ *exists whenever* $g \geq 2h$, $h \geq 0$, *and either* $g \equiv h \equiv 0 \pmod 3$, *or* $g \equiv 1, 2 \pmod 3$ *and* $h \equiv 1, 2 \pmod 3$.

Proof When $g = 2h$, apply Lemma 6.34. If $g - h \equiv 3 \pmod 6$ and $2h + 1 \leq g \leq 2h + 3$, apply Lemma 6.35. If $h = 3$ and $g = 9$, form a 3-HGDD of type $(3 : 5^3)$ by Theorem 6.24, and fill two of the holes using 3-GDDs of type 3^5.

In the remaining cases, Theorem 6.13 gives a $\{3, 4, (h - 1)^*\}$-PBD of order $g - 1$. Giving weight five to each point, using 3-MGDDs of type (3,5) and (4,5) as ingredients, we obtain a 3-HGDD of type $(5 : 1^{g-h}(h - 1)^1)$. Since a 3-IGDD of type $(2 : 1)^5$ exists by Lemma 6.34, we can append a new group of five elements, and place the 3-IGDD of type $(2 : 1)^5$ on each of the $g - h$ holes of size 1, aligning the hole of the IGDD on the five new elements in each. The result is a 3-IGDD of type $(g : h)^5$. □

6.6 3-GDDs with first and second associates

Fu and Rodger (1998), and Fu, Rodger, and Sarvate (1999) consider a more general type of group divisible designs. This more general notion coincides with the 'original' definition of GDDs. A *group divisible design* (GDD) of order v is a v-set V of *elements* (or *points*), together with a partition $\mathcal{G}$ of V into classes called *groups*, and a collection $\mathcal{B}$ of subsets of V called *blocks*. In addition, every pair of elements occurring in a group together occurs in precisely μ blocks, while every pair of elements not occurring together in a group appears in exactly λ blocks. Our usual definition is the same, but requires that $\mu = 0$. The *index* of the GDD is the pair (μ, λ).

Group divisible designs had their origins in the statistical literature, forming a special class of partially balanced incomplete block designs (PBIBDs). In that setting, often an additional constraint was imposed that the group sizes all be equal; in current vernacular, the GDD is *uniform*. Often two elements in the same group are called *first associates*, and two elements from different groups are called *second associates.* Subsequently, GDDs with $\mu = 0$ have played a central role in combinatorial design theory, primarily as a consequence of their use in Wilson's Fundamental Construction. Indeed, the importance of GDDs with $\mu = 0$ has been so great that in most literature in design theory, the term GDD has been identified closely with the restricted case when $\mu = 0$, and among those often just the case when $\lambda = 1$.

Theorem 6.37 *(Fu, Rodger, and Sarvate (1999)) Let g and u be positive integers, and let μ and λ be nonnegative integers. There exists a 3-GDD of type g^u and index (μ, λ) if and only if:*

1. $\mu(g-1) + \lambda g(u-1) \equiv 0 \pmod 2$*;*
2. $\mu ug(g-1) + \lambda g^2u(u-1) \equiv 0 \pmod 6$*;*
3. *if* $g = 2$ *then* $\mu \leq \lambda(u-1)$*;*
4. *if* $u = 2$ *then* $\mu \geq \lambda g/2(g-1)$*; and*
5. *if* $\lambda = 0$ *then* $\mu g(g-1) \equiv 0 \pmod 6$*.*

We give a shorter, but nevertheless technical, proof here. First we establish the necessity of the conditions. In a putative 3-GDD of type g^u and index (μ, λ), every point must occur with $\mu(g-1)+\lambda g(u-1)$ other points in triples; every triple accounts for two of these occurrences, and condition (1) results. Now classify a triple B as *inside* if B lies inside a group, *cross* if it meets exactly two groups, and *transverse* if it meets three groups. Let i, c, and t be the number of inside, cross, and transverse triples, respectively. Then by considering pairs occurring in triples, we obtain a density condition:

$$\mu ug(g-1) = 6i + 2c \quad \text{and} \quad \lambda u(u-1)g^2 = 6t + 4c.$$

Summing these two gives condition (2). Now when $g = 2$ there are no inside triples, when $u = 2$ there are no transverse triples, and when $\lambda = 0$ there are neither transverse nor cross triples. Conditions (3)–(5) follow directly from these.

In view of conditions (3)–(5), when establishing sufficiency for Theorem 6.37, it is natural to handle the 'boundary' cases first. When $g = 1$, the conditions ensure the existence of a triple system of order u and index λ; taking every element as a singleton group gives a 3-GDD of type 1^u and index (μ, λ) (the value of μ is actually of no relevance here). If $u = 1$ or $\lambda = 0$, the conditions suffice for a triple system of order g and index μ to exist. Take u disjoint copies of it to form the desired 3-GDD. If $\mu = 0$, the conditions have been shown to be sufficient by Theorem 3.4. Finally, when $\lambda = \mu$, any TS(gu, λ) can have its elements arbitrarily partitioned into groups of size g to form the required 3-GDD.

Henceforth we consider only cases when $g \geq 2$, $u \geq 2$, $\mu \geq 1$, $\lambda \geq 1$, and $\mu \neq \lambda$. Evidently, the union of a 3-GDD of type g^u and index (μ_1, λ_1) with a 3-GDD of type g^u and index (μ_2, λ_2), in which groups are aligned, forms a 3-GDD of type g^u and index $(\mu_1+\mu_2, \lambda_1+\lambda_2)$. The list (g, u, μ, λ) is an *admissible parameter set* if it meets the necessary conditions of Theorem 6.37; it is *nontrivial* if $\mu + \lambda > 0$. It is *composite* if there are nontrivial admissible parameter sets (g, u, μ_1, λ_1) and (g, u, μ_2, λ_2) satisfying $\mu = \mu_1 + \mu_2$ and $\lambda = \lambda_1 + \lambda_2$, and *primitive* otherwise.

A complete solution rests only on the production of 3-GDDs for primitive parameter sets, since composite ones can be obtained from unions. Naturally, not all isomorphism types of 3-GDDs are produced in this way, but one 3-GDD (at least) for each admissible parameter set results.

6.6.1 *Two groups*

Here we consider 3-GDDs of type g^2 and index (μ, λ). When $g = 2$, condition (3) requires $\mu \leq \lambda$, condition (4) requires $\mu \geq \lambda$; so assume that $g \geq 3$.

Since there are no transverse triples, the number of cross triples is $c = (\lambda g^2)/2$ and the number of inside triples is $i = (g(2\mu(g-1) - \lambda g))/6$. Now condition (1) requires that μ is even when g is even, and that λ is even when g is odd; condition (2) requires that $\lambda \equiv 0 \pmod 3$ when $g \equiv 1 \pmod 3$, and that $\lambda \equiv \mu \pmod 3$ when $g \equiv 2 \pmod 3$ (equivalently, c and i are integral).

A *regular partial triple system* (RPTS$(v, \mu; r)$) is a collection $\mathcal{B}$ of triples on a v-element set V, together with a factorization $\mathcal{F}$ of a multigraph G with vertex set V into 1-factors when v is even and into 2-factors when v is odd, so that every pair of elements occurs in s triples of $\mathcal{B}$ and as an edge $\mu - s$ times in G for some $0 \leq s \leq \mu$; and every element occurs in r triples.

Lemma 6.38 *If an RPTS$(g, \mu; r_0)$ and an RPTS$(g, \mu; r_1)$ exist so that $\mu(g-1) \equiv 2r_i \pmod g$ and $r_0 + r_1 = (2\mu(g-1) - \lambda g)/2$, then a 3-GDD of type g^2 and index (μ, λ) exists.*

Proof We form the 3-GDD on $G \times \{0,1\}$ where $G = \{1, \ldots, g\}$. Place the triples of the first RPTS on $G \times \{0\}$, and of the second on $G \times \{1\}$. Now consider the factorizations $\mathcal{F}_0$ and $\mathcal{F}_1$ of the RPTSs. By construction, the number of 1-factors in each is a multiple of g. So take unions of factors to form an edge decomposition of G_i into g factors $G_{i1}, \ldots, G_{ig}$, each regular of degree $(\mu(g-1) - 2r_i)/g$. Then add triples

$$\{\{k_{2-i}, a, b\} : 1 \leq k \leq g,\ \{a, b\} \in G_{ik},\ i \in \{0, 1\}\}.$$

This completes the construction. □

A simple application of Lemma 6.38 occurs when $r_1 = r_2 = 0$; in this case, the RPTS is just a 1-factorization of μK_v when v is even, or a 2-factorization when v is odd; no triples are present. In that case, a 3-GDD is obtained with $\mu = g$, $\lambda = 2g - 2$ when g is odd, and $\mu = g/2$, $\lambda = g - 1$, when g is even.

Let us consider the situation when $(g, 2, \mu, \lambda)$ is primitive. If $\lambda \geq 2g - 2$ and g is odd, or $\lambda \geq g - 1$ and g is even, $(g, 2, \mu, \lambda)$ is admissible exactly when $(g, 2, \mu - g, \lambda - 2g + 2)$ is (g odd); or exactly when $(g, 2, \mu - (g/2), \lambda - g + 1)$ is (g even). Hence, when $(g, 2, \mu, \lambda)$ is primitive, $\lambda < 2g - 2$ when g is odd, and $\lambda < g - 1$ when g is even. It follows from condition (4) that $\lambda \leq 2\mu - 2$.

Since $(g, 2, \lambda_{\min}(g), 0)$ is admissible when $(g, 2, \mu - \lambda_{\min}(g), \lambda)$ is admissible, $(g, 2, \mu, \lambda)$ is composite. Since $(g, 2, \lambda_{\min}(2g), \lambda_{\min}(2g))$ is admissible when $(g, 2, \mu - \lambda_{\min}(2g), \lambda - \lambda_{\min}(2g))$ is admissible, $(g, 2, \mu, \lambda)$ is composite. When $g \equiv 0, 4 \pmod 6$, $(g, 2, 2, 3)$ is admissible, and hence when $(g, 2, \mu - 2, \lambda - 3)$ is admissible, $(g, 2, \mu, \lambda)$ is composite. When $g \equiv 2 \pmod 6$, $(g, 2, 4, 7)$ is admissible, and hence when $(g, 2, \mu - 4, \lambda - 7)$ is admissible, $(g, 2, \mu, \lambda)$ is composite.

Hence, if the parameters $(g, 2, \mu, \lambda)$ are primitive, then $\lambda = 2\mu - 2$ and g is odd, or $\lambda = 2\mu - 1$ and g is even (so μ is even).

We treat the two cases together. Let $\lambda = 2\mu - \alpha$ for $\alpha \in \{1, 2\}$, and compute $r = (\alpha g)/2 - \mu$. Since $\lambda \leq \alpha(g-1)$ in a primitive parameter set, and $\mu = (\lambda + \alpha)/2$, we find $\mu \leq (\alpha g)/2$, and hence r is nonnegative. It is easily checked that r is integral. Indeed, when $g \not\equiv 0 \pmod 3$, $r \equiv 0 \pmod 3$.

Now write $r = r_1 + r_2$ so that $|r_1 - r_2|$ is minimized subject to the condition that $r_1 \equiv r_2 \equiv 0 \pmod 3$ when $g \not\equiv 0 \pmod 3$. According to Lemma 6.38, we need only produce RPTS$(g, \mu; r_i)$ for $i = 1, 2$ to complete the result for two groups. Since r_1 and r_2 are chosen as equal as possible, it follows that $r_1 = r_2 = (\alpha g - 2\mu)/4$ when $(\alpha g)/2 \equiv \mu \pmod 2$; otherwise, $r_1 = r_2 + 1 = (\alpha g - 2\mu + 2)/4$ if $g \equiv 0 \pmod 3$, and $r_1 = r_2 + 3 = (\alpha g - 2\mu + 6)/4$ if $g \equiv 1, 2 \pmod 3$.

When g is odd, form the required RPTSs as follows. If $g \equiv 3 \pmod 6$, choose r_i parallel classes from a Kirkman TS(g,1); if $g \equiv 1 \pmod 6$, choose $(r_i)/3$ full orbits from a cyclic TS(g,1); and if $g \equiv 5 \pmod 6$, note that $\mu \geq 5$, and choose r_1 full orbits from a cyclic TS(g,3). In each case, a partial triple system is formed whose leave is regular of even degree; hence, by Petersen's theorem (Theorem 1.22), the leave has a 2-factorization and the RPTS is formed.

When g is even, we form the RPTSs as follows. Write $r_i = 3s_i + t_i$, where s_i is an integer, $t_i \in \{0, 1, 2\}$ (actually, $t_i = 0$ unless $g \equiv 0 \pmod 6$). Now choose a set of s_i pairs $\{(a_j, b_j) : 1 \leq j \leq s_i\}$ so that $a_j < b_j$, and $\{a_j, b_j, b_j - a_j : 1 \leq j \leq s_i\}$ contains $3s_i$ distinct elements, all at most $g/2$, and none equal to $g/3$. That this can be done follows from the existence of Skolem sequences. Then form s_i orbits of triples $\{\{0, a_j, b_j\} : 1 \leq j \leq s_i\}$ under the action of the cyclic group of order u. Add t_i copies of the short orbit $\{0, g/3, 2g/3\}$. These form the triples of the RPTS. Now the leave is a circulant multigraph, in which at least one edge at distance $g/2$ is present. Then Theorem 1.19 ensures that the leave has a 1-factorization, and the construction is complete.

6.6.2 *Group size two*

When $g = 2$, μ is even. All primitive parameter sets with $\mu > 0$ have $\lambda = 1$, so we treat only that case here. We find from condition (2) that

$$\mu \equiv \begin{cases} 0 \pmod 2 & \text{if } u \equiv 0 \pmod 3 \\ 0 \pmod 6 & \text{if } u \equiv 1 \pmod 3 \\ 4 \pmod 6 & \text{if } u \equiv 2 \pmod 3 \end{cases}$$

From condition (5), $\mu \leq u - 1$.

We employ a generalization of 3-GDDs with parameters $(2, u, \mu, 1)$. A *rake* of *class* $(\mu_1, \mu_2, \ldots, \mu_u)$ is a set $\{1, \ldots, u\} \times \{0, 1\}$ of $2u$ points, equipped with a collection $\mathcal{B}$ of triples, so that $\{x_i, y_j\}$ is in a unique triple if $x \neq y$, and $\{x_0, x_1\}$ is in exactly μ_x triples for all $x \in \{1, \ldots, n\}$. Some necessary conditions are immediate:

1. $\mu_i \equiv 0 \pmod 2$ for $1 \leq i \leq u$;
2. $\mu_i \leq 2(u-1)$ for $1 \leq i \leq u$, and $\sum_{i=1}^{u} \mu_i \leq u(u-1)$;
3. $\sum_{i=1}^{u} \mu_i = \begin{cases} 0 \pmod 6 & \text{if } u \equiv 1 \pmod 3 \\ 2 \pmod 6 & \text{if } u \equiv 2 \pmod 3 \end{cases}$

It appears plausible that these necessary conditions are sufficient as well. However, a weaker result suffices.

Lemma 6.39 *Let $\delta_1, \ldots, \delta_{u-1}$ satisfy $\delta_i \in \{0, 2\}$, and $\sum_{i=1}^{u-1} \delta_i = \Delta$. If a rake of class $(\mu_1, \ldots, \mu_{u-1})$ exists, then a rake of class $(\mu_1 + \delta_1, \ldots, \mu_{u-1} + \delta_{u-1}, 2u - 2 - \Delta)$ exists.*

Proof Take the rake of class $(\mu_1, \ldots, \mu_{u-1})$. When $\delta_i = 0$, add the triples $\{u_0, u_1, i_0\}$ and $\{u_0, u_1, i_1\}$; when $\delta_i = 2$, instead add triples $\{i_0, i_1, u_0\}$ and $\{i_0, i_1, u_1\}$. □

Now call a rake *semiregular* if $\mu_i - \mu_j \leq 2$ for all $1 \leq i, j \leq u$. Evidently a rake in which $\mu_1 = \cdots = \mu_u$ is equivalent to a 3-GDD with parameters $(2, u, \mu, 1)$, and so these appear among the semiregular rakes.

Theorem 6.40 *Semiregular rakes exist whenever the necessary conditions are met.*

Proof Let $\Lambda = \frac{1}{2} \sum_{i=1}^{u} \mu_i$. If $\Lambda \geq u - 1$, let $\alpha = 2u - 2 - \mu_u$; define δ_i for $1 \leq i \leq u - 1$ so that $\delta_i \geq \delta_j$ whenever $\mu_i \leq \mu_j$, and $\sum_{i=1}^{u-1} \delta_i = \alpha$. In this case, the semiregular rake is produced by Lemma 6.39 from the semiregular rake of class $(\mu_1 - \delta_1, \ldots, \mu_{u-1} - \delta_{u-1})$. So we may assume that $0 < \Lambda < u - 1$, and hence that $\mu_i \in \{0, 2\}$ for $1 \leq i \leq u$.

When u is odd, form a partial triple system $(U, \mathcal{B})$ with $|U| = u$ and index 1, whose leave graph has all vertices of degree 0 or 2, and has exactly Λ edges (this is always possible as a result of the quadratic leave theorem, Theorem 9.15). Now form the rake on $U \times \{0, 1\}$ as follows. For each triple $B \in \mathcal{B}$, place a 3-GDD of type 2^3 (and index (0,1)) on $B \times \{0, 1\}$. Now arbitrarily orient the edges of the leave graph so that every vertex has at most one incoming (and at most one outgoing) arc. When (x, y) appears as an arc in this oriented leave graph, add the triples $\{x_0, x_1, y_0\}$ and $\{x_0, x_1, y_1\}$ to the rake.

When u is even, no cases remain for $u < 6$. Our general strategy is to form a K-GDD on u points where $K = \{k : k \equiv 0, 1 \pmod 3\}$, and then to apply an inflation more general than that used in the odd case. First we form the required K-GDDs. When $u = 6, 8, 16, 18, 20$, we choose a 3-GDD of type 2^3, 2^4, 4^4, 6^3, or $6^3 2^1$, respectively, all having index (0,1); the latter one is from Theorem 6.15. For the remaining values of u, write $u = 3m + a$ where $m \equiv 0, 1 \pmod 3$, $m \neq 6$, and

$0 \leq a < m$. There is a 4-GDD of type m^4 (equivalently, two mutually orthogonal latin squares of order m). So remove $m - a$ points from one group, treat the remaining groups as blocks, and use one deleted point to define groups of type $3^m a^1$ in a $\{3,4,m\}$-GDD. With K-GDDs of order u so produced, let $(U, \mathcal{G}, \mathcal{B})$ be the GDD of order u. Now form the rake on $U \times \{0,1\}$ as follows. For each block $B \in \mathcal{B}$, place a 3-GDD of type $2^{|B|}$ (and index (0,1)) on $B \times \{0,1\}$. For each group $G \in \mathcal{G}$, we place a semiregular rake on $G \times \{0,1\}$; it is now an easy exercise, given the choice of K-GDDs made above, to verify that all possible values of Λ can be realized. □

6.6.3 *The general problem*

Again we consider primitive parameter sets (g, u, μ, λ). If $\mu \geq \lambda_{\min}(g)$, then (g, u, μ, λ) is composite. Hence $0 \leq \mu < \lambda_{\min}(g)$ in a primitive parameter set, and we may take $\mu > 0$. Then no cases remain when $g \equiv 1,3 \pmod 6$; and since μ is even when g is even, no cases remain when $g \equiv 0,4 \pmod 6$. Henceforth, $g \equiv 2 \pmod 3$, and $\mu \in \{1,2\}$ when $g \equiv 5 \pmod 6$, while $\mu \in \{2,4\}$ when $g \equiv 2 \pmod 6$.

When $g \equiv 5 \pmod 6$, (g, u, μ, λ) is composite if a TS$(gu, 1)$ exists and, when $\mu > \lambda$, a TS$(g, \mu - \lambda)$ exists. When $u \equiv 3,5 \pmod 6$, these conditions are met. Now when $u \equiv 1 \pmod 6$, $\mu \equiv 0 \pmod 3$ by condition (2). So, when $g \equiv 5 \pmod 6$, we may assume that u is even (and λ is even). If $\mu = 2$, condition (2) requires $u \equiv 0,2 \pmod 6$. But then a TS(gu,2) exists, and hence (g, u, μ, λ) is composite. Hence we may take $\mu = 1$. Again, condition (2) requires $u \equiv 0,2 \pmod 6$. Since 3-GDDs of type g^u and index $(0, \lambda_{\min}(u))$ exist, the only primitive parameter sets remaining when $g \equiv 5 \pmod 6$ are $(g, u, 1, 2)$ when $u \equiv 0 \pmod 6$, and $(g, u, 1, 4)$ when $u \equiv 2 \pmod 6$. Let $|G| = g$ and $|V| = v = u/2$. We form the required 3-GDDs on $G \times V \times \{0,1\}$.

When $u \equiv 0 \pmod 6$, first place a 3-MGDD of index one and type $(2g, v)$, aligning first groups on $G \times \{y\} \times \{0,1\}$ for $y \in V$, and second groups on $\{x\} \times V \times \{i\}$ for $x \in G$ and $i \in \{0,1\}$. Next place a 3-MGDD of index one and type (g, u), aligning first groups on $G \times \{y\} \times \{i\}$ for $y \in V$ and $i \in \{0,1\}$, and second groups on $\{x\} \times V \times \{0,1\}$. Choose a particular $\widehat{x} \in G$. On $\{\widehat{x}\} \times V \times \{0,1\}$, place a TS$(2g, 2)$. For $x \in G$, $x \neq \widehat{x}$, place on $\{x\} \times V \times \{0,1\}$ a 3-GDD of type 2^v and index (0,2) with groups aligned on $\{x\} \times V \times \{i\}$ for $i = 0,1$. Finally, form a semiregular rake $(\mu_1, \ldots, \mu_g)$ with $\mu_1 = 2$ and all other μ_i zero. Now on $G \times \{y\} \times \{0,1\}$ for $y \in V$, place a copy of this rake, aligning the group for μ_1 on $\{\widehat{x}\} \times \{y\} \times \{0,1\}$, and the remaining groups on $\{x\} \times \{y\} \times \{0,1\}$ for $x \in G \setminus \{\widehat{x}\}$.

When $u \equiv 2 \pmod 6$, start with a 3-MGDD with index 3 of type (g, u), with first groups aligned on $\{\{x\} \times V \times \{0,1\} : x \in G\}$ and second groups aligned on $\{G \times \{y\} \times \{i\} : y \in V, i \in \{0,1\}\}$. Next, place 3-GDDs of type 2^v and index (0,3) on $G \times \{y\} \times \{i\}$ for $y \in V$, $i \in \{0,1\}$. Next, place 3-GDDs of type 2^u and index (4,1) on $\{x\} \times V \times \{i\}$ for $x \in G$, $i \in \{0,1\}$. Finally, place a 3-GDD of type $(2g)^v$ and index (0,1) with groups appearing on $G \times \{y\} \times \{0,1\}$ for $y \in V$.

Now let us turn to $g \equiv 2 \pmod 6$. Condition (2) requires that $u \equiv 0, 2 \pmod 3$ when $\mu \not\equiv 0 \pmod 3$. Hence a TS(gu,2) exists, and we find that when $\lambda > 1$, the parameter set is composite. Hence the only primitive parameter sets are $(g, u, 2, 1)$ when $u \equiv 0 \pmod 3$, and $(g, u, 4, 1)$ when $u \equiv 0, 2 \pmod 3$. Let $G = H \times \{0, 1\}$ where $|H| = h$ and $|G| = 2h = g$. Let $U = \{1, \ldots, u\}$. There is a 3-MGDD of type (h, u) on $H \times U$, with first groups aligned on $\{H \times \{y\} : y \in U\}$, second groups aligned on $\{\{x\} \times U : x \in H\}$, and triple set $\mathcal{B}$. On $H \times U \times \{0, 1\}$, for each block $B \in \mathcal{B}$, place the triples of a 3-GDD of type 2^3 and index (0,1). When $u > 3$ or $\mu = 2$, complete the system by placing a 3-GDD of type 2^h and index $(0, \mu)$ on $H \times \{y\} \times \{0, 1\}$ for each $y \in U$; and place a 3-GDD of type 2^u of index $(\mu, 1)$ on $\{x\} \times U \times \{0, 1\}$ for each $x \in H$. It remains to treat the case when $u = 3$ and $\mu = 4$. A simple adaptation suffices. In place of the 3-GDD of type 2^h, employ a semiregular rake of class $(\mu_1, \ldots, \mu_h)$ with $\mu_i \in \{2, 4\}$, and employ 3-GDDs of type 2^3 and index (0,1) or (2,1) on $\{x\} \times U \times \{0, 1\}$, according to whether $\mu_x = 4$ or 2 in the rake. □

6.7 GDDs and constant weight codes

A $(g + 1)$-ary *constant weight code* of *length* n, *weight* w, and *distance* d is a collection of n-tuples from $\mathbb{Z}_{g+1}^n$ each having w nonzero entries, for which every two n-tuples differ in at least d positions.

Lemma 6.41 *A $(g+1)$-ary constant weight code of length n, weight three, and distance three has at most $g^2 n(n-1)/6$ tuples.*

Proof If the code has more than $g^2 n(n-1)/6$ tuples, counting establishes that two tuples must then agree in two nonzero positions, and the code cannot have distance three as required. □

Etzion (1997) observes that this bound agrees with the number of triples in a 3-GDD of type g^n, and hence 3-GDDs might prove useful in the construction of constant weight codes. Given any 3-GDD of type g^n, we can name the elements x_{ij} for $1 \le i \le g$ and $1 \le j \le n$, so that $\{x_{ij} : 1 \le i \le g\}$ forms a group for each $1 \le j \le n$. Then to construct a code, for each triple B in the 3-GDD, if $x_{ij} \in B$, place $j + 1$ in the ith position of the tuple. All other positions are zero. By construction, this is a constant weight code of length n and weight three. However, in general, its distance may be only two. In particular, its distance is two when the GDD has two blocks which share a common point and intersect the same three groups. Nevertheless, this is the only manner for the GDD to fail to produce a code of minimum distance three. We therefore call a uniform 3-GDD *maximum distance* if every two blocks intersecting the same three groups are disjoint. Etzion (1997) calls such systems *generalized Steiner triple systems*, but we coin a new name here to avoid confusion.

Lemma 6.42 *If a maximum distance 3-GDD of type g^n exists, then $n \ge g + 2$, $g(n-1) \equiv 0 \pmod 2$, and $g^2 n(n-1) \equiv 0 \pmod 3$.*

Proof Theorem 3.4 establishes the necessity of the latter two conditions. Let $(V, \mathcal{G}, \mathcal{B})$ be the maximum distance 3-GDD of type g^n. Choose two groups G and G'. Choose an element $x \in G$. Now consider the elements $Y = \{y : \{x, y, z\} \in \mathcal{B}$ and $z \in G'\}$. The g elements of Y are from different groups, neither of which is G or G'. Thus $n \geq g + 2$ as required. □

Etzion (1997) establishes product constructions for maximum distance 3-GDDs, using a weaker requirement on the 3-GDDs. A 3-GDD is *t-partitionable* if its blocks partition into $t \leq n$ sets $S_1, \ldots, S_t$, so that two triples within any class have distance at least three.

Theorem 6.43 *Let n be a positive integer. Further let a g-partitionable 3-GDD of type 3^m exist.*

1. *If a maximum distance 3-GDD of type g^n exists, then a maximum distance 3-GDD of type g^{nm} exists.*
2. *If a maximum distance 3-GDD of type g^{n+1} exists, then a maximum distance 3-GDD of type g^{nm+1} exists.*

Etzion (1997) uses this construction together with small examples to establish that the conditions of Lemma 6.42 are sufficient when $g = 2$ or 3. Phelps and Yin (1997) give a somewhat different construction when $g = 3$. They then establish sufficiency of the conditions of Lemma 6.42 when $g = 9$, and handle all cases when $g \equiv 3 \pmod 6$ leaving a small number of exceptions for each value of g. Each construction rests on the determination of g-partitionable 3-GDDs of type g^n. Blake-Wilson and Phelps (1999) give a nearly complete characterization:

Lemma 6.44 *There exists a g-partitionable 3-GDD of type g^n when*

1. $g \equiv 0 \pmod 6$ *and* $n \equiv 0, 1, 3, 4, 5 \pmod 6$, $n > 6$;
2. $g \equiv 1, 5 \pmod 6$ *and* $n \equiv 1, 3 \pmod 6$;
3. $g \equiv 2, 4 \pmod 6$ *and* $n \equiv 0, 1, 3, 4 \pmod 6$, $n > 6$;
4. $g \equiv 3 \pmod 6$ *and* $n \equiv 1, 3, 5 \pmod 6$.

They then generalize the constructions of Etzion (1997) and Phelps and Yin (1997) to treat cases for g in general; however, certain sets of disjoint latin squares, some with additional properties, are needed to complete the constructions. Blake-Wilson and Phelps (1999) establish the existence of many of the required sets of squares, but the existence problem in general remains far from completed. Further results appear in Chen, Ge, and Zhu (1999), Chen, Ge, and Zhu (1999a), Ge (1999), Ge and Wu (1999), Phelps and Yin (1999), and Wu, Zhu, and Ge (1999).

6.8 Dimension

The following is easily proved:

Lemma 6.45 *If a triple system TS(v, λ) contains more than one subsystem, any two subsystems intersect in a subsystem.*

Szamkolowicz (1962, 1963, 1964, 1965) initiated a classification of Steiner triple systems according to the structure of certain of its subsystems; the classification was refined by Doyen (1969, 1970c), Hilton (1977), Teirlinck (1979), and Hilton and Teirlinck (1980).

Consider an STS(v) $(V, \mathcal{B})$. Fix $X \subseteq V$, and let $\{(S_1, \mathcal{B}_1)\}, \ldots, (S_r, \mathcal{B}_r)$ be the set of all subsystems for which $X \subseteq S_i$. Then

$$\left(\bigcap_{i=1}^{r} S_i, \bigcap_{i=1}^{r} \mathcal{B}_i\right)$$

is a subsystem, the *closure* of X. X *generates* its closure, and the closure is the minimal subsystem that contains X. The set of all subsystems, ordered by inclusion, is a lattice with minimal element $(\varnothing, \varnothing)$ and maximal element $(V, \mathcal{B})$.

Call a set $\{x_1, \ldots, x_m\}$ *independent* if it generates a subsystem $(Z, \mathcal{D})$, and no set of fewer than m elements generates the same subsystem. All sets of cardinality less than three are independent, and every subset of an independent set is itself independent. A subsystem $(Z, \mathcal{D})$ that is generated by an independent set $\{x_1, \ldots, x_m\}$ has *dimension* $m-1$. The dimension of the empty set is -1, while the dimension of a single point is 0.

A *hyperspace* is a subsystem $(Z, \mathcal{D})$ properly contained in $(V, \mathcal{B})$, but not properly contained in any other subsystem. The dimension of a hyperspace can be at most one smaller than the dimension of $(V, \mathcal{B})$ itself, since by adjoining any element of $V \setminus Z$ to a set X that generates a subsystem $(Z, \mathcal{D})$, we generate a subsystem that properly contains $(Z, \mathcal{D})$. However, the dimension of a hyperspace need not be smaller than the dimension of the STS. In fact, it can be larger! An STS(v) is *nondegenerate* if every hyperspace has dimension smaller than the dimension of the system; it is *degenerate* otherwise.

An STS(v) has dimension -1 if $v = 0$, 0 if $v = 1$, and 1 if $v = 3$; for $v \geq 7$, its dimension must be at least 2. Consider STSs of dimension 2. For such an STS(v) to be nondegenerate, it must have no subsystem of order w, $3 < w < v$. Thus the nondegenerate STSs of dimension 2 are exactly the *subsystem-free* STSs. Doyen (1969, 1970c) proved:

Theorem 6.46 *A subsystem-free STS(v) exists for all* $v \equiv 1, 3 \pmod 6$.

Corollary 6.47 *A nondegenerate STS(v) of dimension 2 exists for all* $v \equiv 1, 3$ $(\mathrm{mod}\ 6)$, $v \geq 7$.

Doyen also determined when degenerate STSs exist:

Theorem 6.48 *A degenerate STS(v) exists if and only if* $v \equiv 1, 3 \pmod 6$, $v \geq 15$.

Steiner triple systems of dimension at least three have the property that no three elements generate the system; such STSs are *Steiner spaces.* Doyen (1970c) and Teirlinck (1976) showed that for $v \leq 37$, the only Steiner spaces are the projective spaces on 15 and 31 elements, and the affine space on 27 elements, and some spaces on 31 elements derived from the projective space. Teirlinck (1979) and Hilton and Teirlinck (1980) establish:

Theorem 6.49 *A Steiner space of order v exists if and only if $v \equiv 1,3 \pmod 6$, $v \geq 15$, $v \notin \{19, 21, 25, 33, 37, 43\}$, except possibly when $v \in \{51, 67, 69, 145\}$.*

Prototypical examples of Steiner triple systems of higher dimension are the projective triple systems (Lemma 2.14) in which d points generate a subsystem of order $2^m - 1$ for some $m \leq d$, and the affine triple systems (Lemma 2.15) in which d points generate a subsystem of order 3^{m-1} for some $m \leq d$. Hilton (1977) established the asymptotic existence of STSs with given dimension d.

One special case has been studied extensively. A *Hall triple system* is a Steiner triple system in which every three points generate either a triple, or a sub-STS(9). Their study was initiated by Hall (1960, 1962, 1965), and more recent work has concentrated on their possible dimensions (see, for example, Bénéteau (1986), Bénéteau and Lacaze (1988), and references therein). Hall triple systems are studied in more detail in §8.7.

6.9 Work points

1. Determine necessary and sufficient conditions for a $\{3\}$-GDD of type $g^t u^r$ to exist.
2. Complete the determination of the list of necessary conditions in Lemma 6.14.
3. Complete the determination of existence for simple ITS$(v, w; \lambda)$.
4. When does a 3-HGDD of type $(g_1^{u_1} \cdots g_s^{u_s})^3$ exist? Equivalently, when is there a latin square of side $\sum_{i=1}^{s} g_i u_i$ having $\sum_{i=1}^{s} u_i$ subsquares sharing no row, column, or symbol, in which u_i of the subsquares have side g_i for $1 \leq i \leq s$?
5. For nonnegative integers u, v, w, and x, when does an STS(v) exist having a sub-STS(w) and a sub-STS(x) whose intersection is a sub-STS(u)?
6. Establish the sufficiency of the conditions of Lemma 6.42 for the existence of maximum distance 3-GDDs.
7. Decide whether there exist Steiner spaces of the four exceptional orders in Theorem 6.49.

7

AUTOMORPHISMS I: SMALL GROUPS

7.1 Orbits and orbit representatives

Let V be a set of v elements, and let Γ be a permutation group acting on V. Γ is a set of permutations containing the identity permutation and closed under composition of permutations. Thus in order to specify the group Γ, it suffices to give a set $\{\pi_1, \ldots, \pi_r\} \subseteq \Gamma$ of *generators*, so that every $\pi \in \Gamma$ can be written as a finite sequence of compositions of the generators.

For a subset $X \subseteq V$, the *orbit* of X under Γ is

$$Orb_\Gamma(X) = \{X\pi : \pi \in \Gamma\}.$$

The number of orbits of k-subsets of V under Γ is typically much smaller than the number of k-subsets of V. Now suppose that Γ is a subgroup of the automorphism group of a TS(v, λ) $(V, \mathcal{B})$. Evidently, if $B \in \mathcal{B}$, $Orb_\Gamma(B) \subseteq \mathcal{B}$. In fact, $\mathcal{B}$ consists of a number of orbits of triples under the action of Γ. If we retain one block from each orbit to form a set $\widehat{\mathcal{B}}$ of orbit representatives, we find that $Orb_\Gamma(\widehat{\mathcal{B}}) = \mathcal{B}$, but the presentation of the triple system is much more succinct. $\widehat{\mathcal{B}}$ is a set of *starter blocks* for the triple system under the action of Γ. Starter blocks are often called *base blocks*.

We are often interested in constructing triple systems with a prescribed set of automorphisms. Consider the orbit of a 3-subset $\{x, y, z\}$ under Γ. The triples in this orbit contain all pairs in the orbits of the pairs $\{x, y\}$, $\{x, z\}$, and $\{y, z\}$. However, while these three pairs are distinct, the three orbits of pairs need not be distinct. We see examples of this in the next section. Our task is to select orbits of triples so that every orbit of pairs occurs exactly λ times; such a selection is a TS(v, λ) whose automorphism group contains Γ.

7.2 Cyclic automorphisms

First we treat one of the easier cases. We examine the cyclic group Cyc_v of order v, choosing $V = \mathbb{Z}_v$ and using the natural generator $\pi : i \mapsto i + 1 \pmod v$ to generate the cyclic group. A triple system admitting the cyclic group of automorphisms is *cyclic*.

Consider a pair $\{x, y\}$. The orbit of this pair is

$$\{\{x + i, y + i\} : 0 \leq i < v\},$$

arithmetic being done modulo v. Now if $\{x+i, y+i\} = \{x+j, y+j\}$, either $i \equiv j \pmod v$ or v is even, $x \equiv y + v/2 \pmod v$, and $i \equiv j + v/2 \pmod v$. Thus orbits of pairs have length v (*full orbits*), except for the orbit of a pair $\{x, x \pm v/2\}$ when v is even; in the latter case, the orbit has $v/2$ pairs (a *half orbit*). In either case, an orbit of pairs is determined completely by specifying the *minimum difference* $\min(a-b, b-a)$, arithmetic mod v, of the pair $\{a, b\}$. Hence a natural set of orbit representatives for pairs is given by $\{\{0, d\} : 1 \leq d \leq \lfloor v/2 \rfloor\}$.

The orbit of a triple $\{x, y, z\}$ is defined similarly. Now if $\{x+i, y+i, z+i\} = \{x+j, y+j, z+j\}$ and $i \not\equiv j \pmod{v}$, a simple case analysis reveals that $v \equiv 0 \pmod{3}$, $\{x, y, z\} = \{x, x+v/3, x+2v/3\}$ and $i \equiv j \pm v/3 \pmod{v}$. Thus an orbit of triples contains v triples (a *full orbit*), unless v is a multiple of 3 and the orbit contains $\{0, v/3, 2v/3\}$ (a *short orbit*). The short orbit of triples contains a single orbit of pairs, while all full orbits of triples contain either three full orbits of pairs, or two full orbits and two copies of the half orbit of pairs.

With each orbit of triples, we can associate the orbits of pairs by specifying the difference representation of the pair orbit. The orbit of triple $\{x, y, z\}$ covers the orbits of pairs with differences $\min(x-y, y-x)$, $\min(x-z, z-x)$, and $\min(y-z, z-y)$, arithmetic mod v. The three differences are not sufficient to determine the orbit of triples, although they determine the pair orbits covered. For each full orbit, we therefore choose as the orbit representative a triple $\{0, a, b\}$ in the orbit that satisfies $a < b$, $a \leq v-b$, and $b-a \leq v-b$. We require that $a < v-b$ unless $b-a = v-b$. Then the *difference triple* of the orbit is $(a, b-a, \min(b, v-b))$. A triple (d_1, d_2, d_3) of integers arises as a difference triple modulo v if and only if each entry is at most $v/2$, and either $d_1+d_2+d_3 \equiv 0 \pmod{v}$ or $d_1+d_2 \equiv d_3 \pmod{v}$. In either case, the orbit of triples corresponding to the difference triple contains $\{0, d_1, d_1+d_2\}$. With the short orbit, we associate a *degenerate difference triple*, containing the entry $v/3$ just once.

Now given a collection of triples that represent the orbits of a cyclic TS(v, λ) once each, every orbit of pairs is represented exactly λ times. Then consider the corresponding collection of difference triples. For v even, the integer $v/2$ occurs $\lambda/2$ times in difference triples; all other integers from 1 to $\lceil v/2 \rceil - 1$ appear λ times each in difference triples. If $v \equiv 0 \pmod{3}$, some of the occurrences of the integer $v/3$ may be in degenerate difference triples.

Example 7.1 Consider the three triples $\{0,1,2\}$, $\{0,2,4\}$, and $\{0,3,6\}$ acted on modulo 7. Each generates a full orbit of 7 triples. The corresponding difference triples $(1,1,2)$, $(2,2,3)$, and $(1,3,3)$ contain all of the required differences three times each, and hence we obtain a (simple) TS$(7,3)$.

Let $\mathcal{D}_{v,\lambda}$ contain the integers from 1 to $\lceil v/2 \rceil - 1$ each λ times; and if v is even, let it also contain the integer $v/2$ $\lambda/2$ times. A cyclic TS(v, λ) is equivalent to a partition of $\mathcal{D}_{v,\lambda}$ into difference triples.

7.2.1 *Direct constructions*

We now examine the existence question for cyclic TS(v, λ)s. In Chapter 2, we developed a number of direct constructions for triple systems, and many yield cyclic triple systems. In particular, the projective spaces (Theorem 2.14), Netto's constructions (Theorems 2.11 and 2.12) and Heffter's construction (Theorem 2.13) all yield cyclic Steiner triple systems. Here we focus on the general existence question for cyclic triple systems of arbitrary index. First we prove a simple necessary condition.

Lemma 7.2 *If* $v \equiv 2 \pmod 4$ *and* $\lambda \equiv 2 \pmod 4$, *there is no cyclic* $TS(v, \lambda)$.

Proof $\mathcal{D}_{v,\lambda}$ contains an odd number of odd differences; each odd difference less than $v/2$ occurs an even number of times, while the odd difference $v/2$ occurs $\lambda/2$ times, which is odd. Every difference triple contains zero or two odd integers, and hence the partition of $\mathcal{D}_{v,\lambda}$ into difference triples cannot exist. □

In order to construct cyclic TS(v, λ)s, we employ Peltesohn's technique, introduced in §2.4. Partitions of $\mathcal{D}_{v,1}$ into difference triples are solutions to Heffter's two difference problems. The solutions presented for Heffter's difference problems in Theorems 2.17 and 2.18 establish:

Theorem 7.3 *There exists a cyclic Steiner triple system of order* v *if and only if* $v \equiv 1, 3 \pmod 6$ *and* $v \neq 9$.

Evidently, the union of a cyclic TS(v, λ_1) and a cyclic TS(v, λ_2) is a cyclic TS$(v, \lambda_1 + \lambda_2)$. Hence for each v, we treat the smallest index for which a cyclic triple system can exist.

We treat congruence classes for v modulo 12, using constructions due to Colbourn and Colbourn (1981b) based on Peltesohn's proof technique.

$(2r+1, 5m-r, 5m+r+1)$	$0 \le r \le m-1$
$(2m-2r-2, 2m+r+1, 4m-r-1)$	$0 \le r \le m-2$
$(2m-2r-1, 2m+r, 4m-r-1)$	$0 \le r \le m-1$
$(2r+2, 5m-r-1, 5m+r+1)$	$0 \le r \le m-2$
$(2m, 3m, 5m)$	
$(4m)$	short; twice

Table 7.1 *Difference triples when* $v \equiv 0 \pmod{12}$

$(2r+2, 3m-r, 3m+r+2)$	$0 \le r \le m-1$	twice
$(2m-2r-1, 4m+r+2, 6m-r+1)$	$0 \le r \le m-1$	twice
$(3m+1, 3m+1, 6m+2)$		

Table 7.2 *Difference triples when* $v \equiv 4 \pmod{12}$

Lemma 7.4 *If* $v \equiv 0 \pmod{12}$, *there is a cyclic* $TS(v, 2)$.

Proof For $v = 12m$, the difference triples in Table 7.1 partition $\mathcal{D}_{v,2}$. □

Lemma 7.5 *If* $v \equiv 4 \pmod{12}$, *there is a cyclic* $TS(v, 2)$.

$(2r+1, 3m-r+1, 3m+r+2)$	$0 \le r \le m-1$
$(2r+2, 5m-r+2, 5m+r+4)$	$0 \le r \le m-2$
$(2r+2, 3m-r-1, 3m+r+1)$	$0 \le r \le m-2$
$(2r+1, 5m-r+2, 5m+r+3)$	$0 \le r \le m-1$
$(2r+2, 3m-r+1, 3m+r+3)$	$0 \le r \le m-2$
$(2r+1, 5m-r+3, 5m+r+4)$	$0 \le r \le m-1$
$(2r+2, 3m-r, 3m+r+2)$	$0 \le r \le m-2$
$(2r+1, 5m-r+2, 5m+r+3)$	$0 \le r \le m-1$
$(2m, 2m+1, 4m+1)$	twice
$(3m+1, 3m+2, 6m+3)$	
$(2m+1, 2m+2, 4m+3)$	
$(2m, 2m, 4m)$	
$(3m, 4m+3, 5m+3)$	
$(4m+2)$	short; four times

Table 7.3 *Difference triples when* $v \equiv 6 \pmod{12}$

$(2r+1, 3m-r+3, 3m+r+4)$	$0 \le r \le m-1$
$(2r+2, 5m-r+4, 5m+r+6)$	$0 \le r \le m-1$
$(2r+1, 3m-r+2, 3m+r+3)$	$0 \le r \le m-1$
$(2r+2, 5m-r+3, 5m+r+5)$	$0 \le r \le m-1$
$(2r+2, 3m-r, 3m+r+2)$	$0 \le r \le m-1$
$(2r+1, 5m-r+5, 5m+r+6)$	$0 \le r \le m-1$
$(2r+2, 3m-r+1, 3m+r+3)$	$0 \le r \le m-1$
$(2r+1, 5m-r+3, 5m+r+4)$	$0 \le r \le m-1$
$(2m+1, 2m+3, 4m+4)$	
$(2m+1, 2m+2, 4m+3)$	
$(4m+2, 4m+3, 4m+5)$	
$(2m+1, 4m+3, 6m+4)$	
$(3m+1, 4m+4, 5m+5)$	
$(2m+2, 3m+2, 5m+4)$	

Table 7.4 *Difference triples when* $v \equiv 10 \pmod{12}$

Proof Write $v = 12m + 4$. The difference triples given in Table 7.2 partition $\mathcal{D}_{v,2}$. □

Lemma 7.6 *If v is odd, there is a cyclic TS$(v, 3)$.*

Proof Write $v = 2m + 1$. Form m difference triples $(i, i, \min(2i, v - 2i))$ for $1 \le i \le m$. When $v \equiv 0 \pmod 3$, replace the 'difference triple' $(v/3, v/3, v/3)$ in this set by three occurrences of the degenerate difference triple $(v/3)$. The resulting collection partitions $\mathcal{D}_{v,3}$. □

Lemma 7.7 *If $v \equiv 6 \pmod{12}$, there exists a cyclic TS$(v, 4)$.*

Proof For $v = 12m+6$, the difference triples in Table 7.3 partition $\mathcal{D}_{v,4}$. □

Lemma 7.8 *If* $v \equiv 10 \pmod{12}$, *there exists a cyclic* $TS(v, 4)$.

Proof For $v = 12m+10$, difference triples are given in Table 7.4 which partition $\mathcal{D}_{v,4}$. □

Lemma 7.9 *If* $v \equiv 0 \pmod 4$, *there exists a cyclic* $TS(v, 6)$.

Proof Write $v = 4m$. The difference triples in Table 7.5 partition $\mathcal{D}_{4m,6}$. □

Lemma 7.10 *If* $v \equiv 2 \pmod{12}$ *and* $v \geq 14$, *there exists a cyclic* $TS(v, 12)$.

Proof Write $v = 12m + 2$, $m \geq 1$. The difference triples in Table 7.6 partition $\mathcal{D}_{v,12}$. □

$(2r-1, 2r, \min(4r-1, 4m-4r+1))$	$1 \leq r \leq m$	twice
$(r, r, \min(2r, 4m-2r))$	$1 \leq r \leq 2m-1$	

Table 7.5 *Difference triples when* $v \equiv 0 \pmod 4$

$(2r+2, 3m-r, 3m+r+2)$	$0 \leq r \leq m-2$	six times each
$(2r+1, 5m-r+1, 5m+r+2)$	$0 \leq r \leq m-1$	six times each
$(2r+1, 3m-r, 3m-r+1)$	$0 \leq r \leq m-1$	twice each
$(2r+2, 5m-r, 5m+r+2)$	$0 \leq r \leq m-2$	twice each
$(2r+1, 3m-r+1, 3m+r+2)$	$0 \leq r \leq m-1$	twice each
$(2r+2, 5m-r, 5m+r+2)$	$0 \leq r \leq m-2$	twice each
$(2r+1, 3m-r, 3m+r+1)$	$0 \leq r \leq m-1$	
$(2r+2, 5m-r, 5m+r+2)$	$0 \leq r \leq m-2$	
$(2r+1, 3m-r-1, 3m+r)$	$0 \leq r \leq m-1$	
$(2r+2, 5m-r-1, 5m+r+1)$	$0 \leq r \leq m-2$	
$(2m, 2m+1, 4m+1)$		seven times
$(2m, 3m+1, 5m+1)$		four times
$(2m+1, 4m+1, 6m)$		
$(3m+1, 4m, 5m+1)$		
$(3m+1, 4m+1, 5m)$		

Table 7.6 *Difference triples when* $v \equiv 2 \pmod 4$

This sequence of lemmas underlies the complete determination of existence of cyclic triple systems:

Theorem 7.11 *There is a cyclic triple system of order* v *and index* λ *if and only if:*

1. $\lambda \equiv 1, 5 \pmod 6$ *and* $v \equiv 1, 3 \pmod 6$, $(v, \lambda) \neq (9, 1)$;
2. $\lambda \equiv 2, 10 \pmod{12}$ *and* $v \equiv 0, 1, 3, 4, 7, 9 \pmod{12}$, $(v, \lambda) \neq (9, 2)$;
3. $\lambda \equiv 3 \pmod 6$ *and* $v \equiv 1 \pmod 2$;

4. $\lambda \equiv 4, 8 \pmod{12}$ *and* $v \equiv 0, 1 \pmod 3$;
5. $\lambda \equiv 6 \pmod{12}$ *and* $v \equiv 0, 1, 3 \pmod 4$; *or*
6. $\lambda \equiv 0 \pmod{12}$ *and* $v \geq 3$.

Proof The necessity of the conditions follow from the existence conditions for TS(v, λ) (Lemma 2.1) and from Lemma 7.2. The nonexistence of a cyclic TS$(9, 1)$ or a cyclic TS$(9, 2)$ follows from a simple hand computation.

For order 9, a cyclic TS$(9, 3)$ is constructed by Lemma 7.6. A cyclic TS$(9, 4)$ has difference triples

$$\{(1,1,2),(1,2,3),(1,3,4),(1,3,4),(1,4,4),(3)\}$$

and a cyclic TS$(9, 5)$ has difference triples

$$\{(1,1,2),(1,2,3),(1,3,4),(1,4,4),(2,2,4),(2,3,4),(3),(3)\}.$$

Cyclic TS$(9, \lambda)$s for all $\lambda \geq 3$ are obtained by taking unions of these.

For $v \neq 9$, we form a cyclic triple system with the smallest admissible λ, produced as follows:

v (mod 12)	Construction	v (mod 12)	Construction
0	Lemma 7.4	1,3,7,9	Theorem 7.3
2	Lemma 7.10	4	Lemma 7.5
5,11	Lemma 7.6	6	Lemma 7.7
8	Lemma 7.9	10	Lemma 7.8

Taking an appropriate multiple yields the desired index. □

7.2.2 *Recursive constructions*

In Chapter 3, we developed a number of recursive constructions for triple systems. For each, we can examine the automorphism group of the triple systems constructed. In the search for cyclic triple systems, the direct product (§3.1) can prove useful:

Lemma 7.12 *The automorphism group of the direct product of two triple systems is the direct product of their automorphism groups.*

When the direct product of a cyclic STS(v) and a cyclic STS(w) is formed, the resulting STS(vw) has automorphism group isomorphic to $Cyc_v \times Cyc_w$. (Strictly speaking, the full automorphism group contains $Cyc_v \times Cyc_w$ as a subgroup.) When v and w are relatively prime, $Cyc_v \times Cyc_w \cong Cyc_{vw}$ and the STS(vw) is again cyclic. However, the restriction on v and w is severe.

Colbourn and Colbourn (1984) describe a simple recursive construction for cyclic TS(v, λ)s:

Theorem 7.13 *Let u be odd.*

1. *If there exist a cyclic TS(v, λ) with no short orbit and a cyclic TS(u, λ), then there exists a cyclic TS(uv, λ).*

2. *If there exist a cyclic STS(3v) and a cyclic STS(3u), then there exists a cyclic STS(3uv).*

Proof For the first statement, consider a collection $\{S_1, \ldots, S_n\}$ of starter blocks for the TS(v, λ), and a collection $\{T_1, \ldots, T_r\}$ of starter blocks for the TS(u, λ). For each $S_j = \{0, s_1, s_2\}$, form u starter blocks $\{\{0, s_1 + iv, s_2 + 2iv\} : 0 \leq i < u\}$, arithmetic modulo uv. For each $T_j = \{0, t_1, t_2\}$, include the starter block $\{0, vt_1, vt_2\}$ in the collection. The result is a collection of starter blocks for a TS(uv, λ).

For the second statement, we let $\{S_1, \ldots, S_n\}$ be the collection of starter blocks from *full* orbits of the STS$(3v)$, and $\{T_1, \ldots, T_r\}$ be the collection of all starter blocks of the TS$(3u, \lambda)$. We then proceed as before. □

Theorem 7.13 generalizes in a natural way. We pursue this next. Bose and Bush (1952) define a (u, t, λ)-*difference matrix* to be a $t \times u\lambda$ array $D = (d_{ij})$ with entries from $\mathbb{Z}_u$, in which for all i, ℓ satisfying $1 \leq i < \ell \leq t$, each element of $\mathbb{Z}_u$ appears exactly λ times among the differences $d_{ij} - d_{\ell j}$, $j = 1, \cdots, \lambda u$. Jungnickel (1978) and Jimbo and Kuriki (1983) develop a construction of difference matrices using mutually orthogonal latin squares and cyclic designs.

Jimbo and Kuriki (1983) then describe a use of difference matrices in constructing cyclic designs. Its restriction to triple systems yields the following:

Theorem 7.14 *If there exist*

1. *a cyclic TS(v, λ_1) having μ short orbits;*
2. *a cyclic TS$(u, (\lambda_1 - \mu)\lambda_2)$;*
3. *a cyclic TS$(3u, \mu\lambda_2)$; and*
4. *a $(u, 3, \lambda_2)$-difference matrix,*

then there exists a cyclic TS$(uv, \lambda_1\lambda_2)$.

The specialization of Theorem 7.14 when $\lambda_1 = \lambda_2 = 1$ can also be obtained from a theorem of Grannell and Griggs (1986) on recursive techniques for prime power block size.

7.2.3 *Cyclic systems with cyclic subsystems*

The cyclic group Cyc_v has a cyclic subgroup Cyc_n whenever $n \mid v$. Write $u = v/n$; u is the index of the subgroup Cyc_n in the group Cyc_v. Consider the action of Cyc_n on $\mathbb{Z}_v$; the elements are partitioned into congruence classes modulo u. For a given factorization $v = n \cdot u$, consider a cyclic TS(v, λ) $(\mathbb{Z}_v, \mathcal{B})$, and define

$$\mathcal{B}_u = \{B \in \mathcal{B} : x \equiv 0 \pmod{u} \ \forall x \in B\}.$$

If $\mathcal{B}_u$ forms a cyclic TS(u, λ) on element set $\{x : x \equiv 0 \pmod{u}, x \in \mathbb{Z}_v\}$, it is a *cyclic subsystem* of order u and index n in $(\mathbb{Z}_v, \mathcal{B})$.

From Theorem 7.11, some necessary conditions for a cyclic STS(v) to have a cyclic sub-STS(u) are immediate: u must divide v, and both must be congruent to 1 or 3 modulo 6, and not equal to 9. These conditions are not sufficient, however. Using a method of Mukerjee, Jimbo, and Kageyama (1987), Phelps, Rosa, and Mendelsohn (1989) establish:

Theorem 7.15 *No cyclic STS(9ℓ) has a cyclic subsystem of order 3ℓ.*

Proof Suppose that such an STS(9ℓ) exists. The sub-STS(3ℓ) accounts for orbits of triples using all pair orbits in which the difference is a multiple of 3. Consider the remaining orbits of triples; there are ℓ such orbits, and each contains a starter block of the form $\{0, 3a+1, 3b+2\}$. Such a starter block covers the differences $\{\pm(3x+1) : x \in \{a, b-a, -b-1\}\}$. The ℓ starter blocks together cover the difference $\pm(3x+1)$ exactly once for each $x \in \{0, \ldots, 3\ell-1\}$. But $a + (b-a) + (-b-1) \equiv -1 \pmod{3\ell}$. Thus $\sum_{x=0}^{3\ell-1} x \equiv -\ell \pmod{3\ell}$. But $\sum_{x=0}^{3\ell-1} x = \binom{3\ell}{2} \equiv 0 \pmod{3\ell}$, which is a contradiction. □

Turning to sufficiency, Phelps, Rosa, and Mendelsohn (1989) employ variants of the constructions in Theorems 7.13 and 7.14. One simple observation is useful to make at the outset:

Lemma 7.16 *If there exists a cyclic STS(u) and a cyclic STS(n) for u and n relatively prime, there exists a cyclic STS($n \cdot u$) having cyclic subsystems of order u and n.*

Proof Apply the direct product from §3.1 to the two systems. Then employ Lemma 7.12. □

Lemma 7.17 *If $n \equiv 1 \pmod 6$ and $u \equiv 1, 3 \pmod 6$, $u \neq 9$, there exists a cyclic STS($n \cdot u$) having a cyclic subsystem of order u.*

Proof If $u \equiv 1 \pmod 6$, a cyclic STS(u) and a cyclic STS(n) both exist. Apply part (1) of Theorem 7.13; the resulting system has a cyclic subsystem of order u. If $u \equiv 3 \pmod 6$, $u \neq 9$, a cyclic STS(u) and a cyclic STS($3n$) both exist; apply part (2) of Theorem 7.13. □

Lemma 7.18 *If $n \equiv 3 \pmod 6$, $n \neq 9$, and $u \equiv 1 \pmod 6$, there exists a cyclic STS($n \cdot u$) having a cyclic subsystem of order u.*

Proof There exist a cyclic STS(n) and a cyclic STS($3u$) containing a cyclic subsystem of order u (Lemma 7.16). Apply part (2) of Theorem 7.13 to obtain a cyclic STS($n \cdot u$) containing the cyclic subsystem of order $3u$, which in turn contains the desired cyclic subsystem of order u. □

Lemma 7.19 *If $n \equiv 3, 5 \pmod 6$, $n \neq 3$, and $u \equiv 3 \pmod 6$, there exists a cyclic STS($n \cdot u$) having a cyclic subsystem of order u.*

Proof Apply statement (2) of Theorem 7.13 to a cyclic STS($3n$) and a cyclic STS(u). □

The simple recursion of Theorem 7.13 fails only in one case that is treated directly by Phelps, Rosa, and Mendelsohn (1989), employing a type of Skolem sequence:

Lemma 7.20 *If* $u \equiv 1 \pmod 6$ *and* $u \geq 7$*, there exists a cyclic* $STS(9u)$ *having a cyclic subsystem of order* u*.*

7.2.4 *Multipliers and multiplier automorphisms*

Let us suppose that $(\mathbb{Z}_v, \mathcal{B})$ is a cyclic triple system. Choose a nonzero integer $\mu \in \mathbb{Z}_v$, and form the collection of blocks

$$\mathcal{B}^\mu = \{\{\mu x_1, \mu x_2, \mu x_3\} : \{x_1, x_2, x_3\} \in \mathcal{B}\},$$

retaining multiple occurrences of triples and performing arithmetic in $\mathbb{Z}_v$. $\mathcal{B}^\mu$ is a cyclic TS(v, λ) if and only if v is relatively prime to μ. An integer $\mu \in \mathbb{Z}_v$ for which $\mathcal{B}^\mu$ is again a triple system of order v is a *multiplier*. When, in addition, $\mathcal{B}^\mu = \mathcal{B}$, μ is a *multiplier automorphism*.

If $(V, \mathcal{B})$ is a TS(v, λ) and μ is a multiplier, the TS(v, λ) $(V, \mathcal{B}^\mu)$ is *(multiplier) equivalent* to $(V, \mathcal{B})$. Multiplier equivalence is a restricted form of isomorphism. Determining equivalence of triple systems, at least with respect to a fixed cyclic presentation, is straightforward. Thus one importance of multipliers is their relationship to isomorphism of cyclic systems.

Bays (1931) and Lambossy (1931) establish the basic theorem in this direction:

Theorem 7.21 *If* $\mathcal{B}_1$ *and* $\mathcal{B}_2$ *are collections of subsets of the same set* V, $|V|$ *is prime, and* $\mathcal{B}_1$ *and* $\mathcal{B}_2$ *admit* $Cyc_{|V|}$ *as automorphisms, then the two collections are isomorphic if and only if they are multiplier equivalent.*

Lambossy (1931) observed that the condition that $|V|$ is prime is essential, by producing isomorphic inequivalent collections of subsets of a 16-set and of a 27-set. However, Bays (1931) and Lambossy (1931) both conjectured that when the collection of subsets is a design, isomorphism and equivalence coincide. Pálfy (1987) establishes a generalization of the Bays–Lambossy theorem:

Theorem 7.22 *If* $\mathcal{B}_1$ *and* $\mathcal{B}_2$ *are families of subsets of the same* n*-set,* n *and* $\phi(n)$ *(Euler's totient function, the number of positive integers less than* n *that are coprime to* n*) are relatively prime, and* $\mathcal{B}_1$ *and* $\mathcal{B}_2$ *admit* Cyc_n *as automorphisms, then the two families are isomorphic if and only if they are multiplier equivalent.*

Naturally, a design can have automorphisms that are not of multiplier type; however, the Bays–Lambossy conjecture remained open for many decades.

Brand (1985, 1985a) first disproved the conjecture by giving classes of isomorphic inequivalent cyclic designs for orders that are powers of two, specifically cyclic TS$(4^n, 2)$ for $n \geq 2$. On 16 points, for example, the three TS$(16, 2)$s generated by the following three sets of starter blocks are isomorphic but inequivalent:

$$\{\{0,1,2\},\{0,2,9\},\{0,3,6\},\{0,4,8\},\{0,5,10\}\}$$
$$\{\{0,1,3\},\{0,1,6\},\{0,2,7\},\{0,3,10\},\{0,4,8\}\}$$
$$\{\{0,1,7\},\{0,1,10\},\{0,2,5\},\{0,2,13\},\{0,4,8\}\}$$

Each of these TS$(16, 2)$s has a full automorphism group of order 128.

Subsequently, Brand (1989) and Phelps (1987) disproved the Bays–Lambossy conjecture even for Steiner triple systems, by producing isomorphic inequivalent STS(p^n) for certain $p \equiv 1 \pmod 6$ and $n \geq 2$. Finally, Phelps (1987a) demonstrated that even when the order is the product of two distinct primes p and q, $p > q$, if $q \mid p-1$ there can exist isomorphic inequivalent cyclic STS(pq)s. In particular, he remarks that whenever $p \equiv 7 \pmod{12}$, and q is a prime divisor of $p-1$ with $q \equiv 1 \pmod 6$, these conditions are met. Thus, for example, there are isomorphic inequivalent cyclic STS($7 \cdot 43$)s.

7.2.5 *Triple systems with multiplier automorphisms*

In this section, we are concerned with determining those orders for which cyclic Steiner triple system having a nontrivial multiplier automorphism.

Examples of Steiner triple systems having nontrivial multiplier automorphisms are provided by the projective spaces, and by Netto's and Heffter's constructions (Theorems 2.11, 2.12, and 2.13). Colbourn and Colbourn (1980) studied small cyclic STS(v)s with multiplier automorphisms, and found that none exist for $v \in \{9, 25, 27, 45, 55, 81\}$. Their results are computational in nature.

Phelps (1987) and Colbourn, Mendelsohn, Praeger, and Tonchev (1992) give some constructions for cyclic STS(v)s with multiplier automorphisms. Netto's construction (Theorem 2.11) generalizes to powers of primes congruent to 1 (mod 6):

Theorem 7.23 *Let $p \equiv 1 \pmod 6$ and $v = p^\beta$, $\beta \geq 1$. There exists a cyclic STS(v) whose multiplier automorphisms include the cube roots of unity.*

Proof The multipliers of Cyc_v form a group that is isomorphic to $Cyc_{p^{\beta-1}(p-1)}$. Thus in Cyc_v there is an element α of (multiplicative) order $p^{\beta-1}(p-1)$. Write $t = (p^{\beta-1}(p-1))/6$. Choose t starter blocks for a cyclic STS(v) by including

$$\{\alpha^i, \alpha^{2t+i}, \alpha^{4t+i}\}$$

for each $i = 0, \ldots, t-1$. If $\beta = 1$, these form the starter blocks of a cyclic STS(p) having the cube roots of unity as multiplier automorphisms (in fact, this is Netto's proof of Theorem 2.11). If $\beta > 1$, add the starter blocks of a cyclic STS($p^{\beta-1}$) on the elements congruent to 0 (mod p) to form the required number of starter blocks. We may choose this STS($p^{\beta-1}$) to have the cube roots of unity in $\mathbb{Z}_{p^{\beta-1}}$ as multipliers by applying the theorem inductively. Now all differences that are multiples of p appear in the subsystem of order $p^{\beta-1}$. Examine the differences accounted for in the first t starter blocks. They are

$$(1+\alpha^t)\{\pm\alpha^{2t+i}, \pm\alpha^{3t+i}, \pm\alpha^{4t+i}\} = (1+\alpha^t)\{\pm\alpha^i, \pm\alpha^{t+i}, \pm\alpha^{2t+i}\}$$

for $i = 0, \ldots, t-1$. This accounts for all differences not congruent to 0 (mod p) provided $\alpha^t \not\equiv -1 \pmod p$. Now $\{\alpha^{jt} : 0 \leq j \leq 5\}$ are the sixth roots of unity in Cyc_v. These are the only sixth roots of unity, since the automorphism group of Cyc_v is itself cyclic. Reducing these sixth roots of unity modulo $p^{\beta-1}$ gives the six sixth roots of unity in $Cyc_{p^{\beta-1}}$ (see, for example, Allenby and Redfern, 1989, p. 111). Thus by an easy induction on β, $\alpha^t \not\equiv -1 \pmod p$.

The system so constructed has the three cube roots of unity as multiplier automorphisms. The three cube roots of unity modulo p^β and the three cube roots of unity modulo $p^{\beta-1}$ are congruent modulo $p^{\beta-1}$, and hence the cube roots modulo p^β act as multiplier automorphisms on the subsystem. Since the three cube roots are $\{\alpha^0, \alpha^{2t}, \alpha^{4t}\}$, and $\alpha^{3t} = -1$, each acts as a multiplier automorphism on the t starter blocks which is outside the subsystem. □

When $p \equiv 7 \pmod{12}$, the construction can yield many more multiplier automorphisms. Taking the starter blocks $\{\alpha^{2i}, \alpha^{2t+2i}, \alpha^{4t+2i}\}$ for $i = 0, \dots, t-1$ ensures that α^2 is a multiplier. This is easy to check, since $\alpha^{3t} \equiv -1 \pmod v$. Now α^2 has order $3t$, and hence a cyclic STS(p^β) can be obtained with $p^{\beta-1}(p-1)/2$ multiplier automorphisms when $p \equiv 7 \pmod{12}$.

Next, prime power orders of the form p^2 are examined for $p \equiv 5 \pmod 6$, p a prime. Let ω be an element of Cyc_{p^2} of multiplicative order $p(p-1)$. A *good triad* is a set $\{\omega^0, \omega^\beta, \omega^\gamma\}$ satisfying $\omega^0 + \omega^\beta + \omega^\gamma \equiv 0 \pmod{p^2}$, for which the remainders β' and γ' of β and γ on division by p satisfy $\beta' \geq (p-2)/3$, $\beta' + (p-2)/3 \leq \gamma' \leq p - (p-2)/3$.

Theorem 7.24 *Let $p \equiv 5 \pmod 6$, p a prime. Let ω be an element of order $p(p-1)$, and $\{\omega^0, \omega^\beta, \omega^\gamma\}$ be a good triad in $\mathbb{Z}_{p^2}$. Then there exists a cyclic STS(p^2) having a multiplier automorphism whose order is the largest odd divisor of $p-1$.*

Proof Write $\mu = \omega^p$. Consider the action of μ on the nonzero elements of $\mathbb{Z}_{p^2}$; its orbits are $p+1$ cycles, each of length $p-1$. One cycle, C^*, contains all nonzero elements that are multiples of p. The p remaining cycles $C_0, \dots, C_{p-1}$ have $C_i = \{\omega^{jp+i} : 0 \leq j < p-1\}$. If $\omega^k \in C_i$, $-\omega^k = \omega^{p(p-1)/2+k} \in C_i$ and so $\pm a$ appear in the same orbit.

Now let v be the largest odd divisor of $p-1$, so that $p-1 = 2^\delta v$. Let $\lambda = \mu^{2^\delta}$. Now use the good triad to form most of the starter blocks, by taking $\{0, \omega^0, \omega^\beta\}\omega^i\mu^j\lambda^k$ for $0 \leq i < (p-2)/3$, $0 \leq j < 2^{\delta-1}$, and $0 \leq k < v$. This covers exactly one of the differences from $\pm a$ for each $a \in \{C_0, \dots, C_{(p-2)/3-1}\} \cup \{C_{\beta'}, \dots, C_{\beta'+(p-2)/3-1}\} \cup \{C_{\gamma'}, \dots, C_{\gamma'+(p-2)/3-1}\}$. Moreover, λ acts as a multiplier automorphism on this set of starter blocks.

There remain three orbits of differences: C^* and two others, say C_r and C_s. Then for each i, C_i contains an element of each nonzero congruence class modulo p. So let $m \in C_r$ have $m \equiv 1 \pmod p$ and $n \in C_s$ have $n \equiv 1 \pmod p$. Then $m - n \in C^*$. So append the starter blocks $\{0, n, m\}\mu^j\lambda^k$ for $0 \leq j < 2^{\delta-1}$ and $0 \leq k < v$. This exhausts all of the remaining differences, and has λ as a multiplier automorphism. □

Theorem 7.24 succeeds in constructing a system for order p^2 whenever a good triad exists, but the multiplier automorphism λ is nontrivial if and only if $p-1$ has a nontrivial odd divisor, that is, p is not a Fermat prime.

A direct product construction also applies:

Theorem 7.25 *If there exists a cyclic STS(v) having a nontrivial multiplier automorphism and a cyclic STS(w) for v and w coprime, there exists a cyclic STS(vw) having a nontrivial multiplier automorphism.*

Proof Form the direct product $\mathcal{B}$ on $\mathbb{Z}_v \times Z_w$ of the STS(v) $\mathcal{B}'$ on $\mathbb{Z}_v$ and the STS(w) $\mathcal{B}''$ on $\mathbb{Z}_w$, placing in $\mathcal{B}$ each triple $\{(a,i),(b,j),(c,k)\}$ with

1. $a = b = c$ and $\{i,j,k\} \in \mathcal{B}''$;
2. $i = j = k$ and $\{a,b,c\} \in \mathcal{B}'$; or
3. $\{a,b,c\} \in \mathcal{B}'$ and $\{i,j,k\} \in \mathcal{B}''$.

Then $\mathcal{B}$ is an STS(vw). In addition, $\phi : (i,j) \mapsto (i+1, j+1) \pmod{(v,w)}$ is an automorphism, which is a single cycle of length vw when v and w are coprime. Hence $\mathcal{B}$ is a cyclic STS(vw).

Now let μ be a multiplier automorphism of $\mathcal{B}'$, and find the unique integer λ in the range $0 \le \lambda < vw$ satisfying $\lambda \equiv 1 \pmod{w}$ and $\lambda \equiv \mu \pmod{v}$. Then λ is a multiplier automorphism of the STS(vw) $\mathcal{B}$, as follows. Consider any triple $\{(a,i),(b,j),(c,k)\} \in \mathcal{B}$, and its image $\lambda\{(a,i),(b,j),(c,k)\}$ under λ. This image is $\{(\mu a,i),(\mu b,j),(\mu c,k)\}$, and hence is also a triple of $\mathcal{B}$. □

Corollary 7.26 *If $v \equiv 1,3 \pmod{6}$ and v has a prime divisor $p \equiv 1 \pmod{6}$, there exists a cyclic STS(v) having a nontrivial multiplier automorphism except possibly when $v = 9p$.*

Proof This follows from Theorems 7.3, 7.23, and 7.25. □

When the order has a prime divisor $p \equiv 1 \pmod{6}$, Corollary 7.26 affords an almost complete solution. When $v = 9p$, the result is conjectured to be affirmative by Colbourn, Mendelsohn, Praeger, and Tonchev (1992). They also establish nonexistence results. To develop these, we can explore the possible multiplier automorphisms.

Lemma 7.27 *If a cyclic STS(v) has -1 as a multiplier automorphism, then $v = 3$.*

Proof Let $\{0,a,b\}$ be a triple. Then $\{0,-a,-b\}$ is also a triple, and thus so also are $\{a,0,a-b\}$ and $\{b,b-a,0\}$ (by adding a and b, respectively). It follows that $b \equiv a-b \pmod{v}$ and that $a \equiv a-b \pmod{v}$, so that $a \equiv -b \pmod{v}$. But then $3a \equiv 3b \equiv 0 \pmod{v}$ for every triple $\{0,a,b\}$, so that $v = 3$. □

Lemma 7.28 *Suppose that $\mathcal{B}$ is a cyclic STS(v) with multiplier $\lambda \neq 1$. Suppose that, for some r, $\gcd(\lambda^r - 1, v) = u$ with $1 < u < v$. If $B_i = \{0, iv/u, z_i\} \in \mathcal{B}$, then v/u divides z_i, and $\{B_i : 1 \le i < u\}$ forms the starter blocks of a subsystem of order u.*

Proof Now $\lambda^r i(v/u) \equiv i(v/u) \pmod{v}$, so $\lambda^r B_i \equiv B_i$. Then $\lambda^r z_i = z_i \pmod{v}$, and (v/u) divides $\frac{\lambda^r - 1}{u} z_i$. Thus v/u divides z_i and the lemma follows. □

Lemma 7.29 *If the prime p divides v with either $v = 9p$ or $p \equiv 5 \pmod 6$, and there exists a cyclic STS(v) having multiplier automorphism $\lambda \neq 1$, then the (multiplicative) order of λ modulo (v/p) is the same as the order of λ modulo v.*

Proof Let r be the order of λ modulo (v/p) for some prime divisor p of v, and consider $C_i = \{0, ip, y_i\} \in \mathcal{B}$. Now $\lambda^r \equiv 1 \pmod{v/p}$, so $\lambda^r ip \equiv ip \pmod v$, whence $\lambda^r C_i = C_i$. Thus $\lambda^r y_i \equiv y_i \pmod v$, and hence v divides $(\lambda^r - 1)y_i$. If v does not divide $\lambda^r - 1$, then since v/p *does* divide $\lambda^r - 1$, p divides y_i for each $1 \leq i < v/p$. Thus the triples $\{C_i : i \in \mathbb{Z}_{v/p}\}$ form a cyclic STS(v/p). This is a contradiction, since either $v/p = 9$ or $v/p \equiv 5 \pmod 6$. Thus v divides $\lambda^r - 1$, i.e. $\lambda^r = 1$. Hence, the order of λ modulo (v/p) agrees with the order of λ modulo v for all prime divisors p of v. □

Next a general nonexistence result for certain prime powers is given:

Theorem 7.30 *Let $p = 2^\alpha + 1$ be a prime. Then there is no cyclic STS(p^β) having a nontrivial multiplier automorphism for $\beta \geq 1$ except when $p = 3$ and $\beta = 1$.*

Proof Multiplier automorphisms of a cyclic STS(v) form a subgroup of the automorphism group of the cyclic group of order v. The automorphism group of Cyc_{p^β} is $Cyc_{p^{\beta-1}(p-1)}$. Hence any multiplier automorphism has order divisible by 2 or by p. The unique multiplier of order two is $-1 \pmod{p^\beta}$, and hence by Lemma 7.27 is ruled out. Then the only nontrivial multipliers must have orders p^γ, and hence there must be a multiplier of order p. But such a multiplier fixes $p^{\beta-1}$ elements, and these must form a subsystem. If $p \equiv 5 \pmod 6$, this subsystem has order 5 (mod 6), a contradiction. So we may suppose that $p = 3$. In this case, the subsystem of order $3^{\beta-1}$ is itself cyclic, since it is on the elements that are multiples of 3. But no cyclic STS($3w$) has a cyclic subsystem of order w by Theorem 7.15, except of course when $w = 1$. □

Products of prime factors congruent to 5 modulo 6 lead to further restrictions:

Theorem 7.31 *Let $\mathcal{B}$ be a cyclic STS(v) with $v = p_1^{\alpha_1} \cdots p_k^{\alpha_k}$, $p_i \equiv 5 \pmod 6$, and $\alpha_i \geq 1$ for $1 \leq i \leq k$. Let $\lambda \neq 1$ be a multiplier automorphism of $\mathcal{B}$. Then if the order of λ modulo (v/p_i) is r, then for each j, $1 \leq j \leq k$, the order of λ modulo v/p_j is also r; and if $\alpha_1 = \ldots = \alpha_k = 1$, then r is odd.*

Proof That the orders of λ modulo (v/p_i) are the same value r follows from Lemma 7.29. Now suppose that r is even, say $r = 2s$. Then v divides $(\lambda^s - 1)(\lambda^s + 1)$. We have shown that for all i, v/p_i does not divide $\lambda^s - 1$, so that each p_i divides $\lambda^s + 1$. Thus when v is square-free, $v = \prod_{i=1}^k p_i$ divides $\lambda^s + 1$, which implies $\lambda^s \equiv -1 \pmod v$. But then $\mathcal{B}$ admits -1 as a multiplier automorphism if v is square-free and r is even, which violates the conclusion of Lemma 7.27. Thus when v is square-free, r is odd. □

To illustrate this, consider square-free orders $v = p_1 p_2 \cdots p_k$ with $p_i \equiv 5 \pmod 6$ for $1 \leq i \leq k$. A nontrivial multiplier is present only if for some $i \neq j$,

$\gcd((p_i-1)/2,(p_j-1)/2)$ has an odd divisor greater than 3. In particular, if at most one p_i is not of the form $2^\alpha+1$, $\alpha \geq 2$, there is no cyclic STS(v) having a nontrivial multiplier automorphism.

Now we change direction somewhat, and examine the main consequence of the nonexistence of a cyclic STS(9):

Theorem 7.32 *If a cyclic STS$(9p)$, $p>3$ a prime, admits a nontrivial multiplier automorphism λ, then $p \equiv 1 \pmod 6$ and either $\lambda^3=1$, or $\lambda^3 \equiv 1 \pmod p$ and $\lambda^3 \equiv -1 \pmod 9$; in the latter case, the STS$(9p)$ has a cyclic subsystem of order p.*

Proof If $v=9p$ then by Lemma 7.29, the order of λ modulo 9 agrees with the order of λ modulo v. Since λ is not trivial, $\lambda \not\equiv 1 \pmod 9$. If $\lambda \equiv -1 \pmod 9$, Lemma 7.27 shows that $\lambda \not\equiv -1 \pmod p$. But $\lambda^2 \equiv 1 \pmod 9$, and to avoid a cyclic STS(9) subsystem, we require that $\lambda \equiv 1 \pmod p$. Then λ fixes the multiples of 9, and hence fixes a cyclic STS(p) subsystem. This cannot happen if $p \equiv 5 \pmod 6$, and when $p \equiv 1 \pmod 6$ every multiplier λ of order divisible by two forces a cyclic subsystem.

It remains to consider the cases when $\lambda \equiv 4,7 \pmod 9$, and $\lambda^3 \equiv 1 \pmod v$. When $p \equiv 5 \pmod 6$, the two cube roots are $3p+1$ and $6p+1$. These multipliers fix all elements that are multiples of three, and hence fix a cyclic subsystem of order $3p$. But no cyclic STS$(9p)$ contains a cyclic subsystem STS$(3p)$ (Theorem 7.15), and hence $p \not\equiv 5 \pmod 6$.

Hence $p \equiv 1 \pmod 6$. If no multiplier $\lambda' \equiv -1 \pmod 9$ is present, $\lambda^3 \equiv 1 \pmod 9$, so λ^3 fixes each $B_i = \{0, ip, z_i\}$. As there is no cyclic STS(9), $\lambda^3 \equiv 1 \pmod v$. □

The determination of orders for which a cyclic STS(v) with nontrivial multiplier automorphism exists remains incomplete.

7.2.6 *Enumerative results*

Numbers of inequivalent cyclic triple systems were first calculated by Kaufmann (1926) and Bays (1931). Colbourn and Mathon (1980) reported numbers of inequivalent cyclic Steiner triple systems. Tables 7.7 and 7.8 give the known information concerning enumeration of inequivalent cyclic triple systems. For each order and index that are admissible, we provide (when known) two quantities: the number of inequivalent cyclic systems is the first entry, and the number of those having no repeated triples is the second entry, when given.

It is clear from Tables 7.7 and 7.8 that the number of cyclic triple systems grows rapidly as a function of the order and the index. How rapidly does it grow? For Steiner triple systems, $v(\frac{1}{6}+o(1))$ starter blocks must be selected. If we consider selecting them sequentially, by choosing a pair not yet in a triple of the system being constructed, and choosing a third element for the triple containing this pair in all possible ways, we find a trivial upper bound, $v^{v(\frac{1}{6}+o(1))}$, on the number of (distinct) cyclic STS(v)s admitting the standard cyclic automorphism on $\mathbb{Z}_v$.

Order	Index								
	1	2	3	4	5	6	7	8	9
5			1			1			1
			1			0			0
6		0		2		0		3	
		0		1		0		0	
7	1	2	3	4	5	7	8	10	12
	1	1	1	1	1	0	0	0	0
8						12			
						1			
9	0	0	4	3	4	24	27	30	108
	0	0	1	1	0	0	1	0	0
10		0		21		0		360	
		0		9		0		1	
11			8			209			2593
			6			6			1
12		4		180		3561		15264	
		0		62		62		0	
13	1	9	47	253	1224	5285	20347	70924	225900
	1	7	23	71	117	117	71	23	7
14						0			
						0			

Table 7.7 *Inequivalent cyclic TS(v, λ)s, $v \leq 14$*

Johnsen and Storer (1974a) established the first lower bound, proving that the number of cyclic STS(v)s is at least $2^{v(\frac{1}{6}+o(1))}$. Abrham (1986) gives an exponential lower bound on the number of Skolem sequences, which leads to an exponential lower bound for cyclic STSs.

We also report numbers of inequivalent cyclic triple systems that have nontrivial multiplier automorphisms in Tables 7.9 and 7.10.

7.3 Rotational automorphisms

A triple system is *k-rotational* if it admits an automorphism consisting of one fixed point and k cycles of the same length. When a triple system TS(v, λ) is $(v-1)/2$-rotational, it is a *reverse* triple system, and its automorphism of order two is an involution fixing one element, called a *reversal.*

7.3.1 *Reverse triple systems*

Doyen (1972), Rosa (1972), and Teirlinck (1973) settled the existence question for reverse Steiner triple systems. We first give a necessary condition:

Lemma 7.33 *A reverse STS(v) exists only if* $(v-1)(v-3) \equiv 0 \pmod{16}$.

	Index					
Order	1	2	3	4	5	6
15	2	9	421	2981	15686	375372
	2	0	237	902	0	6577
16		71		21742		2460490
		67		6916		66418
17			1550			?
			990			204768
18		0		320760		0
		0		109259		0
19	4	201	13316	?	?	?
	4	181	8186	190962	?	?
21	7	175	212968	?	?	?
	7	0	128743	?	0	?
22		0		?		0
		0		?		0
24		2939		?		?
		0		?		?
25	12	19543	?	?	?	?
	12	17899	?	?	?	?
27	8	10841	?	?	?	?
	8	0	?	?	0	?
28		609483		?		?
		560549		?		?
31	80	?	?	?	?	?
33	84	?	?	?	?	?
37	820	?	?	?	?	?
39	798	?	?	?	?	?
43	9508	?	?	?	?	?
45	11616	?	?	?	?	?
49	157340	?	?	?	?	?
51	139828	?	?	?	?	?
55	3027456	?	?	?	?	?
57	2353310	?	?	?	?	?

Table 7.8 *Inequivalent cyclic* $TS(v, \lambda)$*s*, $15 \leq v \leq 57$

Proof Consider the action of a reversal ρ on the element set V. V is partitioned into $(v-1)/2$ orbits of size two, and one element, say ∞, is fixed. From each orbit of size two, choose one representative. Let X be the set of orbit representatives chosen. Let $\overline{X} = \rho(X)$. A pair $\{x, \rho(x)\}$ forming an orbit must appear in exactly one triple, say $\{x, \rho(x), y\}$. Then $\{x, \rho(x), \rho(y)\}$ is also a triple, and hence $y = \rho(y)$ and $y = \infty$. Each two-element orbit of elements under ρ thus forms a triple

	Index									
Order	1	2	3	4	5	6	7	8	9	10
5			1			1			1	
6		0		1		0		1		0
7	1	2	3	4	5	7	8	10	12	14
8						9				
9	0	0	1	0	1	1	2	1	3	2
10		0		1		0		5		0
11			2			5			7	
12		3		13		54		103		267
13	1	6	16	45	102	228	449	874	1558	2738
14						0				

Table 7.9 *Cyclic TS(v, λ)s with multiplier automorphisms I*

with ∞. There remain $(v-1)(v-3)/4$ pairs of elements with one from X and one from $\overline{X}$. Each orbit of triples not containing ∞ contains two triples, and thus accounts for four of these $(v-1)(v-3)/4$ pairs. Thus $(v-1)(v-3) \equiv 0 \pmod{16}$ as required. □

Doyen (1972), Rosa (1972), and Teirlinck (1973) show that these conditions are also sufficient:

Theorem 7.34 *A reverse STS(v) exists if and only if* $v \equiv 1, 3, 9, 19 \pmod{24}$.

One important feature of reverse triple systems arises from a simple observation:

Lemma 7.35 *If there exists a k-rotational TS(v, λ) and α is a divisor of $(v-1)/k$, there is an αk-rotational TS(v, λ). In particular, any k-rotational TS(v, λ) with $v - 1 \equiv 0 \pmod{2k}$ is a reverse TS(v, λ).*

Proof If π is a k-rotational automorphism of a TS(v, λ), π^α is an αk-rotational automorphism of the same system. □

7.3.2 *1-rotational systems*

A 1-rotational TS(v, λ) has an automorphism which is a $(v-1)$-cycle and one fixed point. In general, we employ element set $\mathbb{Z}_{v-1} \cup \{\infty\}$, and let the permutation fixing ∞ and mapping i to $i+1 \pmod{v-1}$ be an automorphism. Naturally, this permutation acts cyclically on the elements of $\mathbb{Z}_{v-1}$. As for cyclic systems, an orbit of pairs on Z_{v-1} is a full orbit (containing $v-1$ distinct pairs), unless v is odd and the orbit contains the pair $\{0, (v-1)/2\}$ in which case the orbit is a half-orbit. Similarly, an orbit of triples on $\mathbb{Z}_{v-1}$ is full unless $v \equiv 1 \pmod 3$ and the orbit contains $\{0, (v-1)/3, 2(v-1)/3\}$ in which case it is a short orbit.

For Steiner triple systems, an examination of the orbit structure of pairs and triples on $\mathbb{Z}_{v-1}$ leads to a strong necessary condition, due to Phelps and Rosa (1981):

	Index					
Order	1	2	3	4	5	6
15	2	8	18	50	104	884
16		7		434		4526
17			1			47
18		0		5		0
19	3	28	203	1212	5924	25442
21	6	54	745	4778	24337	?
22		0		139		0
23			4			?
24		139		?		?
25	0	2	7	276	?	?
27	0	1	123	?	?	?
28		695		?		?
29			14			?
31	17	1495	?	?	?	?
33	6	805	?	?	?	?
36		5872		?		?
37	43	13932	?	?	?	?
39	68	?	?	?	?	?
43	131	?	?	?	?	?
45	0	?	?	?	?	?
49	488	?	?	?	?	?
51	20	?	?	?	?	?
55	0	?	?	?	?	?
57	1951	?	?	?	?	?
61	4694	?	?	?	?	?
63	2268	?	?	?	?	?
67	16688	?	?	?	?	?
69	99	?	?	?	?	?
73	61270	?	?	?	?	?
75	3144	?	?	?	?	?

Table 7.10 *Cyclic $TS(v, \lambda)$s with multiplier automorphisms II*

Lemma 7.36 *A 1-rotational $STS(v)$ exists only if $v \equiv 3, 9 \pmod{24}$.*

Proof First, we consider the triples containing ∞. If $\{\infty, x, y\}$ is a triple, so also is $\{\infty, y, 2y - x\}$. Thus $2x \equiv 2y \pmod{v-1}$, and since $y \neq x$, we have $y = x \pm (v-1)/2$. Thus the pairs in the half-orbit are precisely the pairs occurring in triples with ∞. There remain $(v-3)/2$ orbits of pairs on $\mathbb{Z}_{v-1}$. Second, if there is a short orbit of triples, $v \equiv 1 \pmod 3$ and $(v-5)/2$ orbits of pairs remain; but since these must appear in full orbits of triples, and each such accounts for three pair orbits, we require that $v - 5 \equiv 0 \pmod 6$, which cannot be. Thus there is no short orbit of triples. Since all triple orbits are full, $v - 3 \equiv 0 \pmod{6}$.

Finally, applying Lemma 7.35 ensures that the STS is reverse. Thus by Lemma 7.33, $v \equiv 1, 3, 9, 19 \pmod{24}$. Combining these conditions yields $v \equiv 3, 9 \pmod{24}$. □

Now we turn to direct constructions of 1-rotational triple systems. Each of the constructions given employs either a pure Skolem sequence (Theorem 2.19) or a hooked Skolem sequence (Theorem 2.20).

Phelps and Rosa (1981) settle existence for 1-rotational STSs:

Lemma 7.37 *A 1-rotational STS(v) exists if and only if* $v \equiv 3, 9 \pmod{24}$.

Proof The necessity follows from Lemma 7.36. For sufficiency, write $v = 6k+3$. Since $k \equiv 0, 1 \pmod{4}$, a Skolem sequence of order k, $(a_1, b_1), \ldots, (a_k, b_k)$, exists. Form a set of orbit representatives on $\mathbb{Z}_{v-1} \cup \{\infty\}$ by taking

$$\{\infty, 0, (v-1)/2\} \cup \{\{0, r, b_r\} : 1 \leq r \leq k\}.$$

These orbit representatives generate a 1-rotational STS(v). □

Cho (1982) employs a similar strategy to settle the existence of 1-rotational TS(v, λ)s in general. We pursue this next.

Lemma 7.38 *A 1-rotational TS(v, 2) exists if and only if* $v \equiv 0, 1 \pmod{3}$.

Proof First we treat six small cases:

Order	Starter blocks
3	$\{\infty,0,1\}$, $\{\infty,0,1\}$
4	$\{\infty,0,1\}$, $\{0,1,2\}$
6	$\{\infty,0,1\}$, $\{0,2,4\}$
7	$\{\infty,0,3\}$, $\{\infty,0,3\}$, $\{0,2,4\}$, $\{0,1,2\}$
9	$\{\infty,0,4\}$, $\{\infty,0,4\}$, $\{0,1,2\}$, $\{0,2,5\}$
10	$\{\infty,0,1\}$, $\{0,1,4\}$, $\{0,2,4\}$, $\{0,3,6\}$

Otherwise, write $v = 12t + s$ where $0 \leq s < 12$. In each case, Table 7.11 gives the set of starter blocks, indicating the type (pure or hooked) and order of the Skolem sequence $\{(a_i, b_i)\}$ employed.

This completes the proof of sufficiency. □

Lemma 7.39 *There is a 1-rotational TS(v, 3) if and only if* $v \equiv 1 \pmod{2}$.

Proof Write $v = 2t + 1$. If $t = 1$, the TS(v, 3) with triple $\{\infty, 0, 1\}$ repeated three times is 1-rotational. Otherwise, let $\{(a_r, b_r)\}$ be a Skolem sequence (pure or hooked) of order $t-1$, and form the set of starter blocks $\{\{0, r, b_r\} : 1 \leq r < t\}$. Add the starter block $\{\infty, 0, t\}$. Now if the Skolem sequence is pure, add the starter block $\{\infty, 0, 1\}$; if it is hooked, add instead $\{\infty, 0, 2\}$. □

Lemma 7.40 *There is a 1-rotational TS(v, 6) if and only if* $v \neq 2$.

s	Type	Order	Starter blocks
0	hooked	$4t-1$	$\{\infty,0,8t-2\}$, $\{\{0,r,b_r\}:1\leq r\leq 4t-1\}$
1	hooked	$4t-1$	$\{\infty,0,8t-2\}$, $\{0,4t,8t\}$, $\{\{0,r,b_r\}:1\leq r\leq 4t-1\}$
3	pure	$4t$	$\{\infty,0,4t+1\}$, $\{\{0,r,b_r\}:1\leq r\leq 4t\}$
4	pure	$4t$	$\{\infty,0,4t+2\},\{0,4t+1,8t+2\}$, $\{\{0,r,b_r\}:1\leq r\leq 4t\}$
6	pure	$4t+1$	$\{\infty,0,4t+2\}$, $\{\{0,r,b_r\}:1\leq r\leq 4t+1\}$
7	pure	$4t+1$	$\{\infty,0,4t+3\}$, $\{0,4t+2,8t+4\}$, $\{\{0,r,b_r\}:1\leq r\leq 4t+1\}$
9	hooked	$4t+2$	$\{\infty,0,8t+2\}$, $\{\{0,r,b_r\}:1\leq r\leq 4t+2\}$
10	hooked	$4t+2$	$\{\infty,0,8t\},\{0,4t+3,8t+6\}$, $\{\{0,r,b_r\}:1\leq r\leq 4t+2\}$

Table 7.11 *Sequences for Lemma* 7.38

Proof Appropriate multiples of 1-rotational TS$(v,2)$s and TS$(v,3)$s form 1-rotational TS$(v,6)$s, so it remains only to treat the case when $v\equiv 2 \pmod 6$. Write $v=6t+2$. Let $\{(a_r,b_r)\}$ be a Skolem sequence (pure or hooked) of order $3t-1$. Form the starter blocks $\{\{0,r,b_r\}:1\leq r<3t\}$, and include them twice each. Add to these the starter blocks $\{0,3t,3t+1\}$ and $\{\infty,0,1\}$. Now if the Skolem sequence is pure, add the starter block $\{\infty,0,2\}$ twice; if hooked, add instead the starter block $\{\infty,0,3\}$ twice. □

In summary, we have:

Theorem 7.41 *A 1-rotational TS(v,λ) exists if and only if*

$$\begin{array}{ll} \lambda=1 & \textit{and}\ \ v\equiv 3,9 \pmod{24} \\ \lambda\equiv 1,5 \pmod 6,\ \lambda>1 & \textit{and}\ \ v\equiv 1,3 \pmod 6 \\ \lambda\equiv 2,4 \pmod 6 & \textit{and}\ \ v\equiv 0,1 \pmod 3 \\ \lambda\equiv 3 \pmod 6 & \textit{and}\ \ v\equiv 1 \pmod 2 \\ \lambda\equiv 0 \pmod 6 & \textit{and}\ \ v\neq 2 \end{array}$$

Proof Since multiples of 1-rotational triple systems are 1-rotational, Lemmas 7.37, 7.38, 7.39, and 7.40 settle all cases except when $\lambda\equiv 1,5 \pmod 6$, $\lambda>1$ and $v\not\equiv 3,9 \pmod{24}$. In these cases, however, there exist a 1-rotational TS$(v,\lambda-3)$ and a 1-rotational TS$(v,3)$; their union is the required TS(v,λ). □

Colbourn and Colbourn (1982) and Jimbo (1983) give recursive constructions for 1-rotational TS(v,λ)s that are analogues of Theorems 7.13 and 7.14.

Numbers of inequivalent ($N_{\not\equiv}$) and nonisomorphic ($N_{\not\cong}$) 1-rotational systems are known for some small orders and indices. For index 1, we have:

Order	3	9	27	33	51	57
$N_{\not\equiv}$	1	1	36	242	550528	8660480
$N_{\not\cong}$	1	1	36	242	?	?

For index 2, Chee and Royle (1992) give numbers of nonisomorphic 1-rotational TS$(v, 2)$ for $v \leq 19$. Their table is extended here:

Order	6	7	9	10	12	13	15	16	18	19	21
$N_{\not\equiv}$	1	2	4	1	5	13	53	53	186	465	3017
$N_{\not\cong}$	1	2	3	1	5	13	53	53	186	465	3017

Order	22	24	25	27	28	30	31
$N_{\not\equiv}$	2676	16873	34937	318112	282392	?	?
$N_{\not\cong}$	2676	16873	34910	?	?	?	?

For index 3, known numbers of 1-rotational TS$(v, 3)$s are:

Order	5	7	9	11	13	15	17	19	21
$N_{\not\equiv}$	1	1	6	14	106	376	2599	36442	262446
$N_{\not\cong}$	1	1	5	14	106	376	2587	36442	?

The differences between the numbers of nonisomorphic and inequivalent systems for TS(25,2)s and TS(17,3)s relate to the determination of isomorphic but inequivalent cyclic systems in §7.2.4.

7.3.3 *k-rotational systems*

Phelps and Rosa (1981) apply the $2v+1$ construction to a cyclic STS(v), using the Steiner 1-factorization arising from the same cyclic STS(v) to obtain a 2-rotational STS$(2v+1)$. Thus a 2-rotational STS(v) exists for all $v \equiv 3, 7 \pmod{12}$ except possibly when $v = 19$. They further construct all nonisomorphic 2-rotational STS(19)s (10 of them), and provide constructions using Skolem sequences to establish:

Lemma 7.42 *A 2-rotational STS(v) exists if and only if $v \equiv 1, 3, 7, 9, 15, 19 \pmod{24}$.*

That no 2-rotational STS(v) exists when $v \equiv 13, 21 \pmod{24}$ follows by Lemma 7.35, since such a system would be reverse.

We can generalize the construction of 2-rotational STS$(2v+1)$s from cyclic STS(v)s to obtain:

Lemma 7.43 *Let π be a permutation on v symbols, and for $i \geq 1$ let a_i be the number of i-cycles in π. If there is an STS(v) admitting π as an automorphism, then there is an STS$(2v+1)$ admitting an automorphism σ on $2v+1$ symbols, having $2a_1+1$ fixed points, and $2a_i$ i-cycles for all $i \geq 2$.*

Proof Apply the singular direct product to the unique STS(3) and the STS(v) that admits π (equivalently, apply the $2v+1$ construction to the STS(v) and its Steiner 1-factorization). □

Cho (1982a) proves two results concerning k-rotational STSs:

Theorem 7.44 *A 3-rotational STS(v) exists if and only if* $v \equiv 1, 19 \pmod{24}$. *A 4-rotational STS(v) exists if and only if* $v \equiv 1, 9, 13, 21 \pmod{24}$.

Necessity in both cases follows from Lemmas 7.35 and 7.33, and the obvious condition that a k-rotational system exists only when $v \equiv 1 \pmod{k}$. Sufficiency of the conditions can be established using variants of Skolem sequences.

Applying Lemma 7.35 along with the solution to the 2- and 3-rotational cases yields:

Theorem 7.45 *A 6-rotational STS(v) exists if and only if* $v \equiv 1, 7, 19 \pmod{24}$.

Colbourn and Jiang (1996) develop some recursive constructions for rotational Steiner triple systems with which the spectrum of a k-rotational Steiner triple system of order v is completely determined for each positive integer k. Jiang (1996a) proved:

Lemma 7.46 *Let* v, k *be positive integers such that* $1 \leq k \leq (v-1)/2$. *Then a* k*-rotational STS(v) exists only if*

$$\begin{cases} v \equiv 1, 3 \pmod{6}, & \\ v \equiv 3 \pmod{6} & \text{if } k = 1, \\ v \equiv 1 \pmod{k}, & \\ v \not\equiv 7, 13, 15, 21 \pmod{24} & \text{if } (v-1)/k \text{ is even.} \end{cases} \tag{7.1}$$

Combining Theorems 7.34, 7.41, 7.42, 7.44, 7.45, and the results of Jiang (1998), we obtain:

Theorem 7.47 *Let* v *be an integer. Let* $k = 1, 2, 3, 4, 5, 6, 7, 11,$ *or* $(v-1)/2$. *Then there is a* k*-rotational STS(v) whenever* v *is admissible (as detailed in Table 7.12).*

$k = 1$:	$v \equiv 3, 9 \pmod{24}$;
$k = 2$:	$v \equiv 1, 3, 7, 9, 15, 19 \pmod{24}$;
$k = 3$:	$v \equiv 1, 19 \pmod{24}$;
$k = 4$:	$v \equiv 1, 9, 13, 21 \pmod{24}$;
$k = 5$:	$v \equiv 1, 51, 81, 91 \pmod{120}$;
$k = 6$:	$v \equiv 1, 7, 19 \pmod{24}$;
$k = 7$:	$v \equiv 1, 43, 57, 99 \pmod{168}$;
$k = 11$:	$v \equiv 1, 67, 177, 243 \pmod{264}$;
$k = \frac{v-1}{2}$:	$v \equiv 1, 3, 9, 19 \pmod{24}$.

Table 7.12 *Congruence classes for Theorem 7.47*

Colbourn and Jiang (1996) establish the existence of a k-rotational STS(v) for all admissible orders, showing that the necessary conditions in Lemma 7.46 are also sufficient. We treat their results next.

7.3.4 *Existence for composite orders*

The existence of a k-rotational STS(v) when $k \geq 6$ is composite is examined next.

Lemma 7.48 *Let ℓ, k be positive integers with $\ell | k$. If $(V, \mathcal{T})$ is an ℓ-rotational STS(v), then it is also a k-rotational STS(v) provided $v \equiv 1 \pmod{k}$.*

Proof If π is an ℓ-rotational automorphism of $(V, \mathcal{T})$, π consisting of a single fixed point and ℓ cycles of length $(v-1)/\ell$, then $\pi^{(k/\ell)}$ consists of a fixed point and k cycles of length $(v-1)/k$. Therefore it is a k-rotational permutation on V. Since $\pi^{(k/\ell)}$ is also an automorphism of $(V, \mathcal{T})$, $(V, \mathcal{T})$ is therefore also k-rotational. □

The existence of a 2- or 4-rotational STS(v) for any admissible order essentially solves the existence problem for k-rotational STS(v)s when k is an even composite.

Theorem 7.49 *Let k be a positive even integer. Then there is a k-rotational STS(v) whenever v is admissible.*

Proof If v is an admissible order for a k-rotational Steiner triple system, then $v \equiv 1, 3 \pmod{6}$, $v \equiv 1 \pmod{k}$, and $v \not\equiv 7, 13, 15, 21 \pmod{24}$ if $(v-1)/k$ is even. Now write $k = 2^t h$, where h is odd and $t \geq 1$.

Suppose $t \geq 2$. Then 2 and 4 are both factors of k. If $v \equiv 1, 3, 7, 9, 15, 19 \pmod{24}$, the order v is admissible for a 2-rotational Steiner triple system; if $v \equiv 13, 21 \pmod{24}$, then v is an admissible order for a 4-rotational Steiner triple system. In either case, by Theorem 7.47, and Lemma 7.48 with $\ell = 2$ or $\ell = 4$, there is a k-rotational STS(v).

Suppose $t = 1$. Write $v = 24s + r$, where $r = 1, 3, 7, 9, 13, 15, 19, 21$. Since

$$\frac{v-1}{k} = \frac{24s + r - 1}{2h}$$

is odd when $r = 7, 15$ and it is even when $r = 13, 21$, then $v \equiv 1, 3, 7, 9, 15, 19 \pmod{24}$. Consequently, there is a 2-rotational STS(v) by Theorem 7.47, which implies there is a k-rotational STS(v) by Lemma 7.48 with $\ell = 2$. □

When k is an odd composite, the existence problem of k-rotational STS(v)s depends on that of p-rotational STS(v)s for p an odd prime.

Theorem 7.50 *Let p be an odd prime and k an odd multiple of p. Suppose there is a p-rotational STS(v) whenever v is admissible. Then there is a k-rotational STS(v) whenever v is admissible.*

Proof Assume that v is an admissible order for a k-rotational Steiner triple system, i.e. v satisfies

$$\begin{cases} v \equiv 1, 3 \pmod{6}, \\ v \equiv 1 \pmod{k}, \\ v \not\equiv 7, 13, 15, 21 \pmod{24} \quad \text{if } (v-1)/k \text{ is even.} \end{cases}$$

Let $k = wp$, where w is odd. Then $(v-1)/p = (v-1)w/k$ and consequently $(v-1)/k$ is even if and only if $(v-1)/p$ is even. So v satisfies

$$\begin{cases} v \equiv 1, 3 \pmod{6}, \\ v \equiv 1 \pmod{p}, \\ v \not\equiv 7, 13, 15, 21 \pmod{24} \quad \text{if } (v-1)/p \text{ is even.} \end{cases}$$

Equivalently, v is an admissible order of a p-rotational Steiner triple system. There is a p-rotational STS(v) by assumption. This STS(v) is also k-rotational by Lemma 7.48. □

With the results on 3-, 5-, 7-, and 11-rotational Steiner triple systems in Theorem 7.47, Theorem 7.50 implies

Corollary 7.51 *There is a k-rotational STS(v) of any admissible order v whenever k is an odd multiple of* 3, 5, 7, *or* 11.

Admissible orders of a p-rotational Steiner triple system when p is an odd prime are now considered. The admissible orders of a 3-rotational Steiner triple system are those v such that $v \equiv 1, 19 \pmod{24}$. The admissible orders of a 5-, 7-, and 11-rotational Steiner triple system as shown in Theorem 7.47 reflect general patterns for primes $p \geq 5$. Whenever $p \geq 5$ is a prime, $p \equiv 1, 5, 7, 11 \pmod{12}$.

Lemma 7.52 *Assume p is an odd prime with $p \geq 5$. Then v is an admissible order of a p-rotational Steiner triple system if and only if*

1. $p \equiv 1 \pmod{12} : v \equiv 1, 2p+1, 8p+1, 18p+1 \pmod{24p}$;
2. $p \equiv 5 \pmod{12} : v \equiv 1, 10p+1, 16p+1, 18p+1 \pmod{24p}$;
3. $p \equiv 7 \pmod{12} : v \equiv 1, 6p+1, 8p+1, 14p+1 \pmod{24p}$;
4. $p \equiv 11 \pmod{12} : v \equiv 1, 6p+1, 16p+1, 22p+1 \pmod{24p}$.

Proof We show that the values of v in (1)–(4) are the only solutions of

$$\begin{cases} v \equiv 1, 3 \pmod{6}, \\ v \equiv 1 \pmod{p}, \\ v \not\equiv 7, 13, 15, 21 \pmod{24} \quad \text{if } (v-1)/p \text{ is even.} \end{cases} \tag{7.2}$$

Each of these values is indeed a solution to (7.2). Now let p be a prime with $p \geq 5$. Write $p = 12s + r$, where $s \geq 0$ and $r = 1, 5, 7, 11$. Then $r^2 \equiv 1 \pmod{24}$ and consequently $p^2 \equiv r^2 \equiv 1 \pmod{24}$. Since p is odd, $(v-1)/p$ is always even. So (7.2) is equivalent to

$$\begin{cases} v \equiv 1, 3, 9, 19 \pmod{24}, \\ v \equiv 1 \pmod{p}. \end{cases} \tag{7.3}$$

Suppose that v is a solution to (7.3), or equivalently to (7.2). Then $v \equiv 1 \pmod{p}$. So $v = px + 1$ for some integer x. Substituting $v = px + 1$ into $v \equiv u \pmod{24}$, where $u = 1, 3, 9, 19$, we obtain

$$px + 1 \equiv u \pmod{24},$$

which implies

$$px \equiv u - 1 \pmod{24}.$$

Multiplying both sides by p, we obtain

$$x \equiv (u-1)p \pmod{24}.$$

Substituting x into $v = px + 1$, we find that

$$v \equiv (u-1)p^2 + 1 \pmod{24p}.$$

Since $(u-1)$ is even, $(u-1)p \equiv (u-1)r \pmod{24}$. This implies

$$v \equiv (u-1)rp + 1 \pmod{24p}. \tag{7.4}$$

However,

$$(u-1)r \equiv \begin{cases} 0, 2, 8, 18 \pmod{24} & \text{when } r = 1, \\ 0, 10, 16, 18 \pmod{24} & \text{when } r = 5, \\ 0, 6, 8, 14 \pmod{24} & \text{when } r = 7, \\ 0, 6, 16, 22 \pmod{24} & \text{when } r = 11. \end{cases}$$

Substituting $(u-1)r$ into (7.4), v is one of the forms given. □

The spectra of 5-, 7-, and 11-rotational Steiner triple system are given in Theorem 7.47. The orders of these systems can be obtained from Lemma 7.52. In general, when $p \geq 5$ is a prime, the existence of a p-rotational Steiner triple system for some of the admissible orders follows directly from that of a 1-rotational Steiner triple system.

Lemma 7.53 *Let p be a prime with $p \geq 5$. Then there is a p-rotational $STS(v)$ if v is an admissible order as in the following:*

$$v \equiv \begin{cases} 2p+1, 8p+1 \pmod{24p} & \textit{when } p \equiv 1 \pmod{12}, \\ 10p+1, 16p+1 \pmod{24p} & \textit{when } p \equiv 5 \pmod{12}, \\ 8p+1, 14p+1 \pmod{24p} & \textit{when } p \equiv 7 \pmod{12}, \\ 16p+1, 22p+1 \pmod{24p} & \textit{when } p \equiv 11 \pmod{12}. \end{cases}$$

Proof It is easy to check that $v \equiv 3, 9 \pmod{24}$. Then there is a 1-rotational STS(v) by Theorem 7.47. This STS(v) is also p-rotational by Lemma 7.48. □

We can conclude:

Theorem 7.54 *If there is a p-rotational STS(v) for each prime $p \geq 13$ and of each admissible order v satisfying*

$$v \equiv \begin{cases} 1, 18p+1 \pmod{24p} & \text{when } p \equiv 1,5 \pmod{12}, \\ 1, 6p+1 \pmod{24p} & \text{when } p \equiv 7,11 \pmod{12}, \end{cases}$$

then there is a k-rotational STS(v) of every admissible order v, for k any positive odd composite.

Proof Apply Corollary 7.51 when k is an odd multiple of $p = 3, 5, 7$, or 11 and apply Theorem 7.50, Lemma 7.52, and Lemma 7.53 when k is an odd multiple of a prime $p \geq 13$. □

7.3.5 *Recursive constructions for rotational STSs*

Colbourn and Jiang (1996) also develop some recursive constructions. One main ingredient of these constructions is a group divisible design with a cyclic automorphism. A GDD $(X, \mathcal{G}, \mathcal{A})$ is *cyclic* if it has an automorphism which permutes the elements in each group $G \in \mathcal{G}$ in a $|G|$-cycle. A cyclic k-GDD is denoted by k-CGDD.

Theorem 7.55 *Let $k \geq 1, n \geq 0$ be integers and X be a k-set. Let $(X, \mathcal{G}, \mathcal{A})$ be a group divisible design on X with group set $\mathcal{G}$ and block set $\mathcal{A}$.*

Suppose that there is a $|G|$-rotational Steiner triple system on $|G|n+1$ points for each $G \in \mathcal{G}$, and that there is a cyclic group divisible design with block size 3 of type $n^{|A|}$ for each $A \in \mathcal{A}$. Then there is a k-rotational Steiner triple system on $kn+1$ points.

Proof We construct a k-rotational STS$(kn+1)$ on the point set $(N \times X) \bigcup \{\infty\}$, where N is an n-set and ∞ is a point not in $N \times X$.

Without loss of generality, we assume that $N = \{1, 2, \ldots, n\}$ and denote an element $(i, x) \in N \times X$ by i_x.

For each group $G \in \mathcal{G}$, take a $|G|$-rotational STS$(|G|n+1)$ on the point set $(N \times G) \bigcup \{\infty\}$, with the automorphism fixing the point ∞ and mapping an element i_x to $(i+1)_x$ for $i \in N, x \in G$, where the arithmetic is modulo n. Let $\mathcal{B}_G$ denote the block set of this STS$(|G|n+1)$.

For each block $A \in \mathcal{A}$, take a 3-CGDD on the point set $N \times A$, with groups $\{N \times \{a\} | a \in A\}$, such that the automorphism maps each element i_a to $(i+1)_a$ for $i \in N, a \in A$, where the arithmetic is again modulo n. Let $\mathcal{B}_A$ denote the block set of this 3-CGDD.

Now set $V = (N \times X) \bigcup \{\infty\}$, $\mathcal{B} = (\bigcup_{G \in \mathcal{G}} \mathcal{B}_G) \bigcup (\bigcup_{A \in \mathcal{A}} \mathcal{B}_A)$. We claim that the design $(V, \mathcal{B})$ is the desired STS$(kn+1)$.

The permutation on V fixing the point ∞ and mapping each i_x to $(i+1)_x$ (mod n), for $i \in N, x \in X$, is an automorphism of this design since the action of this permutation on each block $B \in \mathcal{B}$ still results in a block in $\mathcal{B}$, no matter whether $B \in \mathcal{B}_G$ for a $G \in \mathcal{G}$ or $B \in \mathcal{B}_A$ for an $A \in \mathcal{A}$. Since each $N \times \{x\}$

$(x \in X)$ has size n and $|X| = k$, the design is k-rotational. So it remains to show that the design is an STS$(kn+1)$.

It is clear that $|V| = kn+1$ and $|B| = 3$, for each $B \in \mathcal{B}$. We now prove that each pair of distinct points from V appears in exactly one block of $\mathcal{B}$:

Case 1 The given two points are i_x, j_y $(i,j \in N, x, y \in X)$ such that either $x \neq y$ but x, y are in a unique group $G \in \mathcal{G}$, or $x = y \in G \in \mathcal{G}$ but $i \neq j$; or one of the given points is i_x with $i \in N, x \in G \in \mathcal{G}$ and the other is the point ∞. In each of these cases, the two points are contained in $(N \times G) \bigcup \{\infty\}$ and thus are contained in a unique block in $\mathcal{B}_G$ of the STS$(|G|n+1)$ on $(N \times G) \bigcup \{\infty\}$ and no other blocks in $\mathcal{B}_{G'}$, for $G' \neq G \in \mathcal{G}$, or in $\mathcal{B}_A$, for $A \in \mathcal{A}$.

Case 2 The given two points are i_x, j_y $(i,j \in N, x, y \in X)$ such that $x \neq y$ but x, y are in a unique block $A \in \mathcal{A}$. Then i_x, j_y are in $N \times A$ and thus are contained in a unique block in $\mathcal{B}_A$ of the 3-CGDD on $N \times A$ and no other blocks in $\mathcal{B}_{A'}$, for $A' \neq A \in \mathcal{A}$, or in $\mathcal{B}_G$, for $G \in \mathcal{G}$. □

Corollary 7.56 *Let $k = (h-1)g + r$, where $1 \leq r \leq g$, and $n \geq 0$. If there exist a TD(h,g), a g-rotational STS$(gn+1)$, an r-rotational STS$(rn+1)$, and a 3-CGDD of type n^h and one of type n^{h-1}, then there exists a k-rotational STS$(kn+1)$.*

Proof Delete $g-r$ points from a group of the TD(h,g) to obtain an $\{h, h-1\}$-GDD of type $g^{h-1}r^1$. Applying Theorem 7.55 establishes the result. □

The master design of the recursive construction in Theorem 7.55 is a GDD $(X, \mathcal{G}, \mathcal{A})$. A rotational system is employed to fill each of the groups, and a 3-CGDD to fill each of the blocks, to obtain a bigger rotational system. This recursive construction has the flexibility to be generalized. For instance, we may adjoin a new point x to one of the groups, say G_0, in $(X, \mathcal{G}, \mathcal{A})$ and place a $(|G_0|+1)$-rotational subsystem on $G_0 \bigcup \{x\}$ and a *holey* $(|G|+1)$*-rotational* subsystem on each $G \bigcup \{x\}$, $G \in \mathcal{G}, G \neq G_0$. The master design of the construction is a PBD, taking each $G \bigcup \{x\}$, for $G \in \mathcal{G}$, as a block.

Let V be a $(kn+1)$-set and $V = V_0 \bigcup (\bigcup_{i=1}^{k} V_i)$ be a partition of V into $k+1$ subsets such that $|V_0| = 1$ and $|V_i| = n$ $(1 \leq i \leq k)$. Let $\mathcal{B}$ be a set of 3-subsets, or *blocks*, of V. $(V, \mathcal{B})$ is a *holey k-rotational* Steiner triple system of order $kn+1$, or simply k-rotational HS$(kn+1)$, if (1) there is a V_{i_0} $(1 \leq i_0 \leq k)$ such that each pair of points not both from $V_0 \bigcup V_{i_0}$ is contained in exactly one block and no pair of points from $V_0 \bigcup V_{i_0}$ is contained in a block, and (2) the permutation π which fixes the single point in V_0 and permutes the elements in each V_i $(1 \leq i \leq k)$ in a cycle forms an automorphism of $(V, \mathcal{B})$. Equivalently, it has the property that $\pi(B) \in \mathcal{B}$ for each $B \in \mathcal{B}$. The set V_{i_0} is a *hole* of the HS$(kn+1)$. A k-rotational HS(v) is nearly a k-rotational STS(v); it misses only those pairs of points that are from a particular cycle of the rotational automorphism, or one of which is the fixed point and the other from the cycle.

Theorem 7.57 *Let $k \geq 1, n \geq 0$ be integers, X a k-set and $y \in X$. Let $(X, \mathcal{A})$ be a PBD and $A_0, A_1, \ldots, A_h$ be the blocks in $(X, \mathcal{A})$ which contain y.*

Suppose there is

1. *an $|A_0|$-rotational STS($|A_0|n+1$),*
2. *an $|A_j|$-rotational HS($|A_j|n+1$) for each $1 \le j \le h$, and*
3. *a 3-CGDD of type $n^{|A|}$ for each $A \in \mathcal{A} \setminus \{A_0, A_1, \dots, A_h\}$.*

Then there is a k-rotational STS($kn+1$).

Proof Again assume $N = \{1, 2, \dots, n\}$ and denote an element $(i, x) \in N \times X$ by i_x. We construct a k-rotational STS($kn+1$) $(V, \mathcal{B})$, where $V = (N \times X) \bigcup \{\infty\}$ with the automorphism being the permutation which fixes ∞ and maps each $i_a \in N \times X$ to $(i+1)_a$ modulo n.

The block set $\mathcal{B}$ is formed in a similar manner as in Theorem 7.55. For the block A_0, take an $|A_0|$-rotational STS($|A_0|n+1$) $((N \times A_0) \bigcup \{\infty\}, \mathcal{B}_{A_0})$ with the automorphism fixing ∞ and mapping each $i_a \in N \times A_0$ to $(i+1)_a$; For each A_j $(j = 1, 2, \dots, h)$, take an $|A_j|$-rotational HS($|A_j|n+1$) $((N \times A_j) \bigcup \{\infty\}, \mathcal{B}_{A_j})$ with a hole $N \times \{y\}$, with the automorphism fixing ∞ and mapping each $i_a \in N \times A_j$ to $(i+1)_a$. Finally, for each block $A \in \mathcal{A} \setminus \{A_0, A_1, \dots, A_h\}$, take a 3-CGDD $(N \times A, \{N \times \{a\} | a \in A\}, \mathcal{B}_A)$ with the cyclic automorphism mapping each $i_a \in N \times A$ to $(i+1)_a$. Then $\mathcal{B}$ is taken to be

$$\mathcal{B} = \mathcal{B}_{A_0} \quad \cup \quad \left(\bigcup_{j=1}^{h} \mathcal{B}_{A_j} \right) \quad \cup \quad \left(\bigcup_{A \in \mathcal{A} \setminus \{A_0, A_1, \dots, A_h\}} \mathcal{B}_A \right).$$

Checking that $(V, \mathcal{B})$ is a k-rotational STS($kn+1$) is tedious but straightforward as in Theorem 7.55. □

The structure of holey rotational Steiner triple systems is interesting not only because it plays an important role in the above recursive construction, it also suggests an interesting question. For general k, under what conditions on v does a k-rotational HS(v) exist?

Colbourn and Jiang (1996) obtain a partial result on the existence of a 4-rotational HS($4n+1$)s. They use difference methods. Given a $(kn+1)$-set

$$V = (\mathbb{Z}_n \times \{1, 2, \dots, k\}) \bigcup \{\infty\}$$

and a permutation on V $\pi = (\infty)(0_1 1_1 \cdots (n-1)_1) \cdots (0_k 1_k \cdots (n-1)_k)$, a k-rotational HS($kn+1$) on V with a hole $\mathbb{Z}_n \times \{k\}$ and having π as an automorphism can be obtained in the following way. Suppose $\mathcal{D}$ is a set of triples with elements from V and $\mathcal{D}$ satisfies the property that (i) for each $d \in \mathbb{Z}_n$, each $1 \le i, j \le k$ except for $i = j = k$, there is a unique triple in $\mathcal{D}$ containing a unique pair x_i, y_j with $y - x = d$, and (ii) for each $1 \le i \le k-1$ there is a unique triple in $\mathcal{D}$ containing ∞ and an element x_i. Then developing the triples in $\mathcal{D}$ over $\mathbb{Z}_n$ yields the block set $\mathcal{B}$ of the HS($kn+1$). The triples in $\mathcal{D}$ are the *starter triples* of the HS($kn+1$). The correctness of this construction is established by the uniqueness of the triple in $\mathcal{B}$ which contains each pair of points not both from $\{\infty\} \bigcup (\mathbb{Z}_n \times \{k\})$, and by the absence of any pair of points both from $\{\infty\} \bigcup (\mathbb{Z}_n \times \{k\})$.

Theorem 7.58 *If* $n \equiv 0, 2 \pmod 8$, *then there is a 4-rotational HS*$(4n+1)$.

Proof Let $n = 8t + s$, with $s = 0, 2$. Take $V = (\mathbb{Z}_n \times \{1,2,3,4\}) \bigcup \{\infty\}$, $\pi = (\infty)(0_1 1_1 \cdots (n-1)_1) \cdots (0_4 1_4 \cdots (n-1)_4)$. Then a 4-rotational HS$(4n+1)$ on V with a hole $\mathbb{Z}_n \times \{4\}$ and having π as its automorphism has the following starter triples.

Case 1 $n = 8t$, $t \geq 1$:

$$\begin{array}{ll}
\{\infty, 0_i, (4t)_i\} & i = 1,2,3, \\
\{0_1, r_1, (d_r)_4\} & r = 1,2,\ldots,4t-1, \\
\{0_2, r_2, (b_r)_4\} & r = 1,2,\ldots,4t-1, \\
\{0_3, r_3, (b_r)_4\} & r = 1,2,\ldots,4t-1, \\
\{0_1, r_2, (2r+1)_3\} & r = 0,1,\ldots,4t-1, \\
\{0_1, r_2, (2r)_3\} & r = 4t+1, 4t+2, \ldots, 8t-1, \\
\{0_1, (4t)_2, (4t)_4\}, & \{0_2, 0_3, (-2)_4\}, \\
\{0_3, 0_1, 0_4\}, &
\end{array}$$

where (a_r, b_r) $(r = 1,2,\ldots,4t-1)$ is a hooked Skolem sequence, and (c_r, d_r) $(r = 1,2,\ldots,4t-1)$ is a split Skolem sequence.

Case 2 $n = 8t+2$, $t \geq 0$:

$$\begin{array}{ll}
\{\infty, 0_i, (4t+1)_i\} & i = 1,2,3, \\
\{0_1, r_1, (d_r)_4\} & r = 1,2,\ldots,4t \ (t > 1), \\
\{0_2, r_2, (b_r)_4\} & r = 1,2,\ldots,4t \ (t \geq 1), \\
\{0_3, r_3, (b_r)_4\} & r = 1,2,\ldots,4t \ (t \geq 1), \\
\{0_1, r_2, (2r+1)_3\} & r = 0,1,\ldots,4t, \\
\{0_1, r_2, (2r)_3\} & r = 4t+2, 4t+3, \ldots, 8t+1 \ (t \geq 1), \\
\{0_1, (4t+1)_2, (4t+1)_4\}, & \{0_2, 0_3, (-1)_4\}, \\
\{0_3, 0_1, 0_4\} &
\end{array}$$

where (a_r, b_r) $(r = 1,2,\ldots,4t)$ is a pure Skolem sequence and (c_r, d_r) $(r = 1,2,\ldots,4t)$ is a split Skolem sequence. □

7.3.6 *Existence for large prime orders*

Colbourn and Jiang (1996) apply the recursive constructions in §7.3.5 to p-rotational Steiner triple systems with $p \geq 13$ a prime and of orders

$$v \equiv \begin{cases} 1, 18p+1 \pmod{24p} & \text{when } p \equiv 1 \pmod 4, \\ 1, 6p+1 \pmod{24p} & \text{when } p \equiv 3 \pmod 4. \end{cases}$$

These systems require the ingredient 3-CGDDs of the recursive constructions to have a uniform group size that is a multiple of 6. The existence of these 3-CGDDs has been determined by Jiang (1996).

Theorem 7.59 *There is a 3-CGDD of type* $(6g)^u$ *if and only if (i)* $u \geq 4$ *and (ii)* $u \not\equiv 2, 3 \pmod 4$ *when* g *is odd.*

We first deal with primes $p = 13, 17$, and 29, which fall in the congruence class $p \equiv 1 \pmod 4$. The existence of such p-rotational Steiner triple systems of the relevant orders follows from Theorem 7.57.

Theorem 7.60 *Let* $p = 13, 17, 29$. *Then there is a* p*-rotational STS*(v) *when* $v \equiv 1, 18p + 1 \pmod{24p}$ *for each* p.

Proof In general, for $p \in \{13, 17, 29\}$ and for each $v \equiv 1, 18p + 1 \pmod{24p}$, take $n = (v - 1)/p$. Then there is an integer $t \geq 0$ such that $n = 24t + r$, where $r = 0, 18$.

Since $n \equiv 0, 18 \pmod{24}$, we find that $4n + 1 \equiv 1 \pmod{24}$ and $5n + 1 \equiv 1, 91 \pmod{120}$. Then there is a 4-rotational STS$(4n + 1)$ and a 5-rotational STS$(5n + 1)$ by Theorem 7.47. In addition, since $n \equiv 0, 2 \pmod 8$, there is a 4-rotational HS$(4n + 1)$ by Theorem 7.58. Furthermore, since $n \equiv 0 \pmod 6$, there is a 3-CGDD of type n^4 and one of type n^5 by Theorem 7.59. The three values of p are treated as follows.

For $p = 13$, each point in a $(13, 4, 1)$-design is contained in exactly four blocks. Take $(X, \mathcal{A})$ to be a $(13, 4, 1)$-design and A_0, A_1, A_2, A_3 to be the four blocks containing a point $y \in X$. Then the existence of a 4-rotational STS$(4n + 1)$, a 4-rotational HS$(4n + 1)$, and a 3-CGDD of type n^4 implies the existence of a 13-rotational STS$(13n + 1) =$ STS(v) by applying Theorem 7.57.

For $p = 17$, let $(Y, \mathcal{G}, \mathcal{B})$ be a TD$(4, 4)$ on groups $\mathcal{G} = \{G_1, G_2, G_3, G_4\}$, with $y \in G_1$. Then y is contained in four blocks A_1, A_2, A_3, A_4 in $\mathcal{B}$. Adding a new point x to each group G_i in $\mathcal{G}$ to produce a PBD$(17, \{4, 5\}, 1)$ $(X, \mathcal{A})$ with $X = Y \bigcup \{x\}$, $\mathcal{A} = \{G_i \bigcup \{x\} | i = 1, 2, 3, 4\} \bigcup \mathcal{B}$. There are five blocks in the PBD which contain the point y, namely, $A_0 = G_1 \bigcup \{x\}$, of size 5, and A_1, A_2, A_3, A_4, of size 4. With this PBD and the blocks A_i $(i = 0, 1, 2, 3, 4)$ in Theorem 7.57, the existence of a 5-rotational STS$(5n + 1)$, a 4-rotational HS$(4n + 1)$, and a 3-CGDD of type n^4 and one of type n^5 implies the existence of a 17-rotational STS$(17n + 1) =$ STS(v).

For $p = 29$, take a parallel class of a resolvable $(28, 4, 1)$-design, which consists of 7 blocks B_i $(i = 1, 2, \ldots, 7)$, and add a new point x to each of these blocks. This results in a PBD$(29, \{4, 5\}, 1)$ $(X, \mathcal{A})$. Now consider a point $y \in B_1$. There are 9 blocks in the PBD which contain y. One of such blocks is $B_1 \bigcup \{x\}$ and it has size 5. Each of the other 8 blocks containing y has size 4. Assume $A_0 = B_1 \bigcup \{x\}$ and the other 8 blocks containing y are A_i $(i = 1, 2, \ldots, 8)$. Then with this PBD and these blocks A_i $(i = 0, 1, \ldots, 8)$ in Theorem 7.57, the existence of a 29-rotational STS$(29n + 1) =$ STS(v) follows from that of a 5-rotational STS$(5n + 1)$, a 4-rotational HS$(4n + 1)$, a 3-CGDD of type n^4 and one of type n^5. □

To deal with other primes, we need an existence result.

Lemma 7.61 *Let* $n = 24t + r$, *where* $r = 0, 6, 18$ *and* $t \geq 0$.

1. *For $r = 0, 6, 18$, there is a $4m$-rotational STS($(4m)n + 1$) for every integer $m \geq 1$.*
2. *For $r = 0, 6$, there is a $3m$-rotational STS($(3m)n + 1$) for every integer $m \geq 1$ with $m \equiv 1 \pmod 4$.*
3. *For $r = 0, 18$, there is a $3m$-rotational STS($(3m)n + 1$) for every integer $m \geq 3$ with $m \equiv 3 \pmod 4$.*

Proof When $r = 0, 6$, or 18, we find that $4mn+1 \equiv 1 \pmod{4m}$ and $4mn+1 \equiv 1 \pmod{24}$. Therefore, $4mn+1$ is an admissible order of a $4m$-rotational Steiner triple system. The result (1) is then obtained by applying Theorem 7.49.

When $r = 0$ or 6, and $m \equiv 1 \pmod 4$, we find that $3mn + 1 \equiv 1 \pmod{3m}$ and $3mn + 1 \equiv 1, 19 \pmod{24}$. Therefore, $3mn + 1$ is an admissible order of a $3m$-rotational Steiner triple system. The result (2) then follows by Corollary 7.51.

When $r = 0$ or 18, and $m \equiv 3 \pmod 4$, we find that $3mn + 1 \equiv 1 \pmod{3m}$ and $3mn + 1 \equiv 1, 19 \pmod{24}$. Therefore, $3mn + 1$ is an admissible order of a $3m$-rotational Steiner triple system. The result (3) is a consequence of Corollary 7.51. □

Together with the existence results in Lemma 7.61 and those on primes $p = 5, 7, 11, 13, 17$, the existence of a p-rotational Steiner triple system of the relevant orders can be proved by the construction in Corollary 7.56, provided p is large enough.

Theorem 7.62 *Let p be a prime with $p \equiv 1 \pmod 4$. Then there is a p-rotational STS(v), for any $p \geq 97$ and $v \equiv 1, 18p + 1 \pmod{24p}$.*

Proof Since $p \equiv 1 \pmod 4$, then p can be written as $p = 16m + r = 4(4m) + r$, where m, r are integers such that $r = 5, 9, 13, 17$ and $m \geq 0$. Moreover, $p \geq 97$ implies $m \geq 5$ and $4m \geq 20$ ($> r$). So there is a TD($5, 4m$).

Take $n = (v - 1)/p$. Since $v \equiv 1, 18p + 1 \pmod{24p}$, then $n \equiv 0, 18 \pmod{24}$. So n can be written as $n = 24t + s$, where $s = 0, 18$, for some integer $t \geq 0$. There is a $4m$-rotational STS($(4m)n + 1$) for each $m \geq 5$, and a 9-rotational STS($9n + 1$), by Lemma 7.61; there is a 5-rotational STS($5n + 1$) by Theorem 7.47; there is a 13-rotational STS($13n + 1$) and a 17-rotational STS($17n + 1$) by Theorem 7.60. In addition, there is a 3-CGDD of type n^4 and one of type n^5 by Theorem 7.59. Applying Corollary 7.56 establishes the existence of a p-rotational STS($pn + 1$) = STS(v). □

Theorem 7.63 *Let p be a prime with $p \equiv 3 \pmod 4$. Then there is a p-rotational STS(v), for any $p \geq 79$ and $v \equiv 1, 6p + 1 \pmod{24p}$.*

Proof Since $p \equiv 3 \pmod 4$, then p can be written as $p = 16m + r = 4(4m) + r$, where m, r are integers such that $r = 3, 7, 11, 15$ and $m \geq 0$. Moreover, $p \geq 79$ implies $m \geq 4$ and $4m \geq 16$ ($> r$). So there is a TD($5, 4m$).

Take $n = (v - 1)/p$. Since $v \equiv 1, 6p + 1 \pmod{24p}$, then $n \equiv 0, 6 \pmod{24}$. So n can be written as $n = 24t + s$, where $s = 0, 6$, for some integer $t \geq 0$. There is

p	TD	Types of input 3-CGDDs	$\|G\|$-rotational subsystems
$41 = 4 \times 9 + 5$	TD(5, 9)	n^4, n^5	9-, 5-
$61 = 4 \times 13 + 9$	TD(5, 13)	n^4, n^5	13-, 9-
$31 = 4 \times 7 + 3$	TD(5, 7)	n^4, n^5	7-, 3-
$47 = 4 \times 11 + 3$	TD(5, 11)	n^4, n^5	11-, 3-

Table 7.13 *Existence for* $p = 41, 61, 31, 47$ *by Corollary* 7.56

a $4m$-rotational STS$((4m)n+1)$, for each $m \geq 4$, a 3-rotational STS$(3n+1)$ and a 15-rotational STS$(15n+1)$ by Lemma 7.61; there is a 7-rotational STS$(7n+1)$ and a 11-rotational STS$(11n+1)$ by Theorem 7.47. Meanwhile, there is a 3-CGDD of type n^4 and one of type n^5 by Theorem 7.59. Applying Corollary 7.56 establishes the existence of a p-rotational STS$(pn+1) =$ STS(v). □

Thus far, we have treated all primes except for 37, 41, 53, 61, 73, and 89 when $p \equiv 1 \pmod 4$, and for 19, 23, 31, 43, 47, 59, 67, and 71 when $p \equiv 3 \pmod 4$.

The bounds in Theorem 7.62 and Theorem 7.63, namely, 97 and 79, are obtained by only considering 17 in $\{5, 9, 13, 17\}$ and 15 in $\{3, 7, 11, 15\}$. Therefore treating each of the 6 congruence classes for p, namely, $p \equiv 5, 9, 13 \pmod{16}$ or $p \equiv 3, 7, 11 \pmod{16}$, separately, and using the same construction as in the two theorems, we can shorten the list.

Theorem 7.64 *There is a p-rotational STS(v), for* $v \equiv 1, 18p+1 \pmod{24p}$ *if p is in* P_5, P_9 *and for* $v \equiv 1, 6p+1 \pmod{24p}$ *if p is in* P_3, P_7, P_{11}*, where* $P_5 = \{37, 53\}$, $P_9 = \{73, 89\}$, $P_3 = \{19, 67\}$, $P_7 = \{71\}$, $P_{11} = \{59\}$.

Proof For primes in P_5, P_9: $37 = 4 \times 8 + 5$, $53 = 4 \times 12 + 5$, $73 = 4 \times 16 + 9$, $89 = 4 \times 20 + 9$; for primes in P_3, P_7, P_{11}: $19 = 4 \times 4 + 3$, $67 = 4 \times 16 + 3$, $71 = 4 \times 16 + 7$, $59 = 4 \times 12 + 11$. In any case, such a prime p can be written as $p = 4(4m) + r$ such that $4m > r$ (and $4m \neq 2, 3, 6, 10$), where $r = 5, 9$ for p in P_5, P_9, or $r = 3, 7, 11$ for p in P_3, P_7, P_{11}. So the proof of Theorem 7.62 or that of Theorem 7.63 still works in each corresponding case. □

Until the point, only transversal designs TD$(5, 4m)$, with a group size that is a multiple of 4, have been employed. Considering transversal designs with a group size different from $4m$, we can apply Corollary 7.56 to deal with some of the remaining primes. For example, consider primes $p = 41, 61$ (with $p \equiv 1 \pmod 4$) and an order $v \equiv 1, 18p+1 \pmod{24p}$, or $p = 31, 47$ (with $p \equiv 3 \pmod 4$) and an order $v \equiv 1, 6p+1 \pmod{24p}$. The existence of such a p-rotational Steiner triple system of the corresponding orders can be established by applying Corollary 7.56, with the knowledge on 5-, 9-, 13- and 3-, 7-, 11-rotational Steiner triple systems. The proofs are illustrated in Table 7.13. In the table, we take $n = 24t + r$, where $t \geq 0$ is an integer and $r = 0, 18$ for $p = 41, 61$ and $r = 0, 6$ for $p = 31, 47$. The existence of the 3-CGDDs is guaranteed by Theorem 7.59.

p	Master GDD		Types of input 3-CGDDs	$\|G\|$-rotational subsystems
	Block sizes	Type		
$23 = 5 \times 4 + 3$	4, 5	$4^5 3^1$	n^4, n^5	4-, 3-
$43 = 10 \times 4 + 3$	4, 5	$4^{10} 3^1$	n^4, n^5	4-, 3-

Table 7.14 *Existence for* $p = 23, 43$ *by Theorem* 7.55

To deal with the last two primes, 23 and 43, with $p \equiv 3 \pmod 4$, start with a resolvable TD(4, 5) $(Y, \mathcal{H}, \mathcal{B})$, which has 5 parallel classes $\mathcal{B}_i$ $(i = 1, 2, \ldots, 5)$, each of which consists of 5 blocks of size 4. Adjoining a new point ∞_i to all the blocks in $\mathcal{B}_i$, for $i = 1, 2, 3$, and taking $\{\infty_1, \infty_2, \infty_3\}$ and blocks in the parallel class $\mathcal{B}_4$ as new groups we obtain a $\{4, 5\}$-GDD of type $4^5 3^1$ $(X, \mathcal{G}, \mathcal{A})$, where

$$\begin{aligned} X &= Y \bigcup \{\infty_1, \infty_2, \infty_3\}, \\ \mathcal{G} &= \mathcal{B}_4 \bigcup \{\{\infty_1, \infty_2, \infty_3\}\}, \\ \mathcal{A} &= \mathcal{H} \bigcup (\mathcal{B} \setminus \mathcal{B}_1 \bigcup \mathcal{B}_2 \bigcup \mathcal{B}_3 \bigcup \mathcal{B}_4) \bigcup (\textstyle\bigcup_{i=1}^{3} \{B \bigcup \{\infty_i\} \mid B \in \mathcal{B}_i\}). \end{aligned}$$

There is a resolvable $(40, 4, 1)$-design $(Y, \mathcal{B})$ since $40 = 3 \times 13 + 1$ and 13 is a prime congruent to 1 modulo 4. This resolvable block design has 13 parallel classes, each of which consists of 10 blocks of size 4. Suppose $\mathcal{B}_i$, $i = 1, 2, 3, 4$, are four of the parallel classes. Adjoining a new point ∞_i to all the blocks in $\mathcal{B}_i$, for $i = 1, 2, 3$, and taking $\{\infty_1, \infty_2, \infty_3\}$ and blocks in the parallel class $\mathcal{B}_4$ as new groups we obtain a $\{4, 5\}$-GDD of type $4^{10} 3^1$ $(X, \mathcal{G}, \mathcal{A})$, where

$$\begin{aligned} X &= Y \bigcup \{\infty_1, \infty_2, \infty_3\}, \\ \mathcal{G} &= \mathcal{B}_4 \bigcup \{\{\infty_1, \infty_2, \infty_3\}\}, \\ \mathcal{A} &= (\mathcal{B} \setminus \mathcal{B}_1 \bigcup \mathcal{B}_2 \bigcup \mathcal{B}_3 \bigcup \mathcal{B}_4) \bigcup (\textstyle\bigcup_{i=1}^{3} \{B \bigcup \{\infty_i\} \mid B \in \mathcal{B}_i\}). \end{aligned}$$

With a $\{4, 5\}$-GDD of type $4^5 3^1$ and a $\{4, 5\}$-GDD of type $4^{10} 3^1$ as well as the knowledge on 4- and 3-rotational Steiner triple systems, the existence of a p-rotational STS(v), for $p = 23, 43$, of the orders $v \equiv 1, 6p + 1 \pmod{24p}$ can be obtained by applying Theorem 7.55. We outline the proof in Table 7.14, in which $n = 24t$ $(t \geq 0)$ for an order $v \equiv 1 \pmod{24p}$, or $n = 24t + 6$ for an order $v \equiv 6p + 1 \pmod{24p}$. The existence of the 3-CGDDs is guaranteed by Theorem 7.59.

In §7.3.6, we established the existence of a p-rotational STS(v), for a prime $p \geq 13$, of those orders v satisfying

$$v \equiv \begin{cases} 1, 18p + 1 \pmod{24p} & \text{when } p \equiv 1 \pmod 4, \\ 1, 6p + 1 \pmod{24p} & \text{when } p \equiv 3 \pmod 4. \end{cases}$$

In view of Theorems 7.47, 7.49, Lemmas 7.52, 7.53, and Theorem 7.54, the sufficiency of Lemma 7.46 is proved:

Theorem 7.65 *Let v, k be positive integers such that $1 \leq k \leq (v - 1)/2$. Then a k-rotational STS(v) exists if and only if*

$$\begin{cases} v \equiv 1,3 \pmod 6, & \\ v \equiv 3 \pmod 6) & \textit{if } k = 1, \\ v \equiv 1 \pmod k, & \\ v \not\equiv 7,13,15,21 \pmod{24} & \textit{if } (v-1)/k \textit{ is even.} \end{cases} \tag{7.5}$$

7.4 Which permutations are automorphisms?

The existence of cyclic triple systems gives a complete solution to the problem of determining when a permutation with all cycles of the same length arises as an automorphism of an STS(v):

Theorem 7.66 *For positive integers c and ℓ, there exists an STS($c\ell$) admitting an automorphism consisting of c ℓ-cycles if and only if $c\ell \equiv 1,3 \pmod 6$, and $(c,\ell) \neq (1,9)$.*

Proof For $c\ell \equiv 1,3 \pmod 6$, $c\ell \neq 9$, apply Theorem 7.11 to produce a cyclic STS($c\ell$); applying the cyclic automorphism c times yields the desired automorphism. When $c\ell = 9$, the unique STS(9) admits a permutation consisting of 3 cycles of length 3. □

The results on rotational systems give extensive information about the cases when one element is fixed and the remainder are mapped in orbits of equal length.

It is natural to ask a very general question. Which permutations arise as automorphisms of a TS(v, λ)? Of course, the complete triple system TS($v, v-2$) admits Sym_v as its automorphism group, so the problem is most interesting when the index is fixed. We examine the situation for Steiner triple systems primarily.

Colbourn, Magliveras, and Stinson (1992) generate all 172,248 nonisomorphic STS(19)s that admit at least one nontrivial automorphism. The numbers of systems admitting each possible automorphism are tabulated for basic automorphisms in Table 5.13, and for nonbasic automorphisms in Table 5.14.

7.4.1 *Fixed points*

First we prove a simple lemma:

Lemma 7.67 *If $(V, \mathcal{B})$ is an STS(v) admitting π as an automorphism, the elements of V fixed by π induce a subsystem.*

Proof If $x = \pi(x)$ and $y = \pi(y)$, the automorphism π maps the unique triple containing x and y, say $\{x, y, z\}$ to $\{x, y, \pi(z)\}$. Thus $z = \pi(z)$. Hence the fixed points under π induce a subdesign. □

Until this point, we have seen systems in which the subdesign induced on the fixed points contains no triples. Evidently, there can be no more than $(v-1)/2$ fixed points.

Hartman and Hoffman (1987) establish a restriction on the number of fixed points in one particular case:

Lemma 7.68 *Suppose that an STS(v) admits an involution π as an automorphism. When π has f fixed points, and n cycles of length two (so that $v = 2n+f$), the induced subdesign of order f must satisfy $f \le v/3$ when n is odd.*

Proof Let F be the set of points fixed by π, and X the set of remaining points. For a fixed element $\infty \in F$, consider the orbit of a triple $\{\infty, x, y\}$ with $\{x, y\} \subseteq X$; this orbit contains one triple if $x = \pi(y)$, and contains two triples otherwise. There are exactly n triples involving ∞ of this type, and since n is odd, ∞ must appear with an odd number of pairs of the form $\{x, \rho(x)\}$ in triples. Each such pair itself appears in only one triple, and thus $f \le n$. □

Colbourn, Magliveras, and Stinson (1992) establish that when the number of fixed points is $(v-1)/2$, a further necessary condition arises:

Lemma 7.69 *An STS(v) admits an automorphism having exactly $(v-1)/2$ fixed points only if $v = 3$ or $v \equiv 7, 15 \pmod{24}$, and the automorphism is an involution.*

Proof By Lemma 7.67, the STS(v) must have a sub-STS($(v-1)/2$), and thus $v \equiv 3, 7 \pmod{12}$. Now write $n = (v+1)/4$. Let the triple system be $(S \cup T, \mathcal{B})$, where S supports the subsystem of order $(v-1)/2$, and T contains the remaining $(v+1)/2 = 2n$ elements. Let π be the automorphism that fixes each element of S and no element of T. Considering the triples containing one point from S and two from T, each element of S corresponds to a 1-factor on T, and thus we obtain a 1-factorization of K_{2n} with element set T. Since π fixes S, it acts nontrivially on the elements of T. Thus the 1-factorization obtained has a nontrivial automorphism that fixes each 1-factor. Since any automorphism that fixes each element of S and one element of T must fix each element of T, we conclude that π maps the elements of T in cycles of the same length. If π contains a cycle of length three or more on T, say $(t_0, t_1, t_2, \ldots)$, the 1-factor containing $\{t_0, t_1\}$ must also contain $\{t_1, t_2\}$, since factors are fixed. Such a pair repetition cannot occur in a Steiner triple system, and thus π must be an involution with n cycles of length two on T. By Lemma 7.68, we conclude that $n = 1$ or n is even, and hence that $v = 3$ or $v \equiv 7, 15 \pmod{24}$. □

The fact that an automorphism of an STS(v) having $(v-1)/2$ fixed points is an involution was proved by Davies (1987), as a consequence of a more general result on 2-(v,k,1) designs.

7.4.2 *Involutions*

Steiner triple systems admitting an involution with exactly one fixed point are the reverse STSs treated in §7.3.1. Doyen (1973) settles existence of STS(v)s admitting an involution with three fixed points when $v \equiv 3 \pmod 6$. Hartman and Hoffman (1987) prove a substantial generalization of this:

Theorem 7.70 *Let $v = 2n + f \equiv 1, 3 \pmod 6$ where $f \equiv 1, 3 \pmod 6$, $f > 1$ and*

$$v \geq \begin{cases} 2f+1 & \text{if } n \text{ is even} \\ 3f & \text{if } n \text{ is odd} \end{cases}$$

Then there exists an STS(v) having an involution with f fixed points as an automorphism.

7.4.3 *3-cycles*

Kreher, Stinson, and Zhu (1997) prove that the maximum number of fixed points of an STS(v) under an automorphism of order 3 is $(v-3)/2$. They further establish:

Lemma 7.71 *An STS(v) having an automorphism of order* 3 *with* $(v-3)/2$ *fixed points exists if and only if* $v \equiv 9 \pmod{12}$.

7.4.4 *Bicyclic automorphisms*

A *bicyclic* STS(v) is one that admits a permutation consisting of a k-cycle and a $(v-k)$-cycle. Suppose that a bicyclic STS(v) $(V, \mathcal{B})$ exists, with an automorphism π consisting of two cycles. Let k be the length of the shorter cycle, and $\ell = v-k$. Evidently, π^ℓ is an automorphism that fixes at least ℓ elements of V; by Lemma 7.67, all elements must be fixed and thus $k \mid \ell$. Write $\alpha = \ell/k$. Since $v = (\alpha+1)k$, α is even.

Consider the set E of elements in the element orbit of size k under π. For $\{x,y\} \subset E$, the triple $\{x,y,z\} \in \mathcal{B}$ satisfies $\{x,y,z\} = \pi^k(\{x,y,z\})$, so $z = \pi^k(z)$ and thus $z \in E$. Hence the elements of E induce a *cyclic* subsystem of $(V, \mathcal{B})$, and thus $k \equiv 1,3 \pmod 6$ and $k \neq 9$. Next, the pairs of elements on $V \setminus E$ fall into $(\alpha k)/2 - 1$ (full) orbits of length αk, and one short orbit of length $(\alpha k)/2$. Since there are $(\alpha k^2)/2$ triples in $\mathcal{B}$ that contain one element from E and two from $V \setminus E$, there remain $((\alpha-1)k-1)/2$ full orbits of pairs on $V \setminus E$ to form triples. To complete the system, we require that

$$\frac{(\alpha-1)k-1}{2} \equiv \begin{cases} 0 \pmod 3 & \text{if } k \equiv 1 \pmod 6 \\ 0,1 \pmod 3 & \text{if } k \equiv 3 \pmod 6 \end{cases}$$

Thus we find that

$$\alpha \equiv \begin{cases} 2 \pmod 6 & \text{if } k \equiv 1 \pmod 6 \\ 0 \pmod 2 & \text{if } k \equiv 3 \pmod 6 \end{cases}$$

This establishes strong necessary conditions for bicyclic STS(v)s. These conditions were shown to be sufficient for $k = 3$ by Calahan (1990), for $k = 7$ by Gardner (1993), and for all $k > 1$ by Calahan and Gardner (1992):

Theorem 7.72 *There exists an STS(v) admitting an automorphism consisting of a k-cycle and a $(v-k)$-cycle for $k > 1$ if and only if* $v \equiv 1,3 \pmod 6$, $k \mid v$, *and either*

1. $k \equiv 1 \pmod 6$ *and* $3k \mid v$; *or*
2. $k \equiv 3 \pmod 6$ *and* $k \neq 9$.

7.4.5 *Transrotational automorphisms*

A *k-transrotational* STS(v) is an STS(v) admitting an automorphism consisting of one fixed point, one 2-cycle, and k $((v-3)/k)$-cycles. Gardner (1993) establishes:

Theorem 7.73 *A 1-transrotational STS(v) exists if and only if* $v \equiv 1, 7, 9, 15 \pmod{24}$.

Theorem 7.74 *There exists a 2-transrotational STS(v) if and only if* $v \equiv 3, 19 \pmod{24}$.

Das and Rosa (1992) also give a construction for 2-transrotational STS(v)s when $v \equiv 19 \pmod{24}$.

Calahan and Gardner (1991) settle existence for 3-transrotational STS(v)s:

Theorem 7.75 *A 3-transrotational STS(v) exists if and only if* $v \equiv 3, 9, 15 \pmod{24}$.

7.4.6 *Other automorphisms*

A *k-near-rotational* STS(v) is one that admits an automorphism containing three fixed points, and k cycles of equal length $(v-3)/k$.

Gardner (1992) establishes:

Theorem 7.76 *There exists a 2-near-rotational STS(v) if and only if* $v \equiv 1, 3 \pmod{6}$ *and* $v \notin \{13, 21\}$.

Calahan and Hoffman (1990) state:

Theorem 7.77 *If π is a permutation with one fixed point, t 2-cycles, and one cycle of length $2e$, $e > 1$, then an ITS$(2e+2t+1, 2t+1; 1)$ having π as an automorphism for which the $2e$-cycle acts on the points not in the hole exists if and only if*

1. $t \leq \lceil (e-1)/2 \rceil$, *with strict inequality when* $e \equiv 14, 20 \pmod{24}$;
2. $t \equiv \lceil (e-1)/2 \rceil \pmod{2}$; *and*
3. *either* $e \equiv 0 \pmod{3}$, *or* $e \equiv 1 \pmod{3}$ *and* $t \equiv 0 \pmod{3}$.

An STS$(2e+2t+1)$ exists having the same automorphism if and only if, in addition, $2t \equiv 0, 2, 8, 18 \pmod{24}$.

7.4.7 *An application: halving triple systems*

A triple system TS(v, λ) $(V, \mathcal{B})$ can be *halved* if there is a partition $\mathcal{B}_1, \mathcal{B}_2$ so that $(V, \mathcal{B}_1)$ and $(V, \mathcal{B}_2)$ are isomorphic as set systems. Of course, to be halved a TS(v, λ) must have an even number of triples, so $\lambda v(v-1) \equiv 0 \pmod{12}$. When $(V, \mathcal{B})$ has an odd number of triples, we extend the definition to say that $(V, \mathcal{B})$ can be halved if there is a triple $B \in \mathcal{B}$ so that $\mathcal{B} \setminus \{B\}$ has a partition $\mathcal{B}_1, \mathcal{B}_2$ so that $(V, \mathcal{B}_1)$ and $(V, \mathcal{B}_2)$ are isomorphic.

Das and Rosa (1992) examine the question of determining the orders for which a Steiner triple system exist that can be halved. Their approach rests completely on the structure of automorphisms of the STS. When the number of triples is odd, they prove:

Theorem 7.78 *Whenever $v \equiv 3, 7 \pmod{12}$, there exists an STS(v) that can be halved.*

Proof It suffices to show that there exists an STS(v) admitting an automorphism π so that under π, at most one orbit of triples has odd length. For suppose that $\mathcal{O}_1, \ldots, \mathcal{O}_s$ are the orbits, the number of triples in triples in $\mathcal{O}_i$ is ℓ_i, and $T_i \in \mathcal{O}_i$. Then $\mathcal{B}_1 = \{\pi^{2j}(T_i) : 1 \le i \le s, 0 \le j < (\ell_i - 1)/2\}$. The required partition is given by $\mathcal{B}_2 = \pi(\mathcal{B}_1)$.

Choose π to be 1-rotational when $v \equiv 3 \pmod{24}$ (Lemma 7.37), 1-transrotational when $v \equiv 7, 15 \pmod{24}$ (Theorem 7.73), and 2-transrotational when $v \equiv 19 \pmod{24}$ (Theorem 7.74). □

The case when the number of triples is even appears to be more difficult. Consider a class $\mathcal{B}_1$ in a halving, and form the graph G_1 whose edges are the pairs appearing in triples of $\mathcal{B}_1$. G is a self-complementary graph. A self-complementary graph that has an edge-partition into triangles is a *Steiner self-complementary graph.* The existence of an STS(v) with $v \equiv 1, 9 \pmod{12}$ that can be halved is equivalent to the existence of a Steiner self-complementary graph on v vertices.

The complementing permutation τ of G (the isomorphism from G to $\overline{G}$) need not be an automorphism of the STS(v); when it is, however, the STS(v) can be *strongly halved.* Das and Rosa (1992) establish that for an STS(v) to be strongly halved, a necessary condition is that $v \equiv 1, 9 \pmod{24}$. Now if some automorphism π of an STS(v) has all orbits of even length, taking alternate triples from each orbit yields not only a halving, but in fact a strong halving, as π is a complementing permutation. Taking π to be 1-rotational when $v \equiv 1 \pmod{24}$ (Lemma 7.37), and 2-rotational when $v \equiv 9 \pmod{24}$ (Lemma 7.42), we obtain:

Lemma 7.79 *An STS(v) can be strongly halved if and only if $v \equiv 1, 9 \pmod{24}$.*

This leaves the cases when $v \equiv 13, 21 \pmod{24}$. The strategy of choosing alternate triples from each orbit cannot provide halvings in these cases. Nevertheless, the structure of automorphisms provides some information. For example, suppose there exists a cyclic STS(v) admitting a multiplier automorphism λ. If the STS(v) has an even number of orbits under the cyclic automorphism, and if λ maps these orbits in cycles of even length, alternate *orbits* can be selected to form the halving. This works, for example, for prime powers $p^\alpha \equiv 13 \pmod{24}$, using Theorem 7.23. Beyond this, however, determining STS(v)s that can be halved for $v \equiv 13, 21 \pmod{24}$ remains open.

7.5 Automorphism-free triple systems

Lindner and Rosa (1975a) examine the question of determining when a Steiner triple system exists that has no nontrivial automorphisms. A triple system admitting only the identity automorphism is *automorphism-free*. All Steiner triple systems on 3, 7, 9, and 13 elements admit at least one nontrivial automorphism. Lindner and Rosa (1975a) constructed automorphism-free STS(v)s for $v \in \{15, 19, 21, 25, 27, 33\}$; they then employ doubling constructions. First they observe that applying the $2v+1$ construction using an automorphism-free STS(v) and the 1-factorization GK_{v+1}, the resulting STS($2v+1$) is again automorphism-free. Then they establish that applying the $2v+7$ construction of §3.2 to an automorphism-free STS(v) yields an automorphism-free STS($2v + 7$). In summary, they prove:

Theorem 7.80 *An automorphism-free STS(v) exists if and only if $v \equiv 1, 3 \pmod 6$ and $v \geq 15$.*

Babai (1980) proves that almost all Steiner triple systems (as $v \to \infty$) are automorphism-free. The proof rests on the lower bound on the number $s(v)$ of distinct Steiner triple systems of order v (Theorem 5.7), which states that

$$s(v) > v^{v^2\left(\frac{1}{6}+o(1)\right)}.$$

Now let $S(v)$ be the number of nonisomorphic STS(v)s, $a(v)$ the number of distinct automorphism-free STS(v)s, and $A(v)$ the number of nonisomorphic automorphism-free STS(v)s.

Theorem 7.81 *Almost all Steiner triple systems are automorphism-free; more precisely,* $\lim_{n\to\infty} \frac{a(n)}{s(n)} = \lim_{n\to\infty} \frac{A(n)}{S(n)} = 1$. *Moreover,* $\lim_{n\to\infty} \frac{n!S(n)}{s(n)} = 1$.

Proof Let $b(n) = s(n) - a(n)$ and $B(n) = S(n) - A(n)$. Any STS(n) that has a nontrivial automorphism has at least one automorphism of prime order. So fix a set V of n elements, and let Π be the set of all automorphisms of STS(n)s on V of prime order. Let $b(\pi, n)$ be the number of STS(n)s admitting π as an automorphism, for $\pi \in \Pi$. Lemma 7.67 ensures that each $\pi \in \Pi$ fixes at most $(n-1)/2$ points; let f be the number of fixed points under π. Now to form distinct STS(n)s admitting π, there are $S(f)$ ways to choose the subsystem on the fixed points. Suppose that π has order p. We examine the number of ways to select the remaining triples. Whenever a triple is chosen, the entire orbit of triples containing it must be chosen. This orbit has p triples, unless $p = 3$ and the orbit chosen is short. Hence if we examine the number of ways to choose a third element to complete the triple containing each pair disjoint from the fixed points, we find that the number of independent choices to be made does not exceed

$$\frac{1}{3p}\left(\binom{n}{2} - \binom{f}{2}\right)$$

if $p \neq 3$, and does not exceed

$$\frac{n-f}{3} + \frac{1}{9}\left(\binom{n}{2} - \binom{f}{2} + n - f\right)$$

when $p = 3$. Thus

$$\log b(\pi, n) < \log S(f) + (n - f)(n + f + 1) \log n/6p.$$

Since $f \leq (n-1)/2$, we obtain that $\log b(\pi, n) < n^2 \log n(1 + (3/p))/24$. Now $b(n) \leq \sum_{\pi \in \Pi} b(\pi, n)$, so using the (extreme) overestimate of Π containing $n!$ permutations, we obtain

$$\log b(n) < \log n! + \max_p \left(n^2 \log n(1 + (3/p))/24\right),$$

so that

$$\log b(n) < n^2 \log n \left(\frac{5}{48} + \mathrm{o}(1)\right).$$

Now $S(n) \geq s(n)/(n!)$, so applying the bound obtained for $b(n)$ with the bound of Theorem 5.7 for $s(n)$, we obtain that

$$\max\left(\frac{B(n)}{S(n)}, \frac{b(n)}{s(n)}\right) < n!\frac{b(n)}{s(n)} < n^{-n^2\left(\frac{1}{16}+\mathrm{o}(1)\right)},$$

and thus $\lim_{n\to\infty} \frac{a(n)}{s(n)} = \lim_{n\to\infty} \frac{A(n)}{S(n)} = 1$.

Since $n!A(n) = a(n)$, $\lim_{n\to\infty}(n!S(n))/s(n) = 1$. □

7.6 Work points

1. Which permutations are automorphisms of Steiner triple systems?
2. For which orders does a cyclic Steiner triple system having a nontrivial multiplier automorphism exist?
3. Complete the solution of the halving problem: find STS(v)s for $v \equiv 13, 21 \pmod{24}$ that can be halved.
4. Extend the constructive enumeration results in Tables 7.8 and 7.10.

8

AUTOMORPHISMS II: LARGE GROUPS

In this chapter, we study triple systems admitting 'large' automorphism groups. In particular, we examine actions of permutation groups on triple systems that are, in some way, uniform.

8.1 Group actions

Consider a triple system $(V, \mathcal{B})$, and a permutation group Γ acting on V. We can examine the action of Γ in a number of ways. Naturally, the actions on the elements of V, and on the triples of $\mathcal{B}$, are of interest. However, we can also examine actions of Γ on ordered pairs of elements, unordered pairs of elements, flags and antiflags (a pair (x, B) with $x \in V$ and $B \in \mathcal{B}$ is a *flag* if $x \in B$, and an *antiflag* otherwise).

A convenient unification is given by forming the incidence graph of the triple system. Denote the incidence graph by $(V, \mathcal{B}, \mathcal{I})$, with $\mathcal{I}$ giving the element-triple incidence relation. Let $\mathcal{X}$ be a set of (labelled or unlabelled) subgraphs of the incidence graph, so that each $X \in \mathcal{X}$ has the same number of vertices in V, and the same number of vertices in $\mathcal{B}$, and all subgraphs in $\mathcal{X}$ are isomorphic (preserving the element-triple partition, and the vertex labelling when present). A group Γ is *transitive* on $\mathcal{X}$ when the orbit of $X \in \mathcal{X}$ under Γ is $\mathcal{X}$. We are concerned here primarily with transitivity on various subgraphs of the incidence graph, as follows:

Type	Transitive on
transitive or *point-transitive*	elements
block-transitive	triples
2-homogeneous	unordered pairs of elements
2-transitive	ordered pairs of elements
flag-transitive	flags (pairs of incident elements and triples)

An action of a group Γ on a set X is *semiregular* if and only if, for each $x \in X$, the only permutation in Γ fixing x is the identity. In a semiregular action, all orbits have the same length, which is the order of Γ. An action that is both transitive and semiregular is *regular*.

Now consider a transitive action, say Γ acting on X. A subset Δ of X is a *block of imprimitivity* if, for any $\pi \in \Gamma$, $\Delta \cap \Delta\pi = \Delta$ or $\varnothing$. Evidently the empty set, the set X, and all singleton subsets of X are blocks of imprimitivity; these are the *trivial* blocks. An action having a nontrivial block of imprimitivity is *imprimitive*; otherwise it is *primitive*.

Now consider an imprimitive action. Let Δ be a nontrivial block of imprimitivity. Let $x \in \Delta$ and $y \in X \setminus \Delta$. Since Γ acts transitively on X, there is a $\pi \in \Gamma$ for which $x\pi = y$. Then $\Delta\pi$ is disjoint from Δ, and naturally is equicardinal. In this way, we can partition X into blocks of imprimitivity; such a partition is a *system of imprimitivity.* It follows that the size of any block of imprimitivity is a divisor of $|X|$.

In the study of designs in general, and triple systems in particular, the usual objective is to classify all designs and their groups that act transitively, regularly, or primitively on (subsets or tuples of) elements, (subsets or tuples of) blocks, or elements and blocks with prescribed incidence (e.g. flags and antiflags).

8.2 Implications from group theory

Here we are concerned with triple systems, and hence the presentation of general machinery from group theory is somewhat abbreviated; see Camina (1994) and references therein for more general information.

Throughout this section, we let $(V, \mathcal{B})$ be a $\mathrm{TS}(v, \lambda)$, and Γ be its automorphism group or a subgroup thereof. We start with a fundamental theorem of Block (1967):

Theorem 8.1 *The number of orbits of Γ on $\mathcal{B}$ is at least as large as the number of orbits of Γ on V. In particular, if $(V, \mathcal{B})$ is block-transitive, then it is point-transitive.*

Camina and Gagen (1984) relate block- and flag-transitivity; we require some notation to introduce the relationship. The *stabilizer* in Γ of a set or tuple S, denoted Γ_S, is the subgroup of Γ that fixes S (setwise or pointwise, respectively). When Γ' is a subgroup of Γ, the *index* of Γ' in Γ, denoted $[\Gamma : \Gamma']$, is the order of Γ divided by the order of Γ'.

We first establish a simple relation between the indices of some stabilizers of points, blocks, and pairs of points.

Lemma 8.2 *Suppose that Γ acts block-transitively on a Steiner triple system $(V, \mathcal{B})$. Let $B \in \mathcal{B}$ and $x \in B$. Then for all $y \in B \setminus \{x\}$,*

$$[\Gamma_x : \Gamma_{(x,y)}] = (v-1)/6[\Gamma_B : \Gamma_{(x,y)}].$$

Proof By definition,

$$[\Gamma : \Gamma_{(x,y)}] = [\Gamma : \Gamma_x][\Gamma_x : \Gamma_{(x,y)}],$$

and since every pair of points defines a unique block,

$$[\Gamma : \Gamma_{(x,y)}] = [\Gamma : \Gamma_B][\Gamma_B : \Gamma_{(x,y)}].$$

Γ acts block-transitively, so $[\Gamma : \Gamma_B] = (v(v-1))/6$. By Theorem 8.1, Γ also acts point-transitively, so $[\Gamma : \Gamma_x] = v$. Hence we have

$$v[\Gamma_x : \Gamma_{(x,y)}] = (v(v-1))/6[\Gamma_B : \Gamma_{\{x,y\}}].$$

This completes the proof. □

Camina and Gagen (1984) used this result more generally when arbitrary block size is permitted; the case of Steiner triple systems admits some simplifications:

Theorem 8.3 *Let $3|v$ and $\lambda = 1$. Then if Γ acts block-transitively, it also acts flag-transitively.*

Proof By Lemma 8.2, $(v-1)[\Gamma_B : \Gamma_{(x,y)}] \equiv 0 \pmod 6$; by assumption, $v \equiv 3 \pmod 6$ and thus $[\Gamma_B : \Gamma_{(x,y)}] \equiv 0 \pmod 3$. Since $[\Gamma_B : \Gamma_{(x,y)}]$ counts automorphisms that fix B setwise but not pointwise, Γ acts cyclically on the three flags containing the block B. This, together with the block-transitivity of Γ, ensures flag-transitivity. □

Delandtsheer and Doyen (1989) established an important general result about pairwise balanced designs, with the following consequence for triple systems:

Theorem 8.4 *If Γ acts block-transitively, but does not act point-primitively on V, suppose that $\{\Delta_1, \ldots, \Delta_d\}$ is a system of imprimitivity with $|\Delta_i| = c$ for $1 \leq i \leq d$. Then $v = cd$, and there exists a positive integer x and a nonzero integer y for which*

$$c = \frac{3-x}{y}, \qquad d = \frac{3-y}{x},$$

and in particular no such triple system can exist unless $v = 4$.

Proof Fix a block $B \in \mathcal{B}$. Count unordered pairs on B which lie in the same block of imprimitivity; let x denote this (positive) integer. Summing over all blocks, we obtain $\lambda x v(v-1)/6$ as the number of unordered pairs lying in the same block of imprimitivity. This can also be counted directly as $d\lambda\binom{c}{2}$. Hence

$$xv(v-1) = 3dc(c-1).$$

Since $v = cd$, we obtain $x(cd-1) = 3(c-1)$, and thus $c = \frac{3-x}{3-xd}$. (Evidently, $3 - xd \neq 0$ since $3 - xd = 0$ implies $3 - x = 0$, but $d \geq 2$.) Now set y to be the nonzero integer $3 - xd$. Then the required expressions for c and d are obtained.

It is an easy exercise to verify that the only choices for x and y so that $cd = v$ are $x = y = 1$, $c = d = 2$, and $v = 4$. □

Clapham (1976) had previously established Theorem 8.4 for Steiner triple systems. The extension to higher index does require the restriction to $v \neq 4$, as the dihedral group of order 8 acts transitively on the blocks, but imprimitively on the points, of the unique TS(4, 2).

Higman and McLaughlin (1961) proved that if Γ acts flag-transitively, and $(r, \lambda) = 1$, then Γ acts point-primitively. Theorem 8.4 subsumes the Higman–McLaughlin theorem in the case of triple systems.

Summary of implications			
Property	Implied property	Conditions	Authority
2-transitive	2-homogeneous	—	definition
2-transitive	flag-transitive	$\lambda = 1$	definition
2-homogeneous	block-transitive	$\lambda = 1$	definition
flag-transitive	block-transitive	—	definition
block-primitive	block-transitive	—	definition
block-transitive	flag-transitive	$\lambda=1$, $v\equiv 3 \pmod 6$	Theorem 8.3
block-transitive	point-primitive	$v \neq 4$	Theorem 8.4
point-primitive	point-transitive	—	definition

8.3 2-transitive STSs

The only triple systems that admit a 2-homogeneous action on their *blocks* are the unique TS$(3, \lambda)$ for $\lambda \geq 1$, the unique TS$(4, 2)$, and the STS(7) (see Camina, 1994); in each case, the action on blocks of the full automorphism group is 2-transitive.

Turning to 2-transitivity on points, we find a richer structure. The *projective triple systems* from Lemma 2.14 admit the 2-transitive automorphism group $P\Gamma L(n, 2)$. The *affine triple systems* from Lemma 2.15 admit the 2-transitive automorphism group $A\Sigma L(n, 3)$, the affine semilinear group. While there are a number of 2-transitive block designs of index one (see Kantor, 1985), the case for Steiner triple systems is simpler.

Hall (1960, 1962) conjectured that the projective and affine triple systems are the only 2-transitive Steiner triple systems. He established two interesting results in support of this conjecture, in the process defining the Hall triple systems discussed in §8.7.

Lemma 8.5 *(Hall, 1960) A 2-transitive Steiner triple system with more than three elements contains an STS*(7) *or an STS*(9).

Lemma 8.6 *(Hall, 1965) If $(V, \mathcal{B})$ is a Hall triple system having an automorphism group Γ which is 2-transitive on points, then it is an affine triple system.*

A complete resolution of Hall's conjecture awaited the classification of finite simple groups. Key and Shult (1984), as well as Hall (1985) and Kantor (1985), employed the simple group classification to establish:

Theorem 8.7 *If Γ acts 2-transitively on the points of an STS(v) $\mathcal{D}$, then $\mathcal{D}$ is either a projective triple system or an affine triple system.*

8.4 2-homogeneous STSs

There are examples of STS(v)s that admit 2-homogeneous automorphism groups but (for $v > 7$) do not admit 2-transitive groups. These are the Netto triple systems of Lemma 2.12. The automorphism group of the Netto triple system of order p^α, except in the case $p^\alpha = 7$, is the group $A\Gamma^2L(1,p^\alpha)$. In fact, we can give another construction of them using this result. Let

$$\Gamma = \{a^2x\sigma + b \;:\; a, b \in \mathbb{F}_{p^\alpha},\ \sigma \in \mathrm{Aut}(\mathbb{F}_{p^\alpha})\}.$$

Let ε be a primitive sixth root of unity in $\mathbb{F}_{p^\alpha}$. Then the orbit of $\{0, 1, \varepsilon\}$ under the action of Γ is the Netto triple system of order p^α.

Delandtsheer, Doyen, Siemons, and Tamburini (1986) completed the classification for 2-homogeneous STSs:

Theorem 8.8 *If Γ acts 2-homogeneously but not 2-transitively on the points of a Steiner triple system $(V, \mathcal{B})$, then $v = p^\alpha$, where $p \equiv 3 \pmod 4$ is a prime, α is odd, and $(V, \mathcal{B})$ is either*

1. *an affine triple system over a subfield of $\mathbb{F}_{3^\alpha}$; or*
2. *a Netto triple system, and $p \equiv 7 \pmod{12}$.*

8.5 Block-transitive STSs

Clapham (1976), prior to the classification of finite simple groups, undertook the determination of all block-transitive STSs. He established:

Theorem 8.9 *If Γ acts block-transitively but not 2-transitively on points of a Steiner triple system $(V, \mathcal{B})$ of order v, then Γ has odd order and is a subgroup of $A\Gamma L(1, p^\alpha)$ containing the translation group, for some prime p and $\alpha \geq 1$, and one of the following holds:*

1. *$(V, \mathcal{B})$ is an affine triple system of dimension d over $\mathbb{F}_3$, d is odd, and Γ has rank 3 on points;*
2. *$(V, \mathcal{B})$ is a Netto triple system; or*
3. *Γ has rank 7 on points and $p^d \equiv 7 \pmod{12}$.*

In consequence, Clapham (1976) obtained a characterization of flag-transitive STSs originally established by Lüneburg (1963):

Theorem 8.10 *If Γ acts flag-transitively but not 2-transitively on points of a Steiner triple system $(V, \mathcal{B})$ of order v, then Γ has odd order and is a subgroup of $A\Gamma L(1, p^d)$ containing the translation group, for some prime p and $\alpha \geq 1$, and one of the following holds:*

1. *$(V, \mathcal{B})$ is an affine triple system of dimension d over $\mathbb{F}_3$, d is odd, and Γ has rank 3 on points; or*
2. *$(V, \mathcal{B})$ is a Netto triple system.*

The similarity of Theorems 8.8 and 8.10 is no accident; in fact, we have:

Lemma 8.11 *The action of Γ on a Steiner triple system $(V, \mathcal{B})$ of order v is flag-transitive if and only if it is 2-homogeneous on points.*

Proof If Γ is flag-transitive, the order of Γ is a multiple of the number of flags, which is $3(v(v-1))/6 = (v(v-1))/2$. If Γ is 2-homogeneous, the order of Γ is a multiple of the number of unordered pairs of elements $(v(v-1))/2$. Moreover, in either case, Γ acts block-transitively by definition.

Consider the restriction $\widehat{\Gamma}_B$ of Γ_B to the points of $B = \{x, y, z\}$, for some block B. If Γ acts 2-homogeneously, the presence of a permutation mapping $\{x, y\}$ to $\{y, z\}$ ensures that $\widehat{\Gamma}_B$ contains either $(x\ y\ z)$ or $(x\ z)(y)$. Similarly, a permutation mapping $\{x, y\}$ to $\{x, z\}$ requires that $\widehat{\Gamma}_B$ contains either $(x\ z\ y)$ or $(x)(y\ z)$. $\widehat{\Gamma}_B$ either contains one of these 3-cycles, or it contains both 2-cycles; but the composition of the two 2-cycles is itself a 3-cycle. Thus Γ acts (at least) cyclically on the three flags through B; together with block-transitivity, this implies flag-transitivity.

In the other direction, since $\widehat{\Gamma}_B$ contains a permutation mapping the flag (B, x) to the flag (B, y), and another mapping it to the flag (B, z), the same conditions on $\widehat{\Gamma}_B$ as above hold. Then Γ acts cyclically on the three unordered pairs in B; together with block-transitivity, we obtain 2-homogeneity. □

Clapham (1976) also examined when block-transitive STSs admit a regular action on the blocks. Perhaps remarkably, not all Netto triple systems permit such an action. To see this, suppose that Γ acts block-regularly on an STS $(V, \mathcal{B})$. By Theorem 8.9, $V = \mathbb{F}_{p^\alpha}$ and

$$\Gamma = \left\{t \mapsto f^{6k}t + b : b \in \mathbb{F}_{p^\alpha},\ 1 \le k \le \frac{p^\alpha - 1}{6}\right\},$$

where f generates the multiplicative group of $\mathbb{F}_{p^\alpha}$.

Lemma 8.12 *Let $x \in \mathbb{F}_{p^\alpha}$, and let $\mathcal{T}$ be the set of triples which is the orbit of $\{0, 1, x\}$ under Γ. Then $(\mathbb{F}_{p^\alpha}, \mathcal{T})$ is a Steiner triple system if and only if none of x, $x-1$, or $x^{-1}-1$ are cubes in $\mathbb{F}_{p^\alpha}$.*

Proof (Sketch) $(\mathbb{F}_{p^\alpha}, \mathcal{T})$ is a Steiner triple system if and only if the union of those blocks containing 0 is all of $\mathbb{F}_{p^\alpha}$. The blocks containing 0 are

$$\{0, f^{6k}, xf^{6k}\}, \{0, -f^{6k}, (x-1)f^{6k}\}, \{0, -xf^{6k}, (1-x)f^{6k}\}$$

for $1 \le k \le \frac{p^\alpha - 1}{6}\}$. All nonzero quantities here must therefore be distinct, and some manipulations show that this occurs if and only if none of x, $x-1$, or $x^{-1}-1$ are cubes in $\mathbb{F}_{p^\alpha}$. □

Example 8.13 Consider the case when $p = 31$. The choices of $x \in \mathbb{F}_{31} \setminus \{0, 1\}$ for which x, $x-1$, and $x^{-1}-1$ are all noncubes are $\{6, 12, 13, 14, 18, 19, 20, 26\}$. Now $x \in \{6, 26\}$ gives the primitive sixth roots of unity, and hence they lead to the Netto system. The other choices of x give block-regular systems other than the Netto system.

The Netto triple systems are obtained by choosing x to be a primitive sixth root of unity in $\mathbb{F}_{p^\alpha}$. By examining the structure of cubes in $\mathbb{F}_{p^\alpha}$, Clapham (1976) established:

Theorem 8.14 *The Netto triple system of order p^α is block-regular if and only if $p^\alpha \equiv 7, 31 \pmod{36}$.*

8.6 Veblen points and quadrilaterals

Hall (1960) shows that when every three elements not lying in a common block generate an STS(7), the system is necessarily projective. This result falls into a large investigation in projective geometry, characterizing projective spaces via structural or algebraic properties. It also provided the impetus for the study of dimension of Steiner triple systems, discussed in §6.8. For our purposes, their importance is that structural properties can necessitate the presence of certain automorphisms. However, to treat at least the combinatorial aspects of this investigation properly, we examine a related geometric viewpoint.

Veblen and Young (1916) examine geometric consequences of four axioms on a collection of sets:

1. Every two elements appear together in precisely one set in the collection.
2. If p, q, r, and s are distinct elements, and the set containing p and q intersects the set containing r and s, so also does the set containing p and r intersect the set containing q and s.
3. Every set in the collection has at least three elements.
4. There are at least two disjoint sets in the collection.

These are the *Veblen–Young axioms*, used initially to establish that any collection of sets meeting these axioms is a projective space.

If we start with a Steiner triple system, the first and third axioms hold trivially. Indeed, since the number of blocks of an STS(v) intersecting a given block is $(3v-9)/2$, the fourth axiom also holds whenever $v \geq 9$. Thus for STSs, the second axiom is the one of substance. For this reason, an element x of an STS is called a *Veblen point* if whenever $\{x,a,b\}$, $\{x,c,d\}$, and $\{y,a,c\}$ are blocks, so also is $\{y,b,d\}$. The configuration of four triples is a quadrilateral or Pasch configuration.

The second Veblen–Young axiom asserts that every element is a Veblen point. This geometric notion is critical in relating structural properties of the triple system to its automorphism group.

Theorem 8.15 *Let $(V,\mathcal{B})$ be a Steiner triple system of order v, and suppose that $2^n \leq v < 2^{n+1}$. The following six statements are equivalent:*

1. *The system $(V,\mathcal{B})$ is isomorphic to $PG(n+1,2)$, and $v = 2^{n+1}-1$.*
2. *Every three elements of V which do not form a triple of $\mathcal{B}$ generate a sub-STS(7).*
3. *Every element of V is a Veblen point.*
4. *The Steiner 1-factorization of $(V,\mathcal{B})$ has the property that the union of any two of its 1-factors is isomorphic to $v/4$ cycles of length four.*
5. *The Steiner 1-factorization of $(V,\mathcal{B})$ is obtained by doubling a 1-factorization which is isomorphic to boolean 1-factorization $AG(n,2)$.*

6. *The Steiner 1-factorization of* $(V, \mathcal{B})$ *is isomorphic to the boolean 1-factorization* $AG(n+1, 2)$.

Proof The implication of (1) from (6) follows from Lemma 2.16. The implication of (2) from (1) can be seen by taking three binary vectors v_1, v_2, and v_3 of length $n+1$ whose modulo 2 sum is nonzero; then

$$\{v_1, v_2, v_1 \oplus v_2, v_3, v_1 \oplus v_3, v_2 \oplus v_3, v_1 \oplus v_2 \oplus v_3\}$$

supports a sub-STS(7). By the same token, the implication of (3) from (2) follows from the fact that every three points in an STS(7) appear in a quadrilateral.

The implication of (4) from (3) follows from the construction of Steiner 1-factorizations from an STS, and the implication of (6) from (5) is immediate from Lemma 1.20. Hence the substance of the proof is the implication of (5) from (4). Cameron (1976) establishes this statement under the weaker hypothesis that the 1-factorization need not be Steiner. We pursue a similar approach here. Let $F_1, \ldots, F_{v-1}$ be a 1-factorization on X so that for $1 \leq i < j < v$, the 2-factor $F_i \cup F_j$ is a collection of $v/4$ cycles of length four. Form bijections $\phi_i : X \mapsto X$ for $1 \leq i < v$ by setting $\phi_i : x \mapsto y$ if and only if $\{x, y\} \in F_i$. Each mapping ϕ_i is an involution without fixed points. Now if $\{w, x\} \in F_k$ and $k \neq i$, we find $\{w, \phi_i(w)\}$, $\{z, \phi_i(z)\}$, and $\{w, x\}$ in $F_i \cup F_k$, and this must complete to a cycle of length four, so that $\{\phi_i(w), \phi_i(z)\} \in F_k$. It follows that ϕ_i is an automorphism of the 1-factorization.

Whenever $\{\{x, y\}, \{w, z\}\} \subseteq F_i$, it must happen that $\{x, w\}$ and $\{y, z\}$ lie in the same 1-factor F_j; but pairs $\{x, z\}$ and $\{w, y\}$ also lie in the same 1-factor F_k. This groups all 1-factors except for F_i in pairs.

Now choose an arbitrary element x_0 and set $X_0 = \{x_0\}$ and $\overline{X}_0 = \{\phi_1(x_0)\}$. Once X_{i-1} and $\overline{X}_{i-1}$ are defined, if $X_{i-1} \cup \overline{X}_{i-1} \neq X$, let $x_i \in X \setminus (X_{i-1} \cup \overline{X}_{i-1})$ and let $F_{\sigma(i)}$ be the 1-factor containing $\{x_0, x_i\}$. Then define $X_i = X_{i-1} \cup \{y : \{x, y\} \in F_{\sigma(i)}, x \in X_i\}$ and $\overline{X}_i = \{\phi_1(x) : x \in X_i\}$. Let t denote the value of i for which $X_i \cup \overline{X}_i = X$. Using these definitions, it is easily verified that, for each $0 \leq i \leq t$, every 1-factor of the factorization either contains no edges on X_i or induces a 1-factor on X_i. Then X_t carries a 1-factorization in which the union of any two 1-factors consists of cycles of length four, and from which the 1-factorization on X is obtained by Lemma 1.20. A simple induction on n completes the proof. □

8.7 Hall triple systems

In this section, we give a brief treatment of Hall triple systems. More detailed presentations are given in Deza and Sabidussi (1990) and Bénéteau (1996), and references therein.

A *Hall triple system* (HTS) is a Steiner triple system $(V, \mathcal{B})$ such that for each $x \in V$, the automorphism group of the STS has an automorphism which is an involution whose only fixed point is x. Hall (1960) introduced these systems, which were later named after him. He characterized HTSs in the class of all STSs, by examining the structure of their minimal nontrivial subsystems.

Lemma 8.16 *(Hall (1960)) A Steiner triple system $(V, \mathcal{B})$ is an HTS if and only if every three elements not lying in a common block generate an STS(9).*

Hence, affine triple systems are Hall triple systems. In contrast with Theorem 8.15, however, we see that the converse need not hold. The associated Steiner quasigroup $(V, \circ)$ of an HTS $(V, \mathcal{B})$ is distributive, i.e. it satisfies the additional identity $z \circ (x \circ y) = (z \circ x) \circ (z \circ y)$. Conversely, a distributive Steiner quasigroup yields an STS which is a Hall triple system.

An *exponent 3 commutative Moufang loop*, or *exp 3-CML*, is a commutative loop $(L, \oplus)$ satisfying $x^2 \oplus yz = xy \oplus xz$ and $x^3 = \infty$, where ∞ is the loop identity. An exp 3-CML can be obtained from a distributive Steiner quasigroup $(V, \circ)$ by setting $L = V$, fixing an arbitrary $\infty \in V$, and defining $xy = \infty \circ (x \circ y)$. Different choices of ∞ can lead to different loops. Conversely, given an exp 3-CML $(L, \oplus)$, a distributive Steiner quasigroup $(V, \circ)$ can be obtained by setting $V = L$, and $x \circ y = x^2 y^2$. The exp 3-CML so defined is *not* the Steiner loop associated with the STS that is defined in §2.1.1.

Let S be some $(n+1)$-dimensional vector space over $\mathbb{F}_3$, with $n \geq 3$. Let $\{e_0, e_1, \ldots, e_n\}$ be a basis for S. For any two points $x = \sum \alpha^i e_i$ and $y = \sum \beta^i e_i$ set $z = x \circ y$ as defined by: $x + y + z = (\alpha^1 - \beta^1)(\alpha^2 \beta^3 - \alpha^3 \beta^2) e_0$. This defines a binary operation on S. We have either $x = y = z$ or the three points x, y, z are pairwise distinct. The 3-subsets of the form $\{x, y, z\}$ such that $z = x \circ y$ provide S with a structure of an HTS. This particular HTS is referred to as $H(n)$. Any affine space $AG(n, 3)$ over $\mathbb{F}_3$ with the usual lines may be viewed as an HTS. Such an HTS is an *affine* HTS. Hall (1960) essentially proved:

Theorem 8.17 *The cardinality of any HTS is 3^m for some integer $m \geq 2$. Nonaffine HTSs of order 3^m exist for any $m \geq 4$ and do not exist for $m = 2$ or 3.*

The existence of a nonaffine HTS of order $3^m > 3^3$ is provided by $H(m-1)$. For the orders 3^4 and 3^5 there is a unique nonaffine HTS, namely $H(3)$ and $H(4)$, respectively (Hall (1960), and Kepka (1978) and Bénéteau (1980), respectively). The classification of HTSs of order 3^6 is due to Bénéteau (1984), and those of order 3^7 to Bénéteau and Lacaze (1988).

The maximum number of nonisomorphic HTSs of order 3^m for small values of m is:

Order	3^1	3^2	3^3	3^4	3^5	3^6	3^7	3^8	3^9	3^m
Affine HTS	1	1	1	1	1	1	1	1	1	1
Nonaffine HTS	0	0	0	1	1	3	12	>42	?	?

The *rank* r of an HTS is the smallest integer r such that there exists a generator set of r points. Hence the rank is one more than the dimension discussed in §6.8.

Theorem 8.18 *In a Hall triple system of order 3^m the rank r is at most $m+1$, with equality only in the affine case. Moreover any minimal generator subset consists of exactly r distinct points.*

The number of nonisomorphic HTSs of order 3^m of rank r is:

Order	3^4	3^5	3^6	3^7	3^8
Rank 3 HTS	1	0	0	0	0
Rank 4 HTS	1	1	1	1	4
Rank 5 HTS	0	1	2	6	>16
Rank 6 HTS	0	0	1	5	?
Rank 7 HTS	0	0	0	1	?
Rank 8 HTS	0	0	0	0	1

All known classification theorems concerning HTSs have been derived from their description in terms of distributive Steiner quasigroups, or in terms of exponent 3 commutative Moufang loops.

A group Γ is a *Fischer group* (in the restrictive sense) if the set $S = S(\Gamma)$ of order 2 elements generates Γ, if it contains at least two elements, and satisfies order$(xy) = 3$ for any two distinct elements x and y from S. $S(\Gamma)$ may be provided with a structure of a distributive Steiner quasigroup by $x \circ y = xyx$. So the lines of the related HTS are the sets of three involutions from Γ of the form (x, y, xyx). Any HTS arises in this way from various (nonisomorphic) Fischer groups. Among the different Fischer groups associated with some fixed HTS, say $(S, \mathcal{L})$, there is one and only one centreless group (up to isomorphism); it can be described as the group of permutations of S spanned by all the reflections of the form: $x \longrightarrow \sigma \circ x$ with σ ranging in S.

The construction of $H(m)$ may be generalized as follows. Consider some $(n+1)$-dimensional vector space S be over $\mathbb{F}_3$, with $n > 2$, and with some chosen basis, say $e_0, e_1, e_2, \dots, e_n$. For any two points $x = \sum \alpha^i e_i$ and $y = \sum \beta^i e_i$ set $z = x \circ y$ when $x + y + z = \sum_{i<j<k} \Lambda_{ijk}(\alpha^i - \beta^i)(\alpha^j\beta^k - \alpha^k\beta^j)e_0$ where (Λ_{ijk}) for $i < j < k$ form a nonvanishing sequence of elements from $\mathbb{F}_3$. This defines a binary operation of a distributive Steiner quasigroup on S. The corresponding Hall triple system has rank $n+1$.

Theorem 8.19 *Any rank $n+1$ Hall triple system of order 3^{n+1} arises in the manner just described.*

Let V be the n-dimensional vector space spanned by $e_1, e_2, \dots, e_n$ over $\mathbb{F}_3$. Consider again a nonvanishing sequence (Λ_{ijk}), $i < j < k$, of elements from $\mathbb{F}_3$. The equalities $\phi(e_i, e_j, e_k) = \Lambda_{ijk}$ for $i < j < k$ define a unique symplectic trilinear form ϕ from V^3 onto $\mathbb{F}_3$. Any two symplectic trilinear forms ϕ and ϕ' are *equivalent* when $\phi'(x, y, z) = \phi(f(x), f(y), f(z))$ for some automorphism f of V.

Two sequences $(\Lambda_{ijk}{}')$ and (Λ_{ijk}) give rise to equivalent symplectic trilinear forms if their related HTSs are isomorphic. This correspondence between rank $n+1$ HTSs of order 3^{n+1} and nonvanishing symplectic trilinear forms of V

(n-dimensional vector spaces over $\mathbb{F}_3$) is one-to-one up to isomorphism for Hall triple systems and up to equivalence for symplectic trilinear forms.

This correspondence may be extended by considering rank $n+1$ HTSs of order greater than 3^{n+1} and nonvanishing symplectic trilinear mappings. Let V and W be vector spaces over $\mathbb{F}_3$ of dimension n and m respectively. The set $\mathrm{AT}(n,m)$ of symplectic trilinear mappings from V^3 to W whose images have rank m may be provided with a natural notion of equivalence (up to compositions with isomorphisms of V and W). Denote by $\alpha(n,m)$ the number of such equivalence classes. It can be shown that each class is related to an essentially unique HTS (see Bénéteau and Lacaze, 1988). As a byproduct we obtain Theorem 8.20.

Theorem 8.20 *If $m = 1, 2,$ or 3, then $\alpha(n,m)$ coincides with the maximum number $\sigma(n,m)$ of pairwise nonisomorphic rank $n+1$ HTSs of order 3^{n+m}. If $m \geq 4$, then $\sigma(n,m) \geq 3 + \alpha(n,m)$.*

The main significance of Theorem 8.20 is that the classification of the Hall triple systems whose 3-order is not too large with respect to their rank r can be dealt with by counting orbits in $\mathrm{AT}(n,m)$. Among the HTSs whose 3-order is larger than $3+r$, however, there are some systems that are not constructable from symplectic trilinear mappings.

8.8 Abstract groups

Doyen (1973) asked which groups are the automorphism groups of Steiner triple systems. Mendelsohn (1975, 1978a) proved:

Theorem 8.21 *Every abstract group Γ is isomorphic to the group of automorphisms of a finite Steiner triple system. Indeed, when Γ has order γ, there exists an STS(v) whose automorphism group is isomorphic to Γ, for which $v < \gamma^{c\gamma}$ for c a constant.*

We outline Mendelsohn's strategy. A *directed 4-path system* is a set X of elements, together with a set of directed paths on four vertices, i.e. sets of the form $\{(w,x),(x,y),(y,z)\}$ with $\{w,x,y,z\} \subseteq X$ and $|\{w,x,y,z\}| = 4$. It is *amicable* if

1. two 4-paths meet in at most one element;
2. when two 4-paths meet in an element w, it is an endpoint in each of the 4-paths; and
3. every element of X appears on a 4-path.

Lemma 8.22 *The abstract group Γ is the automorphism group of an amicable 4-path system whose number of elements is bounded by $c|\Gamma|\log|\Gamma|$, for c a constant.*

Proof (Sketch) Babai (1974) proved that every abstract group Γ is the automorphism group of a graph having at most $c'|\Gamma|\log|\Gamma|$ edges, for c' a constant. For each edge $e = \{x,y\}$ of this graph, add an element $\widehat{e}$, and add arcs $(x,\widehat{e})$ and $(y,\widehat{e})$, removing edge e. Finally, add a single element ∞, and an arc (∞,x) for each vertex x of the original graph. The result is an oriented graph $(Y, \mathcal{A})$ whose automorphism group is isomorphic to Γ.

Now define $Z = Y \cup (\mathcal{A} \times \{0,1\})$, the set of elements of the 4-path system. The directed 4-paths are obtained by including, for each $a = (x, y) \in \mathcal{A}$, the 4-path $(x, (a, 0), (a, 1), y)$. The resulting directed 4-path system again has automorphism group Γ, and it is amicable provided that the original graph chosen had no isolated vertices. □

In addition to amicable 4-path systems, two more ingredients are needed. The first is the existence of an STS(15), $(W, \mathcal{D})$, which is automorphism-free, and has no subsystem of order 7. (Theorem 7.80 asserts that an automorphism-free STS(15) exists, but to ensure the absence of a subsystem as well, choose STS #23 from Table 5.9, for example.)

The final main ingredient is the projective triple system $(V, \mathcal{B})$ of order $2^n - 1$, from Theorem 2.14. The ingredients are used as follows. Let $(X, \mathcal{P})$ be an amicable directed 4-path system with group Γ. Let $X = \{x_1, \ldots, x_n\}$. Use Theorem 2.14 to form a projective triple system of order $2^n - 1$, whose elements are all nonzero binary vectors of length n. Choosing any four coordinate positions of the n-vectors, we find that the set of 15 vectors in which all entries in the remaining $n - 4$ coordinate positions are zero induce a subsystem of order 15. For every 4-path $(x_a, x_b, x_c, x_d) \in \mathcal{P}$, remove from $\mathcal{B}$ all triples of the STS(15) induced by coordinate positions $\{a, b, c, d\}$. In its place, add a copy of the STS(15) $(W, \mathcal{D})$. The placement of this subsystem is to be fixed by the selection of the mapping to the unit vectors, in the order (a, b, c, d) specified. Since $(X, \mathcal{P})$ is amicable, no two 4-paths share more than one vertex, and hence each of these STS(15) replacements can proceed independently. The result is an STS$(2^n - 1)$ $(V, \widehat{\mathcal{B}})$.

That the automorphism group of $(V, \widehat{\mathcal{B}})$ contains Γ is ensured by the consistent placement of the STS(15)s added. To see that no additional automorphisms are introduced, Mendelsohn (1978) first establishes that the only 'copies' of $(W, \mathcal{D})$ in $(V, \widehat{\mathcal{B}})$ are those introduced in the subsystem replacements (one for each replacement). Every automorphism must map each subsystem to an isomorphic subsystem, and since each of the added subsystems is itself automorphism-free, any automorphism mapping one added STS(15) to another must map the four unit vectors in one, in order, to the four unit vectors of the other. It remains only to verify that every unit vector lies in one of the STS(15)s added, and this is ensured by the requirement that every amicable 4-path system has each element in at least one of its 4-paths.

Mendelsohn's strategy, as we have seen, is to form the projective system large enough to contain all of the automorphisms we want, and many more. Evidently, it contains the symmetric group of order n in its automorphism group. The 4-path system is used in two ways, to destroy unwanted automorphisms, and to ensure that all automorphisms are unique extensions of permutations on the unit vectors. Mendelsohn (1978a, b) extends this, by using amicable directed n-path systems, to prove:

Theorem 8.23 *Let Γ be a finite group. Any finite Steiner triple system can be embedded in a finite Steiner triple system whose automorphism group is isomorphic to Γ.*

Mendelsohn (1978a) also remarks on an interesting consequence of a theorem of Babai (1979):

Theorem 8.24 *Let $\mathcal{L}$ be a finite distributive lattice, and for each $\sigma \in L$ let Γ_σ be a finite group. Then there exists a finite Steiner triple system whose subsystem lattice $\mathcal{S}$ admits a lattice embedding of $\mathcal{L}$, so that for each $\sigma \in \mathcal{L}$ the subsystem corresponding to σ has automorphism group Γ_σ.*

The fact that subsystems of a Steiner triple system form a lattice is almost immediate from Lemma 6.45.

8.9 Graphical triple systems

A TS(v, λ) $(X, \mathcal{A})$ is *graphical* if

1. X is the set of $v = \binom{n}{2}$ edges of the complete graph K_n, and
2. $\mathcal{A}$ is a collection of graphs on three edges closed under isomorphism: whenever $A \in \mathcal{A}$ and σ is an automorphism of K_n, then $\sigma(A) \in \mathcal{A}$.

Such a graphical design has an automorphism group which contains the symmetric group of order n. Chee (1991) proves:

Theorem 8.25 *A simple graphical TS(v, λ) exists if and only if*

$$(v, \lambda) \in \{(10, 4), (15, 1), (28, 6), (28, 10), (55, 25)\}.$$

The graphical STS(15) is defined by the graphs:

A TS(v, λ) is *r-hypergraphical* if

1. X is the set of $v = \binom{n}{r}$ edges of the complete r-hypergraph $K_n^{(r)}$, and
2. $\mathcal{A}$ is closed under isomorphism of hypergraphs: whenever $A \in \mathcal{A}$ and σ is automorphism of $K_n^{(r)}$, then $\sigma(A) \in \mathcal{A}$.

A 3-hypergraphical TS(56,18) is given by:

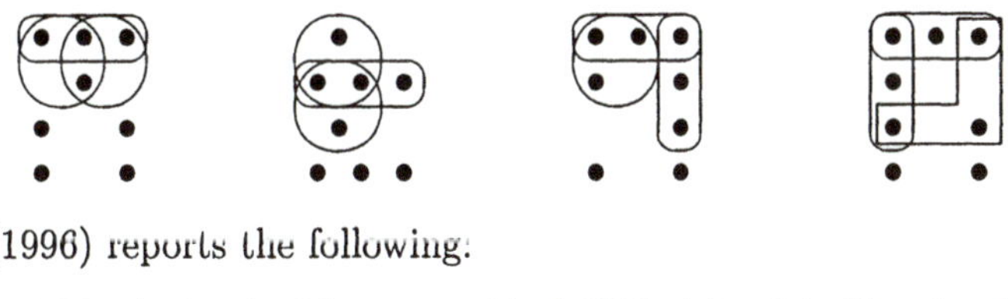

Chee (1996) reports the following:

Theorem 8.26 *A simple 3-hypergraphical TS(v, λ) exists if and only if*

$$(v, \lambda) \in \{(10, 4), (56, 18), (56, 24), (120, 54), (1540, 722)\}.$$

8.10 Work points

1. Prove or disprove: every abstract group Γ is isomorphic to the automorphism group of a Steiner triple system whose order is bounded by a polynomial in the order of Γ.
2. Complete the classification of Hall triple systems of order 3^8.

9

LEAVES AND PARTIAL TRIPLE SYSTEMS

A *partial triple system* of order v and index λ, or PTS(v, λ), is a collection of triples on v elements, so that no pair occurs in more than λ of the triples. Often the parameter λ is omitted when $\lambda = 1$. A PTS is also called a *packing* in the literature. A multigraph on v vertices is a λ-*leave* if and only if there exists a partial triple system of order v and index λ whose leave is the specified multigraph. For example, the 4-vertex multigraph $K_{1,3}$ with edge set $\{\{1,2\},\{1,3\},\{1,4\}\}$ is a 1-leave, but is not a 2-leave. In this chapter, we examine the question of which multigraphs are λ-leaves.

The characterization of λ-leaves is far from complete, even in the case $\lambda = 1$. We first establish some basic necessary conditions, and then establish that recognizing λ-leaves is NP-hard. We then establish the known sufficient conditions for a multigraph to be a λ-leave.

9.1 Necessary conditions

Let $M = (V, E)$ be a multigraph with $|V| = v$ vertices and $|E| = m$ edges. When can M be a λ-leave?

The most basic necessary condition is that M has no edge of multiplicity exceeding λ, and we assume this throughout. Now if M is to be a λ-leave, there must be a partial triple system $(V, \mathcal{B})$ of index λ whose leave is M. Consider the following process. Starting with $G = \lambda K_v$ (the complete graph on v vertices with every edge having multiplicity λ), form a sequence of graphs by removing, for each triple of $\mathcal{B}$, the corresponding triangle of G. At the end of this process, G and M are identical. We derive an elementary necessary condition:

Lemma 9.1 *A λ-leave on v vertices and m edges has all vertex degrees congruent to $\lambda(v-1) \pmod 2$, and has $m \equiv \lambda\binom{v}{2} \pmod 3$.*

Proof λK_v has all vertex degrees equal to $\lambda(v-1)$ and has $\lambda\binom{v}{2}$ edges. The removal of triangles leaves the vertex degrees the same modulo 2, and leaves the number of edges the same modulo 3. □

Lemma 9.1 provides parity and congruence conditions. Not surprisingly, they are insufficient. Consider the multigraph

$$C = 2\{K_3\} = \{\{1,2\},\{1,3\},\{2,3\},\{4,5\},\{4,6\},\{5,6\}\}$$

on the seven vertices $\{1,2,3,4,5,6,7\}$. It has six vertices of degree two, and one of degree 0; it has 6 $\left(\equiv \binom{7}{2} \pmod 3\right)$ edges. But is it a 1-leave?

Consider the partition of the vertex set $V_1 = \{1,2,3\}$ and $V_2 = \{4,5,6,7\}$. Call an edge a *cross edge* if it contains one vertex from each class; otherwise call it an *inside edge.* Then C has no cross edges and six inside edges. Now suppose that C is a 1-leave of the partial triple system $(V, \mathcal{B})$. Each triple of $\mathcal{B}$ can contain pairs corresponding to three inside edges, or to one inside edge and two cross edges. In total, triples of $\mathcal{B}$ must account for 12 cross edges, and for the remaining three inside edges (those not in C). However, since the number of cross edges to deal with is more than twice the number of inside edges, there can be no such partial triple system $\mathcal{B}$.

Graham (see Nash-Williams, 1970) gives an elegant construction for an infinite family of graphs that meet the parity and congruence conditions but are not leaves. We give it in slightly more general form here. Choose a positive index λ, and an order $v = 4t$ for which $\lambda(t-1) \equiv 0 \pmod 2$ and $\lambda t \equiv 0 \pmod 3$. Let G be the multigraph on vertex set $\mathbb{Z}_{4t}$, whose edges are pairs of vertices $\{i,j\}$ with $i - j \equiv 2 \pmod 4$, λ times each. G has all vertex degrees congruent to $\lambda \pmod 2$, and has $4\lambda\binom{t}{2}$ edges; since $4\lambda\binom{t}{2} \equiv \lambda\binom{4t}{2} \pmod 3$, the congruence condition is met. Now consider $\overline{G}$; it contains $4\lambda t^2$ *odd* edges (edges $\{i,j\}$ with $i - j \equiv 1 \pmod 2$), and $4\lambda\binom{t}{2}$ *even* edges (edges with $i - j \equiv 0 \pmod 2$). For G to be a λ-leave, $\overline{G} = \lambda K_{4t} \setminus G$ must have an edge-partition into triangles. But every triangle in $\overline{G}$ contains at least one even edge, and hence we require that $4\lambda t^2 \leq 2 \cdot 4\lambda\binom{t}{2}$ which holds only if $t = 0$.

In general, we obtain the necessary *density condition*:

Lemma 9.2 *Let M be a v-vertex multigraph with m edges. If the vertices of M can be partitioned into sets of size s and $v - s$ so that M has c cross edges, then M is a λ-leave only if*

$$2\left(\lambda\binom{s}{2} + \lambda\binom{v-s}{2} - m + c\right) \geq \lambda s(v-s) - c\,.$$

Proof λK_v has $\lambda\binom{s}{2} + \lambda\binom{v-s}{2}$ inside edges and $\lambda s(v-s)$ cross edges; M accounts for c of the latter and $m - c$ of the former. The number of remaining inside edges must be at least half the number of remaining cross edges. □

Together with the parity and congruence conditions, this density condition is still not strong enough even to eliminate all multigraphs whose complements (with respect to λK_v) have edges appearing in no triangles. An easy computation shows that the density condition cannot eliminate any graph having fewer than $\frac{\lambda}{4}\binom{v}{2} - \frac{3v}{8}$ edges; to see this, observe that if when $s = v - s = v/2$ and $c = 0$ the inequality of Lemma 9.2 holds, then it holds for all $1 \leq s \leq v - 1$ and all $0 \leq c \leq m$. So we may assume that $s = v/2$ and $c = 0$, and extract the inequality on m. Thus, for regular graphs, the density condition cannot be violated if the degree is less than $\lambda(\frac{v}{4} - 1)$. Graham's construction gives a regular graph of degree $\lambda(\frac{v}{4} - 1)$ that does violate the density condition (using the partition into odd and even vertices), and hence is not a 1-leave.

Stinson and Wallis (1987) describe infinite families of graphs meeting the parity, congruence, and density conditions that cannot be realized as leaves for the simple reason that the complement, which must be partitioned into triangles, contains either an edge appearing in no triangle, or a pair of disjoint sets of vertices having no common neighbours, and an odd number of cross edges between the sets.

Colbourn and Mathon (1987) developed a stronger necessary condition that extends the density condition. A pair of subsets X, Y of a multigraph $M = (V, E)$ forms a *fence* for M if X and Y are disjoint and no vertex $z \in V \setminus (X \cup Y)$ is adjacent to vertices both in X and in Y. The *fence-degree* of a vertex $z \in X \cup Y$, $fd(z)$, is the degree of the vertex z in the submultigraph of M induced on $X \cup Y$. The *X-defect* $def_X(X, Y)$ of the fence is the minimum number of edges in a subgraph H on X whose vertex degrees deg_H satisfy $deg_H(x) \equiv fd(x) \pmod 2$ for all $x \in X$. The Y-defect $def_Y(X, Y)$ is defined similarly, and the *defect* $def(X, Y)$ is the sum of the X-defect and the Y-defect.

Finally, for fence (X, Y), let $\epsilon(X, Y)$ equal the number of edges between vertices of X and vertices of Y, and $\iota(Z)$ be the number of edges inside a subset Z of vertices of M. Then we have:

Lemma 9.3 *Let* (X, Y) *be a fence of* $\overline{M} = \lambda K_v - M$. *Then* M *is a* λ*-leave only if* $\epsilon(X, Y)$ *is even, and* $\epsilon(X, Y) \geq 2(\iota(X) + \iota(Y) - def(X, Y))$.

Proof Consider those edges having one endpoint in X and the other in Y. Since (X, Y) is a fence, any triangle containing such an edge has its third endpoint in X or in Y. Thus $\epsilon(X, Y)$ must be even. Moreover, for every two edges between X and Y, either an edge inside X or one inside Y is accounted for. Removing all such triangles leaves the fence-degrees unchanged modulo 2. Hence the number of edges inside X and Y that remain is at least $def(X, Y)$. □

Lemma 9.3 implies Lemma 9.2: take $X \cup Y = V$, which always forms a fence. It further eliminates the trivial case for $\lambda = 1$ when $\overline{G}$ has an edge appearing in no triangle; for if $e = \{x, y\}$ is such an edge, the fence $(\{x\}, \{y\})$ establishes the impossibility. It does not appear to be feasible to check these necessary conditions efficiently, nor are these conditions sufficient for a graph to be a λ-leave.

9.2 Recognizing λ-leaves

In this section, we examine the computational complexity of determining whether a given multigraph is a λ-leave. Colbourn (1983) has shown that the problem is NP-complete, even when $\lambda = 1$.

We require some preliminaries in order to sketch the proof. Let $G = (V, E)$ be a simple undirected r-regular graph on n vertices. A *latin background* for G, LB$[G; m, s]$, is an $s \times s$ array $L = (\ell_{ij})$ which is a symmetric square with symbols from $\{1, 2, \ldots, m\}$ (where $m \geq s$). Every diagonal position ℓ_{ii} contains the symbol m. In the first n rows, each cell of L is either empty or contains a symbol from $\{r + 1, \ldots, m\}$. In the latter $s - n$ rows, every cell is occupied by a symbol from $\{1, \ldots, m\}$. L is a partial latin square: Every symbol appears at most once in each row and at most once in each column. Finally, the $n \times n$ matrix $A(L) = (a_{ij})$ defined by $a_{ij} = 1$ if ℓ_{ij} is an empty cell, and 0 otherwise, is an adjacency matrix of the specified graph G.

Given an arbitrary r-regular n-vertex graph $G = (V, E)$, our first task is to construct a latin background for G. We assume without loss of generality that $V = \{1, 2, \ldots, n\}$. We can form a background $B = (b_{ij})$ of size $m \geq 2n$ by setting $b_{ii} = m$ for all $1 \leq i \leq m$. For $1 \leq i < j \leq m$, if $\{i, j\} \in E$ we leave cells b_{ij} and b_{ji} empty; otherwise, we set $b_{ij} = b_{ji} = ((i + j) \pmod n) + n$. B is then a latin background LB$[G; m, n]$.

Our next task is to extend the latin background L so produced to form an LB$[G; m, m]$. To do this, Colbourn (1983) proves a lemma using a 'bordering' method of Cruse (1974). Let $P_L(i)$ be the number of occurrences of symbol i in L. Let $N_L(i) = P_L(i) + n$ for $1 \leq i \leq r$, and $N_L(i) = P_L(i)$ for $r + 1 \leq i \leq m$.

Lemma 9.4 *Let L be an LB$[G; m, s]$ with m even, for G an r-regular n-vertex graph with n even. If for each $1 \leq i \leq m$, $N_L(i) \geq 2s - m$, there is an LB$[G; m, s + 1]$, L', for which $N_{L'}(i) \geq 2(s + 1) - m$ for all symbols $1 \leq i \leq m$.*

Applying Lemma 9.4 repeatedly to the initial latin background LB$[G; m, n]$ with $m \geq 2n$ yields a latin background LB$[G; m, m]$. Moreover, an explicit construction of the background can be accomplished by using standard bipartite matching algorithms to produce the required 'bordering' in Lemma 9.4.

For such a latin background LB$[G; m, m]$ $B = (b_{ij})$, we define an *idempotent latin background* IB$[G; m - 1, m - 1]$ to be an $m - 1 \times m - 1$ square $L = (\ell_{ij})$ obtained by setting $\ell_{ij} = \ell_{ji} = b_{ij}$ for $1 \leq i < j \leq m - 1$, and setting diagonal cells $\ell_{ii} = b_{im}$.

Now we are in a position to prove the complexity result:

Theorem 9.5 *Deciding whether a graph is a 1-leave is NP-complete.*

Proof Membership in NP follows from the observation that given a partial triple system, it is easy to verify whether or not its leave is a specified graph.

Let G be a cubic graph on vertex set $\{1, \ldots, n\}$. To show that recognizing 1-leaves is NP-hard, we first construct an IB$[G; 2n - 1, 2n - 1]$ *and* an IB$[G; 2n + 1, 2n + 1]$ using the efficient techniques based on bordering. Let $v \in \{2n - 1, 2n + 1\}$ be the order of a Steiner triple system. We construct a partial triple system on $V = \{x_1, \ldots, x_v\} \cup \{y_1, \ldots, y_v\} \cup \{z\}$. On the $\{x_i\}$, we place the triples of a Steiner triple system of order v. Next we include blocks $\{x_i, y_i, z\}$ for $1 \leq i \leq v$. Finally, whenever the (i, j) cell of the idempotent background IB$[G; v, v]$ is nonempty, containing the element k, we add a triple $\{y_i, y_j, x_k\}$. Idempotence ensures that $k \notin \{i, j\}$.

The leave R of the partial triple system so constructed contains all edges of the form $\{x_i, y_j\}$ with $1 \leq i \leq 3$ and $1 \leq j \leq n$, and also all edges of the form $\{y_i, y_j\}$ for $\{i, j\}$ an edge of G. Let $\overline{R}$ denote the complement of R with respect to the complete graph on V. $\overline{R}$ is a 1-leave if and only if R has an edge-partition into triangles. Now R has an edge-partition into triangles if and only if G is 3-edge-colourable. Holyer (1981) has shown that determining whether a cubic graph is 3-edge-colourable is NP-complete. Hence determining whether R has an edge-partition into triangles is NP-hard, and thus determining whether $\overline{R}$ is a 1-leave is NP-hard. □

The proof establishes that for very dense graphs, it is apparently difficult to determine whether or not the graph is a 1-leave. The graphs employed have $4n - 1$ or $4n + 3$ vertices and have minimum degree $3n - 2$ or $3n + 2$. If the edge density is restricted to be 'small', the complexity of recognizing 1-leaves is then open.

9.3 Maximal partial triple systems

In most investigations of partial triple systems, we are concerned with those partial triple systems to which no further triple can be added without forcing at least one pair to appear more than λ times. We call such partial triple systems *maximal*, and denote such a maximal partial triple system of order v and index λ by MPT(v, λ). A partial triple system is maximal if and only if its leave is triangle-free; hence leaves of maximal systems have a structural constraint in addition to the parity, congruence, and fence constraints given before. It remains open whether or not recognizing leaves of MPTs is NP-complete.

In this section, we examine the possible number of edges in a maximal partial triple system. A *maximum* MPT(v, λ), or MMPT(v, λ), is a maximal partial triple system of order v and index λ with the most triples of any MPT(v, λ); a *minimum* MPT(v, λ), or mMPT(v, λ), is analogously one with the fewest triples.

9.3.1 *Maximum partial triple systems*

What is the largest number of triples that a maximal partial triple system of order v and index λ can have? The solution when $\lambda = 1$ is implicit already in Kirkman (1847). Schönheim (1966, 1969) and Spencer (1968) also solved this problem for $\lambda = 1$, and Engel (1981) and Stanton and Rogers (1982) settled it in general.

A maximum partial triple system has the leave with the fewest edges; hence it is natural to examine what the possible leaves are in order to determine MMPT(v, λ). First, if $\lambda \geq \gcd(v - 2, 6)$, we can do no better than to include a triple system of order v and index $\gcd(v-2, 6)$ and an MMPT$(v, \lambda - \gcd(v-2, 6))$. At the other extreme, if $\lambda = 0$, the MMPT has no triples at all. Thus we only need to treat the cases with $0 < \lambda < \gcd(v - 2, 6)$.

Consider the necessary conditions for a leave L. If $\lambda(v - 1) \equiv 0 \pmod 2$, L must have all vertex degrees even; otherwise L has all vertex degrees odd. Similarly, the number of edges of L must be congruent to $\lambda\binom{v}{2} \pmod 3$. Specifically, we have the nontrivial cases given in Table 9.1.

When $v \equiv 0, 2 \pmod 6$ and $\lambda = 1$, the smallest possible leave has all vertices of degree 1. Such a leave is realized by taking a Steiner triple system of order

λ	v (mod 6)			
	0	2	4	5
1	odd deg 0(3) edges	odd deg 1(3) edges	odd deg 0(3) edges	even deg 1(3) edges
2		even deg 2(3) edges		even deg 2(3) edges
3		odd deg 0(3) edges		
4		even deg 1(3) edges		
5		odd deg 2(3) edges		

Table 9.1 *Conditions on leaves*

$v+1$, eliminating all triples containing a fixed element x, and removing the element x.

When $v \equiv 5 \pmod 6$ and $\lambda = 1$, form a pairwise balanced design with one block of size five, and all other blocks of size three. Replace the block $\{u, v, w, x, y\}$ of size five by two triples, $\{u, v, w\}$ and $\{u, x, y\}$. The resulting partial triple system has a leave which is a 4-cycle (v, x, w, y). Since even degrees are required, a leave of one edge is impossible and hence four is the minimum.

For $v \equiv 4 \pmod 6$ and $\lambda = 1$, we proceed similarly and then eliminate the element x and all triples containing it. The resulting partial triple system has a leave consisting of $(v-4)/2$ disjoint edges and a 3-star $\{\{y, u\}, \{y, v\}, \{y, w\}\}$. Hence it has all degrees odd and has $0 \pmod 3$ edges. Since odd degrees are required, at least $v/2$ edges are needed in the leave, and since $v/2 \equiv 2 \pmod 3$, $(v-4)/2+3$ is indeed the minimum possible number of edges in the leave.

When $\lambda = 2$ and $v \equiv 5 \pmod 6$, form MMPT(v)s with leaves

$$\{\{a, b\}, \{b, c\}, \{c, d\}, \{a, d\}\} \text{ and } \{\{a, b\}, \{b, d\}, \{c, d\}, \{a, c\}\}.$$

Take the union of these and add the triples $\{\{a, b, c\}, \{a, b, d\}\}$. The result has leave $\{\{c, d\}, \{c, d\}\}$ and is an MMPT(v, 2).

When $\lambda = 2$ and $v \equiv 2 \pmod 6$, the smallest admissible leave is a pair of parallel edges $\{\{a, b\}, \{a, b\}\}$. To obtain an MPT(v, 2) with such a leave, begin with an MMPT(v) on the set V whose leave L includes $\{a, b\}$ and an STS($v-1$) on the set $V \setminus \{a\}$. Add to their union the set of triples $\{\{x, y, a\} : \{x, y\} \in L \setminus \{a, b\}\}$. The result is an MPT($v$, 2) that leaves the pair $\{a, b\}$ uncovered, but covers each other pair exactly twice.

For the remaining cases with $v \equiv 2 \pmod 6$, we construct the minimum leave as follows. For $\lambda = 3$, take the union of the leaves for $\lambda = 1$ and $\lambda = 2$. For $\lambda = 4$, we take the union of the leaves of two MMPT(v, 2)s. For $\lambda = 5$ we take the union of an MMPT(v) whose leave contains the edges $\{\{a, b\}, \{c, d\}\}$ and an MMPT(v, 4) whose leave contains $\{\{a, c\}, \{a, c\}, \{b, c\}, \{b, c\}\}$ and add the triple

$\{a, b, c\}$; the resulting leave consists of $(v-4)/2$ disjoint edges and a 3-star.

In summary, we have the leave graphs of MMPT(v, λ)s given in Table 9.2. In some cases, leaves need not be isomorphic to the ones given here; however, we present only one possible leave graph.

λ	$v \pmod 6$		
λ	0	1	2
0	$v\{K_1\}$	$v\{K_1\}$	$v\{K_1\}$
1	$\frac{v}{2}\{K_2\}$	$v\{K_1\}$	$\frac{v}{2}\{K_2\}$
2	$v\{K_1\}$	$v\{K_1\}$	$(v-2)\{K_1\} \cup 2K_2$
3	$\frac{v}{2}\{K_2\}$	$v\{K_1\}$	$3K_2 \cup \frac{v-2}{2}\{K_2\}$
4	$v\{K_1\}$	$v\{K_1\}$	$(v-3)\{K_1\} \cup 2K_{1,2}$
5	$\frac{v}{2}\{K_2\}$	$v\{K_1\}$	$K_{1,3} \cup \frac{v-4}{2}\{K_2\}$
	$v \pmod 6$		
λ	3	4	5
0	$v\{K_1\}$	$v\{K_1\}$	$v\{K_1\}$
1	$v\{K_1\}$	$K_{1,3} \cup \frac{v-4}{2}\{K_2\}$	$(v-4)\{K_1\} \cup C_4$
2	$v\{K_1\}$	$v\{K_1\}$	$(v-2)\{K_1\} \cup 2K_2$
3	$v\{K_1\}$	$K_{1,3} \cup \frac{v-4}{2}\{K_2\}$	$v\{K_1\}$
4	$v\{K_1\}$	$v\{K_1\}$	$(v-4)\{K_1\} \cup C_4$
5	$v\{K_1\}$	$K_{1,3} \cup \frac{v-4}{2}\{K_2\}$	$(v-2)\{K_1\} \cup 2K_2$

Table 9.2 *Leaves of MMPT(v, λ)s*

The leaves of MMPT(v, λ)s are not unique in general for $v \equiv 2 \pmod 3$. In summary, we have:

Lemma 9.6 *The number of triples $\mu(v, \lambda)$ in an MMPT(v, λ) satisfies:*

$$\mu(v, \lambda) = \begin{cases} \lfloor \frac{v}{3} \lfloor \frac{\lambda(v-1)}{2} \rfloor \rfloor - 1 & \textit{if } v \equiv 2(6) \textit{ and } \lambda \equiv 4(6) \\ & \textit{or } v \equiv 5(6) \textit{ and } \lambda \equiv 1(3) \\ \lfloor \frac{v}{3} \lfloor \frac{\lambda(v-1)}{2} \rfloor \rfloor & \textit{otherwise} \end{cases}$$

9.3.2 *Minimum maximal partial triple systems*

At the other extreme, we ask for the minimum number of triples in a maximal partial triple system of order v and index λ. Equivalently, what is the largest number of edges a triangle-free λ-leave on v vertices can have?

Novák (1974) determined this minimum in the case $\lambda = 1$.

A well-known theorem in graph theory, Turán's theorem, ensures that the maximum number of edges in a triangle-free multigraph on v vertices with edge-multiplicity at most λ is achieved (uniquely) by $\lambda K_{\lfloor v/2 \rfloor, \lceil v/2 \rceil}$. The following lemma is then immediate:

Lemma 9.7 *Let $v \geq 6$ and $\lambda \geq 1$. Let $v_1 = \lfloor v/2 \rfloor$ and $v_2 = \lceil v/2 \rceil$. Then if $\lambda \equiv 0 \pmod{\gcd(v_1 - 2, 6)}$ and $\lambda \equiv 0 \pmod{\gcd(v_2 - 2, 6)}$, the minimum MPT$(v, \lambda)$ has leave isomorphic to $\lambda K_{v_1, v_2}$.*

Proof By Turán's theorem, we need only show that $\lambda K_{v_1,v_2}$ is a λ-leave under the conditions stated. The congruence conditions on λ ensure that triple systems of index λ and orders v_1 and v_2 exist; taking the union of these on disjoint sets of elements gives the PTS(v, λ) whose leave is $\lambda K_{v_1,v_2}$. □

In general, $\lambda K_{v_1,v_2}$ may have the wrong vertex degrees and may have the wrong number of edges modulo three. Nevertheless, all of the cases are handled by variations of the simplest case described. Novák (1974) shows that a simple procedure constructs an mMPT(v) for all v:

1. Let n_1 be the nearest integer to $v/2$ for which an STS(n_1) exists. Let $n_2 = v - n_1$.
2. Let $\mathcal{B}_1$ be the triples of an STS(n_1) on a set X of n_1 elements.
3. Let $\mathcal{B}_2$ be the triples of an MMPT(n_2) on a set Y of n_2 elements disjoint from X.
4. Let $\mathcal{B}_3$ be a set of triples obtained by adding to each pair on Y in the leave of $\mathcal{B}_2$ an element of X so that no pair with one element from X and one from Y is covered more than once.
5. Then $(X \cup Y, \mathcal{B}_1 \cup \mathcal{B}_2 \cup \mathcal{B}_3)$ is an mMPT(v).

Novák's proof first determines the maximum number of edges in a triangle-free graph with all vertex degrees of the same parity as $n-1$ (a candidate to be a leave), establishing in the process a bound on the maximum number of edges in a leave. Then the construction given is shown, by analyzing the various cases modulo 12, to realize the minimum.

To state Novák's theorem explicitly:

Theorem 9.8 *A minimum maximal partial triple system of order v (i.e. an mMPT(v)) has $(v^2 + d(v))/12$ triples, where*

$$\begin{aligned} d(v) &= -2v + 36 && \textit{if } v \equiv 0, 8 \pmod{12} \\ &= -2v && \textit{if } v \equiv 2, 6 \pmod{12} \\ &= -2v + 4 && \textit{if } v \equiv 4 \pmod{12} \\ &= -2v + 16 && \textit{if } v \equiv 10 \pmod{12} \\ &= -1 && \textit{if } v \equiv 1, 5 \pmod{12} \\ &= 3 && \textit{if } v \equiv 3 \pmod{12} \\ &= 11 && \textit{if } v \equiv 7, 11 \pmod{12} \\ &= 15 && \textit{if } v \equiv 9 \pmod{12}. \end{aligned}$$

9.3.3 *The spectrum of maximal partial triple systems*

At this point, we know the maximum and minimum possible numbers of edges in a λ-leave of a maximal partial triple system. What values between the maximum and the minimum can be realized?

Severn (1984) addressed this question for $\lambda = 1$, and obtained strong partial results. In order to state his theorem, we require some preliminary notation. For $v = 12k + 2t$, we define four functions $m^{(3)}(v)$, $\mu(v)$, $\Omega(v)$, and $M^{(3)}(v)$ as in Table 9.3, writing $\beta = 12k^2$.

t	$m^{(3)}(v)$	$\mu(v)$	$\Omega(v)$	$M^{(3)}(v)$
0	$\beta - 2k + 3$	β	$\beta + 2k$	$2\beta - 4k$
1	$\beta + 2k$	$\beta + 4k + 1$	$\beta + 6k + 1$	$2\beta + 4k$
2	$\beta + 6k + 1$	$\beta + 8k + 2$	$\beta + 10k + 2$	$2\beta + 12k + 1$
3	$\beta + 10k + 2$	$\beta + 12k + 3$	$\beta + 14k + 5$	$2\beta + 20k + 4$
4	$\beta + 14k + 7$	$\beta + 16k + 6$	$\beta + 18k + 8$	$2\beta + 28k + 8$
5	$\beta + 18k + 8$	$\beta + 20k + 9$	$\beta + 22k + 11$	$2\beta + 36k + 13$

Table 9.3 *Functions for MPT(v)s*

For v odd, $M^{(3)}(v)$ is $(v(v-1))/6$ if $v \equiv 1, 3 \pmod 6$, and $(v(v-1))/6 - 8$ if $v \equiv 5 \pmod 6$. In addition, for $v = 12k + 2t + 1$, define $m^{(3)}(v) = 12k^2 + (4k+2) + \beta$, where

$$\beta = \begin{cases} 0 & \text{if } t = 0 \\ 1 & \text{if } t = 1 \\ 2 & \text{if } t = 2 \\ 5 & \text{if } t = 3 \\ 8 & \text{if } t = 4 \\ 11 & \text{if } t = 5 \end{cases}$$

Let $S^{(3)}(v)$ be the spectrum of numbers of triples in maximal partial triple systems on v elements. Let $P(v) = \{s : m^{(3)}(v) \leq s \leq M^{(3)}(v)\}$.

Theorem 9.9 *(Severn, 1984)*

1. *If v is odd, $S^{(3)}(v) = P(v) \setminus \{M^{(3)}(v) - 1\}$ if $v \equiv 1, 3 \pmod 6$, and $S^{(3)}(v) = P(v)$ if $v \equiv 5 \pmod 6$.*
2. *Let v be even. Let*

$$Q(v) = \{s : m^{(3)}(v) < s < \Omega(v) \text{ and } s - \lfloor (v-1)/2 \rfloor \equiv 1 \pmod 2\}, \text{ and}$$

$$Q'(v) = \{s : m^{(3)}(v) < s \leq \mu(v) \text{ and } s - \lfloor (v-1)/2 \rfloor \equiv 1 \pmod 2\}.$$

Then $P(v) \setminus Q(v) \subseteq S^{(3)}(v) \subseteq P(v) \setminus Q'(v)$.

Let $R(v) = Q(v) \setminus Q'(v)$ for v even; this contains the values left in doubt by Severn. Observe that $|R(12k + 2t)|$ is $k - 1$ if $t \in \{0, 1, 2\}$, and is k if $t \in \{3, 4, 5\}$.

Colbourn, Rosa, and Znám (1993) settled these cases, giving the complete determination of $S^{(3)}(v)$.

To state the complete result, we define one further function $\Upsilon(v)$ by:

v	$\Upsilon(v)$
$12k$	$12k^2 + 2k$
$12k + 2$	$12k^2 + 6k + 1$
$12k + 4$	$12k^2 + 10k + 2$
$12k + 6$	$12k^2 + 14k + 5$
$12k + 8$	$12k^2 + 18k + 6$
$12k + 10$	$12k^2 + 22k + 11$

Theorem 9.10 *(Spectrum of MPT(v)s)*

1. *If v is odd, $S^{(3)}(v) = P(v)\setminus\{M^{(3)}(v)-1\}$ if $v \equiv 1, 3 \pmod 6$, and $S^{(3)}(v) = P(v)$ if $v \equiv 5 \pmod 6$.*
2. *Let v be even. Let*

$$Q(v) = \{s : m^{(3)}(v) < s < \Upsilon(v) \text{ and } s - \lfloor (v-1)/2 \rfloor \equiv 1 \pmod 2\}.$$

Then $S^{(3)}(v) = P(v) \setminus Q(v)$.

9.4 Equitable partial triple systems

In a partial triple system $(V, \mathcal{B})$, let $r(v)$ denote the number of triples of $\mathcal{B}$ that contain the element v. A partial triple system is *equitable* if, for every two elements $v, w \in \mathcal{B}$, $|r(v) - r(w)| \leq 1$. Andersen, Hilton, and Mendelsohn (1980) settled the existence question for equitable triple systems with $\lambda = 1$; Rodger and Stubbs (1986) settled it for all λ. We outline their proof here.

The strategy is to make an exchange of triples in an arbitrary partial triple system that is not equitable in such a way that it becomes 'more equitable'. In order to make this exchange, a technical result on edge-colouring multigraphs is useful.

Lemma 9.11 *Let λ be a positive integer. Let G be a multigraph without loops having maximum degree at most 2λ and having no 2λ-regular subgraph that contains an odd number of edges. Then G has an equalized edge-colouring with two colours in which no vertex is incident with more than λ edges of one colour.*

Proof (Sketch) Let G be a multigraph as in the statement of the lemma. Let $G_1, \ldots, G_c$ be the connected components of G. Define a multigraph $\widehat{G}$, having all vertices of G and one additional vertex ∞. $\widehat{G}$ has all edges of G, and in addition has:

1. for each component G_i in which all vertex degrees are even and the number of edges is even, an arbitrary vertex u_i is chosen and two edges connecting ∞ and u_i are added;
2. for each component G_i in which all vertex degrees are even and the number of edges is odd, a vertex u_i is chosen whose degree is at most $2\lambda - 2$, and two edges connecting ∞ and u_i are added;

3. for each vertex v of odd degree in G, an edge $\{\infty, v\}$ is added; and
4. no other edges are added.

Now $\widehat{G}$ is a connected multigraph in which every vertex has even degree. Thus $\widehat{G}$ has an Eulerian circuit. Starting at ∞, colour the edges of $\widehat{G}$ alternately white and black along the Eulerian circuit. Evidently, every vertex of $\widehat{G}$ is incident with the same number of white and black edges, except possibly for ∞. Restricting the edge-colouring so produced to the edges of G yields the desired equalized edge-colouring of G. □

Theorem 9.12 *If a PTS(v, λ) with τ triples exists, then there exists an equitable PTS(v, λ) having τ triples.*

Proof Let $(V, \mathcal{B})$ be a PTS(v, λ) having τ triples. For $v \in V$, let $r(v)$ be the number of triples of $\mathcal{B}$ containing v. Choose $v, w \in V$ so that $|r(v) - r(w)|$ is maximum. If $|r(v) - r(w)| \leq 1$, $(V, \mathcal{B})$ is equitable and the theorem is proved. So suppose that $|r(v) - r(w)| \geq 2$. Form a multigraph G on vertex set $V \setminus \{v, w\}$ whose edges are all pairs on $V \setminus \{v, w\}$ occurring in triples of $\mathcal{B}$ with v or with w (preserving multiplicities as in $\mathcal{B}$). Evidently, G has maximum degree at most 2λ. Moreover, G has an edge-colouring with two colours (v and w) in which each vertex is incident with at most λ edges of each colour, and hence no 2λ-regular subgraph of G can have an odd number of edges. By Lemma 9.11, G has an equalized edge-colouring with two colours, say v and w, in which each vertex is incident with at most λ edges of each colour.

Now form $\mathcal{B}'$ from $\mathcal{B}$ by first removing all triples that contain v or w but not both. Then for each edge $\{a, b\}$ of G, add the triple $\{v, a, b\}$ or $\{w, a, b\}$ according to the colour (v or w) of the edge. The requirement that each vertex of G be incident with at most λ edges of each colour ensures that $(V, \mathcal{B}')$ is a PTS(v, λ). Moreover, the fact that the edge-colouring is equalized ensures that the discrepancy between v and w is reduced, while all other elements occur the same number of times in triples of $\mathcal{B}$ and $\mathcal{B}'$. Repeating this triple exchange a finite number of times results in the required equitable PTS(v, λ). □

9.5 Quadratic leaves

Until this point, for sufficiency we have been concerned primarily with the number of edges in a leave, and have not considered the structure of the leave. Short of characterizing λ-leaves, a natural objective is to identify strong sufficient conditions. We first recall a conjecture of Nash-Williams (1970):

Conjecture 9.13 *Let G be a graph with maximum degree k that has $v \geq 4k+2$ vertices, and that meets the conditions of Lemma 9.1, namely the parity condition and the congruence condition. Then G is a 1-leave.*

The constants chosen are such that the density condition must be met, and Graham's construction shows that they cannot be reduced without violating the density condition.

A weaker formulation of Nash-Williams' conjecture is that there exists an integer n_k such that if G has more than n_k vertices and meets the parity and congruence conditions, it is a 1-leave. Remarkably, little is known about n_3, although Nash-Williams' conjecture would imply $n_3 \leq 13$.

Gustavsson (1991) gave the weaker form of Conjecture 9.13:

Theorem 9.14 *Every n-vertex m-edge graph G having all vertex degrees even, $m \equiv 0 \pmod 3$, and having minimum degree at least $n(1 - 10^{-24})$ has an edge-partition into triangles.*

In other words, $n_k \leq 10^{24}k$. Gustavsson's method employs a key lemma: every partial latin square of order n in which each symbol occurs at least $n(1 - 10^{-7})$ times has a completion to a latin square. The asymptotic result of Theorem 9.14 is a substantial breakthrough, but for small graphs (fewer than 10^{24} vertices), it provides no information.

What can be said about small degree for small orders? Of course, characterizing leaves with maximum degree 1 is trivial; there remains the case of maximum degree 2. Rogers (1985) first considered this case, showing that for order nine, every quadratic graph except $C_4 \cup C_5$ is a 1-leave. Colbourn and Rosa (1986) established the general theorem:

Theorem 9.15 *Let G be a graph on $v \equiv 1 \pmod 2$ vertices with every vertex of degree 0 or 2, and having $0 \pmod 3$ edges if $v \equiv 1, 3 \pmod 6$, and $1 \pmod 3$ edges if $v \equiv 5 \pmod 6$. Then G is a 1-leave unless $v = 7$ and $G = 2\{C_3\} \cup K_1$, or $v = 9$ and $G = C_4 \cup C_5$.*

9.6 Subgraphs of leaves

In view of the apparent difficulty in establishing strong sufficient conditions, one might ask whether certain subgraphs are forbidden from appearing in any leave. Colbourn (1987b) addresses this question, and shows in a strong sense that every graph meeting the basic necessary conditions appears as the subgraph of a leave. In particular, he shows:

Theorem 9.16 *There is a polynomial $p(n)$ such that, for any n-vertex graph G having $0 \pmod 3$ edges:*

1. *if G has all vertex degrees even, there is a number $x \leq p(n)$ for which G together with x isolated vertices is a 1-leave; and*
2. *if G has all vertex degrees odd, there are numbers $x \leq p(n)$ and $y \leq 2$ such that G, together with x disjoint edges and y disjoint 3-stars, is a 1-leave.*

This establishes that there are no forbidden induced subgraphs, or forbidden subgraphs, for leaves.

9.7 An application to group testing

Suppose that there is a population of b *items*. Each item is *defective* or *good*. A *group test* or *pool* is a subset for which we elect to determine whether or not the subset contains a defective. Our objective is to determine which items are defective by examining subsets of the items and asking whether there is at least one defective in the subset. Consequently, define the *feedback* of a test on pool P to be either that at least one item in the pool is defective, or that all items in the pool are good. The *group testing problem*, given a set of items, is to find the number of group tests required to determine exactly which items are defective. Naturally, if the feedback of one pool is known to us, we may choose the second pool based upon the feedback of the first. This leads to *sequential* or *adaptive* group testing, as studied by Dorfman (1943), Sobel and Groll (1959, 1966), and Sobel (1968).

When all group tests are performed independently, and the isolation of defectives carried out afterwards, this is *nonadaptive* group testing. Hwang and Sós (1987) and Weideman and Raghavarao (1987, 1987a) examine such problems, and develop strong connections with combinatorial design theory in the case that at most two items are defective. We explore this connection here, and return to it in §13.4.3.

Let X be a set of b items, and let $\mathcal{G}$ be a set of subsets of X corresponding to the group tests performed. Let $\mathcal{G}_x$, $x \in X$, be the set of group tests in $\mathcal{G}$ that contain x. The pair $(X, \mathcal{G})$ is a solution to the nonadaptive group testing problem if and only if for every four (not necessarily distinct) items $w, x, y, z \in X$, $\mathcal{G}_w \cup \mathcal{G}_x = \mathcal{G}_y \cup \mathcal{G}_z$ occurs precisely when $\{w, x\} = \{y, z\}$. For suppose that $\mathcal{G}_w \cup \mathcal{G}_x = \mathcal{G}_y \cup \mathcal{G}_z$; then the group tests fail to distinguish whether the defective set is $\{w, x\}$ or $\{y, z\}$, and thus if $(X, \mathcal{G})$ solves the group testing problem, we must have $\{w, x\} = \{y, z\}$.

Hwang and Sós (1987) suggest the examination of the dual $(V, \mathcal{B})$ of $(X, \mathcal{G})$; specifically, let $V = \{v_G : G \in \mathcal{G}\}$ be a set of *elements* in one-to-one correspondence with the group tests, and $B_x = \{v_G : x \in G\}$ and $\mathcal{B} = \{B_x : x \in X\}$ be a set of *blocks* in one-to-one correspondence with the items.

For a set V, and collection $\mathcal{B}$ of subsets of V, $(V, \mathcal{B})$ is an *HS-family* if for any four (not necessarily distinct) blocks $B_i, B_j, B_k, B_\ell \in \mathcal{B}$,

$$B_i \cup B_j = B_k \cup B_\ell \iff \{i, j\} = \{k, \ell\}. \tag{9.1}$$

We use the usual notation $v = |V|$ and $b = |\mathcal{B}|$. The central question in nonadaptive group testing (with at most two defectives) is to determine, under various restrictions on the HS-family, how large b can be as a function of v. In other words, how many items can be tested, as a function of the number of group tests? When the HS-family is unrestricted, Erdös, Füredi, and Frankl (1985) and Hwang and Sós (1987) show that b grows exponentially as a function of v.

We treat two natural restrictions here. First, we may be restricted to testing each item no more than some fixed number k of times; this amounts to restricting the block size of an HS-family to be at most k. The resulting problem is *k-restricted.* Vakil, Parnes, and Raghavarao (1990) treat the case when $k = 2$, and Vakil and Parnes (1994) treat the case when $k = 3$. In this latter case, the objects are set systems whose sets are triples.

Second, we may restrict the number of times that two items are tested together to be at most some fixed number μ. This amounts to restricting the intersection of two blocks of an HS-family to be of size at most μ. In the case when $\mu = 1$, Weideman and Raghavarao (1987, 1987a) obtain a nearly complete solution, omitting only the case when $v = 11$. We examine the case when blocks intersect in at most one element here, and obtain their results in a fairly direct way using results on partial triple systems. We call an HS-family a *WR-family* if no two blocks intersect in two or more elements; thus a WR-family is a special type of partial pairwise balanced design. Although two restrictions are outlined here, they can be treated by similar techniques.

9.7.1 *The structure of pairs in an HS-family*

Let $(V, \mathcal{B})$ be an HS-family. Evidently, $\mathcal{B}$ need contain no subsets of size 0 or 1. Let E be the set of pairs in $\mathcal{B}$, and let $\widehat{\mathcal{B}}$ be the set of subsets of size three or greater. We can conveniently view the pairs by forming a *pair graph* $G = (V, E)$ whose vertices are the elements, and whose edges are the pairs. Now define the *square* of the pair graph $G^2 = (V, E^2)$, where $E^2 = E \cup \{\{x, z\} : \exists y \in V \text{ for which } \{x, y\} \in E, \{y, z\} \in E\}$. We prove a simple lemma.

Lemma 9.17 *In an HS-family $(V, \mathcal{B})$, the symmetric difference $B_1 \oplus B_2 = B_1 \cup B_2 \setminus (B_1 \cap B_2)$ of two distinct blocks $B_1, B_2 \in \mathcal{B}$ does not appear as a subset of a block $B_3 \in \mathcal{B}$.*

Proof If $B_1 \oplus B_2 \subseteq B_3$, then $B_1 \cup B_3 = B_2 \cup B_3$, violating (9.1). □

Corollary 9.18 *No edge of the square G^2 of the pair graph $G = (V, E)$ of an HS-family $(V, \mathcal{B})$ appears as a proper subset of a block in $\mathcal{B}$.*

Proof If to the contrary, an edge $\{x, y\}$ of G^2 appears as a proper subset of a block $B \in \mathcal{B}$, then if $\{x, y\} \in E$, $\{x, y\} \cup B = B \cup B$; and if $\{x, y\} \in E^2 \setminus E$, then $\{x, y\}$ is the symmetric difference of two edges in E, and hence by Lemma 9.17, we obtain a contradiction. □

With this in mind, we call the edges of G^2 *forbidden pairs* for the larger blocks. We must determine the number and structure of the forbidden pairs. To do this, we first consider the structure of the pair graph G.

Lemma 9.19 *The pair graph G of an HS-family contains no cycle of length less than five.*

Proof If G has a 3-cycle $\{\{x, y\}, \{x, z\}, \{y, z\}\}$, then $\{x, y\} \cup \{x, z\} = \{x, y\} \cup \{y, z\}$, violating (9.1). Similarly, if G has a 4-cycle $\{\{w, x\}, \{x, y\}, \{y, z\}, \{w, z\}\}$, then $\{w, x\} \cup \{y, z\} = \{w, z\} \cup \{x, y\}$, also violating (9.1). □

Now let $deg(v, G)$ denote the degree of v in G, and $N(v, G)$ denote the set of neighbours of v in G.

Lemma 9.20 *If $G = (V, E)$ is the pair graph of an HS-family, then for each $v \in V$,*

$$deg(v, G^2) = deg(v, G) + \sum_{w \in N(v,G)} (deg(w, G) - 1) = \sum_{w \in N(v,G)} deg(w, G).$$

In addition, G^2 has $(\sum_{v \in V}(deg(v, G))^2)/2$ edges.

Proof By Lemma 9.19, G contains no 3-cycles or 4-cycles. Thus, the number of vertices at distance two from a fixed vertex v in G is $\sum_{w \in N(v,G)}(deg(w, G) - 1)$. Squaring the pair graph makes each vertex at distance two from v adjacent to v.

For the number of edges in G^2, simply sum all of the vertex degrees in G^2, which counts each edge of G^2 twice. □

We summarize as follows. An HS-family has $(\sum_{v \in V} deg(v, G))/2$ pairs, and forbids $(\sum_{v \in V}(deg(v, G))^2)/2$ pairs from appearing in larger blocks.

9.7.2 *Maximum WR-families*

Now we add the WR restriction that every pair occurs as the subset of at most one block. Evidently, some pairs may occur in no blocks at all; a pair that appears in no block and is not forbidden is *unused.*

Theorem 9.21 *In a WR-family $(V, \mathcal{B})$ with pair graph $G = (V, E)$ having degree sequence $d_1, \ldots, d_v$, the number b of blocks satisfies*

$$b \leq \left\lfloor (v(v+1) - \sum_{i=1}^{v}(d_i - 1)(d_i - 2))/6 \right\rfloor$$

with equality only if all blocks have size two or three, and the number of unused pairs is 0, 1, or 2 according to whether $(v(v+1) - \sum_{i=1}^{v}(d_i - 1)(d_i - 2))/2 \equiv 0, 1$ or 2 (mod 3).

Proof There are $\binom{v}{2} - (\sum_{i=1}^{v} d_i^2)/2$ pairs that are not forbidden, by Lemma 9.20. Each such pair occurs in at most one block in $\widehat{\mathcal{B}}$, the blocks of size at least three; since each such block contains at least three pairs, the number of blocks of $\widehat{\mathcal{B}}$ is at most $(v(v-1) - \sum_{i=1}^{v} d_i^2)/6$. Now E contains $(\sum_{i=1}^{v} d_i)/2$ blocks, so the total number of blocks satisfies

$$b \leq (v(v-1) - \sum_{i=1}^{v} d_i(d_i - 3))/6 = (v(v+1) - \sum_{i=1}^{v}(d_i - 1)(d_i - 2))/6.$$

The number of blocks is an integer, and hence

$$b \leq \lfloor (v(v+1) - \sum_{i=1}^{v}(d_i - 1)(d_i - 2))/6 \rfloor.$$

Now consider when equality occurs. If $\widehat{\mathcal{B}}$ contains a block of size $k \geq 4$, the number of blocks in $\widehat{\mathcal{B}}$ is bounded by $1 + (v(v-1) - \sum_{i=1}^{v} d_i^2 - k(k-1))/6$, which reduces the earlier bound by $k(k-1)/6-1$; since $k(k-1)/6-1 \geq 1$ for all $k \geq 4$, equality cannot be met unless all blocks have size two or three.

Now let $\delta \equiv (v(v+1)-\sum_{i=1}^{v}(d_i-1)(d_i-2))/2 \pmod 3$, $\delta \in \{0,1,2\}$. Suppose that the number of unused pairs is γ. If $\gamma \not\equiv \delta \pmod 3$, $\widehat{\mathcal{B}}$ covers a number of pairs which is not divisible by three, which is a contradiction. So $\gamma = \delta$ or $\gamma \geq \delta+3$. If $\gamma \geq \delta+3$, the number of blocks in $\widehat{\mathcal{B}}$ is bounded by $(v(v-1)-2\gamma-\sum_{i=1}^{v} d_i^2)/6$ and thus equality cannot occur. Hence the number of unused pairs must be δ if the WR-family is maximum. □

Corollary 9.22 *A WR-family* $(V, \mathcal{B})$ *has* $b \leq \lfloor v(v+1)/6 \rfloor$, *with equality only if*

1. *every block has size two or three;*
2. *the pair graph* $G = (V, E)$ *satisfies the following:*
   (a) *if* $v \equiv 0, 2 \pmod 3$, *G has all vertices of degree 1 or 2;*
   (b) *if* $v \equiv 1 \pmod 3$, *G either has*
      i. *all vertices of degree 1 or 2, in which case there is one unused edge; or*
      ii. *one exceptional vertex of degree 0 or 3, and all other vertices of degree 1 or 2, in which case there is no unused edge.*

Proof Evidently $\sum_{i=1}^{v}(d_i-1)(d_i-2) \geq 0$, which gives the bound. For equality, we must have $\sum_{i=1}^{v}(d_i-1)(d_i-2) = 0$ when $v \equiv 0, 2 \pmod 3$, or when $v \equiv 1 \pmod 3$ and there is an unused edge. This in turn requires that each vertex degree be 1 or 2. When $v \equiv 1 \pmod 3$ and there is no unused edge, we must have $\sum_{i=1}^{v}(d_i-1)(d_i-2) = 2$. Therefore, one vertex has degree 0 or 3, and all others have degree 1 or 2. □

To proceed further, we make a straightforward observation. In a maximum WR-family $(V, \mathcal{B})$, the triples $(V, \widehat{\mathcal{B}})$ form a partial triple system. Moreover, when there are no unused edges, G^2 is the leave of this partial triple system; and when there is one unused edge e, $G^2 + e$ is the leave. This permits us to characterize possible pair graphs of maximum WR-families. The components that arise turn out to be primarily paths and cycles as one would expect. We let P_k be the path on k vertices, and C_k be the cycle on k vertices. Two other graphs that arise are the *star* $K_{1,3}$, and the *subdivided star*, the unique connected graph with degree sequence (3,2,2,2,1,1,1) and diameter four.

Lemma 9.23 *The pair graph* $G = (V, E)$ *of a WR-family with* $\lfloor v(v+1)/6 \rfloor$ *blocks is one of the following:*

1. *when* $v \equiv 0, 2 \pmod 6$, *it is the union of* $(v/2)$ P_2*s;*
2. *when* $v \equiv 3, 5 \pmod 6$, *it is the union of* ℓ P_3*s and m cycles* $C_{k_1}, \ldots, C_{k_m}$, *where* $3\ell + \sum_{i=1}^{m} k_i = v$, *and* $k_i \geq 5$ *for* $1 \leq i \leq m$*;*
3. *when* $v \equiv 4 \pmod 6$, *it is the union of* $(v-4)/2$ P_2*s and either*
   (a) *a* P_4, *in which case the pair consisting of the two endvertices of the* P_4 *is unused; or*
   (b) *a star, in which case no edge is unused;*

4. *when* $v \equiv 1 \pmod 6$, *it is the union of* ℓ P_3*s*, m *cycles* $C_{k_1}, \ldots, C_{k_m}$ *with* $k_i \geq 5$ *for* $1 \leq i \leq m$, *and one of the following:*
   (a) *one isolated vertex, where* $3\ell + 1 + \sum_{i=1}^{m} k_i = v$, *in which case no edge is unused;*
   (b) *one subdivided star, where* $3\ell + 7 + \sum_{i=1}^{m} k_i = v$, *in which case no edge is unused;*
   (c) *one* P_r *with* $r \geq 6$, *where* $3\ell + r + \sum_{i=1}^{m} k_i = v$, *in which case the two vertices adjacent to the endvertices of the* P_r *form the unused pair.*

Proof This almost follows directly from Corollary 9.22, using Lemma 9.20 to compute the degrees in G^2. First we treat $v \equiv 0, 2 \pmod 6$. By Corollary 9.22, there are no unused edges; thus G has all vertices of degree 1 or 2, and G^2 has all vertices of odd degree. If G had any vertex of degree 2, using Lemma 9.20 we see that if all degrees in G^2 are to be odd, a degree 2 vertex of G must have a neighbour of degree 1. But then this degree 1 vertex of G becomes degree 2 in G^2, which cannot be. Thus G has all vertices of degree 1, and hence is $(v/2)$ copies of P_2.

For $v \equiv 3, 5 \pmod 6$, by a similar argument a degree 2 vertex of G must have either both neighbours of degree 2, or both of degree 1; a degree 1 vertex must have a neighbour of degree 2. Thus every component is a P_3 or a cycle. Since by Lemma 9.17, no C_3 or C_4 is allowed, each such cycle has length at least five.

When $v \equiv 4 \pmod 6$, if we choose to have no unused edge, G must have one exceptional vertex of degree 0 or 3. It cannot have a vertex of degree 0 for then G^2 has a vertex of degree 0 but must have all degrees odd. Thus it must have a vertex of degree 3. The degree 3 vertex has 0 or 2 neighbours of degree 2 in order to have odd degree in G^2. Since a degree 2 vertex must have neighbours whose degrees are of different parity, the degree 3 vertex must have no degree 2 neighbour. Then the degree 3 vertex forms a star, and the remaining components are P_2 as when $v \equiv 0, 2 \pmod 6$. On the other hand, if there is an unused edge $e = \{w, z\}$, we are forced to include the P_4 $\{\{w, x\}, \{x, y\}, \{y, z\}\}$ and the remaining vertices in P_2s.

The last case, $v \equiv 1 \pmod 6$ is slightly more complicated. If there is no unused edge, we may have a vertex of degree 0, in which case all other vertices are in P_3s or cycles of length at least five. Instead we may have a vertex of degree 3; it must have 1 or 3 neighbours of degree 2 since it has even degree in G^2. If it had a neighbour of degree 1, that vertex would have odd degree in G^2, and thus the degree 3 vertex has three neighbours of degree 2; each of these in turn must have a neighbour of degree 1 in order to have even degree in G^2. Thus the degree 3 vertex forces a subdivided star, and the remaining vertices are in P_3s and cycles of length at least five.

Finally, if $v \equiv 1 \pmod 6$ and there is an unused edge $\{w, x\}$, we find that w and x must have degree 2 in G, and each must have one neighbour of degree 1 and one of degree 2. Since every other vertex of degree 2 has either both neighbours of degree 2 or both of degree 1, w and x must be connected in G by a path; the length of this path must be at least four, or the edge $\{w, x\}$ is *forbidden*, not unused. This path together with the degree 1 vertices forms a P_r, $r \geq 6$. □

Lemma 9.23 indicates the possible structures for pairs in a maximum WR-family. We must show that when a partial triple system has the required leave, it forms a maximum WR-family.

Lemma 9.24 *Let $G = (V, E)$ be a possible pair graph from Lemma 9.23; form its square G^2. Then add the unused edge (if any) to G^2 to form $\widehat{G}$. If there exists a partial triple system $(V, \widehat{\mathcal{B}})$ whose leave is $\widehat{G}$, then $(V, E \cup \widehat{\mathcal{B}})$ is a maximum WR-family.*

Proof Evidently, no two blocks intersect in more than one element, and the number of blocks is $\lfloor v(v+1)/6 \rfloor$. So we need only show that $(V, E \cup \widehat{\mathcal{B}})$ is an HS-family. Consider four blocks $B_i, B_j, B_k, B_\ell \in E \cup \widehat{\mathcal{B}}$ with $B_i \cup B_j = B_k \cup B_\ell$. Consider the possibilities for $B_i \cup B_j$. If this union has size two, $B_i = B_j = B_k = B_\ell$. If it has size three, it contains either two edges of E or a triple of $\widehat{\mathcal{B}}$, but not both. If it has size four, and contains two disjoint edges of E, it cannot contain a triple as a result of the two forbidden edges; in addition, it cannot contain two further disjoint edges since G has no C_4. If on the other hand, it has size four but contains a triple and an edge of E, it contains no further triple or edge. If the union has size five, it contains either a triple and a disjoint edge, or two intersecting triples; again, it cannot contain both. If the union has size six, it contains only two disjoint triples. Thus every selection of two sets yields a distinct union. □

With this in hand, we turn to sufficiency.

Theorem 9.25 *For all $v \geq 0$, there is a WR-family with $b = \lfloor v(v+1)/6 \rfloor$ blocks.*

Proof Using Lemma 9.24, we need only choose a pair graph G from the list in Lemma 9.23, construct the leave graph $\widehat{G}$, and produce a partial triple system whose leave is $\widehat{G}$.

For $v \equiv 0, 2 \pmod 6$, $\widehat{G}$ is the union of $(v/2)$ P_2s. A $\{3\}$-GDD of type $2^{(v/2)}$ exists for all $v \equiv 0, 2 \pmod 6$, and is a partial triple system with the required leave. For $v \equiv 4 \pmod 6$, $\widehat{G}$ is (for either the P_4 or star selection) the union of $(v-4)/2$ P_2s and one complete graph on four vertices. A $\{3\}$-GDD of type $2^{((v-4)/2)}4^1$ exists for all $v \equiv 4 \pmod 6$, and is a partial triple system with the required leave.

For $v \equiv 3 \pmod 6$, we choose G to be $v/3$ P_3s, whence $\widehat{G}$ is $v/3$ C_3s. A $\{3\}$-GDD of type $3^{(v/3)}$ exists whenever $v \equiv 3 \pmod 6$ and this is a partial triple system with the required leave. When $v \equiv 5 \pmod 6$, we choose G to be $(v-5)/3$ P_3s and one C_5. Then a $\{3\}$-GDD of type $3^{((v-5)/3)}5^1$ is a partial triple system with the required leave, and one exists for all $v \equiv 5 \pmod 6$ *except* for $v = 11$. We treat the case when $v = 11$ later.

For $v \equiv 1 \pmod 6$, we choose G to be $(v-1)/3$ P_3s and an isolated vertex. A $\{3\}$-GDD of type $3^{((v-1)/3)}1^1$ then has the required leave, and exists for all $v \equiv 1 \pmod 6$, $v \neq 7$. When $v = 7$, the pair graph G must be either a subdivided star or a P_7. In the first case, we obtain

$$\{\{1,2\},\{1,3\},\{1,4\},\{2,5\},\{3,6\},\{4,7\},\{2,6,7\},\{3,5,7\},\{4,5,6\}\},$$

while in the second, the density condition for leaves is not met.

The only remaining case is $v = 11$. By Lemma 9.24, the choices for G are $C_5 \cup C_6$, $P_3 \cup C_8$, or C_{11}. In the first and last cases, $\widehat{G}$ fails the density condition to be a leave. In the middle case, we *can* construct a WR-family on elements $\{a,b,c\} \cup \{0,1,2,3,4,5,6,7\}$ with triples

$$\{\{a,2i,2i+1\},\{b,2i,2i-1\},\{c,i,i+4\} : i \in \{0,1,2,3\}\}$$

and pairs $\{\{i,i+3\} : 0 \leq i < 8\} \cup \{\{a,b\},\{a,c\}\}$. □

9.7.3 *Testing each item at most three times*

In a maximum WR-family, every item is tested either two or three times. Suppose that we relax the restriction that blocks intersect in at most one element, but instead require that all block sizes be at most three; call this a *K3-family.* Evidently, every maximum WR-family is a K3-family; here we show that the maximum number of blocks in a K3-family is the *same* as for a WR-family. We need one further easy lemma.

Lemma 9.26 *If $(V,\mathcal{B})$ is a K3-family, a subset $\{x,y\}$ is the symmetric difference of at most one pair of blocks in $\mathcal{B}$.*

Proof If $B_1, B_2, B_3, B_4 \in \mathcal{B}$ have $B_1 \cup B_2 \neq B_3 \cup B_4$ but

$$(B_1 \cup B_2) \setminus (B_1 \cap B_2) = (B_3 \cup B_4) \setminus (B_3 \cap B_4) = \{x,y\},$$

and $x \in B_1 \cap B_4$, then $B_1 \cup B_3 = B_2 \cup B_4$, in violation of (9.1). □

Now suppose that a pair occurs in s different triples of $\mathcal{B}$. For each of the $\binom{s}{2}$ pairs of such triples, there is a pair that is neither forbidden (as an edge in the square of the pair graph), nor appears in any triple of $\mathcal{B}$ (by Lemma 9.17). All $\binom{s}{2}$ such unused pairs are different, and arise solely because of this s-times repeated pair. Define an equivalence relation on the pairs, in which each equivalence class contains one pair that occurs in $s \geq 1$ triples, and the $\binom{s}{2}$ unused pairs that it accounts for. This is indeed an equivalence relation by Lemma 9.26.

Theorem 9.27 *If $(V,\mathcal{B})$ is a K3-family on v elements with b blocks, $b \leq \lfloor v(v+1)/6 \rfloor$, with equality only if the pair graph meets the conditions in Corollary 9.22.*

Proof Suppose that $(V,\mathcal{B})$ has pair graph $G = (V,E)$ whose square G^2 has ε edges. Then there are ε forbidden pairs, which appear neither in a triple, nor as the symmetric difference of two triples. Thus there remain $\binom{v}{2} - \varepsilon = r$ pairs that are not forbidden. Suppose that t of these r pairs actually appear in triples of $\mathcal{B}$, and that they appear in $s_1, \ldots, s_t$ triples respectively. We have then a total of $(\sum_{i=1}^{t} s_i)/3$ triples; and since each of the t pairs accounts for an equivalence class, we have that

$$\sum_{i=1}^{t}\left(1+\binom{s_i}{2}\right) \leq \binom{v}{2} - \varepsilon.$$

An equivalence class based on an s-times repeated pair contributes s to $\sum_{i=1}^{t} s_i$ but exhausts $1 + \binom{s}{2}$ pairs. Thus the largest ratio of pairs in triples to pairs in the equivalence class that can be achieved, $s/\left(1+\binom{s}{2}\right)$ is 1, and this is realized only when $s \in \{1, 2\}$. Hence there are at most $\left(\binom{v}{2} - \varepsilon\right)/3$ triples; the proof now parallels the WR case.

Theorem 9.25 establishes that there is a K3-family if and only if the number of blocks does not exceed $\lfloor v(v+1)/6 \rfloor$. □

The structure of partial triple systems and the necessary conditions on leaves thus play an important role in applications in nonadaptive group testing.

9.8 Rödl's nibble method

Consider a simple algorithm for constructing a maximal collection of pair-disjoint triples in a set of n points: Repeatedly choose a triple uniformly at random from among all triples which do not share a pair with any previously chosen triple, until there are no more candidate triples. Spencer (1995) and Rödl and Thoma (1996) show that this simple random greedy procedure almost always results in a collection of triples which cover almost all of the pairs. Grable (1997) obtains significantly tighter bounds on the number of uncovered pairs. In particular, he proves:

Theorem 9.28 *For any $\xi > 0$, the random greedy triangle packing algorithm almost always leaves fewer than $O\left(n^{7/4+\xi}\right)$ uncovered pairs.*

Perhaps the bound can be improved; Grable (1997) states a conjecture due to Spencer that the correct bound on the number of uncovered pairs is $O\left(n^{3/2+\xi}\right)$.

The random greedy algorithm constructs a maximal partial triple system in the following way. It starts with an empty partial triple system and the complete list of *candidate* triples. It repeatedly chooses a candidate triple from the list uniformly at random, adds it to the partial triple system, and removes it and all other candidates with which it shares a pair from the list of candidates, until there are no more candidates.

There are two alternate ways to express the algorithm. One, from the point of view of the leave graph, is that the algorithm starts with the complete graph and repeatedly removes random triangles until the graph is triangle-free. Alternatively, all of the randomness takes place in the initial stage of execution. The list of all triples is first randomly ordered and then the triples are considered sequentially in this order. If the triple shares no pair with any previously selected triple (i.e. it can be added to the PTS) then it is added. Otherwise, it is discarded. Grable (1997) analyzes this last procedure.

He uses an 'honest nibble' approach. The nibble method was pioneered by Rödl (1985). The traditional nibble process starts with the empty partial triple system and repeatedly adds a small number of triples. Triples are selected independently at random with a very small probability from all remaining candidates, each selected triple which does not share a pair with any other selected triple is added to the current partial design, and then the list of candidates is updated by removing all selected triples and any triples with which they share a pair. To get the desired theoretical result, it is only necessary to take a large but fixed number of such nibbles, but this process can be iterated until there are no more candidate triples.

To make the connection to the random greedy algorithm, consider the version of the random greedy algorithm where all possible triples are first randomly ordered and then the triples are considered sequentially. The order can be equally well generated in *bursts*. Each of these bursts corresponds roughly to a single nibble.

More specifically, start with an empty partial triple system and the set of all triples. To generate a burst, select triples independently at random with some small probability, randomly order the selected triples, and remove the selected triples from the set of unused triples. The selection probability is a parameter in the method which controls burst size, and can be selected in the analysis to minimize the number of uncovered pairs. After generating each burst, process the triples as in the greedy algorithm, considering each triple in turn in the random order given and adding it to the partial triple system when possible.

In this process, if a selected triple could not be added to the partial triple system at the beginning of the burst, it cannot subsequently be added. In the traditional nibble, such a triple would never have been selected since only candidate triples are selected. Of the remaining selected triples, a triple which is pair-disjoint from all other remaining selected triples is always placed in the partial design. The same thing happens in the traditional nibble. It is only in the case of selected triples which share a pair that the two methods differ. The traditional nibble discards all such triples, but the honest nibble method considers the ordering and mimics the greedy algorithm. This is the only situation where the ordering of the burst matters.

Bursts give well-defined intermediate stopping points in the process, at which the structure of the leave graph of the partial triple system can be determined. Analysis of one burst at a time gives a kind of proof by induction. By taking bursts of the correct size, the leave graph remains fairly regular throughout the entire process. This enables us to predict fairly well how many edges are left near the end. Bursts that are too small, such as taking only one triangle at a time, negate the advantage of the fact that the triangles are presented randomly. The burst affects very few vertices, so the degrees do not remain very regular from one burst to the next. Large bursts make the ordering dependencies among selected triangles sharing pairs very difficult to analyse. Bursts are therefore chosen to have size so that they each contain enough randomly placed triangles that their effects occur in a smooth way over the entire leave graph, but not so many triangles that there are a lot of dependencies.

To obtain the desired result, we also require that at the end, when the leave graph of the partial triple system is triangle-free, there are almost always fewer than $f(n)$ edges, for some function $f(n)$. This is done by carefully analyzing the leave graph after each burst and showing that there is a number t so that after the tth burst, the leave graph already has fewer than $f(n)$ edges. Grable (1997) provides a simplified analysis yielding a weaker bound, and a more detailed analysis to obtain Theorem 9.28.

The nibble method provides an interesting contrast to the hillclimbing technique of Stinson (1985). While Theorem 9.28 shows that nibbling at the right rate gets one close to a Steiner triple system, the hillclimbing method makes no effort to keep the leave graph uniform. However, it has a mechanism for the exchange of a currently selected triple for another, and hence appears to be very successful at avoiding the traps encountered by the random greedy method.

9.9 Work points

1. Is there an efficient algorithm for checking the density condition on leaves?
2. (Nash-Williams' conjecture) If $G = (V, E)$ is an n-vertex m-edge graph with all vertex degrees even, $m \equiv 0 \pmod 3$, and the minimum degree δ_G in G is at least $3(n-1)/4$, then G has an edge-partition into triangles (equivalently, if $\overline{G}$ has all vertices of degree congruent to $n-1 \pmod 2$, maximum degree $\Delta_{\overline{G}} \leq (n-1)/4$, and a number of edges congruent to $\binom{n}{2} \pmod 3$, then $\overline{G}$ is a 1-leave).
3. Is it NP-complete to determine whether a graph is a 1-leave of a *maximal* partial triple system?
4. Extend Novák's theorem to higher index; that is, determine the number of triples in an mMPT(v, λ).
5. Prove Spencer's conjecture that, with high probability, the random greedy method leaves $O\left(n^{3/2+o(1)}\right)$ uncovered pairs.

10

EXCESSES AND COVERINGS

There is a natural duality between packing triples on a set so that every pair appears at most λ times, and covering the pairs by triples so that every pair appears in at least λ triples. A *covering by triples* of order v and index λ, or CT(v, λ), is a set V of v elements, and a collection (multiset) of 3-element subsets of V (*triples* as usual) for which every unordered pair of elements appears in *at least* λ of the triples.

The *excess* of a CT(v, λ) is a multigraph whose vertex set is the element set of the covering, and in which each pair $\{x, y\}$ appears as an edge s times exactly when it appears in $\lambda + s$ triples of the covering.

A multigraph is a λ-*excess* if and only if it is the excess of some CT(v, λ).

A covering by triples CT(v, λ) is *minimal*, and denoted MCT(v, λ) if every triple contains at least one pair that occurs in only λ triples. A minimal covering with the fewest triples is *minimum*, and denoted mMCT(v, λ); one with the most triples is *maximum*, and denoted MMCT(v, λ).

10.1 Necessary conditions

Let M be a v-vertex multigraph. When is M a λ-excess? We examine basic necessary conditions. As with leaves, we obtain parity conditions on degrees, and a congruence condition on the number of edges.

Lemma 10.1 *Let M be a v-vertex multigraph. If M is a λ-excess, all vertices of M have degree congruent to $\lambda(v-1) \pmod 2$, and the number of edges in M congruent to $-\lambda\binom{v}{2} \pmod 3$.*

Proof If M is a λ-excess, the multigraph M' obtained by adding λ additional edges between every pair of vertices of M has an edge-partition into triangles. Thus M' has all vertex degrees even, and since $M = M' \setminus \lambda K_v$, M and λK_v have the same parity of vertex degrees, namely $\lambda(v-1) \pmod 2$. By the same token, M' has a number of edges which is a multiple of 3, and hence the number of edges of M plus the number of edges of λK_v is also a multiple of 3, giving the congruence condition of the lemma. □

When $v \equiv 0, 1 \pmod 3$, the parity and congruence conditions for a graph to be a 1-excess are the same as those for a graph to be a 1-leave. Consider the graph $K_{n,n-2}$ with $n \equiv 3 \pmod 6$. It is a 1-leave. In fact, it is a 1-leave with the maximum number of edges when $2n - 2 = v \equiv 4 \pmod{12}$. Is it a 1-excess?

Suppose that $K_{n,n-2}$ is a 1-excess of a CT$(2n-2, 1)$ $(V, \mathcal{B})$. Partition the elements of the covering into two classes of sizes n and $n-2$ using the bipartition of the putative excess $K_{n,n-2}$. Call an edge *inside* if it has its endpoints in the same class, *cross* otherwise. There are then $\binom{n}{2} + \binom{n-2}{2}$ inside edges of triples of $\mathcal{B}$, and since every cross edge appears in two triples, there are $2n(n-2)$ cross edges. Since any triple can use at most two cross edges and requires at least one inside edge, this requires that

$$\binom{n}{2} + \binom{n-2}{2} \geq n(n-2),$$

which in turn requires that $n \leq 3$.

This argument can be generalized along the same lines as Lemma 9.2 for leaves:

Lemma 10.2 *Let M be a v-vertex multigraph. Suppose that the vertices of M can be partitioned into sets of sizes s and $v-s$ so that c of the m edges of M cross between the two classes of the partition. Then if M is a λ-excess,*

$$2\left[\lambda\binom{s}{2} + \binom{v-s}{2} + m - c\right] \geq \lambda s(v-s) + c.$$

Proof The left-hand side counts twice inside edges in the required covering, and the right-hand side counts cross edges. □

A multigraph is a 0-excess if and only if it has an edge-partition into triangles. Holyer (1981a) shows that deciding whether a graph has an edge-partition into triangles is NP-complete, and hence:

Lemma 10.3 *Deciding whether a multigraph is a 0-excess is NP-complete.*

10.2 Minimum excesses

What is the smallest number of edges that a λ-excess can have? Fort and Hedlund (1958) settled this problem for $\lambda = 1$, and Engel (1981) settled it for all λ. We consider this problem in some detail here.

First, if $\lambda \geq \gcd(v-2, 6)$, the minimum excess of a $\mathrm{CT}(v, \lambda)$ is the same as that of a $\mathrm{CT}(v, \lambda - \gcd(v-2, 6))$, since we can take the union of the covering with smaller index and a triple system of index $\gcd(v-2, 6)$ to obtain the covering with larger index. Moreover, a minimum excess of a $\mathrm{CT}(v, 0)$ is void (has no edges). It remains to treat the cases given in Table 10.1.

First we consider the cases for index one. When $v \equiv 5 \pmod 6$, the minimum excess is a 2-cycle (two copies of an edge); to produce a covering with this excess, use the $\mathrm{CT}(5, 1)$ with triples $\{abc, abd, abe, cde\}$ having excess $\{\{a, b\}, \{a, b\}\}$. A pairwise balanced design with one block of size five and all other blocks having size three exists for all $v \equiv 5 \pmod 6$; see Table 3.1. Replacing the 5-block by a copy of the $\mathrm{CT}(5, 1)$ gives a $\mathrm{CT}(v, 1)$ whose excess is a 2-cycle.

When $v \equiv 2, 4 \pmod 6$, take a triple system on $v-1$ elements with index one containing the triple $\{a, b, c\}$; add a new element ∞. Remove the triple $\{a, b, c\}$,

	$v \pmod 6$			
λ	0	2	4	5
1	odd deg 0(3) edges	odd deg 2(3) edges	odd deg 0(3) edges	even deg 2(3) edges
2		even deg 1(3) edges		even deg 1(3) edges
3		odd deg 0(3) edges		
4		even deg 2(3) edges		
5		odd deg 1(3) edges		

Table 10.1 *Base cases for coverings*

and add triples $\{\infty, a, b\}, \{\infty, a, c\}, \{\infty, b, c\}$. Finally, for all triples $\{a, x, y\}$ with $\{x, y\} \neq \{b, c\}$, add a triple $\{\infty, x, y\}$. The result is a $\mathrm{CT}(v, 1)$ whose excess consists of $(v-4)/2$ disjoint edges and a 3-star, which is easily seen to be minimum.

When $v \equiv 0 \pmod 6$, we proceed in a similar way starting with a pairwise balanced design on $v-1$ elements having a unique block of size 5, and triples otherwise. Let $\{a, b, c, d, e\}$ be the 5-block. Add an element ∞, and for all triples $\{a, x, y\}$ add a triple $\{\infty, x, y\}$. Then remove the 5-block, and on $\{\infty, a, b, c, d, e\}$ place the triples $\{\infty ab, \infty ac, bcd, bce, de\infty, dea\}$. The result has excess consisting of $v/2$ disjoint edges.

Now we turn to cases with higher index. For index two and $v \equiv 5 \pmod 6$, the minimum number of edges is four; hence the union of two $\mathrm{CT}(v, 1)$s with minimum excess give a $\mathrm{CT}(v, 2)$ with minimum excess (although not all $\mathrm{CT}(v, 2)$ with minimum excess are obtained in this way). For index two and $v \equiv 2 \pmod 6$, the minimum excess also has four edges. To produce a $\mathrm{CT}(v, 2)$ whose excess has only four edges, begin with a $\mathrm{CT}(v, 1)$ whose excess contains $(v-4)/2$ disjoint edges and the 3-star $\{\{a, b\}, \{a, c\}, \{a, d\}\}$. Take its union with an $\mathrm{MMPT}(v)$ whose *leave* contains the same $(v-4)/2$ edges and the two further edges $\{\{a, b\}, \{c, d\}\}$. This union has all edges appearing twice *except* that $\{a, c\}$ and $\{a, d\}$ appear three times, and $\{c, d\}$ appears only once. Adding the triple $\{a, c, d\}$ then gives a $\mathrm{CT}(v, 2)$ with excess $\{\{a, c\}, \{a, c\}, \{a, d\}, \{a, d\}\}$.

For index three, $v \equiv 5 \pmod 6$, $v \geq 11$, take the union of

1. a $\mathrm{CT}(v, 1)$ whose excess is the $(v-6)/2$ edges E, the edge $\{e, f\}$, and a 3-star $\{\{a, b\}, \{a, c\}, \{a, d\}\}$,
2. an $\mathrm{MPT}(v)$ with leave E plus the edges $\{\{a, d\}, \{b, c\}, \{e, f\}\}$, and
3. a $\mathrm{CT}(v, 1)$ whose excess consists of E, the edge $\{b, c\}$, and the 3-star $\{\{a, d\}, \{a, e\}, \{a, f\}\}$.

The resulting $\mathrm{CT}(v, 3)$ has excess E and $\{\{a, b\}, \{a, c\}, \{a, d\}, \{a, e\}, \{a, f\}\}$.

λ	$v \pmod 6$: 0	1	2
0	$v\{K_1\}$	$v\{K_1\}$	$v\{K_1\}$
1	$\frac{v}{2}\{K_2\}$	$v\{K_1\}$	$K_{1,3} \cup \frac{v-4}{2}\{K_2\}$
2	$v\{K_1\}$	$v\{K_1\}$	$(v-3)\{K_1\} \cup 2K_{1,2}$
3	$\frac{v}{2}\{K_2\}$	$v\{K_1\}$	$K_{1,5} \cup \frac{v-6}{2}\{K_2\}$
4	$v\{K_1\}$	$v\{K_1\}$	$(v-2)\{K_1\} \cup 2K_2$
5	$\frac{v}{2}\{K_2\}$	$v\{K_1\}$	$\frac{v}{2}\{K_2\}$

λ	$v \pmod 6$: 3	4	5
0	$v\{K_1\}$	$v\{K_1\}$	$v\{K_1\}$
1	$v\{K_1\}$	$K_{1,3} \cup \frac{v-4}{2}\{K_2\}$	$(v-2)\{K_1\} \cup 2K_2$
2	$v\{K_1\}$	$v\{K_1\}$	$(v-3)\{K_1\} \cup 2K_{1,2}$
3	$v\{K_1\}$	$K_{1,3} \cup \frac{v-4}{2}\{K_2\}$	$v\{K_1\}$
4	$v\{K_1\}$	$v\{K_1\}$	$(v-2)\{K_1\} \cup 2K_2$
5	$v\{K_1\}$	$K_{1,3} \cup \frac{v-4}{2}\{K_2\}$	$(v-3)\{K_1\} \cup 2K_{1,2}$

Table 10.2 *Excess graphs for MCT(v)s*

For index four and $v \equiv 5 \pmod 6$, take the union of the CT$(v,3)$ so constructed and an MPT(v) with leave E and $\{\{b,c\},\{a,d\}\}$. The CT$(v,4)$ that results has excess $\{\{a,c\},\{a,c\}\}$. For index five, take the union of the CT$(v,3)$ and an MPT$(v,2)$ whose leave is $\{\{a,c\},\{a,c\}\}$.

In summary, we have the excess graphs of mMCT(v,λ) given in Table 10.2.

10.3 The spectrum for minimal coverings

How many triples can a minimal covering by triples contain? Mendelsohn and Assaf (1987) settle this question for index one (they call minimal coverings by triples *imbrical designs*).

First we determine the maximum possible number of triples. Every triple must contain a pair that appears only in that triple. Define a covering on element set $X \cup \{\infty\}$ with $|X| = v-1$, whose triple set C contains all triples of the form $\{\infty, x, y\}$ for $x, y \in X$; C has $\binom{v-1}{2}$ triples.

Lemma 10.4 *The maximum number of triples in an MCT$(v,1)$ is $\binom{v-1}{2}$.*

Proof Let $\mathcal{B}$ be the triples of any minimal covering, with the maximum number of triples, on $X \cup \{\infty\}$, $|X| = v-1$. If $\mathcal{B} = C$, the covering has $\binom{v-1}{2}$ triples and we are done. Otherwise, there is a triple in $\mathcal{B}$ not containing ∞, say $\{a,b,c\}$.

Since $\mathcal{B}$ is minimal, at least one pair of $\{a,b,c\}$, say $\{a,b\}$, appears only in this triple of $\mathcal{B}$ (and hence there is no triple $\{\infty,a,b\}$ in $\mathcal{B}$ at present). Remove the triple $\{a,b,c\}$ and add the triple $\{\infty,a,b\}$ to form a set $\mathcal{B}'$ of triples. In general, $\mathcal{B}'$ need not be a covering, since $\{a,c\}$ and $\{b,c\}$ may not be covered; if either is not, add a triple containing it together with ∞ to obtain a larger minimal covering. Under the assumption that $\mathcal{C}$ is maximum, therefore, $\mathcal{B}'$ is a maximum minimal covering. $\mathcal{B}'$ shares one more triple containing ∞ with $\mathcal{C}$ than does $\mathcal{B}$. Hence repeating this argument, we transform $\mathcal{B}$ into $\mathcal{C}$, with the result that $\mathcal{B}$ has $\binom{v-1}{2}$ triples if it is maximum. □

To obtain the possible numbers of triples in a minimal covering, we consider a more general *augmentation* operation. Given a minimal covering $\mathcal{B}$ and a fixed element ∞ of the covering, suppose that $\mathcal{B}$ contains a triple $\{a,b,c\}$ not containing ∞. At least one of the pairs in $\{a,b,c\}$ is *essential* in that it appears only in this triple of $\mathcal{B}$ (since $\mathcal{B}$ is minimal); at most all three of the pairs in $\{a,b,c\}$ are essential. Removing the triple $\{a,b,c\}$ and adding triples with ∞ for each essential pair of $\{a,b,c\}$ results in a minimal covering that has zero, one, or two more triples than the original. To determine the possible numbers of triples, it would therefore suffice to show that, for every possible number of triples less than the maximum, there is a minimal covering having a triple not containing ∞ and having two essential pairs; for if such a covering exists, augmentation on the specified triple increases the number of triples by one. More precisely, we have:

Lemma 10.5 *Let $\mathcal{B}$ be an MCT$(v,1)$ having b triples. Let ∞ be an element of the covering. Let $T=\{x,y,z\}$ be a triple of $\mathcal{B}$ not containing ∞, and having exactly two essential pairs, $\{x,z\}$ and $\{y,z\}$. Suppose further that $\{\infty,x,y\}$ is a triple of $\mathcal{B}$. Then for all $b \le b' \le \binom{v-1}{2}$, there is an MCT$(v,1)$ containing b' triples.*

Proof If T is the only triple of $\mathcal{B}$ not containing ∞, $\mathcal{B}$ has $\binom{v-1}{2}-1$ triples, and we apply augmentation to T. Otherwise, there is a triple $T' \neq T$, not containing ∞; applying augmentation to T' gives an MCT$(v,1)$ $\mathcal{B}'$ having b, $b+1$, or $b+2$ triples, and having one more triple containing ∞. Moreover, $\mathcal{B}'$ still contains T, and T still has two essential pairs (no augmentation makes $\{x,y\}$ essential, or makes $\{x,z\}$ or $\{y,z\}$ inessential). Then, applying the lemma repeatedly to $\mathcal{B}'$ handles (at least) the numbers of triples $b+2 \le b' \le \binom{v-1}{2}$. To obtain $b+1$ triples, we apply augmentation to T in $\mathcal{B}$; and to obtain b triples, we simply take $\mathcal{B}$. □

Theorem 10.6 *An MCT$(v,1)$ with b triples exists if and only if*

$$\left\lceil \frac{v}{3}\left\lceil\frac{v-1}{2}\right\rceil\right\rceil \le b \le \binom{v-1}{2},$$

except for $v=7$, $b=8$.

Proof Necessity follows from the characterization of minimum coverings, and from Lemma 10.5 giving the upper bound. Now it is an easy exercise to verify that the minimum coverings for $\lambda = 1$ with $v \equiv 0, 2, 4, 5 \pmod 6$ all have the property that some triple T has exactly two essential pairs. Let T' be a triple intersecting T in exactly two elements (some such T' exists since T has an inessential pair), and let ∞ be the element in $T' \setminus T$. Then apply Lemma 10.5 to obtain the result.

If $v \equiv 1, 3 \pmod 6$, first we handle the cases for $b \geq v(v-1)/6 + 2$. Take any triple system of order v and index 1, and replace some triple $\{x, y, z\}$ by $\{\{w, x, y\}, \{w, x, z\}, \{w, y, z\}\}$ ($w \notin \{x, y, z\}$ is chosen arbitrarily). Then applying Lemma 10.5 using $\infty = y$ and a triple containing $\{w, x\}$ and not containing y gives the required coverings.

For $v \equiv 1, 3 \pmod 6$, the cases when $b = v(v-1)/6$ are realized by triple systems, so it remains only to treat the cases when $b = v(v-1)/6 + 1$. Mendelsohn and Assaf (1987) call these coverings *failed designs*. The necessary conditions for the excess of such a covering specify that the excess is a 2-regular multigraph on 3 edges, for which there is only one choice, namely a triangle, say $\{\{x, y\}, \{x, z\}, \{y, z\}\}$, for which $\{x, y, z\}$ is not a triple.

If $v = 7$, there are four blocks through x, say $\{xy1, xy2, xz3, xz4\}$ (the four elements are distinct since every pair not in the excess occurs in only one triple). There are two additional blocks through y, say $\{yz\alpha, yz\beta\}$. But $\{\alpha, \beta\} \cap \{1, 2, 3, 4\} = \varnothing$ since none of the pairs $\{\{y, 1\}, \{y, 2\}, \{z, 3\}, \{z, 4\}\}$ are repeated. Hence at least nine elements are required, and no minimal covering on seven elements with eight triples exists.

Mendelsohn and Assaf (1987) produce MCT$(v, 1)$s, $v \in \{9, 13, 15\}$, having $b = v(v-1)/6 + 1$ triples. Then since for all $w \geq 2v + 1$, $w \equiv 1, 3 \pmod 6$, there is a Steiner triple system of order w having a subsystem of order v (the Doyen–Wilson theorem, Theorem 6.7), we obtain the required MCT$(w, 1)$ by replacing the subsystem of order v by a failed design of order v. □

The extension to higher index appears not to have been studied.

10.4 Quadratic excesses

A full characterization of excesses that is computationally tractable appears to be unrealistic. It is therefore of interest to ask for strong sufficient conditions for a multigraph to be an excess. In analogy with Nash-Williams' conjecture for leaves, Colbourn (1987a) made a similar conjecture for excesses:

Conjecture 10.7 *For every $d \geq 0$, there is an integer m_d for which any multigraph G of maximum degree d, having more than m_d vertices, and meeting the parity and congruence conditions to be a 1-excess, is a 1-excess.*

Little progress has been made on this conjecture. The cases $d \in \{0, 1\}$ are settled by the existence of minimum coverings. For maximum degree 2, any candidate to be an excess must have all vertex degrees equal to 0 or 2, and the candidate multigraph is *quadratic*. Colbourn and Rosa (1987) show that the necessary conditions for a quadratic multigraph to be a 1-excess are sufficient:

Theorem 10.8 *Every quadratic multigraph on* $v \equiv 1 \pmod 2$ *vertices, and having* e *edges with* $e \equiv 0 \pmod 3$ *when* $v \equiv 1, 3 \pmod 6$ *or with* $e \equiv 2 \pmod 3$ *when* $v \equiv 5 \pmod 6$, *is a* 1*-excess.*

The corresponding problem of determining which quadratic multigraphs are excesses of *minimal* coverings remains open.

10.5 Excesses and leaves: nuclear designs

The similarity between the necessary conditions for a multigraph to be a leave and to be an excess suggests that we may ask for a multigraph to be *both* a leave and an excess. If an n-vertex multigraph is both a λ-leave and a μ-excess, the parity and congruence conditions are equivalent to: $n \equiv 0, 1 \pmod 3$ or $\lambda + \mu \equiv 0 \pmod 3$, and $\lambda \equiv \mu \pmod 2$ when $n \equiv 0 \pmod 2$. These conditions are not sufficient: consider, for example, the 1-leave $K_{6s+1,6s+3}$. We have seen that this is not a 1-excess.

Using techniques of Opencomb (1984), Colbourn, Hamm, and Rosa (1985) established that:

Theorem 10.9 *Every* 1*-leave on an odd number of elements is a* 41*-excess.*

Proof Let G be the 1-leave of an MPT(v), $v \equiv 1 \pmod 2$. Write $v = 8s - t$, $t < 8$. Partition the elements of the MPT into eight sets $S_1, \dots, S_8$, each of size s or $s - 1$. Now let $\mathcal{B}$ be the set of blocks of a Steiner quadruple system of order 8 on the elements $\{S_1, \dots, S_8\}$. Each block of the quadruple system selects four of the sets; construct the PTS containing those triples with all three elements within the union of these four sets. The PTS has at most $4s$ elements, and thus by a result of Colbourn and Hamm (1986) has an embedding into a TS$(v, 3)$. Repeating this for each of the 14 blocks of the Steiner quadruple system, and taking the union of the resulting TS$(v, 3)$ gives a TS$(v, 42)$. Then removing the triples of the MPT$(v, 1)$ with which we started, we obtain a CT$(v, 41)$ whose excess is G. □

A similar technique shows that when the number of elements is even, every 1-leave is an 83-excess. Colbourn, Hamm, and Rosa (1985) make a strong conjecture in this vein:

Conjecture 10.10 *A* 1*-leave on* v *elements is a* μ*-excess for*

$$\begin{array}{ll} \mu = 1 & v \in \{4, 7, 9, 10\} \\ \mu = 2 & v \equiv 1 \pmod 2,\ v \notin \{7, 9\} \\ \mu = 3 & v \equiv 0, 4 \pmod 6,\ v \notin \{4, 10\} \\ \mu = 5 & v \equiv 2 \pmod 6 \end{array}$$

That these are lower bounds follows from considering leaves of mMPT(v)s.

Mendelsohn, Shalaby, and Shen (1991) consider a related problem. A *nuclear partial triple system* NPTS(v, λ) is a PTS(v, λ) $(V, \mathcal{B})$ with the property that there exist a CT(v, λ) $(V, \mathcal{C})$ and a PTS(v, λ) $(V, \mathcal{D})$ for which $\mathcal{B} = \mathcal{C} \cap \mathcal{D}$, and $|\mathcal{B}|$ is the largest possible such intersection. The special case when the nuclear design is the same as the PTS arises precisely when the λ-leave of the PTS is isomorphic to the λ-excess of the covering by triples.

They establish that:

Theorem 10.11 *A nuclear PTS*(v, λ)

1. *is an MMPT*(v, λ) *when* $\lambda(v-1)$ *is even; and*
2. *has* $\lfloor v/6 \rfloor$ *fewer triples than an MMPT*(v, λ) *when* $\lambda(v-1)$ *is odd.*

10.6 Work points

1. Is it NP-complete to decide whether a graph (or multigraph) is a λ-excess of a minimal covering of index λ?
2. Determine the necessary and sufficient conditions for a quadratic multigraph to be the excess of a *minimal* covering by triples of index one.
3. Extend Theorem 10.6 to all values of λ.
4. Prove or disprove Conjecture 10.7.
5. Characterize the graphs for which the density condition for 1-leaves and the density condition for 1-excesses hold simultaneously (and thereby determine a necessary condition for a graph to be both a 1-leave and a 1-excess).
6. Prove or disprove Conjecture 10.10.

11

EMBEDDING AND ITS VARIANTS

In this chapter, we consider embedding (partial) triple systems into triple systems. Given a partial triple system $(V, \mathcal{B})$ of order v and index λ, when is there a triple system $(W, \mathcal{D})$ of order w and index μ such that $V \subseteq W$ and $\mathcal{B}$ is a submultiset of $\mathcal{D}$?

Such a containing system $(W, \mathcal{D})$ is an *enclosing* of $(V, \mathcal{B})$. When $\lambda = \mu$, the enclosing is an *embedding*. When $v = w$, the enclosing is an *immersion*. When both equalities hold, the enclosing is a *completion*.

By way of example, consider the PTS(6) with triples

$$\{\{0,2,4\},\{0,3,5\},\{1,2,5\},\{1,3,4\}\}.$$

This partial system can be enclosed in an STS(7) by adding the triples

$$\{\{0,1,6\},\{2,3,6\},\{4,5,6\}\};$$

this is an embedding. Similarly, adding the blocks of a CT(6, 1) with excess $\{\{0,1\},\{2,3\},\{4,5\}\}$ yields an enclosing in a TS(6, 2); this is an immersion. However, since no STS(6) exists, there can be no completion of the PTS(6).

11.1 Embedding

Not all partial triple systems have completions. Hence it is interesting to ask, for a given PTS(v, λ), $(V, \mathcal{B})$, for the minimum order w so that the partial triple system has an embedding into a TS(w, λ). Denote this minimum by $\nu(V, \mathcal{B})$. Let us then define

$$\kappa(v, \lambda) = \max\{w : w = \nu(V, \mathcal{B}), (V, \mathcal{B}) \text{ a } PTS(v, \lambda)\}.$$

Our first task is to show that $\kappa(v, \lambda)$ is finite.

11.1.1 *Finite embedding theorems*

Treash (1971) examined the case of Steiner triple systems, and proved that $\kappa(v, 1)$ is finite for all v. Her technique employs the $2v + 1$ construction arising from the singular direct product, using extensively the fact that the singular direct product of an STS(v) and an STS(3) contains a subsystem of order seven for every triple of the STS(v).

First a technical lemma is needed:

Lemma 11.1 *If the PTS(v) $(V, \mathcal{B})$ has a finite embedding, and if each of the pairs $\{\{x_i, y_i\} : i = 1, \ldots, n\}$ appears in no triple of $\mathcal{B}$, then $(V, \mathcal{B})$ has a finite* *embedding* $(W, \mathcal{D})$ *in which the triples* $\{\{x_i, y_i, z_i\} \in \mathcal{D} : i = 1, \ldots, n\}$ *have the property that* $\{z_1, \ldots, z_n\}$ *are all distinct, and none belongs to* V.

Proof The proof is by induction on n. If $n = 1$, form a finite embedding $(X, \mathcal{A})$ of $(V, \mathcal{B})$. If the pair $\{x_1, y_1\}$ occurs with third element $z_1 \notin V$ in $\mathcal{A}$, the embedding has the desired property. Otherwise form the singular direct product of $(X, \mathcal{A})$ with the triple $\{0, 1, \infty\}$, which has a subsystem of order 7 on $(\{x_1, y_1, z_1\} \times \{0, 1\}) \cup \{\infty\}$. Replace this subsystem by one containing the triple $\{x_1, y_1, \infty\}$. This completes the base case.

If $n > 1$, inductively embed $(V, \mathcal{B})$ into $(X, \mathcal{A})$ so that the pairs $\{\{x_i, y_i\} : i = 1, \ldots, n-1\}$ have the desired property. Let $\{x_n, y_n, \alpha\}$ be the triple of $\mathcal{A}$ containing $\{x_n, y_n\}$. Form the singular direct product of $(X, \mathcal{A})$ with the triple $\{0, 1, z_n\}$, where z_n is a new symbol, so that there is a subsystem of order 7 on $(\{x_n, y_n, \alpha\} \times \{0, 1\}) \cup \{\infty\}$. Replace this subsystem by one containing the triple $\{x_n, y_n, z_n\}$. Using the singular direct product, the system has a copy of $(X, \mathcal{A})$ missing only the triple $\{x_n, y_n, \alpha\}$, and this copy does not involve the element z_n. Thus the elements $\{z_i : i = 1, \ldots, n\}$ have the required property. □

In order to ensure that an embedding has distinct third elements for triples containing n specified pairs, this lemma may increase the size of the embedding by a factor of $2^{n+1} - 1$; nevertheless, the main observation is that the embedding remains finite.

Theorem 11.2 *A PTS*(v) *has a finite embedding.*

Proof The proof is by induction on v, the number of elements. Any PTS(3) has a finite embedding. So suppose that all PTS(v')s have finite embeddings for $v' < v$, and consider a PTS(v) $(V, \mathcal{B})$. Choose an element $z \in V$ so that z belongs to the fewest triples in $\mathcal{B}$ of any element. Suppose that z occurs in n triples, and let $\mathcal{S} = \{\{x_i, y_i, z\} \in \mathcal{B} : i = 1, \ldots, n\}$. Now the partial triple system $(V \setminus \{z\}, \mathcal{B} \setminus \mathcal{S})$ has fewer elements than $(V, \mathcal{B})$, and hence can be finitely embedded by induction. Then by Lemma 11.1, it has an embedding $(X, \mathcal{A})$ in which the triples $\{x_i, y_i, z_i\}$ for $1 \leq i \leq n$ have the property that the $\{z_i\}$ are all distinct and disjoint from $V \setminus \{z\}$. Apply the singular direct product to $(X, \mathcal{A})$ and a triple $\{0, 1, z\}$, so that for each $1 \leq i \leq n$, there is a subsystem of order 7 on $(\{x_i, y_i, z_i\} \times \{0, 1\}) \cup \{x\}$. These subsystems of order 7 each intersect only in x; hence each can be replaced independently by a subsystem containing the triple $\{x_i, y_i, z\}$. After these n subsystem replacements, the system contains a copy of $(V, \mathcal{B})$, as required. □

While finite, the embedding given by Theorem 11.2 is exponential in the number of elements of the partial system. Lindner and Rosa (1976) gave similar results for $\lambda > 1$, and Lindner and Rosa (1975b) established finite embedding theorems for sets of Steiner triple systems that preserve intersections among the partial systems.

Treash's construction has the property that when we examine all triples, each of whose elements is in the partial system, the triple is itself in the partial system. We call such an embedding *faithful.*

11.1.2 *A lower bound*

Here we examine the basic lower bound on $\kappa(v,\lambda)$. Let $(V,\mathcal{B})$ be a PTS(v,λ) with b triples, and let $(W,\mathcal{D})$ be an embedding of $(V,\mathcal{B})$ on $w = v+s$ elements. $(W,\mathcal{D}\setminus\mathcal{B})$ is a PTS(w,λ); let G be its leave. Since G is a λ-leave, it meets the necessary conditions for leaves. In particular, applying the density condition (Lemma 9.2) to the partition of the vertices of G into two classes, the v vertices of V and the s vertices of $W\setminus V$, we obtain the condition:

$$\lambda vs \leq \lambda v(v-1) + \lambda s(s-1) - 6b.$$

It follows that

$$s \leq \frac{v+1}{2} - \frac{1}{2}\sqrt{(v+1)^2 - 4v(v-1) + \frac{24b}{\lambda}}, \quad \text{or}$$

$$s \geq \frac{v+1}{2} + \frac{1}{2}\sqrt{(v+1)^2 - 4v(v-1) + \frac{24b}{\lambda}}.$$

We use these restrictions on s to prove the lower bound:

Lemma 11.3 *For $v \geq 9$ and $\lambda \geq 1$, $\kappa(v,\lambda) \geq 2v+1$.*

Proof Since s is an integer, if $((v+1)^2 - 4v(v-1) + 24b/\lambda)^{1/2} > v-1$, we obtain $s \leq 0$ or $s \geq v+1$. Hence if there exists a PTS(v,λ) on b triples having nonempty triangle-free leave, with $b > (\lambda v(v-2))/6$, the embedding of this system has $s \geq v+1$ and hence $w \geq 2v+1$. In other words, if there is a PTS(v,λ) whose leave is nonempty, triangle-free, and has fewer than $\lambda v/2$ edges, the lemma holds. We employ this to treat the majority of the cases.

When λ is odd, this restriction on the number of edges in the leave can be weakened. Specifically, if λ is odd, then w is odd, and hence $v \not\equiv s \pmod 2$. In particular, $s \neq v$, so if $s < v+1$, in fact $s \leq v-1$. When $b > (\lambda(v-1)(v-2))/6$, we conclude that

$$s \geq v+1 \text{ or } s \leq \begin{cases} 0 \text{ if } v \text{ odd} \\ 1 \text{ if } v \text{ even} \end{cases}$$

Now $s = 0$ occurs only when the PTS(v,λ) has a leave that can be edge-partitioned into triangles; and $s=1$ only if the leave consists of $\frac{v}{2}\{\lambda K_2\}$ together with a number of triangles. Thus if the PTS(v,λ) has a nonempty, triangle-free leave on fewer than $\lambda(v-1)$ edges, and the leave is not isomorphic to $\frac{v}{2}\{\lambda K_2\}$, then $s \geq v+1$ provided λ is odd.

We apply these observations to obtain the lower bound of the lemma for all $v \geq 9$ as follows. If $v \equiv 1,3 \pmod 6$, use Theorem 9.15 to form a PTS(v) with leave $C_6 \cup (v-6)\{K_1\}$. Form the union of this system with $\lambda - 1$ copies of an

STS(v), to form a PTS(v, λ) with the same leave. The leave has six edges; we require only that $6 < \lambda v/2$ when λ is even, and $6 < \lambda(v-1)$ when λ is odd. When $v \geq 9$, this is immediate.

If $v \equiv 5 \pmod 6$, write $\lambda = 3\alpha + \beta$ with $1 \leq \beta \leq 3$. Form a PTS(v) with leave C_4 by Theorem 9.15 if $\beta = 1$, and form an MMPT($v, 2$) with leave $2K_2$ if $\beta = 2$. When $\beta = 3$, form their union in such a way that the leave contains no triangles. Finally, add the triples of a TS($v, 3$) α times each. In each case, the leave has fewer than $\lambda v/2$ edges.

To handle the cases when v is even, we proceed in each case by forming a $\{3\}$-GDD on v elements, repeating each triple λ times, and filling each group with a PTS. All of the $\{3\}$-GDDs that we employ exist by Theorem 6.15. If $v \equiv 4 \pmod 6$, write $v = 2m + 4$. There is a $\{3\}$-GDD of type $2^m 4^1$. Repeat each triple λ times. Now on the group of size 4, place a PTS($4, \lambda$) whose leave is empty if λ is even, and is $K_{1,3}$ if λ is odd. The resulting PTS(v, λ) has a leave with $\lambda(v-4)/2$ edges if v is even, and $\lambda(v-4)/2 + 3$ edges if λ is odd.

When $v \equiv 0 \pmod 6$, $v \geq 18$, and $\lambda > 1$, use instead a $\{3\}$-GDD of type $2^{(v-6)/2}6^1$, filling the group of size 6 with a PTS($6, \lambda$) having empty leave if λ is even and leave $3\{K_2\}$ if λ is odd. If $\lambda = 1$, form a $\{3\}$-GDD of type $4^{v/4}$ or $4^{(v-6)/4}$, according to whether v is doubly or singly even. On each group of size g, place the triples of an MMPT(g). It remains to treat the case when $v = 12$ and $\lambda > 1$. Employ again a $\{3\}$-GDD of type 4^3; when λ is odd, fill each group with an MMPT($4, \lambda$), and when λ is even fill two groups with TS($4, \lambda$) and one with a PTS($4, \lambda$) having leave $2K_{1,3}$.

Finally, when $v \equiv 2 \pmod 6$, form a $\{3\}$-GDD of type $2^{(v-6)/2}6^1$ and fill the group of size 6 with an MMPT($6, \lambda$).

In each case, we have exhibited a PTS(v, λ) whose leave has a number of edges ensuring that $s \geq v + 1$. □

In Lindner and Evans (1977), Lindner conjectured that, for partial STSs, Lemma 11.3 gives essentially the correct value. The *strong Lindner conjecture* asserts that

$$\kappa(v, 1) \leq \min(w : w \geq 2v + 1 \text{ and } w \equiv 1, 3 \pmod 6).$$

The *weak Lindner conjecture* asserts that

$$\kappa(v, 1) \leq \min(2v' + 1 : v' \geq v \text{ and } v' \equiv 1, 3 \pmod 6).$$

In either case, the conjecture also asserts that an embedding exists whenever $w \geq \kappa(v, 1)$ and $w \equiv 1, 3 \pmod 6$. These two conjectures coincide when $v \equiv 1, 2, 3 \pmod 6$. Lemma 11.3 shows that the strong Lindner conjecture, if true, is best possible.

The strong Lindner conjecture can be stated for all indices:

Conjecture 11.4 *A PTS(v, λ) can always be embedded in a TS(w, λ) whenever $w \geq 2v + 1$ and $\lambda \equiv 0 \pmod{\gcd(w-2, 6)}$.*

11.1.3 *Linear embeddings for index one*

Lindner (1975c) established a much stronger upper bound on $\kappa(v,1)$ than that provided by Theorem 11.2. Although we later see still stronger results, Lindner's method remains elegant and of interest. Essentially, he modified the direct product construction of an STS(v) from a symmetric idempotent latin square, by employing ideas from the Bose construction.

Theorem 11.5 *A PTS(v) can be embedded in an STS(w) for all $w \geq 6v+3$, $w \equiv 3 \pmod 6$.*

Proof Let $w = 6m+3$. Since $w \geq 6v+3$, $m \geq v$. Suppose that $(V, \mathcal{B})$ is a partial STS(v). Let M be a set of m symbols, $M \supseteq V$. Define a partial latin square $L = (\ell_{ij})$ with rows, columns, and symbols indexed by M. For $i \in M$, $\ell_{ii} = i$. If $i,j \in M$, $i \neq j$, and there is a triple $\{i,j,k\} \in \mathcal{B}$, then $\ell_{ij} = \ell_{ji} = k$. Otherwise cell (i,j) of L is empty.

The theorem of Cruse (1974) ensures that L can be embedded in a symmetric idempotent $(2m+1) \times (2m+1)$ latin square $S = (s_{ij})$. Suppose that the rows, columns, and symbols of S are indexed by T, $T \supseteq M$.

Now define a set of triples on element set $T \times \mathbb{Z}_3$, containing:

1. $\{(a,0),(a,1),(a,2)\}$ for each $a \in T$;
2. the triples $\{\{(a,i),(b,j),(\ell_{ab},k)\} : i+j+k \equiv 0 \pmod 3\}$, if $a,b \in V$, $a \neq b$, and the (a,b) entry of L is nonempty,
3. the triples $\{\{(a,i),(b,i),(s_{ab},i+1)\} : i \in \mathbb{Z}_3\}$ otherwise.

A copy of the triples in $\mathcal{B}$ is present on $V \times \{i\}$, for each $i \in \mathbb{Z}_3$. Therefore, the STS($6m+3$) so defined is a (faithful) embedding of $(V,\mathcal{B})$. □

Corollary 11.6 *A PSTS(v) can be embedded in an STS(w) if $w \geq 12v+7$ and $w \equiv 1,3 \pmod 6$.*

Proof Apply Theorem 11.5 to form an embedding in an STS($6m+3$); then apply the Doyen–Wilson Theorem (6.7). □

11.1.4 *Linear embeddings for all indices*

Andersen, Hilton, and Mendelsohn (1980) improved on Lindner's embedding for partial Steiner triple systems:

Theorem 11.7 *A partial Steiner triple system of order v can be embedded in an STS(w) whenever $w \equiv 1,3 \pmod 6$ and $w \geq 4v+1$.*

Subsequently, by transforming problems with higher λ essentially into λ problems with index one, linear embeddings were developed for all λ in a series of papers by Hamm, Lindner, and Rodger (1983), Colbourn, Hamm, and Rodger (1984), and Colbourn, Hamm, Lindner, and Rodger (1986).

However, the best result to date for all indices was obtained by Rodger and Stubbs (1986) generalizing Theorem 11.7 of Andersen, Hilton, and Mendelsohn (1980).

Theorem 11.8 *A PTS(v, λ) can be embedded in a TS(w, λ) whenever $w \geq 4v+1$, w is odd, and w is λ-admissible.*

The general strategy in the proof by Rodger and Stubbs (1986) is to determine a set of sufficient conditions for a PTS(v, λ) to have an embedding in a TS$(2v+1, \lambda)$, and then to apply this by embedding an arbitrary PTS(n, λ) in a PTS$(2n, \lambda)$ that meets this set of sufficient conditions.

One ancillary ingredient is needed. A *partial symmetric quasi-latin square* of order r and index λ on n symbols, or PSQ(n, r, λ), is an $r \times r$ array in which

1. cell (i, j) contains the same multiset of symbols as does cell (j, i) for all $1 \leq i < j \leq r$;
2. cell (i, j) contains at most λ symbols for all $1 \leq i < j \leq r$;
3. each symbol occurs at most λ times in each row, and at most λ times in each column; and
4. each row and each column contain at most $\lambda(r-1)$ symbols.

If each row and column contain *exactly* $\lambda(r-1)$ symbols, the PSQ(n, r, λ) is *filled*. A filled PSQ$(n, n+1, \lambda)$ is a *symmetric quasi-latin square* of index λ (SQ$(n, n+1, \lambda)$), since every symbol occurs exactly λ times in each row and each column.

For a PSQ(n, r, λ) L, let $N_L(\sigma)$ be the number of occurrences of the symbol σ in L. We require an embedding result for partial symmetric quasi-latin squares:

Lemma 11.9 *Let n be a positive integer, and let L be a filled PSQ(n, r, λ) on the symbols $1, \ldots, n$. L can be embedded in an SQ$(n, n+1, \lambda)$ in which cells (i, i) are empty for $r+1 \leq i \leq n+1$ and cells (i, j) contain exactly λ symbols each for $1 \leq i \leq n+1$ and $r+1 \leq j \leq n+1$, $i \neq j$, if and only if for $1 \leq i \leq n$,*

1. *$N_L(i) \equiv \lambda(n+1) \pmod 2$; and*
2. *$N_L(i) \geq \lambda(2r-n-1)$.*

Proof Necessity is straightforward. For sufficiency, it is enough to show that the PSQ(n, r, λ) L can be embedded in a filled PSQ$(n, r+1, \lambda)$ meeting the conditions of the lemma whenever $r \leq n$. If $r = n$, one simply adds a 'border' in which each entry contains all remaining occurrences of symbols for the corresponding row/column. So suppose that $r < n$. Construct a bipartite multigraph G, with one vertex class being $\{1, \ldots, n\}$ and the other $\{\rho_1, \ldots, \rho_r\}$. Connect ρ_j to i in G by x edges if and only if symbol i occurs exactly $\lambda - x$ times in row j. G has maximum degree $\lambda(n+1-r)$, since vertex i has degree $\lambda r - N_L(i)$, and $N_L(i) \geq \lambda(2r-n-1)$ by assumption.

Give G an equitable edge-colouring with $n+1-r$ colours (note that $n+1-r \geq 2$). Let $C_1, \ldots, C_{n+1-r}$ be the colour classes. Now we 'border' the square L by placing the symbol i in the cell $(j, r+1)$ exactly x times if there are exactly x edges $\{i, \rho_j\}$ of G in colour class C_1. Fill cells in row $r+1$ symmetrically, and leave cell $(r+1, r+1)$ empty. The use of an equitable colouring ensures that after bordering, the square is a filled PSQ$(n, r+1, \lambda)$ meeting the conditions of the lemma, and thus the bordering can be repeated, eventually producing an SQ$(n, n+1, \lambda)$. □

In addition to PSQs, we employ the determination of the number $\mu(v, \lambda)$ of triples in an MMPT(v, λ) (Lemma 9.6), the existence of equitable PTS(v, λ) (Theorem 9.12), and results on edge-colourings.

Theorem 11.10 *Let $(V, \mathcal{B})$ be a PTS(v, λ) with $V = \{1, \ldots, v\}$ having leave G. If*

1. *$2v+1$ is λ-admissible;*
2. *G has maximum degree at most $\lambda(\lceil (v+1)/2 \rceil - 1)$;*
3. *G has at most $3\mu(\lceil (v+1)/2 \rceil, \lambda)$ edges; and*
4. *if v is even, then G has minimum degree at least λ,*

then $(V, \mathcal{B})$ can be embedded in a TS$(2v+1, \lambda)$ (possibly having repeated triples).

Proof Let $(V, \mathcal{B})$ be a PTS(v, λ) with $V = \{1, \ldots, v\}$ satisfying the conditions of the theorem. Let the number of edges of the leave G be ε. Since $2v+1$ is λ-admissible, $\varepsilon \equiv 0 \pmod 3$. By Theorem 9.12, there exists an equitable PTS$(\lceil (v+1)/2 \rceil, \lambda)$ on $\varepsilon/3$ triples. Let $W = \{v+1, \ldots, v + \lceil (v+1)/2 \rceil\}$ and let $(W, \mathcal{D})$ be such an equitable PTS in which the number of occurrences $r(i)$ of element i is at least as large as the number of element j whenever $v < i < j \leq v + \lceil (v+1)/2 \rceil$.

Next we observe that since G has maximum degree at most $\lambda(\lceil (v+1)/2 \rceil - 1)$, by a theorem of McDiarmid (1972) and de Werra (1971, 1975) we can give G a proper equalized edge-colouring with $\lambda \lceil (v+1)/2 \rceil$ colours. Let the colour classes be $C_{i,j}$ for $1 \leq i \leq \lambda$ and $1 \leq j \leq \lceil (v+1)/2 \rceil$, with the colour classes indexed so that $|C_{i,j}| \geq |C_{k,\ell}|$ whenever (i,j) precedes (k,ℓ) lexicographically. Now form an edge-colouring using $\lceil (v+1)/2 \rceil$ colours whose colour classes are $C_{v+1}, \ldots, C_{v+\lceil (v+1)/2 \rceil}$, where $C_{v+j} = \bigcup_{i=1}^{\lambda} C_{i,j}$. This edge-colouring is again equalized; moreover, the number of edges of each colour incident at a fixed vertex of G is at most λ, and by construction $|C_{v+1}| \geq |C_{v+2}| \geq \cdots \geq |C_{v+\lceil (v+1)/2 \rceil}|$.

In fact, since $(W, \mathcal{D})$ is equitable, we find that the number of triples in which an element $v+i$ occurs in $(W, \mathcal{D})$ is precisely the number of edges in the colour class C_{v+i}.

We employ the edge-colouring of G to form a filled PSQ$(v, \lceil (v+1)/2 \rceil, \lambda)$ L. To do this, we first form a $\lceil (v+1)/2 \rceil \times \lceil (v+1)/2 \rceil$ square U with rows and columns indexed by the symbols $v+1, \ldots, v + \lceil (v+1)/2 \rceil$. For each vertex w of G, place w in cell $(v+i, v+i)$ x times if and only if x edges of G incident at w are in colour class C_{v+i}. Cell $(n+i, n+i)$ then has $2|C_{n+i}|$ entries. Next, for each triple $\{r, s, t\}$ of $\mathcal{D}$, place a special symbol ∞ in each of the cells (r,s), (r,t), (s,r), (s,t), (t,r), and (t,s) (a cell may contain ∞ many times). Row and column $v+i$ then have $2r(v+i)$ occurrences of ∞. Since $r(v+i) = |C_{v+i}|$, the number of occurrences of ∞ in row $v+i$ (and in column $v+i$) is the same as the number of entries in cell $(v+i, v+i)$. Moreover, if $\{v+i, v+j\}$ occurs in x

triples of $(W, \mathcal{D})$, then cell $(v+i, v+j)$ contains ∞ x times; thus a cell contains ∞ at most λ times.

As long as U contains an offdiagonal cell, say $(n+i, n+j)$, that contains fewer than λ symbols, there is a symbol other than ∞ that occurs fewer than λ times both in row $n+i$ and in column $n+j$, since the number of distinct symbols occurring λ times in one of them is at most $2(\lceil (v+1)/2 \rceil - 1)\lambda + 2(\lambda - 1) < n\lambda$. Thus one of $1, \ldots, n$ can be placed symmetrically in cells $(v+i, v+j)$ and $(v+j, v+i)$. This process can be repeated until all offdiagonal cells have λ entries.

Removing all occurrences of ∞ from U gives a filled PSQ$(v, \lceil (v+1)/2 \rceil, \lambda)$ L. Since G has even degree, every symbol occurs an even number of times on the diagonal of L; and since L is symmetric, every symbol occurs an even number of times off the diagonal as well. Moreover,

$$\lambda(2\lceil (v+1)/2 \rceil - v - 1) = \begin{cases} 0 \text{ if } v \text{ is odd} \\ \lambda \text{ if } v \text{ is even} \end{cases}$$

Since G has minimum degree at least λ when v is even, each symbol occurs at least $\lambda(2\lceil (v+1)/2 \rceil - v - 1)$ times in L. By Lemma 11.9, L can be embedded in a $QS(v, v+1, \lambda)$ on symbols $1, \ldots, v$, and with all diagonal cells of Q but not L empty.

Now we form a collection $\mathcal{A}$ of triples on elements $1, \ldots, 2v+1$:

1. include all triples of $\mathcal{B}$;
2. include all triples of $\mathcal{D}$;
3. for each occurrence of x in cell $(v+i, v+j)$ $(1 \le x \le v,\ 1 \le i < j \le v+1)$, include the triple $\{x, v+i, v+j\}$;
4. for every edge $\{u, w\} \in C_{v+i}$, $1 \le i \le \lceil (v+1)/2 \rceil$, include the triple $\{u, w, v+i\}$.

Evidently, $(\{1, \ldots, 2v+1\}, \mathcal{A})$ contains all triples of $\mathcal{B}$; in fact, it is also a TS$(2v+1, \lambda)$, and hence is the required embedding. To see this, it suffices to verify that every pair of distinct elements from $\{1, \ldots, 2v+1\}$ appears in exactly λ triples of $\mathcal{A}$. We leave this verification to the reader. □

The proof of Theorem 11.8 employs Theorem 11.10 in an essential way. The basic strategy employed by Andersen, Hilton, and Mendelsohn (1980) and by Rodger and Stubbs (1986) is to first embed the PTS(v, λ) in a PTS$((w-1)/2, \lambda)$ whose leave meets the conditions of Theorem 11.10, which then provides the required embedding. A detailed verification is somewhat tedious, so we illustrate the general idea by considering what is perhaps the easiest situation.

Suppose that $w \equiv 3 \pmod 4$. Write $u = (w-1)/2$ and write $p = (u-1)/2$. Since $v \le p$ (as $4v+1 \le w$), any PTS(v, λ) is also a PTS(p, λ). Add additional triples until this partial triple system is a maximal PTS(p, λ) $(P, \mathcal{D})$. Suppose without loss of generality that $P = \{1, \ldots, p\}$. Let the leave of this maximal system be G. Evidently, G has maximum degree at most $\lambda(p-1)$ and by the theorem of Turán (1941), G has at most $\lfloor p^2/4 \rfloor$ edges. Now let $F_1, \ldots, F_p$ be the 1-factors of a 1-factorization of a complete graph K_p whose vertex set W has $W \cap P = \varnothing$. On element set $W \cup P$ form a PTS(u, λ) $(W \cup P, \mathcal{A})$ containing all triples of $\mathcal{D}$, and in addition for each edge $\{x, y\}$ in each 1-factor F_i the triple $\{i, x, y\}$ λ times (this is essentially the $2v+1$ construction). Consider the leave G of $(W \cup P, \mathcal{A})$. It has maximum degree $\lambda(p-1) = \lambda((u-3)/2)$. It has at most $\lfloor (u-1)^2/16 \rfloor$ edges. Thus it meets the conditions for Theorem 11.10 (a tedious verification).

The remaining cases are similar in spirit, but the details are more involved.

Theorem 11.8 is a dramatic advance over previous work on embedding questions, but leaves a number of directions open. First, the embeddings produced by the theorem are not faithful in general. Second, the construction admits repeated triples; simple embeddings of partial triple systems appear to be much more difficult. Third, it leaves open a gap between $2v+1$, the lower bound, and $4v+1$; and for even indices, it leaves open even order embeddings beyond that point.

Hilton and Rodger (1991) took a different approach. By considering a TS(v, λ) as an edge-colouring of λK_v with $\lambda v(v-1)/6$ colour classes, each inducing a triangle, they transform the embedding problem into a specialized problem on edge-colouring. The basic strategy they employ is to consider 'amalgamations' of triple systems, that is, edge-colourings of multigraphs with loops that are obtained by identification(s) of vertices in λK_v. They determine a number of cases in which an edge-colouring corresponding to a TS(v, λ) can be recovered as each identification is reversed. By this method, they prove:

Theorem 11.11 *If w is λ-admissible, $w \geq 2v+1$, and $\lambda \equiv 0 \pmod 4$, then every PTS(v, λ) has a faithful embedding in a TS(w, λ).*

Thus Conjecture 11.4 holds whenever $\lambda \equiv 0 \pmod 4$. Hilton and Rodger (1990) discuss the possible extension of their techniques used in proving Theorem 11.11 to all cases when λ is even, and formulate a conjecture on colourings which would underlie such an extension. Recently, Johansson (1999) established their conjecture, and hence obtained:

Theorem 11.12 *If w is λ-admissible, $w \geq 2v+1$, and λ is even, then every PTS(v, λ) has a faithful embedding in a TS(w, λ).*

11.1.5 *Simple embeddings*

Theorem 11.8 may introduce repeated triples, and at this time no similar bound is known in the event that the embedding is to be simple.

The best current bound for simple embeddings was obtained by Rodger (1986), building on results of Hamm, Lindner, and Rodger (1983), Colbourn, Hamm, and Rodger (1984), and Colbourn, Hamm, Lindner, and Rodger (1986). One key observation concerns the existence of disjoint mutually balanced PTSs with additional properties:

Lemma 11.13 *Let* $(V, \mathcal{B})$ *be a PTS*(v, λ)*, Then there exist two mutually balanced PTS*(w, λ)*s,* $(W, \mathcal{D}_1)$ *and* $(W, \mathcal{D}_2)$*, with* $w \le (3\lambda + 2)v/2$*, so that*

1. $\mathcal{B} \subset \mathcal{D}_1$ *(as multisets);*
2. $\mathcal{D}_2$ *contains no repeated triples;*
3. $\mathcal{D}_2$ *can be partitioned into* λ *PTS*(v)*s; and*
4. $\mathcal{D}_1 \setminus \mathcal{B}$ *contains no repeated triples.*

Since $\mathcal{D}_1$ and $\mathcal{D}_2$ are mutually balanced, any embedding of one yields an embedding of the other immediately. But $\mathcal{D}_2$ has a partition into λ partial Steiner triple systems, and hence techniques for this case can be employed. Combining Theorem 11.7 with Lemma 11.13 thus leads to a bound of $(6\lambda + 4)v + 1$ for embedding a PTS(v, λ). Such an embedding, however, could have repeated triples. Rodger (1986) developed a construction that ensures that each of the λ partial systems is embedded disjointly, to prove:

Theorem 11.14 *A PTS*(v, λ) *with* α *repeated triples can be embedded in a TS*(w, λ) *for some* $w \le 3(2\lfloor \lambda/2 \rfloor + 1)((3\lambda + 2)v + 1)$*, so that the embedding system has the same* α *repeated triples.*

11.2 Enclosing

Colbourn, Hamm, and Rosa (1985) generalized the well-studied notion of embedding to enclosing, when both the index and the order are permitted to increase. A TS(w, μ) $(W, \mathcal{D})$ is an *enclosing* of a PTS(v, λ) $(V, \mathcal{D})$ if $V \subseteq W$ and $\mathcal{B} \subseteq \mathcal{D}$ (as multisets). The enclosing is *faithful* if $\mathcal{D} \setminus \mathcal{B}$ contains no triples entirely on the elements of V, and it is *simple* if $\mathcal{D} \setminus \mathcal{B}$ contains no repeated triples.

Any embedding is an enclosing, but enclosings may exist for partial triple systems that have fewer triples than any embedding. Colbourn, Hamm, and Rosa (1985) make a conjecture for index one:

Conjecture 11.15 *Every PTS*(v) *has an enclosing in a TS*(w, μ) *in which*

$$(w - v) + (\mu - 1) \le \begin{cases} 3 & \textit{if } v \equiv 4 \pmod 6,\ v \ge 16, \\ 2 & \textit{otherwise.} \end{cases}$$

Conjecture 11.15 is best possible, if true, as the minimum MPT(v)s show. There are very few results available on enclosing partial systems. Colbourn and Hamm (1986) employ the correspondence between commutative idempotent quasigroups and triple systems with index three (Lemma 2.8) to prove:

Lemma 11.16 *A partial TS*$(v, 2)$ *can be enclosed in a TS*$(w, 3)$ *for all odd* $w \ge 2v + 1$.

A similar technique was used by Colbourn, Hamm, and Rosa (1985) to establish:

Lemma 11.17 *A partial STS(v) having a set of $\alpha \leq (v/2)$ elements of which no two appear in a triple can be enclosed in a TS(w, 3) for all odd $w \geq 2(v-\alpha)+1$.*

Proof Let $(V, \mathcal{B})$ be an STS(v) on $V = \{1, \ldots, v\}$ having no triples on $\{v - \alpha+1, \ldots, v\}$. Now define a partial symmetric idempotent latin square L of side $v-\alpha$ with rows and columns indexed by $1, \ldots, v-\alpha$ and symbols from V, as follows. In each diagonal cell (i,i), place the symbol i for $1 \leq i \leq v-\alpha$. For each triple $\{x,y,z\} \in \mathcal{B}$ with $x < y < z$, place symbol z in cells (x,y) and (y,x) of L. Now the theorem of Cruse (1974) can be used to embed L in a symmetric idempotent latin square of order w. Apply Lemma 2.8 to this square to obtain a TS(w, 3) that encloses $(V, \mathcal{B})$. □

Enclosings of partial systems have not been seriously studied; in fact, as we see next, even enclosings of triple systems themselves have not been determined. First let us examine necessary conditions. An enclosing of a TS(v, λ) $(V, \mathcal{B})$ in a TS(w, μ) on element set $W \supseteq V$ exists exactly when the leave G of the PTS(w, μ) $(V, \mathcal{B})$ has an edge-partition into triangles. Write $s = w - v$. By the density condition for leaves (Lemma 9.2), we require that

$$(\mu-\lambda)\binom{v}{2} + \mu\binom{s}{2} \geq \frac{\mu v s}{2}.$$

This requires that

$$s \leq \frac{v+1}{2} - \frac{\sqrt{(\frac{4\lambda}{\mu}-3)v^2 + (6-\frac{4\lambda}{\mu})v+1}}{2}$$

or

$$s \geq \frac{v+1}{2} + \frac{\sqrt{(\frac{4\lambda}{\mu}-3)v^2 + (6-\frac{4\lambda}{\mu})v+1}}{2}.$$

Since $\lambda \leq \mu$, if $s \geq v+1$ it satisfies these conditions. However, for $1 \leq s \leq v$, when $3\mu \leq 4\lambda$ some interval of values of s is ruled out by this condition. When μ is 'large' relative to λ, the condition does not eliminate any possibilities.

Beyond this basic necessary condition, enclosings of TS(v, λ) have been studied only with additional constraints. Bigelow and Colbourn (1991, 1992) have studied *faithful* enclosings of TS(v, λ) in TS(w, μ); see §11.1.1. Suppose that a TS(w, μ) $(W, \mathcal{D})$ exists in which a TS(v, λ) $(V, \mathcal{B})$ is faithfully enclosed. Let $X = W \setminus V$ and $\mathcal{C} = \mathcal{D} \setminus \mathcal{B}$. For $0 \leq i \leq 3$, let n_i be the number of triples of $\mathcal{C}$ that intersect X in i points. Evidently $n_0 = 0$ since the enclosing is faithful. Call a pair of elements an *x-pair* if both elements are in X, a *v-pair* if both are in V, and a *cross pair* if one is in X and the other in V. The triples of $\mathcal{C}$ must contain $\mu\binom{w-v}{2}$ x-pairs, $(\mu-\lambda)\binom{v}{2}$ v-pairs, and $\mu v(w-v)$ cross pairs. By considering the pairs in each type of triple we obtain a system of equations:

$$\begin{aligned} n_1 &= (\mu - \lambda)\binom{v}{2} \\ 2n_1 + 2n_2 &= \mu v(w - v) \\ n_2 + 3n_3 &= \mu\binom{w-v}{2} \end{aligned}$$

This has the solution:

$$\begin{aligned} n_1 &= (\mu - \lambda)v(v-1)/2 \\ n_2 &= (\mu v(w - 2v + 1) + \lambda v(v-1))/2 \\ n_3 &= (\mu(w-v)(w-2v) + (\mu-\lambda)v^2 - \mu w + \lambda v)/6 \end{aligned}$$

A nontrivial condition results from the requirement that each n_i be nonnegative. In particular, since $n_2 \geq 0$ it is necessary that $w \geq v + ((\mu - \lambda)/\mu)(v-1)$.

When $\lambda = 0$, a faithful enclosing of a TS(v, λ) in a TS(w, μ) is precisely a TS(w, μ) having an independent set of size v, that is, v elements on which there is no triple. Bigelow and Colbourn (1992) determine the maximum sizes of independent sets in TS(w, μ)s:

Lemma 11.18 *The maximum size α of an independent set in a TS(w, μ) is*

$$\begin{array}{ll} (w+1)/2 & \text{if } w \text{ is odd and } (w-1)/2 \text{ is } \mu\text{-admissible;} \\ w/2 & \text{if } w \text{ is even;} \\ (w-1)/2 & \text{otherwise.} \end{array}$$

Bigelow and Colbourn (1992) use Lemma 11.18, Theorem 6.9, and various doubling constructions to prove:

Theorem 11.19 *A TS(v, λ) has a faithful enclosing in a TS(w, μ) whenever $\mu \geq \lambda$, w is μ-admissible, and $w \geq 2v + 1$.*

This is a natural generalization of Stern's theorem (Theorem 6.9), but it leaves the more difficult cases for faithful enclosing, when $v < w \leq 2v$, open. Bigelow and Colbourn (1991) completed the determination when $\mu = 2\lambda$, showing that:

Theorem 11.20 *A TS(v, λ) has a faithful enclosing in a TS$(w, 2\lambda)$ if and only if $w \geq (3v-1)/2$.*

Both the enclosing and the faithful enclosing problems appear to be far from a solution at this point. The situation is perhaps worse for *simple* faithful enclosings, as even a generalization of Shen's theorem (Theorem 6.22) from embeddings to enclosings appears to be unavailable.

11.3 Immersion and intricacy

In this section, we consider increasing the index while keeping the order fixed. A TS(v, μ) $(V, \mathcal{D})$ is an *immersion* of a TS(v, λ) $(V, \mathcal{B})$ if $\mathcal{B} \subseteq \mathcal{D}$ (as multisets). Colbourn, Hamm, and Rosa (1985) introduced the notion of immersion; Opencomb (1984) introduced a closely related notion. The *intricacy* of a PTS(v, λ) $(V, \mathcal{B})$

is the smallest number μ for which $\mathcal{B}$ has a partition into multisets $\mathcal{B}_1, \ldots, \mathcal{B}_\mu$, with $(V, \mathcal{B}_i)$ being a partial TS(v, λ) that has a completion $(V, \mathcal{D}_i)$. Evidently if $(V, \mathcal{B})$ has intricacy μ, it has an immersion in the TS$(v, \mu\lambda)$ $(V, \bigcup_{i=1}^{\mu} \mathcal{D}_i)$.

Using techniques of Opencomb (1984), we establish a bound on the intricacy that does not depend on v:

Lemma 11.21 *When $v \geq 199$ is odd and v is λ-admissible, the intricacy of a PTS(v, λ) is at most* 130.

Proof Let $(V, \mathcal{B})$ be a PTS(v, λ) and let $V_1, \ldots, V_{26}$ be a partition of V with $||V_i| - |V_j|| \leq 1$ for $1 \leq i, j \leq 26$. Let $(\{1, \ldots, 26\}, \mathcal{A})$ be a 3-(26, 6, 1) design (an inversive plane of order 5). $\mathcal{A}$ contains 130 blocks. For each block $A \in \mathcal{A}$, form

$$\mathcal{B}_A = \left\{ B \in \mathcal{B} : B \subset \bigcup_{i \in A} V_i \right\}.$$

Now $\{\mathcal{B}_A : A \in \mathcal{A}\}$ is a partition of $\mathcal{B}$ into 130 classes. By construction, the triples in each class involve at most $6\lceil \frac{v}{26} \rceil$ elements. Thus the triples in each class can be treated as a PTS(r, λ) where r is odd and $v \geq 4r + 1$, provided that $v \geq 199$. Apply Theorem 11.8 to embed each such PTS in a TS(v, λ). □

Other embedding results with fixed small 3-designs can be employed in a similar manner to establish that the intricacy is fixed. This strategy can also be used directly for immersion, as Colbourn, Hamm, and Rosa (1985) observed:

Lemma 11.22 *A partial STS(v) with v odd can be immersed in a TS$(v, 42)$.*

Proof Let $(V, \mathcal{B})$ be the STS(v), and let $(\{1, \ldots, 8\}, \mathcal{A})$ be a 3-(8, 4, 1) design (a Steiner quadruple system of order 8). For each block $A \in \mathcal{A}$, form

$$\mathcal{B}_A = \left\{ B \in \mathcal{B} : B \subset \bigcup_{i \in A} V_i \right\}.$$

Now $\{\mathcal{B}_A : A \in \mathcal{A}\}$ is a partition of $\mathcal{B}$ into 14 classes. Each class is a partial STS in which the triples involve at most $4\lceil v/8 \rceil$ elements. We can ensure by choosing the partition of V that this partial system is not a subdesign, and hence some pairs of elements do not occur in a triple in the partial system; then Lemma 11.17 establishes that each $\mathcal{B}_A$ can be enclosed in a TS$(v, 3)$. The union of the 14 systems so produced is the required TS$(v, 42)$. □

These bounds for intricacy and immersion are certainly much too large. However, at the present time, little more is known.

11.4 Work points

1. Prove the strong Lindner conjecture, Conjecture 11.4.
2. Establish a linear *faithful* embedding for PTS(v, λ)s.

3. Establish a linear *simple* embedding for simple PTS(v, λ)s.
4. Establish an analogue of Shen's theorem, Theorem 6.22, for simple faithful enclosings of TS(v, λ)s in TS(w, μ)s when $w > 2v$.
5. Determine necessary and sufficient conditions for a TS(w, λ) to have an enclosing in a TS(w, μ).
6. Determine necessary and sufficient conditions for a TS(w, λ) to have a faithful enclosing in a TS(w, μ).
7. Determine the maximum intricacy of a partial TS(v, λ).
8. Establish necessary and sufficient conditions for a TS(w, λ) to have an immersion in a TS(w, μ).

12

NEIGHBOURHOODS

In this chapter, we examine another graph-theoretic problem on triple systems. Let $(V, \mathcal{B})$ be a TS(v, λ), and let $x \in V$. Consider the multiset of triples

$$\mathcal{B}_x = \{B \in \mathcal{B} : x \in B\}.$$

The *neighbourhood* of x, $N_x(\mathcal{B})$, is the *multiset* of pairs

$$\{\{y, z\} : \{x, y, z\} \in \mathcal{B}_x\}.$$

The neighbourhood contains the $v - 1$ elements of $V \setminus \{x\}$.

12.1 Graphs as neighbourhoods

We view the neighbourhood of an element x as a multigraph on the remaining $v - 1$ elements, whose edges are the pairs appearing in triples with x. In general, a multigraph is a λ-*neighbourhood* if it is the neighbourhood of some element in a triple system of index λ. The basic question is to determine which multigraphs are neighbourhoods.

This neighbourhood problem is a special case of the characterization of λ-leaves:

Lemma 12.1 *G is a λ-neighbourhood if and only if G is a λ-regular λ-leave.*

Proof Suppose that G is the neighbourhood of x in a TS(v, λ) $(V, \mathcal{B})$. Consider the partial triple system $(V \setminus \{x\}, \mathcal{B} \setminus \mathcal{B}_x)$; its leave is G.

In the other direction, suppose that $G = (X, E)$ is λ-regular and is the leave of a partial triple system $(X, \mathcal{B})$ of index λ; choose an element $x \notin X$, and let $\mathcal{C}$ be all triples of the form $\{\{x, y, z\} : \{y, z\} \in E\}$. Since G is λ-regular, $\mathcal{C}$ contains every pair of the form $\{x, y\}$ with $y \in X$ λ times; moreover, any triple system of index λ containing the triples of $\mathcal{C}$ has G as the neighbourhood of x. To complete the proof, observe that $(X \cup \{x\}, \mathcal{B} \cup \mathcal{C})$ is a TS$(|X| + 1, \lambda)$. □

Interest in neighbourhoods stems primarily from the effort to explore the structure of triple systems. Moreover, the structure of one neighbourhood can often be used to determine properties of the triple system itself. As an example, if the neighbourhood of some element is a multigraph having no regular nontrivial factor, the triple system is indecomposable (see Chapter 23).

First we examine the necessary conditions. Suppose that G is a λ-regular multigraph on n vertices. If G is to be a neighbourhood, some TS$(n+1, \lambda)$ must exist and hence $\lambda \equiv 0 \pmod{\gcd(n - 1, 6)}$; this is equivalent to the parity and congruence conditions for G to be a λ-leave (Lemma 9.1).

Next we consider the conditions for a λ-regular multigraph to meet the density conditions to be a λ-leave (Lemma 9.2):

Lemma 12.2 *If G is a λ-regular multigraph on $n \geq 8$ vertices, and a $TS(n+1,\lambda)$ exists, G meets the density condition to be a λ-leave.*

Proof Consider any partition of the n vertices of G into classes of sizes s and $n-s$. If the density condition fails, it must fail when all edges of G are inside the classes, and hence we need only check that

$$2\lambda\left(\binom{s}{2}+\binom{n-s}{2}-\frac{n}{2}\right) \geq \lambda s(n-s).$$

Simplifying, we require $3s^2 - 3ns + n^2 - 2n \geq 0$, which holds for all $n \geq 8$, independent of s. □

It is easy to verify that the density condition cannot exclude any multigraph on fewer than four vertices. Thus the density condition only excludes some multigraphs meeting the parity and congruence conditions on 4, 5, 6, and 7 vertices. The only simple graph excluded is the 2-regular 6-vertex graph containing two disjoint triangles.

Colbourn (1987a) states:

Conjecture 12.3 *Let G be a λ-regular n-vertex multigraph with*

$$\lambda \equiv 0 \pmod{\gcd(n-1,6)}.$$

Then if G meets the density condition to be a λ-leave, G is a λ-neighbourhood.

Unlike the analogous problems of characterizing leaves and excesses, progress on proving this conjecture has been substantial, as we see in the remainder of this section.

12.2 Index two

The first nontrivial case of the neighbourhood conjecture is for index two. The density condition rules out only two quadratic multigraphs, namely the disjoint union of a triangle and a double edge, and the disjoint union of two triangles.

Colbourn and Rosa (1987a) show that the neighbourhood conjecture holds for index two:

Theorem 12.4 *Let G be a 2-regular multigraph on $n \equiv 0, 2 \pmod 3$ vertices other than $2K_2 \cup K_3$ and $K_3 \cup K_3$. Then G is a neighbourhood in a $TS(n+1, 2)$.*

The strategy of the proof is quite similar to the quadratic leaves theorem, Theorem 9.15. We outline the construction for the case $n \equiv 0, 2 \pmod 6$, the other case being similar. First we describe the construction when the quadratic graph G contains a 2-cycle. Write $n = 2t+2$, and let Q' be the $(2t)$-vertex graph obtained by removing the vertices of one 2-cycle from G. We construct a TS$(2t{+}3,2)$ on $\{\infty,A,B\}\cup(\mathbb{Z}_t\times\{0,1\})$ in which N_∞ is isomorphic to G. Without loss of generality, suppose that Q' consists of e even length cycles $C_1,\ldots,C_e$ of lengths $2c_1,\ldots,2c_s$, respectively, and $2d$ odd length cycles: $L_1,\ldots,L_d$ of lengths $2\ell_1-1,\ldots,2\ell_t-1$ and $R_1,\ldots,R_d$ of lengths $2r_1{+}1,\ldots,2r_t{+}1$. Let $a_j=\sum_{i=1}^{j-1}c_i$, and $b_j=a_e+\sum_{i=1}^{j-1}(\ell_i+r_i)$.

We then label the vertices of Q' so that each even length cycle C_j is the cycle $((a_j)_0,(a_j{+}1)_0,\ldots,(a_j{+}c_j)_0,(a_j{+}c_j)_1,(a_j{+}c_j{-}1)_1,\ldots,(a_j{+}1)_1,(a_j)_1)$. (When $c_j=2$, this is a 2-cycle on $\{(a_j)_0,(a_j)_1\}$.) Then odd length cycles are mapped in pairs, L_j and R_j, as follows. L_j is mapped onto $((b_j)_0,(b_j+1)_0,\ldots,(b_j+\ell_j)_0,(b_j+\ell_j-1)_1,(b_j+\ell_j-2)_1,\ldots,(b_j)_1)$; for $k=\ell_j+r_j$, R_j is then mapped onto $(k_0,(k-1)_0,\ldots,(k-r_j)_0,(k-r_j-1)_1,(k-r_j)_1,\ldots,k_1)$. Having labelled vertices in this way, we take $N_\infty=Q'\cup\{\{A,B\},\{A,B\}\}$.

Now we choose neighbourhoods for A and B. Form a collection of edges E on $\mathbb{Z}_t\times\{0,1\}$ by taking each edge of the form $\{i_0,i_1\}$ twice, and edges of the form $\{i_0,(i+1)_0\}$, $\{i_1,(i+1)_1\}$, and $\{i_1,(i+1)_0\}$ once each; then remove from this collection the edges in the mapping of Q'. What remains is a 3-regular multigraph R. Now R can be factored into a 1-factor and a 2-factor; to get the 1-factor, take all edges of the form $\{i_1,(i+1)_0\}$ that remain in R, along with edges of the form $\{i_j,(i+1)_j\}$ required to complete the 1-factor F. Now let the neighbourhood of A, N_A, be $R\setminus F$ together with a 2-cycle on $\{\infty,B\}$. Now N_B is to contain the edges of a 1-factor F, and the edges of an arbitrary 1-factor on $\mathbb{Z}_t\times\{0,1\}$ containing no edge of E.

The key observation to make at this point is that no matter what cycle lengths Q' contains, it remains only to partition all edges on the *fixed* multigraph M containing all edges on $\mathbb{Z}_t\times\{0,1\}$ *less* those in E into one 1-factor F' and a collection of triangles. The neighbourhood of B then contains $F\cup F'$, and the triangles complete the required TS$(2t+3,2)$. The required partition of M is a problem in mixed differences, since M has the automorphism $i_j\mapsto(i+1)_j$; Colbourn and Rosa (1987a) give solutions for the difference problem.

To complete the case when $n\equiv 0,2 \pmod 6$, it remains to handle the case when G has no 2-cycle. Suppose that G contains an h-cycle with $h\geq 4$; form G' by replacing the h-cycle by a 2-cycle $\{A,B\}$ and an $(h-2)$-cycle. Now proceed as before, ensuring that some edge of the $(h-2)$-cycle is $\{i_0,i_1\}$. Then a *pullup* of the resulting TS$(2t+3,2)$ is the replacement of the triples

$$\{\infty,A,B\},\ \{\infty,i_0,i_1\},\ \{A,i_0,(i+1)_0\},\text{ and }\{B,i_1,(i+1)_0\}$$

by the triples

$$\{\infty,A,i_0\},\ \{\infty,B,i_1\},\ \{A,B,i_1\},\text{ and }\{i_0,i_1,(i+1)_0\}.$$

After pullup, the neighbourhood of ∞ is isomorphic to G.

The final case arises when G consists of $n/3$ disjoint triangles, and this is easily handled using pairwise balanced designs with the single exception when $G=2\{K_3\}$.

When $n\equiv 3,5 \pmod 6$, the strategy is quite similar; as a result of the change in parity, we remove a triangle from G (when one is present), and proceed as above to map the remaining cycles. Pullup can be used in a very similar way to handle the case when the shortest odd cycle (of which there must be at least one) has length at least five.

Theorem 12.4 has an important corollary for larger indices:

Corollary 12.5 *Let G be a λ-regular multigraph with λ even, on $n \equiv 0, 2 \pmod 3$ vertices. Then G is a neighbourhood in a TS$(n+1, \lambda)$.*

Proof Let $s = \lambda/2$. By Petersen's theorem (Theorem 1.22), a 2-factorization of G exists, into s 2-factors $G_1, \ldots, G_s$. By Theorem 12.4, each is the neighbourhood of an element in a TS$(n+1, 2)$; their union is a TS$(n+1, \lambda)$ with an element having neighbourhood G. □

12.3 Index three

For index three, there are three small cubic multigraphs that are eliminated by the density conditions, namely the disjoint union of two triply repeated edges, a 4-cycle with two opposite edges duplicated, and two disjoint triangles with a superimposed 1-factor.

Colbourn and McKay (1987) showed:

Theorem 12.6 *Every n-vertex cubic multigraph meeting the density condition is a neighbourhood in a TS$(n+1, 3)$.*

Their proof relies on partitioning cubic multigraphs into almost regular factors. A *nontrivial path factor* in a graph is a spanning subgraph in which each component is a path of length at least one. We outline the construction for the easiest case, when $n \equiv 0, 2 \pmod 6$.

Lemma 12.7 *Let $G = (V, E)$ be a cubic multigraph on $n \equiv 0, 2 \pmod 6$ vertices. Then if $n \geq 8$, G is a 3-neighbourhood.*

Proof Colbourn and McKay show that every cubic multigraph G has a nontrivial path factor H containing at least one path of length one $\{x, y\}$. Let H' be the complement of H with respect to G; H' has maximum degree 2. Form Q' by adding an arbitrary matching on the odd degree vertices of H'. Q' may be a multigraph. Using the solution for index two, form a PTS$(n+1, 2)$ $\mathcal{B}_1$ in which the neighbourhood of the new element, ∞, is Q'.

Form Q from $H - \{x, y\}$ by adding a matching on the odd degree vertices of H, so that every cycle formed has even length (which is possible since $n-2$ is even). Q is a 2-regular simple graph on $V \setminus \{x, y\}$; in addition, Q has a 1-factorization F_1, F_2 since every cycle length is even. Using the quadratic leaves theorem (Theorem 9.15), form a PTS$(n-1, 1)$ with triples $\mathcal{D}$ on $V \setminus \{x\}$ whose leave is Q; form a collection of triples $\mathcal{B}_2$ as the union of $\mathcal{D}$, $\{\{x, y, z\} : \{y, z\} \in F_1\}$, $\{\{\infty, y, z\} : \{y, z\} \in F_2\}$, and the triple $\{\infty, x, y\}$.

The union $\mathcal{B} = \mathcal{B}_1 \cup \mathcal{B}_2$ has $F_2 \cup Q'$ as the neighbourhood of ∞, but the important fact is that all edges of G that are *not* at present in the neighbourhood of ∞ are in the neighbourhood of x. Let $\mathcal{I}$ be the edges in the neighbourhood of ∞ but not in G, and $\mathcal{O}$ be the edges in G but not the neighbourhood of ∞. $\mathcal{I}$ and $\mathcal{O}$ are matchings on the same vertex set. Thus, replacing all triples of the form $\{\infty, a, b\}$ for $\{a, b\} \in \mathcal{I}$ by $\{x, a, b\}$, and all triples of the form $\{x, c, d\}$ for $\{c, d\} \in \mathcal{O}$ by $\{\infty, c, d\}$, gives a triple system of order $n+1$ in which element ∞ has neighbourhood G. □

The case when $n \equiv 4 \pmod 6$ is more complicated. The order is not admissible for $\lambda = 1$ or $\lambda = 2$, so neither the solution for index two nor the quadratic leaves theorem can be used in the same way. Nevertheless, Colbourn and McKay (1987) obtain results that are analogous to, but more restricted than, the quadratic leaves theorem and the neighbourhood theorem for index two (Theorems 9.15 and 12.4); then using certain factorizations of cubic multigraphs into nontrivial path factors, they combine these results in a manner quite similar to the easier case outlined here.

12.4 Simple neighbourhoods

Using the solutions for indices two and three, and the quadratic leaves theorem, Colbourn (1989) obtained:

Theorem 12.8 *Let G be an n-vertex λ-regular simple graph, other than two disjoint triangles. Then if $\lambda \equiv 0 \pmod{\gcd(n-1, 6)}$, G is a λ-neighbourhood.*

We indicate the main ideas of the proof, showing in the process a stronger result: for $n \geq 8$, $n \equiv 0, 2 \pmod 3$, every λ-regular n-vertex *multigraph* with $\lambda \equiv 0 \pmod{\gcd(n-1, 6)}$ is a λ-neighbourhood. We may assume that $\lambda \geq 4$ using the results of previous sections.

Now suppose that the index is even. If $n \equiv 0, 2 \pmod 3$, the result is obtained (more generally for multigraphs) in Corollary 12.5. This leaves only $n \equiv 1, 4 \pmod 6$. For even index, we have $\lambda \equiv 0 \pmod 6$.

Lemma 12.9 *Let G be a 6-regular simple multigraph on $n \equiv 1 \pmod 6$ edges. Then G is the neighbourhood in a TS$(n+1, 6)$.*

Proof By Petersen's theorem (Theorem 1.22), G has a 2-factorization with three 2-factors Q_1, Q_2, Q_3. In addition, since $n \equiv 1 \pmod 3$, there is an element x that does not appear in a triangle in Q_1. For $i = 1, 2, 3$, let $\{y_i, z_i\}$ be the neighbours of x in Q_i. Let R_i be the result of replacing the edges $\{\{x, y_i\}, \{x, z_i\}\}$ in Q_i by the single edge $\{y_i, z_i\}$. Since x is not in a triangle of Q_1, R_1 is simple; R_2 and R_3 may each have a multiple edge. Now since R_1 is a simple quadratic graph with $n - 1 \equiv 0 \pmod 3$ edges, there is a PTS$(n, 1)$ whose leave is R_1; let $\mathcal{B}_1$ be the triples of this system. For R_2, proceed as follows. If R_2 has no multiple edge, form a PTS$(v, 2)$ $\mathcal{B}_2$ by taking the union of a PTS$(n, 1)$ with leave R_2 and a TS$(n, 1)$. Otherwise, if R_2 has a repeated edge at $\{y_2, z_2\}$, form S_2 by choosing any other edge $\{a, b\}$ of R_2, removing the edge $\{a, b\}$ and one copy of the edge $\{y_2, z_2\}$, and adding the edges $\{\{a, y_2\}, \{b, z_2\}\}$ (this cannot introduce repeated edges). Now take the union of a PTS$(n,1)$ with leave S_2 and a TS$(n,1)$ in which the neighbourhood of x contains $\{\{a,b\},\{y_2,z_2\}\}$; produce $\mathcal{B}_2$ by replacing the triples $\{\{x,a,b\},\{x,y_2,z_2\}\}$ by $\{\{x,a,y_2\},\{x,b,z_2\}\}$ in this union. $\mathcal{B}_2$ is a PTS$(v,2)$ whose leave is R_2. Proceed similarly to produce $\mathcal{B}_3$ from R_3. Finally, let $\mathcal{B}_4$ be the triples of a TS$(n,1)$ having the triples $\mathcal{U} = \{\{x,y_1,z_3\},\{x,y_2,z_1\},\{x,y_3,z_2\},\{y_1,y_2,y_3\}\}$. Now the system with triples $\mathcal{C} = \mathcal{B}_1 \cup \mathcal{B}_2 \cup \mathcal{B}_3 \cup \mathcal{B}_4$ is a PTS$(n,6)$ with leave $R_1 \cup R_2 \cup R_3$. Then $\mathcal{B} = \mathcal{C} \setminus \mathcal{U} \cup \{\{y_1,z_1,y_3\},\{y_2,z_2,y_1\},\{y_3,z_3,y_2\}\}$ is the collection of triples of a PTS$(n,6)$ with leave $G = Q_1 \cup Q_2 \cup Q_3$. □

Repeated application of Petersen's theorem to factorize the graph then handles all simple graphs with $n \equiv 1 \pmod 6$. A similar strategy handles the case when $n \equiv 4 \pmod 6$.

For odd index, the strategy is somewhat different. When the index is even, Petersen's theorem ensures that the graph is 2-factorable; for any odd degree λ, however, there are graphs with no regular factor of degree less than λ. Hence we cannot partition the graph as before into regular factors. Nevertheless, we can adapt the ideas used in the case when $\lambda = 3$. We require a graph-theoretic result:

Lemma 12.10 *Let G be a λ-regular n-vertex graph G with λ odd. Then G has a nontrivial path factor containing a path of length one.*

Proof Add an arbitrary 1-factor F to G to form G', a $(\lambda+1)$-regular graph. Petersen's theorem ensures that G' has a 2-factor Q. Then deleting one edge from each cycle in $Q \setminus F$ is a nontrivial path factor H. If H has any path of length one, we are done. Otherwise, if H has a path of length $\ell > 2$, we can delete one further edge to split the path into paths of lengths 1 and $\ell - 2$, and we are done. Otherwise H has all paths of length 2. Hence it remains only to handle the case when $n \equiv 0 \pmod 6$. For some 2-factorization $Q_1, \ldots, Q_{(\lambda+1)/2}$ of G', consider the edge-partition of G into subgraphs $G_i = Q_i \setminus F$. If any one of the $\{G_i\}$ has a cycle other than a triangle, or a path of length different from two, it is easily seen that the desired path factor exists. Now G has $\lambda n/2$ edges, and hence one of the subgraphs, say G_j, has at least $(\lambda n)/(\lambda+1)$ edges. Thus if we have not yet obtained the required path factor, G_j has at most $n/(\lambda+1)$ paths of length two.

First, whenever G_j contains a 2-path in which the two endvertices are adjacent in G, we add the edge in G_j, turning the 2-path into a triangle. Now consider the neighbours of the vertex on a triangle. If adjacent to any vertex of another triangle, or to the endpoint of a 2-path, the required factor is easily produced. Thus each vertex on a triangle is adjacent only to other vertices on the triangle, and to centres of 2-paths. Similarly, an endvertex of a 2-path can be adjacent only to centres of other 2-paths; otherwise, we can combine two 2-paths with the additional edge to form a 5-path which can in turn be repartitioned as a 1-path and a 3-path.

Since the degree is odd, every triangle has at least one edge connecting a vertex of the triangle to a vertex not on the triangle. Hence, there must remain some 2-paths. Moreover, since the 2-paths account for twice as many edges at the centres of 2-paths as at the leaves, an elementary counting argument shows that the number of edges required to account for the degrees in G of the endvertices of the paths in G_j exceeds the degree remaining to be accounted for at the centres of the paths. □

With Lemma 12.10 in hand, we now address the neighbourhood problem with $n \equiv 0, 2 \pmod 6$ and all odd λ:

Lemma 12.11 *Let G be a λ-regular multigraph on $n \equiv 0, 2 \pmod 6$ vertices, $n \geq 8$, with $\lambda \geq 5$ odd. Then G is a λ-neighbourhood.*

Proof (Sketch) This parallels Lemma 12.7 closely. Find a nontrivial path factor H of G having a path $\{x, y\}$ of length 1. Let $R = G - H$. Choose an arbitrary matching M on the vertices of degree $\lambda - 2$ in R, so that $R + M$ is $\lambda - 1$-regular; then apply Corollary 12.5 to obtain a TS$(n+1, \lambda-1)$ whose additional point, ∞, has neighbourhood $R+M$. Next form a simple quadratic graph Q from $H \setminus \{x, y\}$ by adding a matching M' on the odd degree vertices so as to form only even length cycles. Let F_1, F_2 be a 1-factorization of Q, and form a TS$(n + 1, 1)$ in which the neighbourhood of ∞ is $F_2 \cup \{x, y\}$ and the neighbourhood of x is $F_1 \cup \{\infty, y\}$. Take the union of the two triple systems to form a TS$(n + 1, \lambda)$; then interchange the role of x and ∞ in the triples involving the pairs in $N_\infty \setminus G$ and in $G \setminus N_\infty$. □

It remains only to consider the case when $n \equiv 4 \pmod 6$; here, Colbourn (1989) uses a decomposition of the graph into a nontrivial path factor with at least two paths of length one, a spanning subgraph of maximum degree 2, and a spanning subgraph with all vertex degrees from $\{\lambda - 4, \lambda - 3\}$; the ingredients are more complicated, but the basic strategy remains the same.

12.5 Neighbourhood uniform triple systems

Until this point, we have been concerned with a single neighbourhood in a triple system. We might instead ask that every neighbourhood be isomorphic; we call such a triple system *neighbourhood uniform.* Such triple systems have been studied under the name *G-triple systems*, where G is the neighbourhood multigraph of each element. Naturally, every triple system whose automorphism group acts transitively on its elements is neighbourhood uniform; however, the structure of possible neighbourhoods in transitive systems appears not to have been studied.

When one asks for a specific graph to be the neighbourhood in a neighbourhood uniform triple system, little is known. Ducrocq and Sterboul (1978, 1980) observe:

Theorem 12.12 *For every $n \equiv 0, 1 \pmod 3$, there exists a simple TS$(n, 2)$ in which the neighbourhood of every element is an $(n - 1)$-cycle.*

Proof Ringel (1975) establishes that for every $n \equiv 0, 1 \pmod 3$, there is a nonorientable surface which K_n triangulates (this is the Map Colour Theorem). The triples of the simple $\mathrm{TS}(v, 2)$ are just the faces of this embedding. Now consider the faces incident at a vertex v of K_n; they are necessarily in a cyclic sequence about v in the embedding, and hence the neighbourhood of v is a cycle. Since every vertex other than v is adjacent to v, the neighbourhood is an $(n-1)$-cycle. □

12.6 Double neighbourhoods for index one

The neighbourhood problem is trivial for $\lambda = 1$, since for each order n (necessarily even), the only isomorphism type of 1-regular multigraph is a 1-factor. It is the neighbourhood of any element in any Steiner triple system of order $n + 1$.

Colbourn, Colbourn, and Rosa (1983) examined a nontrivial extension of the neighbourhood problem in the case $\lambda = 1$. A *double star* of triples based on a pair $\{x, y\}$ of a v-set V is a partial triple system on V so that every triple contains at least one of $\{x, y\}$, and every element $z \in V \setminus \{x, y\}$ appears in exactly one triple with x and exactly one triple with y. Equivalently, we can think of a double star as a simultaneous specification of the neighbourhoods of x and y, and hence as a *double neighbourhood.*

Colbourn, Colbourn, and Rosa (1983) established that every double star is completable to a Steiner triple system; we give a simpler proof using more recent results:

Theorem 12.13 *Every double star on* $n \equiv 1, 3 \pmod 6$ *elements can be completed to a Steiner triple system of order* n.

Proof Let $\mathcal{D}$ be the triples of the double star on $V \cup \{x, y\}$ based at $\{x, y\}$. $\mathcal{D}$ contains a triple $\{x, y, z\}$ containing $\{x, y\}$. Let $\mathcal{C} = \mathcal{D} \setminus \{\{x, y, z\}\}$, and form the set of pairs $E = \{\{b, c\} : \{a, b, c\} \in \mathcal{D}, a \in \{x, y\}\}$. Then the graph $G = (V, E)$ is a quadratic graph with $n - 2$ vertices and $n - 3$ edges; moreover, every cycle of G has even length. Thus G is the leave of some partial triple system $(V, \mathcal{B})$ of order $n - 2$ and index 1 by Theorem 9.15. Then $(V \cup \{x, y\}, \mathcal{B} \cup \mathcal{D})$ is a Steiner triple system of order n containing the double star $\mathcal{D}$. □

The quadratic graph in the double star has been widely studied, primarily as an isomorphism invariant. The isomorphism type of such a graph is determined simply by the number of cycles of each length, and hence the double neighbourhood graph is typically encoded as a list of cycle lengths; this is the cycle structure of the pair $\{x, y\}$. The fact that every double star can be completed is equivalent to the fact that, for $n \equiv 1, 3 \pmod 6$, every partition of $n - 3$ into even parts each at least four is a cycle structure for some pair in a Steiner triple system of order n.

12.7 Uniform and perfect triple systems

Rather than prescribing the double neighbourhood at a single pair, one could ask that *all* pairs of vertices have isomorphic double neighbourhood graphs. We call such a Steiner triple system *uniform*. If in addition each double neighbourhood consists of a single cycle, we call the triple system *perfect* (in analogy with 'perfect 1-factorizations', in which the union of any two factors forms a hamiltonian cycle).

Two classes of uniform Steiner triple systems arise from the infinite families of 2-transitive systems, the projective systems and the affine systems. In a projective system of order n, each double neighbourhood is a set of $(n-3)/4$ 4-cycles, since any three noncollinear elements generate a subsystem of order seven. In an affine system of order n, every double neighbourhood is a set of $(n-3)/6$ 6-cycles, since every three noncollinear elements generate a subsystem of order nine. More generally, the Hall triple systems (from §8.7), although not 2-homogeneous in general, have the same double neighbourhoods as the affine triple systems; see Bénéteau (1983, 1984).

A further class of uniform Steiner triple systems arises from 2-homogeneous, but not 2-transitive, systems. These are the Netto systems that exist for $q \equiv 7 \pmod{12}$, q a prime power; see Delandtsheer, Doyen, Siemons, and Tamburini (1986). Necessarily, all pairs of elements have isomorphic double neighbourhoods; however, the structure of the double neighbourhood is only partially understood. Robinson (1975) established that the Netto triple system of p^α has a double neighbourhood containing a 4-cycle whenever $p \equiv 7 \pmod{24}$, and having no 4-cycle whenever $p \equiv 19 \pmod{24}$. He also examines the distribution of numbers of 4- and 6-cycles in the double neighbourhood.

Beyond these classes for which uniformity is forced by the structure of the automorphism group, little is known. It follows from the classification of 2-homogeneous triple systems and our remarks above that the only perfect 2-homogeneous systems are the obvious cases for orders seven and nine.

Perfect Steiner triple systems are known to exist for only 13 orders: 7, 9, 25, 33, 79, 139, 367, 811, 1531, 25771, 50923, 61339, and 69991. The known perfect Steiner triple system of order 25 is a transitive design acted on by the group $\mathbb{Z}_5 \times \mathbb{Z}_5$; orbit representatives are:

$$\{(0,0),(0,1),(1,0)\}, \quad \{(0,0),(0,2),(2,1)\},$$
$$\{(0,0),(1,1),(2,3)\}, \quad \{(0,0),(1,3),(3,3)\}.$$

The perfect Steiner triple system of order 33 is also transitive; in fact, it is cyclic with starter blocks: $\{0,1,7\}$, $\{0,2,21\}$, $\{0,3,20\}$, $\{0,4,28\}$, $\{0,8,18\}$, and $\{0,11,22\}$. The nine larger orders are given in Grannell, Griggs, and Murphy (1999). These systems are each generated from a block $\{0,1,\alpha\}$ under the action of the mappings $z \mapsto z+1 \pmod{v}$ and $z \mapsto \omega^6 z \pmod{v}$ where ω is a primitive root of v.

Grannell, Griggs, and Murphy (1999) also found three additional uniform Steiner triple systems, of orders 43, 13063, and 34303. In each case the double neighbourhood consists of one 4-cycle and one $(v-7)$-cycle.

At the present time, no infinite family of perfect Steiner triple systems is known; nor is any recursive construction known that preserves perfection.

12.8 Work points

1. Complete the proof of Conjecture 12.3 for $n \geq 8$; it remains to handle multigraphs with at least one repeated edge on $n \equiv 1 \pmod 3$ vertices.
2. Which graphs are neighbourhoods in *simple* $\mathrm{TS}(v, \lambda)$s? This is open for all $\lambda \geq 2$.
3. Find an infinite family of perfect Steiner triple systems, or prove that there are only finitely many.
4. Characterize the quadratic multigraphs for which some triple system of index two exists with every neighbourhood isomorphic to the specified one.
5. Determine the possible double neighbourhoods in a triple system of index two.

13

CONFIGURATIONS

By a configuration we mean a PTS(v, λ), typically having a fixed small number of triples. The triples are sometimes called *lines* here to conform with geometric terminology (in the same vein, elements are sometimes called *points* here). The *degree* of a point is the number of lines containing the point.

Evidently, there are configurations which must occur in every nontrivial triple system, while others may be avoided altogether. This leads naturally to questions about ubiquity, as well as questions about avoidance and decompositions. A configuration whose number of occurrences in a TS(v, λ) depends only upon v and λ is *constant* (for these parameters). Otherwise, it is *variable.*

13.1 Constant and variable configurations

In this section, we restrict to configurations in Steiner triple systems. A (k, ℓ)-*configuration* in an STS(v) is simply a set of ℓ lines whose union contains precisely k points.

Each configuration with at most three lines is constant. Indeed, these are easily counted. The number of 1-line configurations is just the number of triples, $v(v-1)/6$. There are two 2-line configurations (Figure 13.1): A_1 is a pair of disjoint triples, while A_2 is a pair of intersecting triples. Their numbers are $a_1 = v(v-1)(v-3)/72$ and $a_2 = v(v-1)(v-3)/8$, respectively. The five 3-line configurations B_1 (3-PPC), B_2 ('hut'), B_3 ('star'), B_4 ('3-path' or '3-snake'), and B_5 ('triangle') are shown in Figure 13.1. Grannell, Griggs, and Mendelsohn (1995) establish that the corresponding numbers are:

$$\begin{aligned}
b_1 &= n_v(v-7)(v^2-19v+96)/1296 \\
b_2 &= n_v(v-7)(v-9)/48 \\
b_3 &= n_v(v-4)/48 \\
b_4 &= n_v(v-7)/8 \\
b_5 &= n_v/6 = 4a_2/3
\end{aligned}$$

where $n_v = v(v-1)(v-3)$.

For each $n \geq 4$, there exist both variable and constant n-line configurations. Grannell, Griggs, and Mendelsohn (1995) show that of the 16 4-line configurations (Figure 13.1), five are constant and 11 are variable. The five constant configurations are C_4, C_7, C_8, C_{11}, and C_{15}; their numbers are:

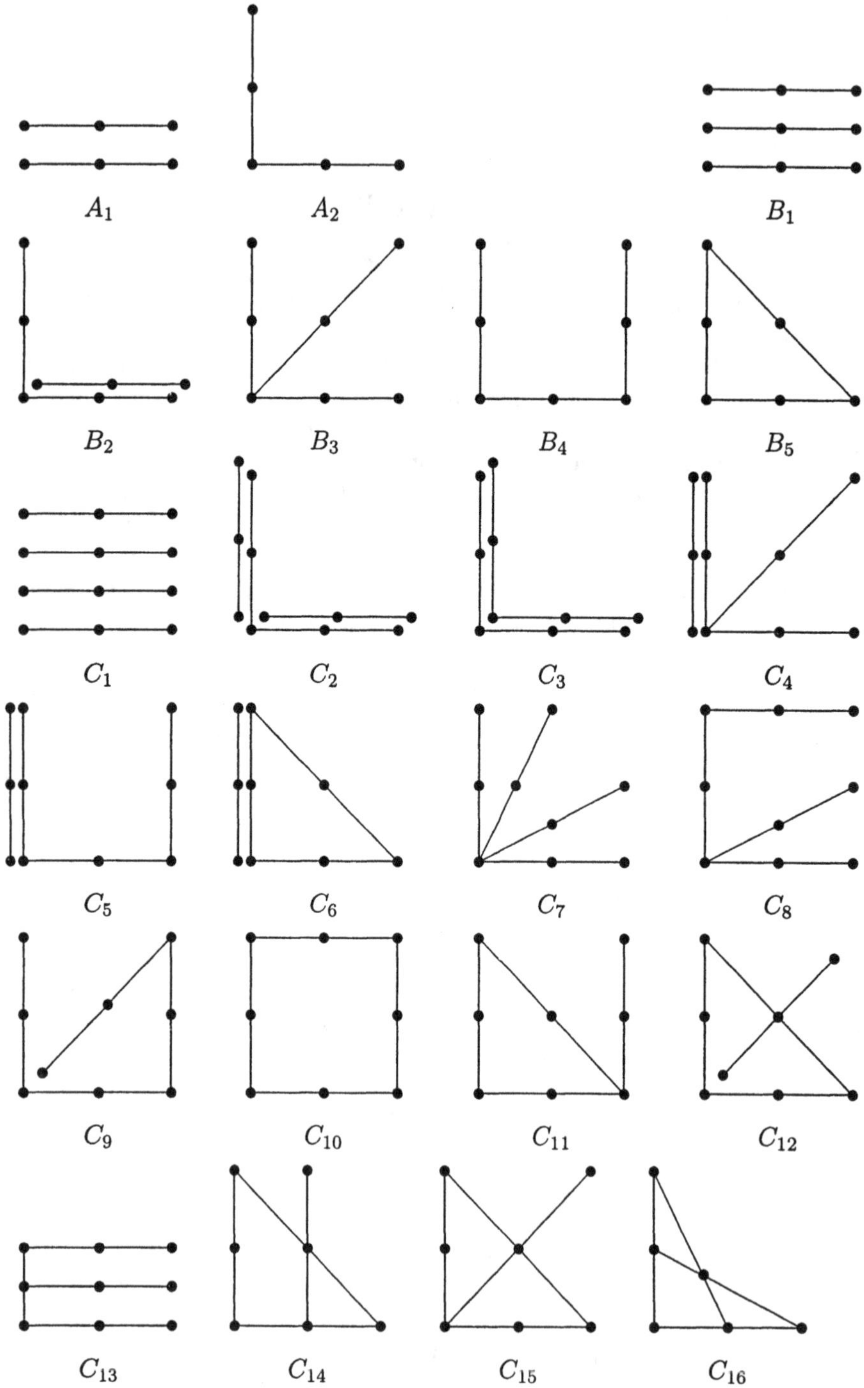

FIG. 13.1. Small configurations

$$
\begin{aligned}
c_4 &= n_v(v-7)(v-9)(v-11)/288 \\
c_7 &= n_v(v-5)(v-7)/384 \\
c_8 &= n_v(v-7)(v-9)/16 \\
c_{11} &= n_v(v-7)/4 \\
c_{15} &= n_v/6
\end{aligned}
$$

The number of occurrences of each variable 4-line configuration is determined by the number of occurrences of any single one of them. Consider, for instance, the *Pasch configuration* C_{16}. Let c_i denote the number of occurrences of the configuration C_i. As noted by Grannell, Griggs, and Mendelsohn (1995), we find:

$$
\begin{aligned}
c_1 &= n_v(v-9)(v-10)(v-13)(v^2-22v+141)/31104 + c_{16} \\
c_2 &= n_v(v-9)(v-10)(v^2-22v+129)/576 - 6c_{16} \\
c_3 &= n_v(v-9)^2(v-11)/128 + 3c_{16} \\
c_5 &= n_v(v-9)(v^2-20v+103)/48 + 12c_{16} \\
c_6 &= n_v(v-9)(v-10)/36 - 4c_{16} \\
c_9 &= n_v(v-9)^2/8 - 12c_{16} \\
c_{10} &= n_v(v-8)/8 + 3c_{16} \\
c_{12} &= n_v(v-9)/4 + 12c_{16} \\
c_{13} &= n_v(v^2-18v+85)/48 - 4c_{16} \\
c_{14} &= n_v/4 - 6c_{16}
\end{aligned}
$$

We can obtain these equations by considering extensions of 3-line configurations to 4-line configurations in all possible ways. For example, consider in B_4 the two lines containing only one point of degree 2. On each of the two lines choose one of the points of degree 1 and add the line joining them. This can be done in four ways. The resulting configuration is either C_{10} or C_{14}. Now each configuration C_{10} can be obtained from four distinct B_4 configurations, while each C_{14} can be obtained from two. This gives the equation $4c_{10} + 2c_{14} = 4b_4$. Similar considerations lead to a number of equations which reduce, in conjunction with the constant values for the b_is, to the relationships given.

In general, a set M of m-line configurations, $1 \leq m \leq n$, is a *generating set* for n-line configurations if, for each admissible order v, the number of occurrences of any n-line configuration can be expressed as a linear combination of the number of occurrences of the configurations in M. To treat all values of v simultaneously, these numbers of occurrences are expressed as polynomials in v. A minimal generating set is a *basis.* Thus, for 4-line configurations, any constant configuration together with the Pasch configuration forms a basis.

The *mitre* configuration (in Figure 13.4) is a 5-line configuration isomorphic to one with lines abc, ade, afg, bdf, and ceg. Horák, Phillips, Wallis, and Yucas (1997) establish that any constant configuration, together with the Pasch and mitre configurations, forms a basis for the 5-line configurations. This is a consequence of a more general theorem:

Theorem 13.1 *Any constant n-line configuration, together with all m-line configurations for $m \leq n$ having all points of degree at least 2, forms a generating set for the n-line configurations.*

Indeed the Pasch and mitre configurations are the only two configurations with five or fewer lines having all points of degree at least 2. The generating set consisting of any constant configuration, the Pasch, and the mitre, is easily seen to be minimal for 5-line configurations. Danziger, Grannell, Griggs, and Mendelsohn (1996) give explicit formulas for the numbers of each of the 56 5-line configurations.

Horák, Phillips, Wallis, and Yucas (1997) state:

Conjecture 13.2 *The generating set of Theorem 13.1 forms a basis for the n-line configurations.*

They verify this conjecture for $n = 6$, and Urland (1999) verified it for $n = 7$.

One open problem concerns the characterization of constant configurations. Let S_n, T_n, U_n, V_n, W_n be five n-line configurations obtained from the $(n-1)$-star by adding a line. Figure 13.2 shows the star S_n in which n lines meet at a single point, obtained by adding a line to the $(n-1)$-star which meets the remaining lines at their common point. Also shown are T_n, U_n, V_n, W_n obtained by adding a line which meets 0, 1, 2, or 3 other points of the $(n-1)$-star.

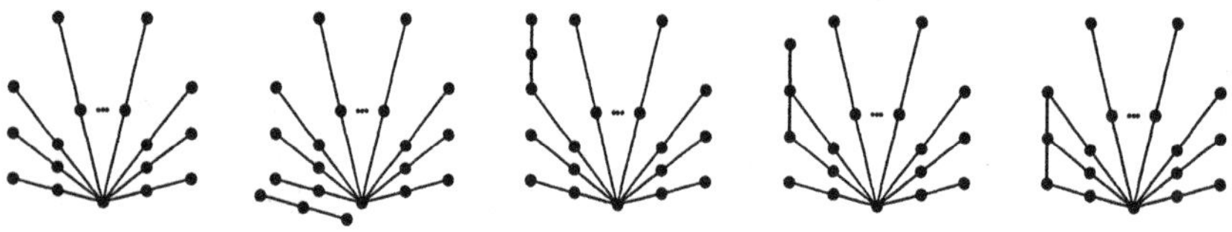

S_n T_n U_n V_n W_n

FIG. 13.2. Adding a line to a star

Theorem 13.3 *(Horák, Phillips, Wallis, and Yucas, 1997) For each $n \geq 4$, the configurations S_n, T_n, U_n, V_n, and W_n are constant.*

Proof We have seen that the statement holds for $n = 4$. Let $n \geq 5$, and let $(V, \mathcal{B})$ be an STS(v). Let F_n be one of the five n-line configurations, and F_{n-1} be the same type of configuration on $n-1$ lines. Let x be the centre of the star of F_{n-1}. For each configuration F_{n-1} of $(V, \mathcal{B})$ there are $(v - s - k)/2$ lines $\mathcal{L} \subset \mathcal{B}$ such that $F_{n-1} \cup L$ forms a configuration F_n when $L \in \mathcal{L}$, where s is the total number of points in F_{n-1} and k is the number of points of F_{n-1} not collinear with x. Any configuration F_n is obtained in this way from t different configurations F_{n-1} where t is the number of pendant lines of F_n containing x. Thus, if f_n is the number of occurrences of F_n, we find $(v - s - k)f_{n-1} = 2tf_n$. The result now follows by induction. □

From the proof of Theorem 13.3, the number of occurrences of these constant configurations can be computed explicitly:

$$\begin{aligned}
s_n &= n_v(v-5)(v-7)\cdots(v-2n+1)/(2^n n!)\\
t_n &= n_v(v-7)(v-9)\cdots(v-2n-3)/(3\cdot 2^n(n-1)!)\\
u_n &= n_v(v-7)(v-9)\cdots(v-2n-1)/(2^{n-1}(n-2)!)\\
v_n &= n_v(v-7)(v-9)\cdots(v-2n+1)/(2^{n-1}(n-3)!)\\
w_n &= n_v(v-7)(v-9)\cdots(v-2n+3)/(3\cdot 2^{n-3}(n-4)!)
\end{aligned}$$

Horák, Phillips, Wallis, and Yucas (1997) state:

Conjecture 13.4 *The five n-line configurations described here are the only constant configurations in Steiner triple systems.*

The maximum and minimum numbers of occurrences of each variable 4-line configuration is completely determined by the extreme numbers of possible Pasch configurations in an STS(v). Stinson and Wei (1993) have determined that the number of Pasch configurations cannot exceed $v(v-1)(v-3)/4$, and equality is attained if and only if the Steiner triple system is a projective space PG$(n-1,2)$ (and hence $v = 2^n - 1$). Equality in this bound therefore cannot be attained in general, and indeed the maximum value for other orders is not determined at present. See Gray and Ramsay (1998) for asymptotic results. The minimum number of Pasch configurations is zero; such quadrilateral-free STS(v)s are the subject of §13.2.

We can obtain a result for mitres that is similar to the result of Stinson and Wei (1993) for Pasch configurations:

Theorem 13.5 *The number of mitres in an STS(v) is at most $v(v-1)(v-3)(v-7)/24$, and equality is attained if and only if v is a power of 3 and the STS(v) is a Hall triple system.*

Proof In any STS(v) there are exactly $v(v-1)(v-3)(v-7)/72$ distinct pairs of disjoint lines (this is a constant configuration). Each pair of disjoint lines can appear in at most three distinct mitres, so we obtain the upper bound. Conversely, when equality is attained, every pair of disjoint lines must lie in three distinct mitres. So let $\{a,b,c\}$ and $\{u,v,w\}$ be two disjoint lines, and the three mitres in which they occur contain in addition the lines $\{a,u,x\},\{b,w,x\},\{c,v,x\}$; $\{a,w,y\},\{b,v,y\},\{c,u,y\}$; and $\{a,v,z\},\{b,u,z\},\{c.w.z\}$. Consider next the two disjoint lines $\{a,u,x\}$ and $\{b,v,y\}$. Some mitre contains these two lines and the lines $\{a,v,z\}$ and $\{b,u,z\}$ only if there is also a line $\{x,y,z\}$. Thus any two disjoint lines generate an STS(9), and so the system is indeed a Hall triple system or affine space. □

Just as in the case of Pasch configurations, the maximum number of mitres for other orders has not been determined yet. At the other extreme, a Steiner triple system may be *anti-mitre* (have no mitres); this is the topic of §13.3.

13.2 Avoidance: Pasch configurations

When an STS contains no Pasch configurations, it is *quadrilateral-free* (or *anti-Pasch*), and is denoted QFSTS(v).

Erdös (1976) conjectured that for every r there is an integer $v_0(r)$ so that for every $v > v_0(r)$, $v \equiv 1,3 \pmod 6$, there is a Steiner triple system on v elements with the property that for $2 \le j \le r$, no $j+2$ points carry j triples. Such an STS(v) is *r-sparse.* For $r = 4$, Erdös's conjecture states that for all $v > v_0(4)$ with $v \equiv 1, 3 \pmod 6$, a QFSTS(v) exists.

The unique STS(7), and both nonisomorphic STS(13)s, contain at least one quadrilateral, but Brouwer (1977) conjectures:

Conjecture 13.6 *For $v \equiv 1, 3 \pmod 6$, a QFSTS(v) exists unless $v \in \{7, 13\}$.*

Brouwer's conjecture remains far from settled, although substantial partial results are available which we develop here. Erdös's more general conjecture apparently remains completely open for $r \ge 6$. For $r = 5$, Brouwer (1977) purports to give one infinite family of QFSTS in which no seven points carry five triples, but there is an error in the construction; see Hartman and Phelps (1992). In §13.3.1, positive results for $r = 5$ are discussed.

For quadrilateral-free STSs, using affine triple systems and Robinson's characterization of quadrilaterals in Netto triple systems, we have:

Lemma 13.7 *A QFSTS(p^α) exists for $p = 3$ or p a prime with $p \equiv 19 \pmod{24}$, and α a nonnegative integer.*

A more general construction for prime powers was obtained by Brouwer (1977). For prime power order q, a TS(q, 1) $\mathcal{S}_C$ can be formed as follows. Let x be a primitive element of $\mathbb{F}_q$, $t = (q-1)/6$ and $y = x^{2t}$ so that $y^2 + y + 1 = 0$. Let C be a subset of $\mathbb{F}_q$ of size t with $0 \notin C$ and if $x^\alpha, x^\beta \in C$, then $\alpha \not\equiv \beta \pmod t$. The system $\mathcal{S}_C$ contains the triples $\{c\{1, y, y^2\} + i : c \in C, i \in \mathbb{F}_q\}$.

Brouwer (1977) proves that $\mathcal{S}_C$ contains a quadrilateral whenever, for some triple T, $\mathcal{S}_C$ contains the triple $2T$, and in addition $|3T| = 3$. Using this observation, he proved:

Theorem 13.8 *Let $q \equiv 1 \pmod 6$, $q = p^\alpha$ and p prime. Let C be as above. Then*

1. *if $p \in \{7, 13\}$, $\mathcal{S}_C$ contains a quadrilateral;*
2. *if $p \notin \{7, 13\}$, $\mathcal{S}_C$ contains a quadrilateral if and only if for some $S \in \mathcal{S}_C$, $2S \in \mathcal{S}_C$, where $S \ne 2S$.*

Corollary 13.9 *For $q \equiv 1 \pmod 6$, $q = p^\alpha$, $p \notin \{7, 13\}$ a prime, there is a QFSTS(q) whenever $p \equiv 1, 3 \pmod 8$ or $\alpha \equiv 0 \pmod 2$.*

Proof (Sketch) The conditions on p and α are equivalent to the statement that -2 is not a square in $\mathbb{F}_q$. Choose a set C so that $\mathcal{S}_C$ is invariant under multiplication by -2. □

Brouwer (1977) and Doyen (1981) obtained the first results on QFSTSs for orders other than prime powers. First, we recall the Bose construction for Steiner triple systems, Lemma 2.3. Let Γ be a commutative idempotent quasigroup of order $g = 2s + 1$ with binary operation $\odot$. Let $V = \Gamma \times \mathbb{Z}_3$, and write (x, i) as x_i. Let $\mathcal{B}$ contain the triples

1. for $x \in \Gamma$, include the triple $\{x_0, x_1, x_2\}$;

2. for $x, y \in \Gamma$, $x \neq y$, include the triples $\{x_i, y_i, (x \odot y)_{i+1}\}$ for $i \in \mathbb{Z}_3$ (reducing subscripts modulo 3 as necessary).

Then $(V, \mathcal{B})$ is a Steiner triple system. Brouwer (1977) and Doyen (1981) established that:

Theorem 13.10 *Whenever g is odd and $g \not\equiv 0 \pmod 7$, a QFSTS($3g$) exists.*

Proof (Sketch) Form a triple system $(V, \mathcal{B})$ using the Bose construction just described, based on the commutative idempotent quasigroup on $\mathbb{Z}_g$ with $x \odot y = (x+y)/2$. A quadrilateral arises in the Bose construction whenever the latin square of the quasigroup contains a subsquare of order two; since g is odd, none of this type is present. The second type of quadrilateral requires that $x \odot (x \odot (x \odot y)) = y$ for distinct symbols x, y. By construction, this requires $\frac{7}{8}x + \frac{1}{8}y = y$, or $7(x-y) = 0$; but $g \not\equiv 0 \pmod 7$, and hence no quadrilaterals of this second type are present. □

Brouwer (1977) and Griggs, Murphy, and Phelan (1990) extended this result:

Theorem 13.11 *For all $v \equiv 3 \pmod 6$, there exists a QFSTS(v) with a parallel class of triples.*

Proof If $v \not\equiv 0 \pmod 7$, the Bose construction yields the desired QFSTS with triples of the form $\{x_0, x_1, x_2\}$ giving a parallel class. There is a QFSTS(21) that has a parallel class of triples; one such is given by the starter blocks

$$\{\{0,1,3\}, \{0,4,12\}, \{0,6,11\}, \{0,7,14\}\}.$$

The parallel class is the 'short orbit' $\{0,7,14\} \pmod{21}$.

Now suppose that $v \equiv 3 \pmod 6$, and that $v = 7u$. Then proceeding inductively there is a QFSTS(u) $\mathcal{B}$ having a parallel class $\mathcal{P}$ on a u-set U. Let $V = \mathbb{Z}_7 \times U$. For each $T = \{a, b, c\} \in \mathcal{S} \setminus \mathcal{P}$, $i, j \in \mathbb{Z}_7$, include the triple $\{(i,a), (j,b), (i+j,c)\}$ (arithmetic mod 7). For each $T \in \mathcal{P}$, take the triples of the QFSTS(21) with parallel class on $\mathbb{Z}_7 \times T$. It is then easy to check that the result is a QFSTS(v) with a parallel class. □

Grannell, Griggs, and Phelan (1988) established that the Schreiber–Wilson construction also yields QFSTSs under certain restrictions. First we recall a restricted form of the construction from §2.6 for the case when for every prime divisor p of $n = v-2$ the order of $-2 \pmod p$ is singly even. To form an STS(v), write $n = v-2$ and choose an abelian group Γ of order n with binary operation $\oplus$ and identity 0. Since $n \equiv 1, 5 \pmod 6$, Γ has no elements of order two or three. Now consider all triples of group elements $\{a, b, c\}$ with $a \oplus b \oplus c = 0$. Let $\mathcal{D}$ be the set of such triples in which all three group elements are distinct, and let $\mathcal{T}$ be those in which there are two distinct elements. Finally, let $\mathcal{S}$ be those containing a single element; in Γ, $a \oplus a \oplus a = 0$ has the unique solution $a = 0$, since Γ has no elements of order three. Thus $\mathcal{S}$ contains the single triple $\{0,0,0\}$.

Now to form a triple system $(\Gamma \cup \{\alpha, \beta\}, \mathcal{B})$,

Order	Number	Quadrilateral-free	Anti-mitre	5-Sparse
7	1	0	1	0
13	1	0	1	0
15	2	1	1	0
19	4	2	2	1
21	7	1	2	0
25	12	3	3	0
27	8	1	2	0
31	80	13	15	0
33	84	15	6	1
37	820	157	51	4
39	798	112	46	1
43	9508	1279	257	28
45	11616	1485	130	13
49	157340	13936	2430	155
51	139828	12429	1597	104
55	3027456	241971	21877	1197
57	2353310	165527	17523	843

Table 13.1 *Summary for small cyclic STS(v)s*

1. include all triples of $\mathcal{D}$ in $\mathcal{B}$;
2. first replace every unordered pair $\{a,b\}$ by the ordered pair (a,b) if $a \oplus a \oplus b = 0$, and by (b,a) otherwise. Partition the pairs of $\mathcal{T}$ into orbits under the mapping $\phi : z \mapsto z'$ where $z \oplus z' \oplus z' = 0$ (if $(a,b) \in \mathcal{T}$, $a = \phi(b)$). Now consider an orbit of pairs of $\mathcal{T}$ under ϕ. Under the stated restriction that for every prime divisor p of n, the order of $-2 \pmod p$ is singly even, each orbit of elements from $\mathcal{T}$ is of even length. Suppose that the orbit has length $2k$, and that the pair (a,b) is in the orbit, add to $\mathcal{B}$ the triples $\{\{\alpha, \phi^{2i}(a), \phi^{2i+1}(a)\}, \{\beta, \phi^{2i+1}(a), \phi^{2i+2}(a)\}\}$ for $0 \le i < k$.
3. include the triple $\{\alpha, \beta, 0\}$ in $\mathcal{B}$.

It is a somewhat tedious case analysis to verify that this system has no quadrilaterals, to obtain the theorem of Grannell, Griggs, and Phelan (1988):

Theorem 13.12 *Let $p_1, \ldots, p_s$ be primes so that for $1 \le i \le s$, -2 has singly even order* $\pmod{p_i}$. *Then there exists a QFSTS*$(2 + \prod_{i=1}^{s} p_i)$.

Griggs, Murphy, and Phelan (1990) and Stinson and Wei (1992) observe that the direct product construction for triple systems does not introduce quadrilaterals:

Theorem 13.13 *If there exists a QFSTS(u) and a QFSTS(v), then there exists a QFSTS(uv).*

Chee and Lim (1989) give cyclic QFSTS(v)s for $v \in \{31, 37\}$, and subsequently Colbourn, Mendelsohn, Rosa, and Širáň (1994) did an exhaustive search of inequivalent cyclic STSs for small orders. Table 13.1 summarizes their results. In the process, they also proved:

Theorem 13.14 *For $19 \leq v \leq 99$ and $v \equiv 1, 3 \pmod 6$, there is an STS(v) which is 5-sparse (and hence quadrilateral-free and anti-mitre) except possibly when $v \in \{21, 25, 27, 31\}$. In these cases, there is an STS(v) which is quadrilateral-free but not anti-mitre, and an STS(v) which is anti-mitre but not quadrilateral-free.*

13.2.1 *Stinson and Wei's construction*

Ling, Colbourn, Grannell, and Griggs (1999) extend a singular direct product construction of Stinson and Wei (1993); we examine the construction here. It employs a certain type of N_2-latin square (see Theorem 1.5).

We need N_2-latin squares with additional properties, similar to (but weaker than) the 'special' latin squares used by Stinson and Wei (1993). A *Kotzig square* of order $n = 2w$ is an N_2-latin square L of order $2w$ with rows, columns, and symbols indexed by $\{0, 1, \ldots, n-1\}$, and enjoying three properties:

1. $\{L(2i, s), L(2i+1, s)\} \neq \{2j, 2j+1\}$ for $0 \leq i, j < w$;
2. $\{L(s, 2i), L(s, 2i+1)\} \neq \{2j, 2j+1\}$ for $0 \leq i, j < w$;
3. $L(2i, 2j) \neq L(2i+1, 2j+1)$, $L(2i, 2j+1) \neq L(2i+1, 2j)$ for $0 \leq i, j < w$.

Stinson and Wei (1993) used similar N_2-latin squares to prove:

Theorem 13.15 *If there exists a QFSTS(u) and $u \equiv 1 \pmod 4$, and $u-1$ has an odd divisor exceeding three, then there exists a QFSTS$(3(u-1)+1)$.*

Ling, Colbourn, Grannell, and Griggs (1999) extend Theorem 13.15 by relaxing the condition that $u \equiv 1 \pmod 4$, and the condition on divisors. Kotzig squares are used in this extension, so we state their existence result:

Lemma 13.16 *There exists a Kotzig square of order $2w$ whenever $w \geq 4$.*

Theorem 13.17 *If there exist a QFSTS(v) and a QFSTS(u), and $u > 3$, then there exists a QFSTS$(v(u-1)+1)$.*

Proof Take a QFSTS$(v) = (V, \mathcal{B})$ and let $I_{u-1} = \{0, 1, \ldots, u-2\}$. We construct a QFSTS$(v(u-1)+1)$ on the set $\{\infty\} \cup (V \times I_{u-1})$. For each $a \in V$, put a QFSTS(u) on $V_a = \{\infty\} \cup (\{a\} \times I_{u-1})$. Without loss of generality, we can stipulate that this STS contains the blocks $\{\infty, a_{2i}, a_{2i+1}\}$ for $0 \leq i < (u-1)/2$. For each block $\{a, b, c\} \in \mathcal{B}$, fix the order of the elements and put the TD$(3, u-1)$ which arises from a Kotzig square of order $u-1$ on $\{a, b, c\} \times I_{u-1}$. We claim that this produces a QFSTS$(v(u-1)+1)$. It is easy to see that we obtain an STS$(v(u-1)+1)$, so we concentrate on proving it is quadrilateral-free.

First of all, if there is a Pasch configuration in the STS which involves the point ∞, then for some $a, b \in V$ two of the blocks must come from V_a, V_b. If $a = b$ then these and the other two blocks come from the the same QFSTS(u), a contradiction. So $a \neq b$ and the other two blocks come from the TD associated with the block in $\mathcal{B}$ which contains the pair $\{a, b\}$. A case analysis in Ling, Colbourn, Grannell, and Griggs (1999) establishes that this supposition produces a contradiction.

Hence any possible Pasch configuration contains at most one block from any of the V_as. If it contained one such block, then it would necessarily have the form

$$\{a_i, a_j, a_k\},\ \{a_i, b_l, c_m\},\ \{a_j, b_l, D\},\ \{a_k, D, c_m\}.$$

From the third block we have $D = c_f$ and from the fourth $D = b_g$, a contradiction.

Thus any Pasch configuration must have all four blocks from the TDs. It therefore has the form

$$\{a_i, b_j, c_k\},\ \{a_i, y_m, z_n\},\ \{x_l, b_j, z_n\},\ \{x_l, y_m, c_k\},$$

where the blocks obtained from these by deleting the subscripts lie in $\mathcal{B}$. Since there is no quadrilateral in $\mathcal{B}$, at least two (and consequently all four) of these resulting blocks are identical; indeed $x = a$, $y = b$, and $z = c$. But then the supposed quadrilateral comprises

$$\{a_i, b_j, c_k\},\ \{a_i, b_m, c_n\},\ \{a_l, b_j, c_n\},\ \{a_l, b_m, c_k\},$$

and consideration of the subscripts establishes that the underlying latin square has a subsquare of order 2, a final contradiction. □

13.2.2 *Bose-type constructions*

Ling, Colbourn, Grannell, and Griggs (1999) establish:

Theorem 13.18 *Let* $2s + 1$ *be an odd positive integer that is not divisible by* 3 *or* 7. *If there exists a* QFSTS$(6n+1)$, *then there exists a* QFSTS$(6n(2s+1)+1)$.

Proof Take $2s+1$ copies of the QFSTS$(6n+1)$, one on each of the point sets V_i where

$$V_i = \{\infty, (0, i_0), (0, i_1), (0, i_2), (1, i_0), (1, i_1), (1, i_2), \ldots$$
$$\ldots, (2n-1, i_0), (2n-1, i_1), (2n-1, i_2)\},$$

$0 \le i \le 2s$. The corresponding sets of blocks C_i are chosen so that the triples containing ∞ are

$$\{\infty, (0, i_0), (0, i_1)\}, \{\infty, (0, i_2), (1, i_0)\}, \{\infty, (1, i_1), (1, i_2)\},$$

$$\{\infty, (2, i_0), (2, i_1)\}, \ldots, \{\infty, (2n-1, i_1), (2n-1, i_2)\}.$$

Next take an abelian group $(G, \circ)$ of order $2s+1$ represented on $G = \{0, 1, \ldots, 2s\}$ with identity I and define $x * y$ for $x, y \in G$ to be the unique $z \in G$ for which $z \circ z = x \circ y$. Note that $x, y, x * y$ are either all equal or all distinct.

Let N be an N_2-latin square of order $2n$ represented on $\{0, 1, \ldots, 2n-1\}$; such squares exist for all $n \ge 3$ by Theorem 1.5, a condition which is required for the existence of a QFSTS$(6n+1)$.

Let $\mathcal{L}$ be the set of triples of the form $\{(a, i_\alpha), (b, j_\alpha), (c, (i * j)_{\alpha+1})\}$ where $a, b, c \in \{0, 1, \ldots, 2n-1\}, 0 \leq i < j \leq 2s$, subscript arithmetic is modulo 3 and $c = N(a, b)$.

$$\text{Put} \quad V = \bigcup_{i=0}^{2s} V_i \quad \text{and} \quad \mathcal{B} = \left(\bigcup_{i=0}^{2s} C_i\right) \cup \mathcal{L}.$$

We claim that $(V, \mathcal{B})$ is a QFSTS$(6n(2s+1)+1)$. To verify this,

$$|C_i| = n(6n+1) \quad \text{and} \quad |\mathcal{L}| = 4n^2 \cdot 3 \cdot s(2s+1).$$

Hence $|\mathcal{B}| = n(2s+1)[6n(2s+1)+1]$, the number of blocks required to form an STS$(6n(2s+1)+1)$. Moreover, any pair of points from V which contains ∞ lies in a block in one of the C_is as does any pair of the form $\{(a, i_\alpha), (b, i_\beta)\}$. All remaining pairs are of the form $\{(a, i_\alpha), (b, j_\beta)\}$ with $i < j$. We now identify the third element (c, k_γ) of the triple from $\mathcal{L}$ which contains this pair.

1. If $\beta = \alpha$, then $\gamma = \alpha + 1$, $k = i * j$ and $c = N(a, b)$.
2. If $\beta = \alpha + 1$, then $\gamma = \alpha$ and k is chosen so that $i * k = j$. If $k > i$ then c is chosen so that $b = N(a, c)$; if $k < i$ then c is chosen so that $b = N(c, a)$.
3. If $\beta = \alpha + 2$, then $\gamma = \alpha + 2$ and k is chosen so that $j * k = i$. If $k > j$ then c is chosen so that $a = N(b, c)$; if $k < j$ then c is chosen so that $a = N(c, b)$.

It remains to prove that $\mathcal{B}$ is quadrilateral-free. Since none of the C_is contains a quadrilateral, any possible quadrilateral in $\mathcal{B}$ must comprise either:

(a) a triple from C_i, another from C_j, $i \neq j$, plus two from $\mathcal{L}$, or
(b) a triple from C_i, plus three from $\mathcal{L}$, or
(c) four triples from $\mathcal{L}$.

Ling, Colbourn, Grannell, and Griggs (1999) establish a contradiction in each case. □

Theorem 13.19 *If there exist a QFSTS$(6n+1)$, $n \geq 4$, and a QFSTS$(6s+3)$, the latter having a parallel class, then there exists a QFSTS$(6n(2s+1)+1)$.*

Proof Take $2s+1$ copies of the QFSTS$(6n+1)$, one on each of the point sets V_i where

$$V_i = \{\infty, (0, i_0), (0, i_1), (0, i_2), (1, i_0), (1, i_1), (1, i_2), \ldots \\ \ldots, (2n-1, i_0), (2n-1, i_1), (2n-1, i_2)\},$$

$0 \leq i \leq 2s$. The corresponding sets of blocks C_i are chosen so that the triples containing ∞ are

$$\{\infty, (0, i_0), (1, i_0)\}, \{\infty, (0, i_1), (1, i_1)\}, \\ \{\infty, (0, i_2), (1, i_2)\}, \{\infty, (2, i_0), (3, i_0)\}, \\ \{\infty, (2, i_1), (3, i_1)\}, \{\infty, (2, i_2), (3, i_2)\}, \\ \vdots \\ \{\infty, (2n-2, i_0), (2n-1, i_0)\}, \{\infty, (2n-2, i_1), (2n-1, i_1)\}, \\ \{\infty, (2n-2, i_2), (2n-1, i_2)\}.$$

Take the QFSTS$(6s+3)$ on the points $\{i_\alpha : 0 \le i \le 2s, \quad \alpha = 0, 1, \text{ or } 2\}$ with the parallel class $\mathcal{P} = \{\{i_0, i_1, i_2\} : 0 \le i \le 2s\}$. Let $\mathcal{B}_s$ denote the set of blocks forming this system.

Let L be a Kotzig square of side $2n$ represented on $\{0, 1, \ldots, 2n-1\}$, and take $\mathcal{L}$ to be the set of triples of the form $\{(a, i_\alpha), (b, j_\beta), (c, k_\gamma)\}$ where $\{i_\alpha, j_\beta, k_\gamma\} \in \mathcal{B}_s \backslash \mathcal{P}, i < j < k, \quad a, b, c \in \{0, 1, \ldots, 2n-1\}$ and $c = L(a, b)$.

$$\text{Put} \quad V = \bigcup_{i=0}^{2s} V_i \quad \text{and} \quad \mathcal{D} = \left(\bigcup_{i=0}^{2s} \mathcal{C}_i\right) \cup \mathcal{L}.$$

The verification that $(V, \mathcal{D})$ is a QFSTS$(6n(2s+1)+1)$ is given by Ling, Colbourn, Grannell, and Griggs (1999). □

Using also Theorem 13.11, we may state:

Corollary 13.20 *If there exists a QFSTS$(6n+1)$, $n \ge 4$, then for any odd integer m there exists a QFSTS$(6nm+1)$.*

13.2.3 *Lu's construction*

We employ a variant of a construction of Lu (1983), from Ling, Colbourn, Grannell, and Griggs (1999):

Theorem 13.21 *If there exist a QFSTS$(m+2)$ and a QFSTS$(n+2)$, then there exists a QFSTS$(mn+2)$.*

Proof Let $(Z_m \cup \{a, b\}, \mathcal{A})$ be a QFSTS$(m+2)$ with $\{a, b, 0\} \in \mathcal{A}$, and let $(Z_n \cup \{a, b\}, \mathcal{B})$ be a QFSTS$(n+2)$. To avoid trivialities we may assume that $m, n > 1$. Let $N_{ab} = \{\{x_i, x_j\} : x_i, x_j \in Z_m \backslash \{0\}, \ell \in \{a, b\}$ and $\{\ell, x_i, x_j\} \in \mathcal{A}\}$. N_{ab} is a set of pairs on $Z_m \backslash \{0\}$ with every element appearing in two pairs. Each pair can then be ordered so that each element is the first element of one pair, and the second element of another; call this set of ordered pairs Q_{ab}. Define a permutation π on $Z_m \backslash \{0\}$ by setting $\pi(i) = j$ whenever $(i, j) \in Q_{ab}$. Subsequently, it is crucial that since $\{a, b, 0\}$, $\{a, i, \pi(i)\}$, and $\{b, \pi(i), \pi^2(i)\}$ appear in $\mathcal{A}$ (or the three blocks obtained by interchanging a and b appear in $\mathcal{A}$), no block of the form $\{0, i, \pi^2(i)\}$ can appear in $\mathcal{A}$ since it is quadrilateral-free.

We construct an STS$(mn+2)$ on the point set $(Z_m \times Z_n) \cup \{a, b\}$ with triples of the following forms where $x_1, x_2, x_3 \in Z_m$ and $y_1, y_2, y_3 \in Z_n$.

(i) $\{(0, y_1), (0, y_2), (0, y_3)\}$ if $\{y_1, y_2, y_3\} \in \mathcal{B}$
$\{\ell, (0, y_2), (0, y_3)\}$ if $\{\ell, y_2, y_3\} \in \mathcal{B}$ and $\ell \in \{a, b\}$
$\{a, b, (0, y_3)\}$ if $\{a, b, y_3\} \in \mathcal{B}$

(ii) $\{(x_1, y_1), (x_1, y_2), (x_2, y_3)\}$ where $(x_1, x_2) \in Q_{ab}$ and $y_1 + y_2 \equiv 2y_3 \pmod{n}$.

(iii) $\{\ell, (x_1, y_1), (x_2, y_1)\}$ where $\ell \in \{a, b\}$ and $\{\ell, x_1, x_2\} \in \mathcal{A}$.

(iv) $\{(x_1, y_1), (x_2, y_2), (x_3, y_3)\}$ where $\{x_1, x_2, x_3\} \in \mathcal{A}$, $x_1 < x_2 < x_3$ and $y_1 + y_2 + y_3 \equiv 0 \pmod{n}$.

First of all, we prove that the construction gives an STS($mn+2$). The number of type (i) blocks is $(n+2)(n+1)/6$. The number of type (ii) blocks is $(m-1)n(n-1)/2$. The number of type (iii) blocks is $(m-1)n$. The number of type (iv) blocks is $[((m+2)(m+1)/6)-m)]n^2$. So the total number of blocks is $(mn+2)(mn+1)/6$ as expected. Therefore, it suffices to show that every pair of points is contained in a triple. All possibilities are exhausted as follows:

(1) Pairs $\{a,b\}$, $\{a,(0,y_1)\}$, $\{b,(0,y_2)\}$, and $\{(0,y_1),(0,y_2)\}$ are contained in some type (i) triple.
(2) When $x \in Z_m \backslash \{0\}$, $\{(x,y_1),(x,y_2)\}$ is contained in some type (ii) triple, since x is the first element of some pair in Q_{ab} and, n being odd, the equation $y_1+y_2 \equiv 2y_3 \pmod n$ has a unique solution for y_3.
(3) Pairs $\{a,(x,y)\}$ and $\{b,(x,y)\}$ for $x \in Z_m \backslash \{0\}$ are contained in some type (iii) triple.
(4) If $\{x,x'\} \in N_{ab}$, then $\{(x,y_1),(x',y_2)\}$ for $x \neq x'$ is contained in a type (ii) triple when $y_1 \neq y_2$, or a type (iii) triple when $y_1 = y_2$. If $\{x,x'\} \notin N_{ab}$, then $\{(x,y_1),(x',y_2)\}$ for $x \neq x'$ is contained in some type (iv) triple.

To complete the proof, Ling, Colbourn, Grannell, and Griggs (1999) show that the STS($mn+2$) is quadrilateral-free by a lengthy case analysis. □

13.2.4 *GDD constructions*

Ling, Colbourn, Grannell, and Griggs (1999) explore further constructions involving GDDs. A TD$(3,n)$ without any sub-TD$(3,2)$ is equivalent to an N_2-latin square of order n. We call such a TD$(3,n)$ an N_2-TD$(3,n)$.

Theorem 13.22 *If there exist a QFSTS($2v+1$) and a QFSTS($2n+1$), and $n>4$, then there exists a QFSTS($2vn+1$).*

Proof This is an application of the Fundamental Construction of Wilson (Theorem 3.3). Delete a point from the QFSTS($2v+1$) to form a 3-GDD of type 2^v. Replace each remaining point x by n points $x_1, x_2, \ldots, x_n$ and then use an N_2-TD$(3,n)$ to produce a 3-GDD of type $(2n)^v$. Add one new point ∞ and, on each group together with ∞, place a copy of the QFSTS($2n+1$) so that when $\{\infty, a_i, b_j\}$ is a triple, then $a \neq b$. Call the triples of the 3-GDD of type $(2n)^v$ *vertical*, and the triples of the STS($2n+1$)s *horizontal*. Points and blocks from the same copy of the STS($2n+1$) are said to be on the same *level*. The resulting set of triples forms an STS($2vn+1$), which we prove is quadrilateral-free. Suppose to the contrary that a Pasch configuration P is present.

Suppose first that P contains ∞. Then P contains at least two horizontal blocks. Since the STS($2n+1$) used is quadrilateral-free these two horizontal blocks cannot be from the same level. Hence P must contain two horizontal and two vertical blocks. But then the STS($2v+1$) would have to contain a Pasch configuration. Hence P cannot contain ∞.

Suppose next that P contains two horizontal blocks which do not contain ∞. These must intersect in a common point and consequently they and the remaining two blocks must be from the same level. But then the STS$(2n+1)$ would have to contain a Pasch configuration. Hence P contains at most one horizontal block.

Suppose now that P contains precisely one horizontal block. Without loss of generality this block has the form $\{a_i, a_j, b_k\}$ (where b may or may not equal a) and there are vertical blocks $\{a_i, c_l, d_m\}$ and $\{a_j, c_l, e_p\}$. But then d_m and e_p come from the same level and so the fourth block must be horizontal, a contradiction.

Finally, suppose that P contains four vertical triples. If all six points come from different levels then the STS$(2v+1)$ would contain a Pasch configuration. This would also be the case if just two of the six points came from the same level (with the remaining four from different levels). If two points come from one level and a further two from another level then the third pair must also share a common level, which is impossible since the STS$(2v+1)$ used was quadrilateral-free and the TD used was an N_2-TD. □

Ling, Colbourn, Grannell, and Griggs (1999) give one more recursive construction using GDDs which we state here without proof.

Theorem 13.23 *If there exist a QFSTS$(2v+1)$, a QFSTS$(4n+1)$, and a QFSTS$(2n(v-1)+1)$, and if $n, v > 4$, then there exists a QFSTS$(2n(3v-1)+1)$.*

13.2.5 *The current status*

Using the constructions given, we can evidently produce QFSTS(v)s for a wide range of admissible values v. Such systems are already known to exist for all $v \equiv 3 \pmod 6$ and so interest naturally focuses on $v \equiv 1 \pmod 6$. We show first how to cover the residue classes $v \equiv 1$ or $7 \pmod{18}$, $v \neq 7$. Subdivision of the remaining residue class, $v \equiv 13 \pmod{18}$, yields further progress. However, the existing constructions for QFSTS(v)s do not appear to be capable of establishing Conjecture 13.6. We show that the existence of QFSTS(v)s in the special cases $v = 6p+1, p \equiv 5 \pmod 6$, a prime, and $v = 12p+1, p \equiv 1 \pmod 6$, a prime, would suffice to complete a proof of the conjecture. We make use of some specific QFSTS(v)s for $v < 100$, all from Colbourn, Mendelsohn, Rosa, and Širáň (1994).

Theorem 13.24 *If $v \equiv 1$ or $7 \pmod{18}$ and $v \neq 7$ then a QFSTS(v) exists.*

Proof We treat two cases, according to the congruence class of v modulo 18.
Case 1 v ≡ 7 (mod 18). Suppose that $v = 18s+7$ and $s \geq 1$. Since there exists a QFSTS$(6s+3)$, an application of Theorem 13.17 with $v = 3$ gives a QFSTS$(18s+7)$.
Case 2 v ≡ 1 (mod 18). A QFSTS$(54s+19)$ for $s \geq 1$ is obtained by reapplying Theorem 13.17 with $v = 3$. Since there exists a QFSTS(19), we may assert the existence of QFSTS$(54s+19)$s for $s \geq 0$. Next, taking a QFSTS(37) and using Theorem 13.22 with $n = 18$ and $v = 3s+1$ we obtain a QFSTS$(108s+37)$ for $s \geq 0$. Taking a QFSTS(55), we may apply Theorem 13.19 with $n = 9$ to obtain a QFSTS$(108s+55)$ for $s \geq 0$. Again taking a QFSTS(19) and using Theorem 13.18

with $n = 3$ and $s = 3t + 2$ ($t \geq 0$), we obtain a QFSTS($108s + 91$) except in the case when $7|(6t+5)$, i.e. when $t = 7u+5$ (corresponding to a QFSTS($756u+631$)). To deal with this exception we may use Theorem 13.19 with $n = 21$; this requires a QFSTS(127) which may be obtained from Theorem 13.17 with $v = 3$ using a QFSTS(43). We obtain from Theorem 13.19 a QFSTS($252s+127$) ($s \geq 0$) which covers the case of QFSTS($756u + 631$).

At this point it is appropriate to summarize the residue classes modulo 108 which are covered. These are 19, 37, 55, 73, and 91. To deal with the remaining case, $v \equiv 1 \pmod{108}$, we use induction. First take a QFSTS(37) and apply Theorem 13.17 with $v = 3$ to obtain a QFSTS(109). Next assume that QFSTS(v)s exist for all $v \equiv 1 \pmod{18}$ satisfying $19 \leq v \leq v_0$. (We have established this for $v_0 = 109$.) Now choose $v' \equiv 1 \pmod{18}$ satisfying $19 \leq v' \leq 3v_0$. If $v' \equiv$ 19, 37, 55, 73, or 91 (mod 108), we already have a QFSTS(v'), so suppose that $v' = 108s + 1$. Put $u = (v' + 2)/3 = 36s + 1$. Then $u \leq v_0$ and $u \equiv 1$ (mod 18) and so, by assumption, there exists a QFSTS(u). Theorem 13.17 with $v = 3$ then gives a QFSTS($3u - 2$), i.e. a QFSTS(v'). It now follows by induction that a QFSTS(v) exists for all $v \equiv 1 \pmod{18}$, $v \geq 19$. □

Theorem 13.25 *If* $v \equiv 49 \pmod{72}$ *then there exists a QFSTS*(v).

Proof Write $v-1 = 24(3s+2)$ for some integer s. Thus $v-1 = 3 \cdot 2^{a+3} \cdot (2t+1)$ where $a, t \geq 0$. Suppose first that $t = 0$. The cases $a = 0$ and $a = 1$ then correspond to a QFSTS(25) and a QFSTS(49) respectively, which are known to exist. Next apply Theorem 13.22 with ingredients QFSTS(9) and QFSTS($3 \cdot 2^{b+3} + 1$) (noting $3 \cdot 2^{b+3} > 4$ if $b \geq 0$) to form a QFSTS($3 \cdot 2^{b+5} + 1$). Proceeding inductively from the cases $b = 0$ and $b = 1$, we obtain QFSTS($3 \cdot 2^{b+3} + 1$)s for all $b \geq 0$. Now suppose that $t > 0$. Applying Corollary 13.20 with $n = 2^{a+2}$ we obtain a QFSTS($6n(2t + 1) + 1$), i.e. a QFSTS(v). □

Theorem 13.26 *If* $v \equiv 31 \pmod{36}$ *and* v *is not of the form* $v = 6p + 1$, *where* p *is a prime, then there exists a QFSTS*(v).

Proof If $v \equiv 31 \pmod{36}$ then $v - 1 = 6p_1p_2 \cdots p_k$, where each p_i is an odd prime, at least 5, and an odd number of these primes satisfy $p_i \equiv 5 \pmod 6$. From the assumption in the statement of the theorem, $k \geq 2$, and we may also assume that $p_k \equiv 5 \pmod 6$. But then $3p_k \equiv 3 \pmod 6$ and $2p_1p_2 \cdots p_{k-1}+1 \equiv 3 \pmod 6$. Applying Theorem 13.17, we obtain a QFSTS(v). □

Theorem 13.27 *If* $v \equiv 13 \pmod{72}$ *and* $v - 1$ *has a factor of the form* $6t + 5$, *then there exists a QFSTS*(v).

Proof If $v \equiv 13 \pmod{72}$ then $v - 1 = 12p_1p_2 \cdots p_k$. Since $v \neq 13$, each p_i is an odd prime, at least 5, and the number of these primes satisfying $p_i \equiv 5$ (mod 6) is even. From the assumption in the statement of the theorem, $k \geq 2$ and one of the primes is congruent to 5 (mod 6), say $p_k \equiv 5 \pmod 6$. But then $3p_k \equiv 3 \pmod 6$ and $4p_1p_2 \cdots p_{k-1} + 1 \equiv 3 \pmod 6$. Applying Theorem 13.17 we obtain a QFSTS(v). □

Theorem 13.28 *(Ling, Colbourn, Grannell, and Griggs, 1999)*

1. *Suppose that* $v \equiv 1, 3 \pmod 6$ *and* $v \not\equiv 13, 31, 67 \pmod{72}$. *Then there exists a QFSTS(v) provided that* $v \notin \{7, 13\}$.
2. *Suppose that it could be shown that there exists a QFSTS(v) for all* $v = 6p+1$ *where* $p \equiv 5 \pmod 6$ *is a prime, and for all* $v = 12p+1$ *where* $p \equiv 1 \pmod 6$ *is a prime. Then it would follow that there exists a QFSTS(v) if and only if* $v \equiv 1$ *or* $3 \pmod 6$ *and* $v \neq 7$ *or* 13.

Proof Theorems 13.24, 13.25, 13.26, and 13.27 establish the first statement. Now consider the second statement. When $v \equiv 31 \pmod{36}$, a QFSTS(v) exists either as a consequence of the assumption in the statement of the theorem or as a consequence of Theorem 13.26.

To deal with the case $v \equiv 13 \pmod{72}$, $v \neq 13$, Theorem 13.27 produces a QFSTS(v) unless $v - 1$ has no factor of the form $6t + 5$. But then $v - 1 = 12p_1p_2 \cdots p_k$ where each p_i is an odd prime, at least 7, and all of these primes are congruent to 1 (mod 6). If $k = 1$ then the assumption in the statement of the theorem gives a QFSTS(v). If $k \geq 2$ then take a QFSTS($12p_1 + 1$) and apply Corollary 13.20 with $n = 2p_1$ and $m = p_2p_3 \cdots p_k$. This produces a QFSTS($12p_1p_2 \cdots p_k + 1$), i.e. a QFSTS(v). □

Theorem 13.28 essentially investigates the factorization of $v - 1$. It is possible to obtain similar (but more complicated) criteria which would ensure the existence of QFSTS(v) by investigating either the factorization of v itself or of $v - 2$. In the case of factoring v, Theorem 13.13 is used. In the case of factoring $v - 2$, Theorem 13.21 is used. It is also possible to combine these criteria with those of Theorem 13.28 and with other constructions for QFSTS(v). We may thereby narrow the range of values of v, needed as the basis of an inductive argument, from the $6p + 1$ and $12p + 1$ of Theorem 13.28. However, this range still (apparently) remains infinite.

13.2.6 *Erasure codes*

Hellerstein, Gibson, Karp, Katz, and Patterson (1994) describe a family of codes for handling erasures in a redundant array of inexpensive disks (RAID). The codes are linear, systematic, constant weight codes which have sparse parity-check matrices (typically, small constant row and column sums). The essential defining property for such a code is that the modulo 2 sum of any t columns be nonzero, for small values of t. When rows are indexed by the elements of a design and columns form the characteristic vectors of blocks of size three, a modulo 2 sum of zero arises from two identical triples. If the design contains a Pasch configuration, this also leads to a zero sum (involving the columns for the four triples). In this way, the construction of erasure codes employs the existence of quadrilateral-free STSs. Hellerstein, Gibson, Karp, Katz, and Patterson (1994) employ the affine STSs. Subsequently, Chee, Colbourn, and Ling (1999) made the connection with QFSTSs. See Lefmann, Pudlák, and Rödl (1997) for related work.

13.3 Avoidance of mitres

The mitre is one of two (7,5) configurations. The other (7,5) configuration, the *mia*, is obtained by joining two noncollinear points in a Pasch configuration. In particular, then, a quadrilateral-free system cannot contain a mia configuration either.

The unique STS(7) contains no mitre. Nor does the cyclic STS(13), nor do four of the 80 STS(15)s (Nos. 1, 2, 3, and 16 in Tables 5.8–5.12). On the other hand, the unique STS(9) contains 36 mitres. Nevertheless, in light of Theorem 13.14, this appears to be the only exceptional order; indeed, Colbourn, Mendelsohn, Rosa, and Širáň (1994) state:

Conjecture 13.29 *For* $v \equiv 1, 3 \pmod 6$, *an antimitre STS*(v) *exists unless* $v = 9$.

They also give a doubling construction:

Theorem 13.30 *If there exists an anti-mitre STS of order* v, *then there also exists an anti-mitre STS of order* $2v + 1$.

Proof Let $S = (V, \mathcal{B})$ be an anti-mitre STS of order v. Apply the $2v + 1$ construction from §3.2, using as the 1-factorization in the construction the Steiner 1-factorization from S on elements V'. This gives an STS S' of order $2v + 1$ on the set $V \cup V'$ with the triple set $\mathcal{B}' = (\cup_{x \in V} T_x) \cup \mathcal{B}$. It remains to prove that the anti-mitre property of $S = (V, \mathcal{B})$ implies that $S' = (V \cup V', \mathcal{B}')$ is anti-mitre as well.

Assume the contrary and let M be a mitre in S'. We concentrate on the 'parallel' triples of M, i.e. the unique pair of triples of M with no common element. If they both belong to $\mathcal{B}$ then the whole mitre would have to be a subset of $\mathcal{B}$, contrary to the anti-mitre property of S. If exactly one of the parallel triples of M is in $\mathcal{B}$ then, by inspection, the central element of M can be neither in V nor in V'. It remains to consider the case when both parallel triples are in $\mathcal{B}' \setminus \mathcal{B}$.

If neither of the two parallel triples of M contains ∞, then (without loss of generality) the mitre has the form abe, $ac'f'$, $ad'g'$, $bc'd'$, and $ef'g'$, where $a, b, c, d, e, f, g \in V$ are seven distinct elements. But then, by our construction, acf, adg, bcd, and efg are triples from $\mathcal{B}$ that, together with abe, form a mitre in the original system S, a contradiction. Finally, if one of the two parallel triples of M contains the symbol ∞, then (up to a possible renaming of elements) the triples of the mitre are abe, $ac'f'$, $ad'\infty$, $bc'd'$, and $ef'\infty$. Now, the definition of $\mathcal{B}'$ implies that $d' = a'$, and hence $d = a$. Since $bc'd' \in \mathcal{B}'$, we have $bcd = bca \in \mathcal{B}$ and also $abe \in \mathcal{B}$, whence $c = e$. Moreover, $e = f$ since $ef'\infty$ is a triple. Thus $c = e = f$, a contradiction. This completes the proof. □

Colbourn, Mendelsohn, Rosa, and Širáň (1994) also adapt Bose's construction (Lemma 2.3) to produce anti-mitre STS(v):

Theorem 13.31 *For every* $v \equiv 3$ *or* $9 \pmod{18}$, $v \geq 27$ *there exists an anti-mitre STS of order* v.

Proof Let $v \geq 27$ be a given number, $v \equiv 3$ or $9 \pmod{18}$. Then $v = 3n$ where $n \geq 9$ is an admissible order of an STS. Let W be a set of cardinality n. There exist three disjoint STSs on W by Theorem 15.15. More precisely, there exist STSs $S_i = (W, \mathcal{B}_i)$, $i \in \{0,1,2\} = \mathbb{Z}_3$, such that $\mathcal{B}_i \cap \mathcal{B}_j = \varnothing$ if $i \neq j$. For each $i \in \mathbb{Z}_3$ let $W_i = \{a_i;\ a \in W\}$; we assume that the three sets W_i are pairwise disjoint. We introduce a new triple system on the set $V = W_0 \cup W_1 \cup W_2$ as follows. If $i \in \mathbb{Z}_3$, we let $\mathcal{B}'_i$ be the set of all triples of the form $a_i b_i c_{i+1}$ where (a,b,c) runs through all ordered 3-tuples of elements of W such that $abc \in \mathcal{B}_i$. In addition, let $\mathcal{B}^* = \{a_0a_1a_2;\ a \in W\}$, and $\mathcal{B} = \mathcal{B}'_0 \cup \mathcal{B}'_1 \cup \mathcal{B}'_2 \cup \mathcal{B}^*$. It is routine to check that the system $S = (V, \mathcal{B})$ is indeed an STS(v). Now we show that S is anti-mitre.

Assume the contrary and let M be a mitre in S. Again, the proof depends on the position of the unique pair of parallel triples of the mitre within $\mathcal{B}$. It is easy to see that if both parallel triples of M belong to $\mathcal{B}'_i$ for some $i \in \mathbb{Z}_3$, then there is no way of placing the central point of M in V. The same is true if exactly one of the two parallel triples of the mitre belongs to $\mathcal{B}^*$.

If both parallel triples of M are in $\mathcal{B}^*$, then we may assume without loss of generality that the central vertex of M is in W_0, and that the mitre consists of the triples $a_0b_0c_1$, $a_0c_0b_1$, $b_2c_2a_0$, and the two parallel triples $b_0b_1b_2$ and $c_0c_1c_2$. According to the construction of S we see that $a_0b_0c_1 \in \mathcal{B}'_0$, and therefore $abc \in \mathcal{B}_0$. Similarly, $b_2c_2a_0$ is in $\mathcal{B}'_2$, which implies that $bca \in \mathcal{B}_2$. But then the STS S_0 and S_2 are not disjoint, a contradiction.

It remains to consider the case when the two parallel triples of the mitre belong to different sets $\mathcal{B}'_i$, say, $\mathcal{B}'_0$ and $\mathcal{B}'_1$. Let $b_0c_0d_1$ and $e_1f_1g_2$ be the parallel triples of M. Clearly, the central vertex of M cannot lie in W_1. Let the central vertex of the mitre be $a_0 \in W_0$ (the case when the central vertex is in W_2 is analogous and left to the reader). Then, without loss of generality, the remaining three triples of M are $a_0b_0e_1$, $a_0c_0f_1$, and $a_0d_1g_2$. The only triples in $\mathcal{B}$ that use three different subscripts are those in $\mathcal{B}^*$; therefore $a_0d_1g_2 = a_0a_1a_2$ and hence $a = d = g$. Now, $b_0c_0d_1 = b_0c_0a_1 \in \mathcal{B}'_0$ and also $a_0c_0f_1 \in \mathcal{B}'_0$, implying that $bca \in \mathcal{B}_0$ and $acf \in \mathcal{B}_0$, which is possible only if $b = f$. Similarly, $a_0b_0e_1 \in \mathcal{B}'_0$ implies that $abe \in \mathcal{B}_0$ and thus $c = e$. Finally, we have $b_0c_0d_1 = b_0c_0a_1 \in \mathcal{B}'_0$ and $f_1e_1g_2 = b_1c_1a_2 \in \mathcal{B}'_1$. Hence $bca \in \mathcal{B}_0$ and also $bca \in \mathcal{B}_1$, contrary to the fact that the STSs S_0 and S_1 are disjoint. □

Theorems 13.30 and 13.31 together show that anti-mitre STS(v)s exist whenever $v \equiv 3, 7, 9, 19, 21, 27 \pmod{36}$. Applying Theorem 13.30 two more times yields anti-mitre STS(v) for 27 of the 48 admissible residue classes modulo 144. Thus the arithmetic density of the set of orders for which anti-mitre STS(v)s are known to exist is currently 9/16.

13.3.1 *5-sparse STSs*

It can happen that an STS(v) is both quadrilateral-free and anti-mitre; in this case, the system is in fact 5-sparse. The Netto triple systems of orders congruent to 19 (mod 24) are quadrilateral-free (Lemma 13.7), and Colbourn, Mendelsohn,

Rosa, and Širáň (1994) establish that all Netto triple systems are anti-mitre. This provides an infinite class of 5-sparse triple systems. Theorem 13.14 provides many more small examples. However, only one recursive construction is known, due to Ling (1997a). We examine his construction next.

Let $S = (V, \mathcal{B})$ be a transitive 5-sparse triple system on $\mathbb{Z}_v$ or $\mathbb{F}_v$, where $|V| = v$. Let $\mathcal{C} = \{\{-a, -b, -c\} : \{a, b, c\} \in \mathcal{B}\}$. Let $S' = (V, \mathcal{C})$. S' and S are isomorphic, and hence S' is also a transitive 5-sparse triple system. When $v \equiv 1 \pmod 6$, S and S' are block-disjoint.

Theorem 13.32 *If $v \equiv 1 \pmod 6$ and a transitive 5-sparse STS(v) exists, then a 5-sparse STS$(3v)$ exists.*

Proof (Sketch) We construct an STS$(3v)$ on $V \times \{0, 1, 2\}$:

1. For $\{a, b, c\} \in \mathcal{B}$, construct two blocks of the form $\{(a, i), (b, i), (c, i)\}$ for $i = 0, 1$.
2. For $\{a, b, c\} \in \mathcal{C}$, construct a block $\{(a, 2), (b, 2), (c, 2)\}$.
3. Construct the blocks $\{(i, 0), (j, 1), (i + j, 2)\}$ for $i, j \in \mathbb{Z}_v$ or $\mathbb{F}_v$.

The rest of the proof requires that the system be shown to be anti-mitre and quadrilateral-free; the latter verification uses the method of Griggs, Murphy, and Phelan (1990). □

Ling (1997a) also gives an extension of Theorem 13.32 to a more general direct product construction. As expected, with neither quadrilateral-free nor anti-mitre STS(v)s having their existence question settled, less is known about 5-sparse STS(v)s.

For r-sparse STS(v)s with $r \geq 6$, not a single example is known at the present time, and none exists among the cyclic STS(v)s with $v \leq 57$.

13.4 Avoidance: weakly union-free systems

Chee, Colbourn, and Ling (1997) settle a problem of Frankl and Füredi (1984) to determine the maximum number of hyperedges in a 3-uniform hypergraph in which no two pairs of distinct hyperedges have the same union (compare with §9.7). The extremal case corresponds to the existence of weakly union-free twofold triple systems, which is settled with seven definite and three possible exceptions. An application to group testing is also given. A 3-GDD $(X, \mathcal{G}, \mathcal{B})$ is a *weakly union-free* GDD (*wuf* GDD) if

1. whenever $\{\{a, b, x\}, \{a, b, y\}\} \subseteq \mathcal{B}$, the points x and y are in different groups; and
2. whenever four distinct blocks B_1, B_2, B_3, B_4 are chosen from $\mathcal{B}$, it does not happen that $B_1 \cup B_2 = B_3 \cup B_4$.

An examination of every isomorphism type of 3-uniform hypergraph having four blocks makes the second condition more explicit. There cannot exist four blocks of any of the following four forms:

$$\begin{aligned}
&\text{C1}: \{\{a,b,c\},\{a,b,d\},\{a,c,d\},\{b,c,d\}\},\\
&\text{C2}: \{\{x,a,b\},\{x,a,c\},\{x,b,d\},\{x,c,d\}\},\\
&\text{C3}: \{\{x,a,b\},\{x,a,c\},\{x,b,d\},\{a,c,d\}\},\\
&\text{C4}: \{\{x,a,b\},\{x,c,d\},\{y,a,b\},\{y,c,d\}\}.
\end{aligned}$$

Our interest is in the construction of wuf 3-GDDs, and in particular those of type 1^n and index two. Frankl and Füredi (1984) began the study of wuf TTS(n)s in the study of an old problem of Erdös. Erdös (1977) asked how many hyperedges a k-uniform hypergraph can have, so that whenever four hyperedges A, B, C, D satisfy $A \cup B = C \cup D$, we find $\{A, B\} = \{C, D\}$. Such a family is *union-free.* Frankl and Füredi (1984) settled this question when $k = 3$, showing that the Steiner triple systems realize the maximum.

They also addressed the related question of enforcing the union-free condition only for sets of four distinct blocks A, B, C, D. This gives the notion of weakly union-free, already defined. Frankl and Füredi (1984) established an important bound, and showed that it is realized infinitely often:

Theorem 13.33 *A weakly union-free 3-uniform hypergraph on n vertices has at most $\lfloor n(n-1)/3 \rfloor$ hyperedges. Equality occurs when all, or all but one, pairs of vertices occur in two hyperedges each.*

They established that this bound is met whenever $n \equiv 1 \pmod 6$, and either n is a prime power at least 13 or n is sufficiently large. Chee, Colbourn, and Ling (1997) establish that equality is met for all $n \equiv 0, 1 \pmod 3$, with a small number of definite, and a small number of possible, exceptions. Colbourn (1999a) establishes the analogous result for $n \equiv 2 \pmod 3$.

The difficulty of this problem appears initially to be that, while catalogues of twofold triple systems for small orders are available, no TTS(n) is weakly union-free when $n \in \{3, 4, 6, 7, 9, 10\}$. Moreover, when a wuf 3-GDD of type T can be decomposed into two 3-GDDs of index one and type T, condition (1) together with the exclusion of C4 ensures that these two index one 3-GDDs are orthogonal (see §14.4.1). The existence of orthogonal uniform 3-GDDs with group size 1, the orthogonal Steiner triple systems, remained open for 30 years. The exclusion of further configurations adds to the difficulty of the problem for wuf TTSs.

13.4.1 *Direct constructions*

Chee, Colbourn, and Ling (1997) develop a direct construction technique that is used to construct both wuf TTS(n)s and, more generally, wuf 3-GDDs of index two. The general framework follows. We aim to construct a 3-GDD of index two on point set $\mathbb{Z}_{gu} \cup \{\infty_1, \ldots, \infty_k\}$, whose type is $g^u k^1$. Naturally, we chose $\mathbb{Z}_{gu}$ for a portion of the point set to suggest the cyclic action of the cyclic group on these points. Indeed our goal is to construct 3-GDDs that have $\mathbb{Z}_{gu}$ as an automorphism group.

Let $X = \mathbb{Z}_{gu} \cup \{\infty_1, \ldots, \infty_k\}$, and let σ be a permutation mapping $i \mapsto i+1 \bmod gu$ for $i \in \mathbb{Z}_{gu}$, and fixing $\{\infty_1, \ldots, \infty_k\}$. Let $\mathcal{B}$ be the blocks of a 3-GDD of type $g^u k^1$ on X that admits σ as an automorphism. The action of σ partitions $\mathcal{B}$ into orbits of size gu or, when $gu \equiv 0 \pmod 3$, possibly $gu/3$. A set of representatives of these orbits forms a set of starter blocks for the 3-GDD. Starter blocks of the form $\{0, a, b\} \subset \mathbb{Z}_{gu}$ may generate orbits of length gu under σ, in which case the starter block is said to *cover* the differences $\pm a$, $\pm b$, $\pm(b-a)$ with arithmetic in $\mathbb{Z}_{gu}$ (if repetitions occur, such differences are covered the number of times that they occur). When $gu \equiv 0 \pmod 3$, a starter block of the form $\{0, gu/3, 2gu/3\}$ generates only $gu/3$ distinct blocks (a *short orbit*), and is said therefore to *cover* the differences $\pm gu/3$ once each. Finally, a starter block may have the form $\{\infty_i, 0, d_i\}$; again, gu blocks appear in the orbit generated, but here only the differences $\pm d_i$ are covered, once each.

A set $\mathcal{D}$ is a set of starter blocks for a 3-GDD of index two and type $g^u k^1$ (under the action of σ) if

1. for $1 \le i \le k$, there is exactly one starter block containing ∞_i; and
2. each $d \in \mathbb{Z}_{gu}$ is covered twice as a difference, unless $d \equiv 0 \pmod u$, in which case the difference is not covered.

The reader can quickly verify that these conditions on starter blocks are equivalent to the existence of a 3-GDD of index two and type $g^u k^1$ admitting σ.

In order to be a wuf 3-GDD, further conditions are imposed. Suppose that $\mathcal{D}$ is the set of starter blocks for a 3-GDD of index two and type $g^u k^1$. Partition $\mathcal{D}$ into the blocks $\mathcal{A}$ which contain one of the infinite points, and the blocks $\mathcal{B}$ which do not. Evidently, $\mathcal{A}$ contains exactly k blocks, one for each of the infinite points. In addition, in order to meet the first wuf condition, we have:

1. If $\{\infty_i, 0, a\}, \{\infty_j, 0, b\} \in \mathcal{A}$, then $a \not\equiv \pm b \pmod{gu}$.

Call a difference *external* if it is covered once in $\mathcal{A}$ and once in $\mathcal{B}$, and *internal* if it is covered twice in $\mathcal{B}$. For each external difference d, define $\alpha(d) = \min(\pm 2d)$. For each internal difference d, when blocks $\{0, d, x\}$ and $\{0, d, y\}$ appear in the orbits of blocks of $\mathcal{B}$, define $\alpha(d) = \min(\pm(x-y))$.

First we examine constraints resulting from prohibiting the appearance of one of the infinite points in one of the configurations C1, C2, C3, or C4. In order to ensure that no infinite point occurs in a C1 configuration, we require that

2. If $gu \equiv 0 \pmod 3$ and $gu/3$ is an external difference, then $\mathcal{B}$ does not contain $\{0, gu/3, 2gu/3\}$.

In order to ensure that no infinite point occurs in a C2 or C3 configuration, we require that

3. If d is an external difference, then $4d \not\equiv 0 \pmod{gu}$.
4. If d is an external difference and $\{0, d, x\}$ is a block in an orbit of a starter block of $\mathcal{B}$, then $2x \not\equiv d \pmod{gu}$ and none of $\{0, d, 3d\}$, $\{0, 2d, 3d\}$, $\{0, 2d, d+x\}$, $\{0, 2d, x\}$, or $\{0, d, gu/2\}$ when $gu \equiv 0 \pmod 2$, appears in the orbits of the starter blocks in $\mathcal{B}$.

In order to ensure that no infinite point occurs in a C4 configuration, we require that

5. If d and d' are external differences, or if d is external and d' is internal, then $\alpha(d) = \alpha(d')$ only if $d = d'$.

Once conditions (1)–(5) are met, any violation of the wuf conditions occurs entirely among the blocks on $\mathbb{Z}_{gu}$.

In order to check that none of the conditions are violated on the blocks involving no infinite points, we first observe that the first wuf condition is equivalent to:

6. If d is an internal difference then $\alpha(d) \not\equiv 0 \pmod{u}$.

To check that the four configurations are missed, we first form the restricted neighbourhood of the point 0, defined by $N_0 = \{\{x, y\} : \{0, x, y\}$ is in the orbit of a block of $\mathcal{B}\}$. This is a graph on the vertex set consisting of $\mathbb{Z}_{gu}$, except integers congruent to 0 modulo u. In this graph, every vertex has degree 1 or 2, and so N_0 consists of disjoint paths and cycles.

To avoid C1, we require:

7. If (a, b, c) forms a 3-cycle in N_0, then $\{a, b, c\}$ does not appear in the orbit of any block in $\mathcal{B}$.

To avoid C2, we require:

8. N_0 does not contain a 4-cycle.

To avoid C3, we require:

9. For every 3-edge path (a, b, c, d) in N_0, no block of the form $\{a, c, d\}$ or $\{a, b, d\}$ appears in the orbit of a block of $\mathcal{B}$.

To avoid a C4 configuration, we require that

10. If d and d' are internal differences, then $\alpha(d) = \alpha(d')$ only if $d = d'$.

This is apparently an extensive list of conditions, but each condition is easily checked. In fact, Frankl and Füredi (1984) establish the existence of some GDDs of index 2 and type 1^q when $q \equiv 1 \pmod{6}$. They show that if ω is a primitive element of the finite field $\mathbb{F}_q$ and $q \equiv 1 \pmod{6}$, $q \neq 7$, then $\{\{0, \omega^i, \omega^{2t+i} + \omega^i\} : 0 \leq i < 2t\}$ is a set of starter blocks for a wuf TTS(q). Hence we have:

Lemma 13.34 *A 3-GDD of index two and type* 1^q *exists whenever* $q \equiv 1$ (mod 6) *is a prime power, except when* $q = 7$.

Ingredients for other congruence classes modulo 6 are needed. Chee, Colbourn, and Ling (1997) employed a combination of backtracking and hillclimbing techniques to produce a large number of wuf GDDs.

Numerous 3-GDDs of type $1^u x^1$ over $\mathbb{Z}_u$ are needed. To present them concisely, we define a *presentation* of a wuf 3-GDD of type $g^u t^1$ to be a set of pairs of elements from $\mathbb{Z}_{gu}$. For each pair $\{a, b\}$, $\{0, a, b\}$ is a starter block. Then consider the collection of unused differences modulo gu which are not multiples of u. Then for t of these remaining differences, a starter block is formed using the pairs having that difference together with one of the t infinite points; the differences are selected so as to leave the difference $u/3$ in the case that there are $t+1$ unused differences. In addition, if a further unused difference remains (in which case we require that $u \equiv 0 \pmod 3$ and $x \equiv 1 \pmod 3$), then $\{0, u/3, 2u/3\}$ is a starter block.

Lemma 13.35 *A wuf TTS(n) exists for* n = 21, 24, 27, 28, 30, 33, 34, 36, 39, 40, 42, 45, *and* 46.

Proof Table 13.2 gives presentations for the required designs. □

GDD	Presentation
$1^{20}1^1$	{1,7} {1,9} {2,4} {3,8} {3,13} {4,9}
$1^{23}1^1$	{1,6} {2,13} {2,16} {3,12} {3,18} {4,8} {6,16}
$1^{26}1^1$	{1,6} {2,12} {2,23} {3,19} {4,13} {4,18} {6,17} {7,18}
$1^{28}0^1$	{1,2} {2,13} {3,7} {3,12} {4,12} {5,11} {5,19} {6,13} {8,18}
$1^{29}1^1$	{1,7} {2,15} {2,18} {3,8} {3,12} {4,22} {4,23} {5,19} {8,17}
$1^{32}1^1$	{1,3} {1,8} {3,10} {4,19} {4,20} {5,18} {5,26} {6,23} {8,18} {9,21}
$1^{33}1^1$	{1,4} {2,8} {2,21} {3,16} {4,26} {5,15} {5,24} {6,24} {7,23} {8,21}
$1^{35}1^1$	{1,4} {2,6} {2,19} {3,20} {5,12} {5,29} {7,18} {8,16} {9,22} {9,23} {10,20}
$1^{38}1^1$	{1,4} {2,7} {2,17} {3,15} {4,18} {5,13} {6,12} {7,27} {8,22} {9,22} {9,28} {10,27}
$1^{40}0^1$	{1,2} {2,5} {3,7} {4,27} {5,15} {6,24} {7,29} {8,21} {8,28} {9,21} {9,26} {11,26}
$1^{41}1^1$	{1,4} {2,6} {2,14} {3,30} {5,24} {5,31} {6,21} {7,18} {7,32} {8,20} {8,25} {9,28} {10,28}
$1^{44}1^1$	{1,2} {2,5} {3,7} {5,13} {6,20} {6,33} {7,23} {8,24} {9,26} {9,19} {10,34} {11,31} {12,30} {13,28}
$1^{45}1^1$	{1,4} {2,6} {2,7} {3,29} {5,35} {6,34} {7,25} {8,22} {8,29} {9,20} {9,31} {10,27} {12,24} {13,26}

Table 13.2 *Small weakly union-free systems*

Chee, Colbourn, and Ling (1997) give direct constructions for a number of further orders:

Lemma 13.36 *A wuf TTS(n) exists for* n = 48, 51, 52, 54, 55, 57, 58, 60, 63, 64, 66, 69, 70, 72, 75, 76, 78, 81, 82, 84, 85, 87, 88, 90, 91, 93, 94, 96, 99, 100, 102, 105, 108, 111, 112, 114, 115, 117, 118, 120, 123, 124, 126, 129, 130, 132, 133, 135, 136, 138, 141, 142, 144, 145, 148, 150, 154, 156, 159, 160, 161, 165, 166, 171, 177, 178, 184, 195, 201, 207, 213, 219, *and* 243.

The remaining small values do not appear to be able to be handled by this general approach. However, Chee, Colbourn, and Ling (1997) have succeeded in one more case:

Lemma 13.37 *A wuf TTS(16) exists.*

Proof Let $X = \mathbb{Z}_8 \times \{0,1\}$. For succinctness, we write $(x,i) \in X$ as x_i. Let $\sigma : X \to X$ be the permutation such that $\sigma : x_i \mapsto (x+1 \bmod 8)_i$. Developing the set of starter blocks

$$\{0_0,1_0,3_1\}\ \{0_0,4_0,0_1\}\ \{0_0,2_0,5_0\}\ \{0_0,2_0,1_1\}\ \{3_0,0_1,1_1\}$$
$$\{0_0,1_1,3_1\}\ \{0_0,1_0,5_1\}\ \{0_1,2_1,5_1\}\ \{0_0,2_1,6_1\}\ \{0_0,0_1,7_1\}$$

by σ gives a wuf TTS(16) on X. □

13.4.2 *Recursive constructions*

Chee, Colbourn, and Ling (1997) employ two well known constructions, the Fundamental Construction of Wilson (Theorem 3.3), and Filling In Groups (§3.4). We leave as an easy exercise that when all of the ingredient GDDs are wuf, so also is the GDD constructed. Filling in groups preserves the wuf property primarily as a consequence of the first requirement, since none of the forbidden configurations can have both a block from the wuf GDD of type $g_1g_2\cdots g_n$ and one from a wuf GDD of type $1^{g_i}h^i$.

Now we give some applications of Theorem 3.3.

Lemma 13.38 *If a TD(6, n) exists, then a wuf 3-GDD of type* $(3n)^5(6n)^1$ *exists. Moreover, there exist wuf TTSs of orders* 106, 147, 168, 189, *and* 231.

Proof A wuf 3-GDD of type $3^5 6^1$ with presentation $\{\{1,12\},\{2,9\}\}$ exists. Use the TD$(6,n)$ as a master design and the 3-GDD of type $3^5 6^1$ as an ingredient design in Theorem 3.3. Apply with $n = 5,7,8,9,11$ and fill in holes using wuf 3-GDDs of types $1^{15}1^1$ and $1^{30}1^1$ when $n = 5$, and of types 1^{3n} and 1^{6n} for the remaining values of n. □

Lemma 13.39 *If a TD(7, n) exists, then a wuf 3-GDD of type* $(2n)^7$ *exists. Hence wuf TTSs of orders* 112, 183, *and* 225 *exist.*

Proof The presentation $\{\{1,4\},\{1,6\},\{2,6\},\{2,11\}\}$ gives one wuf 3-GDD of type 2^7; Theorem 3.3 gives the wuf 3-GDD of type $(2n)^7$. Applying with $n = 8,13,16$, and filling holes with wuf 3-GDDs of types 1^{16}, $1^{26}1^1$, and $1^{32}1^1$ gives the required consequences. □

Lemma 13.40 *If a TD(8, n) exists, and* $0 \le x \le n$, *then a wuf 3-GDD of type* $(3n)^7(3n+6x)^1$ *exists. Hence there exist wuf TTSs of orders* 174, 180, 186, 192, 198, 204, 210, 216, 222, 228, *and* 187.

Proof A wuf 3-GDD of type $3^8 = 3^7 3^1$ exists over $\mathbb{Z}_{24}$ with presentation

$$\{\{1,3\},\{1,20\},\{2,12\},\{3,10\},\{4,11\},\{5,18\},\{6,15\}\}.$$

A wuf 3-GDD of type $3^7 9^1$ with presentation $\{\{1,13\},\{2,5\},\{4,10\}\}$ exists. Apply Theorem 3.3 using weight 3 in seven groups and weights 3 or 9 in the eighth, to produce a wuf 3-GDD of type $(3n)^7(3n+6x)^1$. Apply with $n = 7,8$ and fill in holes to obtain the stated consequences. For the final value, apply with $n = 7$ and employ a wuf 3-GDD of type $1^{21}7^1$ to fill holes. It has presentation $\{\{1,4\},\{1,6\},\{2,9\},\{2,13\}\}$. □

Lemma 13.41 *If a TD(14, q) exists, and* $0 \leq x \leq 6q$ *satisfies* $x \equiv 0 \pmod 3$, *then a wuf 3-GDD of type* $q^{13}x^1$ *exists. If, in addition, a wuf 3-GDD of type* $1^q h^1$ *exists, so also does a wuf GDD of type* $1^{13q}(x+h)^1$.

Proof Use as ingredient wuf 3-GDDs the ones of type $1^{13}0^1$ from Lemma 13.34, of type $1^{13}3^1$ presented as $\{\{1,4\},\{3,4\},\{2,8\}\}$, and the one of type $1^{13}6^1$ presented as $\{\{1,4\},\{2,8\}\}$. Give all points in 13 groups of the TD$(14, q)$ weight 1, and points in the final group weights 0, 3, or 6 so that the total weight in the final group is x. Theorem 3.3 then gives a wuf 3-GDD of type $q^{13}x^1$. Filling in holes with a 3-GDD of type $1^q h^1$ (when one exists) gives a wuf GDD of type $1^{13q}(x+h)^1$. □

Corollary 13.42 *A wuf TTS(n) exists whenever* $n = 169$ *or* $190 \leq n \leq 304$ *and* $n \equiv 1 \pmod 3$.

Proof If $n \leq 247$, write $n = 13 \cdot 13 + x$; then $0 \leq x \leq 6 \cdot 13$ and a wuf TTS(x) has been previously given. Apply Theorem 13.41 with a TD$(14, 13)$. If instead $n > 247$, write $n = 13 \cdot 16 + x$ and proceed similarly with a TD$(14, 16)$. □

Corollary 13.43 *A wuf TTS(n) exists whenever* $n \equiv 0 \pmod 3$ *and* $234 \leq n \leq 327$, *except possibly when* $n = 243$.

Proof Write $n = 13 \cdot 17 + 4 + x$. A wuf TTS$(x+4)$ has been presented, and a wuf 3-GDD of type $1^{17}4^1$ has presentation $\{\{1,3\},\{1,5\},\{2,12\},\{3,11\}\}$. Apply Lemma 13.41 with $h = 4$. □

Lemma 13.44 *If a TD(k, q) exists for* $k \geq 14$, *a wuf TTS(q) exists, and* $0 \leq x \leq 6(k-13)$ *satisfies* $x \equiv 0 \pmod 3$, *then a wuf 3-GDD of type* $1^{13(q-1)}(x+13)^1$ *exists. Hence wuf TTSs of orders* 172 *and* 175 *exist.*

Proof Use the same ingredients as Lemma 13.41. Give all points in 13 groups of the TD$(14, q)$ weight 1, and choose a single block B of length k. Assign all points in the remaining $k - 13$ levels weight 0 if they are *not* on B, and weight 0, 3, or 6 if they are on B so that the total weight of such points is x. When $q = 13$ and $x = 3, 6$, wuf 3-GDDs of types $1^{156}16^1$ and $1^{156}19^1$ are obtained. □

Theorem 13.45 *Let* $n \equiv 0, 1 \pmod 3$. *If wuf TTSs exist for all orders n satisfying* $24 \leq n \leq 304$, *then wuf TTSs exist for all orders* $n \geq 24$.

Proof Form two infinite sequences of integers $(r_i : i \geq 0)$ and $(s_i : i \geq 0)$ for which

1. $r_0 = 19$ and $s_0 = 23$;
2. $r_{i+1} > r_i$ and $s_{i+1} > s_i$;
3. $13r_{i+1} + 21 \leq 19r_i$ and $13s_{i+1} + 25 \geq 19s_i + 1$;
4. $r_i \equiv 1 \pmod 3$ and $s_i \equiv 2 \pmod 3$; and
5. TD$(14, r_i)$ and TD$(14, s_i)$ exist.

A TD$(14, n)$ exists whenever n is relatively prime to 2, 3, 5, 7, and 11 by MacNeish's theorem; see Abel, Brouwer, Colbourn, and Dinitz (1996). Among the integers congruent to 1 modulo 3, considering the sequence of those relatively prime to 2, 3, 5, 7, and 11, we find a largest difference between consecutive values of 24. Choose the r_is to be the sequence of numbers congruent to 1 modulo 3 and relatively prime to 2, 3, 5, 7, and 11, beginning with 19, in addition to the number 25. It is now an easy verification that we have the specified properties. In the same way, the s_is are the sequence of numbers congruent to 2 modulo 3 and relatively prime to 2, 3, 5, 7, and 11, beginning with 23, in addition to the number 32.

To prove the theorem, proceed inductively. In general, suppose that wuf TTSs have been produced for all orders smaller than n, where $n \equiv 0, 1 \pmod 3$ and $n \geq 24$. Then we establish that a wuf TTS(n) exists. By assumption, wuf TTS(n)s exist whenever $24 \leq n \leq 304$. Now if $n \equiv 1 \pmod 3$, find the largest i for which $13r_i + 24 \leq n \leq 19r_i$; such a choice exists by the definition of the sequence. Then a TD$(14, r_i)$ exists. Wuf 3-GDDs of type 1^{r_i} and 1^{n-13r_i} exist by the inductive hypothesis. Apply Theorem 13.41 to obtain the wuf TTS(n). In the same way, if $n \equiv 0 \pmod 3$, find the largest s_i for which $13s_i+25 \leq n \leq 19s_i+1$; such a choice exists by the definition of the sequence. Then a TD$(14, s_i)$ exists. Wuf 3-GDDs of type $1^{s_i}1^1$ and 1^{n-13s_i} exist by the inductive hypothesis. Apply Theorem 13.41 to obtain the wuf TTS(n). □

Now we can prove the existence theorem.

Theorem 13.46 *A wuf TTS(n) exists whenever* $n \equiv 0, 1 \pmod 3$ *except when* $n \in \{3, 4, 6, 7, 9, 10, 12\}$ *and possibly when* $n \in \{15, 18, 22\}$.

Proof The definite exceptions can all be verified by an exhaustive search. Denny (1998) completed the search for wuf TTS(12)s, showing nonexistence, and established that there are two nonisomorphic wuf TTS(13)s. Now if n is a prime or prime power, apply Lemma 13.34. Otherwise, apply Lemmas 13.35, 13.36, and 13.37 to treat most small orders, and Lemmas 13.38, 13.39, 13.40, 13.44 and Corollaries 13.42 and 13.43 to treat $n = 21$ and all remaining values satisfying $24 \leq n \leq 304$. Then apply Theorem 13.45. □

Using similar techniques, Colbourn (1999a) proves:

Theorem 13.47 *A wuf 3-GDD of type* $1^{n-2}2^1$ *exists whenever* $n \equiv 2 \pmod 3$ *and* $n \geq 17$, *except possibly when* $n = 23$.

13.4.3 *Group testing revisited*

Chee, Colbourn, and Ling (1997) employ Theorem 13.46 in an application to group testing. We employ the same scenario as in §9.7. For simplicity, we denote the k-restricted group testing problem, with at most r defective items, by GTP$_k(r)$.

An algorithm for the group testing problem is an *α-approximation algorithm* if it returns a set S of at most αr items, so that S contains all defective items of Ω. The *complexity* of a group testing algorithm is defined to be the number of tests conducted (hence, also the number of pools). The best adaptive algorithm has a complexity no higher than any nonadaptive algorithm. However, the advent of massively parallel computers prompted Hwang and Sós (1987) to make a case for the study of nonadaptive algorithms.

Our focus is on nonadaptive (3/2)-approximation algorithms for $\text{GTP}_3(2)$. Any nonadaptive algorithm $\mathcal{A}$ for $\text{GTP}_3(2)$ corresponds to a 3-uniform hypergraph $\mathcal{H}(\mathcal{A}) = (X, \mathcal{B})$ as follows:

1. $X = \{x_P : P \text{ is a pool of } \mathcal{A}\}$.
2. $\mathcal{B} = \{B_\omega : \omega \in \Omega\}$.
3. $x_P \in B_\omega$ if and only if $\omega \in P$.

We call $\mathcal{H}(\mathcal{A})$ the *hypergraph of* $\mathcal{A}$. We make a useful observation concerning $\mathcal{H}(\mathcal{A})$. Let $\mathcal{O}$ be the set of all defective items in Ω. Then $x_P \in \bigcup_{\omega\in\mathcal{O}} B_\omega$ if and only if P is a pool of $\mathcal{A}$ whose feedback is defective. Hence, if we know that one of $\mathcal{O}$ or $\mathcal{O}'$ contains the set of all defective items in Ω, then a necessary and sufficient condition which allows us to distinguish them is

$$\bigcup_{\omega\in\mathcal{O}} B_\omega \neq \bigcup_{\omega\in\mathcal{O}'} B_\omega.$$

Lemma 13.48 *Suppose that $\mathcal{A}$ is a nonadaptive (3/2)-approximation algorithm for $GTP_3(2)$. Then $\mathcal{H}(\mathcal{A}) = (X, \mathcal{B})$ is weakly union-free.*

Proof Assume to the contrary that there are four distinct hyperedges $B_{\omega_i} \in \mathcal{B}$, $1 \leq i \leq 4$, such that $B_{\omega_1} \cup B_{\omega_2} = B_{\omega_3} \cup B_{\omega_4}$. If one of $\{\omega_1, \omega_2\}$ or $\{\omega_3, \omega_4\}$ is the pair of defective items, then $\mathcal{A}$ cannot distinguish them. The best $\mathcal{A}$ can do is then to conclude that $\{\omega_1, \omega_2, \omega_3, \omega_4\}$ contains all defective items of Ω. But this violates the condition that $\mathcal{A}$ is a (3/2)-approximation algorithm. □

Lemma 13.49 *Any wuf PTS$(n, 2)$ is the hypergraph of a nonadaptive (3/2)-approximation algorithm for $GTP_3(2)$.*

Proof Let $\mathcal{A}$ be the nonadaptive algorithm given by a wuf PTS$(n, 2)$, $\mathcal{H}(\mathcal{A}) = (X, \mathcal{B})$. Let $\omega_1, \omega_2, \omega_3 \in \Omega$ be any three distinct items. Then $B_{\omega_1} \neq B_{\omega_2}$ since $H(\mathcal{A})$ contains no repeated hyperedges, and $B_{\omega_1} \neq B_{\omega_2} \cup B_{\omega_3}$ since the union of two distinct hyperedges contains at least four vertices. Hence if Ω contains only one defective item, then $\mathcal{A}$ can identify that item precisely. We are thus left with the task of considering the case with two defective items.

It suffices to show that for any three distinct hyperedges $B_{\omega_1}, B_{\omega_2}, B_{\omega_3} \in \mathcal{B}$ such that $B_{\omega_1} \cup B_{\omega_2} = B_{\omega_1} \cup B_{\omega_3} = F$, we have $\{B, B'\} \subseteq \{B_{\omega_1}, B_{\omega_2}, B_{\omega_3}\}$ whenever $B \cup B' = F$. So let $B \cup B' = F$. Suppose that at least one of B or B' is not B_{ω_1}, B_{ω_2}, or B_{ω_3}, for otherwise we are done. Therefore we must have $\{B, B'\} = \{B_{\omega_1}, B_{\omega_4}\}$, for some $\omega_4 \notin \{\omega_1, \omega_2, \omega_3\}$ since $(X, \mathcal{B})$ is weakly union-free. Now $|B_{\omega_1} \cap B_{\omega_2}| \neq 0$ or 3 because $\mathcal{B}$ contains no repeated hyperedges. If $|B_{\omega_1} \cap B_{\omega_2}| = 2$, then $|F| = 4$, implying that $\{B_{\omega_1}, B_{\omega_2}, B_{\omega_3}, B_{\omega_4}\}$ is the complete 3-uniform hypergraph on four vertices, which is not weakly union-free. Hence $|B_{\omega_1} \cap B_{\omega_2}| = 1$. But then $B_{\omega_2} \setminus B_{\omega_1}$ is a 2-subset that must also be contained in the blocks B_{ω_3} and B_{ω_4}. This contradicts the assumption that $(X, \mathcal{B})$ is a PTS$(n, 2)$. □

To summarize:

Theorem 13.50 *For any $n > 23$, there exists a nonadaptive $(3/2)$-approximation algorithm of complexity n for $GTP_3(2)$ with a population of $\lfloor n(n-1)/3 \rfloor$ items. This complexity is optimal.*

13.5 Avoidance in general

For a given configuration C and index λ, the *avoidance set* $\Omega(C, \lambda)$ is the set of orders v for which a TS(v, λ) exists which avoids C. For the Pasch configuration P, for example, we have $\{v : v \equiv 3 \pmod 6\} \subset \Omega(P, 1)$ by Theorem 13.11, while Conjecture 13.6 would imply $\Omega(P, 1) = B(1) \setminus \{7, 13\}$. (Here and throughout this section, $B(\lambda)$ denotes the set of λ-admissible orders.) Similarly, Conjecture 13.29 would imply that for the mitre M, $\Omega(M, 1) = B(1) \setminus \{9\}$. The avoidance set for C_{14} is known: $\Omega(C_{14}, 1) = \{2^n - 1 : n > 1\}$, as the only Steiner triple systems avoiding C_{14} are projective spaces.

When $v \geq 13$, all configurations with at most three lines are unavoidable in an STS(v). Griggs and Rosa (1994) determine the avoidance sets for almost all 2- and 3-line configurations in which lines are permitted to meet in any number of points. One immediate observation with general applicability is that if C' is a subconfiguration of a configuration C, then $\Omega(C', \lambda) \subseteq \Omega(C, \lambda)$.

By way of example, let A_i be the 2-line configuration in which the two lines share $i - 1$ points. Then

$$\begin{aligned} \Omega(A_1, \lambda) &= B(\lambda) \cap \{3, 4, 5, 6, 7\} \\ \Omega(A_2, \lambda) &= B(\lambda) \cap \{3, 4\} \\ \Omega(A_3, \lambda) &= B(1) \\ \Omega(A_4, \lambda) &= B(\lambda) \cap \{v : v \geq \lambda + 2\} \end{aligned}$$

In order to determine the avoidance sets for 3-line configurations, use:

Lemma 13.51 *If C is a PTS(v, λ) without A_4 (i.e. it is simple), then $v \in \Omega(C, \lambda)$ implies $v \in \Omega(C, m\lambda)$ for each integer $m \geq 1$.*

Proof Replicate the triples m times each in a TS(v, λ) avoiding C. □

Figure 13.3 depicts the 2- and 3-line configurations in which some pair of elements occurs on two or more lines; these supplement the configurations of Figure 13.1 which are partial triple systems of index one. The avoidance sets $\Omega(B_i, \lambda)$ are finite when $i \in \{1, 2, 3, 4, 5\}$, but are infinite for $6 \leq i \leq 16$. Griggs and Rosa (1994) determined the latter, given in Table 13.3.

Let $\mathcal{L}(k, \ell)$ be the set of all configurations of ℓ triples on k points. Brown, Erdös, and Sós (1971) showed:

Theorem 13.52 *For every k and ℓ, there is a constant c_k for which a set of at least $c_k n^{(3\ell - k)/(\ell - 1)}$ triples on n points exist having no configuration of ℓ triples on k points.*

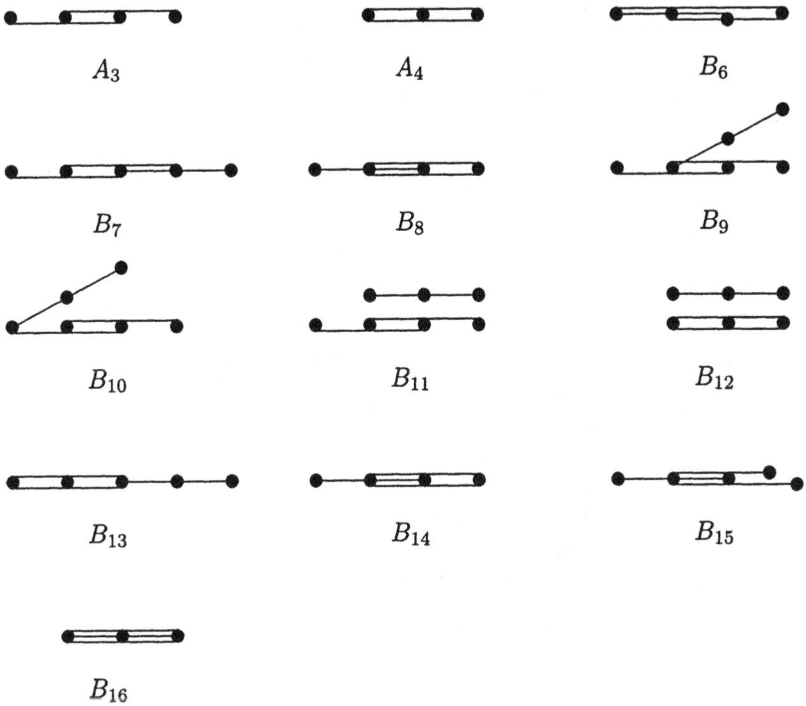

FIG. 13.3. More small configurations

Sós, Erdös, and Brown (1973) establish lower bounds for some specific configurations. They show that the largest set of triples on n elements avoiding the triangle configuration has at least $cn^{1.5}$ triples; they conjectured that any triple system containing cn^2 triples for n sufficiently large must contain a triangle. Their conjecture was proved by Ruzsa and Szemerédi (1978) who also improved the lower bound to $cr_3(n)n$ where $r_3(n)$ is the largest integer m such there exist m positive integers smaller than n, no three forming an arithmetic progression.

Lefmann, Phelps, and Rödl (1993) examine (6,4) configurations. Among other results, they establish:

Theorem 13.53 *A largest set of triples on n elements which avoids the Pasch configuration has at least $c_1 n^{5/2}$ triples, and at most $c_2 n^{11/4}$ triples, for constants c_1 and c_2 independent of n.*

Proof The upper bound follows from Theorem 13.52. For the lower bound, on $m = \lfloor 2n/3 \rfloor$ vertices, form a simple graph without 4-cycles. There exists such a graph $G = (V, E)$ having at least $cm^{3/2}$ edges, for constant c. Now add $n - m$ further elements X, and whenever $\{v, w\} \in E$ and $x \in X$, form a triple $\{x, v, w\}$. The resulting set of triples contains no Pasch configuration, since exactly two

$$\Omega(B_6,\lambda) = \begin{cases} B(1) & \text{if } \lambda \equiv 1,5 \pmod 6 \\ B(2)\setminus\{4\} & \text{if } \lambda \equiv 2,4 \pmod 6 \\ B(3)\setminus\{5\} & \text{if } \lambda \equiv 3 \pmod 6 \\ B(6)\setminus X_6 & \text{if } \lambda \equiv 0 \pmod 6 \\ & \text{where } \{4,5\} \subseteq X_6 \subseteq \{4,5,8,14,20\} \end{cases}$$

$$\left.\begin{array}{r}\Omega(B_7,\lambda)\\ \Omega(B_8,\lambda)\end{array}\right\} = \begin{cases} B(1) & \text{if } \lambda \equiv 1 \pmod 2 \\ B(2)\setminus\{6\} & \text{if } \lambda \equiv 0 \pmod 2 \end{cases}$$

$$\Omega(B_9,\lambda) = \begin{cases} B(1) & \text{if } \lambda \equiv 1,5 \pmod 6 \\ B(1)\cup\{4\} & \text{if } \lambda \equiv 2,4 \pmod 6 \\ B(1)\cup\{5\} & \text{if } \lambda \equiv 3 \pmod 6 \\ B(1)\cup\{4,5\} & \text{if } \lambda \equiv 0 \pmod 6 \end{cases}$$

$$\left.\begin{array}{r}\Omega(B_{10},\lambda)\\ \Omega(B_{11},\lambda)\end{array}\right\} = \begin{cases} B(1) & \text{if } \lambda \equiv 1,5 \pmod 6 \\ B(1)\cup\{4,6\} & \text{if } \lambda \equiv 2,4 \pmod 6 \\ B(1)\cup\{5\} & \text{if } \lambda \equiv 3 \pmod 6 \\ B(1)\cup\{4,5,6\} & \text{if } \lambda \equiv 0 \pmod 6 \end{cases}$$

$$\Omega(B_{12},\lambda) = (B(\lambda)\cap\{v : v \geq \lambda+2\})\cup\{3,4,5,6,7\}$$

$$\Omega(B_{13},\lambda) = (B(\lambda)\cap\{v : v \geq \lambda+2\})\cup\{3,4\}$$

$$\Omega(B_{14},\lambda) = (B(\lambda)\cap\{v : v \geq \lambda+2\})\cup\{3\}$$

$$\Omega(B_{15},\lambda) = \begin{cases} B(1) & \text{if } \lambda \equiv 1 \pmod 2 \\ B(2) & \text{if } \lambda \equiv 0 \pmod 2 \end{cases}$$

$$\Omega(B_{16},\lambda) = B(\lambda)\cap\{v : \lambda \leq 2v-4\}$$

Table 13.3 *Avoidance sets for 4-line configurations*

elements from X would be required, which would induce a 4-cycle in G. The number of triples chosen is $(n-m)\cdot cm^{3/2}$, which is bounded below by $c_1 n^{5/2}$ for c_1 a constant. □

Akbari, Khosrovshahi, Maysoori, and Shariari (1997) also examine this question. Theorem 13.53 indicates the substantial difference in avoiding configurations in a simple hypergraph, and avoiding the same configurations in a design. While avoiding Pasch configurations, for example, it is possible to choose a number of triples which would correspond, in some sense, to a triple system of index $c\sqrt{n}$. However, even the problem of quadrilateral-free designs of index one is challenging.

13.6 Block intersection graphs

Often results on configurations in triple systems are proved by establishing graph-theoretic properties of block intersection graphs. For $\mathcal{I} \subset \{0,1,2,3\}$, the *$\mathcal{I}$-block intersection graph* of a collection $\mathcal{B}$ of triples is the graph whose vertex set is the triples in $\mathcal{B}$, so that two vertices B and $\widehat{B}$ are adjacent if the corresponding triples share exactly ℓ elements and $\ell \in \mathcal{I}$. When $\mathcal{I} = \{i\}$, the shorter notation of i-block intersection graph is used. When $\mathcal{I} = \{1,2,3\}$, the $\mathcal{I}$-block intersection graph is simply the block intersection graph.

Alspach, Heinrich, and Mohar (1990) attribute to Graham the conjecture that the block intersection graph of a Steiner triple system is always hamiltonian. In this section, we establish a much stronger result using a technique of Horák, Pike, and Raines (1999). The stronger version applies to 1-block intersection graphs; for STSs, the 1-block intersection graph and the block intersection graph coincide. Some preliminary results are needed.

Theorem 13.54 *The 1-block intersection graph of a TS(v,λ) has no independent set with more than $\lambda v/2$ vertices.*

Proof Let $S = (V, \mathcal{B})$ be a TS(v, λ). Let $\mathcal{N} \subset \mathcal{B}$ be an independent set in the 1-block intersection graph of S. We show that each element $x \in V$ occurs in at most $3\lambda/2$ triples of $\mathcal{N}$. Suppose a triple $\{x,y,z\} \in \mathcal{N}$. If all triples in $\mathcal{N}$ containing the element x also contain y, then x occurs in at most λ triples of $\mathcal{N}$. Otherwise, there exists an element $t \in V$ so that any triple in $\mathcal{N}$ containing the element x has the form $\{x,y,z\}$, $\{x,y,t\}$, or $\{x,z,t\}$. Then denote by c_i for $i \in \{1,2,3\}$ the number of occurrences of the triples $\{x,y,z\}$, $\{x,y,t\}$, and $\{x,z,t\}$ in $\mathcal{N}$, respectively. As each of the pairs $\{x,y\}$, $\{x,z\}$, and $\{x,t\}$ occurs in exactly λ triples of S we find that $c_i + c_j \leq \lambda$ for any $1 \leq i < j \leq 3$, and hence $c_1 + c_2 + c_3 \leq 3\lambda/2$. Each triple contains three elements, and thus $\mathcal{N}$ contains at most $\lambda v/2$ triples. □

Theorem 13.54 is tight: equality occurs in a TS$(4n, \lambda)$ that contains n disjoint sub-TS$(4,\lambda)$s.

Theorem 13.55 *When $v \geq 12$, the 1-block intersection graph of a TS(v,λ) has vertex connectivity of at least $\lambda v/2$.*

Proof Let $(V, \mathcal{B})$ be a TS(v, λ) and let $G = (\mathcal{B}, E)$ be its 1-block intersection graph. Let $\mathcal{C} \subset \mathcal{B}$ be a vertex cutset of G. Let $\mathcal{A}$ be the vertex set of one component of the graph obtained from G by deleting all vertices of $\mathcal{C}$, and let $\mathcal{D} = \mathcal{B} \setminus (\mathcal{A} \cup \mathcal{C})$. Then define

$$\begin{aligned} V_A &= \{x : \exists T \in \mathcal{A} \text{ such that } x \in T\}, \\ V_D &= \{x : \exists T \in \mathcal{D} \text{ such that } x \in T\}, \\ V_C &= V \setminus (V_A \cup V_D). \end{aligned}$$

Since $\mathcal{A}$ and $\mathcal{D}$ each contain at least one triple, $|V_A| \geq 3$ and $|V_D| \geq 3$. If $|V_C| \geq 2$, then $\mathcal{C}$ contains at least $2\lambda(v-1)/2 - \lambda = \lambda(v-2) \geq \lambda v/2$ triples. So we suppose henceforth that $|V_C| \leq 1$.

If $V_A \cap V_D = \varnothing$, then whenever $x \in V_A$ and $y \in V_D$, all λ triples containing the pair $\{x,y\}$ appear in $\mathcal{C}$. So $\mathcal{C}$ contains at least $\min(\frac{\lambda}{2}x(v - |V_C| - x) : 3 \leq x \leq v - |V_C| - 3)$ triples. Since $|V_C| \leq 1$, $|\mathcal{C}| \geq (3/2)\lambda(v-4)$, which is at least $\lambda\frac{v}{2}$ since $v \geq 7$.

It remains to treat cases in which $V_A \cap V_D \neq \varnothing$. So suppose that $x \in V_A \cap V_D$, and consider a triple $T = \{x,y,z\} \in \mathcal{A}$. If $T' \in \mathcal{D}$ and $x \in T'$, we may suppose without loss of generality that $y \in T'$. If, in addition, $z \in T'$, then let γ denote the number of occurrences of the triple $\{x, y, z\}$ in $\mathcal{B}$. The number of triples sharing at least two elements with T and T' is then $\gamma + 3(\lambda - \gamma)$. Since each of x, y, and z is in $\lambda(v-1)/2$ triples of $\mathcal{B}$, there are $3\lambda(v-1)/2 - 3\lambda + \gamma = 3\lambda(v-3)/2 + \gamma$ triples of $\mathcal{B}$ that contain at least one of x, y, and z. Hence there are $3\lambda(v-3)/2 - 3\lambda + 3\gamma = 3\lambda(v-5)/2 + 3\gamma$ triples that contain exactly one of x, y, and z. Each such triple is in $\mathcal{C}$. Since $\gamma \geq 2$, $\mathcal{C}$ is large enough provided that $3\lambda(v-5)/2 + 6 \geq \lambda\frac{v}{2}$, which occurs when $v \geq 8$.

Finally, we suppose that the third element of T' is w, and $w \neq z$. There are $2\lambda(v-1)/2 - \lambda = 2\lambda(v-2)/2$ triples containing at least one of x or y. Let α be the number of occurrences of the triple $\{x, z, w\}$ and β be the number of occurrences of the triple $\{y, z, w\}$ in $\mathcal{B}$. At most $5\lambda - \alpha - \beta$ triples of $\mathcal{B}$ contain one of the pairs $\{x, y\}$, $\{x, z\}$, $\{x, w\}$, $\{y, z\}$, and $\{y, w\}$. But then at least $\lambda(v-2) - 5\lambda + \alpha + \beta$ triples in $\mathcal{B}$ contain exactly one of x or y and neither of w or z; and $\lambda - \alpha - \beta$ triples of $\mathcal{B}$ contain w and z but neither x nor y. Both types of triples must be in $\mathcal{C}$, so that $|\mathcal{C}| \geq \lambda(v-6) \geq \lambda\frac{v}{2}$. □

Theorem 13.56 *When $v \geq 12$, the 1-block intersection graph of a TS(v, λ) is hamiltonian.*

Proof Chvátal and Erdös (1972) prove that any graph whose vertex connectivity is at least as large as the maximum size of its independent sets is hamiltonian. Apply Theorems 13.54 and 13.55. □

Corollary 13.57 *The block intersection graph of an STS(v) is hamiltonian.*

Proof We need only verify the statement when $v \in \{7, 9\}$, and apply Theorem 13.56 for larger values of v. □

Corollary 13.57 was proved by similar techniques by Horák and Rosa (1988), and by somewhat different methods by Alspach, Heinrich, and Mohar (1990).

Theorem 13.56 has applications to decomposing triple systems into configurations. Hamiltonicity of other block intersection graphs is also of interest. Colbourn and Johnstone (1984) examine the question of when the blocks of a simple TS(v, λ) can be cyclically ordered so that every pair of consecutive blocks shares two elements. When such a cyclic ordering exists, the blocks can be listed with a *minimum change* (one replacement of a symbol) to get from each block to the next. These *minimum change* triple systems are precisely those whose 2-block intersection graph is hamiltonian. Colbourn and Johnstone (1984) establish that every component of the 2-block intersection graph of a simple TS$(v, 2)$ is 3-connected, but in general need not be hamiltonian. They establish that minimum change TS$(v, 2)$s do exist for all orders $v \equiv 0, 1 \pmod 3$, $v \geq 7$.

13.7 Decompositions

The study of decompositions originated with a question of Füredi. Suppose that the number of triples in a Steiner triple system is divisible by three. Can the set of triples of this STS be partitioned into triangles? A *triangle* of triples is the configuration B_5, a set of three lines that intersect pairwise but whose intersection is empty.

Füredi's problem remains open. More generally, let C be an ℓ-line configuration and let $(V, \mathcal{B})$ be an STS(v). Write $|\mathcal{B}| = b = q\ell + s$ for $0 \leq s < \ell$. Any q line-disjoint copies of C constitute a *packing of* $(V, \mathcal{B})$ *with* C. When $s = 0$, the packing with C is a *decomposition.*

Two basic questions arise here. First, for which orders v does there exist a TS(v, λ) which admits a packing or decomposition with a given configuration C? Second, for a given configuration C, does every STS(v) for sufficiently large v admit a packing (decomposition) with C?

We consider the latter question first. An ℓ-star (or a *Dutch ℓ-windmill*) is an ℓ-line configuration in which every pair of lines intersects in one and the same point. Horák and Rosa (1988) prove:

Theorem 13.58 *Let $(V, \mathcal{B})$ be an arbitrary STS(v) with $v > 6\ell$. Then there exists a packing of $\mathcal{B}$ with ℓ-stars.*

Proof Let $\mathcal{A} = \{A_i : i \in V\}$ where $A_i = \{B : i \in B \in \mathcal{B}\}$. Let $p = \max\{r : r \leq (v-1)/6 \text{ and } \ell \mid r\}$. Let $\alpha = (v-1)/6 - p$ (so that $0 \leq \alpha < \ell$). Let $c = \lfloor \alpha v/\ell \rfloor$. Horák and Rosa (1988) prove that there exist sets $X_i \subseteq A_i$ for $i \in V$ so that $|X_i| = p+\ell$ for $1 \leq i \leq c$, $|X_i| = p$ for $c < i \leq v$, and $X_i \cap X_j = \varnothing$ for all $i, j \in V$ with $i \neq j$.

Now writing $|\mathcal{B}| = v(v-1)/6 = q\ell + s$ where $0 \leq s < \ell$, we employ the sets $\{X_i\}$ and partition each into ℓ-stars. It can be checked that in this way exactly s triples are not chosen for inclusion in one of the stars. □

An *ℓ-PPC* (partial parallel class) is a configuration of ℓ pairwise disjoint lines. In contrast to the ℓ-star which is constant, the ℓ-PPC is variable. Nevertheless, an analogue of Theorem 13.58 can be established:

Theorem 13.59 *Let $(V, \mathcal{B})$ be an arbitrary STS(v) with $v > 9\ell$. Then there exists a packing of $\mathcal{B}$ with ℓ-PPCs.*

Proof Any ℓ-PPC corresponds to an independent set in the block intersection graph G of $\mathcal{B}$. A packing of $\mathcal{B}$ by ℓ-PPCs is equivalent to the existence in the complement $\overline{G}$ of a factor whose components are ℓ-cliques, except for at most $\ell-1$ vertices appearing in smaller components. A theorem of Hajnal and Szemerédi (1970) guarantees the existence of such a factor when $v > 9\ell$. □

Now we turn to packings with specific configurations with few lines.

Theorem 13.60 *The following statements hold:*

1. *Any STS(v) admits a packing with 2-stars.*
2. *For $v \neq 7, 9$, any STS(v) admits a packing with 2-PPCs.*

Proof Horák and Rosa (1988) show that the block intersection graph G of any STS(v) is hamiltonian; see Corollary 13.57. By a theorem of Dirac (1952), the complement of the block intersection graph is also hamiltonian whenever $v \geq 13$.

A packing with 2-stars (with 2-PPCs) corresponds to the presence of a 1-factor or near 1-factor in G (in $\overline{G}$, respectively). □

Theorem 13.61 *The following statements hold:*

1. *Every STS(v) with $v \geq 13$ admits a packing with 3-stars.*
2. *Every STS(v) with $v \geq 27$ admits a packing with 3-PPCs.*
3. *Every STS(v) with $v \geq 55$ admits a packing with huts (B_2 configurations).*
4. *Every cyclic STS(v) with $v \geq 13$ admits a packing with 3-snakes (B_4 configurations).*

Proof For $v \geq 19$, (1) follows from Theorem 13.58. The cases when $v = 13$ and $v = 15$ can be verified exhaustively. For $v > 27$, (2) follows from Theorem 13.59. When $v = 27$, the block intersection graph has a complement with 117 vertices which is regular of degree 80; a theorem of Corradi and Hajnal (1963) establishes that it has a triangle-factor. In addition, statement (2) holds when $v \in \{9, 13, 15\}$, but remains open when $v \in \{19, 21, 25\}$. For statements (3) and (4), see Horák and Rosa (1988). □

Packing with triangles, the subject of Füredi's original question, is much less understood. Mullin, Poplove, and Zhu (1987) prove that there is an STS(v) admitting a packing with triangles for every admissible order v. Horák and Rosa (1988) show that when $v \equiv 1, 19 \pmod{72}$, the number of STS(v)s admitting a packing with triangles tends to infinity with v. All STS(15)s admit a packing with triangles. It therefore appears reasonable to expect an affirmative answer to Füredi's original question:

Conjecture 13.62 *When $v \geq 13$, every STS(v) admits a packing with each of the five 3-line configurations.*

Lindner and Rodger (1989) prove that, for $v \geq 6$ and $v \equiv 0, 1 \pmod 3$, there exists a TS(v, 2) admitting a packing with triangles.

13.7.1 *Four-line configurations*

By Theorem 13.58, every STS(v) with $v \geq 25$ admits a packing with 4-stars (configuration C_7). By Theorem 13.59, every STS(v) with $v \geq 37$ admits a packing with 4-PPCs (configuration C_1). For an STS(v) to admit a decomposition into a given 4-line configuration, it is necessary that $v \equiv 1, 9 \pmod{24}$. Griggs, de Resmini, and Rosa (1992) show that every STS(v) satisfying this condition can be decomposed into copies of C_3 when $v \geq 57$, and into copies of C_4 when $v \geq 169$.

For some of the 4-line configurations, it is certainly untrue that any Steiner triple system of sufficiently large order can be decomposed into copies of the configuration. In §13.2, for example, numerous systems are given that contain no Pasch configurations whatsoever. Similarly, the projective spaces avoid the configuration C_{14}. However, the remaining configurations are unavoidable, and

Griggs, de Resmini, and Rosa (1992) conjecture that for each unavoidable configuration C_i, there exists an integer $v_0(C_i)$ for which every STS(v) with $v \equiv 1, 9$ (mod 24) and $v > v_0(C_i)$ admits a decomposition into copies of C_i. Progress has only been made when $i \in \{3, 4\}$.

For a configuration C, the *spectrum* of C, denoted $S(C)$, is the set of orders v for which an STS(v) exists which is decomposable into copies of C. Griggs, de Resmini, and Rosa (1992) show:

Theorem 13.63 *For* $1 \le i \le 16$, $S(C_i) = \{v : v \equiv 1, 9 \pmod{24}, v \ge 25\} \cup \Delta_i$, *where* $\Delta_i = \{9\}$ *when* $i \in \{11, 12, 13, 15\}$, *and* $\Delta_i = \varnothing$ *otherwise.*

The main tool in proving Theorem 13.63 is:

Lemma 13.64 *Suppose there exists a* K*-GDD of type* $g_1^{u_1} \cdots g_s^{u_s}$ *with* $v = \sum_{i=1}^{s} u_i g_i$. *Further suppose that for each* $k \in K$, *there exists a 3-GDD of type* w^k *whose blocks can be partitioned into copies of* C_i. *Finally suppose that for* $1 \le i \le s$, *we have* $wg_i + 1 \in S(C_i)$. *Then* $wv + 1 \in S(C_i)$.

Proof This is a standard application of Wilson's Fundamental Construction, giving weight w to the K-GDD of type $g_1^{u_1} \cdots g_s^{u_s}$ using the 3-GDDs of type w^k, and then filling in groups with the systems of order $wg_i + 1$ for $1 \le i \le s$ appending in the process one point at infinity. □

The proof of Theorem 13.63 is obtained by applying Lemma 13.64 to some small ingredient GDDs each decomposable into copies of C_i; these appear in Griggs, de Resmini, and Rosa (1992).

Adams, Billington, and Rodger (1994) examine decompositions into Pasch configurations, and establish:

Theorem 13.65 *There is a TS*(v, λ) *which has a partition into Pasch configurations if and only if* $(v, \lambda) \ne (9, 1)$, $v \ge 6$, *and* v *and* λ *satisfy the conditions:*

v	λ (mod 12)
1, 9 (mod 24)	1, 5, 7, 11
1, 9 (mod 12)	2, 10
1 (mod 8)	3, 9
0, 1 (mod 3)	4, 8
1 (mod 4)	6
all	0

13.7.2 *Simultaneous decompositions*

Let $\mathcal{C}$ be a set of configurations, not necessarily each having the same number of lines. The *admissible set* for $\mathcal{C}$, denoted $A(\mathcal{C})$, is the set of orders v for which $v(v-1)/6$ is evenly divisible by $|C|$ for each $C \in \mathcal{C}$. Griggs, Mendelsohn, and Rosa (1994) consider *simultaneous* decompositions into STSs into configurations from $\mathcal{C}$. The *spectrum* for $\mathcal{C}$, denoted $S(\mathcal{C})$, is the set of orders v for which there exists an STS(v) which is decomposable into copies of C for each $C \in \mathcal{C}$.

Let $\mathcal{C}_n$ be the set of all n-line configurations. By a simple computation, $S(\mathcal{C}_2) = \{v : v \equiv 1, 9 \pmod{12}, v \geq 13\}$. Now $A(\mathcal{C}_3) = \{v : v \equiv 1, 9 \pmod{18}\}$, and Griggs, Mendelsohn, and Rosa (1994) show that $S(\mathcal{C}_3) = \{v : v \equiv 1, 9 \pmod{18}, v \geq 19\}$. Naturally, the truth of Conjecture 13.62 would also imply this.

The spectrum $S(\mathcal{C}_4)$ is not determined at present. Griggs, Mendelsohn, and Rosa (1994) show that $\{v : v \equiv 1 \pmod{24}, v \neq 97\} \cup \{v : v \equiv 33 \pmod{96}\} \subseteq S(\mathcal{C}_4)$. For the general situation, they show:

Lemma 13.66 *If C, C' are arbitrary configurations, then the following statements are equivalent.*

1. $S(\{C, C'\}) \neq \varnothing$.
2. *There exists an integer v such that there exists a partial STS(v) whose set of lines is simultaneously decomposable into copies of C and of C'.*
3. *There exists an integer $v_0(C, C')$ such that $\{v : v \in A(\{C, C'\})$ and $v > v_0(C, C')\} \subseteq S(\{C, C'\})$.*

Proof The only nontrivial implication is that (2) implies (3). This follows from the theorem of Wilson (1974) concerning the asymptotic existence of graph decompositions. □

By Lemma 13.66, in order to show that $S(\{C, C'\})$ contains all sufficiently large orders in $A(\{C, C'\})$, it suffices to produce a single partial STS having decompositions into copies of C and into copies of C'. Lemma 13.67 demonstrates this for a large class of configurations, using the concept of strong colouring from Chapter 18.

Lemma 13.67 *If C and C' are strongly 3-colourable configurations, then there exists a partial STS which has decompositions into copies of C and into copies of C'.*

The proof in Griggs, Mendelsohn, and Rosa (1994) consists in showing that the weak product of C and C' is simultaneously decomposable into copies of C and C'. Lemma 13.67 applies to a wide class of configurations, including all 4-line configurations except for C_{15}. In addition, all path-like, cycle-like, tree-like, and bipartite-like configurations are strongly 3-colourable. Taking the weak product of a strongly 3-colourable configuration with an arbitrary configuration gives a strongly 3-colourable configuration.

Lemma 13.68 *For any set $\mathcal{C}$ of strongly 3-colourable configurations, there exists a constant $v_0(\mathcal{C})$ for which $A(\mathcal{C}) \cap \{v : v > v_0(\mathcal{C}) \subseteq S(\mathcal{C})\}$.*

Griggs, Mendelsohn, and Rosa (1994) state:

Conjecture 13.69 *For two arbitrary configurations C and C', there exists a partial STS which is simultaneously decomposable into copies of C and C'.*

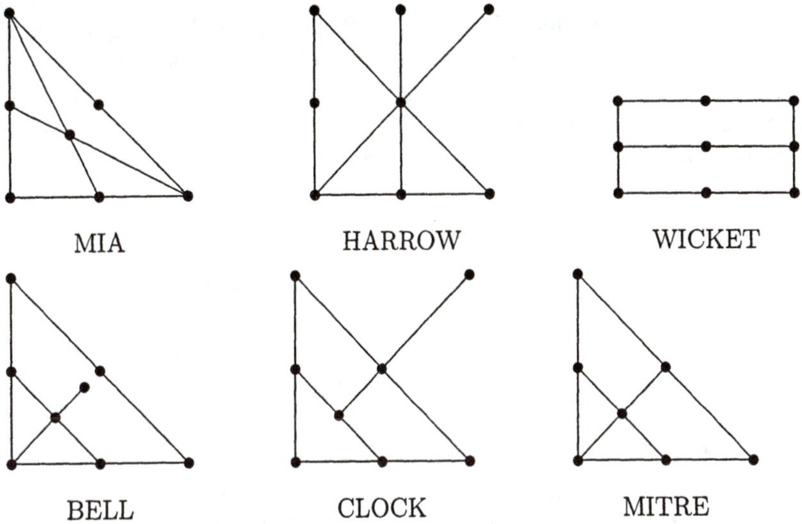

FIG. 13.4. Some 5-line configurations

13.7.3 *Ubiquity*

An STS $(V, \mathcal{B})$ is *C-ubiquitous* if every line of $\mathcal{B}$ is contained in at least one copy of the configuration C. Similarly, $(V, \mathcal{B})$ is *$\mathcal{C}$-ubiquitous* for a set $\mathcal{C}$ of configurations when it is C-ubiquitous for each $C \in \mathcal{C}$; and it is *n-ubiquitous* when it is $\mathcal{C}_n$-ubiquitous.

If C is a configuration such that every sufficiently large STS is C-ubiquitous, then evidently C is an unavoidable configuration. Mendelsohn and Rosa (1997) conjecture that the converse is also true, and verify it for each configuration with at most six lines.

The *ubiquity spectrum* $U(C)$ for a configuration C is the set of orders for which a C-ubiquitous STS exists; $U(\mathcal{C})$ and $U(n)$ are defined analogously. By our remarks, $U(2) = \{v : v \equiv 1, 3 \pmod 6, v \geq 9\}$ and $U(3) = \{v : v \equiv 1, 3 \pmod 6, v \geq 13\}$. The case when $n = 4$ is more complicated as $\mathcal{C}_4$ contains two avoidable configurations. Nevertheless, Mendelsohn and Rosa (1997) show that $U(4) = \{v : v \equiv 1, 3 \pmod 6, v \geq 13\}$. The proof is obtained by utilizing a construction that is similar to Lemma 13.64, as well as the 'Pendant Star' Lemma 13.70.

A *pendant star* in a configuration C is a subconfiguration D which is a star for which no triple of $C \setminus D$ contains the centre of the star D, and every line of D has at least one element of degree 1.

Lemma 13.70 *(Pendant Star) Let D be a pendant star in a configuration C, and let k be the degree of the centre of the star. Let q be the number of pairs of points that do not occur on a line of $C \setminus D$. Then $v > |C \setminus D| + 2k + 1 + q$ implies that a $(C \setminus D)$-ubiquitous STS(v) is also C-ubiquitous.*

Lemma 13.70 establishes ubiquity for all but six of the 56 5-line configurations. The six 5-line configurations not having a pendant star are the mia, harrow, wicket, bell, clock, and mitre shown in Figure 13.4.

Mendelsohn and Rosa (1997) establish that the ubiquity spectra for the mia, bell, and clock are each $\{v : v \equiv 1, 3 \pmod 6), v \geq 13\}$, while for the harrow, wicket, and mitre, they are each $\{v : v \equiv 1, 3 \pmod 6), v \geq 9\}$. Determining the ubiquity spectrum for the mitre appears by far the most difficult.

To determine $U(5)$ appears more difficult yet. There does exist a 5-ubiquitous STS(19), the cyclic system with base blocks {0,1,4}, {0,5,11}, and {0,2,9}. However, not much else is known.

One interesting case for ubiquity occurs when the configuration is itself an STS.

13.8 Work points

1. Prove Conjecture 13.2, that the set consisting of any constant configuration and all m-line configurations with $m \leq n$ having all points of degree at least 2 forms a basis for n-line configurations in Steiner triple systems.
2. Prove Conjecture 13.4, that the only constant n-line configurations are those obtained by adding a single line to an $(n-1)$-star.
3. Prove that a quadrilateral-free Steiner triple system of order v exists if and only if $v \equiv 1, 3 \pmod 6$ and $v \notin \{7, 13\}$.
4. Prove that an anti-mitre Steiner triple system of order v exists if and only if $v \equiv 1, 3 \pmod 6$ and $v \neq 9$.
5. Determine the spectrum for 5-sparse Steiner triple systems.
6. Prove or disprove Erdös's conjecture: for every r, there exists an integer $v_0(r)$ such that if $v > v_0(r)$ and $v \equiv 1, 3 \pmod 6$, there is an STS(v) in which no set of $j+2$ points carries j triples for $2 \leq j \leq r$.
7. Prove that every STS(v) with $v \in \{19, 21, 25\}$ admits a packing with 3-PPCs.
8. Prove that every STS(v) with $v \geq 9$ admits a packing with 3-snakes.
9. (Füredi's problem) Prove that every STS(v) admits a packing with triangles.
10. Show that $\{v : v \equiv 9 \pmod{24}\} \setminus \{9\} \subseteq S(C_4)$.
11. Prove Conjecture 13.69.

14

INTERSECTIONS

In this chapter, we examine the intersections of two triple systems on the same set of elements. We then consider the existence of disjoint Steiner triple systems that exhibit a stronger orthogonality property.

14.1 Making two STSs disjoint

Given two STS(v)s $(V, \mathcal{B}_1)$ and $(V, \mathcal{B}_2)$, we examine the *intersection size* $|\mathcal{B}_1 \cap \mathcal{B}_2|$. Two STS($v$)s are *disjoint* if their intersection size is 0. Teirlinck (1977) proved the remarkable theorem that every two STS(v)s can be made disjoint:

Theorem 14.1 *If $(V, \mathcal{B}_1)$ and $(V, \mathcal{B}_2)$ are STS(v)s, $v \geq 7$, there exists an STS(v) $(V, \mathcal{B}_3)$ so that $\mathcal{B}_1 \cap \mathcal{B}_3 = \varnothing$ and $(V, \mathcal{B}_2) \cong (V, \mathcal{B}_3)$.*

Proof If $\mathcal{B}_1 \cap \mathcal{B}_2 = \varnothing$, we are done. Otherwise, we determine a relabelling of the elements of $(V, \mathcal{B}_2)$ that reduces the number of triples in common with $\mathcal{B}_1$. Fix a triple $\{1, 2, 3\} \in \mathcal{B}_1 \cap \mathcal{B}_2$. Define the *spread* σ of element 3 to be the set

$$\sigma = \{1, 2, 3\} \cup \Gamma \cup \Delta,$$

where

$$\Gamma = \{a : \{z, w, a\} \in \mathcal{B}_1 \setminus \{\{1, 2, 3\}\} \text{ and } \{z, w, 3\} \in \mathcal{B}_2\} \text{ and}$$

$$\Delta = \{b : \{x, y, b\} \in \mathcal{B}_2 \setminus \{\{1, 2, 3\}\} \text{ and } \{x, y, 3\} \in \mathcal{B}_1\}.$$

Now if the spread σ has $|\sigma| < v$, let $\theta \in V \setminus \sigma$. Permuting the elements of $(V, \mathcal{B}_2)$ by interchanging 3 and θ, the number of triples in common with $(V, \mathcal{B}_1)$ is necessarily reduced by at least one.

It remains then to consider the case when the spread of each element in each triple of $\mathcal{B}_1 \cap \mathcal{B}_2$ contains all v elements. In this case, any transposition as used above leaves the number of triples in common unchanged (providing that $\theta \in \Gamma \cup \Delta$). Then consider a triple $\{1, 2, 3\}$ in both systems. Select a further triple $\{3, x, y\} \in \mathcal{B}_2$, and let $\{x, y, c\} \in \mathcal{B}_1$. Since the spread of element 3 has v elements, $c \notin \{1, 2, 3\}$. Next examine $\{3, c, e\} \in \mathcal{B}_2$, and finally examine $\{c, d, e\} \in \mathcal{B}_1$. Again, $\{c, d, e\} \cap \{1, 2, 3\} = \varnothing$ since the spread of 3 has v elements. Now interchange 3 and d in $(V, \mathcal{B}_2)$, so that the triple $\{1, 2, 3\}$ is no longer in common, but the triple $\{c, d, e\}$ now is shared. However, with respect to this triple, elements c and d have spread smaller than v, and hence a second transposition can be applied to reduce the number of triples in common.

Applying these repeatedly makes the systems disjoint. □

No such result is known for making three or more systems disjoint. In fact, Cayley (1850) showed that there are only two mutually disjoint STS(7)s, and hence an extension to three or more systems, if found, must employ additional information.

14.2 Intersections of STSs

Two triple systems $(V, \mathcal{B}_1)$ and $(V, \mathcal{B}_2)$ on the same set are said to *intersect in k triples* if $|\mathcal{B}_1 \cap \mathcal{B}_2| = k$; they are *disjoint* if $k = 0$ and *almost disjoint* if $k = 1$. Define

$$\text{Int}(v, \lambda) = \{k \ : \ \exists \text{ two simple } TS(v, \lambda)\text{s intersecting in } k \text{ triples}\}.$$

The *intersection problem* for TS(v, λ)s is to determine Int(v, λ). We begin with Steiner triple systems, and write Int(v) for Int$(v, 1)$.

Examination of the 30 distinct STS(7)s on the same set reveals that for any fixed STS(7), there are exactly eight STS(7)s that are disjoint from it, exactly 14 that are almost disjoint, and exactly seven that intersect it in three triples. Hence Int$(7) = \{0, 1, 3, 7\}$.

Kramer and Mesner (1974) have determined completely the intersection sizes for $v = 9$, obtaining Int$(9) = \{0, 1, 2, 3, 4, 6, 12\}$. For any given STS(9), there are 192 other STS(9)s disjoint from it, 216 that are almost disjoint, 216 that intersect it in two triples, 152 that intersect it in three, 27 that intersect it in four, and 36 that intersect it in six. Any two pairs of STS(9)s intersecting in k triples are isomorphic, except when $k = 0$ or 3, in which cases there are two nonisomorphic pairs. The 192 pairs of disjoint systems fall into two orbits of sizes 148 and 44, and the 152 pairs intersecting in three triples fall into two orbits of sizes 144 and 8.

Two partial triple systems $(V, \mathcal{D}_1)$ and $(V, \mathcal{D}_2)$ are *mutually balanced* if any 2-subset of V is contained in a triple of $\mathcal{D}_1$ if and only if it is contained in a triple of $\mathcal{D}_2$. Two partial triple systems are *disjoint* if they have no triples in common.

For example, two PTSs with four triples on the set $\{1, 2, 3, 4, 5, 6\}$ are disjoint and mutually balanced exist:

$$\mathcal{D}_1 = \{134, 156, 235, 246\} \qquad \mathcal{D}_2 = \{135, 146, 234, 256\}.$$

However, the following is easily checked:

Lemma 14.2 *There exists no pair of disjoint mutually balanced PTSs each having one, two, three, or five triples.*

As a consequence, we define $b_v = v(v-1)/6$, and

$$\mathcal{I}_v = \{0, 1, \ldots, v\} \setminus \{b_v - 1, b_v - 2, b_v - 3, b_v - 5\},$$

and obtain:

Lemma 14.3 *For all* $v \equiv 1, 3 \pmod 6$, *Int*$(v) \subseteq \mathcal{I}_v$.

Proof The statement that $b_v - i \in \text{Int}(v)$ is equivalent to saying that there exists a pair of disjoint mutually balanced PTSs, each having i triples. □

Lizzio and Milici (1983, 1985) have undertaken a fairly extensive study of pairs of disjoint mutually balanced PTSs.

Lindner and Rosa (1975c) completely determined the set Int(v). The main tools in the proof of Theorem 14.6 are the $2v+1$ and $2v+7$ constructions of §3.2. In Lemmas 14.4 and 14.5, these constructions are adapted for use in the intersection problem.

Lemma 14.4 *For* $v \geq 13$, *Int*$(v) = \mathcal{I}_v$ *implies Int*$(2v+1) = \mathcal{I}_{2v+1}$.

Proof Let $(V, \mathcal{B}_1)$ and $(V, \mathcal{B}_2)$ be two STS(v)s intersecting in k triples, and let V, X, and $\mathcal{F} = \{F_1, \ldots, F_v\}$ be as in §3.2; here $\mathcal{F}$ is a 1-factorization. Let α be any permutation of V fixing exactly s elements. Let $\mathcal{C} = \bigcup_{i \in V} i \cdot F_i$, $\mathcal{C}_\alpha = \bigcup_{i \in V} i\alpha \cdot F_i$. Then $(V \cup X, \mathcal{B}_1 \cup \mathcal{C})$ and $(V \cup X, \mathcal{B}_2 \cup \mathcal{C}_\alpha)$ are both STS$(2v+1)$s. Each 1-factor of $\mathcal{F}$ contains $(v+1)/2$ edges, and thus $\mathcal{C}$ and $\mathcal{C}_\alpha$ have exactly $s(v+1)/2$ triples in common. Thus $k + s(v+1)/2 \in \text{Int}(2v+1)$ whenever $k \in \text{Int}(v)$. Choosing consecutively $s = 0, 1, 2, \ldots, v-2$, we get $\{0, 1, 2, \ldots, b_{2v+1} - (v+7)\} \subset \text{Int}(2v+1)$, while $s = v$ gives

$$\{b_{2v+1} - b_v, b_{2v+1} - b_v + 1, \ldots, b_{2v+1} - 6, b_{2v+1} - 4, b_{2v+1}\} \subset \text{Int}(2v+1).$$

Since $b_v \geq 7$ for $v \geq 13$, we have $b_{2v+1} - (v+7) \geq b_{2v+1} - b_v$, which completes the proof. □

Lemma 14.5 *For* $v \geq 15$, *Int*$(v) = \mathcal{I}_v$ *implies Int*$(2v+7) = \mathcal{I}_{2v+7}$.

Proof Let $(V, \mathcal{B}_1)$ and $(V, \mathcal{B}_2)$ be two STS(v)s intersecting in k triples, and let V, A, B, G, and $\mathcal{H} = \{H_1, \ldots, H_v\}$ be as in the $2v+7$ construction of §3.2. Let $(B, \mathcal{D}_1)$ and $(B, \mathcal{D}_2)$ be two STS(7)s with $|\mathcal{D}_1 \cap \mathcal{D}_2| = \gamma$ for some $\gamma \in \text{Int}(7)$. Let $\mathcal{C} = \{\{a_i, a_{i+1}, a_{i+3}\} : i \in V\}$, and let $\mathcal{C}_\delta$ be $\mathcal{C}$ if $\delta = 1$, or $\{\{a_i, a_{i+2}, a_{i+3}\} : i \in V\}$ if $\delta = 0$. Further, let α be any permutation of V fixing exactly s elements. Now let

$$\mathcal{E} = \bigcup_{i \in V} i \cdot H_i \quad \text{and} \quad \mathcal{E}_\alpha = \bigcup_{i \in V} i\alpha \cdot H_i.$$

Then $(V \cup A \cup B, \mathcal{B}_1 \cup \mathcal{D}_1 \cup \mathcal{E} \cup \mathcal{C})$ and $(V \cup A \cup B, \mathcal{B}_2 \cup \mathcal{D}_2 \cup \mathcal{E}_\alpha \cup \mathcal{C}_\delta)$ are both STS$(2v+7)$s. Now $|\mathcal{E} \cap \mathcal{E}_\alpha| = s(v+7)/2$ and $|\mathcal{C} \cap \mathcal{C}_\delta| = \delta v$. Thus $k + s(v+7)/2 + \delta v + \gamma \in \text{Int}(2v+7)$ whenever $k \in \text{Int}(v)$.

Choosing $s = 0, 1, \ldots, v-2$ consecutively, we get $\{0, 1, \ldots, b_{2v+7} - (v+19)\} \subset \text{Int}(2v+7)$. Then choosing $s = v$ we obtain

$$\{b_{2v+7} - b_v, b_{2v+7} - b_v + 1, \ldots, b_{2v+7} - 6, b_{2v+7} - 4, b_{2v+7}\} \subset \text{Int}(2v+7).$$

Since $b_v \geq v + 19$ for $v \geq 15$, we obtain the required result. □

Theorem 14.6 *For every $v \geq 13$, $Int(v) = \mathcal{I}_v$.*

Proof For $v \in \{13, 15, 19, 21, 25, 33\}$, Lindner and Rosa (1975) prove the statement directly. They employ the proofs of Lemmas 14.4 and 14.5, and explicitly provide pairs of STS(v)s for certain cases. When $v \in \{27, 31\}$, Lemmas 14.4 and 14.5 apply. So suppose that $v \geq 37$. Assume inductively that $\text{Int}(w) = \mathcal{I}_w$ for all admissible w satisfying $15 \leq w < v$. If $v \equiv 3, 7 \pmod{12}$, then $(v-1)/2 \equiv 1, 3 \pmod 6$ and $15 \leq (v-1)/2 < v$; hence by Lemma 14.4, $\text{Int}(v) = \mathcal{I}_v$. If instead $v \equiv 1, 9 \pmod{12}$, then $(v-7)/2 \equiv 1, 3 \pmod 6$ and $15 \leq (v-7)/2 < v$; hence by Lemma 14.5, $\text{Int}(v) = \mathcal{I}_v$. □

The pairs of STS(v)s in Theorem 14.6 may or may not be isomorphic. Lindner and Rosa (1975) posed the problem of determining the set

$$\text{Int}^\star(v) = \{k : \exists \text{ isomorphic } STS(v) \text{ intersecting in } k \text{ triples}\},$$

but stated incorrectly that $\text{Int}^\star(v) \neq \text{Int}(v)$ when $v > 13$. Müller (1982) was the first to address the problem seriously. A complete answer was given by Koszarek (1983):

Theorem 14.7 *For all $v \equiv 1, 3 \pmod 6$, $Int^\star(v) = Int(v)$.*

Another extension of the problem is to cases with higher index. Define $b_{v,\lambda} = \lambda\binom{v}{2}/3$, and define

$$\mathcal{I}_{v,\lambda} = \{0, 1, \ldots, v\} \setminus \{b_{v,\lambda} - 1, b_{v,\lambda} - 2, b_{v,\lambda} - 3, b_{v,\lambda} - 5\}.$$

The sets $\text{Int}(v, \lambda)$ for $\lambda = \lambda_{\min}$ have been completely determined by Ajoodani-Namini and Khosrovshahi (1995); Guo (1999) independently handled the case when $\lambda = 2$. We have:

Lemma 14.8 *For $\lambda = 2$,*

1. *$Int(4, 2) = \{4\}$;*
2. *$Int(6, 2) = \{0, 4, 6, 10\}$;*
3. *$Int(7, 2) = \{2, 5, 8, 14\}$;*
4. *for $v \equiv 0, 1 \pmod 3$, $v \geq 9$, $Int(v, 2) = \mathcal{I}_{v,2}$.*

Lemma 14.9 *For all $v \equiv 5 \pmod 6$, $Int(v, 3) = \mathcal{I}_{v,3}$.*

Lemma 14.10 *For $\lambda = 6$,*

1. *$Int(8, 6) = \{56\}$;*
2. *$Int(14, 6) = \{0, 4, 6, 7, \ldots, 175, 176, 178, 182\}$;*
3. *for all $v \equiv 2 \pmod 6$, $v > 14$, $Int(v, 6) = \mathcal{I}_{v,6}$.*

14.3 Intersections of 3-GDDs

Butler and Hoffman (1992) considered the generalization of intersection problems to 3-GDDs of type g^t and index one. Two 3-GDDs $(V, \mathcal{G}, \mathcal{B}_1)$ and $(V, \mathcal{G}, \mathcal{B}_2)$ (on the same set, with the same groups) *intersect in k triples* if $|\mathcal{B}_1 \cap \mathcal{B}_2| = k$. Let $b_{g^t} = g^2t(t-1)/6$ (the number of blocks in a 3-GDD of type g^t), and denote by $\mathcal{I}_{g^t}$ the set $\{0, 1, \ldots, b_{g^t}\} \setminus \{b_{g^t}-5, b_{g^t}-3, b_{g^t}-2, b_{g^t}-1\}$. Let $\text{Int}(g^t)$ denote the numbers of blocks in which two 3-GDDs of type g^t intersect. The same argument as in Lemma 14.3 establishes that $\text{Int}(g^t) \subseteq \mathcal{I}_{g^t}$.

Since 3-GDDs of type 1^t are Steiner triple systems, the intersection problem when $g = 1$ is settled in Theorem 14.6. A 3-GDD of type 2^t is obtained from an STS$(2t+1)$ by deleting one point and taking the resulting blocks of size two as groups. In this case, the 3-GDD intersection problem has been called the *flower intersection problem* for STSs (the set of all blocks through a point being colourfully called a 'flower'), and has been solved completely by Hoffman and Lindner (1987).

Another special case is that of 3-GDDs of type g^3, which are equivalent to latin squares of side g. In that setting, Fu (1980) solved the intersection problem. These special cases are all incorporated in a theorem due to Butler and Hoffman (1992):

Theorem 14.11 *Let g and t be positive integers satisfying the conditions of Theorem 3.4. Then $Int(g^t) = \mathcal{I}_{g^t}$ except that:*

1. $Int(1^9) = \mathcal{I}_{1^9} \setminus \{5, 8\}$;
2. $Int(2^4) = \mathcal{I}_{2^4} \setminus \{1, 4\}$;
3. $Int(3^3) = \mathcal{I}_{3^3} \setminus \{1, 2, 5\}$;
4. $Int(4^3) = \mathcal{I}_{4^3} \setminus \{5, 7, 10\}$.

14.4 Orthogonal Steiner triple systems

Two STS(v)s on the same set of elements, say $(V, \mathcal{A})$ and $(V, \mathcal{B})$, are *orthogonal* if $\mathcal{A} \cap \mathcal{B} = \varnothing$, and if

$$\{\{a, b, w\}, \{x, y, w\}\} \subset \mathcal{A} \text{ and } \{\{a, b, s\}, \{x, y, t\}\} \subset \mathcal{B}$$

then $s \neq t$. We denote a pair of orthogonal Steiner triple systems of order v as OSTS(v).

Orthogonal Steiner triple systems were introduced as a means of constructing Room squares (defined in §1.3.6) by O'Shaughnessy (1968). To recover a Room square from orthogonal STS(v)s $(V, \mathcal{A})$ and $(V, \mathcal{B})$, define a $v \times v$ square with rows and columns indexed by the elements of V. Whenever $\{a, b, w\} \in \mathcal{A}$ and $\{a, b, x\} \in \mathcal{B}$, place the pair $\{a, b\}$ in cell (w, x). The result is a Room square of side v. The existence question for Room squares and many restricted types of Room squares was settled long in advance of the solution of the existence problem for OSTSs. Dinitz and Stinson (1992) survey current knowledge about Room squares. Nevertheless, OSTSs have been investigated largely independent of this connection.

O'Shaughnessy (1968) constructed OSTS(v) for $v \in \{7, 13, 19\}$. He conjectured that OSTS(v)s exist whenever $v \equiv 1 \pmod 6$, and further conjectured that none exist when $v \equiv 3 \pmod 6$. It is trivial that no OSTS(3) can exist. Mullin and Nemeth (1969, 1970) established that no OSTS(9) exists. They further established:

Lemma 14.12 *An OSTS(v) exists whenever v is a prime power congruent to* 1 *modulo* 6.

However, Rosa (1974) disproved O'Shaughnessy's conjecture by establishing the existence of OSTS(27). Subsequently, Gibbons (1987) found OSTSs of order 15, and in fact enumerated all nonisomorphic OSTS(15)s.

The set $\mathcal{OSTS} = \{v : \exists OSTS(v)\}$ is PBD-closed. Thus Wilson's techniques (Theorem 3.2) ensure the existence of a finite v_0 so that if $v \geq v_0$ and $v \equiv 1, 3 \pmod 6$, an OSTS(v) exists. A major step in determining the spectrum for OSTSs of orders congruent to 1 modulo 6 came when Mullin and Stinson (1987) examined the spectrum for pairwise balanced designs whose block sizes are prime powers congruent to 1 modulo 6. Their results ensure that an OSTS(v) exists for $v \equiv 1 \pmod 6$ for all $v \geq 1927$, and for all but 31 values less than 1927.

A similar closure result was obtained for the $v \equiv 3 \pmod 6$ class by examining PBDs that permit blocks of sizes 15 and 27 in addition to prime powers congruent to 1 modulo 6. Stinson and Zhu (1991) showed that an OSTS(v) exists for all $v \equiv 3 \pmod 6$, $v \geq 27369$, and for all but at most 917 orders in the range $15 \leq v \leq 27363$. In this class, the smallest unresolved case was that of OSTS(21).

Greig (1999) improved the PBD result for the 1 (mod 6) class; he produces PBDs whose block sizes are prime powers congruent to 1 (mod 6) for orders 295, 655, 1243, 1255, 1795, 1819, and 1921. Stinson and Zhu (1991) eliminated two further cases for OSTSs.

Colbourn, Gibbons, Mathon, Mullin, and Rosa (1994) completed the solution for both classes. Their solution strategy involved two main directions. First, Gibbons and Mathon (1993) developed a powerful hill-climbing technique for finding OSTS(v)s. Then Colbourn, Gibbons, Mathon, Mullin, and Rosa (1994) develop new recursions. We examine these techniques in some detail here, in part because it illustrates a more complicated set of applications of 'Wilson-type' techniques than we have yet treated.

The key ingredient used repeatedly is a generalization of orthogonal Steiner triple systems suggested by Stinson and Zhu (1991). An *orthogonal group divisible design* (OGDD) $(X, \mathcal{G}, \mathcal{B}_1, \mathcal{B}_2)$ is a set X and a partition $\mathcal{G}$ of X into classes (again called *groups*), and two disjoint sets $\mathcal{B}_1$ and $\mathcal{B}_2$ of 3-subsets of X, so that each pair $\{x, y\}$ of elements of X appears once in a 3-subset of $\mathcal{B}_1$ and once in a 3-subset of $\mathcal{B}_2$ if x and y are from different groups, and does not appear in a 3-subset of either if x and y are from the same group. Moreover, if $\{x, y, a\} \in \mathcal{B}_1$ and $\{x, y, b\} \in \mathcal{B}_2$, then a and b are in different groups; and for two distinct intersecting triples $\{x, y, z\}$ and $\{u, v, z\}$ of $\mathcal{B}_1$, the triples $\{x, y, a\}$ and $\{u, v, b\}$ of $\mathcal{B}_2$ satisfy $a \neq b$. We abuse the notation somewhat by referring to the pair of 3-GDDs $(X, \mathcal{G}, \mathcal{B}_1)$ and $(X, \mathcal{G}, \mathcal{B}_2)$ as an OGDD, and also saying that one of the 3-GDDs is *orthogonal* to the other. An OSTS(v) is precisely the same as an OGDD in which there are v groups, each having a single element. Adopting usual notation, an OGDD has *type* $(g_1)^{u_1}\dots(g_s)^{u_s}$ if the OGDD has u_i groups of size g_i for $1 \le i \le s$, and no other groups. Thus an OSTS(v) is the same as an OGDD of type 1^v.

We also use conjugate orthogonal quasigroups (COQs) in a very standard way. Two quasigroups are *conjugate orthogonal* if each conjugate of one is orthogonal to each conjugate of the other. We refer the reader to Stinson and Zhu (1991) for more details concerning COQs.

Lemmas 14.13, 14.14, 14.15, and 14.16 summarize basic constructions from Stinson and Zhu (1991).

Lemma 14.13 *Suppose that there exists an OGDD of type g^u. Further suppose that $v=1$ or that a COQ(v) exists. Then there exists an OGDD of type $(gv)^u$. If in addition there exists an OSTS(gv) (OSTS($gv+1$)) then there exists an OSTS(guv) (OSTS($guv+1$), respectively).*

Lemma 14.14 *(Filling In Groups) If there exists an OGDD of type $u^g v^h$ and there exists an OSTS($u+1$) and an OSTS($v+1$), then there exists an OSTS($gu+hv+1$).*

Lemma 14.15 *(Direct Product) If there exists an OSTS(u) and a COQ(v), then there exists an OGDD of type v^u. If, in addition, there exists an OSTS(v), there exists an OSTS(uv).*

Lemma 14.16 *(Singular Direct Product) If an OSTS(u), a COQ($v-1$), and an OSTS(v) all exist, then there exists an OSTS($u(v-1)+1$).*

To apply these results, COQs are needed. These are produced by the Lemmas 14.17 and 14.18 without further comment:

Lemma 14.17 *If v is a prime power and $v \notin \{2,3\}$, then there exists a COQ(v).*

Lemma 14.18 *If there exists a COQ(u) and a COQ(v), then there exists a COQ(uv).*

14.4.1 *Constructions and uses of OGDDs*

In this section, we first develop some constructions for OGDDs, and then describe applications of Wilson's Fundamental Construction to produce OSTSs from the OGDDs found.

Let G be a finite abelian group of order v and let H be a subgroup of G. Then a (v,k,λ)-*relative difference set* based on G and H is a triple (G,H,F) where F is a family of k-subsets of G which has the property that every element of $G\setminus H$ occurs exactly λ times as a difference of elements in the subsets of F, and no member of H appears as such a difference. Further, let d be a fixed element of $G\setminus H$. Then a (v,k,λ)-*near relative difference set* based on G, H, and d is a quadruple (G,H,F,d) where F is a family of k-subsets which has the property that every element of $G \setminus (H \cup \{\pm d\})$ occurs exactly λ times as a difference of elements in the subsets of F, and no member of $H \cup \{\pm d\}$ occurs as such a difference.

Let $T = \{x, y, z\}$ be a triple of elements of an abelian group G. Then the three pairs $\{x - z, y - z\}$, $\{x - y, z - y\}$, and $\{y - x, z - x\}$ are the *fundamental pairs* of T. Let F be the set of triples of a $(v, 3, 1)$-relative difference set or near relative difference set. The multiset of $3|F|$ fundamental pairs corresponding to the triples of F is the *set of fundamental pairs* of F.

Now let $S_1 = (G, H, F_1)$ and $S_2 = (G, H, F_2)$ be a pair of $(v, 3, 1)$-relative difference sets based on G and H. Then for each fundamental pair $P = \{x, y\}$ of S_1 there is a unique triple $T(P)$ in S_2 having a translate $T'(P)$ containing P, say $T'(P) = \{x, y, z(P)\}$. The multiset $\mathcal{L} = \{z(P) : P \text{ is a fundamental pair of } S_1\}$ is an *orthogonality certificate* for the (ordered) pair (S_1, S_2) of relative difference sets if all members of $\mathcal{L}$ are distinct and no member of $\mathcal{L}$ lies in H.

In an similar way, let $S_1 = (G, H, F_1, d_1)$ and $S_2 = (G, H, F_2, d_2)$ be $(v, 3, 1)$-near relative difference sets. The multiset $\mathcal{L} = \{z(P) : P \text{ is a fundamental pair of } S_1\} \cup \{\pm d_1 \pm d_2\}$ is an *orthogonality certificate* for the ordered pair (S_1, S_2) if all members of $\mathcal{L}$ are distinct and no member of $\mathcal{L}$ lies in H.

The significance of orthogonality certificates for pairs of relative difference sets is given in the Lemmas 14.19 and 14.20.

Lemma 14.19 *Let $S_1 = (G, H, F_1)$ and $S_2 = (G, H, F_2)$ be a pair of $(v, 3, 1)$-relative difference sets. Let $h = |H|$, $v = |G|$, and $n = v/h$. If there exists an orthogonality certificate for the pair (S_1, S_2) then there exists an OGDD of type h^n.*

Proof We consider the members of G to be the elements of the OGDD, and the cosets of H in G to be its groups. The $v|F_1|$ triples $\{T + g : T \in F_1, g \in G\}$ form a $\{3\}$-GDD on G, and similarly the $v|F_2|$ triples $\{T + g : T \in F_2, g \in G\}$ form a $\{3\}$-GDD on G. The second 3-GDD is orthogonal to the first, as can be verified using the definition of the orthogonality certificate. □

The situation for near relative difference sets is somewhat more complicated. Let $S = (G, H, F, d)$ be a $(v, 3, 1)$-near relative difference set. Then S is *partitionable* if d and $-d$ are distinct in G, and there exists a subset P of G containing $|G|/2$ elements and having the property that $\{\{0, d\} + g : g \in G\} = \{\{0, d\} + p : p \in P\} \cup \{\{0, -d\} + p : p \in P\}$. The set P is a *partitioning set.* We write $S = (G, H, F, d, P)$ if (G, H, F, d) is a near relative difference set with partitioning set P.

Lemma 14.20 *Let $S_1 = (G, H, F_1, d_1, P_1)$ and $S_2 = (G, H, F_2, d_2, P_2)$ be a pair of partitionable $(v, 3, 1)$-near relative difference sets. Let $h = |H|$, $v = |G|$, and $n = v/h$. If there exists a orthogonality certificate for (S_1, S_2) then there exists an OGDD of type $h^n 2^1$.*

Proof We take the elements of G, together with two new elements ∞_1 and ∞_2, to be the points of the OGDD. The cosets of H in G form the n groups each of size h, and $\{\infty_1, \infty_2\}$ forms a group of size 2. The set of $v|F_1|$ triples $\{T + g : T \in F_1, g \in G\}$ together with the triples $\{\{\infty_1, 0, d_1\} + p : p \in P_1\} \cup \{\{\infty_2, 0, -d_1\} + p : p \in P_1\}$ form the triples of a 3-GDD. (Here we adopt the usual convention that $\infty_i + g = \infty_i$.) A second 3-GDD is defined similarly using S_2. The second is orthogonal to the first, as can be verified using the definition of the orthogonality certificate. □

Colbourn, Gibbons, Mathon, Mullin, and Rosa (1994) employ Lemmas 14.19 and 14.20 to construct a number of OGDDs, in each case taking G to be the cyclic group of integers modulo v, and the subgroup H is uniquely determined by its order. In order that a near relative difference set be partitionable, it is sufficient that $v/\gcd(v, d)$ be even and that $d \neq v/2$.

In addition to OGDDs produced using relative difference sets, Colbourn, Gibbons, Mathon, Mullin, and Rosa (1994) employ three more OGDDs by employing smaller groups of automorphisms. On $\mathbb{Z}_7 \times \{0, 1\}$, define a 3-GDD of type 2^7 by taking $\{\{i \times \{0, 1\}\} : i \in \mathbb{Z}_7\}$ to form the seven groups, and develop the starter blocks

$$\{\{0_0, 1_0, 4_1\}, \{0_0, 2_0, 1_1\}, \{0_0, 3_0, 5_1\}, \{0_1, 1_1, 3_1\}\}$$

modulo $(7, -)$ to form the triples. Form a second 3-GDD by developing

$$\{\{0_0, 3_1, 4_1\}, \{0_0, 1_1, 6_1\}, \{0_0, 2_1, 5_1\}, \{0_0, 1_0, 3_0\}\}.$$

It is easily verified that the two 3-GDDs are orthogonal.

Similarly, on $\mathbb{Z}_9 \times \{0, 1\}$, form two 3-GDDs by developing the following starter blocks modulo $(9, -)$:

3-GDD # 1	3-GDD # 2
$0_0, 2_0, 8_1$	$0_1, 2_1, 8_0$
$0_0, 4_0, 7_1$	$0_1, 4_1, 7_0$
$0_0, 1_0, 5_1$	$0_1, 1_1, 5_0$
$0_0, 1_1, 2_1$	$0_1, 1_0, 2_0$
$0_1, 2_1, 5_1$	$0_0, 2_0, 5_0$
$0_0, 3_0, 6_0$	$0_1, 3_1, 6_1$

Each of the first five starter blocks develops into nine blocks, while the sixth generates only three distinct blocks. It is an easy exercise to verify that these two 3-GDDs form an OGDD of type 2^9.

Finally, on $\mathbb{Z}_5 \times \{0, 1, 2, 3\}$, we present OGDDs of type 2^{10} having groups $\{\{i \times \{0, 1\}\} : i \in \mathbb{Z}_5\} \cup \{\{i \times \{2, 3\}\} : i \in \mathbb{Z}_5\}$.

The starter blocks for the two 3-GDDs to be developed modulo $(5, -)$ are:

GDD # 1		GDD # 2	
$0_0, 3_0, 4_1$	$0_0, 4_0, 2_1$	$0_0, 3_0, 2_2$	$0_0, 1_0, 1_3$
$0_0, 1_2, 4_2$	$0_0, 2_2, 3_2$	$0_0, 3_1, 2_1$	$0_0, 1_1, 3_2$
$0_0, 0_2, 2_3$	$0_0, 1_3, 3_3$	$0_0, 4_1, 3_3$	$0_0, 0_2, 1_2$
$0_0, 4_3, 0_3$	$0_1, 2_1, 4_2$	$0_0, 2_3, 4_3$	$0_1, 3_1, 0_3$
$0_1, 1_1, 2_3$	$0_1, 3_2, 4_3$	$0_1, 3_2, 1_2$	$0_1, 4_2, 3_3$
$0_1, 1_2, 0_3$	$0_1, 0_2, 3_3$	$0_1, 0_2, 1_3$	$0_2, 2_3, 3_3$

Using these methods, Colbourn, Gibbons, Mathon, Mullin, and Rosa (1994) establish:

Lemma 14.21 *There exists an OGDD of type*

1. 2^n *for* $n \in \{7, 9, 10, 12, 13, 15, 16\}$*;*
2. 3^n *for* $n \in \{11, 17\}$*;*
3. 4^{10} *and* $4^9 2^1$*;*
4. 6^n *for* $n \in \{8, 9\}$.

Now we turn to constructions that use OGDDs to produce OSTSs. The main construction is an application of Wilson's Fundamental Construction:

Lemma 14.22 *If there exist an OGDD of type $g^n u^1$, an OGDD of type $g^n v^1$, and a TD$(n+1, m)$, then there exists an OGDD of type $(mg)^n((m-t)u+tv)^1$ for all $0 \le t \le m$.*

Proof Let $G_1, \ldots, G_{n+1}$ be the groups of the TD$(n+1, m)$. Apply Wilson's Fundamental Construction, giving each point of groups $G_1, \ldots, G_n$ weight g, $m-t$ points of G_{n+1} weight u and the remaining t points of G_{n+1} weight v. The result is a pair of 3-GDDs of type $(mg)^n((m-t)u+tv)^1$ whose orthogonality can be verified easily using the definition of OGDD. □

Next we consider some applications of Lemma 14.22.

Lemma 14.23 *Let $X = \{18, 24, 30\}$. Let $x \in X$ and suppose that there exists a TD$((x/2)+1, m)$, an OSTS$(2m+1)$ and an OSTS$(2t+1)$, where $0 \le t \le m$. Then there exists an OSTS$(xm+2t+1)$.*

Proof For each $x \in X$, there exist OGDDs of type $2^{x/2}$ and of type $2^{(x/2)+1}$, by Lemma 14.21. Apply Lemma 14.22 with $u = 0$ and $v = 2$ to obtain OGDDs of type $(2m)^{x/2}(2t)^1$. Apply Lemma 14.14 to complete the proof. □

Lemma 14.24 *If a TD$(10, m)$, an OSTS$(4m+1)$, and an OSTS$(2m+2t+1)$ where $0 \le t \le m$ all exist, then an OSTS$(38m+2t+1)$ exists.*

Proof There exists an OGDD of type $4^9 2^1$ and of type 4^{10}. Apply Lemma 14.22 with $u = 2$ and $v = 4$, and then apply Lemma 14.14. □

In order to apply Lemma 14.23, we require that $m \equiv 0, 1 \pmod 3$ and that $t \equiv 0, 1 \pmod 3$, $t \notin \{1, 4\}$. In order to apply Lemma 14.24, we require that $m \equiv 0, 2 \pmod 3$ and that $t \equiv 0, 1 \pmod 3$; moreover, when $m \equiv 2 \pmod 3$, we require that $t \equiv 1 \pmod 3$ for an OSTS$(2m+2t+1)$ to exist. In particular, the smallest order obtained when $m \equiv 2 \pmod 3$ is $38m+3$.

For dealing with the case when $v \equiv 1 \pmod 6$, one further construction of this type is useful:

Lemma 14.25 *Suppose that a TD$(9, m)$, an OSTS$(6m+1)$, and an OSTS$(6t+1)$ where $0 \leq t \leq m$ all exist. Then there exists an OSTS$(48m+6t+1)$.*

Proof There exist OGDDs of types 6^8 and 6^9 by Lemma 14.21. Apply Lemma 14.22 with $u = 0$ and $v = 6$, and then apply Lemma 14.14. □

Each of Lemmas 14.23, 14.24, and 14.25 require transversal designs. Fortunately, all of the transversal designs that we require are produced by a single classical construction due to MacNeish (1922):

Lemma 14.26 *Let $n \geq 2$ be an integer. Suppose that $n = {p_1}^{\alpha_1}{p_2}^{\alpha_2}\cdots{p_s}^{\alpha_s}$ where $p_1, \ldots, p_s$ are distinct primes. Then there exists a TD(m, n) for all $2 \leq m \leq \min_i({p_i}^{\alpha_i}) + 1$.*

Lemma 14.26 is used here to produce all of the needed transversal designs.

14.4.2 *OSTSs with $v \equiv 1 \pmod 6$*

In this section, we show that if $v \equiv 1 \pmod 6$, there is an OSTS(v). In the process, we consider a number of smaller cases for the class $v \equiv 3 \pmod 6$, but for the most part we treat the two classes separately.

We require two direct constructions of Gibbons and Mathon (1993):

Lemma 14.27 *There exists an OSTS(v) for $v \in \{115, 145\}$.*

Mullin and Stinson (1987) show that if $v \equiv 1 \pmod 6$ and $v \geq 1927$, there exists an OSTS(v). Some of the remaining cases were handled by Stinson and Zhu (1991), who show:

Lemma 14.28 *For $v \equiv 1 \pmod 6$, there exists an OSTS(v) with the possible exception of $v \in \mathcal{A}$, where $\mathcal{A} = \{55, 115, 145, 205, 235, 265, 295, 319, 355, 391, 415, 445, 451, 493, 649, 655, 679, 697, 745, 781, 799, 805, 1243, 1255, 1315, 1585, 1795, 1819, 1921\}$.*

Lemma 14.29 *For $v \equiv 1 \pmod 6$, $v \leq 307$, there exists an OSTS(v).*

Proof By Lemma 14.28 and Lemma 14.27, it suffices to consider

$$v \in \{55, 205, 235, 265, 295\}.$$

For these values, apply Lemmas 14.15 and 14.13 using:

Order	OGDD	COQ	OSTS
55	6^9	–	7
205	3^{17}	4	13
235	2^{13}	9	19
265	3^{11}	8	25

Greig (1999) gives a PBD on 295 points having a unique block of size 49 and all other blocks of size 7, producing an OSTS(295). □

Lemma 14.30 *For* $v \equiv 3 \pmod 6$, $15 \le v \le 201$, *there exists an OSTS*(v).

Proof Rosa (1974) constructed an OSTS(27), and Gibbons (1987) produced an OSTS(15). Stinson and Zhu (1991) constructed OSTS(v)s for $v \in \{105, 189, 195\}$. Gibbons and Mathon (1993) have constructed OSTS(v)s for $v \in \{21, 33, 39, 45, 51, 57, 63, 69, 75, 81, 87, 93, 99, 111, 117, 123, 129, 135, 153, 159, 171\}$. This leaves the values $v \in \{141, 147, 165, 177, 183, 201\}$. The values $v \in \{141, 147\}$ are handled by Lemmas 14.15 and 14.16 using:

Order	OSTS	COQ	Construction
141	7	20	Lemma 14.16
147	7	21	Lemma 14.15

The values $v \in \{165, 183, 201\}$ are handled by Lemma 14.13 using:

Order	OGDD	COQ	OSTS
165	3^{11}	5	15
183	2^{13}	7	15
201	4^{10}	5	21

For $v = 177$, apply Lemma 14.23 with $x = 18$, $m = 9$, and $t = 7$ to form an OGDD of type $(18)^9(14)^1$, and apply Lemma 14.14 using OSTS(19) and OSTS(15). □

Lemma 14.31 *Suppose that* $v \equiv 1, 3 \pmod 6$ *and that* v *satisfies one of* $301 \le v \le 381$; $397 \le v \le 441$; $463 \le v \le 541$; $571 \le v \le 741$; $757 \le v \le 865$; $877 \le v \le 993$; $1027 \le v \le 1743$; *or* $1759 \le v \le 1941$. *Then there exists an OSTS*(v).

Proof Applications of Lemmas 14.23 and 14.24 are given in Tables 14.1 and 14.2, respectively. In Table 14.1, the OSTS($2m+1$) exists as a consequence of Lemmas 14.30 and 14.29. Similarly, for Table 14.2, an OSTS($4m+1$) exists by Lemmas 14.30 and 14.29. □

Now we are in a position to state the definitive result for $v \equiv 1 \pmod 6$.

Theorem 14.32 *If* $v \equiv 1 \pmod 6$, *there exists an OSTS*(v).

Proof In view of Lemma 14.29, we need only consider $v \ge 313$, and in view of Lemma 14.28, we need only consider $v \le 1921$. By Lemma 14.31, if $301 \le v \le 1941$ and $v \equiv 1, 3 \pmod 6$, then there exists an OSTS(v) except possibly when $385 \le v \le 393$; $445 \le v \le 459$; $543 \le v \le 567$; $867 \le v \le 873$; $997 \le v \le 1023$; or $1747 \le v \le 1755$.

Upon considering the set $\mathcal{A}$ of Lemma 14.28, it suffices to consider values in the first two of these intervals. We cover these intervals using Lemma 14.25 with $m = 8$ (so $48m+1 = 385$ and $54m+1 = 433$) and with $m = 9$ (so $48m+1 = 433$ and $54m+1 = 487$). This completes the proof of the theorem. □

m	x	$mx+13$	$m(x+2)+1$	m	x	$mx+13$	$m(x+2)+1$
16	18	301	321	13	24	325	339
19	18	355	381	16	24	397	417
25	18	463	501	27	18	499	541
31	18	571	621	25	24	613	651
37	18	679	741	31	24	757	807
43	18	787	861	27	30	823	865
49	18	895	981	31	30	943	993
43	24	1045	1119	61	18	1113	1221
67	18	1219	1341	73	18	1327	1461
79	18	1435	1581	64	24	1549	1665
67	24	1621	1743	97	18	1759	1941

Table 14.1 *Applications of Lemma* 14.23

m	$38m+1$	$38m+3$	$40m+1$
9	343		361
11		421	441
17		649	681
23		877	921
27	1027		1081

Table 14.2 *Applications of Lemma* 14.24

14.4.3 *OSTSs with* $v \equiv 3 \pmod 6$

In this section, we complete the proof of Theorem 14.37, showing that if $v \equiv 3 \pmod 6$ and $v \geq 15$, there is an OSTS(v).

Lemma 14.33 *If* $v \equiv 3 \pmod 6$ *and* $15 \leq v \leq 297$, *there exists an OSTS*(v).

Proof In view of Lemma 14.30, we need only consider values in the range $207 \leq v \leq 297$. Using constructions of Gibbons and Mathon (1993), we need only consider

$$v \in \{225, 231, 249, 255, 261, 267, 273, 285, 297\}.$$

Applying Lemma 14.23 with $x = 18$, $m = 13$, and $t \in \{7, 10, 13\}$ yields OSTS(v) for $v \in \{249, 255, 261\}$. Then applying Lemmas 14.15 and 14.16 gives OSTS(v) for $v \in \{225, 231, 273, 285\}$, using ingredients as follows:

Order	OSTS	COQ	Construction
225	7	32	Lemma 14.16
231	7	33	Lemma 14.15
273	21	13	Lemma 14.15
285	15	19	Lemma 14.15

The final value, $v = 297$, is handled by Lemma 14.13 using an OGDD of type 3^{11}, a COQ of order 9, and an OSTS(27). This completes the proof. □

Now we extend the interval of orders for which OSTSs are known:

Lemma 14.34 *If* $v \equiv 3 \pmod 6$ *and* $15 \le v \le 1941$, *there exists an OSTS*(v).

Proof By Lemma 14.33, we need only consider $v \ge 303$. The following list of intervals is what remains to consider after applying Lemma 14.31 (as summarized in the proof of Theorem 14.32), restricting the intervals to those values congruent to 3 (mod 6): $387 \le v \le 393$; $447 \le v \le 459$; $543 \le v \le 567$; $867 \le v \le 873$; $999 \le v \le 1023$; or $1749 \le v \le 1755$. Again using constructions of Gibbons and Mathon (1993), what remains is

$$v \in \{393, 459, 555, 561, 567, 867, 999, 1005, 1011, 1023, 1749, 1755\}.$$

For $v \in \{459, 867\}$, we apply Lemma 14.13 using an OGDD of type 3^{17}. When $v = 459$, an OSTS(27) and a COQ of order 9 are used; when $v = 867$, an OSTS(51) and a COQ of order 17 are used.

For the remaining cases, we apply Lemmas 14.15 and 14.16, using the ingredients specified in Table 14.3. □

Order	OSTS	COQ	Construction
393	7	56	Lemma 14.16
555	15	37	Lemma 14.15
561	7	80	Lemma 14.16
567	7	81	Lemma 14.15
999	37	27	Lemma 14.15
1005	15	67	Lemma 14.15
1023	33	31	Lemma 14.15
1749	135	13	Lemma 14.15
1755	19	92	Lemma 14.16

Table 14.3 *Ingredients for Lemma* 14.34

Although all cases when $v \equiv 1 \pmod 6$ have already been settled, in the remainder classes $v \equiv 1, 3 \pmod 6$ are both treated for convenience.

Lemma 14.35 *Let* $(m_1, m_2, \ldots, m_u)$ *be a sequence of positive integers, and let* s *be a positive integer, satisfying*

1. $m_i \equiv 1 \pmod 6$ *for* $1 \le i \le u$;
2. *there exists a TD*$(10, m_i)$ *for* $1 \le i \le u$;
3. $0 < m_{i+1} - m_i \le 6s$ *for* $1 \le i < u$; *and*
4. $m_i \ge 54s + 6$.

Suppose further that if $v \equiv 1, 3 \pmod 6$ *and* $13 \le v \le 18m_1 + 9$, *there exists an OSTS*(v).

Then there exists an OSTS(v) *for* $v \equiv 1, 3 \pmod 6$ *and* $13 \le v \le 20m_u + 1$.

Proof Apply a straightforward induction based on Lemma 14.23 with $x = 18$. The fact that $m \geq 54s + 6$ implies $20m + 1 \geq 18(m + 6s) + 13$ ensures that all values are produced. □

Corollary 14.36 *Let $(m_1, m_2, \ldots)$ be an infinite sequence of positive integers, and let s be a positive integer, satisfying*

1. $m_i \equiv 1 \pmod 6$ *for* $1 \leq i \leq u$;
2. *there exists a TD*$(10, m_i)$ *for* $1 \leq i \leq u$;
3. $0 < m_{i+1} - m_i \leq 6s$ *for* $1 \leq i < u$; *and*
4. $m_i \geq 54s + 6$.

Suppose further that if $v \equiv 1, 3 \pmod 6$ *and* $13 \leq v \leq 18m_1 + 9$, *there exists an OSTS*(v).

Then there exists an OSTS(v) *for* $v \equiv 1, 3 \pmod 6$, $v \geq 13$.

Corollary 14.36 provides the vehicle to complete the solution.

Theorem 14.37 *If* $v \equiv 1, 3 \pmod 6$, $v \geq 7$, *and* $v \neq 9$, *then there exists an OSTS*(v).

Proof If $m \equiv 1 \pmod 6$ and $(m, 35) = 1$, there exists a TD$(10, m)$ by Lemma 14.26. Since at least one of $6m$, $6m+6$, and $6m+12$ is relatively prime to 35, we can apply Corollary 14.36 with $s = 3$ to the sequence whose entries are elements of $M = \{m : m \geq 169, m \equiv 1 \pmod 6), (m, 35) = 1\}$, provided that an OSTS$(v)$ exists for all $v \equiv 1, 3 \pmod 6$, $13 \leq v \leq 3051$. We begin by applying Lemma 14.35 with $s = 1$ to the sequence (103,109), noting that $18 \cdot 103 + 9 = 1863$ and that $20 \cdot 109 + 1 = 2181$. Thus, together with Lemma 14.34 and Theorem 14.32, we have the result for $13 \leq v \leq 2181$. Extend this interval to include 2185 and 2187 by noting that $2185 \equiv 1 \pmod 6$, and $2187 = 3^7$, so we can apply Lemma 14.15 to an OSTS(27) and a COQ(81) to obtain an OGDD of type 81^{27}, and thus an OSTS(2187).

Apply Lemma 14.35 ($s = 2$) to the sequence (121, 127, 139, 151, 157), noting that $121 \cdot 18 + 9 = 2187$ and $157 \cdot 20 + 1 = 3141$. Since $3141 > 3041$, we can now apply the Corollary as stated to complete the proof. □

14.5 More OGDDs

Colbourn and Gibbons (1998) established more detailed information about the existence of OGDDs. They generalize the definitions of COQs and OGDDs to obtain a stronger product construction. Let ψ be a relation from Sym_3 to itself. Then call a pair of (not necessarily distinct) quasigroups Q_1 and Q_2 a ψ-*COQ* if, whenever $\pi, \pi' \in Sym_3$ and $\pi' \in \psi(\pi)$, the π-conjugate of Q_1 is orthogonal to the π'-conjugate of Q_2. Choosing ψ so that all permutations are related recovers the definition of conjugate orthogonal quasigroups. Choosing ψ instead so that no permutations are related, an arbitrary pair of quasigroups forms a ψ-COQ. But the useful choices lie between these two extremes.

We have described a way to weaken the definition of COQs; now we need a corresponding way to strengthen the definition of OGDDs. For every block

$\{x, y, z\}$ in both 3-GDDs forming the OGDD, we impose some total order on the elements of the block (any of six possible orders can be chosen independently for each block). Let (x_1, x_2, x_3) be an ordered block, and let $\pi \in Sym_3$; then

$$\pi((x_1, x_2, x_3)) = (x_{\pi(1)}, x_{\pi(2)}, x_{\pi(3)}).$$

Now consider any pair x, y of elements. These appear as elements (x_i, x_j) in some (ordered) block of the first 3-GDD, and as elements (y_a, y_b) in some (ordered) block of the second 3-GDD in an OGDD. Let π be the permutation in Sym_3 that maps $(1, 2)$ to (i, j), and let π' be the permutation in Sym_3 that maps $(1, 2)$ to (a, b). If for every choice of elements x and y, we find $\pi' \in \psi(\pi)$, the (ordered) OGDD is a *ψ-OGDD.*

Restricting to a ψ-compatible ordering typically strengthens the requirements for an OGDD to exist. Now we establish that for various choices of ψ, the weakened notion of ψ-COQs and strengthened one of ψ-OGDDs provide the correct match.

Theorem 14.38 generalizes Lemma 14.13:

Theorem 14.38 *Let ψ be a relation from Sym_3 to itself. Suppose that there is a ψ-OGDD of type $g_1^{u_1} \cdots g_s^{u_s}$. Let k be the order of a ψ-COQ. Then there is an OGDD of type $(k \cdot g_1)^{u_1} \cdots (k \cdot g_s)^{u_s}$.*

Proof Let $(Q, \odot_1)$ and $(Q, \odot_2)$ be the first and second quasigroups of order k in a ψ-COQ. Let $(V, \mathcal{G}, \mathcal{B}_1)$ and $(V, \mathcal{G}, \mathcal{B}_2)$ be the first and second (ordered) 3-GDDs in a ψ-OGDD of type $g_1^{u_1} \cdots g_s^{u_s}$. We form the OGDD on $\widehat{V} = V \times Q$, with groups $\widehat{\mathcal{G}} = \{G \times Q : G \in \mathcal{G}\}$. Define

$$\mathcal{D}_i = \bigcup_{(x,y,z) \in \mathcal{B}_i} \bigcup_{u,v \in Q} \{\{(x, u), (y, v), (z, (u \odot_i v))\}\}.$$

Two 3-GDDs of type $(k \cdot g_1)^{u_1} \cdots (k \cdot g_s)^{u_s}$ are $(\widehat{V}, \widehat{\mathcal{G}}, \mathcal{D}_i$ for $i \in \{1, 2\}$; our task is to show that they are orthogonal. We check the two conditions in turn. Consider two points (x, u) and (y, v) in different groups of $\widehat{V}$. They appear in a block with a third element (z_i, w_i) in $\mathcal{D}_i$ for $i \in \{1, 2\}$. Now z_1 and z_2 are in different groups since the triples $\{x, y, z_1\}$ and $\{x, y, z_2\}$ arise from the two orthogonal 3-GDDs (the ordering is immaterial for this).

To check the second requirement, suppose to the contrary that, for $i, j \in \{1, 2\}$, $\{(x_i, u_i), (y_i, v_i), (z_j, w_j)\} \in \mathcal{D}_j$. Immediately, $z_1 \neq z_2$ since $\{x_1, y_1, z_j\} \in \mathcal{B}_j$ for $j \in \{1, 2\}$, but these two 3-GDDs are orthogonal. Moreover, $\{x_1, y_1\} = \{x_2, y_2\}$ by the orthogonality of $\mathcal{B}_j$ for $j \in \{1, 2\}$, so without loss of generality, we take $x = x_1 = x_2$ and $y = y_1 = y_2$. Hence $u_1 \neq u_2$ and $v_1 \neq v_2$. If in the ordered blocks of $\mathcal{B}_j$ $(j \in \{1, 2\})$, we find x in position α and y in position β, define the permutation π_j to be the one with $\pi_j(1) = \alpha$ and $\pi_j(2) = \beta$. Since $\mathcal{B}_j$ for $j \in \{1, 2\}$ form a ψ-OGDD, we have $\pi_2 \in \psi(\pi_1)$. So let $(Q, \oplus)$ be the π_1-conjugate of $(Q, \odot_1)$, and let $(Q, \otimes)$ be the π_2-conjugate of $(Q, \odot_2)$; the property of the ψ-COQ ensures that $(Q, \oplus)$ and $(Q, \otimes)$ are orthogonal. But by construction, $w_1 = u_1 \oplus v_1 = u_2 \oplus v_2$ and $w_2 = u_1 \otimes v_1 = u_2 \otimes v_2$, which is a contradiction. □

One case of interest occurs when ψ is the universal relation; here we recover our usual notions of COQ and OGDD, and the theorem is then Lemma 14.13. Zhang (1996) explored the use of orderings defined by the relation $\psi(abc) = \{bca, cab\}$ for every $abc \in Sym_3$. Unfortunately, in this case relatively few ψ-OGDDs are known.

We examine a relation that is less strict than Stinson and Zhu's, but stricter than that of Zhang. Partition Sym_3 into three classes: $\{123, 213\}$, $\{132, 312\}$, and $\{231, 321\}$. Then $\pi \in \phi(\pi')$ if and only if π and π' are in different classes.

The key observation is:

Theorem 14.39 *Every OGDD underlies a ϕ-OGDD.*

Proof Suitable orderings of the two 3-GDDs $(X, \mathcal{G}, \mathcal{B}_1)$ and $(X, \mathcal{G}, \mathcal{B}_2)$ forming the OGDD must be found. To do this, form a graph H whose vertex set is $\mathcal{B}_1 \cup \mathcal{B}_2$. For $B_1 \in \mathcal{B}_1$ and $B_2 \in \mathcal{B}_2$, place an edge between B_1 and B_2 if $|B_1 \cap B_2| \geq 2$ (since $\mathcal{B}_1$ and $\mathcal{B}_2$ are disjoint, this requires $|B_1 \cap B_2| = 2$). By construction, H is a cubic, bipartite graph; hence its edges can be properly coloured using three colours, say 0, 1, and 2.

Now for $i = 0, 1$ and for $B = \{x_0, x_1, x_2\} \in \mathcal{B}_{i+1}$, suppose that the three edges incident with B in H receive colours as follows: colour $\alpha + i$ to $\{x'_0, x_1, x_2\}$, colour $\beta + i$ to $\{x_0, x'_1, x_2\}$, and colour $\gamma + i$ to $\{x_0, x_1, x'_2\}$. Colour numbers are computed modulo 3. Then order the triple as $(x_\alpha, x_\beta, x_\gamma)$. Now consider a pair $\{x, y\}$ of elements, and find $\{x, y, z_j\}$ in $\mathcal{B}_j$ for $j = 1, 2$. These two blocks are connected by an edge in H, so suppose that that edge was coloured δ. Then $\{x, y\}$ occupy positions $\{\delta - 1, \delta + 1\}$ in the ordered block of the first system, but they occupy positions $\{\delta - 2, \delta\}$ in the ordered block of the second. □

The importance of Theorem 14.39 is that, while the restrictions on the OGDD seem to have been strengthened, in practice they have not. But the prerequisites on COQs have indeed been weakened. First the existence of some ingredient ϕ-COQs is established:

Lemma 14.40 *There is an idempotent ϕ-COQ of order q whenever $q \geq 7$ and q is a prime power. There are ϕ-COQs of order 4 and 5.*

Then a basic PBD construction establishes a fairly complete existence spectrum.

Lemma 14.41 *Let $(V, \mathcal{G}, \mathcal{B})$ be a K-GDD of type $g_1^{u_1} \cdots g_2^{u_s}$, for which*

1. *for each $k \in K$, there is an idempotent ϕ-COQ of order k;*
2. *for every $1 \leq i \leq s$, there is a ϕ-COQ of order g_i.*

Then there is a ϕ-COQ of order $v = |V|$.

Theorem 14.42 *If m is a positive integer and $m \notin \{2, 3, 6, 10, 12, 14, 18, 26, 30, 38, 42\}$, then there is a ϕ-COQ of order m.*

Theorem 14.43 *If m is a positive integer and $m \notin \{2, 3, 6, 10, 12, 14, 18, 26, 30, 38, 42\}$, and there is an OGDD of type g^u, then there is an OGDD of type $(mg)^u$.*

Proof Apply Theorem 14.38 using $\psi = \phi$, and Theorem 14.42 to provide the ϕ-COQ. Theorem 14.39 orders the given OGDD as a ϕ-OGDD. □

Using this machinery, Wilson's fundamental construction, and computational constructions, Colbourn and Gibbons (1998) show:

Theorem 14.44 *(Existence of OGDDs)*

1. *There is an OGDD of type $(6m)^u$ whenever $m \geq 1$ and $u \geq 5$, except possibly when*
   (a) $m = 1$ *and* $u \in \{6, 10, 14, 18, 22, 26, 30, 38, 42, 46\}$;
   (b) $m = 3$ *and* $u \in \{6, 22, 23, 34\}$;
   (c) $m = 9$ *and* $u \in \{6, 14, 22, 26\}$;
   (d) $m \equiv 1, 5 \pmod 6$, $m \geq 5$, *and* $u \in \{6, 14, 22, 26, 38\}$; *and*
   (e) $m \equiv 3 \pmod 6$, $m \geq 15$ *and* $u \in \{6, 22\}$.
2. *Let $g = 2\gamma$ and $\gamma \equiv 1, 2 \pmod 3$. Then an OGDD of type g^u exists whenever $u \equiv 0, 1 \pmod 3$, $u \geq 6$, except when $g = 2$ and $u = 6$, and possibly when*
   (a) $u \in \{6, 22, 34, 58, 94, 142\}$, *and* $\gamma \in \{4, 20, 28, 52, 76\}$ *or* γ *is odd;*
   (b) $u \in \{51, 69, 87, 93, 123\}$ *and* $\gamma \in \{1, 5, 7, 13, 19\}$.
3. *Let $g \equiv 1, 5 \pmod 6$ and $u \equiv 1, 3 \pmod 6$, $u \geq 7$. Then there is an OGDD of type g^u except possibly when $u = 9$.*
4. *There is an OGDD of type $(3\gamma)^u$ whenever $u \geq 7$, and $u \equiv \gamma \equiv 1 \pmod 2$.*

There is no OGDD of type 1^9, 2^4, 2^6, or 3^5, and the lack of these account in part for the gaps in the characterization given. Colbourn and Gibbons (1998) treat some of the infinite classes of possible exceptions that remain in Theorem 14.44 by generalizing this notion of orthogonality to modified 3-GDDs. The notion of orthogonality for 3-GDDs extends naturally to 3-MGDDs, by requiring in the first orthogonality condition that if $\{x, y, z\} \in \mathcal{B}$ and $\{x, y, w\} \in \mathcal{D}$, then z and w are in different first groups *and* in different second groups. Two orthogonal 3-MGDDs of type (u, v) form an OMGDD of type (u, v). The relevance of OMGDDs is that one of type (u, v) may exist despite the nonexistence of OGDDs of type u^v and v^u. It is easily verified that when an OMGDD of type (u, v) and an OSTS(v) both exist, then an OGDD of type u^v exists.

An important use of OMGDDs follows:

Lemma 14.45 *Suppose there is a K-GDD of order v and type $g_1 \cdots g_t$ for which an OMGDD of type (k, u) exists for every $k \in K$, and an OGDD of type g_i^u exists for $1 \leq u \leq t$. Then there is an OGDD of type v^u.*

Proof Form the K-GDD on V with groups $G_1, \ldots, G_t$. The OGDD formed has point set $V \times \{1, \ldots, u\}$, with groups $\{V \times \{i\} : 1 \leq i \leq u\}$. For every block B

in the K-GDD, place an OMGDD of type $(|B|, u)$, aligning the first groups on $\{x \times \{1, \ldots, u\} : x \in B\}$ and the second groups on $\{B \times \{i\} : 1 \le i \le u\}$. For every group G_i, place an OGDD of type $|G_i|^u$, aligning its groups on $\{G_i \times \{i\} : 1 \le i \le u\}$. Verification that the result is an OGDD is straightforward. □

Lemma 14.45 can be used to treat cases with a fixed number of large groups.

14.6 Intersections of resolvable STSs

Rosa (1980) posed the problem of determining

$$\mathrm{Int}_R(v) = \{\ell : \exists \text{ two resolvable STS}(v)\text{s sharing } \ell \text{ triples}\}.$$

Since the only STS(9) is resolvable, it is easily verified that

$$\mathrm{Int}_R(9) = \{0, 1, 2, 3, 4, 6, 12\}.$$

Lo Faro (1987), Shen (1999a), and Chang and Lo Faro (1999) establish that $\mathcal{I}_{15} \setminus \{26, 29\} \subseteq \mathrm{Int}_R(15)$. Lo Faro (1987) and Chang and Lo Faro (1999) establish that $\mathcal{I}_{27} \setminus \{54, 61, 62\} \subseteq \mathrm{Int}_R(27)$.

To treat cases in general, Chang and Lo Faro (1999) proceed as follows. There exist two Kirkman frames of type 2^4 that have groups the same and intersect in s triples for $s \in \{0, 2, 8\}$. Using the Fundamental Frame Construction (Theorem 19.17) with these Kirkman frames and {4,7}-GDDs of type 3^n, two Kirkman frames of type 6^n can be produced which share exactly ℓ triples whenever ℓ is even, $0 \le \ell \le 6n(n-1) - 6$ or $\ell = 6n(n-1)$, $n \ge 4$, and $n \ne 6$.

Taking these two Kirkman frames, and adding three infinite points, we can place two STS(9)s on a group together with the infinite points so that the infinite points form a triple. Placing the STS(9)s differently on the groups of the two frames produces two (resolvable) STS$(6n + 3)$s sharing any number of triples in $\mathcal{I}_{6n+3}$, *except* for numbers 4, 7, or 13 less than the maximum intersection size. Chang and Lo Faro (1999) also extend this construction to employ Kirkman frames of type 6^6, and hence to obtain the same result when $v = 39$. Then Theorem 19.68 can be used to treat the three remaining cases, to show:

Theorem 14.46 *$Int_R(6n + 3) = \mathcal{I}_{6n+3}$ for all $n \ge 7$.*

Chang and Lo Faro (1999) also establish that the intersection size 13 less than the maximum can be realized by two resolvable STS(27)s, and hence leave 13 possible exceptions:

v	$\in \mathrm{Int}_R(v)$?
15	$b_{15} - 9$, $b_{15} - 6$
21	$b_{21} - 16$, $b_{21} - 9$, $b_{21} - 8$
27	$b_{27} - 7$, $b_{27} - 4$
33	$b_{33} - 13$, $b_{33} - 7$, $b_{33} - 4$
39	$b_{39} - 13$, $b_{39} - 7$, $b_{39} - 4$

14.7 Work points

1. Under what conditions can three arbitrary STS(v)s be made disjoint?
2. Under what conditions on v, λ, and μ can an arbitrary simple TS(v, λ) and an arbitrary simple TS(v, μ) be made disjoint?
3. Complete the determination of existence of OGDDs.
4. In particular, does any OGDD with four groups exist? OMGDDs with one set of groups of size four are given in Colbourn and Gibbons (1998).
5. Complete the intersection problem for resolvable STSs.
6. Two Steiner triple systems $S_1 = (V, \mathcal{B}_1)$ and $S_2 = (V, \mathcal{B}_2)$ are *skew-orthogonal* if they are orthogonal, and satisfy the restriction that whenever $\{u, v\} \neq \{x, y\}$, $\{\{u, v, w\}, \{x, y, s\}\} \subset \mathcal{B}_1$, and $\{\{u, v, t\}, \{x, y, w\}\} \subset \mathcal{B}_2$, we find that $s \neq t$. Dukes and Mendelsohn (1999) establish that skew-orthogonal STSs lead to Room squares which have a transversal on the main diagonal (they are *standardized*) and for $i \neq j$, whenever cell (i, j) is occupied by a symbol, cell (j, i) is empty (they are *skew*). Dinitz and Stinson (1992) survey current knowledge about skew Room squares. Dukes and Mendelsohn (1999) establish using the Mullin–Nemeth proof of Lemma 14.12 that skew-orthogonal STSs exist whenever the order v is a prime power and $v \equiv 1 \pmod 6$. They also exhibit a solution of $v = 27$, and thereby establish asymptotic existence. Does a skew-orthogonal pair of STSs exist for all orders except for 3, 9, and 15?

15

LARGE SETS AND PARTITIONS

Two triple systems are *disjoint* when they have no triples in common. A set of s mutually disjoint simple triple systems of order v, and of indices $\lambda_1, \ldots, \lambda_s$, forms a *partition* of the complete design of order v if $v - 2 = \sum_{i=1}^{s} \lambda_i$. When $\lambda = \lambda_1 = \cdots = \lambda_s$, such a partition is a *large set* of triple systems, denoted LTS(v, λ). When $\lambda = 1$, this is often abbreviated to LTS(v).

Elementary necessary conditions for an LTS(v, λ) to exist are that a TS(v, λ) exists (so that $\lambda \equiv 0 \pmod{\gcd(v-2, 6)}$), and that $\lambda \mid v - 2$. From this, it can be verified that the basic necessary conditions are met when $\lambda = \lambda_{\min}(v)$ (using the notation from Theorem 2.10). Cayley (1850) first showed that these conditions are not sufficient. For order 7 and index 1, there are at most two pairwise disjoint STS(7)s. This motivated the determination of the maximum number $D(v, \lambda)$ of mutually disjoint simple TS(v, λ)s. The maximum possible value, $D(v, \lambda) = (v-2)/\lambda$, occurs if and only if a large set LTS(v, λ) exists.

Kirkman (1850) exhibited the first large set of triple systems, an LTS(9,1). Sylvester (1861, 1893) later studied the same case, as did Bays (1917) and Emch (1929). To obtain one LTS(9,1), seven disjoint STS(9)s are required on the same set of elements; to obtain all seven systems, apply the cyclic group $\mathbb{Z}_7$ to the base system:

$$\begin{array}{cccc} \{\infty, \infty', 0\} & \{0,1,6\} & \{0,2,5\} & \{0,3,4\} \\ \{1,2,4\} & \{3,5,6\} & \{\infty,1,5\} & \{\infty,2,3\} \\ \{\infty,4,6\} & \{\infty',1,3\} & \{\infty',2,6\} & \{\infty',4,5\} \end{array}$$

15.1 Large sets: even index

Schreiber (1974) established the existence of large sets with index two and orders 0 and 4 modulo 12:

Lemma 15.1 *When* $v \equiv 0, 4 \pmod{12}$, *there is an LTS*$(v, 2)$.

Proof Write $v = 2t+2$, so that $t \equiv 1, 5 \pmod 6$. We form a base system S_0 on $(\mathbb{Z}_t \times \{0,1\}) \cup \{\infty_1, \infty_2\}$. If $a, b, c \in \mathbb{Z}_t$ and $i, j, k \in \{0,1\}$, whenever $a+b+c \equiv 0 \pmod t$ and a, b, c are distinct, place the triple $\{a_i, b_j, c_k\}$ in S_0. For each pair (a, b) with $a \neq 0$ and $b \equiv -2a \pmod t$, place the triples $\{a_0, a_1, b_j\}$ for $j = 0, 1$ in S_0. Then include the four triples on elements $\{\infty_1, \infty_2, 0_0, 0_1\}$. Now consider all pairs (a, b) with $a \neq 0$ and $b \equiv -2a \pmod t$. Colour these ordered pairs with two colours, say white and black, so that if (a, b) gets one colour, then $(-a, -b)$ gets the opposite colour. Now for each white pair (a, b), add triples $\{\infty_1, a_0, b_0\}$, $\{\infty_1, a_1, b_1\}$, $\{\infty_2, a_0, b_1\}$, and $\{\infty_2, a_1, b_0\}$; for each black pair (a, b), add triples $\{\infty_2, a_0, b_0\}$, $\{\infty_2, a_1, b_1\}$, $\{\infty_1, a_0, b_1\}$, and $\{\infty_1, a_1, b_0\}$.

Form altogether t disjoint systems from S_0 by adding $(i, -)$ modulo t to the triples in S_0. □

Subsequently, Teirlinck (1975b) settled existence of large sets with even index. First, he proved:

Lemma 15.2 *When* $v \equiv 8 \pmod{12}$, *there is an LTS*$(v, 6)$.

Proof Write $v = 6t + 2$, so that $t \equiv 1 \pmod 2$. We form the large set on $(\mathbb{Z}_{3t} \times \{0,1\}) \cup \{\infty_1, \infty_2\}$. An auxiliary directed graph is needed, which is obtained from a complete graph by choosing arbitrarily a direction for each of its edges; this is a *tournament* (see Bondy and Murty, 1976). Place an arbitrary tournament T on $\mathbb{Z}_{3t}$. For $1 \le g \le t$, follow the prescription:

1. all 3-subsets of $\{\infty_1, \infty_2, (g+\alpha t)_0, (g+\alpha t)_1\}$ for $\alpha \in \{0,1,2\}$;
2. $\{x_i, y_j, z_k\}$ where $x+y+z \in \{3g, 3g+1, 3g+2\} \pmod{3t}$ and $x, y, z \in \mathbb{Z}_{3t}$;
3. $\{\infty_1, u_i, (3g-2u)_i\}$ and $\{\infty_2, u_i, (3g-2u)_{i+1}\}$ where $u \in \mathbb{Z}_{3t} \setminus \{g, g+t, g+2t\}$, $i \in \mathbb{Z}_2$, and $(u, 3g-2u) \in T$;
4. $\{\infty_1, u_i, (3g-2u)_{i+1}\}$ and $\{\infty_2, u_i, (3g-2u)_i\}$ where $u \in \mathbb{Z}_{3t} \setminus \{g, g+t, g+2t\}$, $i \in \mathbb{Z}_2$, and $(u, 3g-2u) \notin T$;
5. $\{\infty_1, u_i, (3g-2u+r)_i\}$ and $\{\infty_2, u_i, (3g-2u+r)_{i+1}\}$ where $u \in \mathbb{Z}_{3t}$, $i \in \mathbb{Z}_2$, $r \in \{1,2\}$, and $(u, 3g-2u+r) \in T$;
6. $\{\infty_2, u_i, (3g-2u+r)_i\}$ and $\{\infty_1, u_i, (3g-2u+r)_{i+1}\}$ where $u \in \mathbb{Z}_{3t}$, $i \in \mathbb{Z}_2$, $r \in \{1,2\}$, and $(u, 3g-2u+r) \notin T$;
7. $\{u_0, u_1, (3g-2u)_i\}$ where $u \in \mathbb{Z}_{3t} \setminus \{g, g+t, g+2t\}$ and $i \in \mathbb{Z}_2$;
8. $\{u_0, u_1, (3g-2u+r)_i\}$ where $u \in \mathbb{Z}_{3t}$, $i \in \mathbb{Z}_2$, and $r \in \{1,2\}$.

to obtain the blocks of S_g. □

Teirlinck (1975b) then established a type of doubling construction:

Lemma 15.3 *If an LTS*(v, λ) *exists and* λ *is even, then an LTS*$(2v-2, \lambda)$ *exists.*

Proof Let $S_1, \ldots, S_m$ be the systems in an LTS(v, λ) on element set $X \cup \{\infty\}$ (so that $|X| = v-1$). We form the LTS$(2v-2, \lambda)$ on $X \times \{0,1\}$. The neighbourhood of ∞ in each system S_i is a λ-regular graph G_i. By Petersen's theorem (Theorem 1.22), G_i can be partitioned into 2-factors since λ is even. Hence all edges of G_i can be oriented to form a regular digraph D_i; its converse (obtained by reversing all directions) is also a regular digraph. Now from each system S_i, we form two TS$(2v-2, \lambda)$s, T_{ij} for $j = 0, 1$, as follows. When $\{a, b, c\}$ is a block of S_i not containing ∞, place in T_{ij} the blocks $\{a_k, b_m, c_n\}$ with $k + m + n \equiv j \pmod 2$. When $(x, y) \in D_i$, place the blocks $\{x_0, x_1, y_r\}$ for $r = 0, 1$ in T_{i0} and the blocks $\{y_0, y_1, x_r\}$ for $r = 0, 1$ in T_{i1}. □

Theorem 15.4 *(Large sets for even index)*

1. *There is a large set of TS*$(v, 2)$*s if and only if* $v \equiv 0, 4 \pmod 6$.
2. *There is a large set of TS*$(v, 6)$*s whenever* $v \equiv 2 \pmod 6$.

Proof Write $v = 2^\ell t + 2$. If $\ell > 1$, we can apply Lemma 15.3 inductively. So it suffices to treat cases with $\ell = 1$. Then if $t \equiv 1, 5 \pmod 6$, apply Lemma 15.1, and if $t \equiv 3 \pmod 6$ apply Lemma 15.2. □

15.2 Large sets: index three

Teirlinck (1984) settled the existence of LTS$(v, 3)$s in the course of solving a more difficult question. A *quadruple system* QS(v, λ) is a v-set V equipped with a collection $\mathcal{B}$ of 4-subsets of V, so that every 3-subset occurs in precisely λ 4-subsets of $\mathcal{B}$. Teirlinck (1984) establishes:

Theorem 15.5 *When $v \equiv 0 \pmod 6$, the set of all 4-subsets of a v-set can be partitioned into disjoint QS$(v, 3)$s, each having no repeated blocks. Hence, there is a large set of TS$(v, 3)$s whenever $v \equiv 5 \pmod 6$.*

Proof Write $v = 3w + 3$. Let $\preceq$ be a total order on $\mathbb{Z}_w$. Form w sets of 4-tuples with elements from $\mathbb{Z}_w$, defined by

$$\mathcal{L}_a = \left\{ (x_1, x_2, x_3, x_4) : \sum_{i=1}^{4} x_i = a \right\}, a \in \mathbb{Z}_w.$$

All possible 4-tuples with elements from $\mathbb{Z}_w$ are partitioned by $\{\mathcal{L}_a : a \in \mathbb{Z}_w\}$.

We construct the large set of w QS$(v, 3)$s on $(\mathbb{Z}_w \cup \{\infty\}) \times \mathbb{Z}_3$, with blocks in the rth system as follows:

1. $\{x_i, y_j, z_k, v_\ell\}$, where $(x, y, z, v) \in \mathcal{L}_r$, $|\{x, y, z, v\}| = 4$, and $i, j, k, \ell \in \mathbb{Z}_3$;
2. $\{x_i, y_j, z_k, z_\ell\}$ for $(x, y, z, z) \in \mathcal{L}_r$, $|\{x, y, z\}| = 3$, and $i, j, k, \ell \in \mathbb{Z}_3$ but $k \neq \ell$;
3. $\{\infty_i, x_j, y_k, z_\ell\}$, where $(x, y, z, z) \in \mathcal{L}_r$, $|\{x, y, z\}| = 3$, $i, j, k, \ell \in \mathbb{Z}_3$, and

$$i + j + k + \ell = \begin{cases} 0 & \text{if } x \preceq z \text{ and } y \preceq z \\ 1 & \text{if } x \preceq z \preceq y \text{ or } y \preceq z \preceq x \\ 2 & \text{if } z \preceq x \text{ and } z \preceq y \end{cases}$$

4. all 4-subsets of $\{x, y\} \times \mathbb{Z}_3$ when $(x, x, y, y) \in \mathcal{L}_r$, and $x \neq y$;
5. $\{\infty_i, \infty_j, x_k, y_\ell\}$ when $(x, x, y, y) \in \mathcal{L}_r$, $x \neq y$, $i, j, k, \ell \in \mathbb{Z}_3$, and $i \neq j$;
6. $\{\infty_i, x_j, x_k, y_\ell\}$ when $(x, x, x, y) \in \mathcal{L}_r$, $x \neq y$, $i, j, k, \ell \in \mathbb{Z}_3$, and $j \neq k$;
7. all 4-subsets of $\{\infty, x\} \times \mathbb{Z}_3$ when $(x, x, x, x) \in \mathcal{L}_r$.

The conclusion is obtained by taking the derived systems with respect to any element, i.e. by selecting all quadruples containing a fixed element and deleting that element from each. □

15.3 Large sets: index one

The most challenging case is that of Steiner triple systems. Doyen (1972) apparently first suggested that an LTS(v) could exist except when $v = 7$, and in support of this he showed that $D(6v + 1, 1) \geq 2v + 1$. Teirlinck (1973) then proved:

Theorem 15.6 *For v odd, $D(3v) \geq 2v + D(v)$. In particular, if an LTS(v) exists, so also does an LTS($3v$).*

Proof Lemma 16.25 establishes that there exist $2v$ STS($3v$)s, and $v-2$ 3-GDDs of type v^3 each having the same three groups, all of which are disjoint, and none contain any triple on the three groups. Filling the groups with $D(v)$ disjoint STS(v)s establishes the desired result. □

At the same time, Schreiber (1973) and Wilson (1974a) proved, using an extension of the construction in §2.6, that:

Lemma 15.7 *An LTS(v) exists whenever, for every prime divisor p of $v-2$, the multiplicative order of -2 modulo p is singly even.*

Denniston (1974) settled a large number of small cases by direct constructions. Rosa (1975) developed a doubling construction:

Theorem 15.8 *If $v > 3$ and $v \equiv 1, 3 \pmod 6$, then $D(2v+1, 1) \geq v + 1 + D(v, 1)$. In particular, if an LTS(v) exists, an LTS($2v+1$) also exists.*

Proof Write $n = (v+1)/2$, $N' = \{1, \ldots, 2n\}$, and $X = \{x_1, \ldots, x_v\}$, with $N' \cap X = \emptyset$. Let $\mathcal{F} = \{F_1, \ldots, F_{2n-1}\}$ be a 1-factorization of the complete graph on vertex set N'. We choose $\mathcal{F}$ so that $F_1 \cup F_2 \cup F_3$ forms a cubic bipartite graph on N'. Such 'good' 1-factorizations are constructed by Rosa (1975). Use $\mathcal{F}$ to define a latin square $N^\star = (n_{ij})$. First set $n_{r,2n} = r$ for $1 \leq r \leq 2n$. Next, for $2 \leq k \leq n-1$, form the 2-factor $T_k = F_{2k} \cup F_{2k+1}$, and orient each edge so that the resulting directed graph $T_k^\star$ has every vertex of indegree and outdegree 1; for $(r,s) \in T_k^\star$, set $n_{r,2k} = s$ and $n_{s,2k+1} = r$. It remains to produce the first three columns of $N^\star$. Form three 2-factors $A_1 = F_2 \cup F_3$, $A_2 = F_1 \cup F_3$, and $A_3 = F_1 \cup F_2$. Orient the edges in each to form directed graphs $A_i^\star$ so that each has all vertices of indegree and outdegree 1, and no directed edge is in common between two of them (the bipartiteness of $F_1 \cup F_2 \cup F_3$ ensures that this can be done). To complete the definition of $N^\star$, set $n_{r,i} = s$ for $(r,s) \in A_i^\star$ and $i \in \{1,2,3\}$.

Next let $\mathcal{H} = \{H_1, \ldots, H_{2n-1}\}$ be a 1-factorization on $X \cup \{\infty\}$, where ∞ is a new symbol and $\{\infty, i\} \in H_i$ for $1 \leq i \leq 2n-1$. For $i \in N'$, the 1-factorization $\mathcal{H}^i = \{H_1^i, \ldots, H_{2n-1}^i\}$ is obtained from $\mathcal{H}$ by replacing each occurrence of the symbol ∞ by the symbol i. Next let $L = (\ell_{ij})$ be a latin square of side $2n-1$ on the set X, whose last two rows are

$$\begin{array}{cccccccccc} x_3 & x_1 & x_2 & x_4 & x_5 & x_6 & x_7 & \cdots & x_{2n-2} & x_{2n-1} \\ x_2 & x_3 & x_1 & x_5 & x_4 & x_7 & x_6 & \cdots & x_{2n-1} & x_{2n-2} \end{array}$$

Any $2 \times v$ latin rectangle can be completed to a latin square, and hence such an L exists.

Finally, let $(N', \mathcal{B})$ be a Steiner quadruple system of order $2n$, and let $(N' \setminus \{i\}, \mathcal{B}_i)$ be the derived Steiner triple system from element $i \in N'$.

Now we turn to the construction. For $i \in N'$ and $1 \leq j \leq 2n-1$, define a set of triples $C_{ij} = \{\{u, w, n_{ij}\} : \{u, w\} \in H_j^i\}$. Then set $\mathcal{C}_i = \bigcup_{j=1}^{2n-1} C_{ij}$. Then

$\mathcal{B}_i \cup \mathcal{C}_i$ forms, for each $i \in N'$, a Steiner triple system of order $2v+1$ on $X \cup N'$. These $2n = v+1$ STSs are all disjoint (an analysis of cases verifies this).

It remains to construct $D(v)$ further systems. In fact, we construct $v-2$ incomplete triple systems, each of order $2v+1$ and index one, and each having a hole of size v on the set X. By ensuring that these $v-2$ incomplete systems share no triples with the $v+1$ disjoint STSs already constructed, and observing that no triple already used lies entirely on the set X, we can obtain the conclusion of the theorem by selecting any $D(v)$ of the incomplete systems and filling the hole in each using a different STS(v) from a set of $D(v)$ disjoint systems. Now we turn to the construction of the $v-2$ incomplete systems.

For $1 \le i \le v-2$ and $1 \le j \le 2n$, set $E_{ij} = \{\{u, w, \ell_{ij}\} : \{u, w\} \in F_j\}$, and put $\mathcal{E}_i = \bigcup_{j=1}^{2n} E_{ij}$. For each $1 \le i \le v-2$, $\mathcal{E}_i$ contains the triples of an incomplete triple system of order $2v+1$ with a hole of size v on X. Since $\mathcal{F}$ is a 1-factorization on N', the disjointness of $\mathcal{E}_i$ and $\mathcal{E}_j$ for $i \neq j$ is immediate.

It remains to verify that $\mathcal{C}_i$ and $\mathcal{E}_j$ are disjoint for all choices of i and j; see Rosa (1975). □

The major breakthrough came with the remarkable work of Lu (1983, 1983a, b, 1984, 1984a, b, c). Through an intricate combination of recursive constructions, he succeeded in settling the existence of LTS(v)s with only six possible exceptions. Indeed, the final paper (Lu, 1984c) addressed these six remaining cases, but remained incomplete because of Lu's untimely death. Subsequently, Teirlinck (1991) simplified Lu's proof and in the process, completed the six remaining cases. Kang (1995) and Sharry and Street (1996) give complete, simplified treatments of Lu's strategy and proof, and therefore a more abbreviated description is given here.

One of Lu's main tactics, introduced in Lu (1983b), involved the use of an auxiliary design denoted LD(n). A *Lu design* LD(n) on an n-set X is $(\mathcal{A}^1, \mathcal{A}^2, \{\mathcal{L}_x : x \in X\})$, where $\mathcal{A}^i$ for $i \in \{1,2\}$ is a *resolvable* 3-GDD on $X \times \{0,1,2\}$ with groups on $X \times \{j\}$ for $j = 0,1,2$, and $\mathcal{L}_x$ for $x \in X$ is a 3-GDD on $(X \setminus \{x\}) \times \{0,1,2\}$ with groups on $(X \setminus \{x\})$ for $j = 0,1,2$. In addition, the $n+2$ systems are disjoint except that $\mathcal{A}^1 \cap \mathcal{A}^2 = \{\{x_0, x_1, x_2\} : x \in X\}$.

When an LD(n) satisfies the additional property that, for $w, x \in X$ we find exactly one solution for y and z to the requirement that $\{x_0, y_1, z_2\} \in \mathcal{A}^1$ and $\{w_0, y_1, z_2\} \in \mathcal{A}^2$, the Lu design is denoted by LD$^\star(n)$.

Teirlinck (1991) made an important observation:

Theorem 15.9 *The set of orders for which an LD$^\star(n)$ exists is PBD-closed.*

Lu (1983b, 1984a) developed both direct and recursive constructions for LD$^\star(n)$s, and in the process produced solutions for orders 4, 5, 7, 8, 11, and 19, among many others. Kang (1995) gives a solution of Chen's for LD$^\star(9)$. These small examples have powerful consequences:

Theorem 15.10 *An LD$^\star(n)$ exists except when $n \in \{2,3,6\}$, and possibly when $n \in \{10,12,14,15,18,23,26,27,30\}$.*

Proof Apply Theorem 15.9, using the determination of {4,5,7,8,9}-PBDs in Bennett, Gronau, Ling, and Mullin (1996). This handles all values except the stated omissions and in addition the orders 11, 19, and 51. Lu (1984b) handles 11 and 19. A {4,5,11}-PBD of order 51 handles the last case. □

Kang (1987) proved that whenever an LD(n) exists, an LD($3n$) also exists. Consequently, we have:

Theorem 15.11 *An LD(n) exists except when $n \in \{2,3,6\}$, and possibly when $n \in \{10,14,18,23,26,30\}$.*

Proof First employ Theorem 15.10. When $n = 12, 15, 27$, there is an LD*($n/3$); apply Kang's tripling construction. □

The importance of Lu designs is the recursive construction from Lu (1983b, 1984a):

Theorem 15.12 *If an LD(n) and an LTS($n+2$) both exist, then an LTS($3n$) exists.*

Proof Let $(\mathcal{A}^1, \mathcal{A}^2, \{\mathcal{L}_x : 0 \le x < n\})$ be an LD(n) on elements $\{0,\dots,n-1\} \times \{0,1,2\}$. Let $(\mathcal{B}_i : 0 \le x < n)$ be an LTS($n+2$) on elements $\{0,\dots,n-1\} \cup \{\infty_1, \infty_2\}$, indexed so that $\{\infty_1, \infty_2, i\} \in \mathcal{B}_i$. Assume without loss of generality that $\{\{x_0, x_1, x_2\} : 0 \le x < n\}$ forms parallel class $P_0^0 = P_0^1$ of $\mathcal{A}^1$ and of $\mathcal{A}^2$. For $j \in \{0,1\}$, let the remaining parallel classes of $\mathcal{A}^j$ be P_i^j for $1 \le i < n$.

Now we form the LTS($3n$) on the same elements as the LD(n). First, for $0 \le i < n$, we form an STS($3n$) S_i with triples:

1. $\{i_0, i_1, i_2\}$;
2. all triples of $\mathcal{L}_i$;
3. $\{u_k, v_k, w_k\}$ when $k \in \{0,1,2\}$ and $\{u,v,w\} \in \mathcal{B}_i$;
4. $\{i_k, v_{k+j}, w_{k+j}\}$ when $k \in \{0,1,2\}$, $0 \le v, w < n$, $v \ne i \ne w$, and $\{\infty_j, v, w\} \in \mathcal{B}_i$.

Then, for $1 \le i < n$ and $j \in \{1,2\}$, we form an STS($3n$) T_{ij} with triples:

1. all triples of P_i^j;
2. $\{u_k, v_{k+j}, w_{k+j}\}$ when $k \in \{0,1,2\}$, $0 \le t < n$, $\{\infty_j, v, w\} \in \mathcal{B}_t$, $v \ne t \ne w$, $\{\alpha_0, \beta_1, \gamma_2\} \in P_i^j$, where

$$u = \begin{cases} \alpha \text{ if } k = 0 \\ \beta \text{ if } k = 1 \\ \gamma \text{ if } k = 2 \end{cases}$$

and

$$t = \begin{cases} \alpha \text{ if } k+j \equiv 0 \pmod 3 \\ \beta \text{ if } k+j \equiv 1 \pmod 3 \\ \gamma \text{ if } k+j \equiv 2 \pmod 3 \end{cases}$$

Then $\{S_i : 0 \le i < n\} \cup \{T_{ij} : 0 \le i < n, j \in \{1,2\}\}$ forms the required LTS($3n$). □

Theorem 15.11 then establishes that whenever an LTS($n+2$) exists and $n \notin \{1, 23\}$, an LTS($3n$) exists. Let us summarize progress until this point. Theorem 15.8 can handle orders $v \equiv 3, 7 \pmod{12}$; Theorem 15.6 in addition those orders $v \equiv 9, 21 \pmod{36}$; and Theorem 15.12 those orders $v \equiv 33 \pmod{36}$. Hence it remains to treat orders $v \equiv 1 \pmod{12}$. The main attack of Lu (1983a) on this class is through a restricted and intricate hextupling construction:

Theorem 15.13 *If there is an LTS*($4n+3$) *and a TD*($14, 2n+1$), *then there is an LTS*($24n+13$).

This theorem can treat (eventually) all orders $v \equiv 13, 37 \pmod{72}$, since for n sufficiently large a TD($14, 2n+1$) exists (here, $2n+1 > 3565$ or $\gcd(n, 3{\cdot}5{\cdot}7{\cdot}11) = 1$ suffice; see Colbourn and Dinitz, 1996). To complete the solution, Lu (1984) gives a family of recursive constructions as follows:

Theorem 15.14 *If there is an LTS*($4n+1$), *then*

1. *there is an LTS*($mn+1$) *for* $m = 12, 24, 60$;
2. *if* $n \equiv 0 \pmod 3$, *there is an LTS*($mn+1$) *for* $m = 28, 44$.

The application with $m = 12$ handles orders $v \equiv 1, 25 \pmod{36}$, and application with $m = 24$ handles the last remaining case, orders $v \equiv 49 \pmod{72}$. The constructions introduced can, in principle, handle all congruence classes eventually. However, the requirement for a TD($14, 2n+1$) in Theorem 15.13 is the only situation in which a large number of applications of the recursion are not possible. This is the motivation for Lu handling the additional cases of $m = 60, 28, 44$ in Theorem 15.14. Indeed the resulting simplification is dramatic:

Theorem 15.15 *A large set of Steiner triple systems of order* v, *or LTS*(v), *exists if and only if* $v \equiv 1, 3 \pmod 6$ *and* $v \neq 7$.

Proof Direct constructions for LTS(v)s for $v = 3$, 13, 15, and 69 are given by Denniston (1974a). We proceed inductively on v. Suppose that $v \geq 9$, $v \notin \{13, 15, 69\}$, and for each admissible order w smaller than v, with $w \neq 7$, there is an LTS(w).

If $v \equiv 3, 7 \pmod{12}$, there is an LTS($(v-1)/2$) (recall that $v \neq 15$), so apply Theorem 15.8. If $v \equiv 9, 21 \pmod{36}$, apply Theorem 15.6 using LTS($v/3$). If $v \equiv 33 \pmod{36}$, apply Theorem 15.12 using LD($v/3$) and LTS($(v/3)+2$); recall that $v \neq 69$ so the LD exists by Theorem 15.11.

In the cases that remain, $v \equiv 1 \pmod{12}$. If $v \equiv 1, 25 \pmod{36}$, apply Theorem 15.14 with $m = 12$ and $n = (v-1)/12$. If $v \equiv 49 \pmod{72}$, apply Theorem 15.14 with $m = 24$ and $n = (v-1)/24$.

In the final cases, $v \equiv 13 \pmod{72}$ so write $v = 12\widehat{v}+1$ where $\widehat{v} \equiv 1 \pmod 6$. If $5 \mid \widehat{v}$, apply Theorem 15.14 with $m = 60$ and $n = (v-1)/60 = \widehat{v}/5$. Similarly, if $7 \mid \widehat{v}$, apply Theorem 15.14 with $m = 28$ and $n = (v-1)/28 = 3\widehat{v}/7$. Also, if $11 \mid \widehat{v}$, apply Theorem 15.14 with $m = 44$ and $n = (v-1)/44 = 3\widehat{v}/11$. If none of these apply, $\widehat{v}$ is relatively prime to 3, 5, 7, and 11, and hence a TD($14, \widehat{v}$) exists. Apply Theorem 15.13. □

Since order 7 is the only exceptional value in the characterization, Griggs and Rosa (1995) showed that there are six Steiner triple systems of order 7, among which each of the 35 possible triples appears at least once:

123	145	167	247	256	346	357
123	147	156	246	257	345	367
124	135	167	236	257	347	456
125	137	146	234	267	356	457
126	134	157	237	245	356	467
127	136	145	235	246	347	567

Each row is an STS(7). An analogous structure with only five rows (and hence containing every triple exactly once) would be a large set of STS(7)s. Colbourn (1996) gives an application of this covering by STSs to the design of a lottery scheme.

15.4 Large sets of 3-GDDs

Two 3-GDDs on the same set of groups are disjoint when they share no triples. Using 3-GDDs of type $g_1 \cdots g_s$ and index 1 with $g_1 \geq g_2 \geq \cdots \geq g_s$ leads to a bound on the number of mutually disjoint 3-GDDs of this type by considering an element from the group of size g_1 and one from the group of size g_2. Since there are $\sum_{i=3}^{s} g_i$ distinct triples involving the chosen points and not intersecting a group in two elements, this provides a trivial upper bound on the maximum number of mutually disjoint 3-GDDs. For this reason, a *large set of* 3-*GDDs* of type $g_1 \cdots g_s$ is a set of $\sum_{i=3}^{s} g_i$ 3-GDDs having the same groups, but pairwise disjoint. When the 3-GDDs are uniform, such a large set is a partition of all triples not intersecting a group in two or more elements.

Stinson and Vanstone (1988) established the motivation for studying large sets of 3-GDDs, by giving a connection to the design of threshold schemes (discussed in §15.4.1). Subsequently, Schellenberg and Stinson (1988) produced a number of useful small examples, including a large set of 3-GDDs of type 2^4. Chen and Stinson (1990) provided more examples, and subsequently Chen and Stinson (1993) undertook a systematic study of large sets of 3-GDDs of type 2^u. They proved:

Theorem 15.16 *If a large set of* 3-*GDDs of type* 2^{u+2} *exists, and* $v \equiv 1,3 \pmod 6$, *then a large set of* 3-*GDDs of type* $2^{u(v-2)+2}$ *exists.*

Chen and Stinson (1993) provide a number of related constructions to obtain a partial existence result, that a large set of 3-GDDs of type 2^u exists whenever $u \equiv 0,1 \pmod 3$ but $u \not\equiv 10,18 \pmod{24}$. Their result was then improved by Chen, Lindner, and Stinson (1992):

Theorem 15.17 *If large sets of* 3-*GDDs of types* t^{2u} *and* t^6 *exist, and* $u \notin \{2,6\}$, *then a large set of* 3-*GDDs of type* t^{6u} *exists.*

Using this tripling construction primarily, Chen, Lindner, and Stinson (1992) obtained a result on uniform 3-GDDs with group size two:

Theorem 15.18 *A large set of* 3*-GDDs of type* 2^u *exists whenever* $u \equiv 0, 1 \pmod 3$ *but* $u \not\equiv 130, 258 \pmod{384}$.

Some more sporadic results for uniform 3-GDDs with larger group size, and for 3-GDDs of type $g^u t^1$, were found by Chen and Stinson (1993) and Chen, Lindner, and Stinson (1992). However, the real breakthough came in Teirlinck (1993). He introduced a novel technique that we describe next.

A *large set with holes*, LS$(\lambda, 1; t, (k, K), v)$, for λ a positive integer, $t \le k \le \min(b : b \in K)$, is a collection of r t-wise balanced designs $(X, \mathcal{B}_i)$, $1 \le i \le r$. It is required that for every block $B \in \bigcup_{i=1}^r \mathcal{B}_i$, there are exactly $\lambda\binom{|B|-t}{k-t}$ solutions for the integer i in $B \in \mathcal{B}_i$. Hence an LS$(\lambda, 1; t, (k, K), v)$ must have $r = \lambda\binom{v-t}{k-t}$.

The crucial observation of Teirlinck (1993) is:

Lemma 15.19 *If an* $LS(m, 1; 2, (3, K), v)$ *exists and a large set of* 3*-GDDs of type* m^k *exists for each* $k \in K$*, then a large set of* 3*-GDDs of type* m^v *exists.*

In order to apply this lemma, Teirlinck (1993) established the existence of numerous large sets with holes:

Theorem 15.20 *A large set with holes* $LS(m, 1; 2, (3, K), v)$ *exists when*

1. $K = \{3\}$ *and* $v \equiv 1, 3 \pmod 6$, $v \ge 3$, $m \ge 1$, *and* $(m, v) \ne (1, 7)$;
2. $K = \{3, 4, 6\}$ *and* $v \equiv 0, 4 \pmod 6$, $v \ge 3$, $m \ge 1$;
3. $K = \{3, 5\}$ *and* $v \equiv 5 \pmod 6$, $v \ge 3$, $m \ge 1$;
4. $K = \{3, 4, 6, 8, 14, 26, 50\}$ *and* $v \equiv 2 \pmod 6$, $v \ge 3$, $m \ge 1$.

Teirlinck used these to settle all cases with $v \not\equiv 2 \pmod 6$ in the existence question for large sets of 3-GDDs of type m^v. Lei (1997) produced the ingredients needed for this final case, establishing:

Theorem 15.21 *A large set of* 3*-GDDs of type* m^v *exists if and only if* $v(v-1)m^2 \equiv 0 \pmod 6$, $(v-1)m \equiv 0 \pmod 2$, *and* $(m, v) \ne (1, 7)$.

15.4.1 *Threshold schemes*

Let X be a set of points, and $\mathcal{B}$ be a set of (distinct) triples on the points of X. Let K be a set of m elements, and let $\Phi : \mathcal{B} \mapsto K$. The pair $(\mathcal{B}, \Phi)$ is a *threshold scheme* provided that, for every 2-subset $P \subset X$, we have $|\{\Phi(B) : P \subset B \in \mathcal{B}\}| \ne 1$. The terminology arises from an application. Suppose that there are m keys which we wish to distribute. However, since a key permits access to sensitive information, we want to ensure that cooperation among authorized parties is needed to validate or determine a key. For example, suppose that we wish to distribute *shadows* of a key to three participants, so that if all three cooperate they can determine the key. However, if only two cooperate, they cannot. This is sometimes called a (3,3)-threshold scheme. The elements of X play the role of the shadows, and the keys are the elements of K. To distribute shadows of a key $k \in K$, select a triple $B \in \mathcal{B}$ with $\Phi(B) = k$, and give the three elements of B (one each) to the three people. They can then establish that collectively they hold key k, but no two of them can establish this without the third.

Security in such a system is of paramount importance. *Perfect security* requires that no two of the participants can deduce *any* information about the key from the shadows that they hold. Let $K = \{k_1, \ldots, k_m\}$, and consider the partition of $\mathcal{B}$ induced by Φ, with classes $\mathcal{B}_i = \{B \in \mathcal{B} : \Phi(B) = k_i\}$ for $1 \leq i \leq m$. Consider a pair $P \subset X$. If P appears in no triples of $\mathcal{B}_i$, knowledge of the pair P of shadows permits the exclusion of key k_i from consideration. More generally, if pair P occurs more frequently in $\mathcal{B}_i$ than in $\mathcal{B}_j$, knowledge of the pair P of shadows supports the conclusion that it is more likely that the key is k_i. This is made precise using probability distributions by Stinson and Vanstone (1988). With this in mind, a *perfect* threshold scheme is a threshold scheme in which every pair occurs equally often in triples of each class of the partition induced by Φ. In a perfect threshold scheme, therefore, if there is some pair occurring λ times in one of the $\mathcal{B}_i$, there can be at most $m \leq \lfloor (|X| - 2|)/\lambda \rfloor$ keys. For specified choices of X, the number of keys in a perfect threshold scheme can be expected is maximized when $\mathcal{B}_i$ to be a partial triple system of index one, for $1 \leq i \leq m$.

When $v = |X| \equiv 1, 3 \pmod 6$, as the reader must expect by now, we maximize the number of keys by employing an LTS$(v, 1)$ to produce the partition. When $v \equiv 0, 2 \pmod 6$, employ a large set of 3-GDDs of type $2^{v/2}$. When $v \equiv 4 \pmod 6$, some partial results appear in Chen and Stinson (1990, 1993). When $v \equiv 5 \pmod 6$, partial results are given in Chen, Stanton, and Stinson (1991).

15.5 Mutually almost disjoint systems

Two triple systems are *almost disjoint* if they share exactly one triple. Lindner and Rosa (1975) first studied the existence of sets of Steiner triple systems of order v that are mutually almost disjoint (or MAD). A *large set* of MAD STS(v)s is a set in which every triple is contained in at least one of the STS(v)s. Lindner and Rosa prove:

Theorem 15.22 *A large set of MAD STS(v)s contains $v-1$, v, or $v+1$ STS(v)s when $v > 13$.*

Proof Let $\mathcal{B}_1, \ldots, \mathcal{B}_s$ be a large set of MAD STS(v)s. Let k_i be the number of triples appearing in precisely i of the systems in the large set. Then we have the equations:

$$\sum_{i=1}^{s} k_i = \binom{v}{3}$$

$$\sum_{i=1}^{s} i k_i = \frac{sv(v-1)}{6}$$

$$\sum_{i=1}^{s} i(i-1) k_i = \frac{s(s-1)}{2}$$

Now subtracting the first from the second equation gives:

$$\sum_{i=1}^{s}(i-1)k_i = (s-(v-2))v(v-1)/6$$

and hence we find that $\sum_{i=1}^{s}(i-1)k_i > \sum_{i=1}^{s} i(i-1)k_i$ when $s \geq v+2$ and $v > 13$. But this cannot be, and hence $s \leq v+1$ when $v > 13$. □

Evidently, any triple $\{a, b, c\}$ can appear in at most $s-(v-3)$ of the systems, since the remaining $v-3$ triples containing $\{a, b\}$ must each appear in at least one system. Using this fact, Webb (1977) subsequently eliminated the possibility of a large set of $v-1$ MAD STS(v), observing that when $s = v-1$, the equations require that $k_2 = v(v-1)/6$ and also that $k_2 = (v-1)(v-2)/2$; these can be simultaneously satisfied only when $v \in \{1, 3\}$.

Lindner and Rosa (1975) established that large sets of v MAD STS(v)s do exist, however:

Theorem 15.23 *A large set of v MAD STS(v)s exists if and only if $v \equiv 1, 3$ (mod 6).*

Proof Form a Steiner quadruple system of order $v+1$, and choose one element r. Form the set of v derived Steiner triple systems (see §22.4), and replace r in each by the deleted element. □

The remaining case, large sets of $v+1$ MAD systems, has proved more difficult. Griggs and Rosa (1995) found a solution for orders 13 and 15, and they conjecture that such large sets exist for all $v \geq 13$ with $v \equiv 1, 3$ (mod 6).

15.6 Large sets of Kirkman triple systems

Cayley (1850) mentions a problem of Sylvester, namely to partition all triples on 15 elements into 13 Kirkman triple systems. Sylvester (1861c) shows that a large set of KTS(9)s exists, and Sylvester (1893) reports that a large set of Kirkman triple systems of order 3^m exists for all $m \geq 1$. The existence of large sets of Kirkman triple systems has often been called *Sylvester's problem.*

Denniston (1974b) found a large set of KTS(15)s. Denniston (1979) gave the basic recursive construction. A KTS(v) is *transitive*, and denoted TKTS(v), if there is a group G acting sharply transitively on the points of the KTS(v).

Theorem 15.24 *(Denniston, 1979) If a TKTS(v) and a large set of KTS(v)s both exist, then there is a large set of KTS($3v$)s.*

Theorem 15.25 *(Wu, 1990) When $x \in \{1, 5, 11, 17, 25, 35, 43\}$ and $m \geq 1$, a large set of KTS($3^m x$)s exists.*

Theorem 15.26 *(Lei, 1996) A TKTS(v) exists whenever $v \equiv 9$ (mod 18), and the prime divisors of v greater than three are each either in the class $\{x : x \equiv 1$ (mod 6)$\}$, or in the set $\{5, 11, 17\}$.*

Theorem 15.27 *(Lei, 1999) If a TKTS(v) and a large set of KTS(v)s both exist, then there is a large set of KTS($(2 \cdot 13^n + 1)sv$) for each $n \geq 0$.*

15.7 Overlarge sets

We can also ask for partitions of all triples on w elements into triple systems on v elements. One particular case of interest arises when $w = v+1$. An *overlarge set* of TS(v, λ)s is a partition of all triples on $v+1$ elements into TS(v, λ)s. Unlike the large set problem, the basic question here is easily answered:

Lemma 15.28 *Whenever* $v \equiv 1, 3 \pmod 6$, *there is an overlarge set of STS*(v)*s.*

Proof Take the v derived Steiner triple systems of a Steiner quadruple system of order $v+1$. □

Sharry and Street (1988) remark that not all overlarge sets are derived in this manner from Steiner quadruple systems, showing that there are 11 nonisomorphic overlarge sets of STS(7)s, while there is a unique overlarge set derived from a Steiner quadruple system of order eight. Sharry and Street (1988a) show that there are exactly 77 nonisomorphic overlarge sets of STS(9)s, and Sharry and Street (1993) show that there are 70 *cyclic* overlarge sets of STS(13)s. Using a doubling construction of overlarge sets of STS$(2v+1)$s from an overlarge set of STS(v)s, Sharry and Street (1991b) establish:

Theorem 15.29 *For every* $v \equiv 1, 3 \pmod 6$, $v \geq 7$, *there is an overlarge set of STS*(v)*s which is not derived from a Steiner quadruple system of order* $v+1$.

15.8 Work points

1. Complete the determination of the largest number m of keys in a (3,3)-threshold scheme which is perfectly secure and has $6k+4$ or $6k+5$ shadows.
2. Let v be an integer, and write $\binom{v}{3} = mq + r$ for integers m, q, and r. When can the set of $\binom{v}{3}$ distinct triples on v elements be partitioned into $q+1$ classes, so that q contain exactly m triples, and one contains exactly r triples, in such a way that every class forms a partial triple system of index one?
3. Under the same hypotheses, when can we in addition form an $m \times q$ array, each entry of which contains a triple, so that the triples in each row form a PTS, the triples in each column form a PTS, and the r remaining triples form a PTS?
4. When can the $\binom{v}{3}$ triples on a v-set be partitioned into Steiner triple systems of order w? Evidently, $w(w-1) \mid v(v-1)(v-2)$ and $w-1 \mid (v-1)(v-2)$ are necessary conditions. When $v = w$, this asks for large sets, and when $v = w+1$, for overlarge sets.
5. Prove that a large set of $v+1$ MAD STS(v)s exists for all $v \geq 13$, $v \equiv 1, 3 \pmod 6$.
6. Prove that a large set of KTS(v)s exists for all $v \equiv 3 \pmod 6$.

16

SUPPORT SIZES

Triple systems are permitted to have repeated triples, triples that are identical as 3-subsets. A theme throughout studies of triple systems is the goal to produce *simple* triple systems having no repeated blocks. However, in some applications, the presence of repeated triples poses no difficulty. In fact, for certain applications in experimental design (Wynn, 1977; Foody and Hedayat, 1977), it is desirable to have many triples repeated.

In this chapter, we examine the possible numbers of distinct triples in a TS(v, λ). Let $(V, \mathcal{B})$ be a TS(v, λ). Naturally, $\mathcal{B}$ contains $\lambda v(v-1)/6$ triples, but in general some may be repeated. The *support* of $\mathcal{B}$, denoted $\mathcal{B}^*$, is the set of distinct triples in $\mathcal{B}$. The *support size* of the triple system $(V, \mathcal{B})$ is the number of distinct triples, or $|\mathcal{B}^*|$.

We are interested in the spectrum of possible support sizes of a TS(v, λ), and denote the set of support sizes by SS(v, λ). The main subject of this chapter is the essentially complete determination of SS(v, λ).

The solution employs a number of ingredients that we have already developed, in particular results on the intersection problem (§14.2), the large set problem (§15), and the necessary conditions for leaves (§9.1). In addition, we employ one further ingredient in the support size problem, namely 'null designs' or 'trades'. We treat this topic next.

16.1 Trades: the linear algebra of designs

Fix an element set V of cardinality $v \geq 3$. Let $\mathcal{P} = (P_1, \ldots, P_{\binom{v}{2}})$ be a list of all 2-subsets of V. Let $\mathcal{T} = (T_1, \ldots, T_{\binom{v}{3}})$ be a list of all 3-subsets of V. Let A_{23} be the $\binom{v}{2} \times \binom{v}{3}$ (0,1)-matrix in which

$$A_{23}(i,j) = \begin{cases} 1 & \text{if } P_i \subset T_j \\ 0 & \text{otherwise.} \end{cases}$$

A theorem of Wilson (1973) has as a corollary that:

Lemma 16.1 *For $v \geq 6$, A_{23} has rank $\binom{v}{2}$, which is the full rank.*

Let $\mathbf{u}$ be a nonnegative integral vector of length $\binom{v}{3}$, $\mathbf{1}$ be a vector of length $\binom{v}{2}$ with all entries equal to one, and $\lambda > 0$ be an integer. Consider the matrix equation:

$$A_{23}\mathbf{u} = \lambda\mathbf{1}. \tag{16.1}$$

Lemma 16.2 *There exists a bijection between nonnegative integer solutions for* $\mathbf{u}$ *in (16.1) and TS*(v,λ)*s. Under this bijection,* $\mathbf{u}$ *is a* $(0,1)$ *vector if and only if the corresponding TS*(v,λ) *is simple.*

Proof Consider a TS(v,λ) $(V,\mathcal{B})$. Form a vector

$$\mathbf{u} = (u_1, \ldots, u_{\binom{v}{3}})$$

by setting u_i equal to the number of occurrences of T_i in $\mathcal{B}$. Since $\mathcal{B}$ covers each pair of $\mathcal{P}$ exactly λ times, $\mathbf{u}$ solves the matrix equation (16.1). Moreover, if the triple system is simple, $\mathbf{u}$ is a (0,1) vector by construction.

In the other direction, consider an integer solution $\mathbf{u}$ of (16.1). Form a collection $\mathcal{B}$ of triples that contains each triple $T_i \in \mathcal{T}$ precisely u_i times. Consider a pair $P_j \in \mathcal{P}$. This indexes row j of A_{23}; denote this row $\mathbf{a}_j$. Consider the inner product of $\mathbf{u}$ and $\mathbf{a}_j$; since $\mathbf{u}$ solves (16.1), this product is λ, and hence the pair P_j appears in precisely λ triples of $\mathcal{B}$. This holds for every pair, and hence $\mathcal{B}$ forms a TS(v,λ). When $\mathbf{u}$ is a (0,1)-vector, the triple system is simple by construction. □

The vector $\mathbf{u}$ corresponding to a triple system is the *characteristic vector* of the triple system. This linear algebraic framework was introduced more generally for t-designs by Wilson (1973). In this context, the support of a triple system is the same as the (linear algebraic) support of its characteristic vector $\mathbf{u}$, the set of nonzero positions in $\mathbf{u}$.

Consider two TS(v,λ)s $(V,\mathcal{B})$ and $(V,\mathcal{D})$, with characteristic vectors $\mathbf{u}_1$ and $\mathbf{u}_2$. Now it is easy to verify that $\mathbf{u}_1 - \mathbf{u}_2$ is a solution to (16.1) with $\lambda = 0$; that is, $\mathbf{u} = \mathbf{u}_1 - \mathbf{u}_2$ is an integer solution to

$$A_{23}\mathbf{u} = \mathbf{0}. \tag{16.2}$$

The set of integer vectors solving (16.2) is the *kernel* of A_{23}; a vector in the kernel is a *trade*, or a *null design*. The study of supports of triple systems can be viewed in large part as the study of trades: given the characteristic vector $\mathbf{u}$ of any TS(v,λ), the characteristic vectors of all TS(v,λ)s form precisely the set of positive integral vectors of

$$\{\mathbf{u} + \mathbf{t} : \mathbf{t} \in \ker A_{23}\}.$$

Now consider a trade $\mathbf{t} = (t_i)$. Form two vectors $\mathbf{t}^+ = (t_i^+)$ and $\mathbf{t}^- = (t_i^-)$ as follows:

$$t_i^+ = \begin{cases} t_i & \text{if } t_i \geq 0 \\ 0 & \text{otherwise} \end{cases}, \quad t_i^- = \begin{cases} 0 & \text{if } t_i \geq 0 \\ -t_i & \text{otherwise} \end{cases}.$$

Now $\mathbf{t}^+$ is the characteristic vector of a collection of triples $\mathcal{N}^+$, and $\mathbf{t}^-$ is the characteristic vector of a (disjoint) collection of triples $\mathcal{N}^-$. A trade is then a pair of collections $(\mathcal{N}^+,\mathcal{N}^-)$ of triples, with the additional property that $\mathcal{N}^+$ and $\mathcal{N}^-$ are *mutually balanced*, i.e. each pair is covered the same number of times in each. The collection $\mathcal{N}^+$ is the *positive* part of the trade, and $\mathcal{N}^-$ the *negative part*. Since the parts are mutually balanced, the parts have the same number of triples; this number is the *volume* of the trade.

The volume of trades dictates how 'close' two TS(v, λ)s can be, and hence we are concerned with possible volumes of trades. Hedayat and Li (1979, 1980) prove:

Lemma 16.3 *There is no trade of volume* 1, 2, 3 *or* 5.

Proof Suppose that $(\mathcal{N}^+, \mathcal{N}^-)$ is a trade of volume ℓ. Form a multigraph M from this trade on vertex set V, containing the edge $\{a, b\}$ the same number of times that the pair $\{a, b\}$ appears in triples of $\mathcal{N}^+$. M is a multigraph with even vertex degrees having 3ℓ edges, and having two disjoint edge-partitions into triangles (from $\mathcal{N}^+$ and $\mathcal{N}^-$). If M is a simple graph, the statement follows from Lemma 14.2. So suppose that M is a multigraph. When a vertex v has nonzero degree in V, it must appear in at least two distinct triples of $\mathcal{N}^+$ and hence must have degree at least 4. Thus the number of vertices of nonzero degree in M cannot exceed $3\ell/2$. Moreover, if a vertex v has degree equal to 4, it must have at least three distinct neighbours in M in order to belong to two distinct triangles. But since v must belong to a disjoint set of two distinct triangles in $\mathcal{N}^-$, it must have four distinct neighbours. Hence since M is not simple, it must contain at least one vertex of degree 6, and all vertices of degree 4 must have four distinct neighbours.

These elementary considerations eliminate $\ell = 1$, 2, and 3. It remains to consider $\ell = 5$. Now consider a vertex v of degree $d \geq 6$. Vertex v requires at least four neighbours to belong to disjoint triangles in $\mathcal{N}^+$ and $\mathcal{N}^-$. Hence we need only consider 15-edge multigraphs on 5, 6, and 7 vertices in which every vertex has at least four distinct neighbours. It is now a tedious exercise to generate these graphs and verify that none has two disjoint partitions into triangles. □

There is a trade of volume four:

$$\begin{array}{ll} \mathcal{N}^+ & \{\{a,c,e\},\{a,d,f\},\{b,c,f\},\{b,d,e\}\} \\ \mathcal{N}^- & \{\{b,c,e\},\{b,d,f\},\{a,c,f\},\{a,d,e\}\} \end{array}$$

Both parts of this trade are quadrilaterals (Pasch configurations). This trade is a *minimal trade*. Next we explore a basis for the trades. Graham, Li, and Li (1980) prove the following:

Theorem 16.4 *The dimension of the kernel of* A_{23} *is* $\binom{v}{3} - \binom{v}{2}$ *for* $v \geq 6$, *and it is zero for* $1 \leq v \leq 5$.

A basis for the trades is therefore a set of $\binom{v}{3} - \binom{v}{2}$ linearly independent trades. Following Graver and Jurkat (1973), Graham, Li, and Li (1980) produced a basis for trades that contains only minimal trades. We outline their approach here. Suppose that $V = \{1, \ldots, v\}$. Associate with each element $i \in V$

a variable x_i. Then a triple of $\mathcal{B}$ is a monomial of the form $\{x_{i_1}x_{i_2}x_{i_3}\}$. A minimal trade can then be succinctly written as $(x_{i_1} - x_{j_1})(x_{i_2} - x_{j_2})(x_{i_3} - x_{j_3})$, where $\{i_1, i_2, i_3, j_1, j_2, j_3\}$ are all distinct. Graham, Li, and Li (1980) use this monomial representation to prove that $\ker A_{23}$ is generated by such minimal trades, and then explicitly eliminate dependencies to find a basis. Hedayat and Hwang (1983) outline an implementation of their method. Subsequently, Khosrovshahi and Mahmoodian (1987a) improved upon this procedure. Khosrovshahi and Ajoodani-Namini (1990) establish that there is a basis for trades with a remarkably simple structure. We pursue their development here.

Let $\mathbf{u} \in \ker A_{23}$ be a minimal trade. The *leading index* of $\mathbf{u} = (u_1, \ldots, u_{\binom{v}{3}})$ is the smallest index j for which $u_j \neq 0$. Now consider s trades $\mathbf{u}_1, \ldots, \mathbf{u}_s$ having leading indices $\ell_1, \ldots, \ell_s$, respectively. If $1 \leq \ell_1 < \ell_2 < \cdots < \ell_s$, the set of trades is *semitriangular*, and the s trades are linearly independent. Hence Khosrovshahi and Ajoodani-Namini (1990) conclude that we need only determine the number of possible leading indices for a trade. For triple systems, this number is easy to determine. Assume that the 3-subsets are listed in lexicographic order. Using the monomial representation, we must choose six distinct indices $\{i_1, i_2, i_3, j_1, j_2, j_3\}$, each at most v. Assuming that $i_1 < j_1$, $i_2 < j_2$, and $i_3 < j_3$, the first triple in the trade is $\{i_1, i_2, i_3\}$. Thus we need only determine the number of ways to choose a triple $\{i_1, i_2, i_3\}$ so that j_1, j_2, j_3 can be suitably chosen. Thus we may

1. choose all three among the first $v - 5$ elements in $\binom{v-5}{3}$ ways;
2. choose two among the first $v - 5$ in $\binom{v-5}{2}$ ways, and choose the last independently in four ways; or
3. choose one among the first $v-5$, and take the last two to be $\{v-4, v-3\}$, $\{v-4, v-2\}$, $\{v-4, v-1\}$, $\{v-3, v-2\}$, or $\{v-3, v-1\}$.

This gives a total of

$$\binom{v-5}{3} + 4 \cdot \binom{v-5}{2} + 5(v-5) = \binom{v}{3} - \binom{v}{2}$$

choices for a leading triple in a trade. Thus choosing arbitrarily one minimal trade with each possible leading triple, we obtain the special case for triples of the result of Khosrovshahi and Ajoodani-Namini (1990):

Theorem 16.5 *$\ker A_{23}$ has a semitriangular basis of $\binom{v}{3} - \binom{v}{2}$ minimal trades for $v \geq 5$.*

A consequence of this characterization of the basis of trades is that from *any* TS(v, λ), every other TS(v, λ) can be produced by a finite number of additions of minimal trades. Unfortunately, this does *not* require that, after some number of additions of minimal trades, the result is a TS(v, λ). Triples may appear with negative multiplicity. For triple systems, quadrilateral-free triple systems (§13.2) are precisely those to which the addition of a minimal trade *must* yield a triple of negative multiplicity.

An interesting question for triple systems had been whether every TS(v, λ) *containing a quadrilateral* can be transformed to any other by addition of minimal trades, subject to the constraint that after each addition no triples of negative multiplicity are required. This holds for Steiner triple systems of order 13 and 15 (§7.2.3.2 of Gibbons, 1976), but Grannell, Griggs, and Murphy (1997) showed that it does not hold for Steiner triple systems of order 19.

16.2 Support sizes: necessity

Now we turn to the main subject of this chapter, the determination of SS(v, λ), the spectrum of support sizes of TS(v, λ)s.

Colbourn and Mahmoodian (1990) determine an extensive set of necessary conditions on membership in SS(v, λ); we pursue their strategy here.

Let $(V, \mathcal{B})$ be a TS(v, λ). Now for each $1 \leq i \leq \lambda$, let $\mathcal{B}_i$ be the *set* of triples each of which appears precisely i times in $\mathcal{B}$. Then $\mathcal{B}^* = \cup_{i=1}^{\lambda} \mathcal{B}_i$. Let $b^* = |\mathcal{B}^*|$.

How large can b^* be? Since every one of the distinct triples of $\mathcal{B}^*$ is a triple of $\mathcal{B}$, we have

$$b^* \leq \frac{\lambda v(v-1)}{6}.$$

In addition, since every triple is one of the distinct 3-subsets of the element set,

$$b^* \leq \binom{v}{3}.$$

Combining these, we obtain one basic necessary condition:

$$b^* \leq M_{v,\lambda} = \max\left(\frac{\lambda v(v-1)}{6}, \binom{v}{3}\right). \tag{16.3}$$

When $\lambda = v-2$, the only simple TS(v, λ) is the complete design; in this case, we obtain a further necessary condition:

Lemma 16.6 *There is no TS$(v, v-2)$ with support size $\binom{v}{3} - s$ for $s \in \{1, 2, 3, 5\}$.*

Proof Consider a TS$(v, v-2)$, $(V, \mathcal{B})$, with support size $\binom{v}{3} - s$. Let $\mathbf{z}$ be the characteristic vector of this triple system. Then $\mathbf{z} - \mathbf{1}$ is a trade of volume s. Lemma 16.3 completes the proof. □

How small can b^* be? This is a much more difficult question! We start with an easy observation. Every pair of elements must appear in at least one triple, and each triple contains three distinct pairs. Thus

$$b^* \geq \left\lceil \frac{\binom{v}{2}}{3} \right\rceil = \left\lceil \frac{v(v-1)}{6} \right\rceil = m_v. \tag{16.4}$$

Equality in (16.4) is achieved when $v \equiv 1, 3 \pmod 6$ by repeating each triple of an STS(v) λ times. In the remaining congruence classes, equality is not possible. To see this, $\mathcal{B}_\lambda$ is a partial triple system of index one; hence we can examine its leave G. We call this graph the *surplus*. Lemma 9.1 then ensures that G has all vertices of degree congruent to $\lambda(v-1)$ (mod 2), and the number of edges in G is congruent to $\lambda\binom{v}{2}$ (mod 3). In addition, the collection of triples $\cup_{i=1}^{\lambda-1} i\mathcal{B}_i$ is an edge-partition of λG into triangles, in which each edge of G belongs to at least two distinct triangles.

The determination of triple systems with support size s is thereby reduced to finding graphs G with t edges, for which

1. G is a 1-leave; and
2. λG has an edge-partition into triangles with no triangle repeated λ times, containing $s - m_v + \lceil \frac{t}{3} \rceil$ distinct triangles.

The second condition can be used to determine in part the structure of G:

Lemma 16.7 *G has no vertices of degree 1 or 2. Moreover, the subgraph induced on the neighbourhood of a vertex of G has minimum degree at least 2.*

Proof If G has a degree 1 vertex, λG has no edge-partition into triangles. If G has a degree 2 vertex x with neighbours y and z, every edge-partition of λG into triangles contains $\{x, y, z\}$ λ times. If G has a vertex v whose neighbourhood induces a subgraph containing a vertex w of degree 1, G contains a unique triangle containing the edge $\{v, w\}$, and hence every edge-partition of λG contains a λ-times repeated triple. □

One important consequence is that every degree 3 vertex of G appears in a complete subgraph of order four.

The determination now partitions naturally along the same lines as the necessary conditions for leaves (§9.1). We treat first the case $v \equiv 1, 3 \pmod 6$. The minimum support size is realized with $G = \overline{K_v}$. However, as with Lemma 14.2 for intersections, and Lemma 16.3 for trades, we find that we cannot realize 1, 2, 3, or 5 more than the minimum. This is not simply a consequence of earlier results, however, since in this case the edge-partition of λG into triangles need not be decomposable into mutually balanced parts. Despite this, Colbourn and Mahmoodian (1990) (see also Rosa and Hoffman, 1986) prove:

Lemma 16.8 *For $v \equiv 1, 3 \pmod 6$, there is no $TS(v, \lambda)$ with support size $m_v + s$ for $s \in \{1, 2, 3, 5\}$.*

For $v \equiv 5 \pmod 6$, the situation is somewhat different. Here G is required to have all vertex degrees even, but to have 1 (mod 3) edges. By Lemma 16.7, G then has minimum degree 4, and thus has at least five vertices of nonzero degree. If G has only five such vertices, they induce K_5. In this case, the unique partition of λK_5 is obtained by taking $\lambda/3$ copies of the unique TS(5, 3). This has ten distinct triples, leading to $b^* = m_v + 6$.

So suppose that G has at least six vertices of nonzero degree. It follows that G has at least 13 edges, and that the partition of λG contains at least 26/3 distinct triples. Thus b^* must be at least $m_v + 4$.

At the other extreme, if G has more than 11 vertices of nonzero degree, it has at least 25 edges; in this case, λG requires at least 17 distinct triples, and then $b^* \geq m_v + 8$. Colbourn and Mahmoodian (1990) examine all graphs on 13, 16, 19, and 22 edges meeting the necessary conditions, and they prove:

Lemma 16.9 *For* $v \equiv 5 \pmod 6$, *there is no* $TS(v, \lambda)$ *with support size* $m_v + t$ *for* $t \in \{0, 1, 2, 3, 4, 5, 7, 8\}$.

The problem is complicated for v even by the fact that all vertices of G must have *odd* degree. Each vertex of G must have degree at least 3 for λG to have an edge-partition into triangles. Let $s_v = \lceil v(v+2)/6 \rceil$. Then we have

Lemma 16.10 *For* v *even, every* $TS(v, \lambda)$ *has support size at least* s_v.

Proof Consider the surplus G of a $\text{TS}(v, \lambda)$ with v even. G has $t \geq 3v/2$ edges, since its minimum degree is at least 3. Now the number of λ-times repeated triples in the triple system is $(v(v-1) - 2t)/6$, and the number of distinct triples in the edge-partition of λG into triangles is at least $\lceil 2t/3 \rceil$, for a total of $\lceil (v(v-1) + 2t)/6 \rceil$ distinct triples. This is minimized when t is as small as possible, subject to $t \geq 3v/2$. So $s_v = \lceil v(v+2)/6 \rceil$ triples are required. □

This can be refined to eliminate some further small values:

Lemma 16.11 *For* $v \equiv 0, 4 \pmod{12}$, *there is no* $TS(v, \lambda)$ *with support size* $s_v + 1$, *and for* $v \equiv 6, 10 \pmod{12}$ *there is no* $TS(v, \lambda)$ *with support size* s_v.

Proof By Lemma 16.7, every vertex of degree 3 in the surplus of the triple system forms a complete subgraph of order four with its neighbours. If the support size equals s_v, the surplus has all vertices of degree 3, and hence must consist of $v/4$ disjoint complete subgraphs of order four; then for $b^* = s_v$, $v \equiv 0 \pmod 4$.

For $b^* = s_v + 1$, the surplus must have degree sequence $9^1 3^{v-1}$, $7^1 5^1 3^{v-2}$, or $5^3 3^{v-3}$. Only the first of these leads to a graph satisfying Lemma 16.7 for every vertex, namely the graph obtained by identifying three vertices, one from each of three distinct K_4s, in a graph consisting of $(v+2)/4$ disjoint K_4s. Thus for $b^* = s_v + 1$, it must be the case that $v \equiv 2 \pmod 4$. □

The remaining cases, when $v \equiv 2 \pmod 6$, are the most complicated. The surplus has all vertex degrees odd and at least 3, and in this case has $1 \pmod 3$ edges. Colbourn and Mahmoodian (1993) examined the possible surplus graphs in this case as well. Unlike the cases that arise when v is odd, however, the minimum degree is 3. Using the fact that degree 3 vertices participate only as vertices of a K_4, however, permits us to consider only that portion of the surplus graph on the vertices of larger degree. This permits the examination of all possible cases in a manner that is independent of the order v. As a result of extensive computations, Colbourn and Mahmoodian (1993) prove:

Lemma 16.12 *For* $v \equiv 2 \pmod{12}$, *there is no* $TS(v, \lambda)$ *of support size smaller than* $s_v + 7$. *For* $v \equiv 8 \pmod{12}$, *there is no* $TS(v, \lambda)$ *of support size smaller than* $s_v + 6$.

They also conjecture that there is no TS(v, λ) with $v \equiv 8 \pmod{12}$ having support size $s_v + 7$, but the computational effort to establish this appears to be prohibitive.

Define PS$(v, \lambda) = \varnothing$ if $\lambda \not\equiv 0 \pmod{\gcd(v-2, 6)}$. On the other hand, when the order v is λ-admissible, and in addition $v - 2 \neq \lambda$, PS(v, λ) is defined by:

v (mod 12)	PS(v, λ)
0,4	$s_v, s_v + 2, s_v + 3, \ldots, M_{v,\lambda}$
1,3,7,9	$m_v, m_v + 4, m_v + 6, m_v + 7, \ldots, M_{v,\lambda}$
2	$s_v + 7, \ldots, M_{v,\lambda}$
5,11	$m_v + 6, m_v + 9, m_v + 10, \ldots, M_{v,\lambda}$
6,10	$s_v + 1, s_v + 2, \ldots, M_{v,\lambda}$
8	$s_v + 6, s_v + 8, s_v + 9, \ldots, M_{v,\lambda}$

When $\lambda = v - 2$, PS(v, λ) is defined similarly, but with the omission of $\{M_{v,\lambda} - 5, M_{v,\lambda} - 3, M_{v,\lambda} - 2, M_{v,\lambda} - 1\}$.

Summarizing the necessary conditions, we have:

Theorem 16.13 *For $v \geq 3$ and $\lambda \geq 1$, $SS(v, \lambda) \subseteq PS(v, \lambda)$.*

16.3 Support sizes: sufficiency

Next we describe some of the ingredients used in establishing sufficient conditions for membership in SS(v, λ). The basic strategy for sufficiency is a common one. First, the solution is obtained directly by computation for small orders. Then techniques are developed that increase the order and the index while keeping the support size near the minimum. Next techniques are developed that, for fixed index, increase the order and the support sizes obtained. These underlie the determination for all orders for small fixed indices. Finally, with the cases for small orders and small indices both solved, constructions are introduced to increase the order, index, and support size at the same time.

16.3.1 *Small orders*

For $v \in \{3, 4, 5\}$, we can easily verify that SS$(v, \lambda) = \left\{\binom{v}{3}\right\}$. For example, when $V = \{a, b, c, d, e\}$, consider the partition $\{\{a, b, c\}, \{d, e\}\}$; the triple $\{a, b, c\}$ must be included, for if it is not, each pair 'inside' a class occurs with two pairs that 'cross' between the classes, and there are four inside pairs and six cross pairs. Thus every possible triple must be included at least once.

When $v \geq 6$, however, different support sizes are possible. Table 16.1 summarizes the known results for small orders.

Trades have been used extensively in this determination for small values. A local optimization method, starting with a fixed (typically simple) system, and applying trades to reduce support, has proved very effective in the solution for small orders.

v	Index λ	SS(v, λ)	Reference
6	2	$\{10\}$	
6	4	$\{10, 14, 16, 20\}$	Hedayat and
6	6	$\{10, 14, 16, 17, 19, 20\}$	Khosrovshahi (1981)
6	$2\mu \geq 8$	$\{10, 14, 16-20\}$	
7	1	$\{7\}$	
7	2	PS$(7, 2)$	
7	3	PS$(7, 3) \setminus \{16\}$	Foody and
7	4	PS$(7, 4) \setminus \{16, 24, 27\}$	Hedayat (1977)
7	5	PS$(7, 5) \setminus \{16\}$	
7	6	PS$(7, 6) \setminus \{16, 34\}$	
7	$\lambda \geq 7$	PS$(7, \lambda) \setminus \{16\}$	
8	$6\mu \geq 6$	PS$(8, 6\mu) \setminus \{20\}$	Hedayat and Li (1979)
9	$\lambda \geq 2$	PS$(9, \lambda) \setminus \{16, 19\}$	Khosrovshahi and Mahmoodian (1988)
10	$2\mu \geq 2$	PS$(10, 2\mu) \setminus \{22\}$	Khosrovshahi and Mahmoodian (1987)
11	$3\mu \geq 3$	PS$(11, 3\mu)$	Khosrovshahi and Mahmoodian (1987)
12	$2\mu \geq 2$	PS$(12, 2\mu) \setminus \{30\}$	Colbourn and Mahmoodian (1988)

Table 16.1 *SS(v, λ) for $v \leq 12$*

16.3.2 *Intersections and support*

The intersection theorem for Steiner triple systems (Theorem 14.6) provided the impetus for recursive methods in the solution of the support size problem. If two STS(v)s intersect in s triples, their union has support size $v(v-1)/3 - s$. Adding further copies of either system does not increase this support size. Thus we obtain:

Lemma 16.14 *For $v \equiv 1, 3 \pmod 6$, $v \geq 13$, and $\lambda \geq 2$, $PS(v, 2) \subseteq SS(v, \lambda)$.*

With the complete solution for small orders in hand, recursive methods are now used to determine the spectrum of support sizes for larger orders and larger indices. We first examine the solution for support sizes at or near the minimum value.

Lemma 16.15 *If there exists a $\{3\}$-GDD with group sizes $g_1, \ldots, g_\ell$ for which $\sum_{i=1}^{\ell} g_i = v$, and $s_i \in SS(g_i, \lambda)$ for $1 \leq i \leq \ell$, then*

$$\frac{1}{3}\left(\binom{v}{2} - \sum_{i=1}^{\ell}\binom{g_i}{2}\right) + \sum_{i=1}^{\ell} s_i \in SS(v, \lambda).$$

Proof Repeat each block of the GDD λ times, and place on the ith group a TS(g_i, λ) having support size s_i. □

This underlies a near complete determininination of the minimum support size:

Theorem 16.16 *The minimum support size of a TS(v, λ), $v \notin \{2, 6, 8, 14\}$, is m_v if $v \equiv 1, 3 \pmod 6$, $m_v + 6$ if $v \equiv 5 \pmod 6$, s_v if $v \equiv 0, 4 \pmod{12}$, $s_v + 1$ if $v \equiv 6, 10 \pmod{12}$, $s_v + 6$ if $v \equiv 8 \pmod{12}$, and $s_v + 7$ if $v \equiv 2 \pmod{12}$.*

Proof By Theorem 16.13, no TS(v, λ)s can have smaller support size. Now using Theorem 6.15, form a $\{3\}$-GDD with the following group type, depending on the congruence class of v:

Order	Group type	Restriction
$v \equiv 0, 4 \pmod{12}$	$4^{(v/4)}$	$v \geq 12$
$v \equiv 1, 3 \pmod 6$	1^v	$v \geq 3$
$v \equiv 2 \pmod{12}$	$4^{(v-26)/4}26^1$	$v \geq 62$
$v \equiv 5 \pmod 6$	$1^{(v-5)}5^1$	$v > 11$
$v \equiv 6, 10 \pmod{12}$	$4^{(v-10)/4}10^1$	$v \geq 28$
$v \equiv 8 \pmod{12}$	$4^{(v-20)/4}20^1$	$v \geq 44$

To complete the proof using Lemma 16.15, we must establish the statement for $v \in \{18, 20, 22, 26, 32, 38, 50\}$. Rosa and Hoffman (1986) give examples for $v \in \{18, 22\}$, and Colbourn and Mahmoodian (1993) give the rest. □

Near the minimum, Lemma 16.15 can also be employed with larger groups. For example, using a $\{3\}$-GDD of type $1^{u+1}u^1$ for $u \equiv 1, 3 \pmod 6$, we obtain that for $s \in SS(u, \lambda)$, $\binom{u+1}{2} + s \in SS(2u + 1, \lambda)$.

In the same vein, every block of a TS(v, λ) can be replaced by ℓ copies of the block, to obtain that SS$(v, \lambda) \subseteq SS(v, \lambda\ell)$ for all $\ell \geq 1$.

The determination of intersection sizes for $\{3\}$-GDDs in Theorem 14.11 can be used in a similar manner. Taking the union of a number of copies of each GDD, and filling groups in the result, yields support sizes near the minimum.

16.3.3 *Recursive constructions with fixed index*

Next we require constructions that increase the order, while keeping the index fixed. Of course, essentially all recursive constructions for triple systems have this property; we therefore need only examine ways to control the support size of the resulting system.

First we examine the basic $v \to 2v + 1$ construction from §3.2. Let $(V, \mathcal{B})$ be a TS(v, λ), and let X be a $(v+1)$-set of elements disjoint from V. Let $F_1, \ldots, F_v$ be a λ-factorization of λK_{v+1} with vertices indexed by X. Let $V = \{x_1, \ldots, x_v\}$. Now define $x_i \cdot F_i = \{\{x_i, a, b\} : \{a, b\} \in F_i\}$. Then $\mathcal{B} \cup \bigcup_{i=1}^{v} x_i \cdot F_i$ form the triples of a TS$(2v + 1, \lambda)$ on $V \cup X$. What is the support size of this triple system? For $1 \leq i \leq v$, let s_i be the number of distinct edges in F_i. Let s be the number of distinct triples in $\mathcal{B}$. Then the support size of the resulting system is $s + \sum_{i=1}^{v} s_i$.

For a factorization $F_1, \ldots, F_\ell$, if factor F_i contains s_i distinct edges for $1 \leq i \leq \ell$, then the *support size* of the factorization is $\sum_{i=1}^{\ell} s_i$. We denote by FS(n, λ) the spectrum of support sizes of λ-factorizations of λK_n. Then the $2v+1$ construction yields:

Lemma 16.17 *Let $s \in SS(v, \lambda)$ and $t \in FS(v+1, \lambda)$. Then $s+t \in SS(2v+1, \lambda)$.*

The use of Lemma 16.17 requires information concerning FS$(v + 1, \lambda)$. Colbourn and Rosa (1991) prove:

Theorem 16.18 *For $v \geq 8$ even, let*

$$m = v(v-1)/2 \quad \textit{and} \quad M = \min(\lambda, v-1) \cdot v(v-1)/2.$$

Then $FS(v, \lambda)$ is

$$\begin{cases} \{m, \ldots, M\} \setminus \{m+1, m+2, m+3, m+5\} & \textit{if } \lambda \neq v-1 \\ \{m, \ldots, M\} \setminus \{m+i, M-i : i \in \{1,2,3,5\}\} & \textit{if } \lambda = v-1 \end{cases}$$

Theorem 16.19 *For $v \geq 7$ odd, let*

$$m = v(v+3)/2 \quad \textit{and} \quad M = \min(\lambda, v-1) \cdot v(v-1)/2.$$

Then

$$FS(v, \lambda) = \begin{cases} \{m, \ldots, M\} & \textit{if } \lambda \neq v-1 \\ \{m, \ldots, M\} \setminus \{M-1, M-2, M-3, M-5\} & \textit{if } \lambda = v-1 \end{cases}$$

When v is odd, Lemma 16.17 does not yield support sizes near the minimum; however, filling in groups handles the majority of these. In addition, Lemma 16.17 is only effective when a TS(v, λ) exists; for example, we obtain no information about SS$(17, 3)$ since no TS$(8, 3)$ exists.

A large number of other doubling constructions have been applied to the support size problem; see Rosa and Hoffman (1986), Colbourn and Mahmoodian (1990, 1993), and Colbourn and Lindner (1992). We describe here just one of the general doubling constructions, essentially from Colbourn and Mahmoodian (1993). The idea is to provide a $v \to 2v + s$ construction for $1 \leq s \leq 7$. In these constructions, we employ λ-factorizations of graphs other than λK_n. In general, we do not require a complete determination of the spectrum of support sizes of factorizations for each such graph, but we do require a partial determination. To do this, we adapt an approach from Colbourn and Rosa (1991a).

Define a (v, λ, s)-*selection* to be a set of v multisets of size λ, whose union contains the numbers $\{7, \ldots, v+s-1\}$ each λ times, and the numbers $\{1, \ldots, 6\}$ each $(7-s)\lambda/6$ times. The support size of a selection is taken to be the sum over the v multisets of the number of distinct numbers in the multiset. Let ES(v, λ, λ, s) be the spectrum of support sizes of (v, λ, s)-selections.

Lemma 16.20 *For $v > \lambda \geq 2$,*

1. $ES(v, \lambda, 1) = ES(v, \lambda, 7) = \{v, \ldots, \lambda v\} \setminus \{v+1\}$, *and*
2. $ES(v, \lambda, 2) = ES(v, \lambda, 6) = \{v+5, \ldots, \lambda v\}$,
3. $ES(v, \lambda, 3) = ES(v, \lambda, 5) = \{v+4, \ldots, \lambda v\} \setminus \{v+5\}$,
4. $ES(v, \lambda, 4) = \{v+3, \ldots, \lambda v\} \setminus \{v+4\}$.

One application of Lemma 16.20 follows. Let G be a $(v+s-1)$-regular graph on $2n$ vertices, having a 1-factorization $F_1, \ldots, F_{s+v-1}$. Let G' be the multigraph obtained by repeating each edge in factor $F_1, \ldots, F_6$ $(7-s)\lambda/6$ times, and each other edge λ times. Then G' has a collection of v λ-factors with support size ℓn for every $\ell \in ES(v, \lambda, s)$.

When the number of vertices is odd, Lemma 16.20 can be applied instead to a 2-factorization, using a $(v, \lambda/2, s)$-selection. Now we turn to doubling constructions that allow us to control the congruence class modulo 6.

Lemma 16.21 *Let v and m be integers, $v > \lambda$, $v+m$ even, and $1 \leq m \leq 7$. Let $s \in SS(v, \lambda)$ and $r \in ES(v, \lambda, m)$. Let δ be 0 if $m = 1$, 1 if $m = 2$, and $\delta \in \{1, 2\}$ if $m \geq 3$. Then $s + r(v+m)/2 + \delta(v+m) \in SS(2v+m, \lambda)$.*

Proof To form the triples of the required triple system, use elements

$$\{x_1, \ldots, x_v\} \cup \{0, \ldots, v+m-1\}.$$

On the $\{x_i\}$s, take the triples of a TS(v, λ) with support size s. Now on the $v+m$ points, we employ triples using differences 1, 2, and 3 as follows. If $\delta = 0$, we take no such triples. If $\delta = 1$, we take the $v+m$ triples $\{i, i+1, i+3\}$ for $i = 0, \ldots, v+m-1$, each $(m-1)\lambda/6$ times. If $\delta = 2$, we take the triples $\{i, i+1, i+3\}$ for $i = 0, \ldots, v+m-1$, each once, and triples $\{i, i+2, i+3\}$ each $(m-1)\lambda/6 - 1$ times.

Now form a 1-factorization on the $v+m$ points, so that the first six factors cover edges of differences 1, 2, and 3, and the remaining $v+m-7$ factors cover the remaining differences. It is well known that since these graphs are cyclic, and each has a good difference, they are 1-factorable by Theorem 1.16. Finally, use a (v, λ, m)-selection with support size r, and form v λ-factors from the pairs left on the $v+m$ points, having support size $r(v+m)/2$. Form triples by associating each of the $\{x_i\}$ with one of these λ-factors. □

When $v+m$ is odd, the situation is slightly more complicated, but is similar nonetheless.

Lemma 16.22 *Let v and m be integers with $v > \lambda$, v even, m odd, and $1 \leq m \leq 7$. Let $s \in SS(v, 6)$ and $r \in ES(v, \lambda/2, m)$. Let $\delta = 0$ if $m = 1$, and $\delta \in \{1, 2\}$ if $m = 3, 5, 7$. Then $s + r(v+m) + \delta(v+m) \in SS(2v+m, \lambda)$.*

Proof We proceed as in the proof of Lemma 16.21 to form triples on $\{x_i\}$ and to form $\delta(v+m)$ triples on $\{0, \ldots, v+m-1\}$. At this point, we instead form 2-factors from the K_{v+m} on $\{0, \ldots, v+m-1\}$, again by 2-factoring the graph on differences 1, 2, and 3 and then 2-factoring the graph on the remaining differences. Using a $(v, \lambda/2, m)$-selection with support size r to form triples results in each x_i appearing with each element j λ times as required. □

The last case requires a small modification.

Lemma 16.23 *Let v and m be integers, $m \in \{2,4,6\}$, v odd, λ even and $v > \lambda$. Let $s \in SS(v, \lambda)$ and $r \in ES(v-1, \lambda/2, m+1)$. Let δ be as in Lemma 16.21. Then $s + r(v+m) + \lambda(v+m)/2 + \delta(v+m) \in SS(2v+m, \lambda)$.*

Proof Proceed as in the proof of Lemma 16.22. In the last step, only $v-1$ λ-factors are constructed, and the edges of difference 1, 2, and 3 are only covered $\lambda - 1$ times in total. Taking all remaining edges forms a (simple) λ-factor. This provides the last λ-factor, and adds $\lambda(v+m)/2$ to the support size. □

We have presented doubling constructions that are both sufficiently general, and not unnecessarily complicated. However, there is much latitude in selecting the triples on the points $\{0, \ldots, v+m-1\}$.

The second main class of recursive constructions involves tripling. We outline one of the easier constructions here, and examine general tripling constructions in the next section.

Lemma 16.24 *Let $s_1, s_2, s_3 \in SS(v, \lambda)$ and $s \in FS(v, \lambda)$. Then*

$$s_1 + s_2 + s_3 + 2s + v \in SS(3v, \lambda).$$

Proof We form the required TS$(3v, \lambda)$ on $\mathbb{Z}_v \times \mathbb{Z}_3$. First, on $\mathbb{Z}_v \times \{i\}$, place the triples of a TS(v, λ) with support size s_i. Now let $F_1, \ldots, F_{v-1}$ be a λ-factorization of λK_v having support size s. For each $\{x, y\} \in F_i$, let $\{i_0, x_1, y_2\}$ and $\{i_0, x_2, y_1\}$ be triples. Finally, let $\{0_0, a_1, a_2\}$ be a triple λ times for each $a \in \mathbb{Z}_v$. □

16.3.4 *Partitions with subsystems*

Khosrovshahi and Ajoodani-Namini (1991) first applied a construction to the support size problem that increases both the order and the index. Theirs is a $v \rightarrow 2v-2$ construction that doubles the index; using it, they settled the support size problem when $\lambda \equiv 0 \pmod 4$.

Colbourn and Lindner (1992) also develop constructions that increase the order and the index simultaneously. The basic idea is to form a partition of the set of all triples on v points into triple systems so that for a fixed set of $w < v$ points, each triple system either contains no triple on the w points, or contains a subsystem on the w points.

The $3v$ and $3v-2$ constructions

We describe recursive constructions for determining SS$(3v, \lambda)$ and SS$(3v-2, \lambda)$ from SS(v, μ). We form a number of collections of sets of triples on $\mathbb{Z}_n \times \mathbb{Z}_3$, in which the sets of the collection are disjoint, or nearly disjoint.

Let $\mathcal{T}$ be the triples of a resolvable TD$(3, n)$ on $\mathbb{Z}_n \times \mathbb{Z}_3$, with groups aligned on $\mathbb{Z}_n \times \{i\}$ for $i \in \mathbb{Z}_3$. We use the subscript notation x_i for $(x, i) \in \mathbb{Z}_n \times$

$\mathbb{Z}_3$. $\mathcal{T}$ has a partition into parallel classes of triples $P_0, \ldots, P_{n-1}$. Without loss of generality P_0 contains $\{x_0, x_1, x_2\}$ for all $x \in \mathbb{Z}_n$. A resolvable TD$(3,n)$ determines $6n$ permutations on $\mathbb{Z}_n \times \mathbb{Z}_3$ as follows. For all $x \in \mathbb{Z}_n$ and all $i \in \mathbb{Z}_3$, permutation π_j^i fixes x_i and x_{i-1} and maps x_{i+1} to y_{i+1} if $\{x_i, y_{i+1}\}$ appears in a triple of P_j; ψ_j^i fixes x_i and x_{i+1} and maps x_{i-1} to y_{i-1} if $\{x_i, y_{i-1}\}$ appears in a triple of P_j. We use another permutation on $\mathbb{Z}_n \times \mathbb{Z}_3$: σ fixes x_0 and x_1 and maps x_2 to $(x+1)_2$ for $x \in \mathbb{Z}_n$.

Let L be an $n \times n$ commutative idempotent latin square, for n odd. Let

$$T_1 = \left\{\{x_i, y_i, L(x,y)_{i+1}\} \mid x < y, x, y \in \mathbb{Z}_n, i \in \mathbb{Z}_3\right\} \cup P_0$$

$$T_2 = \left\{\{x_i, y_i, L(x,y)_{i-1}\} \mid x < y, x, y \in \mathbb{Z}_n, i \in \mathbb{Z}_3\right\} \cup P_0.$$

Then for $j \in \mathbb{Z}_n$ set

$$T_1^j = \{\pi_j^i(\{x_i, y_i, L(x,y)_{i+1}\}) \mid x, y \in \mathbb{Z}_n, x < y; i \in \mathbb{Z}_3\} \cup P_j$$

and

$$T_2^j = \{\psi_j^i(\{x_i, y_i, L(x,y)_{i-1}\}) \mid x, y \in \mathbb{Z}_n, x < y; i \in \mathbb{Z}_3\} \cup P_j.$$

Lemma 16.25 *For n odd, the sets of triples*

$$\sigma^2(\mathcal{T}), \ldots, \sigma^{n-1}(\mathcal{T}), T_1^0, \ldots, T_1^{n-1}, \sigma(T_2^0), \ldots, \sigma(T_2^{n-1})$$

are pairwise disjoint.

Proof That $\sigma^i(\mathcal{T})$ and $\sigma^j(\mathcal{T})$ are disjoint for $i \neq j$ follows from the fact that $\mathcal{T}$ contains a unique triple containing $\{x_0, y_1\}$ for $x, y \in \mathbb{Z}_n$. Since T_1^i contains only those triples $\{x_0, y_1, z_2\}$ that appear in $\mathcal{T}$, and $\sigma(T_2^i)$ contains only those in $\sigma(\mathcal{T})$, the sets $\sigma^i(\mathcal{T})$ are disjoint from the rest.

Moreover, sets obtained from T_1 are disjoint from those obtained from T_2 since the former contain triples of the form $\{x_i, y_i, z_{i+1}\}$ while the latter contain triples of the form $\{x_i, y_i, z_{i-1}\}$. We need only verify that T_1^j and T_1^k are disjoint for $j \neq k$, the case for T_2 being similar. This disjointness is a result of the fact that $\pi_j^i(z_{i+1}) = \pi_k^i(z_{i+1})$ only if $j = k$. □

Similarly, we have:

Lemma 16.26 *For n odd, the sets of triples*

$$\sigma^1(\mathcal{T}) \ldots, \sigma^{n-1}(\mathcal{T}), T_1^0, \ldots, T_1^{n-1}, T_2^0, \ldots, T_2^{n-1}$$

have the property that $T_1^i \cap T_2^i = P_i$ for $i \in \mathbb{Z}_n$, and the sets are otherwise pairwise disjoint.

These lemmas are used as the basis for one pair of $3v$ and $3v-2$ constructions.

Lemma 16.27 *Let v be odd. Let $\lambda = \lambda_1 + \lambda_2$ with $\lambda_1 \equiv 0 \pmod 3$ if $v \equiv 5 \pmod 6$. Let $s_0, s_1, s_2 \in SS(v, \lambda)$. Let τ be 0 if $\lambda_1 = 0$, and $1 \leq \tau \leq \min(v-2, \lambda_1)$ otherwise. Let μ be 0 if $\lambda_2 = 0$, and $1 \leq \mu \leq \min(2v, \lambda_2)$ otherwise. Then $s_0 + s_1 + s_2 + \tau v^2 + \mu v(3v-1)/2 \in SS(3v, \lambda)$.*

Proof Form the required triple system on $\mathbb{Z}_v \times \mathbb{Z}_3$ with blocks $\mathcal{B}$. First, on $\mathbb{Z}_v \times \{i\}$ place the triples of a TS(v, λ_1) having s_i distinct blocks, for each $i \in \mathbb{Z}_3$. Now use a resolvable TD$(3, v)$ $\mathcal{T}$ to form permutations, and any commutative idempotent latin square L to form the sets T_1 and T_2 as before.

Choose any τ of the sets $\sigma^i(\mathcal{T})$ for $2 \leq i \leq n-1$; if τ is not zero, place $\tau - 1$ of the chosen sets in $\mathcal{B}$ once each, and place the triples of one of the chosen $\sigma^i(\mathcal{T})$ in $\mathcal{B}$ $\lambda_1 - \tau + 1$ times. Finally, choose any μ of the sets T_1^i and $\sigma(T_2^i)$; if $\mu > 0$, place the blocks of $\mu - 1$ of these in $\mathcal{B}$ once each, and place the blocks of the remaining one in $\mathcal{B}$ $\lambda_2 - \mu + 1$ times. Using Lemma 16.25, the result is a TS$(3v, \lambda)$ with the required support size. □

Lemma 16.28 *Let v be even. Let $\lambda = \lambda_1 + 2\lambda_2$ where $\lambda_1 \equiv 0 \pmod 2$ for $v \equiv 0, 4 \pmod 6$, and $\lambda_1 \equiv 0 \pmod 6$ if $v \equiv 2 \pmod 6$. Let $s_0, s_1, s_2 \in SS(v, \lambda_1)$. Let τ be 0 if $\lambda_1 = 0$, and $1 \leq \tau \leq min(\lambda_1, v-2)$ otherwise. Let μ be 0 if $\lambda_2 = 0$, and $1 \leq \mu \leq min(\lambda_2, v-1)$ otherwise. Then*

$$s_0 + s_1 + s_2 + \tau(v-1)^2 + \mu(v-1)(3v-2) \in SS(3v-2, \lambda).$$

Proof Use a resolvable TD$(3, v-1)$ $\mathcal{T}$ on $\mathbb{Z}_{v-1} \times \mathbb{Z}_3$ to form the permutations, and a commutative idempotent latin square of side $v-1$ to form the sets T_1 and T_2. Add a further point ∞. On $(\mathbb{Z}_{v-1} \times i) \cup \{\infty\}$, place a TS$(v, \lambda_i)$ having s_i distinct blocks for each $i \in \mathbb{Z}_3$.

Choose any τ of the sets $\sigma^i(\mathcal{T})$ for $1 \leq i \leq v-2$; if $\tau > 0$, include any $\tau - 1$ of these once each, and include one more $\lambda_1 - \tau + 1$ times.

If $\mu > 0$, include $T_1^i \cup T_2^i$ once for each $1 \leq i < \mu$; recall that T_1 and T_2 intersect in P_0, so retain only one copy of each such triple in their union. Then include $T_1 \cup T_2$ $\lambda_2 - \mu + 1$ times.

Finally, for each triple $\{x, y, z\}$ of parallel classes $P_1, \ldots, P_{\mu-1}$, include the triples $\{\infty, x, y\}$, $\{\infty, x, z\}$, and $\{\infty, y, z\}$. Include the same triples for each triple of P_0 $\lambda_2 - \mu + 1$ times each. Lemma 16.26 ensures that the result is a TS$(3v-2, \lambda)$ with the stated support size. □

We now follow the same strategy as Colbourn and Lindner (1992), but repair in the process an oversight in their solution.

Lemma 16.29 *When $n \equiv 0, 1, 3 \pmod 4$, there is a set*

$$G_1, \ldots, G_{n-1}, U_0, \ldots, U_{n-1}$$

of disjoint sets of triples on $X \times \{0, 1, 2\}$ with $|X| = n$, with the properties that

1. *each G_ℓ contains only transverse triples (of the form $\{a_0, b_1, c_2\}$), and covers every pair of the form $\{a_i, b_j\}$ with $i \neq j$ exactly once;*
2. *each U_ℓ contains n disjoint transverse triples (a genuine parallel class), and covers all pairs of $X \times \{0, 1, 2\}$ twice each, except for $3n$ pairs which together would form a set of disjoint transverse triples (a shadow parallel class); the latter pairs are covered once each by U_ℓ; and*

3. *genuine and shadow parallel classes contain triples not appearing in any* G_ℓ.

Proof For n odd, apply Lemma 16.26 and let $G_i = \sigma^i(\mathcal{T})$ and $U_i = T_1^i \cup T_2^i$. In U_i, the parallel class P_i appears only once as the genuine parallel class, and its shadow parallel class is also P_i.

Next we treat the case when $n = 2s = 2^\alpha$ for $\alpha \geq 2$. Let ω be a primitive element of $\mathbb{F}_{2^\alpha}$. Let X be $\mathbb{F}_{2^\alpha}$. The pairs of elements in X form a complete graph, whose edges can be factored into s hamiltonian paths; in such a factorization, every vertex appears as the end of precisely one path by simple counting. So, by appropriately renaming symbols, we can choose a decomposition for any nonzero $\beta \in \mathbb{F}_{2^\alpha}$ of the pairs on $\mathbb{F}_{2^\alpha}$ into hamiltonian paths $B_{a_1}^\beta, \ldots, B_{a_s}^\beta$ where $a_i \notin \{a_j, a_j + \beta\}$ unless $i = j$, and the endpoints of path $B_{a_i}^\beta$ are a_i and $a_i + \beta$. Now for $Y \subset \mathbb{F}_{2^\alpha}$, and A a set of unordered pairs on $\mathbb{F}_{2^\alpha}$, define

$$Y \star_i A = \{\{y_i, a_{i+1}, b_{i+1}\} : \{a, b\} \in A, y \in Y\}.$$

Subscripts are reduced modulo 3 as needed. Similarly, define

$$Y \circ_i A = \{\{y_i, a_{i-1}, b_{i-1}\} : \{a, b\} \in A, y \in Y\}.$$

With this notation, choose $a_1, \ldots, a_s$ in $\mathbb{F}_{2^\alpha}$ so that $a_i \notin \{a_j, a_j + 1\}$ unless $i = j$. Form parallel classes $P_\beta = \{\{j_0, (j+\beta)_1, (j+\omega\beta)_2\} : j \in \mathbb{F}_{2^\alpha}\}$ when $\beta \in \mathbb{F}_{2^\alpha}$. Then in the system U_{2i-2}, take triples

$$(\{c, c+1\} \circ_0 B_{a_i+c}^1) \cup (\{c, c+\omega+1\} \circ_1 B_{a_i+c}^{\omega+1}) \cup (\{c, c+\omega\} \circ_2 B_{a_i+c}^\omega)$$

for $c \in \mathbb{F}_{2^\alpha}$, together with P_{a_i}, with genuine parallel class P_{a_i} and shadow parallel class P_{a_i+1}. Form the system U_{2i-1} by replacing $\circ_i$ by $\star_{i-1}$, and interchanging P_{a_i} and P_{a_i+1} in the definition of U_{2i-2}. Finally, to form the G_is, let $n_1, \ldots, n_{2s-1}$ be the nonzero elements of $\mathbb{F}_{2^\alpha}$, and form G_i as

$$\{\{a_0, b_1, (c+n_i)_2\} : \{a, b, c\} \in P_\beta, \beta \in \mathbb{F}_{2^\alpha}\}.$$

It remains to treat the case when $n = n_1 n_2$ with n_1 odd, and $n_2 = 2^\alpha$ for some $\alpha \geq 2$. Form a solution $\widehat{G}_1, \ldots, \widehat{G}_{n_1-1}, \widehat{U}_0, \ldots, \widehat{U}_{n_1-1}$ as above on the set $\widehat{X} \times \{0,1,2\}$. Choose a solution in which genuine and shadow parallel classes coincide. Form the required solution on $X \times \{0,1,2\}$ with $X = \widehat{X} \times \mathbb{F}_{2^\alpha}$. The technique is a standard Wilson-type inflation. First consider inflating one of the $\widehat{U}_i$s. For each non-transverse triple $\{a_i, b_i, c_j\}$, form on the groups $\{a\} \times \mathbb{F}_{2^\alpha} \times \{i\}$, $\{b\} \times \mathbb{F}_{2^\alpha} \times \{i\}$, and $\{c\} \times \mathbb{F}_{2^\alpha} \times \{j\}$ altogether 2^α disjoint TD$(3, 2^\alpha)$s. For the genuine parallel classes, inflate using the 2^α systems $\tilde{U}_i$ in a solution of order 2^α as produced before. Simply ignore the shadow parallel classes. Now to inflate the classes $\widehat{G}_i$, use the same technique for each transverse triple as was used for the nontransverse triples of $\widehat{U}_i$. This gives thus far $2^\alpha \cdot (n_1 - 1)$ of the required G_is. The remaining $2^\alpha - 1$ G_is are obtained by employing the (as yet unused) triples in the systems $\tilde{G}_i$ from the solutions of order 2^α. □

This underlies the second $3v-2$ construction:

Lemma 16.30 *Let v be odd, $v \notin \{3,7\}$. Let $\lambda = \lambda_1 + 2\lambda_2$, with $\lambda_1 \equiv 0 \pmod 3$ if $v \equiv 5 \pmod 6$. Let $s_0, s_1, s_2 \in SS(v, \lambda_1)$. Let $\tau = 0$ if $\lambda_1 = 0$, and $1 \le \tau \le min(\lambda_1, v-2)$ otherwise. Let $\mu = 0$ if $\lambda_2 = 0$, and $1 \le \mu \le min(\lambda_2, v-1)$ otherwise. Then $s_0 + s_1 + s_2 + \tau(v-1)^2 + \mu(v-1)(3v-2) \in SS(3v-2, \lambda)$.*

Proof If $v \equiv 1 \pmod 4$, form the required blocks $\mathcal{B}$ on $(\mathbb{Z}_{v-1} \times \mathbb{Z}_3) \cup \{\infty\}$. For $i \in \mathbb{Z}_3$, include in $\mathcal{B}$ the blocks of a TS(v, λ_1) on $(\mathbb{Z}_{v-1} \times \{i\}) \cup \{\infty\}$. Apply Lemma 16.29. If $\tau > 0$, place in $\mathcal{B}$ the blocks of G_{j-1} once each for $2 \le j \le \tau$, and $\lambda_1 - \tau + 1$ times each for $j = 1$. If $\mu > 0$, place the blocks of U_j in $\mathcal{B}$ once each for $1 \le j < \mu$, and $\lambda_2 - \mu + 1$ times each for $j = 0$. Finally, if $\{x, y, z\}$ is in the shadow parallel class of U_j, place the blocks $\{\infty, x, y\}$, $\{\infty, x, z\}$, and $\{\infty, y, z\}$ in $\mathcal{B}$ once each for $1 \le j < \mu$, and $\lambda_2 - \mu + 1$ times each for $j = 0$.

If $v \equiv 3 \pmod 4$, write $v - 1 = n = 2s$, so that s is odd. Use Lemma 16.29 to form a solution on $X \times \{0,1,2\}$, namely $G_1, \ldots, G_{s-1}, U_0, \ldots, U_{s-1}$ in which genuine and shadow parallel classes in the U_is coincide. Now form some sets on $(X \times \{0,1\}) \times \{0,1,2\}$. For each U_i, form two sets of triples, $\overline{U}_{2i}$ and $\overline{U}_{2i+1}$, as follows. For each non-transverse triple of U_i, inflate it using two disjoint TD(3,2)s. When $\{a_0, b_1, c_2\}$ is a triple in the genuine parallel class of U_i, place the triples obtained from $\{(a,0)_0, (a,1)_0, (b,0)_1\}$ and $\{(a,0)_0, (a,1)_0, (c,1)_2\}$ under the automorphism $((a,0)_0\ (b,0)_1\ (c,0)_2)((a,1)_0\ (b,1)_1\ (c,1)_2)$ in $\overline{U}_{2i}$, and the triples obtained from these by negating the second coordinate in $\overline{U}_{2i+1}$. For each G_i, form systems $\overline{G}_{2i-1}$ and $\overline{G}_{2i}$ by inflating each triple to two disjoint TD(3,2)s. Now we complete the proof. Choose τ and μ as in the proof of Lemma 2.6. In each case, include the blocks of $\overline{G}_j$ once each for $2 \le j \le \min(\tau, v-3)$, and $\lambda_1 - \tau + 1$ times each for $j = 1$. Writing $\mu = 2t + x$, $x \in \{0,1\}$, include the blocks of $\overline{U}_j$ once each for $1 \le j < \mu$, and $\lambda_2 - \mu + 1$ times for $j = 0$. Further, include the blocks $\{\infty, (a,e)_i, (b,f)_j\}$ once each when $1 \le j < t$, $e, f \in \{0,1\}$, and $\{a_i, b_j\}$ appears in a triple T of the shadow parallel class of U_j, along with the four triples of a TD(3,2) on the inflation of T by 2. Now if $\mu \equiv \lambda_2 \pmod 2$, add triples with ∞ and the triples of a TD(3,2) each $(\lambda_2 - \mu + 2)/2$ times for the triples of the shadow parallel class of U_0. If $\mu \not\equiv \lambda_2 \pmod 2$, for each block $\{a_0, b_1, c_2\}$ in the shadow parallel class of U_0, include the blocks $\{(a,h)_0, (b,h)_1, (c,h)_2\}$ for $h = 0, 1$. Edges in these blocks then are added to each form $(\lambda_2 - \mu + 1)/2$ triples each with ∞, and the other edges are placed in $(\lambda_2 - \mu + 3)/2$ edges with ∞. Then the blocks of a TD(3,2) are placed $(\lambda_2 - \mu + 1)/2$ times for the inflation of each triple in the shadow parallel class of U_0. Finally, when $x = 1$, for each block $\{a_0, b_1, c_2\}$ in the shadow parallel class of U_t, include the blocks $\{(a,h)_0, (b,h)_1, (c,h)_2\}$ for $h = 0, 1$ and form triples by inserting ∞ in the pairs $\{(a,h)_0, (b, 1-h)_1\}$, $\{(a,h)_0, (c, 1-h)_1\}$, and $\{(b,h)_0, (c, 1-h)_1\}$ for $h = 0, 1$. This completes the construction when $\tau < v - 2$.

Now we treat the final case when $\lambda_1 = v - 2 = 2s - 1$. For all shadow parallel classes except for those of U_0 when $\lambda_2 \not\equiv \mu \pmod 2$ and of U_t when $x = 1$, simply include the blocks of the TD(3,2) *disjoint from* the one chosen in inflating the shadow parallel class before. Under the stated conditions, blocks from the shadow parallel classes of U_0 and/or U_t are treated differently. When necessary, follow the construction given, but for each triple $\{a_0, b_1, c_2\}$ in the shadow parallel class, remove the blocks $\{(a,h)_0, (b,h)_1, (c,h)_2\}$ for $h = 0, 1$ and the triples formed by inserting ∞ in the pairs $\{(a,h)_0, (b, 1-h)_1\}$, $\{(a,h)_0, (c, 1-h)_1\}$, and $\{(b,h)_0, (c, 1-h)_1\}$ for $h = 0, 1$. Replace them with the triples $\{(a,i)_0, (b,j)_1, (c,j)_2\}$ for all $i, j, k \in \{0, 1\}$ with two but not three of $\{i, j, k\}$ equal, along with the triples obtained by inserting ∞ in the pairs $\{(a,h)_0, (b,h)_1\}$, $\{(a,h)_0, (c,h)_1\}$, and $\{(b,h)_0, (c,h)_1\}$ for $h = 0, 1$. □

The $3v$ construction when v is even is similar. When $v \equiv 0, 2, 6 \pmod 8$, inflate by 2 the solution from Lemma 16.29 for $v/2$, using two disjoint TD(3,2)s for each nontransverse and each shadow transverse triple in the U_is, and for each transverse triple in the G_is. Transverse triples in genuine parallel classes are treated differently. When $\{a_0, b_1, c_2\}$ is a triple in the genuine parallel class of U_i, placc the triples obtained from

$$\{(a,0)_0, (a,1)_0, (b,0)_1\} \quad \text{and} \quad \{(a,0)_0, (a,1)_0, (c,1)_2\}$$

under the automorphism $((a,0)_0\ (b,0)_1\ (c,0)_1)((a,1)_0\ (b,1)_1\ (c,1)_1)$ in the first system, and the triples obtained from these by negating the second coordinate in the second. This yields a set of v TS$(v,2)$s and a set of $v-2$ collections of transverse triples, all of which are disjoint and contain no triple of $X \times \{0,1,2\}$ with all second coordinates equal. It remains to treat the case when $v \equiv 4 \pmod 8$. Since there is a set of four TS(12,2)s and two sets of transverse triples, partitioning all triples on $X \times \{0,1,2\}$ except those with three second coordinates equal, it is easy to produce a similar set for any odd multiple of 4, as follows. Write $n = 4s$, s odd. Use Lemma 16.29 to produce a solution $G_1, \ldots, G_{s-1}, U_0, \ldots, U_{s-1}$. Now inflate every element to four elements, using four disjoint TD(3,4)s on nontransverse triples, and all triples in the G_is. Assume that genuine and shadow parallel classes coincide. For the triples in genuine parallel classes, use the four disjoint TS(12,2)s. The remaining two sets of transverse triples from the solution on 12 points, taken over all triples in genuine parallel classes, can be used to form two additional sets of transverse triples to complete the partition into $4s$ TS$(12s, 2)$s and $4s-2$ sets of transverse triples.

Lemma 16.31 *Let v be even, $v \notin \{4, 12\}$. Let $\lambda = \lambda_1 + 2\lambda_2$, where $\lambda_1 \equiv 0 \pmod 2$ if $v \equiv 0, 4 \pmod 6$, and $\lambda_1 \equiv 0 \pmod 6$ if $v \equiv 2 \pmod 6$. Let $\tau = 0$ if $\lambda_1 = 0$, and $1 \le \tau \le min(\lambda_1, v-2)$ otherwise. Let $\mu = 0$ if $\lambda_2 = 0$, and $1 \le \mu \le min(\lambda_2, v)$ otherwise. Let $s_0, s_1, s_2 \in SS(v, \lambda_1)$. Then*

$$s_0 + s_1 + s_2 + \tau v^2 + \mu v(3v-1) \in SS(3v, \lambda).$$

Proof The sets of triples defined above are used. For $i \in \mathbb{Z}_3$, place on $\mathbb{Z}_2 \times \mathbb{Z}_{v/2} \times \{i\}$ a TS(v, λ_1) having s_i distinct triples. Next include any τ of the sets of transverse triples; when $\tau > 0$, include the triples of one of them $\lambda_1 - \tau + 1$ times, and the triples for the remaining $\tau - 1$ sets once each. Finally, if $\mu > 0$, choose any μ of the TS$(v, 2)$s, and include the triples of one $\lambda_2 - \mu + 1$ times, and of the remaining $\mu - 1$ sets once each. □

The $3v - 4$ constructions

The $3v - 4$ constructions pose a different type of problem. In the $3v$ and $3v - 2$ constructions, the recursive constructions form three systems that intersect in zero or one common element. The natural extension for the $3v - 4$ construction is to form three systems on $(\mathbb{Z}_{v-2} \times \mathbb{Z}_3) \cup \{\infty_1, \infty_2\}$ that intersect in the two common elements $\{\infty_1, \infty_2\}$. However, the three triple systems cannot intersect in a subsystem as before, since there is no triple system on two elements. Nevertheless, Colbourn and Lindner (1992) adapt the same basic strategy to obtain the $3v - 4$ construction.

In developing the $3v$ and $3v - 2$ constructions, Colbourn and Lindner (1992) amalgamated two well-known constructions: the direct product, and the Bose construction. The first task is to develop analogous basic constructions for the $3v - 4$ case. First, an analogue of the Bose construction for v odd is derived.

Let L_1, L_2, L_3 be commutative idempotent $(v-2) \times (v-2)$ latin squares. Let π_1, π_2, π_3 be permutations on $\mathbb{Z}_{v-2}$ for which $\pi_1(0) = \pi_2(0) = \pi_3(0) = 0$, and let π_1, π_2, π_3 permute the nonzero elements of $\mathbb{Z}_{v-2}$ in cycles of even length. Then for $j \in \{1, 2, 3\}$, form $D_j = \left\{\{x_i, y_i, (\pi_j(z))_{i+1}\} \mid x < y, L_j(x, y) = z, i \in \mathbb{Z}_3\right\}$. D_j covers all pairs on $\mathbb{Z}_{v-2} \times \mathbb{Z}_3$ once each *except* those of the form $\{x_i, (\pi_j(x))_{i+1}\}$ for $x \in \mathbb{Z}_{v-2}$ and $i \in \mathbb{Z}_3$. Now colour the nonzero elements of $\mathbb{Z}_{v-2}$ in two colours, say white and black, so that $\pi_j(x)$ is coloured oppositely from x (this can be done since π_j consists of *even* length cycles). Then if x is white, add the triple $\{\infty_1, x_i, (\pi_j(x))_{i+1}\}$ to D_j for $i \in \mathbb{Z}_3$; if x is black, add $\{\infty_2, x_i, (\pi_j(x))_{i+1}\}$ instead.

D_j now covers all pairs on $(\mathbb{Z}_{v-2} \times \mathbb{Z}_3) \cup \{\infty_1, \infty_2\}$ once each, except precisely those on $\{\infty_1, \infty_2, 0_0, 0_1, 0_2\}$. Hence $D_1 \cup D_2 \cup D_3$, together with the blocks of a TS$(5, 3)$ on $\{\infty_1, \infty_2, 0_0, 0_1, 0_2\}$ gives a TS$(3v - 4, 3)$. This construction has much flexibility, since the permutations and the latin squares can all be chosen independently; it is an easy exercise to check that choosing two of the permutations to be the same makes it unnecessary for the common permutation π to consist of even length cycles, since in that case both $\{\infty_1, x_i, (\pi(x))_{i+1}\}$ and $\{\infty_2, x_i, (\pi(x))_{i+1}\}$ can be included for all nonzero x and $i \in \mathbb{Z}_3$.

To achieve the full spectrum of support sizes, Colbourn and Lindner (1992) also employ a construction that uses transverse triples (of the form $\{x_0, y_1, z_2\}$). Hence a second $3v - 4$ construction is developed. Suppose that $v \equiv 1, 3 \pmod 6$, and let M be an arbitrary $(v-2) \times (v-2)$ latin square. Let L_1, L_2 be commutative idempotent $(v - 2) \times (v - 2)$ latin squares, and choose an arbitrary permutation π on $\mathbb{Z}_{v-2}$.

Using L_1 and L_2, form D_j for $j \in \{1, 2\}$ on $\mathbb{Z}_{v-2} \times \mathbb{Z}_3$. Then add the triples $\{\infty_1, x_i, (\pi(x))_{i+1}\}$ and $\{\infty_2, x_i, (\pi(x))_{i+1}\}$ for *all* $x \in \mathbb{Z}_{v-2}$ and $i \in \mathbb{Z}_3$. At this point, all pairs on $(\mathbb{Z}_{v-2} \times \mathbb{Z}_3) \cup \{\infty_1, \infty_2\}$ have been covered twice, except the pair $\{\infty_1, \infty_2\}$ that is uncovered. Add the triples $\{\{x_0, y_1, z_2\} \mid M(x,y) = z\}$ and finally, for $i \in \mathbb{Z}_3$, place an STS(v) on $(\mathbb{Z}_{v-2} \times \{i\}) \cup \{\infty_1, \infty_2\}$. The pair $\{\infty_1, \infty_2\}$ is covered once in each of the three STS(v)s, and hence appears in three triples in total as required. Hence the result is a TS($3v-4, 3$).

The strategy is an amalgamation of these two basic constructions. The easiest case is treated first:

Lemma 16.32 *Let* $\lambda = 3\lambda_1$, *where* $\lambda_1 \equiv 0 \pmod{\gcd(v-2, 6)}$. *Let* $s_0, s_1, s_2 \in SS(v, \lambda)$. *Let* $1 \leq \tau, \mu \leq min(v-2, \lambda_1)$. *Then*

$$s_0 + s_1 + s_2 + \tau(v-2)^2 + \mu(v-1)(3v-6) \in SS(3v-4, \lambda).$$

Proof Form the blocks $\mathcal{B}$ of the required TS($3v-4, \lambda$) on $(\mathbb{Z}_{v-2} \times \mathbb{Z}_3) \cup \{\infty_1, \infty_2\}$. First, for $i \in \mathbb{Z}_3$, place on $(\mathbb{Z}_{v-2} \times \{i\}) \cup \{\infty_1, \infty_2\}$, a TS($v, \lambda_1$) having s_i distinct triples.

Choose an arbitrary $(v-2) \times (v-2)$ latin square M, and define

$$E_j = \{\{x_0, y_1, (z+j)_2\} \mid M(x,y) = z\}$$

for $j \in \mathbb{Z}_{v-2}$. Add to $\mathcal{B}$ the triples of E_j once each for $1 \leq j < \tau$, and add the triples of E_τ $\lambda_1 - \tau + 1$ times each.

Choose an arbitrary $(v-2) \times (v-2)$ idempotent latin square L, and an arbitrary permutation π on $\mathbb{Z}_{v-2}$. For $j \in \mathbb{Z}_{v-2}$, define

$$A_j = \{\{x_0, y_0, (z+j)_1\}, \{x_1, y_1, (z+j)_2\}, \{x_2, y_2, (z-2j)_0\} : .$$
$$x < y;\ z = \pi(L(x,y))\}$$

and

$$B_j = \{\{x_0, y_0, (z+2j)_2\}, \{x_2, y_2, (z-j)_1\}, \{x_1, y_1, (z-j)_0\} :$$
$$x > y;\ z = \pi^{-1}(L(x,y))\}.$$

Then form C_j to be the union of A_j, B_j, and

$$\{\{\iota, x_0, (\pi(x)+j)_1\}, \{\iota, x_1, (\pi(x)+j)_2\}, \{\iota, x_2, (\pi(x)-2j)_0\} :$$
$$\iota \in \{\infty_1, \infty_2\}, x \in \mathbb{Z}_{v-2}\}$$

Then include in $\mathcal{B}$ the triples of C_j once each for $0 \leq j < \mu - 1$, and $\lambda_1 - \mu + 1$ times each for $j = \mu - 1$. □

In this proof, the usual ordering $<$ in $\mathbb{Z}_{v-2}$ could in fact be replaced by an arbitrary ordering $\prec$ on the unordered pairs of $\mathbb{Z}_{v-2}$.

Lemma 16.32 gives us the necessary tool to handle all cases when $v \not\equiv 2 \pmod 3$, but yields only $\lambda \equiv 0 \pmod 9$ when $v \equiv 5 \pmod 6$, and only $\lambda \equiv 0 \pmod{18}$ when $v \equiv 2 \pmod 6$. Hence the construction is modified to treat these anomalous cases.

Lemma 16.33 *Let $v \equiv 2 \pmod 3$, $v \geq 5$. Let $\lambda = 3\lambda_1 + 6$, $\lambda_1 > 0$, $\lambda_1 \equiv 0 \pmod{\gcd(v-2, 6)}$. Let $s_0, s_1, s_2 \in SS(v, \lambda_1)$. Let $1 \leq \tau \leq min(v-2, \lambda_1)$. Let $1 \leq \mu \leq min(v-2, \lambda_1 + 3)$. Let $\alpha = 0$ if every pair of a TS(v, λ_1) with support size s_i appears in at least $v-3$ distinct triples, and $\alpha \in \{0, 6\}$ otherwise. Let $\beta = 0$ if $\mu = 1$, $\beta = -6$ if $\mu = \lambda_1 + 3$, and $\beta \in \{0, -6\}$ otherwise. Let $\gamma = 3$ if $\tau = 1$, $\gamma = 0$ if $\tau = v-2$, and $\gamma \in \{0, 3\}$ otherwise. Let $\delta = 0$ if $\tau = 1$, $\delta = -1$ if $\tau = \lambda_1$, and $\delta \in \{0, -1\}$ otherwise; moreover, if $\delta = -1$ and $\gamma = 0$, $\tau \geq 2$. Then $s_0 + s_1 + s_2 + \tau(v-2)^2 + \mu(v-1)(3v-6) + \alpha + \beta + \gamma + \delta \in SS(3v-4, \lambda)$.*

Proof The proof of Lemma 16.32 is followed closely. Choose a $(v-2) \times (v-2)$ idempotent latin square L arbitrarily, and choose an arbitrary permutation π that satisfies $\pi(0) = 1$ and $\pi(1) = 0$. The choice of L and π defines sets C_j for $j \in \mathbb{Z}_{v-2}$ as in the proof of Lemma 16.32. Next choose an arbitrary $(v-2) \times (v-2)$ latin square M that satisfies $M(0,0) = M(1,1) = 0$ and $M(0,1) = M(1,0) = 1$ (that is, M has a 2×2 subsquare on $\{0, 1\}$). Use M to form sets E_j for $j \in \mathbb{Z}_{v-2}$ as in the proof of Lemma 16.32.

Define $\hat{C}_0$ to be C_0 *omitting* the six triples consisting of $\{\infty_1, 0_i, 1_{i+1}\}$ and $\{\infty_2, 0_i, 1_{i-1}\}$ for $i \in \mathbb{Z}_3$.

Now the required TS$(3v-4, \lambda)$ is constructed. For $i \in \mathbb{Z}_3$, place the triples of a TS(v, λ_1) on $(\mathbb{Z}_{v-2} \times \{i\}) \cup \{\infty_1, \infty_2\}$ in such a way that $\{\infty_1, \infty_2, 0_i\}$ and $\{\infty_1, \infty_2, 1_i\}$ *are not* in the TS(v, λ_1) if $\alpha = 6$, and *are* in the TS(v, λ_1) if $\alpha = 0$. In the first case, we can always choose to place the system in this way since there is some pair appearing in fewer than $v-3$ distinct triples. In the second, since no TS(v, λ_1) has all blocks λ_1-times repeated, some pair appears with two different elements in triples.

Now if $\beta = -6$, include the triples of $\hat{C}_0$ once, those of C_j for $1 \leq j \leq \mu - 2$ once, and those of $C_{\mu-1}$ each $\lambda_1 - \mu + 1$ times. If $\beta = 0$, include instead the triples of $\hat{C}_0$ once, those of C_0 $\lambda_1 - \mu$ times, and those of C_j once each for $1 \leq j < \mu$.

If $\gamma = 3$ and $\delta = -1$, include the triples of E_j once for $1 \leq j < \tau$, and $\lambda_1 - \tau + 1$ times for $j = \tau$. If $\gamma = 3$ and $\delta = 0$, include instead the triples of E_j once each for $2 \leq j \leq \tau$, and $\lambda_1 - \tau + 1$ times for $j = 1$. If $\gamma = 0$ and $\delta = -1$, include instead the triples of E_j once each for $0 \leq j \leq \tau - 2$, and $\lambda_1 - \tau + 1$ times for $j = \tau - 1$. Finally, if $\gamma = 0$ and $\delta = 0$, include the triples of E_j once each for $j = 0$ and $2 \leq j < \tau$, and $\lambda_1 - \tau + 1$ times for $j = 1$. By construction, E_1 contains the triple $\{1_0, 1_1, 1_2\}$; omit *one* copy of this triple.

The collection of triples $\mathcal{B}$ so constructed is 'nearly' a TS$(3v-4, \lambda)$. All pairs are covered precisely λ times each, *except* that

- $\{\infty_1, \infty_2\}$ appears only $3\lambda_1$ times,
- $\{\iota, x_i\}$ appears only $3\lambda_1 + 5$ times for $\iota \in \{\infty_1, \infty_2\}$, $x \in \{0, 1\}$, $i \in \mathbb{Z}_3$,
- $\{0_i, 1_j\}$ appears only $3\lambda_1 + 5$ times for $i \neq j$, $i, j \in \mathbb{Z}_3$, and
- $\{1_i, 1_j\}$ appears only $3\lambda_1 + 5$ times.

The omission of the six triples in $\hat{C}_0$ reduced the number of distinct triples by 6 if $\beta = -6$, and left it unchanged when $\beta = 0$. Similarly, the omission of the triple $\{1_0, 1_1, 1_2\}$ reduced the number of distinct triples by one when $\delta = -1$, and left it unchanged for $\delta = 0$.

Now complete $\mathcal{B}$ to a TS$(3v-4, \lambda)$ by adding the triples

- $\{\infty_1, \infty_2, x_i\}$ for $x \in \{0,1\}$, $i \in \mathbb{Z}_3$;
- $\{0_0, 1_1, 1_2\}$, $\{1_0, 0_1, 1_2\}$, and $\{1_0, 1_1, 0_2\}$.

The six triples containing $\{\infty_1, \infty_2\}$ add six distinct triples when $\alpha = 6$, and duplicate existing triples when $\alpha = 0$. The three remaining triples all appear in E_0 (they arise from the 2×2 subsquare of M), and hence add three distinct triples when $\gamma = 3$ and duplicate existing triples when $\gamma = 0$. □

The necessity to exchange some triples for others in the proof makes the construction somewhat complicated, but it also affords the flexibility to handle the wide variety of cases that arise.

For the remaining case, a similar proof can be used, aligning a second 2-cycle in π with a disjoint 2×2 subsquare of M to perform the triple exchange twice. However, when $v \equiv 5 \pmod 6$, and $\lambda = 3v - 3$, the ingredients would then be TS$(v, v-5)$s and we would fail to cover the entire spectrum needed. The triple exchange in the proof is instead reversed.

Lemma 16.34 *Let $v \equiv 2 \pmod 3$, $v \geq 5$. Let $\lambda = 3\lambda_1 - 6$, $\lambda_1 \geq 6$, and $\lambda_1 \equiv 0 \pmod{\gcd(v-2, 6)}$. Let $s_0, s_1, s_2 \in SS(v, \lambda_1)$. Let $1 \leq \tau \leq min(v-2, \lambda_1)$. Let $1 \leq \mu \leq min(v-2, \lambda_1 - 3)$. Let $\alpha \in \{-6, 0\}$; $\alpha = -6$ only if there are $TS(v, \lambda_1)$s with support sizes s_0, s_1, s_2 each having some pair of elements appearing in two singly repeated blocks, and $\alpha = 0$ only if there is some pair in two multiply repeated blocks. Let $\beta = 0$ if $\mu = v - 2$, and $\beta \in \{0, 6\}$ otherwise. Let $\gamma = 0$ if $\tau = \lambda_1 - 3$, and $\gamma \in \{-3, 0\}$ otherwise. Let $\delta = 0$ if $\tau = v - 2$, and $\delta \in \{0, 1\}$ otherwise. If $\delta = 1$ and $\gamma = -3$, suppose that $\tau \geq 2$. Then*

$$s_0 + s_1 + s_2 + \tau(v-2)^2 + \mu(v-1)(3v-6) + \alpha + \beta + \gamma + \delta \in SS(3v-4, \lambda).$$

Proof **(Sketch)** We proceed by reversing the triple exchange in the proof of Lemma 16.33. □

The essential observation is that if $s_i > \lambda_1 v(v-1)/12$, some pair of elements in a TS(v, λ_1) having support size s_i must appear in more than $\lambda_1/2$ distinct blocks, and hence must appear in at least two singly repeated blocks. Align the pair with $\{\infty_1, \infty_2\}$ so that the two singly repeated blocks appear with 0_i and with 1_i. This ensures a loss of six triples in the exchange. On the other hand, if $s_i < (\lambda_1 - 1)v(v-1)/6$, some pair must appear in two multiply repeated blocks (for $v \equiv 2 \pmod 3$).

16.3.5 *Putting the pieces together*

Rosa and Hoffman (1986) employ GDD constructions, and doubling and tripling constructions, to determine SS$(v, 2)$ for all v. Colbourn and Mahmoodian (1990) employ doubling constructions to determine SS$(v, 3)$, and Colbourn and Mahmoodian (1993) employ a general doubling construction for SS$(v, 6)$. Colbourn and Milici (1989) complete the determination for indices 4, 5, 7, and 8. The constructions based on partitions with subdesigns then underlie the proof of the Support Size Theorem:

Theorem 16.35 *For* $v \notin \{6, 7, 8, 9, 10, 12, 14\}$, *and* $\lambda \geq 1$, $SS(v, \lambda) = PS(v, \lambda)$, *when* $v \not\equiv 8 \pmod{12}$. *When* $v \equiv 8 \pmod{12}$, $PS(v, \lambda) \subseteq SS(v, \lambda) \subseteq PS(v, \lambda) \cup \{s_v + 7\}$.

With the results on small orders given earlier, this leaves only the case when $v = 14$ to consider. Colbourn and Mahmoodian (1990) establish that $\{49, \ldots, 182\} \subseteq SS(14, 6)$; but the minimum remains in doubt here. The necessary conditions (Theorem 16.13) would permit 45, 46, 47, and 48. However, no TS$(14, 6)$ with one of these support sizes is known.

16.4 Fine structure

A number of variants of the support size problem have been studied. Milici and Quattrocchi (1988) determine the possible numbers of triply repeated blocks in a TS$(v, 3)$. Milici (1991a) examines TS$(v, 3)$s without doubly repeated blocks. Quattrocchi (1991) considers TS$(v, 3)$s without triply repeated blocks.

A common generalization is obtained as follows. For a TS(v, λ), let c_i be the number of distinct blocks that are each i-times repeated in the system. The *fine structure* of the system is $(c_1, c_2, \ldots, c_\lambda)$. Evidently,

$$\sum_{i=1}^{\lambda} ic_i = b = \lambda v(v-1)/6 \quad \text{and} \quad \sum_{i=1}^{\lambda} c_i = b^*.$$

The fine structure thus indicates not just the support size, but the way in which it is achieved.

Now suppose that $(c_1, \ldots, c_\lambda)$ is the fine structure of a TS(v, λ). Let $f = \lfloor(\lambda - 1)/2\rfloor$, and consider all triples occurring at least $\lambda - f$ times. The number of such triples is $\sum_{i=\lambda-f}^{\lambda} c_i$, and since each pair occurs in at most one such triple, this sum does not exceed $v(v-1)/6$. Moreover, if a triple occurs at least $\lambda - f$ times but does not occur λ times, each pair of the triple must occur in other triples that themselves occur fewer than $\lambda - f$ times. In fact, we obtain that $\sum_{i=1}^{f} ic_i \geq \sum_{i=1}^{f}(\lambda - i)c_{\lambda-i}$. When $\lambda = 3$, this states that $c_1 \geq c_2$.

The number of possible fine structures is unwieldy for large λ. However, for $\lambda = 3$, Colbourn, Mathon, Rosa, and Shalaby (1991) give a complete solution for $v \equiv 1, 3 \pmod 6$ and $v \geq 19$; and Colbourn, Mathon, and Shalaby (1991) give a complete solution for $v \equiv 5 \pmod 6$ and $v \geq 17$. Since in this case, any two of $\{c_1, c_2, c_3\}$ determine the third, we denote the fine structure by a pair (t, s) where $c_2 = t$ and $c_3 = \lfloor v(v-1)/6 \rfloor - s$. Then $c_1 = 3s - 2t$ if $v \equiv 1, 3 \pmod 6$, and $c_1 = 3s - 2t + 1$ if $v \equiv 5 \pmod 6$. We simply state the results here:

Theorem 16.36 *For $v \equiv 1, 3 \pmod 6$, $v \geq 19$, (t, s) is the fine structure of a TS$(v, 3)$ if and only if*

1. $0 \leq t \leq s \leq v(v-1)/6$;
2. $s \notin \{1, 2, 3, 5\}$; *and*
3. $(t, s) \notin \{(1,4), (2,4), (3,4), (1,6), (2,6), (3,6), (5,6), (2,7), (5,7), (1,8), (3,8), (5,8)\}$.

Theorem 16.37 *For $v \equiv 5 \pmod 6$, $v \geq 17$, (t, s) is the fine structure of a TS$(v, 3)$ if and only if*

1. $0 \leq t < s \leq \lfloor v(v-1)/6 \rfloor$;
2. $s \notin \{1, 2, 4, 5\}$; *and*
3. $(t, s) \notin \{(1,3), (2,3), (0,6), (1,6), (2,6), (3,6), (0,7), (1,7), (2,7), (3,7), (5,7), (0,8), (1,8), (2,8), (6,8), (7,8), (1,9)\}$.

16.5 Defining sets

A set of blocks which is a subset of a unique t-(v, k, λ_t) design $D = (V, \mathcal{B})$ is a *defining set* of that design, denoted by $d(t\text{-}(v, k, \lambda_t))$ or dD. A defining set is *minimal* ($d_m D$) if it does not properly contain a defining set of D, and *smallest* ($d_s D$) if no defining set of D has fewer blocks; $|d_s D|$ is denoted by s.

Examples for the STS(9) in Table 2.1 are:

1. A smallest defining set: $S = \{456, 789, 258, 369\}$.
2. A minimal defining set: $S = \{123, 456, 147, 249, 159\}$.

Exact sizes of smallest defining sets of TS(v, λ)s are known for some small orders. For each order and index, all nonisomorphic triple systems are included, and x^y means that y nonisomorphic designs with the given parameters have a smallest defining set of size x.

v	λ	Defining sets
7	1	3
9	1	4
13	1	8,9
15	1	$(\leq 11)^{25}$, $(\leq 12)^{43}$, $(\leq 13)^{5}$, 13^3, 14^3, 16^1
6	2	3
7	2	6^4
9	2	10^5, 9^{21}, 8^{10}
7	3	7, 9^9

Street (1995, 1996) provides further details. Few general results are known. Gower (1997) proves:

Theorem 16.38 *In the projective geometry $PG(d, 2)$, with $d \geq 2$, there exists a set of $d+1$ hyperplanes such that no point of $PG(d, 2)$ is incident with all of them. The hyperplanes of such a set cover a total of* $(\frac{2}{3})4^d - (\frac{1}{2})3^d - 2^d + (\frac{5}{6})$

lines which form a minimal defining set of the point-line design of $PG(d,2)$, that is, an $STS(2^{d+1}-1)$.

Theorem 16.39 *In the affine geometry $AG(d,3)$, there is a set of $2d$ hyperplanes, two from each of d parallel classes, such that no line of $AG(d,3)$ is incident with any d of them, and there is a point of $AG(d,3)$ incident with none of them. The hyperplanes of such a set cover a total of $[3^d(3^d-1)-7^d+1]/6$ lines which form a minimal defining set of the point-line design of $AG(d,3)$, that is, an $STS(3^d)$.*

16.6 Work points

1. Is there a $\mathrm{TS}(v,\lambda)$ with $v \equiv 8 \pmod{12}$ having $s_v + 7$ distinct triples?
2. What is the minimum support size of (i) a $\mathrm{TS}(14,6)$? (ii) a $\mathrm{TS}(14,12)$?
3. Let $2w+1 \leq v$ and $\gcd(w-2,6) | \gcd(v-2,6)$. Let V be a v-set and $W \subset V$ be a w-set. When can the triples $\binom{V}{3} \setminus \binom{W}{3}$ be partitioned into $(v-w)/\gcd(v-2,6)$ $\mathrm{TS}(v,\gcd(v-2,6))$ and $(w-2)/\gcd(w-2,6)$ partial $\mathrm{TS}(v,\gcd(w-2,6))$ each with a hole on W? This is a *large set of triple systems with a hole.*
4. What is the size of (i) a smallest defining set in an $\mathrm{STS}(v)$? (ii) a largest minimal defining set?
5. What is the computational complexity of determining the size of a smallest defining set in a particular $\mathrm{STS}(v)$?

17

INDEPENDENT SETS

A subset $S \subseteq V$ in a TS(v, λ) $(V, \mathcal{B})$ is *independent* if $B \not\subseteq S$ for all $B \in \mathcal{B}$. An independent set is *maximal* if, for all $x \in V \setminus S$, $S \cup \{x\}$ is not independent (that is, it is maximal with respect to set inclusion). It is *maximum* if it has the largest cardinality of any independent set in the design. Independent sets are often called *arcs*, and maximal independent sets are called *complete arcs*.

17.1 The independence number

Define the *independence number* $\beta(D)$ of a design D,

$$\beta(D) = \max(|S| : S \text{ is an independent set of } D).$$

Further define

$$\beta_{\min}(v, \lambda) = \min(\beta(D) : D \text{ a } TS(v, \lambda)),$$
$$\beta_{\max}(v, \lambda) = \max(\beta(D) : D \text{ a } TS(v, \lambda)), \text{ and}$$
$$B(v, \lambda) = \{\beta : \exists TS(v, \lambda) D \text{ with } \beta(D) = \beta\}.$$

The numbers $\beta'_{\min}(v, \lambda)$, $\beta'_{\max}(v, \lambda)$, and $B'(v, \lambda)$ are defined analogously for *partial* TS(v, λ)s.

An early result of Sauer and Schönheim (1969) determines the size of a maximum independent set in an STS:

Theorem 17.1

$$\beta_{\max}(v, 1) = \begin{cases} (v+1)/2 \text{ } \textit{if } v \equiv 3, 7 \pmod{12} \\ (v-1)/2 \text{ } \textit{if } v \equiv 1, 9 \pmod{12} \end{cases}$$

Bigelow and Colbourn (1992) extended this to triple systems of all indices:

Theorem 17.2

$$\beta_{\max}(v, \lambda) = \begin{cases} (v+1)/2 & \textit{if } v \equiv 3 \pmod 4 \\ & \textit{or } v \equiv 1 \pmod 4 \textit{ and } \lambda \equiv 0 \pmod 2 \\ \frac{v}{2} & \textit{if } v \equiv 0 \pmod 2 \\ (v-1)/2 & \textit{if } v \equiv 1 \pmod 4 \textit{ and } \lambda \equiv 1 \pmod 2 \end{cases}$$

Lenz and Zeitler (1982) called independent sets of size $(v-1)/2$ *ovals*, and those of size $(v+1)/2$ *hyperovals*. Determining $\beta_{\min}(v, \lambda)$ appears to be much harder. Erdös and Hajnal (1966) proved:

Theorem 17.3 $\beta'_{\min}(v,1) \geq \lfloor\sqrt{2v}\rfloor$.

Proof Let $(V,\mathcal{B})$ be a partial STS(v), and let $X \subset V$ be a maximal independent set. Since X is maximal, each $x \in V \setminus X$ belongs to a triple $T_x \in \mathcal{B}$ with $T_x \setminus \{x\} \subseteq X$. Since $|T_x \cap T_y| \leq 1$ for $x, y \in V \setminus X$, $x \neq y$, the sets $\{T_x \cap X\}$ have cardinality two and are pairwise distinct. Thus $|V \setminus X| = v - |X| \leq \binom{|X|}{2}$, which implies that $|X| \geq \lfloor\sqrt{2v}\rfloor$. □

On the other hand, the affine triple systems provide examples of STS(v)s without independent sets of size εv for any $\varepsilon > 0$ and v sufficiently large, as shown by Brown and Buhler (1982).

A probabilistic method has been used by de Brandes and Rödl (1984) and by Phelps and Rödl (1986) to prove:

Theorem 17.4 *$c_1\sqrt{v\ln v} \leq \beta_{\min}(v,1) \leq c_2\sqrt{v\ln v}$ for all $v \geq v_0(c_2)$, where c_1 is an absolute constant, $c_2 > 4$ is any constant, and $v_0(c_2)$ depends only on c_2.*

Theorem 17.5 *$c_1\sqrt{v\ln v} \leq \beta'_{\min}(v,1) \leq c_3\sqrt{v\ln v}$ for all $v \geq v_0$, where c_1 and c_3 are absolute constants.*

Grable, Phelps, and Rödl (1995) extend Theorem 17.4 to TS(v,λ)s for all λ.

For small orders, results from Sauer and Schönheim (1969), de Brandes and Rödl (1984), Zeitler and Lenz (1985), Bigelow and Colbourn (1992), and Colbourn and Ling (1998a) apply. We give some previously unpublished results in addition. In Table 17.1, we present starter blocks for cyclic STS(v)s, omitting the short orbit when $v \equiv 3 \pmod 6$, for some small orders, along with their independence number β.

Systems with larger independence numbers are somewhat easier to find, although not necessarily more numerous. To find an STS(v) with independence number β, we modified Stinson's hill-climbing algorithm so that it never selects a triple on the first β elements, ensuring that they form an independent set. Then a simple backtracking technique checks that the independence number is indeed at most β, so that it equals β. In this way, systems have been produced to yield the results in Tables 17.2 and 17.3. Beyond affirmative results in these tables, it is known that $7 \notin B(25,1)$.

Colbourn and Ling (1998a) calculated the size of a maximum independent set in each STS of order $v \leq 39$ that admits a cyclic automorphism. For each order, the possible maxima, and in parentheses the number of STS(v)s realizing each maximum, are given in Table 17.4.

17.2 Complete arcs, spanning and scattering sets

A subset S of the set X of elements of an STS is *spanning* if $|B \cap S| \geq 1$ for each $B \in \mathcal{B}$. A subset which is both independent and spanning is precisely a maximal independent set, or complete arc. For a set S of elements, we say that an element x of $X \setminus S$ is *spanned* by a pair $\{s,t\} \in S$ if $\{s,t,x\} \in \mathcal{B}$. A subset S of elements is *scattering* if it is independent and, for any two blocks $B, C \in \mathcal{B}$

v	Base blocks	β
13	{0,1,4}, {0,2,7}	6
15	{0,1,4}, {0,2,9}	6
15	{0,1,4}, {0,2,8}	8
19	{0,1,4}, {0,2,12}, {0,5,13}	7
19	{0,1,8}, {0,2,5}, {0,4,10}	8
21	{0,1,5}, {0,2,10}, {0,3,9}	8
21	{0,1,5}, {0,2,10}, {0,3,15}	9
25	{0,1,3}, {0,4,11}, {0,5,13}, {0,6,15}	9
25	{0,1,3}, {0,4,11}, {0,5,13}, {0,6,16}	10
25	{0,1,3}, {0,4,11}, {0,5,17}, {0,6,16}	11
27	{0,1,3}, {0,4,11}, {0,5,17}, {0,6,19}	10
27	{0,1,3}, {0,4,14}, {0,5,11}, {0,7,15}	11
31	{0,1,3}, {0,4,11}, {0,5,15}, {0,6,18}, {0,8,17}	11
31	{0,1,3}, {0,4,11}, {0,5,15}, {0,6,18}, {0,8,22}	12
31	{0,1,12}, {0,2,24}, {0,3,8}, {0,4,17}, {0,6,16}	16
33	{0,1,3}, {0,4,23}, {0,5,25}, {0,6,18}, {0,7,16}	12
33	{0,1,3}, {0,4,23}, {0,5,25}, {0,6,18}, {0,7,24}	13
37	{0,1,3}, {0,4,9}, {0,6,21}, {0,7,18}, {0,8,25}, {0,10,23}	13
37	{0,1,3}, {0,4,9}, {0,6,21}, {0,7,18}, {0,8,25}, {0,10,24}	14
39	{0,1,3}, {0,4,18}, {0,5,27}, {0,6,29}, {0,7,15}, {0,9,20}	13
39	{0,1,3}, {0,4,18}, {0,5,27}, {0,6,29}, {0,7,15}, {0,9,28}	14
39	{0,1,3}, {0,4,18}, {0,5,27}, {0,6,29}, {0,7,31}, {0,9,20}	15

Table 17.1 *Independence numbers of cyclic STSs*

v	λ	$B(v,\lambda)$
7	1	$\{4\}$
9	1	$\{4\}$
13	1	$\{6\}$
15	1	$\{6,7,8\}$
19	1	$\{7,8,9,10\}$
21	1	$\supseteq \{8,9,10\}$
25	1	$\supseteq \{9,10,11,12\}$
27	1	$\supseteq \{9,10,11,12,13,14\}$
31	1	$\supseteq \{11,12,13,14,15,16\}$
33	1	$\supseteq \{12,13,14,15,16\}$
37	1	$\supseteq \{13,14,15,16,17,18\}$
39	1	$\supseteq \{13,14,15,16,17,18,19,20\}$
43	1	$\supseteq \{14,15,16,17,18,19,20,21,22\}$
45	1	$\supseteq \{15,16,17,18,19,20,21,22\}$

Table 17.2 *Independence in STSs*

v	λ	$B(v,\lambda)$	v	λ	$B(v,\lambda)$
6	2	$\{3\}$	7	2	$\{3,4\}$
9	2	$\{4,5\}$	10	2	$\{4,5\}$
12	2	$\supseteq \{5,6\}$	13	2	$\supseteq \{5,6,7\}$
15	2	$\supseteq \{6,7,8\}$	16	2	$\supseteq \{6,7,8\}$
18	2	$\supseteq \{6,7,8,9\}$	19	2	$\supseteq \{7,8,9,10\}$
5	3	$\{2\}$	7	3	$\{3,4\}$
9	3	$\{3,4\}$	11	3	$\supseteq \{4,5,6\}$
13	3	$\supseteq \{4,5,6\}$	15	3	$\supseteq \{5,6,7,8\}$
17	3	$\supseteq \{5,6,7,8\}$	19	3	$\supseteq \{6,7,8,9,10\}$
6	4	$\{2,3\}$	7	4	$\{3,4\}$
9	4	$\{3,4,5\}$	10	4	$\{3,4,5\}$
12	4	$\supseteq \{4,5,6\}$	13	4	$\supseteq \{4,5,6,7\}$
15	4	$\supseteq \{5,6,7,8\}$	16	4	$\supseteq \{5,6,7,8\}$
18	4	$\supseteq \{5,6,7,8,9\}$	19	4	$\supseteq \{6,7,8,9,10\}$
8	6	$\{2,3,4\}$	9	6	$\{3,4,5\}$
10	6	$\{3,4,5\}$	11	6	$\{3,4,5,6\}$
12	6	$\{3,4,5,6\}$	13	6	$\supseteq \{4,5,6,7\}$

Table 17.3 *Independence in TS(v,λ)s, $\lambda \geq 2$*

v	Maximum(#)	v	Maximum(#)
7	4(1)		
13	6(1)	15	6(1) 8(1)
19	7(2) 8(2)	21	8(5) 9(2)
25	9(3) 10(8) 11(1)	27	10(6) 11(2)
31	11(48) 12(31) 16(1)	33	12(77) 13(7)
37	13(733) 14(87)	39	13(262) 14(529) 15(7)

Table 17.4 *Maximum independent sets in cyclic STS(v)s*

with $|B \cap S| = |C \cap S| = 2$, we find $B \setminus S \neq C \setminus S$. In other words, all elements spanned by pairs in S are distinct.

A subset of s elements in an STS(v) spans at most $s + \binom{s}{2} = \binom{s+1}{2}$ elements. Hence a spanning set of size s must have $\binom{s+1}{2} \geq v$, while a scattering set must have $\binom{s+1}{2} \leq v$. Solving for s in terms of v, we find an upper bound $L(v)$ on the size of a scattering set, and a lower bound $U(v)$ on the size of a spanning set:

$$L(v) = \left\lfloor \tfrac{1}{2}(\sqrt{8v+1} - 1) \right\rfloor,$$

$$U(v) = \left\lceil \tfrac{1}{2}(\sqrt{8v+1} - 1) \right\rceil.$$

Colbourn, Dinitz, and Stinson (1992) establish that these bounds are met with equality:

Theorem 17.6 *Let* $v \equiv 1, 3 \pmod 6$.

1. *There is an STS*(v) *having a scattering set of size* $L(v)$. *When* $v = 15$ *or* $v \geq 31$, *this STS has in addition a subsystem of order* 7 *containing two elements of the scattering set, and for which four elements of the subsystem are spanned by pairs in the scattering set.*
2. *There is an STS*(v) *having a spanning set of size* $U(v)$. *Moreover, this spanning set can be chosen to be independent.*

Since the spanning sets produced are independent, they form complete arcs. This determines the minimum size $U(v)$ and the maximum size $\beta_{\max}(v)$ of complete arcs. In fact, Colbourn and Dinitz (1997) treat all intervening values, and show that:

Theorem 17.7 *The spectrum of sizes of complete arcs in STS*(v)*s is the interval* $[U(v), \beta_{\max}(v)]$.

In the remainder of this section, we examine the proof in some detail. The singular direct product is used repeatedly. Given an STS(v) with a sub-STS(w), the singular direct product produces an STS($3(v - w) + w$) having subsystems of order v and w (and perhaps others). The construction is introduced in §3.1, but repeated here to provide some notation. Let $X = V \setminus W$. Form the STS on $(X \times \{1, 2, 3\}) \cup W$, writing x_i for $(x, i) \in X \times \{1, 2, 3\}$. Take a latin square L of side $v - w$ with rows, columns, and symbols indexed by the elements of X. When $L(a, b) = c$, form a triple $\{a_1, b_2, c_3\}$ (these are *latin square triples*). Now, for $i = 1, 2, 3$, place a copy of the STS(v) on $(X \times \{i\}) \cup W$, so that the subsystem of order w appears on W (these are i-subsystem triples). Suppress the second and third copy of the triples on W. The result is an STS($3(v - w) + w$) with a sub-STS(w) on W, and sub-STS(v)s on $(X \times \{i\}) \cup W$ for $i = 1, 2, 3$. In addition, if the STS(v) had other subsystems, these appear again in the STS($3(v-w)+w$). For convenience, we always take $V = \{1, \ldots, v\}$ and $W = \{v - w + 1, \ldots, v\}$.

Since the effective application of the singular direct product requires the presence of appropriate subsystems, it is important to limit the size of the subsystems that we require. If $3(v - w) + w$, v, and w are all to be admissible orders for an STS, and we want to write every admissible order in the form $3(v - w) + w$, we note that $w = 0$ works for $v \equiv 3, 9 \pmod{18}$, $w = 1$ works for $v \equiv 1, 7 \pmod{18}$, and $w = 3$ works for $v \equiv 15 \pmod{18}$. But an STS(v) always has subsystems of order 0, 1, and 3 when $v \geq 3$, and the singular direct product can always be applied with $w \in \{0, 1, 3\}$. Unfortunately, one congruence class remains: $v \equiv 13 \pmod{18}$. In this case, $w = 7$ works; however, in an inductive strategy it is then necessary to ensure that the STSs constructed have sub-STS(7)s.

Colbourn and Dinitz (1997) therefore establish a stronger characterization of the spectrum for complete arcs:

Theorem 17.8 *If* $x \equiv 1, 3 \pmod 6$ *and* $U(x) \leq s \leq \beta_{\max}(x)$, *there exists an STS*($x$) *with a complete s-arc. When* $x = 7$ *or* $x \geq 15$, *this STS*(x) *has a sub-STS*(7) *and this sub-STS*(7) *contains at least two elements of the complete arc.*

When $x \geq 15$, for the complete arc and sub-STS(7) chosen, there is an element of the sub-STS(7) that is spanned by a pair within the complete arc but not within the sub-STS(7).

The constructions proceed inductively on v. Table 17.5 gives the minima and maxima for some small orders.

v	$U(v)$	$\beta_{\max}(v)$	v	$U(v)$	$\beta_{\max}(v)$
3	2	2	7	4	4
9	4	4	13	5	6
15	5	8	19	6	10
21	6	10	25	7	12
27	7	14	31	8	16
33	8	16	37	9	18

Table 17.5 *Minima and maxima for complete arcs*

Theorem 17.7 holds for $x \in \{3, 7, 9, 13\}$, since in these cases there are no sizes strictly between the minimum and the maximum.

The singular direct product can be applied to obtain different sizes of complete arcs by using both the flexibility in selecting the latin square, and in selecting the ingredient STSs. To produce a number of useful latin squares, employ Ryser's theorem (1.6).

Lemma 17.9 *Let n be an integer and $m \leq n/2$. Let σ be an integer satisfying $m \leq \sigma \leq \min(m^2, n)$. There exists a latin square of order n in which, in the subarray in rows $1, \ldots, m$ and columns $1, \ldots, m$, the set of distinct symbols that appear is precisely $\{n - \sigma + 1, \ldots, n\}$.*

Proof Form a latin square of side m on symbols $1, \ldots, m$. Now for $i = n - \sigma + 1, \ldots, n - m$, select arbitrarily some symbol occurring two or more times in the square, and replace one of the occurrences by the symbol i. This can be done since $\sigma \leq m^2$, and the resulting partial $m \times m$ square has exactly $\alpha \leq n$ symbols in it. Apply Theorem 1.6 to obtain the conclusion. □

Theorem 17.10 *Let $x \equiv 1, 3 \pmod 6$. Let s be an integer satisfying*

$$\begin{array}{ll} 2U(x/3) \leq s \leq (x/3) - 1 + \beta_{\max}(x/3) & \textit{if } x \equiv 3, 9 \pmod{18} \\ 2U((x+2)/3) - 1 \leq s \leq (x-1)/2 & \textit{if } x \equiv 1, 7 \pmod{18} \\ 2U((x+6)/3) - 1 \leq s \leq (x-1)/2 & \textit{if } x \equiv 15 \pmod{18} \\ 2U((x+14)/3) - 2 \leq s \leq (x-9)/2 & \textit{if } x \equiv 13 \pmod{18} \end{array}$$

Then there is an STS(x) having a complete s-arc, and when $x = 7$ or $x \geq 15$ also having a sub-STS(7).

Proof When $x \equiv 3, 9 \pmod{18}$, set $v = x/3$ and $w = 0$. When $s \geq 3U(v)$, write $s = 2q + \beta$ for $\beta \in \{q - 1, q, q + 1\}$; otherwise, write $q = U(v)$ and $s = 2q + \beta$. Now $q < v/2$, so apply Lemma 17.9 with $m = q$ and $\sigma = v - \beta$

to form a latin square L. Use L to form the latin square triples in the singular direct product. For $i = 1, 2$, the i-subsystem triples are formed by an STS(v) having a complete q-arc on $\{1, \ldots, q\}$. The 3-subsystem triples are formed by an STS(v) having an independent set on $\{1, \ldots, \beta\}$. The first and second systems exist since $q \geq U(v)$. Since $\beta \leq q + 1$ when $q \leq (v-3)/2$, and $\beta \leq \beta_{\max}(v)$ when $q = (v-1)/2$, the third STS(v) with an independent set of size β also exists. The singular direct product produces an STS(x) having a complete s-arc on $(\{1, \ldots, q\} \times \{1, 2\}) \cup (\{1, \ldots, \beta\} \times \{3\})$. Independence is immediate. Every element in the first two subsystems is spanned by the complete q-arcs, while all elements of the third ingredient STS that are not in the arc are spanned by the latin square triples. Inductively, select the first STS(v) to have a sub-STS(7) when $v = 7$ or $v \geq 15$. It remains to treat $x \in \{27, 39\}$.

Now when $x \equiv 1, 7 \pmod{18}$, write $v = (x+2)/3$ and $w = 1$. When $s \geq 3U(v)$, write $s = 2q + \beta$ for $\beta \in \{q-1, q, q+1\}$; otherwise when $q \geq 2U(v)$, write $q = U(v)$ and $s = 2q + \beta$. Now proceed as above, ensuring that the point v appears in neither of the complete arcs, nor the independent set of the third system. If $q = 2U(v) - 1$, set $q = U(v) - 1$, $\beta = 0$, and include the single point of the subsystem in the complete arc. To ensure the presence of a sub-STS(7), it remains to treat $x \in \{25, 37\}$.

When $x \equiv 15 \pmod{18}$, if $v = 15$ see Lemma 17.12. Otherwise, write $v = (x+6)/3$ and $w = 3$. When $s \geq 3(U(v) - 1) + 1$, write $s = 2q + \beta + 1$ for $\beta \in \{q-1, q, q+1\}$; otherwise, write $q = U(v) - 1$ and $s = 2q + \beta + 1$. Apply Lemma 17.9 with $m = q$ and $\sigma = (x-3)/3 - \beta$ to form a latin square L of side $v - w$, yielding the latin square triples. Choose the first two STS(v)s with complete q-arcs, arranged in such a way that $v - 2$ is a point of the q-arc, but $v - 1$ and v are not; this can always be done since $q \leq (v-1)/2$. Choose the last STS(v) so that $\{1, \ldots, \beta\} \cup \{v+2\}$ is an independent set. The complete s-arc is on $(\{1, \ldots, q\} \times \{1, 2\}) \cup (\{1, \ldots, \beta\} \times \{3\}) \cup \{v-2\}$. To ensure the presence of a sub-STS(7) inductively, we must treat the case $x = 33$. The parameters used in this case also handle $x = 39$, left from the first case.

When $x \equiv 13 \pmod{18}$, write $v = (x+14)/3$ and $w = 7$. Although we have required that the systems constructed have sub-STS(7)s, we have exercised no control over the size of the intersection of various complete arcs with the subsystem. As a result, we obtain a weaker result here. Suppose inductively that an STS(v) exists having a complete t-arc for $U(v) \leq t \leq \beta_{\max}(v)$; further suppose that in the STS(v) with the complete t-arc, we find exactly δ_t of the points of the arc in the sub-STS(7). Evidently, $2 \leq \delta_t \leq 4$. Write $s = 2q + \beta - \delta_q$ so that $q \geq U(v)$, and $0 \leq \beta \leq \beta_{\max}(v) - 4$. When this can be done, choosing latin square triples as before, then placing STS(v)s having a complete q-arc on elements $\{1, \ldots, q - \delta_q\} \cup \{v-6, \ldots, v-7+\delta_q\}$, and finally placing the third STS with an independent set on $\{1, \ldots, \beta\} \cup \{v-6, v-5, v-4, v-3\}$ yields the desired result. Following the prescription for the singular direct product blindly, this third STS would require an independent set meeting a sub-STS(7) in precisely four points, and our induction hypothesis does not support this.

Instead, omit the sub-STS(7)s when placing the first two STS(v)s, and place the third STS(v) on $(\{1,\ldots,v-7\}\times\{3\})\cup\{v-6,\ldots,v\}$ so that its independent set of size $\beta+4$ appears on the desired points. This may, of course, fail to place an STS(7) on the points $\{v-6,\ldots,v\}$. Nevertheless, provided the STS(v) employed indeed has a sub-STS(7), it remains in the STS(x).

Since the $\{\delta_q\}$s are unknown, we must be careful to see what values can be handled. For each value of q, by selecting $0\leq\beta\leq\beta_{\max}((u+14)/3)-4$, we handle an interval of values, and δ_q contributes only some small uncertainty about which interval is covered. However, the interval for q and the interval for $q+1$ must meet (or overlap) if $\beta_{\max}((u+14)/3)-4\geq 3$, which always holds.

Colbourn and Dinitz (1997) handle the cases when $v\in\{25,27,33,37\}$. □

Theorem 17.10 misses a narrow but growing interval of values near the minimum, and only a few values near the maximum. We treat the missed values near the maximum first. The maximum value itself, $\beta_{\max}(v)$, has been settled in Theorem 17.1, but it is necessary to ensure the presence of the required sub-STS(7).

Here the $2v+1$ construction can be employed. If an STS(v) on elements V exists, and a 1-factorization of the complete graph of order $v+1$ on a disjoint set X of elements exists, we produce an STS($2v+1$) on $V\cup X$. Evidently, X always forms an independent set.

Typically, the STS($2v+1$) constructed has many more complete arcs. We examine this next. A set D of vertices in a graph is *independent* if it induces a void subgraph, and it is *dominating* if every vertex not in D has at least one neighbour in D. Consider a 1-factorization $\mathcal{F}=\{F_1,\ldots,F_v\}$ of K_{v+1}, and let $G_r=\bigcup_{i=1}^{r}F_i$. Now choose an STS(v) for which $\{z_1,\ldots,z_r\}$ is a complete arc, and choose a 1-factorization $\mathcal{F}$ of K_{v+1} so that D is an independent dominating set of size s in G_r. Then applying the $v\to 2v+1$ construction, it is easily verified that $\{z_1,\ldots,z_r\}\cup D$ is a complete $(r+s)$-arc in the STS($2v+1$).

It is easy to produce 1-factorizations of K_{v+1} in which the first $(v-1)/2$ 1-factors form a bipartite graph with an independent, dominating set of size $(v+1)/2$ — this is just a class of the bipartition. In fact, the same independent, dominating set is present in G_r for every r satisfying $1\leq r\leq(v-1)/2$. Hence we conclude:

Theorem 17.11 *If an STS(v) having a complete s-arc exists, then an STS($2v+1$) having a complete $(s+(v+1)/2)$-arc also exists. If the STS(v) has a sub-STS(7), so also does the STS($2v+1$).*

Inductively, Theorem 17.11 fails to produce the required sub-STS(7) within the sub-STS(v) when $2v+1\in\{19,27\}$. However, in both cases, choose the 1-factorization to have a sub-1-factorization of order 4. Then attaching the three 1-factors of the subfactorization to three elements of a triple in the sub-STS(v) yields the required sub-STS(7).

The structure of the independent dominating sets can also be exploited:

Lemma 17.12 *There is an STS(15) having a complete s-arc in which δ_s arc elements lie in a sub-STS(7) when $(s,\delta_s)\in\{(8,4),(7,4),(6,4),(5,3)\}$.*

Proof For (8,4), apply Theorem 17.11. For (7,4), consider the 1-factorization of K_8:

1,4	1,5	1,6	1,7	1,8	1,2	1,3
2,8	2,4	2,5	2,6	2,3	3,6	2,7
3,7	3,8	3,4	3,5	4,6	4,7	4,5
5,6	6,7	7,8	4,8	5,7	5,8	6,8

Each column forms a 1-factor, and the first four 1-factors yield a graph with {1,2,3} as an independent dominating set. Employ these in the doubling construction using an STS(7) having a complete 4-arc, attaching the first four 1-factors to the arc elements. The 4-arc, together with elements {1,2,3} from the 1-factorization, gives the complete 7-arc.

For (6,4), choose a 1-factorization in which the first three 1-factors form two disjoint complete graphs on four elements each. Then the first four 1-factors induce a graph with an independent dominating set of size two.

For (5,3), employ the same 1-factorization, noting that the first three 1-factors also induce a graph with an independent dominating set of size two. Suppose without loss of generality that the edge joining the two elements in the independent dominating set appears in the seventh 1-factor. Then choose the STS(7) to be used to have triples $\{x_1, x_2, x_4\}$, $\{x_1, x_3, x_5\}$, and $\{x_1, x_4, x_6\}$, without loss of generality. Then $\{x_1, x_2, x_3\}$ is not a complete arc, but it fails to span only x_7, and it is spanned by the two elements of the independent dominating set. □

Now we turn to cases when $x \equiv 1, 9 \pmod{12}$. When $x \equiv 1, 3, 7, 15 \pmod{18}$, Theorem 17.10 reaches the maximum, $(x-1)/2$. Indeed when $x \equiv 9 \pmod{18}$ and $x \equiv 9 \pmod{12}$, we find $x \equiv 9 \pmod{36}$ so that $x/3 \equiv 3 \pmod{12}$ and again Theorem 17.10 reaches the maximum. Therefore the only remaining case is when $x \equiv 13 \pmod{36}$.

Theorem 17.13 *Let $v = 36t + 13$, $t \geq 1$. When $15t + 5 \leq s \leq 18t + 6$, there exists an STS($v$) having a complete s-arc. In addition, it contains a sub-STS(7).*

Proof Form an STS($6t+3$) on elements $(X \times \{0,1\}) \cup \{\infty\}$ in which $\{\infty, x_0, x_1\}$ is a triple for each $x \in X$. Let $\mathcal{D}$ be the set of triples not containing ∞. On $X \times \{0,1\} \times \{0,1,2\} \times \{0,1\}$, we form a set of triples, writing the 4-tuple (x, i, a, p) as (x_i, a_p). When $\{x_i, y_j, z_k\} \in \mathcal{D}$, we form the 36 triples obtained by taking $\{(x_i, a_p), (y_j, b_q), (z_k, c_r)\}$ whenever $a + b + c \equiv 0 \pmod 3$ and $p + q + r \equiv 1 \pmod 2$, and of course $\{p, q, r\} \subseteq \{0,1\}$ and $\{a, b, c\} \subseteq \{0,1,2\}$. Call the resulting set of triples $\mathcal{B}$. Now no triple contains more than two elements of $X \times \{0,1\} \times \{0,1,2\} \times \{0\}$, since in each triple chosen we required that $p+q+r \equiv 1 \pmod 2$. However, for every point of $X \times \{0,1\} \times \{0,1,2\} \times \{1\}$, there are two points in $X \times \{0,1\} \times \{0,1\} \times \{0\}$ with which it forms a triple of $\mathcal{B}$. To verify this, consider a point (z_k, c_1). Find a triple of $\mathcal{D}$, say $\{x_i, y_j, z_k\}$ of $\mathcal{D}$, that contains z. If $c = 2$, select $a = 0$ and $b = 1$; otherwise select $a = b = c$. In either case, choose

$p = q = 0$. Then $\{(x_i, a_p), (y_j, b_q), (z_k, c_1)\}$ is indeed in $\mathcal{B}$, and $\{(x_i, a_p), (y_j, b_q)\}$ appear in the required set.

Let $\widehat{s} = s - (15t + 5)$. For any $\widehat{s}$ different selections of $x \in X$, place on $\{x\} \times \{0,1\} \times \{0,1,2\} \times \{0,1\}$ together with a new point ∞ an STS(13) with a complete 5-arc on

$$\{(x_i, a_p) : i \in \{0,1\}, p = 0, \text{ and } a \in \{0,1\} \text{ or } (a,i) = (2,1)\}.$$

For the remaining $3t + 1 - \widehat{s}$ choices of $x \in X$, on $\{x\} \times \{0,1\} \times \{0,1,2\} \times \{0,1\}$ together with the new point ∞, place an STS(13) with a complete 6-arc on $\{x\} \times \{0,1\} \times \{0,1,2\} \times \{0\}$. In both cases, we ensure (without loss of generality) that $\{\infty, (x_i, 0_0), (x_i, 0_1)\}$ is a triple for $i = 0, 1$.

The union of the complete arcs in the STS(13)s so placed is a complete s-arc as required. When $\{x_i, y_j, z_k\} \in \mathcal{D}$, a sub-STS(7) on $(\{(x,i),(y,j),(z,k)\} \times \{0\} \times \{0,1\}) \cup \{\infty\}$ is present. □

Theorem 17.10 treats all values from approximately $(2\sqrt{2}/\sqrt{3})\sqrt{x}$ upwards, whereas $U(x)$ is approximately $\sqrt{2}\sqrt{x}$. To continue with the singular direct product, we must then abandon the assumption that within the first two STS(v)s the elements included form a complete arc of the subsystem. In this case, the machinery developed by Colbourn, Dinitz, and Stinson (1992) can be used.

A latin square of side r is (a,b)-*scattered* if

1. The $a \times a$ subarray indexed by rows $1, \ldots, a$ and columns $1, \ldots, a$ contains a^2 distinct symbols, none from the set $\{1, \ldots, b\}$.
2. In rows $1, \ldots, a$, symbols $1, \ldots, b$ appear in ab distinct columns.
3. In columns $1, \ldots, a$, symbols $1, \ldots, b$ appear in ab distinct rows.

Colbourn, Dinitz, and Stinson (1992) prove that

Lemma 17.14 *If $b \le a$ and $r \ge \max(a^2 + b, ab + a + b)$, then an (a,b)-scattered latin square of side r exists.*

Colbourn and Dinitz (1997) then follow the prescription by Colbourn, Dinitz, and Stinson (1992) closely.

Theorem 17.15 *Let $x = 3r + w$, with $w = 0$ when $x \equiv 3, 9 \pmod{18}$, $w = 1$ if $x \equiv 1, 7 \pmod{18}$, $w = 3$ when $x \equiv 15 \pmod{18}$, and $w = 7$ when $x \equiv 13 \pmod{18}$. Let $e = w$ when $w \in \{0,1\}$, $e = 2$ when $w = 7$, and $e = U(x) \bmod 3$ when $w = 3$. Let a_1, a_2, a_3 satisfy $\lceil (U(x)-e)/3 \rceil \ge a_1 \ge a_2 \ge a_3 \ge \lfloor (U(x)-e)/3 \rfloor$ and $a_1 + a_2 + a_3 + e = U(x)$. Then whenever $U(x) \le s \le a_1 + a_2 + e + (r - a_1 a_2)$, there is an STS($x$) having a complete s-arc.*

Proof Form the STS(x) on $(\{1, \ldots, r\} \times \{1,2,3\}) \cup \{r+1, \ldots, r+w\}$. First we form an (a_1, a_3)-scattered latin square of side r; in Colbourn, Dinitz, and Stinson (1992), the tedious verification is done to ensure that a_1, a_3, and r meet the requirements of Lemma 17.14 (more precisely, in the cases when the condition of the lemma fails, the square is constructed there explicitly). For concreteness, suppose that the latin square L constructed has symbols $\{r - a_1 a_2 + 1, \ldots, r\}$

in the leading $a_1 \times a_2$ subarray, that it has symbols $1, \ldots, a_3$ in rows $1, \ldots, a_1$ appearing in columns $r - a_1a_3 + 1, \ldots, r$, and that it has symbols $1, \ldots, a_3$ in columns $1, \ldots, a_2$ appearing in rows $r - a_2a_3 + 1, \ldots, r$. Our objective is to form a complete arc that contains (at least) the points $P = \{r+1, \ldots, r+e\} \cup \bigcup_{i=1}^{3}(\{1, \ldots, a_i\} \times \{i\})$. If we use L to form latin square triples, then immediately we find that no element x_i can be added to P while keeping the set independent if $x > a_j a_k$ and $\{i, j, k\} = \{1, 2, 3\}$. We can, however, for fixed $i \in \{1, 2, 3\}$, add all elements $\{x_i : a_i < x \leq r - a_j a_k, \{i, j, k\} = \{1, 2, 3\}\}$ to P while retaining independence among the latin square triples. In general, it is not possible to add elements with different subscripts and retain independence.

The subsystem triples must now be selected. Choose the first system to be an STS$(r+w)$ with a sub-STS(w), having a scattering set of size $a_1 + e$ with e of the scattering set elements in the sub-STS(w). In general, we further prescribe the number d_1 of elements of the sub-STS(w) spanned by the scattering set. Place the STS with scattering set on $(\{1, \ldots, a_1\} \times \{1\}) \cup \{r+1, \ldots, r+e\}$, so that the $\binom{a_1+e}{2}$ triples having two elements in the scattering set have third elements (which are all different) equal to $\{a_1 + 1, \ldots, a_1 + \binom{a_1+e}{2} - d_1\}$ outside of the subsystem, and to $\{r+e+1, \ldots, r+e+d_1\}$ within the subsystem. Evidently, when $a_1 + \binom{a_1+e}{2} - d_1 \geq r - a_2a_3$, the 1-subsystem triples, together with the latin square triples, prevent the addition of any element x_1 with $x > a_1$ to P, while retaining independence. The verification requires that we specify d_1; always take $d_1 = 0$ when $x \equiv 1, 3, 7, 9 \pmod{18}$. Take $d_1 = 1$ when $x \equiv 15 \pmod{18}$, and take $d_1 = 4, 3, 3$ when $U(x) \equiv 0, 1, 2 \pmod 3$, respectively. Colbourn, Dinitz, and Stinson (1992) verify that the required inequality is met by these choices of parameters.

The second subsystem is placed similarly, with third elements in the sub-STS(w) from $\{r+e+1, \ldots, r+e+\binom{e}{2}\} \cup \{r+e+d_1+1, \ldots, r+e+d_1+d_2-\binom{e}{2}\}$. When $r+e+d_1+d_2-\binom{e}{2}$ exceeds $r+w$, for the latter set take $\{r+w-d_2+\binom{e}{2}+1, \ldots, r+w\}$. Again the verification requires that we specify d_2. Take $d_2 = d_1$ except when $x \equiv 13 \pmod{18}$ and $U(x) \equiv 0 \pmod 3$, in which case we take $d_2 = 2$. Colbourn, Dinitz, and Stinson (1992) verify that $a_2 + \binom{a_2+e}{2} - d_2 \geq r - a_1a_3$, so that no element of the form x_2 can be added to P while retaining independence, once the 2-subsystem triples are present.

If we were to place the third subsystem also similarly, this would in fact realize the minimum size $U(x)$ of the complete arc. However, we vary the prescription for this last ingredient. The basic fact upon which we rely is that having placed latin square, 1-subsystem, and 2-subsystem triples, the only candidates to add to P while retaining independence are $\{x_3 : a_3 < x \leq r - a_1a_2\} \cup \{r+e+d_1+d_2-\binom{e}{2}+1, \ldots, r+w\}$. Some computation shows that with the specified choices of e, d_1, and d_2 for each choice of w, the latter set (within the subsystem) is empty except when $x \equiv 13, 15 \pmod{18}$, in which cases it contains at most one element. This limitation depends in no way on the selection of the third subsystem to place.

So write $\tilde{s} = s - U(x)$, the amount by which the size of the desired complete arc exceeds the size of a minimum complete arc. Form an STS$(r+w)$ on $\{1,\ldots,r+w\}$ with a sub-STS(w) on $\{r+1,\ldots,r+w\}$ having an independent set on $\{1,\ldots,a_3+\tilde{s}\}\cup\{r+1,\ldots,r+e\}$ by which all elements $\{a_3+\tilde{s}+1,\ldots,r-a_1a_2\}\cup\{r+e+d_1+d_2-\binom{e}{2}+1,\ldots,r+w\}$ are spanned. The size of the latter set (within the subsystem) is at most one. When $w \in \{0,1,3\}$, this can be ensured for any STS$(r+w)$. When $w = 7$, the induction hypothesis provides the required system (except when $r + w \leq 13$, which cannot hold).

Indeed we can choose $0 \leq \tilde{s} \leq r - a_1a_2 - a_3$ as we like, since a_1a_2 comprises approximately two-thirds of the elements. That $\tilde{s}$ can always be chosen to be the maximum without exceeding the size of an independent set in the STS$(r+w)$ is easy but tedious. The more serious issue is whether $\tilde{s}$ can be chosen to be small. However, Colbourn, Dinitz, and Stinson (1992) verify that when $\tilde{s} = 0$ we have $a_3 + \binom{a_3+e}{2} - d_3 \geq r - a_1a_2$, so the result is always a complete arc. □

Colbourn and Dinitz (1997) establish that these three constructions treat all of the relevant values, to complete the proof of Theorem 17.7.

17.3 Partitions into complete arcs

We have seen that a complete s-arc in an STS(v) exists only if $\binom{s+1}{2} \geq v$, and that equality can hold only when $s \equiv 1,2 \pmod 4$. When $s \equiv 1 \pmod 4$ and equality holds, an intriguing possibility arises. In this case, it is numerically possible for the STS(v) to admit a partition of its elements into $(s+1)/2$ classes, each a complete s-arc. Colbourn, Phelps, de Resmini, and Rosa (1991) prove two theorems about such partitions:

Theorem 17.16 *If $s \equiv 1,5 \pmod{12}$, there exists a cyclic STS$(\binom{s+1}{2})$ admitting a partition into complete s-arcs.*

Theorem 17.17 *If $s \equiv 9 \pmod{12}$ and there exists a KTS(s) for which the $(s-1)/2$ parallel classes can be paired to form $(s-1)/4$ isomorphic pairs, then there exists an STS$(\binom{s+1}{2})$ admitting a partition into complete s-arcs. In particular, such a system exists when $s \equiv 9 \pmod{24}$, and when $s \equiv 21 \pmod{120}$.*

They also establish that all of the STS(15)s in Tables 5.8–5.12 have a complete 5-arc except for #80, but that only nine of the systems (numbers 1, 2, 4, 5, 7, 9, 10, 70, and 76) admit a partition into three complete 5-arcs. System #1 admits 336 such partitions, while systems #70 and #76 each admit only one.

For general TS(v,λ)s, a complete s-arc exists only when $\lambda\binom{s}{2}+s \geq v$. Equality can hold for any s when λ is even, and numerically a partition into complete s-arcs is possible for every positive integer s. When λ is odd, a partition into complete s-arcs can be present only when $s \equiv 1 \pmod 4$.

Phelps and de Resmini (1992) considered partitions into complete arcs when $\lambda = 2$. They treat the cases $s \equiv 0,1,2 \pmod 3$ separately. We examine their solution when $s \equiv 0 \pmod 3$.

Lemma 17.18 *There exists a TTS(s^2) with a partition into complete s-arcs whenever $s \equiv 0 \pmod 3$, except possibly when $s = 6$.*

Proof Let $V = \mathbb{Z}_s \times \mathbb{Z}_s$. Let $(\mathbb{Z}_s, \mathcal{B})$ and $(\mathbb{Z}_s, \mathcal{D})$ be resolvable TTS(s)s, with parallel classes $\{T_1, \ldots, T_{s-1}\}$ and $\{U_1, \ldots, U_{s-1}\}$, respectively. Give each triple in both systems an arbitrary but fixed total order; for example, for $\{a, b, c\}$ we could choose (a, b, c). Let L be an idempotent latin square of side $s/3$ on the symbols $\mathbb{Z}_{s/3}$. Replace each cell containing α in L by the 3×3 subsquare

$$\begin{pmatrix} \alpha & \alpha + s/3 & \alpha + 2s/3 \\ \alpha + 2s/3 & \alpha & \alpha + s/3 \\ \alpha + s/3 & \alpha + 2s/3 & \alpha \end{pmatrix}$$

and remove the 3×3 subsquares arising from the diagonal cells of L. The result is an incomplete latin square M of side s having 3×3 holes on the diagonal.

For $1 \leq i \leq s-1$, let $U_i = \{(a_j, b_j, c_j) : j = 0, \ldots, (s/3) - 1\}$, and define a permutation π_i on $\mathbb{Z}_s$ mapping j to a_j, $j + s/3$ to b_j, and $j + 2s/3$ to c_j for $0 \leq j < s/3$. Let M_i be the incomplete latin square $\pi_i(M)$; M_i has its 3×3 holes on the triples of U_i. Now on $\mathbb{Z}_s \times \mathbb{Z}_s$, form a set $\mathcal{T}$ of triples by including, for each $1 \leq i < s$ and each $(a, b, c) \in T_i$, the triples:

1. $\{\{(a, f), (b, g), (c, h)\} : M_i(f, g) = h,\ f, g \in \mathbb{Z}_s\}$;
2. triples $\{(a, k), (a, \ell), (b, m)\}$, $\{b, k), (b, \ell), (c, m)\}$, and $\{(c, k), (c, \ell), (a, m)\}$ when (k, ℓ, m), (ℓ, m, k), or (m, k, ℓ) is in U_i; and
3. $\{\{(a, x), (b, x), (c, x)\} : x \in \mathbb{Z}_s\}$.

The system $(\mathbb{Z}_s \times \mathbb{Z}_s, \mathcal{T})$ is a TTS(s^2), and for each $a \in \mathbb{Z}_s$, the set $\{a\} \times \mathbb{Z}_s$ forms a complete s-arc. The verification is primarily that this arc is, in fact, complete. Consider the arc $\{a\} \times \mathbb{Z}_s$, and an element (b, x) not in the arc. Since $\{a, b\}$ appears in a unique triple, say (a, b, c), in $\mathcal{B}$, we may suppose that $(a, b, c) \in T_i$. There is a triple containing x, say $\{x, y, z\}$, in U_i and hence there is a triple $\{(a, y), (a, z), (b, x)\}$ in $\mathcal{T}$. Hence $\{a\} \times \mathbb{Z}_s$ is complete. □

Phelps and de Resmini (1992) treat the exceptional case when $s = 6$, and then develop fundamentally similar constructions to treat the remaining cases when $s \equiv 1, 2 \pmod 3$. In summary, they prove:

Theorem 17.19 *For $s \geq 1$, there is a TTS(s^2) admitting a partition into s complete s-arcs.*

17.4 Generating sets

A k-generating set generalizes the notion of a complete arc. Consider the Steiner quasigroup $(V, \oplus)$ of a Steiner triple system. Let $W \subset V$. Then some elements of $V \setminus W$ can be written in the form $w \oplus \widehat{w}$ for $w, \widehat{w} \in W$. In turn, further elements can be written in the form $\tilde{w} \oplus (w \oplus \widehat{w})$, and so on. If every element of V can be written using elements of W and at most $k-1$ applications of $\oplus$, then W is called a *k-generating set.* When $k = 2$, such a generating set is a spanning or dominating set, and these have been constructed in the course of constructing STSs with complete arcs.

Colbourn and Dinitz (1999) examine the case $k = 3$, focusing on the situation that every element of $V \setminus W$ can be written in *exactly one* way as a word with at most two occurrences of $\oplus$, and using the elements of W. When $x = |W|$, we find that $|V| \le x + \binom{x}{2} + x \cdot \binom{x-1}{2}$ for a 3-generating set, and we examine when equality can hold.

When an STS(v) with Steiner quasigroup $(V, \oplus)$ has a 3-generating set of size x, there are

1. x *primary* elements in the generating set;
2. at most $\binom{x}{2}$ *secondary* elements defined as $w \oplus w'$ for w, w' primary; and
3. at most $x \cdot \binom{x-1}{2}$ *tertiary* elements defined as $w \oplus z$ where w is primary and z is secondary.

Thus, $|V| \le x + \binom{x}{2} + x \cdot \binom{x-1}{2}$. We are interested here in cases when equality holds. In principle, this can occur when $x \equiv 1, 2, 3 \pmod 4$, since in these cases we find that $x + \binom{x}{2} + x \cdot \binom{x-1}{2} \equiv 1, 3 \pmod 6$. When $x \equiv 0 \pmod 4$, the threshold quantity is even and hence not the order of a Steiner triple system.

The existence of STS$(x + \binom{x}{2} + x \cdot \binom{x-1}{2})$s with a 3-generating set of size x for all $x \equiv 1, 2, 3 \pmod 4$ is settled by Colbourn and Dinitz (1999). Here we examine the case when x is odd.

Let $x = 2t + 1$ be an odd integer, and $v = x + xt + xt(x - 2)$. Let $X = \{\infty\} \cup \{i : 1 \le i \le t\} \cup \{(i, j) : 1 \le i \le t, j \in \mathbb{Z}_x, j \not\equiv \pm i \pmod x\}$. We form a Steiner triple system on $X \times \mathbb{Z}_x$ as follows. First we specify the triples containing ∞_0 (this subscript notation is used for $(\infty, 0) \in X \times \mathbb{Z}_x$. They are:

1. $\{\infty_0, \infty_{2k}, k_k\}$ for $1 \le k \le t$;
2. $\{\infty_0, k_j, (k, j)_0\}$ for $1 \le k \le t$, $j \in \mathbb{Z}_x \setminus \{\pm k \bmod x\}$;
3. $\{\infty_0, (k, j)_y, (k, j)_{-y}\}$ for $1 \le k \le t$, $j \in \mathbb{Z}_x \setminus \{\pm k \bmod x\}$, and $1 \le y \le t$.

Develop these triples in $X \times \mathbb{Z}_x$ mod $(-, x)$ to obtain all triples containing an element of $\{\infty\} \times \mathbb{Z}_x$. If we produce a Steiner triple system of order v containing these triples, it is readily verified that $\{\infty\} \times \mathbb{Z}_x$ is a 3-generating set.

Consider the structure of the pairs that remain. No pair on $\{i : 1 \le i \le t\} \times \mathbb{Z}_x$ has been chosen. For fixed $1 \le k \le t$ and $j \in \mathbb{Z}_x \setminus \{\pm k \bmod x\}$, all pairs of the form $\{(k, j)_y, (k, j)_z\}$ for $y, z \in \mathbb{Z}_x$, $y \ne z$, are used. The only other used pairs form, for fixed $1 \le k \le t$ and $j \in \mathbb{Z}_x \setminus \{\pm k \bmod x\}$, a 1-factor between the elements $\{k\} \times \mathbb{Z}_x$ and the elements $\{(k, j)\} \times \mathbb{Z}_x$. Each element of $\{(k, j)\} \times \mathbb{Z}_x$ occurs in exactly one such pair, while each element of $\{k\} \times \mathbb{Z}_x$ occurs in $x - 2$, one for each selection of j. Our task reduces to showing that this is a leave of a partial STS.

Case 1 $x \equiv 3 \pmod 6$, and hence $t \equiv 1 \pmod 3$. For $1 \le k \le t$, place an STS(x) on $\{k\} \times \mathbb{Z}_x$. Next, for $1 \le k \le t$, place a 3-MGDD of type $(x - 1, x)$ on $(\{k\} \cup \{(k, j) : j \in \mathbb{Z}_x, j \not\equiv \pm k \pmod x\}) \times \mathbb{Z}_x$, with first groups defined by constant last coordinates, and second groups defined by constant first coordinates. Now place the triples of an STS$(x-2)$ on $(\{(k,j) : j \in \mathbb{Z}_x, j \not\equiv \pm k \pmod{x}\}) \times \{\ell\}$ for each $\ell \in \mathbb{Z}_x$ and each $1 \leq k \leq t$.

Pairs used now account for all those on $(\{k\} \cup \{(k,j) : j \in \mathbb{Z}_x, j \not\equiv \pm k \pmod{x}\}) \times \mathbb{Z}_x$ for each $1 \leq k \leq t$, and for no others. Then form a 3-GDD of type $(x(x-1))^t$, with groups aligned on $(\{k\} \cup \{(k,j) : j \in \mathbb{Z}_x, j \not\equiv \pm k \pmod{x}\}) \times \mathbb{Z}_x$ for $1 \leq k \leq t$. This completes the case $x \equiv 3 \pmod 6$.

Case 2 $x \equiv 1 \pmod 6$, and hence $t \equiv 0 \pmod 3$. Write $s = t/3$. For $1 \leq k \leq t$, place an STS(x) on $\{k\} \times \mathbb{Z}_x$. Now partition $\{1, \ldots, t\}$ into s disjoint classes $C_1, \ldots, C_s$, each of size 3. Place a 3-GDD of type $(3x(x-1))^s$ whose groups are $(\{i : i \in C_k\} \cup \{(i,j) : i \in C_k, j \in \mathbb{Z}_x, j \not\equiv \pm i \pmod{x}\}) \times \mathbb{Z}_x$, for $1 \leq k \leq s$. Next, on each of these s groups, we place a 3-HGDD of type $(x-1 : 3^x)$ with the groups of size $3(x-1)$ on $(\{i : i \in C_k\} \cup \{(i,j) : i \in C_k, j \in \mathbb{Z}_x, j \not\equiv \pm i \pmod{x}\}) \times \{\ell\}$ for each $\ell \in \mathbb{Z}_x$. The groups of size $3x$ are on $\{i : i \in C_k\} \times \mathbb{Z}_x$ and $\{(i,j) : i \in C_k\} \times \mathbb{Z}_x$ for $j \in \mathbb{Z}_x, j \not\equiv \pm i \pmod{x}$. These two sets of groups intersect in sets of size three, namely $(\{i\} \times \{\ell\}$ and $\{(i,j)\}) \times \{\ell\}$ for each $\ell \in \mathbb{Z}_x$, $i \in C_k$, and $j \in \mathbb{Z}_x, j \not\equiv \pm i \pmod{x}$.

On each of the groups of size $3x$, examining pairs already used, we find that within three disjoint sets of size x, all pairs are used, and that no other pairs are used. Place a 3-MGDD of type $(3 : x)$ on each such group, so that the groups of size x align on these three previously used sets, and the groups of size 3 align within the groups of size $3(x-1)$ of the HGDD selected earlier.

It remains only to fill the groups of size $3(x-1)$. Again, considering pairs already used within such a group, we find that in a group $(\{i : i \in C_k\} \cup \{(i,j) : i \in C_k, j \in \mathbb{Z}_x, j \not\equiv \pm i \pmod{x}\}) \times \{\ell\}$, the pairs used thus far are $\{\{i_\ell, (i,j)_\ell\} : i \in C_k, j \in \mathbb{Z}_x, j \not\equiv \pm i \pmod{x}\}$, and no others. To cover the remaining pairs, we select a partial triple system of order $3(x-2)$ on $\{(i,j) : i \in C_k, j \in \mathbb{Z}_x, j \not\equiv \pm i \pmod{x}\} \times \{\ell\}$, whose leave is a $3(x-2)$-cycle Y. Such a partial triple system exists by Theorem 9.15. When $C_k = \{a,b,c\}$, we place the partial triple system so that, reading the first coordinate of the points sequentially along the cycle forming the leave, we always encounter b after a, c after b, and a after c. This can be done since the cycle is hamiltonian and has a number of edges which is a multiple of three. To complete filling the hole of size $3(x-1)$, whenever $\{(b,j)_\ell, (c,j')_\ell\}$ is an unused pair, the triple $\{a_\ell, (b,j)_\ell, (c,j')_\ell\}$ is added; this process is repeated for (b,c,a) and (c,a,b) in place of (a,b,c). This completes the proof for $x \equiv 1 \pmod 6$.

Case 3 $x \equiv 11 \pmod{12}$, and hence $t \equiv 5 \pmod 6$. Place an STS$(x-2)$ on $\{(k,j)_i : j \in \mathbb{Z}_x, j \neq \pm k\}$ for $1 \leq k \leq t$ and $i \in \mathbb{Z}_x$. Next place a 3-MGDD of type $(x, x-1)$ on $(\{k\} \cup \{(k,j) : j \in \mathbb{Z}_x, j \not\equiv \pm k \pmod{x}\}) \times \mathbb{Z}_x$, with first groups defined by constant last coordinates, and second groups defined by constant first coordinates. Then place the triples of a 3-IGDD of type $(x(x-1) : x)^t$ with the hole on $\{1, \ldots, t\} \times \mathbb{Z}_x$ and the groups on $(\{k\} \cup \{(k,j) : j \in \mathbb{Z}_x, j \not\equiv \pm k \pmod{x}\}) \times \mathbb{Z}_x$ for $1 \leq k \leq t$. Finally place the triples of an STS(xt) on $\{1, \ldots, t\} \times \mathbb{Z}_x$.

Case 4 $x \equiv 5 \pmod{12}$, and hence $t \equiv 2 \pmod 6$. Place a 3-GDD of type $(2x(x-1))^{t/2}$ with groups $\{\{k_i\} \cup \{(k,j)_i\} : i \in \mathbb{Z}_x, j \in \mathbb{Z}_x, j \neq \pm k, k \in \{2\ell-1, 2\ell\}\}$ for $1 \leq \ell \leq t/2$. Evidently all as yet uncovered pairs now lie within a group of this 3-GDD. We describe how to handle the pairs when $\ell = 1$, observing that the same method treats all other values $2 \leq \ell \leq t/2$.

Next we place a 3-GDD of type 2^{x-2} on $\{\{(k,j)_i\} : i \in \mathbb{Z}_x, j \in \mathbb{Z}_x, j \neq \pm k, k \in \{1,2\}\}$, aligning the groups on $\{(k,j)_i, (k,-j)_i\}$ for $i \in \mathbb{Z}_x$, $k \in \{1,2\}$, and $j \in \mathbb{Z}_x \setminus \{0, \pm k\}$. Now add triples $\{(3-k)_i, (k,j)_i, (k,-j)_i\}$ for $i \in \mathbb{Z}_x$, $k \in \{1,2\}$, and $j \in \mathbb{Z}_x \setminus \{0, \pm k\}$. It is easily verified that, among elements with the same subscript i, all pairs are covered except $\{1_i, 2_i\}$, $\{1_i, (2,0)_i\}$, $\{2_i, (1,0)_i\}$, and $\{(1,0)_i, (2,0)_i\}$ — i.e. a 4-cycle remains uncovered.

Now since $2x - 7 \equiv 3 \pmod{24}$, there is a Kirkman triple system of order $2x-7$ by Theorem 19.9. To each of five resolution classes, add an infinite point to each block in the class to form a {3,4}-GDD of type $1^{2x-7}5^1$. Using 3-MGDDs of type $(3,x)$ and $(4,x)$ to give weight x to the {3,4}-GDD, produce a 3-HGDD of type $(x : 1^{2x-7}5^1)$. Place this on the points $\{\{k_i\} \cup \{(k,j)_i\} : i \in \mathbb{Z}_x, j \in \mathbb{Z}_x, j \neq \pm k, k \in \{1,2\}\}$ so that the first groups align on points with the same subscript and the second groups align on $\{1, 2, (1,0), (2,0), (1,1)\} \times \mathbb{Z}_x$ (for the group of size $5x$) and on groups of the form $\{(k,j)\} \times \mathbb{Z}_x$ (for the groups of size x).

There remain pairs to be covered on $\{1, 2, (1,0), (2,0), (1,1)\} \times \mathbb{Z}_x$. All pairs with the same entry from $\mathbb{Z}_x$ are already covered except those in the 4-cycle $(1_i, (2,0)_i, (1,0)_i, 2_i)$. Pairs with the same entry from $\{1, 2, (1,0), (2,0), (1,1)\}$ are already covered.

Form a 3-GDD of type $3^{(x-5)/3}5^1$ on $\mathbb{Z}_x$. For each block B of this 3-GDD, form a 3-MGDD of type (3,5) on $\{1, 2, (1,0), (2,0), (1,1)\} \times B$, with groups aligned on elements with constant first, or constant second, coordinate. To handle pairs on $\{1, 2, (1,0), (2,0), (1,1)\} \times G$ where G is a group of size three or five, the two ingredients needed have 15 and 25 points. See Colbourn and Dinitz (1999) for details.

17.5 Covering sets

Related to independence is the notion of a covering set. A subset $F \subseteq V$ is a *covering set* of a TS(v,λ) $D = (V, \mathcal{B})$ if $B \cap F \neq \varnothing$ for all $B \in \mathcal{B}$. A covering set need not be independent. The *domination number* of D, dom(D), is the minimum cardinality of a covering set of D. Let $f(v,\lambda) = \min(\mathrm{dom}(D) : D \text{ a } TS(v,\lambda))$. Hanani (1964) determined $f(v,\lambda)$ completely:

Theorem 17.20 *If a TS(v,λ) exists, then*

$$f(v,\lambda) = \begin{cases} (v-1)/2 & \textit{if } v \equiv 3 \pmod 4 \\ & \quad \textit{or } v \equiv 1 \pmod 4 \textit{ and } \lambda \equiv 0 \pmod 2 \\ (v+1)/2 & \textit{if } v \equiv 1 \pmod 4 \textit{ and } \lambda \equiv 1 \pmod 2 \\ \frac{v}{2} & \textit{if } v \equiv 0 \pmod 2 \end{cases}$$

On the other hand, determining the domination number of affine STSs is a computationally difficult problem; see Fulkerson, Nemhauser, and Trotter (1974), Avis (1980), and Peter-Orth (1987). Related work appears in Feo and Resende (1989), Karmarkar, Ramakrishnan, and Resende (1991), Huang, Kao, and Horng (1994), Mannino and Sassano (1994, 1995), Lorena and de Souza Lopes (1997), and Odijk and van Maaren (1998).

Compare Theorems 17.20 and 17.2. These are in fact the same theorem, since the elements *not* in a covering set form an independent set, and hence maximum independent sets are complements of minimum covering sets.

17.6 An application: point codes

Baartmans, Landjev, and Tonchev (1996) study certain codes associated with Steiner triple systems. We introduce necessary definitions next. Let $(V, \mathcal{B})$ be an STS(v). There are $b = v(v-1)/6$ triples, every point lying in $r = (v-1)/2$ of them. Form a $v \times b$ matrix M, whose rows are indexed by V and columns indexed by $\mathcal{B}$. In the entry indexed by $x \in V$ and $B \in \mathcal{B}$, there is a '1' if $x \in B$, a '0' otherwise. The *point code* of the STS is the binary linear code $C(M)$ spanned by the rows of M.

Baartmans, Landjev, and Tonchev (1996) ask which Steiner triple systems have the property that their point code uniquely determines them; that is, when does the point code of an STS contain only one isomorphism type of STSs of maximal 2-rank among its vectors of weight r? They remark that one way to ensure that this occurs is to establish that the point code of the STS has exactly v vectors of weight r, which then necessarily correspond to the points of the STS. Using connections with ovals in Steiner triple systems, and other arguments, they establish the existence of such a *point code minimum* STS for orders v satisfying $v \equiv 3, 9 \pmod{18}$.

Colbourn and Ling (1998a) study the existence of point code minimum STSs; we examine their results next. Choosing any subset $X \subseteq V$, examine the modulo 2 sum of the rows indexed by the elements in X. We are interested in when this sum of rows has weight r, in particular when $|X| \geq 2$, since no such X exists precisely when the STS is point code minimum. We derive some simple necessary conditions. Choose a set X of size s, and let n_i be the number of triples in $\mathcal{B}$ which intersect X in precisely i points. We have

$$\begin{aligned} n_2 + 3n_3 &= \tbinom{s}{2} \\ 2n_1 + 2n_2 &= s(v-s) \\ 3n_0 + n_1 &= \tbinom{v-s}{2} \end{aligned}$$

by considering pairs intersecting X in 2, 1, or 0 points. Now specifying that $n_1 + n_3 = r$, we can solve for n_0, n_1, n_2, and n_3 as in Table 17.6.

Since $n_1 < 0$ when $s \geq r+4$, Table 17.6 presents all of the cases permitted. Now consider the last case, when $s = r+3$. The triples counted by n_0 all lie on the $r-2$ elements not in X. Since n_1 is an integer, $r \equiv 2 \pmod 4$ so that $r - 2 \equiv 0 \pmod 4$. The maximum number of triples in a partial triple system on

s	n_3	n_2	n_1	n_0
s	$\frac{(s-r)(s-1)}{4}$	$\frac{(s-1)(3r-s)}{4}$	$r - \frac{(s-r)(s-1)}{4}$	$\frac{8r^2+3s^2-9rs+r-3s}{12}$
r	0	$\binom{r}{2}$	r	$\frac{1}{3}\binom{r}{2}$
$r+1$	$\frac{r}{4}$	$\frac{r(2r-1)}{4}$	$\frac{3r}{4}$	$\frac{r(2r-5)}{12}$
$r+2$	$\frac{r+1}{2}$	$\frac{(r-1)(r+1)}{2}$	$\frac{r-1}{2}$	$\frac{(r-1)(r-3)}{6}$
$r+3$	$\frac{3r+6}{4}$	$\frac{(r+2)(2r-3)}{4}$	$\frac{r-6}{4}$	$\frac{2r^2-11r+18}{12}$

Table 17.6 *Solutions for* n_0, n_1, n_2, *and* n_3

$r-2$ points, when r is even, does not exceed $(r-2)(r-4)/6$. But the required value of n_0 exceeds the maximum permitted by $(r+2)/12$, so the case when $s = r+3$ is prohibited by simple structural reasons.

When $s = r+2$, a similar analysis shows that r is odd, and on the $r-1$ elements not in X, there can be at most $(r-1)(r-3)/6$ triples. This is the required value for n_0. We conclude that the triples entirely on $V \setminus X$ form a 3-GDD of type $2^{(r-1)/2}$. In other words, every element of $V \setminus X$ lies in precisely one triple counted by n_1.

When $s = r+1$, we require that $r \equiv 0 \pmod 4$.

Lemma 17.21 *For* $v \geq 37$, $v \equiv 13, 21 \pmod{24}$, *there exists a point code minimum STS*(v).

Proof Write $v = 18x + 2w + 1$ with $w \in \{0, 1, 3, 4, 6, 7\}$. Form a 3-GDD of type $9^{2x}(2w+1)^1$ which exists by Theorem 6.15. Let $G_1, \ldots, G_{2x}$ be the groups of size 9, and H be the group of size $2w+1$. Let $V = H \cup \bigcup_{i=1}^{2x} G_i$. Since each group has size which is the order of a Steiner triple system, we can place the blocks of an STS on each to form an STS(v) on V, having a subsystem on each of $G_1, \ldots, G_{2x}$, and H. We claim that this STS meets the requirements of the lemma.

By the earlier discussion, we need only consider sets of r rows, since $r \equiv 2 \pmod 4$. For r rows to have modulo 2 sum of weight r, the corresponding elements form an independent set (i.e. no triple of the STS lies on the set). But an STS(9) has maximum independent set of size 4, and the STS($2w+1$) has a maximum independent set of size at most $w+1$. Hence if X is independent in the STS(v) constructed, it contains no more than $8x + w + 1$ elements. Since $x \geq 2$, this is fewer than r and hence the selection of r rows is ruled out. □

Next we treat the cases when r is odd.

Lemma 17.22 *For* $v \geq 39$, $v \equiv 3, 7 \pmod{12}$, *there exists a point code minimum STS*(v).

Proof If $v \geq 61$, write $v = 30x+2w+1$ with $0 \leq w \leq 13$. Form a 3-GDD of type $15^{2x}(2w+1)^1$ using Theorem 6.15. Fill the group of size $2w+1$ using an arbitrary STS($2w+1$). Use the unique quadrilateral-free STS(15) (#80 from Table 5.12) to fill the groups of size 15. This STS(15) has its maximum independent set of size 6.

We must consider the selection of $\xi \in \{r, r+2\}$ rows, since r is odd. For $\xi = r$, our STS(v) has no independent set of size larger than $6 \cdot 2x + w + 1$, but $12x + w + 1 < 15x + w = r$, so no subset of r rows has modulo 2 sum with weight r. Then suppose that $\xi = r + 2$. The complementary $r - 1$ points in our STS(v) must induce a 3-GDD of type $2^{(r-1)/2}$. However, since the STS(15) used is quadrilateral-free, this requires that among the ξ points selected, the number lying within a particular STS(15) is 0, 1, 12, 13, 14, or 15. The absence of a configuration of six elements carrying four triples prohibits the selection of 8 or 9 elements, which would otherwise be allowed. It is now an easy computation that some of the STS(15)s must have 12 or more points selected, but then the number of triples on the selected points exceeds $(r+1)/2$ in each case. This guarantees that no suitable selection of $r + 2$ points is present.

For $v \in \{39, 43, 51, 55\}$, use 3-GDDs of type $15^1 9^1 3^5$, $15^1 9^3 1^1$, $15^1 9^4$, and $15^1 13^1 9^3$, respectively, all from Colbourn (1993). Fill each hole of size 15 with the STS(15) #80 as above. In each case, consider a point of the STS(15). It appears in $(v - 15)/2$ triples of the GDD; hence, if the point of the STS(15) is included in an independent set, then at most $(v - 15)/2$ elements not in the STS(15) can also be included. However $(v - 15)/2 + 6 < r = (v - 1)/2$, so if any point of the STS(15) appears in an independent set, its size is less than r. If no such point appears in the independent set, its size is bounded by 14, 13, 16, and 18 in the four cases for v, and hence in each case is again less than r. Therefore, no r rows have modulo 2 sum with weight r.

Consider the case with $r + 2$ rows. If 12 or more points are selected from the STS(15), at least 16 triples lie on the chosen set, but $16 > (r + 1)/2$ in the four cases at issue. So at most one point can be chosen from the STS(15). Now consider the complementary $r - 1$ points. They induce a 3-GDD of type $2^{(r-1)/2}$, which must itself contain either an STS(15) or a 3-GDD of type 2^7. The former requires that $(r - 1)/2 \geq 15$, which does not hold. Simple counting shows that the latter requires that $(r - 1)/2 \geq 14$ or $(r - 1)/2 = 7$, and neither holds. Therefore, no $r + 2$ rows have modulo 2 sum with weight r, and the four 'small' cases are completed. □

Finally we turn to the cases when $r \equiv 0 \pmod 4$.

Lemma 17.23 *For $v \geq 97$, $v \equiv 1, 9 \pmod{24}$, there exists a point code minimum STS(v).*

Proof Our earlier computations show that, for the statement of the lemma to fail for a particular STS(v), it has either an independent set of size $(v-1)/2$, or a set of $(v+1)/2$ points carrying exactly $\frac{v-1}{8}$ triples. In the latter case, if we greedily select and delete a point lying in a triple, we can delete at most $\frac{v-1}{8}$ elements to find an independent set of size $3\frac{v-1}{8} + 1$. Hence, in order to establish the statement of the lemma for a particular order v, it suffices to produce an STS(v) whose largest independent set has size no larger than $3\frac{v-1}{8}$. When $v \geq 97$, write

$v = 24n+2w+1$, with $w \in \{12, 16\}$, and produce a 3-GDD of type $24^n(2w-2)^1$ using Theorem 6.15. Add three 'infinite' points, and fill the holes of size 24, with the three infinite points, using the STS(27) whose largest independent set has size 9, omitting in the process a triple on the three infinite points. Fill the hole of size $2w-2$, and the three infinite points, with an STS$(2w+1)$, having a maximum independent set of size 9 when $w = 12$, size 12 when $w = 16$. The largest independent set has size no more than $9n + \frac{3w}{4}$, which is less than $\frac{3}{4}(12n + w) + 1$. □

In order to establish the existence of point code minimum STSs for every order at least 15, it remains only to treat some small orders; see Colbourn and Ling (1998a).

In summary,

Theorem 17.24 *For $v \geq 15$, $v \equiv 1, 3 \pmod 6$, there exists a point code minimum STS(v).*

The role of independent sets is of particular interest here. Indeed, Colbourn (1999) pursues this connection in examining minimum weights of point codes. He proves:

Theorem 17.25 *There is a Steiner triple system of order $v = 2r + 1$ whose point code contains a vector of weight $w = r - \varepsilon$, with $\varepsilon > 0$, which is the modulo 2 sum of s rows if and only if one of:*

1. $s = r + 1$, $w \equiv 0 \pmod 4$, *and*

$$w \geq \begin{cases} 0 & \text{if } r \equiv 1, 3 \pmod 6 \\ \frac{2r}{3} & \text{if } r \equiv 0 \pmod 6 \\ \frac{2r+4}{3} & \text{if } r \equiv 4 \pmod 6 \end{cases}$$

2. $s = r + 2$, $r \equiv 0, 4 \pmod 6$, $w \equiv r + 2 \pmod 4$, *and*

$$w \geq \begin{cases} \frac{r+2}{3} & \text{if } r \equiv 4 \pmod 6 \\ \frac{r+18}{3} & \text{if } r \equiv 0 \pmod 6 \end{cases}$$

3. $s = r + 3$, $r \equiv 1, 3 \pmod 6$, $w \equiv 0 \pmod 4$, *and*

$$w \geq \begin{cases} \frac{2r+6}{3} & \text{if } r \equiv 3 \pmod 6 \\ \frac{2r+22}{3} & \text{if } r \equiv 1 \pmod 6 \end{cases}$$

Baartmans, Landjev, and Tonchev (1996) indicate a number of interesting coding-theoretic questions as well. In particular, they give examples of STSs which are not point code minimum, and indeed since an STS(v) with an oval exists for every order $v \equiv 1, 3 \pmod 6$, we cannot hope that every STS(v) is point code minimum. Nevertheless, it appears plausible that, although not point code minimum, every STS is determined by its point code. Tonchev and Weishaar (1997) show that the 80 STS(15)s are determined by their point codes, although only two are point code minimum. Indeed, they also show that the point code of an STS(15) may contain more than one isomorphism type of STS among its vectors of weight $r = 7$; however, in each case (when $v = 15$) the point code does not contain two or more nonisomorphic STS(15)s, each of the largest 2-rank in the point code.

17.7 Work points

1. Determine $B(v, 1)$, in particular for $v \in \{21, 25, 27\}$.
2. Determine $B(v, 2)$, in particular for $v \in \{12, 13\}$.
3. Is $B(v, \lambda)$ an interval?
4. Extend Theorem 17.17 to the cases when $s \equiv 45, 69, 93, 117 \pmod{120}$. One way to do this is to establish the existence of 4-rotational KTS(s)s when $s \equiv 21 \pmod{24}$.
5. Is every STS determined by its point code? In other words, in the point code of an STS, can another (nonisomorphic) STS of larger or equal 2-rank be found among the vectors of weight r?

18

CHROMATIC NUMBER

Erdös and Hajnal (1966) and Lovász (1968, 1968a) were apparently the first to consider (weak) vertex-colourings of hypergraphs. Somewhat later, Berge (1973) formalized the notions of weak and strong chromatic numbers of hypergraphs. Both are natural generalizations of the chromatic number of a graph: view a proper vertex-colouring of a graph either as one in which no edge is monochromatic (*weak colouring*) or as one in which the vertices incident upon a given edge all receive different colours (*strong colouring*).

Methods used to establish results on colourings of designs cover a wide spectrum. In particular, colourings of designs form one of the few subjects concerning triple systems in which probabilistic methods find effective application. Implicit in many of the arguments are often deep results in number theory, such as Szemerédi's theorem on the density of arithmetic progression-free sets, or van der Waerden's permanent theorem.

18.1 Colourings

Let $S = (V, \mathcal{B})$ be a partial triple system of order v and index λ. A *colouring* of S is a mapping $\phi : V \mapsto C$. The elements of C are *colours*; if $|C| = k$, we have a *k-colouring* of S. For each $c \in C$, the set $\phi^{-1}(c) = \{x \in V : \phi(x) = c\}$ is a *colour class.* A triple all of whose elements have the same colour is *monochromatic.* For $U \subseteq V$, we write $\phi(U) = \{\phi(x) : x \in U\}$. A colouring of S is *weak* if for all $B \in \mathcal{B}$, $|\phi(B)| > 1$, it is *strong* if for all $B \in \mathcal{B}$, $|\phi(B)| = 3$. Each colour class of a weak or strong colouring is an independent set. In a weak colouring, no triple is monochromatic; while in a strong colouring, the elements of each triple receive three distinct colours.

The *weak chromatic number* of S, $\chi_w(S)$, is the smallest k for which S admits a weak k-colouring. Similarly, the *strong chromatic number* of S, $\chi_s(S)$, is the smallest k for which S admits a strong k-colouring. A design S is *weakly (strongly) k-colourable* if it admits a weak (strong) k-colouring; it is *weakly (strongly) k-chromatic* if it has weak (strong) chromatic number k.

We consider weak 2-colourings first.

Lemma 18.1 *Let $(V, \mathcal{B})$ be a weakly 2-chromatic partial TS(v, λ). Then $|\mathcal{B}| \leq \lambda v^2/8$.*

Proof If t elements are coloured green, and $v - t$ red, and no triple is monochromatic, then $|\mathcal{B}| \leq \lambda t(v-t)/2 \leq \lambda v^2/8$. □

Thus the only weakly 2-chromatic TS(v, λ)s are the TS$(3, \lambda)$s for all λ, and the TS$(4, \lambda)$s for all even λ.

In the case of colour classes of nearly equal size, we can say more about the number of monochromatic triples in a 2-colouring. Phelps (1984) essentially proved:

Lemma 18.2 *When a TS*(v, λ) *has a colouring with colour classes of size* $\lfloor v/2 \rfloor$ *and* $\lceil v/2 \rceil$, *the number of monochromatic triples does not depend on the particular triple system. It equals* $\lambda \lfloor v/2 \rfloor (\lceil v/2 \rceil - 2)/6$.

Proof We give the proof for v odd; the proof when v is even is similar. Let $v = 2t+1$, and suppose that there are t green elements and $t+1$ red elements. Let b_g be the number of monochromatic green triples and b_r the number of monochromatic red triples. Then there are $\lambda\binom{t}{2} - 3b_g$ triples with two green elements and one red element; and there are $\lambda\binom{t+1}{2} - 3b_r$ triples with two red elements and one green element. The total number of triples is $\lambda\binom{2t+1}{2}/3$. Thus

$$b_g + b_r + \lambda\binom{t}{2} - 3b_g + \lambda\binom{t+1}{2} - 3b_r = \lambda\binom{2t+1}{2}/3,$$

whence $b_g + b + r = \lambda t(t-1)/6 = \lambda \lfloor v/2 \rfloor (\lceil v/2 \rceil - 2)/6$. □

Hence there exist weakly 2-chromatic partial TS(v, λ)s with exactly b triples whenever $1 \leq b \leq \lfloor \lambda v^2/8 \rfloor$. Indeed, take any TS$(v, \lambda)$ and partition its elements into two colour classes of nearly equal size. Delete the monochromatic triples, of which Lemma 18.2 ensures there are $\lambda \lfloor v/2 \rfloor (\lceil v/2 \rceil - 2)/6$. There remain $\lfloor \lambda v^2/8 \rfloor$ triples, and any further $\lfloor \lambda v^2/8 \rfloor - b$ triples can be deleted to obtain the partial system required.

We have seen, as a consequence of Lemma 18.1, that the weak chromatic number of any TS(v, λ) with $v \geq 5$ is at least three. But do three colours suffice? Not always: for example, the simple TS$(8, 6)$ has a maximum independent set of size two, and hence four colours are needed. A weakly 3-chromatic TS$(8, 6)$ does exist, having two disjoint independent sets of three elements each. One example of such a TS$(8, 6)$ is:

Example 18.3 A weakly 3-chromatic TS$(8, 6)$.
The triples are: 125, 125, 126, 126, 127, 128, 135, 135, 136, 137, 137, 138, 145, 146, 146, 147, 147, 148, 158, 168, 178, 235, 236, 236, 237, 237, 238, 245, 245, 246, 247, 247, 248, 258, 268, 278, 345, 345, 346, 346, 347, 348, 358, 368, 378, 458, 468, 478, 567, 567, 567, 567, 567, 567, 568, 578, and 678.

In fact, three colours *do* suffice, in the following sense:

Theorem 18.4 *For every* $v \geq 5$ *and every admissible* λ, *there exists a weakly 3-chromatic TS*(v, λ).

Proof It suffices to prove the statement for $\lambda = \lambda_{\min}(v)$ (see §2.1.2). When $\lambda = 1$ and $v \equiv 3 \pmod 6$, use the commutative $3v$ construction (Lemma 2.3); colour the elements of the set Q_i with colour i, for $1 \leq i \leq 3$. When $\lambda = 1$ and

$v \equiv 1 \pmod 6$, use the commutative $3v+1$ construction (Lemma 2.5); colour the elements of Q_i with i, and the element ∞ with any of the three colours.

When $\lambda = 2$ and $v \equiv 0, 4 \pmod 6$, use the noncommutative $3v$ or $3v+1$ construction (Lemma 2.4) when $v \geq 10$, and colour elements as above. The unique TS(6, 2) is obviously 3-chromatic. When $\lambda = 3$ and $v \equiv 5 \pmod 6$, use the special $3v+2$ construction (Lemma 2.6). Finally, when $\lambda = 6$ and $v \equiv 2 \pmod 6$, use the general $3v+2$ construction when $v > 8$ (Lemma 2.7); Example 18.3 gives a 3-chromatic TS(8, 6). □

For the remainder of this section, we assume that $\lambda = 1$.

Erdös and Hajnal (1966) were the first to establish that for any $m \geq 2$, there exist weakly m-chromatic *partial* Steiner triple systems. The probabilistic method was used by de Brandes, Phelps, and Rödl (1984), and by Phelps and Rödl (1986), to establish:

Theorem 18.5 *Let u_m be the smallest order of a weakly m-chromatic partial STS. Then*

$$C_1 m^2 \log m < u_m < C_2 m^2 \log m,$$

where C_1 and C_2 are absolute constants.

The lower bound in Theorem 18.5 is obtained as a corollary of an inequality due to Phelps and Rödl (1986):

$$\chi_w(S) \leq C\sqrt{v/\ln v}$$

for any partial STS(v) S, where C is an absolute constant. This inequality is, in turn, a consequence of the upper bound on $\beta'_{\min}$ from Theorem 17.4.

Returning to Steiner triple systems, all STS(v)s with $v \leq 15$ are weakly 3-chromatic (see Mathon, Phelps, and Rosa, 1983). Nevertheless, the existence of weakly m-chromatic partial STS(v), coupled with the finite embedding theorem for partial STSs (Chapter 11), shows immediately the existence of Steiner triple systems whose weak chromatic number is at least m, for any $m \geq 3$. More is true:

Theorem 18.6 *For all $m \geq 3$, there exists v_m so that for every admissible $v > v_m$, there exists a weakly m-chromatic STS(v).*

Proof Theorem 18.5 gives a weakly m-chromatic PTS(u_m, 1) $(V, \mathcal{B})$. Construct from $(V, \mathcal{B})$ a partial idempotent quasigroup $(V, \oplus)$, and embed it into an idempotent (or half-idempotent) commutative quasigroup $(T, \otimes)$ of order $2t+1$ (or $2t$, respectively) for every $t \geq u_m$. Applying the commutative $3v$ construction (the commutative $3v+1$ construction, respectively) gives an STS($6t+3$) (an STS($6t+1$), respectively) S. In either case, the proof of Theorem 18.4 establishes that S is 3-chromatic. For every triple $t \in \mathcal{B}$ of the original PTS, there is a sub-STS(9) on the set $t \times \{1, 2, 3\}$. Let $T_1, \ldots, T_b$ be the triples in $\mathcal{B}$, and define a sequence S_i, $0 \leq i \leq b$ of STS as follows. $S_0 = S$. We form S_{i+1} from S_i by replacing in S_{i-1} the triples of the sub-STS(9) on $T_i \times \{1, 2, 3\}$ by the triples of an STS(9) that include the four triples $\{x_1, x_2, x_3\}$ for $x \in T_i$, and $T_i \times \{1\}$, among others.

Evidently, S_b has chromatic number at least m, since a copy of the triples $\mathcal{B}$ of the original PTS are on $V \times \{1\}$ in S_b. Since S_0 has chromatic number 3, and S_b has chromatic number at least m, it suffices to show that the chromatic number of S_i is at most one larger than the chromatic number of S_{i-1}.

Suppose then that S_{i-1} is q-chromatic, $q < m$, and consider the system S_i. Now form a weak q-colouring of S_{i-1}. When S_i is coloured in the same way, the only monochromatic triples must lie in the sub-STS(9) on $T_i \times \{1,2,3\}$. Now any STS(9) is weakly 3-chromatic, and in any 3-colouring a colour class has at most four elements. It follows that there are at most three monochromatic triples in S_i, and an element can be chosen from each monochromatic triple to form an independent set. Assigning colour $q+1$ to the elements of this independent set gives a weak $q+1$-colouring of S_i. Thus for every q, $3 \le q \le m$, at least one of the triple systems $\{S_i : 0 \le i \le b\}$ is q-chromatic. □

Theorem 18.7 is very useful in obtaining reasonably good upper bounds on v_m for small $m \ge 4$:

Theorem 18.7 *Let* $m \ge 4$. *If, for some* $v \equiv 3 \pmod 6$, *there exists a weakly* m*-chromatic STS*(v) *having an* m*-colouring in which three colour classes contain at most* $v/3$ *elements each, then there exists a weakly* m*-chromatic STS*(w) *for all* $w \ge 2v+1$.

Proof Let $(V, \mathcal{B})$ be an STS(v) with $v = 6t+3$. Let $C_1, \ldots, C_m$ be the colour classes of a weak m-colouring $\mathcal{C}$ of $(V, \mathcal{B})$, where $|C_i| \le 2t+1$ for $i = 1,2,3$. Let $w \ge 2v+1$ be admissible. First we treat the case when $w \equiv 3 \pmod 6$; say $w = 6u+3$. By a theorem of Cruse (1974), there is a commutative idempotent quasigroup $(U, \odot)$ of order $2u+1$ containing a commutative idempotent subquasigroup of order $2t+1$. Apply the commutative $3v$ construction (Lemma 2.3) to $(U, \odot)$ to form a weakly 3-chromatic STS($6u+3$) with a weakly 3-chromatic sub-STS(v), $(V', \mathcal{B}')$. Now replace the sub-STS(v) with the weakly m-chromatic STS(v), $(V, \mathcal{B})$ in such a way that $C_i \subseteq U \times \{i\}$, for $i = 1,2,3$. The result is a weakly m-chromatic STS(w): it suffices to colour the elements of $(U \setminus V) \times \{i\}$ with colour i ($i = 1,2,3$), and the remaining elements as in $\mathcal{C}$.

When $w \equiv 1 \pmod 6$, write $w = 6u+1$. Use the theorem of Lindner (1980) to produce a commutative half-idempotent quasigroup $(U, \odot)$ containing a commutative idempotent subquasigroup of order $2t+1$. Next apply the commutative $3v+1$ construction (Lemma 2.5) to form an STS($6u+1$) that is weakly 3-chromatic and has a sub-STS(v). The subsystem can be replaced by $(V, \mathcal{B})$ as before, assigning the extra element ∞ any of the m colours. A weakly m-chromatic STS(w) results. □

Corollary 18.8 *If* $m \ge 5$, *and for some* $v \equiv 3 \pmod 6$ *there exists an* m*-chromatic STS*(v), *then there exists an* m*-chromatic STS*(w) *for all admissible* $w \ge 2v+1$.

Proof When $m \geq 5$, there must exist three colour classes whose size is bounded by $v/3$. □

Corollary 18.9 $v_4 \leq 43$ *and* $v_5 \leq 127$.

Proof A weakly 4-chromatic STS(21) was constructed by Haddad (1999a). If $(V, \mathcal{B})$ is the cyclic STS(7) with $V = \{0, \ldots, 6\}$ and starter block $\{0, 1, 3\}$, let $W = V \times \{1, 2, 3\}$, and let $\mathcal{C}$ contain, for each triple $\{i, j, k\} \in \mathcal{B}$, the 12 triples

$$\{i_1, i_2, i_3\},\ \{j_1, j_2, j_3\},\ \{k_1, k_2, k_3\},\ \{i_1, j_1, k_1\},\ \{i_2, j_1, k_2\},\ \{i_3, j_1, k_3\},$$
$$\{i_2, j_3, k_1\},\ \{i_2, j_2, k_3\},\ \{i_3, j_3, k_2\},\ \{i_1, j_3, k_3\},\ \{i_3, j_2, k_1\},\ \text{and } \{i_1, j_2, k_2\}.$$

Then $(W, \mathcal{C})$ is an STS(21) that does not admit a weak 3-colouring. A weak 4-colouring of this system has

C_1: $\{0_1, 1_1, 2_1, 0_2, 1_2, 2_2\}$
C_2: $\{3_1, 4_1, 3_2, 4_2\}$
C_3: $\{5_1, 5_2, 6_1, 6_2\}$
C_4: $V \times \{3\}$

This satisfies the conditions of Theorem 18.7, and thus $v_4 \leq 43$.

Fugère, Haddad, and Wehlau (1994) have shown that the projective Steiner triple system on 63 elements has chromatic number 5, and hence by Corollary 18.8, we have $v_5 \leq 127$. □

The bound for v_4 can be improved upon:

Theorem 18.10 $15 \leq v_4 \leq 19$.

Proof As mentioned earlier, every STS(v) with $7 \leq v \leq 15$ is weakly 3-chromatic. Rosa (1970a) shows that the projective Steiner triple system on 31 elements is weakly 4-chromatic. In view of Corollary 18.9, it suffices to exhibit weakly 4-chromatic STS(v)s for $v = 25$, 27, 33, 37, and 39. These were found by computer; the first four are from de Brandes, Phelps, and Rödl (1982). Each is cyclic, and so we present starter blocks. The solution for $v = 39$ was found by Haddad (1999a).

$v = 25$:

$\{\{1,2,4\}, \{1,5,14\}, \{1,6,12\}, \{1,8,18\}\}$
C_1: $\{1,2,3,6,7,8,11\}$
C_2: $\{5,9,10,13,14,15,19\}$
C_3: $\{12,16,17,18,21,22\}$
C_4: $\{4,20,23,24,25\}$

$v = 27$:

$\{\{1,2,4\}, \{1,5,12\}, \{1,6,18\}, \{1,7,15\}, \{1,10,19\}\}$
C_1: $\{16,21,27\}$
C_2: $\{4,5,9,17,26\}$
C_3: $\{10,13,14,15,18,19,20,23,24\}$
C_4: $\{1,2,3,6,7,8,11,12,22,25\}$

$v = 33$:

$\{\{1,2,4\},\{1,5,15\},\{1,6,14\},\{1,7,19\},\{1,8,17\},\{1,12,23\}\}$

C_1: $\{1,2,3,6,7,8,12,13,30\}$

C_2: $\{14,15,16,19,20,21,24,25,26,31\}$

C_3: $\{4,5,11,27,28,29,32,33\}$

C_4: $\{9,10,17,18,22,23\}$

$v = 37$:

$\{\{1,2,4\},\{1,5,15\},\{1,6,14\},\{1,7,22\},\{1,8,20\},\{1,10,21\}\}$

C_1: $\{1,2,3,6,7,8,11,12,13,18,32\}$

C_2: $\{14,15,16,19,20,21,24,25,26,37\}$

C_3: $\{4,9,23,29,30,31,34,35,36\}$

C_4: $\{5,10,17,22,27,28,33\}$

For $v = 39$, let $(V, \mathcal{B})$ be the cyclic STS(13) with starter blocks $\{0,1,4\}$ and $\{0,2,7\}$. Construct an STS(39) on $V \times \{1,2,3\}$ by replacing each triple of the STS(13) by 12 triples in the STS(39), as in the proof of Corollary 18.9. This STS(39) does not admit a weak 3-colouring, but does admit a weak 4-colouring with colour classes $\{0,1,2,3,9\} \times \{1,2\}$, $\{4,5,6,7\} \times \{1,2\}$, $\{8,10,11,12\} \times \{1,2\}$ and $V \times \{3\}$. □

The projective triple systems have already been used to provide examples of 4- and 5-chromatic STSs. In general, their chromatic numbers are only partially determined:

Theorem 18.11 *The projective triple system S_n^* of order $2^{n+1}-1$, arising from the points and lines of $PG(n,2)$, has weak chromatic number 3 for $n \in \{2,3\}$, 4 for $n = 4$, and 5 for $n = 5$. In general, $\chi_w(S_n^*) \leq \chi_w(S_{n+1}^*) \leq 1 + \chi_w(S_n^*)$.*

Proof Rosa (1970a) settled the cases $n = 2,3,4$. Fugère, Haddad, and Wehlau (1994) settled the case $n = 5$, and their argument is already quite involved. For the final inequality, in S_{n+1}^*, give a weak colouring to any fixed projective subsystem isomorphic to S_n^* using $\chi_w(S_n^*)$ colours. Colour all remaining elements using a new colour. The result is a weak colouring of S_{n+1}^*. □

18.2 Equitable weak colourings

We might ask for a colouring with additional properties, such as balanced sizes of colour classes. A weak colouring of a TS(v,λ) is *equitable* if the cardinalities of the colour classes differ by at most one. For example, each of the 80 STS(15)s has an equitable weak 3-colouring; see Mathon, Phelps, and Rosa (1983). Moreover, every 3-colouring of S_3^* is equitable (Pelikán, 1970), and the colouring is unique up to automorphism (Fugère, Haddad, and Wehlau, 1994). This suggests that perhaps every STS(v) admits an equitable χ_w-colouring. Haddad and Rödl (1994) support this suggestion for 3-chromatic STSs:

Theorem 18.12 *Let $S = (V, \mathcal{B})$ be a weakly 3-chromatic STS(v), and let V_i, $i = 1, 2, 3$, be the colour classes of a weak 3-colouring of S. Let $v_i = |V_i|$. Then*

$$v = v_1 + v_2 + v_3 \geq \frac{1}{2}\left[(v_1 - v_2)^2 + (v_1 - v_3)^2 + (v_2 - v_3)^2\right].$$

Thus, for sufficiently large v, the colour classes must have approximately the same size.

Proof For $i, j \in \{1, 2, 3\}$, $i \neq j$, let x_{ij} be the number of triples in $\mathcal{B}$ that intersect V_i in two elements and V_j in one element. Let x be the number of triples that intersect each colour class in one element. We have:

$$\begin{aligned}
x + \sum x_{ij} &= v(v-1)/6 \\
x_{12} + x_{13} &= \tbinom{v_1}{2} \\
x_{21} + x_{23} &= \tbinom{v_2}{2} \\
x_{31} + x_{32} &= \tbinom{v_3}{2} \\
2x_{12} + 2x_{21} + x &= v_1 v_2 \\
2x_{23} + 2x_{32} + x &= v_2 v_3 \\
2x_{13} + 2x_{31} + x &= v_1 v_3
\end{aligned}$$

Adding the second and the third, and subtracting the fifth gives

$$x_{13} + x_{23} - x_{21} - x_{12} = \binom{v_1}{2} + \binom{v_2}{2} - v_1 v_2 + x \tag{18.1}$$

so that

$$x_{13} + x_{23} - x_{21} - x_{12} \geq \frac{(v_1 - v_2)^2 - (v_1 + v_2)}{2} \tag{18.2}$$

and similarly we get

$$x_{32} + x_{12} - x_{31} - x_{13} \geq \frac{(v_1 - v_3)^2 - (v_1 + v_3)}{2} \tag{18.3}$$

$$x_{31} + x_{21} - x_{23} - x_{32} \geq \frac{(v_2 - v_3)^2 - (v_2 + v_3)}{2} \tag{18.4}$$

Adding inequalities (18.2), (18.3), and (18.4) and rearranging terms gives the theorem. □

On the other hand, a second theorem of Haddad and Rödl (1994) refutes the existence of equitable weak m-colourings in a very strong sense:

Theorem 18.13 *For every $0 < \varepsilon < 1$, $m \geq 6$ and $t \geq m$, there exists a weakly m-chromatic STS(v) S such that for every t-colouring of S, there are $m - 3$ colour classes whose union contain at most εv elements.*

Corollary 18.14 *For every $m \geq 6$ there exists a weakly m-chromatic STS that does not admit an equitable m-colouring.*

18.3 Equitable colourings with many colours

Colbourn, Haddad, and Linek (1996) show that when the order v is large enough with respect to the number r of colours, and $v \equiv 1, 3 \pmod 6$, an equitably r-coloured r-chromatic Steiner triple system of order v exists. We examine their method here.

If X is a nonempty set, an *r-partition* of X is a partition $(C_1, C_2, \ldots, C_r)$ of X; equivalently it is a mapping $f : X \to \{1, 2, \ldots, r\}$, where $C_r = f^{-1}(r)$. The C_i are the *cells* of the partition. If $S = (V, \mathcal{B})$ is an STS, then an r-partition of V is an r-colouring of S if and only if no triple of $\mathcal{B}$ is entirely contained in any cell C_i, $i = 1, \ldots, r$. The r-partition $(C_1, C_2, \ldots, C_r)$ of X is *equitable* if $\big||C_i| - |C_j|\big| \leq 1$ holds for every $i, j \in \{1, 2, \ldots, r\}$.

To give an example of the kind of colouring problems encountered in Lemmas 18.17 and 18.18, let $X = I_5 \times I_3$ have the 4-partition

$$\mathbf{c} = (\{0_0, 0_1, 1_0, 2_1, 3_2, 4_1\}, \{2_0, 3_0, 4_0\}, \{0_2, 1_2, 2_2\}, \{1_1, 3_1, 4_2\}).$$

Is it possible to construct a 3-GDD $(X, \mathcal{G}, \mathcal{B})$ such that $\mathcal{G} = \{\{i_0, i_1, i_2\} : 0 \leq i \leq 4\}$ and no block of $\mathcal{B}$ is contained in any cell of $\mathbf{c}$? To see that the answer is affirmative, we use Bose's construction. Define a commutative, idempotent quasigroup

$\circ$	1	2	3	4	5
1	1	4	2	5	3
2	4	2	5	3	1
3	2	5	3	1	4
4	5	3	1	4	2
5	3	1	4	2	5

and let

$$\mathcal{B} = \{\{x_i, y_i, (x \circ y)_{i+1}\} : x \neq y,\ 1 \leq x, y \leq 5,\ i \in I_3\}, \tag{18.5}$$

where addition is performed modulo 3. Then $(X, \mathcal{G}', \mathcal{B})$ with $\mathcal{G}' = \{\{i\} \times I_3 : 1 \leq i \leq 5\}$ is a 3-GDD. To solve the original problem define a bijection $f : \{1, \ldots, 5\} \times \mathbb{Z}_3 \to X$ by

$$\begin{array}{ccc} 0_0\ 1_0\ 2_1\ 3_2\ 4_0 & & 1_0\ 2_0\ 3_0\ 4_0\ 5_0 \\ 0_2\ 1_1\ 2_0\ 3_1\ 4_2 & \xrightarrow{f} & 1_1\ 2_1\ 3_1\ 4_1\ 5_1 \\ 0_1\ 1_2\ 2_2\ 3_0\ 4_1 & & 1_2\ 2_2\ 3_2\ 4_2\ 5_2 \end{array}$$

and take the 3-GDD $(X, \mathcal{G}, f(\mathcal{B}))$ where

$$f(\mathcal{B}) = \{f(\{x, y, z\}) : \{x, y, z\} \in \mathcal{B}\}.$$

Then $(X, \mathcal{G}, f(\mathcal{B}))$ is the required GDD. Modifying the partition $\mathbf{c}$ by first freely permuting the contents of the groups $\{i_0, i_1, i_2\}$, $i = 0, 1, \ldots, 4$, a new partition $\mathbf{c}'$ is produced, for which again the problem is solved. Indeed, permuting the elements of X in any manner that leaves the *collection* of groups intact still ensures that the partition $\mathbf{c}'$ (obtained from $\mathbf{c}$ by this permutation) is a colouring for some GDD on X with the required groups. As the number of partitions of X to consider is large, this strategy of considering orbits of partitions under the action of the largest group that preserves the set of GDDs on X is employed.

In general, write $X = \{1,\ldots,5\} \times I_3$, put $G_i = \{i\} \times I_3$, $1 \le i \le 5$, and let $\mathcal{G} = (G_1,G_2,\ldots,G_5)$. We also consider the same problem for $X = \{1,2,3\} \times I_3$, $G_i = \{i\} \times \mathbb{Z}_3$, $i = 1,2,3$, and $\mathcal{G} = (G_1, G_2, G_3)$, but our efforts concentrate on the former case.

Denote by $\mathrm{Sym}(X)$ the symmetric group of permutations of X, and let $\Gamma(X,\mathcal{G})$ be the subgroup of $\mathrm{Sym}(X)$ consisting of those permutations of X that induce permutations of the members of $\mathcal{G}$. In other words, $\Gamma(X,\mathcal{G}) = \{\gamma \in \mathrm{Sym}(X){:}\, \gamma(G_i) \in \mathcal{G}$ for each $i\}$, so $|\Gamma(X,\mathcal{G})| = 5! \cdot 6^5$. Thus $\Gamma(X,\mathcal{G})$ is (up to isomorphism) the wreath product of $\mathrm{Sym}(I_3)$ and $\mathrm{Sym}(\{1,\ldots,5\})$. If $(X,\mathcal{G},\mathcal{B})$ is a 3-GDD then $(X,\mathcal{G},\gamma(\mathcal{B}))$ is also a 3-GDD for any $\gamma \in \Gamma(X,\mathcal{G})$. Let $\mathfrak{S}$ be the collection of all r-partitions $\mathbf{c} = (C_1,C_2,\ldots,C_r)$ of X such that $|C_i \cap G_j| \le 1$ fails to hold for at most one pair of indices (i,j), in which case $|C_i \cap G_j| = 2$ holds for the exceptional pair of indices (i,j).

Lemma 18.15 *With notation as above, for any* $\mathbf{c} \in \mathfrak{S}$ *there exists a 3-GDD* $(X,\mathcal{G},\mathcal{B})$ *in which no block lies in any cell of* $\mathbf{c}$.

Proof The case $|X| = 9$ is left to the reader, so consider the case where $|X| = 15$. Consider first $\mathbf{c} = (C_1,C_2,\ldots,C_r)$, $|C_1| \le 2$, and $r > 3$. In this case it is possible to redistribute the points of C_1 among the remaining cells to obtain an $(r-1)$-partition $\mathbf{c}' = (C'_2, C'_3,\ldots,C'_r) \in \mathfrak{S}$. Then any 3-GDD $(X,\mathcal{G},\mathcal{B})$ that admits $\mathbf{c}'$ as a colouring also admits $\mathbf{c}$ as a colouring. It therefore suffices to consider only those r-partitions $\mathbf{c}$ with $|C_i| \ge 3$ for each i. Furthermore, if $|C_1| \le 3$ and $|C_1 \cap G_j| = 2$ for some j (the exceptional case), then we may again redistribute the points of C_1 among the other cells to obtain $\mathbf{c}' = (C'_2, C'_3,\ldots,C'_r) \in \mathfrak{S}$. Then we need only consider $\mathbf{c}'$.

Given $\mathbf{c} \in \mathfrak{S}$, the notation $g^* g_1^{t_1} g_2^{t_2} \cdots g_r^{t_r}$ indicates that $\mathbf{c}$ has a cell of size g that intersects a group in exactly two points and in addition $\mathbf{c}$ has t_i cells of size g_i for each i. In light of the preceding observations, it suffices to consider only the types $6^*\cdot 5\cdot 4$, $6^*\cdot 3^3$, $5^*\cdot 4\cdot 3^2$, $4^*\cdot 5\cdot 3^2$, $4^*\cdot 4^2\cdot 3$, 5^3, $3\cdot 4^3$, $3^2\cdot 4\cdot 5$, and 3^5.

Now $\Gamma(X,\mathcal{G})$ induces an action on $\mathfrak{S}$ that preserves the (possibly starred) type of any partition $\mathbf{c} \in \mathfrak{S}$, so the partitions of a given type fall into orbits under the action of $\Gamma(X,\mathcal{G})$. It is easily verified that each of the types 5^3, $3\cdot 4^3$, $3^2\cdot 4\cdot 5$, $6^*\cdot 5\cdot 4$, $6^*\cdot 3^3$, and $4^*\cdot 5\cdot 3^2$ determine a single orbit, while the types 3^5, $5^*\cdot 4\cdot 3^2$, and $4^*\cdot 4^2\cdot 3$ each split into exactly two orbits. It suffices then to exhibit for each orbit a representative partition and a 3-GDD which admits that orbit representative as a colouring.

Given $(X,\mathcal{G},\mathcal{B})$ we write $\overline{\mathcal{B}} = \{A \in X{:}\, |A| = 3,\ A \notin \mathcal{B}$, and $|A \cap G_i| \le 1$ for each $G_i \in \mathcal{G}\}$. For $A, A' \in \overline{\mathcal{B}}$, $A \cap A' = \varnothing$ and $\mathbf{c} = (C_1,C_2,\ldots,C_r) \in \mathfrak{S}$ we define $\mathbf{c} - A$ to be $(A,\, C_1 - A,\, C_2 - A,\ldots,\, C_r - A)$ and $\mathbf{c} - A - A'$ to be $(A,\, A',$

$C_1 - A - A'$, $C_2 - A - A', \ldots, C_r - A - A'$). Let $(X, \mathcal{G}, \mathcal{B})$ be the 3-GDD with blocks given in (18.5), the required r-partitions are:

(5^3)	$C_1 = \{1,\ldots,5\}\times\{0\}$, $C_2 = \{1,\ldots,5\}\times\{1\}$, $C_3 = \{1,\ldots,5\}\times\{2\}$, $\mathbf{c} = (C_1, C_2, C_3)$
$(3 \cdot 4^3)$	$\mathbf{c} - \{1_0, 2_1, 3_2\}$
$(3^2 \cdot 4 \cdot 5)$	$\mathbf{c} - \{1_1, 2_2, 3_2\}$
$(6^* \cdot 5 \cdot 4)$	take $\mathbf{c}' = (C_1 \cup \{1_2\}, C_2, C_3 - \{1_2\})$
$(6^* \cdot 3^3)$	$\mathbf{c}' - \{2_2, 3_2, 4_1\} - \{3_1, 4_2, 5_2\}$
$(4^* \cdot 5 \cdot 3^2)$	$\mathbf{c}' - \{2_0, 3_0, 4_2\}$
(3^5)	$\mathbf{c} - \{2_2, 4_0, 5_0\} - \{3_2, 4_1, 5_1\}$ and $\mathbf{c} - \{1_0, 2_1, 3_2\} - \{3_0, 4_1, 5_2\}$
$(5^* \cdot 4 \cdot 3^2)$	$\mathbf{c}' - \{1_1, 2_0, 3_2\}$ and $\mathbf{c}' - \{2_1, 3_0, 4_2\}$
$(4^* \cdot 4^2 \cdot 3)$	$\mathbf{c}' - \{1_1, 4_0, 5_0\}$ and $\mathbf{c}' - \{3_1, 4_0, 5_0\}$

This completes the proof. □

As an example of the kind of colouring problem encountered in Lemma 18.18, consider the following. Let $X = \{\infty\} \cup I_3 \times I_4$ have the 5-partition $\mathbf{c} = (\{\infty, 0_0, 1_0, 3_0, 3_1\}, \{0_1, 1_1, 2_1\}, \{1_2, 2_2, 3_2\}, \{0_2\}, \{2_0\})$. Is it possible to construct an STS(13) on X such that $\mathbf{c}$ is a 5-colouring and such that $\{3_0, 3_1, 3_2\}$ is a block of the STS(13)? To see that the answer is affirmative, it suffices to produce an STS(13) that contains three mutually disjoint independent sets of sizes 5, 3, and 3 and a block, b, that intersects the set of size 5 in two points and intersects one of the sets of size 3 in one point. Such a configuration exists in both STS(13)s.

In general, let $|X| = 13$ and let $X = \{\infty\} \cup G_1 \cup G_2 \cup G_3 \cup G_4$ where the G_i are mutually disjoint sets of size 3. Consider the collection of r-partitions $\mathbf{c} = (C_1, C_2, \ldots, C_r)$ such that for each i we have $|C_i \cap G_j| \le 1$, $1 \le j \le 3$ and $|C_i \cap C_4| \le 2$. Our problem is to construct an STS(13) on X such that $\mathbf{c}$ is an r-colouring and G_4 is a block. In the case $|X| = 7$ we need something similar. If $X = \{\infty\} \cup G_1 \cup G_2$ where G_1 and G_2 are disjoint sets of size 3 and $\mathbf{c} = (C_1, C_2, \ldots, C_r)$ satisfies $|C_i \cap G_1| \le 1$ and $|C_i \cap G_2| \le 2$ for each i, then, is it possible to construct an STS(7) on X such that G_2 is a block? To give an affirmative answer to each of these questions we need a bit more terminology.

Given v let $P(g_1^{e_1} \cdots g_m^{e_m})$ denote the proposition that there exists an STS(v) that contains m mutually disjoint independent sets of sizes $g_1, g_2, \ldots, g_m$ and which contains a block, b, that intersects the set of size g_i in e_i points (we do not insist that the g_is partition the integer v). A technical lemma for Lemma 18.18 is repeated here from Colbourn, Haddad, and Linek (1996) without proof.

Lemma 18.16 *The following propositions all hold.*

1. *The propositions $P(4^2)$, $P(3^2, 3^1)$, $P(3^2, 3^0)$, and $P(3^1, 3^1)$ all hold in the unique STS(7).*
2. *Let $g_1^{e_1} \cdots g_m^{e_m}$, $3 \le e_1 \le e_2 \le \cdots \le e_m$ satisfy*

(1) $3 \leq g_i \leq 6$ *and* $\sum g_i \leq 13$
(2) $g_1 + g_2 \leq 11$
(3) $0 \leq e_i \leq 2$ *and* $\sum e_i \leq 3$
(4) $g_1^{e_1} \cdots g_m^{e_m} \neq 6^1 5^1$ *or* $6^1 5^0$

then $P(g_1^{e_1} \cdots g_m^{e_m})$ *holds in some STS*(13).

A *partition sequence* of an r-chromatic STS(v) is a sequence of integers $t_1 \geq t_2 \geq \cdots \geq t_r$ that correspond to the sizes of colour classes in some r-colouring of the r-chromatic STS(v).

Lemma 18.17 *Suppose there exists an r-chromatic STS(w) with $w \equiv 3 \pmod 6$ and partition sequence $t_1 \geq t_2 \geq \cdots \geq t_r$. If v satisfies*

$$v \equiv 3 \pmod 6 \tag{18.6}$$

$$v \geq \max(2w, rt_1) \tag{18.7}$$

$$\lceil v/r \rceil - t_r \leq \frac{1}{3}(v - w) \tag{18.8}$$

then there exists an equitably r-coloured, r-chromatic STS(v).

Proof Write $v = 3u$, $w = 3k$, $n = u - k$, and let $X = (I_k \cup I'_n) \times I_3$. Let $s_1 \leq s_2 \leq \cdots \leq s_r$ be an equitable partition of v and let

$$u_1 = s_1 - t_1$$
$$u_2 = s_2 - t_2$$
$$\cdots$$
$$u_r = s_r - t_r.$$

Inequality (18.7) ensures that each u_i is nonnegative. Put $t_0 = u_0 = 0$ and let $\tau_m = \sum_{i=0}^{m} t_i$, $\mu_m = \sum_{i=0}^{m} u_i$ for $0 \leq m \leq r$. Now form the following partition of $I_w \cup I'_{3n}$:

$$C_1 = (I_{\tau_1} - I_{\tau_0}) \cup (I'_{\mu_1} - I'_{\mu_0})$$
$$C_2 = (I_{\tau_2} - I_{\tau_1}) \cup (I'_{\mu_2} - I'_{\mu_1})$$
$$\cdots$$
$$C_r = (I_{\tau_r} - I_{\tau_{r-1}}) \cup (I'_{\mu_r} - I'_{\mu_{r-1}}).$$

By construction, $(C_1, C_2, \ldots, C_r)$ is an equitable r-partition of the v element set $I_w \cup I'_{3n}$. Define the bijection $f: I_w \cup I'_{3n} \to X$ by the equations

$$f(z) = (z \bmod k, \lfloor z/k \rfloor)$$
$$f(z') = ((z \bmod n)', \lfloor z/n \rfloor).$$

(all residues taken in the appropriate set) and for each i put $\overline{C}_i = f(C_i)$. Then $(\overline{C}_1, \overline{C}_2, \ldots, \overline{C}_r)$ is an equitable r-partition of X.

Informally, the idea here is to take an r-coloured STS(w) (on I_w), arrange the colour classes 'linearly' and then stack them in three layers (f does this). If the colouring is not equitable to begin with, then just enough coloured points are added to make an equitable colouring (those in I'_{3n}), and the new points are stacked in a similar way so that they do not 'mix' with the points of the original r-chromatic STS(w).

In order to use Lemma 18.15 for the construction of an STS(v) on X, by Lemma 17.1 at most two points of each set $\{j_0, j_1, j_2\}$, $j \in I_w$, can lie in some $\overline{C}_i$. Furthermore, by inequality (18.8), no pair of points from $\{j'_0, j'_1, j'_2\}$, $j' \in I'_n$, is in any $\overline{C}_i$; that is, these three points occur in three *distinct* $\overline{C}_i$s.

Using Theorem 6.12, construct an $(\{\mathcal{O} \cup \{k^*\})$-PBD on $I_k \cup I'_n$ such that I_k is the unique block of size k. Form a set of blocks, $\mathcal{B}$, on X as follows:

1. Take the blocks of an r-coloured, r-chromatic STS(w) on $I_k \times I_3$.
2. For each block B in the PBD with $|B| = 3$ or 5 take the blocks of a 3-GDD on $B \times I_3$ so that the groups are $\{b_0, b_1, b_2\}$, $b \in B$ and such that none of the blocks constructed lies in any $\overline{C}_i$. This is possible by Lemma 18.15.
3. For each $j' \in I'_n$ take the block $\{j'_0, j'_1, j'_2\}$.

By construction, $(X, \mathcal{B})$ is an STS(v) that admits an equitable r-colouring $(\overline{C}_1, \overline{C}_2, \ldots, \overline{C}_r)$ and it has an r-chromatic STS(w) as a subsystem, so it is an equitably r-coloured, r-chromatic STS(v). □

Lemma 18.18 treats the case where $v \equiv 1 \pmod 6$.

Lemma 18.18 *Suppose there exists an r-chromatic STS(w) with $w \equiv 3 \pmod 6$ and partition sequence $t_1 \geq t_2 \geq \cdots \geq t_r$. If v satisfies*

$$v \equiv 1 \pmod 6 \tag{18.9}$$
$$v > \max(2w, rt_1) \tag{18.10}$$
$$\lceil \frac{v}{r} \rceil - t_r \leq \frac{1}{3}(v - w - 1) \tag{18.11}$$

then there exists an equitably r-coloured, r-chromatic STS(v).

Proof Write $v - 1 = 3u$, $w = 3k$, $n = u - k$, and let $X = \{\infty\} \cup (I_k \cup I'_n) \times I_3$. Let $s_1 \leq s_2 \leq \cdots \leq s_r$ be an equitable partition of $v - 1$ and let

$$\begin{aligned} u_1 &= s_1 - t_1 \\ u_2 &= s_2 - t_2 \\ &\cdots \\ u_r &= s_r - t_r. \end{aligned}$$

Inequality (18.10) ensures that each u_i is nonnegative. Put $t_0 = u_0 = 0$ and let $\tau_m = \sum_{i=0}^{m} t_i$, $\mu_m = \sum_{i=0}^{m} u_i$ for $0 \leq m \leq r$. Now form the following partition of $I_w \cup I'_{3n}$:

$$\begin{aligned} C_1 &= (i\tau_1 - i\tau_0) \cup (I'_{\mu_1} - I'_{\mu_0}) \\ C_2 &= (i\tau_2 - i\tau_1) \cup (I'_{\mu_2} - I'_{\mu_1}) \\ &\cdots \\ C_r &= (i\tau_r - i\tau_{r-1}) \cup (I'_{\mu_r} - I'_{\mu_{r-1}}). \end{aligned}$$

By construction, $(C_1,C_2,\ldots,C_r)$ is an equitable r-partition of the $(v-1)$ element set $I_w \cup I'_{3n}$. We transfer this partition to the $(v-1)$ element set $(I_k \cup I'_n) \times I_3$ by means of the bijection $f\colon I_w \cup I'_{3n} \to X\backslash\{\infty\}$ given by

$$f(z) = (z \bmod k, \lfloor z/k \rfloor)$$
$$f(z') = ((z \bmod n)', \lfloor z/n \rfloor).$$

We now take the r-partition $\overline{C}_1 = \{\infty\} \cup f(C_1)$, $\overline{C}_2 = f(C_2), \ldots, \overline{C}_r = f(C_r)$ of X.

By Lemma 17.1, for each $j \in I_k$ at most two points of $\{j_0, j_1, j_2\}$ can lie in the same cell of $(\overline{C}_1,\overline{C}_2,\ldots,\overline{C}_r)$, whereas inequality (18.11) implies that for each $j' \in I'_n$ the points of $\{j'_0, j'_1, j'_2\}$ must lie in three distinct cells of $(\overline{C}_1,\overline{C}_2,\ldots,\overline{C}_r)$.

By Theorem 6.12 construct an $(\{\mathcal{O} \cup \{k^*\})$-PBD on $I_k \cup I'_n \cup \{a\}$ such that I_k is the unique block of size k. Form a set of blocks, $\mathcal{B}$, on X as follows:

1. Take the blocks of an r-coloured, r-chromatic STS(w) on $I_k \times I_3$.
2. For each block B of the PBD such that $a \notin B$ and $|B| = 3$ or 5 take the blocks of a 3-GDD on $B \times I_3$ so that the groups are $\{b_0, b_1, b_2\}$, $b \in B$, and such that none of the blocks constructed lies in any $\overline{C}_i$. This is possible by Lemma 18.15.
3. For each block B of the PBD such that $a \in B$, $B \cap I_k = \varnothing$, and $|B| = 3$ or 5 take the blocks of an STS(7) or STS(13) on $\{\infty\} \cup (B - \{a\}) \times I_3$ constructed in such a way that no block is entirely contained in any $\overline{C}_i$. This is possible by Lemma 18.16.
4. For each block B of the PBD such that $a \in B$, $B \cap I_k = \{j\}$, and $|B| = 3$ or 5 we construct the blocks of an STS(7) or STS(13) on $\{\infty\} \cup (B - \{a\}) \times I_3$ in such a way that $\{j_0, j_1, j_2\}$ is a block, and in such a way that no block constructed is entirely contained in any $\overline{C}_i$. This is possible by Lemma 18.16. We then take all the blocks *except* $\{j_0, j_1, j_2\}$, which we delete.

By construction, $(X, \mathcal{B})$ is an STS(v) that admits an equitable r-colouring $(\overline{C}_1,\overline{C}_2,\ldots,\overline{C}_r)$. It has an equitably r-coloured, r-chromatic STS(w) as a subsystem, so it is an equitably r-coloured, r-chromatic STS(v). □

In the special case where we start with an STS(w) that admits an equitable r-colouring, Lemma 18.17 and Lemma 18.18 combine to give:

Theorem 18.19 *If there exists an equitably r-coloured, r-chromatic STS(w) with $w \equiv 3 \pmod 6$ and $r \geq 4$, then there exists an equitably r-coloured, r-chromatic STS(v) for each admissible $v \geq 2w$.*

In the case $w \equiv 1 \pmod 6$, first triple an r-chromatic STS(w) to get into the right congruence class:

Lemma 18.20 *Any r-chromatic STS(w) with $r \geq 4$ can be embedded in an r-chromatic STS($3w$). In addition, if the r-chromatic STS(w) is equitably r-coloured, then it can be embedded in a equitably r-coloured, r-chromatic STS($3w$).*

Proof Let $(X, \mathcal{B})$ be an r-chromatic STS(w) with r-colouring

$$\mathbf{c}: X \to \{1, 2, \ldots, r\}.$$

Define a new colouring $\mathbf{c}': X \times I_3 \to \{1, 2, \ldots, r\}$ by the rule $\mathbf{c}'(x, j) = \mathbf{c}(x) + j \pmod{r}$, where the residue is computed in $\{1, 2, \ldots, r\}$. It is not hard to check that for each block $\{a, b, c\} \in \mathcal{B}$ it is possible to construct a 3-GDD on $\{a, b, c\} \times I_3$ such that (1) $\{a_0, b_0, c_0\}$ is a block, (2) the groups are $\{a\} \times I_3$, $\{b\} \times I_3$ and $\{c\} \times I_3$, and (3) no block of the 3-GDD receives a single colour under $\mathbf{c}'$. The construction is finished by taking the blocks $\{x_0, x_1, x_2\}$.

This gives an r-coloured STS($3w$) on the point set $X \times I_3$ with a sub-STS(w) which is r-chromatic on the point set $X \times \{0\}$; so we have constructed an r-chromatic STS($3w$). (If $\mathbf{c}$ is an equitable colouring, then by suitably arranging the colour classes, $\mathbf{c}'$ can be made to be equitable.) □

Combining Lemma 18.20 with Theorem 18.19 gives

Theorem 18.21 *If there exists an equitably r-coloured, r-chromatic STS(w) with $w \equiv 1 \pmod 6$ and $r \geq 4$, then there exists an equitably r-coloured, r-chromatic STS(v) for each admissible $v \geq 6w + 1$.*

Theorem 18.21 is not as sharp as Theorem 18.19. However, Colbourn, Haddad, and Linek (1996) give a sharp embedding in the case $v \equiv w \equiv 1 \pmod 6$:

Lemma 18.22 *Suppose there exists an r-chromatic STS(w) with $w \equiv 1 \pmod 6$ and partition sequence $t_1 \geq t_2 \geq \cdots \geq t_r$. If v satisfies*

$$v \equiv 1 \pmod 6 \tag{18.12}$$

$$v > \max(2w, rt_1) \tag{18.13}$$

$$\lceil \frac{v}{r} \rceil - t_r \leq \frac{1}{3}(v - w) \tag{18.14}$$

then there exists an equitably r-coloured, r-chromatic STS(v).

18.4 Balanced colourings

Colbourn, Haddad, and Linek (1997) consider the stronger requirement that *every* r-colouring be equitable, and there is at least one r-colouring. If every r-colouring of a triple system is equitable then the system is *r-balanced.* An r-balanced system is necessarily r-chromatic.

Theorem 18.23 *With the possible exceptions of v in {19, 21, 37, 49, 55, 57, 67, 69, 85, 109, 139}, for all $v \equiv 1, 3 \pmod 6$ and $v \geq 15$ there exists a 3-balanced Steiner triple system of order v.*

18.5 Strong colouring

Trivially, the strong chromatic number χ_s of any TS(v, λ) equals v. Indeed, since every pair of elements is contained in $\lambda \geq 1$ triples, v colours are both necessary and sufficient.

However, the situation is quite different for partial triple systems. Consider a strong colouring of a partial TS(v, λ); any colour class consists of elements, no two of which appear together in a triple. Consider a graph G whose vertices are the elements of the PTS, with two vertices connected by an edge when the corresponding pair appears in no triple of the PTS; G is the *strong leave* of the PTS. Every colour class in a strong colouring of the PTS induces a clique in G, and it follows that:

Lemma 18.24 *A strong colouring of a PTS is equivalent to finding a (vertex) partition of its strong leave into cliques.*

Thus the strong chromatic number of a PTS equals the minimum number of cliques in a clique partition of its strong leave. When the partial triple system is maximal, every clique in such a partition is a K_1 or a K_2. Thus, any partition of the strong leave of a maximal PTS(v, λ) into k cliques consists of $v - k$ edges (K_2) and $2k - v$ isolated vertices (K_1). Therefore the partition of the strong leave into the fewest cliques corresponds to a maximum matching of the strong leave.

Colbourn, Jungnickel, and Rosa (1988) use these simple observations to determine the spectrum of strong chromatic numbers for maximal PTS(v, λ)s, that is, the set

$$S^{(\lambda)}(v) = \{k : \exists PTS(v, \lambda) D \text{ with } \chi_s(D) = k\}.$$

Obviously, $v/2 \leq \chi_s \leq v$ for any maximal PTS(v, λ). Let $J(v) = \{v/2, \ldots, v\}$.

Theorem 18.25 *(Strong chromatic numbers)*

1. (a) $S^{(\lambda)}(3) = \{3\}$

   (b) $S^{(\lambda)}(4) = \begin{cases} \{3\} & \text{if } \lambda = 1 \\ \{3, 4\} & \text{if } \lambda \geq 2 \end{cases}$

   (c) $S^{(\lambda)}(5) = \begin{cases} \{3\} & \text{if } \lambda = 1 \\ \{3, 4\} & \text{if } \lambda = 2 \\ \{3, 4, 5\} & \text{if } \lambda \geq 3 \end{cases}$

   (d) $S^{(\lambda)}(6) = \begin{cases} \{3\} & \text{if } \lambda = 1 \\ \{3, 5, 6\} & \text{if } \lambda \geq 2 \end{cases}$

2. *for* $\lambda = 1$ *and* $v > 6$,

$$S^{(\lambda)}(v) = \begin{cases} J(v) \setminus \{v-2, v-1\} & \text{if } v \equiv 1, 3 \pmod 6 \\ J(v) \setminus \{v-1, v\} & \text{if } v \equiv 5 \pmod 6 \\ J(v) \setminus \{v-2, v-1, v\} & \text{if } v \equiv 0 \pmod 6 \\ J(v) \setminus \{v\} & \text{if } v \equiv 2, 4 \pmod 6 \end{cases}$$

3. *for* $\lambda \geq 2$ *and* $v > 6$, $S^{(\lambda)}(v) = J(v)$.

We do not present a complete proof, but illustrate the methods in one case. Let v be odd. In this case, to show that $v-t \in S^{(1)}(v)$, it suffices to show that there is a maximal PTS$(v,1)$ having a leave with a maximum matching of size t. When $v \equiv 1,3 \pmod 6$, Theorem 9.15 guarantees that the following are leaves:

$$\begin{array}{ll} C_{2t} & t \equiv 0 \pmod 3 \\ C_{2t+1} & t \equiv 1 \pmod 3,\ t \geq 4 \\ C_{2t-3} \cup C_5 & t \equiv 2 \pmod 3,\ t \geq 5 \end{array}$$

In each case, the maximum matching in the leave has size t. This leaves only $t \in \{1,2\}$; but no leave can have a maximum matching with one or two edges in this case.

When $v \equiv 5 \pmod 6$, we proceed similarly, again using Theorem 9.15, with leaves

$$\begin{array}{ll} C_{2t+1} & t \equiv 0 \pmod 3,\ t \geq 3 \\ C_{2t-3} \cup C_5 & t \equiv 1 \pmod 3,\ t \geq 4 \\ C_{2t} & t \equiv 2 \pmod 3 \end{array}$$

Again, $t \in \{0,1\}$ are impossible as sizes of maximum matchings in this case.

Similar constructions handle the remaining cases.

18.6 Strict colourings and the upper chromatic number

An interesting colouring concept was suggested by Voloshin (1995) via mixed hypergraphs. Restricted to Steiner triple systems, it amounts to the following. If $(V,\mathcal{B})$ is an STS, we have $\mathcal{B} = \mathcal{E} \cup \mathcal{A}$ where $\mathcal{E}$ is the set of *edges* and $\mathcal{A}$ is the set of *co-edges*. The sets $\mathcal{E}$ and $\mathcal{A}$ are not necessarily disjoint, and one of them may be empty. A *strict colouring* of $(V,\mathcal{E},\mathcal{A})$ is a colouring of the elements such that no edge is monochromatic and no co-edge is multicoloured (each co-edge has at least two elements of the same colour). The *upper (lower) chromatic number* $\overline{\chi}(\chi)$ of an STS is the maximum (minimum) number of colours in a strict colouring of $(V,\mathcal{E},\mathcal{A})$.

If all triples are edges and there are no co-edges (equivalently, $\mathcal{E} = \mathcal{B}$ and $\mathcal{A} = \varnothing$), the lower chromatic number of an STS(v) is the usual weak chromatic number while the upper chromatic number trivially equals v.

Two special cases are of interest. The first arises when all triples are co-edges, and there are no edges (equivalently, $\mathcal{E} = \varnothing$ and $\mathcal{A} = \mathcal{B}$). In this case we have a *co-STS*. The second arises when all triples are edges and also co-edges (equivalently, $\mathcal{E} = \mathcal{B}$ and $\mathcal{A} = \mathcal{B}$). In this case we have a *bi-STS*.

For co-STSs, monochromatic triples are allowed, so a strict colouring (say, with one colour) always exists. The lower chromatic number of a co-STS thus trivially equals one. Milazzo and Tuza (1997) obtain the following upper bound on the upper chromatic number of co-STSs.

Theorem 18.26 *In a co-STS(v) with $v \leq 2^k - 1$, k a positive integer, the upper chromatic number does not exceed k.*

This bound is best possible in the sense that there exist co-STSs of order $2^k - 1$ whose upper chromatic number equals k. Indeed, consider the projective STS$(2^k - 1)$ represented on the set of nonzero binary vectors of length k. Number the coordinates of these vectors from 1 to k, from right to left. Now assign an element colour i if the rightmost coordinate in which it has a '1' entry is the ith. This is a strict colouring of the co-STS from the projective system.

For bi-STSs, Example 18.27 displays a bi-STS(25) with a strict 3-colouring.

Example 18.27 A bi-STS(25) on $\mathbb{Z}_5 \times \{0,1,2,3,4\}$, with strict 3-colouring $\mathbb{Z} \times A_i$ $(i = 1,2,3)$, where $A_1 = \{0\}$, $A_2 = \{1,2\}$, and $A_3 = \{3,4\}$. Starter blocks are given, to be developed modulo 5.

$$\begin{array}{llll} 0_0,1_0,0_1 & 0_0,2_0,0_2 & 0_0,0_3,4_4 & 0_0,1_3,3_4 \\ 0_0,2_3,2_4 & 0_0,3_3,1_4 & 0_0,4_3,0_4 & 0_0,1_1,1_2 \\ 0_0,2_1,4_2 & 0_0,3_1,2_2 & 0_1,1_1,3_4 & 0_1,0_3,2_3 \\ 0_1,1_2,4_4 & 0_1,0_4,4_4 & 0_1,3_2,4_3 & 0_1,2_1,3_3 \\ 0_2,1_2,3_3 & 0_2,2_2,1_4 & 0_2,0_3,4_3 & 0_2,0_4,2_4 \end{array}$$

Milazzo and Tuza (1997) (see also Milazzo and Tuza, 1998) show that a strict colouring does not always exist. Such a bi-STS is *uncolourable.* A strong condition establishes the existence of numerous uncolourable bi-STSs:

Theorem 18.28 *Let $S = (V, \mathcal{B})$ be an STS(v) whose independence number satisfies $\beta(S) \le v/3$. Then the corresponding bi-STS is uncolourable.*

Proof Suppose there exists a strict colouring of our bi-STS(v) with k colours in which the colour classes have sizes $n_1, n_2, \ldots, n_k$ where $k \ge 3$, $\sum_{i=1}^{k} n_i = v$, and $n_i \le v/3$. Since each triple must contain two elements of the same colour but cannot be monochromatic, the number of triples must equal $\sum_{i=1}^{k} \binom{n_i}{2}$. When $n = 6t + 3$, the maximum value of this expression is attained when $k = 3$ and $n_1 = n_2 = n_3 = 2t + 1$. But then $\sum_{i=1}^{k} \binom{n_i}{2} = 3\binom{2t+1}{2}$ which is strictly less than $(3t + 1)(2t + 1)$, the number of triples in any STS$(6t + 3)$. The case $v = 6t + 1$ is handled similarly. □

The necessary condition for colourability of bi-STSs provided by Theorem 18.28 is not sufficient. For example, a bi-STS(15) is colourable if and only if it contains a sub-STS(7) (and the upper chromatic number of each colourable bi-STS(15) equals 4).

If we put $\overline{\chi} = 0$ for an uncolourable STS, then the upper bound of Theorem 18.26 is valid also for bi-STSs (or for any STS with $\mathcal{A} = \mathcal{B}$, for that matter).

Many further results on strict colourings and the upper chromatic number of Steiner triple systems can be found in Milazzo (1997) and Milazzo and Tuza (1997, 1998).

18.7 Achromatic number

A *complete m-colouring* of a (partial) design $\mathcal{D} = (V, \mathcal{B})$ is a weak m-colouring C of V with the additional property that for every two distinct colour classes C_i, C_j

there exists a block $B \in \mathcal{B}$ such that $B \subseteq C_i \cup C_j$. Alternatively, a weak colouring is complete if merging any two colour classes would force a monochromatic block. The maximum possible number of colours of a complete colouring of $\mathcal{D}$ is the *achromatic number* of $\mathcal{D}$, denoted $\psi(\mathcal{D})$.

An interpolation theorem for complete m-colourings is proved by Cockayne, Miller, and Prins (1972) (see also Tarski, 1975):

Theorem 18.29 *Given a design $\mathcal{D}$, there exists a complete m-colouring of $\mathcal{D}$ for all m, $\chi_w(\mathcal{D}) \leq m \leq \psi(\mathcal{D})$.*

Nešetřil, Phelps, and Rödl (1983) have established bounds for the maximum value $\psi_{\max}(v,k)$ of the achromatic number of a partial Steiner system $S(2,k,v)$:

Theorem 18.30 $C_1(k)v^{2/3} \leq \psi_{\max}(v,k) \leq C_2(k)v^{2/3}$.

For $k = 3$, the lower bound is attained by STSs, for v sufficiently large. Let $\overline{\psi}_{\min}(v,k)$ be the minimum value of the achromatic number for *maximal* partial $S(2,k,v)$s.

Theorem 18.31 $\overline{\psi}_{\min}(v,3) \geq c\sqrt{v}$.

Nešetřil, Phelps, and Rödl (1983) remark that a proof similar to theirs can be obtained to show $\overline{\psi}_{\min}(v,k) \to \infty$ as $n \to \infty$ for arbitrary k.

18.8 The complexity of strong and weak colouring

For a general discussion of algorithmic aspects of combinatorial design theory, we refer the reader to an excellent survey by Marlene Colbourn (1985). Part of her survey (§4) is devoted to colourings of block designs.

Phelps and Rödl (1984) prove:

Theorem 18.32 *Deciding whether a partial $S(2,k,v)$ is weakly m-colourable is NP-complete for any $m \geq 3$.*

This was proved earlier for partial triple systems by Colbourn, Colbourn, Phelps, and Rödl (1984). The proof of Theorem 18.32 is based on two lemmas by Phelps and Rödl (1984).

Lemma 18.33 *For each $m \geq 2$, there exists an m-chromatic partial $S(2,k,v)$ for which any m-colouring assigns the same colour to all elements of a particular fixed block.*

Lemma 18.34 *For each $m \geq 2$, there exists an m-chromatic partial $S(2,k,v)$ such that for some fixed pair of vertices x, x' any weak m-colouring assigns different colours to x and x'.*

What follows was one of the first examples of an algorithmic problem in design theory proved to be NP-complete, by Colbourn, Colbourn, Phelps, and Rödl (1982).

Theorem 18.35 *Deciding whether a (v,k,λ)-design is weakly m-colourable is NP-complete for all $m \geq 9$.*

The proof uses the result of Theorem 18.32, together with Theorem 11.5; from a partial STS(v), in polynomial time a TS($18tv+3, 12tv+3$) is constructed which is $3m$-colourable if and only if the partial STS(v) is m-colourable.

The result of Theorem 18.35 is established for designs with relatively large λ, and hence has no consequences for designs with small λ. However, the following theorem shows that a restriction to small fixed λ does not necessarily result in an improvement.

Theorem 18.36 *(Phelps and Rödl, 1984) Deciding whether an STS is weakly 14-colourable is NP-complete.*

Theorem 18.37 *Determining the weak chromatic number of STSs is NP-hard.*

Turning to strong colouring, Colbourn, Jungnickel, and Rosa (1988) prove:

Theorem 18.38 *Deciding whether a partial STS has a strong m-colouring is NP-complete for any fixed $m \geq 3$.*

The proof uses a reduction from the problem of m-edge-colouring for ordinary graphs (which Holyer (1981) shows is NP-complete) to the problem of strong m-colouring for partial STSs.

On the other hand, Colbourn, Jungnickel, and Rosa (1988) show:

Theorem 18.39 *The strong chromatic number of maximal partial triple systems can be computed in polynomial time.*

Indeed, a strong m-colouring of a PTS corresponds to a partition of its leave into m cliques. If the PTS is maximal, every clique in such a partition is either K_1 or K_2. The smallest partition into cliques therefore corresponds to a maximum matching in the leave, and hence can be found efficiently (see, for example, Lovász and Plummer, 1986).

18.9 Work points

1. Does there exist a weakly 4-chromatic STS(19)?
2. Prove or disprove: $\chi_w(S_n^*) = n$ for $n \geq 3$.
3. Let $C(v) = \{m : \exists$ weakly m-chromatic STS$(v)\}$. Is $C(v)$ an interval?
4. Does every weakly 3-chromatic STS(v) admit an equitable weak 3-colouring?
5. Characterize the uncolourable bi-STSs.

19

CHROMATIC INDEX AND RESOLVABILITY

19.1 Partial parallel classes

In a triple system TS(v, λ) $(V, \mathcal{B})$, any set of pairwise disjoint triples is a *partial parallel class* (PPC). A partial parallel class that partitions V is a *parallel class* (PC), and a partial parallel class that partitions $V \setminus \{x\}$ for some element x is an *almost parallel class* (APC). A PPC $\mathcal{P}$ is *maximal* if for any triple $B \in \mathcal{B} \setminus \mathcal{P}$, there exists a triple $B' \in \mathcal{P}$ for which $B \cap B' \neq \varnothing$. A PPC $\mathcal{P}$ is *maximum* if $|\mathcal{P}| \geq |\mathcal{Q}|$ for any PPC $\mathcal{Q}$; thus a maximum PPC is a PPC having the largest possible number of triples for the triple system.

Evidently, $|\mathcal{P}| \leq \lfloor v/3 \rfloor$ for any TS(v, λ). For example, Bose's construction (Lemma 2.3) always produces an STS having a parallel class. In fact, the following is not difficult to prove:

Theorem 19.1 *A TS(v, λ) having a maximum PPC of size $\lfloor v/3 \rfloor$ exists for every admissible pair (v, λ) except for $(6, 2)$ and $(7, 1)$.*

Proof This holds trivially when $v \leq 5$. Any two triples of the unique TS$(6, 2)$ intersect, and hence the maximum PPC contains only one triple. The same statement holds for the unique TS$(7, 1)$. On the other hand, it is easy to find TS$(6, \lambda)$s with $\lambda \geq 4$, and TS$(7, \lambda)$s with $\lambda \geq 2$, having two disjoint triples. For $v \geq 8$, it suffices to consider TS(v, λ)s with $\lambda = \lambda_{\min}$. The direct constructions given in Lemmas 2.3, 2.4, and 2.9 produce TS(v, λ)s with a PPC of size $\lfloor v/3 \rfloor$ when $v \equiv 3 \pmod 6$ and $\lambda = 1$, $v \equiv 0, 4 \pmod 6$ and $\lambda = 2$, $v \equiv 5 \pmod 6$ and $\lambda = 3$, and $v \equiv 2 \pmod 6$ and $\lambda = 6$. Theorem 19.44 (on Hanani triple systems) yields such an STS(v) when $v \equiv 1 \pmod 6$. The proof is completed by observing that both of the nonisomorphic STS(13)s contain an APC (of size four). □

Determining the size of a maximum PPC in an arbitrary TS(v, λ) appears to be a more difficult question. The first nontrivial bound on the size of a maximum PPC in STSs was established by Lindner and Phelps (1978); it says that every STS(v) contains a PPC of size at least $(v-1)/4$ for v sufficiently large. This was improved upon by Woolbright (1980); for STSs, his result gives a bound of $(3v/10) - 7$. Brouwer (1981) made a significant advance:

Theorem 19.2 *Any STS(v) with v sufficiently large contains a PPC of size* $(v - 5v^{2/3})/3$.

Asymptotically, the best current bound on the size of a maximum PPC in an STS(v) is due to Alon, Kim, and Spencer (1997):

Theorem 19.3 *Any STS(v) contains a PPC of size* $(v/3) - cv^{1/2}(\ln v)^{3/2}$ *for c an absolute constant.*

Brouwer (1981) conjectures that every STS(v) contains a PPC of size $(v-c)/3$ for c an absolute constant (based on information currently known, c could be as small as 4). Examples of STS(v)s without a single parallel class are known for $v = 15$ and 21. Ten of the 80 STS(15)s have no parallel class (see Tables 5.8–5.12); for an example of an STS(21) with no parallel class, see Mathon and Rosa (1981). Bennett and others have stated:

Conjecture 19.4 *For all* $v \equiv 3 \pmod 6$ *with* $v \geq 15$, *an STS(v) without a parallel class exists.*

Conjecture 19.4 is evidently open. Similarly, examples of STS(v)s with $v \equiv 1 \pmod 6$ with no APC are known to exist for $v = 7$ and 19 (see Lo Faro, 1985). Rosa and Colbourn (1982) state:

Conjecture 19.5 *An STS with no APC exists for all* $v \equiv 1 \pmod 6$ *except for* $v = 13$.

Conjecture 19.5 is also open. The only general result in this direction is due to Wilson (see Rosa and Colbourn, 1992), and gives an infinite family of STS(v)s with no APC.

Theorem 19.6 *The projective triple system of order* $2^{n+1} - 1$, *arising from the points and lines of an even-dimensional projective space* $PG(n, 2)$, *contains no APC.*

Proof The number of elements satisfies $2^{n+1} - 1 \equiv 1 \pmod 6$ only when n is even. The main observation is that in any STS($2v+1$) having a sub-STS(v), no APC can miss an element x not in the subsystem. Indeed, every triple contains either zero or two elements not in the subsystem, and hence any PPC misses an even number of elements not in the subsystem. Since an APC misses only one element in total, it must contain all elements not in the subsystem. In the projective triple system on $2^{n+1} - 1$ elements, for every element x, there is a projective subsystem on $2^n - 1$ elements that misses x. Since no element can be missed by an APC, there is no APC. □

Remarkably, all known examples of STSs whose maximum PPC has size less than $\lfloor v/3 \rfloor$ do have PPCs of size $\lfloor v/3 \rfloor - 1$.

19.2 Resolvable triple systems

A TS(v, λ) $(V, \mathcal{B})$ is *resolvable* if there exists a partition of the triples of $\mathcal{B}$ into parallel classes. Any such partition is a *resolution*. Since $v \equiv 0 \pmod 3$ is a necessary condition for the presence of a parallel class in a TS(v, λ), in order to consider the existence of resolvable TS(v, λ)s, it suffices to consider the case $\lambda = 1$ and $v \equiv 3 \pmod 6$ and the case $\lambda = 2$ and $v \equiv 0 \pmod 6$.

If $\mathcal{R} = \{R_1, \ldots, R_r\}$ is a resolution of an STS(v) $(V, \mathcal{B})$ then the triple $(V, \mathcal{B}, \mathcal{R})$ is a *Kirkman triple system.* Thus a Kirkman triple system (KTS) is a Steiner triple system together with some resolution of it; an STS may admit several resolutions, and hence be the underlying system of many KTSs.

Again, it was the Reverend Kirkman who was the first to study what we today call Kirkman triple systems. In 1850, the following problem, posed by Kirkman, appeared as Query 6 in the *Lady's and Gentleman's Diary*:

Fifteen young ladies in a school walk out three abreast for seven days in succession: it is required to arrange them daily, so that no two will walk twice abreast.

A solution is provided, of course, by a Kirkman triple system of order 15. Earliest solutions presented were by both Kirkman (1850) and Cayley (1850). The solution given in Example 19.7 is that of Cayley.

Example 19.7 A Kirkman triple system

$\mathcal{R}$:	abc	d35	e17	f28	g46
	ade	b26	c48	f15	g37
	afg	b13	c57	d68	e24
	bdf	a47	c16	e38	g25
	beg	a58	c23	d14	f67
	cdg	a12	b78	e56	f34
	cef	a36	b45	d27	g18

Example 19.8 A Kirkman triple system of order 21. Let $V = \mathbb{Z}_7 \times \mathbb{Z}_3$. Form a base parallel class

$$\{1_0, 2_0, 4_0\}, \{3_0, 1_1, 0_2\}, \{5_0, 2_1, 4_2\}, \{6_0, 6_1, 6_2\},$$

$$\{0_0, 3_1, 1_2\}, \{0_1, 4_1, 5_1\}, \{2_2, 3_2, 5_2\}$$

and develop modulo $(7, -)$ to produce seven parallel classes. Develop $\{1_0, 2_1, 3_2\}$ modulo $(7, -)$ to produce a parallel class, $\{1_0, 3_1, 6_2\}$ to produce another, and $\{1_0, 0_1, 4_2\}$ to produce the last.

Of course, we encountered an example of a KTS(9) already in Figure 2.1 (the unique STS(9) shown there has a unique resolution).

An effort, at times extensive, spanning 120 years culminated in the proof of the existence of KTS(v) for all $v \equiv 3 \pmod 6$ by Ray-Chaudhuri and Wilson (1971). Remarkably, in relative isolation, Lu (1965) had also established this theorem, but it was not published for a variety of reasons until his collected works were published posthumously in 1990. A good account of the early history of Kirkman triple systems is given by Ahrens (1901), and a good bibliography of early works can be found in Eckenstein (1912) (and again in Ray-Chaudhuri and Wilson, 1971).

Theorem 19.9 *A KTS(v) exists if and only if* $v \equiv 3 \pmod 6$.

Two direct constructions for KTSs, due to Ray-Chaudhuri and Wilson (1971), apply only to prime powers and thus do not suffice to prove Theorem 19.9. Nevertheless, they are both important and elegant:

Theorem 19.10 *If $q = p^\alpha \equiv 1 \pmod 6$ is a prime power, then there exists a KTS$(3q)$ and a KTS$(2q+1)$.*

Proof Let $t = (q-1)/6$. To construct a KTS$(3q)$, take as elements $\mathbb{F}_q \times \{1,2,3\}$. Let ω be a primitive element in $\mathbb{F}_q$, and let $\mathcal{B}$ consist of triples:

1. $C = \{0_1, 0_2, 0_3\}$;
2. $B_{ij} = \{\omega^i_j, \omega^{i+2t}_j, \omega^{i+4t}_j\}$, $0 \le i < t$, $j \in \{1,2,3\}$;
3. $A_i = \{\omega^i_1, \omega^{i+2t}_2, \omega^{i+4t}_3\}$, $0 \le i < t$.

It is readily verified that each of the (nonzero) pure and mixed differences occurs exactly once in triples of $\mathcal{B}$, and thus $\mathcal{B}$ is the set of starter triples for an STS$(3q)$. This STS$(3q)$ is resolvable. Indeed, $R_0 = C \cup \{B_{ij} : 0 \le i < t, j \in \{1,2,3\}\} \cup \{A_{jt+i} : j \in \{1,3,5\}, 0 \le i < t\}$ forms a parallel class; when developed modulo $6t+1$, it yields a further $6t$ parallel classes. Each A_i, when developed modulo $6t+1$, also yields a parallel class; taking those parallel classes only from A_{jt+i} with $j \in \{0,2,4\}$ and $0 \le i < t\}$ thus yields a further $3t$ parallel classes, for a total of $9t+1$ forming the resolution.

To construct a KTS$(2q+1)$, take as elements $(\mathbb{F}_q \times \{1,2\}) \cup \{\infty\}$. Let ω be a primitive element of $\mathbb{F}_q$, and let m satisfy $2\omega^m = \omega^t + 1$. Let $\mathcal{B}$ consist of triples

1. $C = \{0_1, 0_2, \infty\}$;
2. $B_i = \{\omega^i_1, \omega_1 i + t, \omega^{i+m}_2\}$, $0 \le i < t$, $2t \le i < 3t$, or $4t \le i < 5t$;
3. $A_i = \{\omega^{i+m}_2, \omega^{i+m+3t}_2, \omega^{i+m+5t}_2\}$, $0 \le i < t$.

Again, it is straightforward that every pure and every mixed difference occurs exactly once and hence $\mathcal{B}$ is a set of starter triples for an STS$(2q+1)$. But $\mathcal{B}$ itself forms a parallel class, whose development modulo q yields the required q parallel classes for the KTS$(2q+1)$. □

The proof of Theorem 19.9 rests on a simple but important construction:

Lemma 19.11 *If $(V, \mathcal{B})$ is a PBD$(v, K, 1)$ such that for each $k \in K$ there exists a KTS$(2k+1)$, then there exists a KTS$(2v+1)$.*

Proof Put $V^\star = (V \times \{1,2\}) \cup \{\infty\}$. For each $B \in \mathcal{B}$, let $V_B = (B \times \{1,2\}) \cup \{\infty\}$. For each $B \in \mathcal{B}$, let $(V_B, \mathcal{C}_B, \mathcal{R}_B)$ be a KTS$(2|B|+1)$ such that $B_a = \{\infty, a_1, a_2\} \in \mathcal{C}_B$ for each $a \in B$. Let $\mathcal{R}_B(a)$ be the unique parallel class of $\mathcal{R}_B$ that contains a. Let now $\mathcal{B}^\star = \bigcup_{a \in V, B \in \mathcal{B}} \mathcal{R}_B(a)$, $\mathcal{R}^\star(a) = \bigcup_{B \ni a} \mathcal{R}_B(a)$, and $\mathcal{R}^\star = \{\mathcal{R}^\star(a) : a \in V\}$. Then $(V^\star, \mathcal{B}^\star, \mathcal{R}^\star)$ is a KTS$(2v+1)$. □

Proof of Theorem 19.9: Hanani (1961) establishes the existence of an $S(2,4,v)$ for all $v \equiv 1, 4 \pmod{12}$, and Brouwer (1979) establishes the existence of a $\{4, 7^*\}$-PBD for all $v \equiv 7, 10 \pmod{12}$ except for $v = 10$ or 19. Since a KTS(9) and a KTS(15) exist, Lemma 19.11 yields a KTS(v) for all $v \equiv 3 \pmod 6$ with the exception of $v = 21$ and $v = 39$. Both are constructed by Theorem 19.10 (a KTS(21) is also given in Example 19.8). □

Other proofs of Theorem 19.9 are given in Ray-Chaudhuri and Wilson (1971), Beth, Jungnickel, and Lenz (1985), Street and Street (1987), Wallis (1988), Anderson (1990), and Stinson (1991).

The existence problem for resolvable TS$(v, 2)$s with $v \equiv 0 \pmod 6$ was settled by Hanani (1974):

Theorem 19.12 *A resolvable TS$(v, 2)$ exists if and only if $v \equiv 0 \pmod 3$.*

Theorem 19.12 follows also from Theorem 25.33 on the existence of resolvable Mendelsohn triple systems.

19.3 Kirkman frames and resolvable 3-GDDs

While Kirkman triple systems have a long history, the generalization to Kirkman frames is a subject more recently explored. Stinson (1987) introduces them explicitly, although they are implicit in earlier work by Ray-Chaudhuri and Wilson (1971) and Hanani (1976). Kirkman frames have been used widely in the construction of KTSs with subsystems, and in the generalization of KTSs to resolvable 3-GDDs.

19.3.1 *Frames*

A *holey parallel class* or *frame parallel class* in a 3-GDD $(V, \mathcal{G}, \mathcal{B})$ is a partial parallel class that partitions the set $V \setminus G$ for some $G \in \mathcal{G}$; G is the *hole* missing from the partial parallel class. If the set $\mathcal{B}$ of triples can be partitioned into a set $\mathcal{P}$ of holey parallel classes, then $(V, \mathcal{G}, \mathcal{B}, \mathcal{P})$ is a *3-frame* or a *Kirkman frame* (or simply a *frame*). Its *type* is the type of the underlying 3-GDD. A frame of type t^u (that is, one having all groups of the same size) is *uniform*. These definitions extend in the natural way to 3-GDDs of index λ; in this case, the frame is a $(3, \lambda)$*-frame*.

Example 19.13 A uniform Kirkman frame of type 2^4.
Holes: $\{\{1,5\},\{2,4\},\{3,6\},\{7,8\}\}$;
Holey parallel classes: 267 348; 168 357; 147 258; 123 456.

Example 19.14 A uniform Kirkman frame of type 4^4.
Holes: $\{\{1,2,3,4\},\{5,6,7,8\},\{9,10,11,12\},\{13,14,15,16\}\}$;
Holey parallel classes:

5 9 14	6 12 16	7 10 15	8 11 13
5 11 15	6 10 13	7 12 14	8 9 16
1 9 15	2 11 16	3 12 13	4 10 14
1 10 16	2 12 15	3 11 14	4 9 13
1 5 13	2 8 14	3 6 15	4 7 16
1 6 14	2 7 13	3 5 16	4 8 15
1 7 11	2 6 9	3 8 10	4 5 12
1 8 12	2 5 10	3 7 9	4 6 11

An obvious necessary condition for the existence of a frame is that $|V\backslash G| \equiv 0 \pmod 3$ for each $G \in \mathcal{G}$. Another, perhaps less obvious, necessary condition can be established (as in Stinson, 1987):

Theorem 19.15 *If $(V, \mathcal{G}, \mathcal{B}, \mathcal{P})$ is a Kirkman frame, then for any $G \in \mathcal{G}$ there are exactly $|G|/2$ frame parallel classes in $\mathcal{P}$ for which G is the hole.*

Corollary 19.16 *A uniform Kirkman frame of type t^u exists only if t is even, $u \geq 4$, and $t(u-1) \equiv 0 \pmod 3$ (that is, $u \geq 4$ and if $t \equiv 2, 4 \pmod 6$ then $u \equiv 1 \pmod 3$, and if $t \equiv 0 \pmod 6$ then there is no further restriction on u).*

In order to prove that the necessary condition for the existence of Kirkman frames is also sufficient, a basic construction is quite useful:

Theorem 19.17 *(Fundamental Frame Construction) Let $(V, \mathcal{G}, \mathcal{B})$ be a GDD, and let $w : V \mapsto N$ be a weight function from elements to nonnegative integers. If for every block $T \in \mathcal{B}$, there exists a frame of type $\{w(x) : x \in T\}$, then there exists a frame of type $\{\sum_{x \in G} w(x) : G \in \mathcal{G}\}$.*

Another simple recursive construction for frames is given by the following.

Theorem 19.18 *If there is a frame of type T, and there exists a pair of mutually orthogonal latin squares of side n, then there exists a frame of type $\{nt : t \in T\}$.*

Stinson (1987) proved the following:

Theorem 19.19 *A uniform Kirkman frame of type t^u exists if and only if t is even, $u \geq 4$, and $t(u-1) \equiv 0 \pmod 3$.*

Proof Necessity is established by Corollary 19.16. For sufficiency, first observe that deleting a single element from a Kirkman triple system of order $2v+1$ gives a Kirkman frame of type 2^v, and hence a Kirkman frame of type 2^u exists for all $u \equiv 1 \pmod 3$. Further, for every $u \equiv 1 \pmod 3$, we obtain a Kirkman frame of type 4^u as follows. Brouwer, Schrijver, and Hanani (1977) establish that a 4-GDD of type 2^u exists for $u \equiv 1 \pmod 3$, $u \geq 7$. Apply the Fundamental Frame Construction, giving each element weight $w = 2$ and using the frame of type 2^4 in Example 19.13.

Proceeding now to frames of type 6^u, Stinson (1987) gives a frame of type 6^6. If $u \neq 6$, form a GDD of type 3^u having all blocks of size 4, or all blocks of size 4 and one of size 7 (these are obtained by removing a point from an $S(2, 4, 3u+1)$ if $u \equiv 0, 1 \pmod 4$, or removing a point not on the 7-block in a $\{4, 7^*\}$-PBD on $3u+1$ points if $u \equiv 2, 3 \pmod 4$). Now apply the Fundamental Frame Construction giving every element weight $w = 2$ and using the frame of type 2^4. Frames of type 12^u for $u \neq 6$ are produced similarly using weight 4, and the frame of type 4^4 in Example 19.14. A frame of type 12^6 is obtained by applying the Fundamental Frame Construction using weight 2 to a 4-GDD of type 6^6, constructed by Brouwer, Schrijver, and Hanani (1977).

To complete the proof, let t be even, $t \notin \{2,4,6,12\}$. If $t \equiv 0 \pmod 6$, write either $t = 6t_1$ or $12t_1$ so that $t_1 \neq 2,6$; if $t \equiv 2,4 \pmod 6$, write $t = 2t_1$ (again, $t_1 \neq 2,6$). In either case, apply Theorem 19.17 with $n = t_1$. □

Assaf and Hartman (1989) proved the analogous result for higher indices:

Theorem 19.20 *A uniform Kirkman frame of type t^u and index λ exists if and only if $t\lambda$ is even, $t(u-1) \equiv 0 \pmod 3$, and $u \geq 4$.*

19.3.2 *Reverse frames and an application*

Kreher, Stinson, and Zhu (1997) explore the maximum number of fixed points that an automorphism of prime order of a $(v,k,1)$-design can have. Camina and Mischke (1996) give a short proof, establishing that a $(3r+1,4,1)$-design having an automorphism of order 2 can have at most $r+1$ fixed points under such an automorphism. Kreher, Stinson, and Zhu (1997) determine the necessary conditions that $v \equiv 1,4,13,28 \pmod{36}$ (i.e. $r \equiv 0,1,4,9 \pmod{12}$) in this case. They further establish that when $v \equiv 4,13 \pmod{36}$, the design exists. We consider a somewhat different proof here.

Wilson (unpublished) observed that the complete directed graph on $r \equiv 1,4 \pmod{12}$ vertices, with one red arc, one white arc, and one blue arc from each vertex to each other vertex can be partitioned into subdigraphs, each isomorphic to one of the following:

digraph	# 1	# 2
red arcs	(1,2), (1,3), (1,4)	(1,2), (1,3), (1,4)
white arcs	(2,3), (2,4), (3,4), (3,2), (4,2), (4,3)	(3,4), (4,3)
blue arcs		(2,3), (2,4), (3,2), (4,2)

Since a partition is easily produced when $r = 4$, one employs an $(r,4,1)$-design and places the solution for $r = 4$ on each block. This in turn yields a frame:

Lemma 19.21 *When $n \equiv 1,4 \pmod{12}$, there is a frame of type 2^r which has a reverse automorphism (which exchanges the two elements within each group).*

Proof Form the partition of the edge-coloured directed complete graph on vertex set X into copies of coloured digraphs # 1 and # 2. The frame is formed as follows on $X \times \{0,1\}$. Whenever a coloured directed graph on $\{a,b,c,d\} \subset X$ is contained in the partition which is isomorphic to graph # 1 under the mapping $a \mapsto 1$, $b \mapsto 2$, $c \mapsto 3$, and $d \mapsto 4$, then place the triples $\{\{b_0,c_0,d_0\},\{b_1,c_1,d_1\}\}$ in the frame parallel class for group $\{a_0,a_1\}$. Similarly, for each coloured directed graph in the partition which is isomorphic to graph # 2, place the triples $\{\{b_1,c_0,d_0\},\{b_0,c_1,d_1\}\}$ in the frame parallel class for group $\{a_0,a_1\}$. □

The reverse frames from Lemma 19.21 allow one to construct the desired $(3r+1,4,1)$-design:

Theorem 19.22 *A $(3r+1, 4, 1)$-design having an automorphism of order 2 with $r+1$ fixed points exists whenever $r \equiv 1, 4 \pmod{12}$.*

Proof Form a reverse frame on $X \times \{0, 1\}$ using Lemma 19.21. Append the points $X \times \{2\}$, adding the point x_2 to the group containing x_0 and x_1. For each triple of the frame, append the element y_2 to the triple if it belongs to the frame parallel class for the group $\{y_0, y_1\}$. Next place an $(r, 4, 1)$-design on the points $X \times \{2\}$ arbitrarily. Finally, add a single infinite point and adjoin it to each group to form blocks. The automorphism of order 2 exchanges x_0 and x_1 for each $x \in X$, and leaves all other elements fixed. □

This provides a useful application of frames to the construction of designs with larger block size.

19.3.3 *Nearly Kirkman triple systems*

Kotzig and Rosa (1974) proposed an analogue of the Kirkman problem for orders which are multiples of 6. Essentially, they asked when a 3-GDD of type 2^{3n} exists whose triples can be partitioned into $3n-1$ parallel classes. This object is a *nearly Kirkman triple system*, or *NKTS*. No NKTS of type 2^3 or 2^6 exists, but Kotzig and Rosa (1974) produced an NKTS of type 2^9. Its eight parallel classes can be produced on the elements $\{1, \ldots, 9, \hat{1}, \ldots, \hat{9}\}$. For each column in the table

$$\begin{array}{cccccccc}
147 & 159 & 12\hat{6} & 13\hat{5} & 16\hat{3} & 18\hat{7} & 89\hat{1} & 24\hat{1} \\
258 & 267 & 45\hat{9} & 46\hat{8} & 29\hat{8} & 35\hat{2} & 23\hat{4} & 68\hat{5} \\
369 & 348 & 78\hat{3} & 79\hat{2} & 57\hat{4} & 49\hat{6} & 56\hat{7} & 37\hat{9}
\end{array}$$

form a parallel class by adjoining the three triples obtained by applying the automorphism interchanging i and $\hat{i}$ for $1 \le i \le 9$.

Kotzig and Rosa (1974) gave a general construction of NKTSs from KTSs:

Theorem 19.23 *If $r \equiv 1 \pmod 3$, $r \ge 4$, and $s \equiv 1 \pmod 2$, then there exists an NKTS of type 2^{3rs}.*

Proof Let $r = 3p+1$ and $s = 2q-1$. Let $R = \{1, \ldots, 2r\}$ and $S = \{1, \ldots, 3s\}$. We form the NKTS on $R \times S$. On $R \cup \{\infty\}$, there exists a Kirkman triple system with parallel classes $P_1, \ldots, P_r$. For each parallel class, let $\{a_i, b_i\}$ be the pair for which $\{\infty, a_i, b_i\} \in P_i$, and let $\tilde{P}_i = P_i \setminus \{\infty, a_i, b_i\}$.

Now form a Kirkman triple system of order $3s$ on S, and give it weight $2r$ to obtain a resolvable 3-GDD of type $(2r)^{3s}$. This can be done so that $P = \{\{(x, j), (x, j+s), (x, j+2s)\} : x \in R, 1 \le j \le s\}$ forms one of the parallel classes. Now retain all $r(3s-1)-1$ parallel classes of the RGDD except for P. Form a further parallel class H_i by placing the triples of $\tilde{P}_i$ on $R \times \{j\}$ for each $j \in S$, and adjoining the triples $\{\{(x, j), (x, j+s), (x, j+2s)\} : x \in \{a_i, b_i\}, 1 \le j \le s\}$. Forming H_i for $1 \le i \le r$ gives $3rs-1$ parallel classes altogether, completing the NKTS. □

Corollary 19.24 *An NKTS of type 2^{9x+3} exists for all $x \ge 1$.*

Proof Apply Theorem 19.23 with $s = 1$. □

Baker and Wilson (1977) settled existence of NKTSs with three possible exceptions: 2^{42}, 2^{51}, and 2^{87}. Their main construction follows:

Lemma 19.25 *If there exists a K-GDD of type* $g_1^{u_1} \cdots g_s^{u_s}$ *so that*

1. *an NKTS of type* 2^{g_i+1} *exists for* $1 \le i \le s$, *and*
2. *a KTS*$(2k+1)$ *exists for every* $k \in K$,

then an NKTS of type $2^{1+\sum_{i=1}^{s} u_i g_i}$ *exists.*

Proof Let X be the point set of the K-GDD. We form the NKTS on $V = (X \cup \{\infty\}) \times \{0,1\}$, with groups $\{i\} \times \{0,1\}$ for $i \in X \cup \{\infty\}$. For each group G of the K-GDD, place an NKTS of type $2^{|G|+1}$ on $(G \cup \{\infty\}) \times \{0,1\}$, aligning the groups of the smaller NKTS on those of the larger. Index parallel classes of the NKTS of type $2^{|G|+1}$ arbitrarily using the elements $y \in G$. For each block B of the K-GDD, form a KTS$(2|B|+1)$ on $(B \times \{0,1\}) \cup \{\sigma\}$, where $\sigma \notin B$, so that $\{\sigma, (y,0), (y,1)\}$ is a block of the KTS chosen, for each $y \in B$. The parallel class containing triple $\{\sigma, (y,0), (y,1)\}$ is used to form a parallel class of triples on $(B \setminus \{y\}) \times \{0,1\}$, indexed by y and obtained by deleting the triple $\{\sigma, (y,0), (y,1)\}$.

The resulting set of triples is a 3-GDD of type $2^{1+\sum_{i=1}^{s} u_i g_i}$, and its resolution is obtained by forming the union of all parallel classes in the ingredients which are indexed by y to obtain a parallel class, with each element y of the K-GDD yielding one parallel class. □

Baker and Wilson (1977) apply this lemma by using a TD$(10, m)$ to form a $\{4,10\}$-GDD on $10m + 3m' + 3m''$ points with group sizes from $\{m, m+3m', m+3m''\}$. When $m \equiv 2 \pmod 3$, the lemma can then be applied. They provide further small examples needed in this recursive method. Brouwer (1978) produced a 3-GDD of type $2^6 6^1$ having six parallel classes and two frame parallel classes missing the group of size 6. Together with frames of type 12^x, this establishes:

Theorem 19.26 *An NKTS of type* 2^{6x+3} *exists for all* $x \ge 4$.

The last possible exception, type 2^{42}, was settled by Rees and Stinson (1987), to complete the proof of the existence result:

Theorem 19.27 *An NKTS of type* 2^n *exists if and only if* $n \equiv 0 \pmod 3$ *and* $n \ge 9$.

The use of parallel classes and frame parallel classes in the lemma of Baker and Wilson (1977) deserves particular mention as an elegant example of filling in holes to produce resolvable designs.

19.3.4 *Resolvable 3-GDDs*

Let us now turn to resolvable 3-GDDs (or 3-RGDDs) in general. In such a design, all groups must have the same size and so the type is of the form g^u. Necessarily, $u \ge 3$ and $g(u-1) \equiv 0 \pmod 2$ for the 3-GDD to exist; and for the design to be resolvable, $gu \equiv 0 \pmod 3$. Summarizing these conditions, we have:

Lemma 19.28 *If a 3-RGDD of type g^u exists then $u \geq 3$ and*

1. *if $g \equiv 1, 5 \pmod 6$ then $u \equiv 3 \pmod 6$;*
2. *if $g \equiv 3 \pmod 6$ then $u \equiv 1 \pmod 2$;*
3. *if $g \equiv 2, 4 \pmod 6$ then $u \equiv 0 \pmod 3$; and*
4. *if $g \equiv 0 \pmod 6$ then there are no further restrictions on u.*

Theorem 19.9 settles the existence of 3-RGDDs with group size 1, and Theorem 19.27 settles group size 2. Rees and Stinson (1987) undertook the first systematic study of RGDDs for arbitrary group sizes. Their main tool is a frame construction:

Theorem 19.29 *If there exists a Kirkman frame of type $t_1^{u_1} \cdots t_j^{u_j}$ where $t_i \equiv 0 \pmod t$ for $1 \leq i \leq j$, and there exists a 3-RGDD of type $t^{1+(t_i/t)}$ for $1 \leq i \leq j$, then there exists a 3-RGDD of type $t^{1+\sum_{i=1}^{j} t_i u_i / t}$.*

In addition, there are two simple inflation constructions:

Theorem 19.30 *If there exist 3-RGDDs of type g^u and h^3, then there exists a 3-RGDD of type $(gh)^u$.*

Theorem 19.31 *If there exist 3-RGDDs of type g^u, w^3, and w^g, then there exists a 3-RGDD of type $(3w)^{gu/3}$.*

One basic ingredient in these constructions consists of 3-RGDDs of type g^3:

Theorem 19.32 *For g a positive integer, a 3-RGDD of type g^3 exists if and only if $g \notin \{2, 6\}$.*

Proof A 3-RGDD of type g^3 is equivalent to a TD(4,g). □

It follows from Theorems 19.9 and 19.30 that a 3-RGDD of type g^u with $g = 1, 5 \pmod 6$ exists if and only if $u \equiv 3 \pmod 6$. Similarly, from Theorems 19.27 and 19.30, a 3-RGDD of type g^u exists whenever $g \equiv 2, 4 \pmod 6$, $u \equiv 0 \pmod 3$, and $u \geq 9$, except when $g = 4$. When $g = 4$, if $u \equiv 3 \pmod 6$, we can apply Theorems 19.9 and 19.30. Rees and Stinson (1987) exhibit 3-RGDDs of types 4^6, 4^{12}, and 4^{18}, and use Theorem 19.17 to treat the remaining cases with group size 4. What remains is then cases with $g \equiv 2, 10 \pmod{12}$ and $u = 6$.

For $g \equiv 3 \pmod 6$, first consider $g = 3$. If we take the blocks of a KTS of order $3n$ as groups, we obtain a 3-RGDD of type 3^n, and hence (by Theorem 19.9) these exist for all odd values of n. Applying Theorem 19.30 handles all larger group sizes.

Group sizes which are multiples of six pose the largest subproblem. First consider $g = 6$. If $u \equiv 0 \pmod 3$ and $u \geq 9$, use Theorem 19.30 to give weight 3 to a 3-RGDD of type 2^u. Rees and Stinson (1987) give a 3-RGDD of type 6^6, to complete the cases when $u \equiv 0 \pmod 3$ (a solution in the case when $u = 3$ does not exist). They also give a 3-RGDD of type 6^4. Using a frame of type $18^{(u-1)/3}$ and this 3-RGDD in Theorem 19.29 handles all cases with $u \equiv 1 \pmod 3$ for $u \geq 13$. They handle the remaining cases ($u = 7, 10$) when $u \equiv 1 \pmod 3$ by direct constructions. It remains to treat cases when $u \equiv 2 \pmod 3$. These are settled by Rees and Stinson (1987) and Assaf and Hartman (1989).

Rees (1993) settled all of the remaining cases, treating the case of six groups. In summary, the existence statement for 3-RGDDs of index one is:

Theorem 19.33 *A 3-RGDD of type g^u exists if and only if $u \geq 3$ and*

1. $g \equiv 1, 5 \pmod 6$ *and* $u \equiv 3 \pmod 6$*;*
2. $g \equiv 3 \pmod 6$ *and* $u \equiv 1 \pmod 2$*;*
3. $g \equiv 2, 4 \pmod 6$ *and* $u \equiv 0 \pmod 3$*, except for* $g^u \in \{2^3, 2^6\}$*; and*
4. $g \equiv 0 \pmod 6$*, except for* $g^u = 6^3$*.*

Assaf and Hartman (1989) further considered the existence of 3-RGDDs with higher index, settling all cases except some with exactly six groups. Together with results of Rees (1993), this more general existence result is also completed:

Theorem 19.34 *A 3-RGDD of type g^u and index λ exists if and only if $u \geq 3$, $\lambda g(u-1) \equiv 0 \pmod 2$, and $gu \equiv 0 \pmod 3$, except when $\lambda = 1$ and $g^u \in \{2^6, 6^3\}$, when $\lambda \equiv 1 \pmod 2$ and $g^u = 2^3$, and when $\lambda \equiv 2 \pmod 4$ and $g^u = 1^6$.*

19.3.5 *Kirkman school project designs*

A *partial PBD* of order v with block sizes K, or *K-PPBD* of order v, is a set of v points together with a set of blocks with the property that no two blocks intersect in two or more points. The notion of resolvability extends in the natural way to partial PBDs, and we employ the notation K-RPPBD(v, c) to denote a resolvable PPBD on v elements having c parallel classes.

A $\{2, 3, 4\}$-RPPBD(v, c) in which every parallel class consists of $\lfloor (v-2)/3 \rfloor$ blocks of size three and one block of size $v - 3\lfloor (v-2)/3 \rfloor$ is a *school project design*, denoted SP(v, c). These designs were introduced by Černý, Horák, and Wallis (1997). They showed that

$$c \leq \begin{cases} (v-2)/2 & \text{if} \quad v \equiv 0, 2 \pmod 6, v \geq 6 \\ (v-3)/2 & \text{if} \quad v \equiv 1 \pmod 6 \\ (v-1)/2 & \text{if} \quad v \equiv 3, 5 \pmod 6 \\ (v-4)/2 & \text{if} \quad v \equiv 4 \pmod 6, v \geq 10 \\ 1 & \text{if} \quad v \in \{2, 4\} \end{cases}$$

When c attains this maximum value, the SP(v, c) is a *Kirkman school project design*, denoted KSP(v).

Černý, Horák, and Wallis (1997) settle the existence question for KSP(v)s when $v \equiv 0, 2 \pmod 3$. When $v \equiv 3 \pmod 6$, these are Kirkman triple systems. When $v \equiv 0 \pmod 6$, they are nearly Kirkman triple systems. When $v \equiv 2 \pmod 6$, they are 3-frames of type $2^{v/2}$ (add each group as a block of size two to its parallel class). When $v \equiv 5 \pmod 6$, delete a point from an NKTS$(v+1)$; the $(v-1)/2$ groups form uncovered pairs, while the pairs obtained from triples that contained the point deleted are those appearing in the parallel classes. Within the class $v \equiv 1 \pmod 3$, Černý, Horák, and Wallis (1997) obtain solutions for two small orders (22 and 25), and some infinite classes that result. Colbourn and Ling (1999) settle existence completely when $v \equiv 4 \pmod 6$, and give a solution with few possible exceptions when $v \equiv 1 \pmod 6$. We examine the case when $v \equiv 4 \pmod 6$ in some detail here.

An *incomplete KSP* of order $v \equiv 1 \pmod 3$ with a hole of size $h \equiv v \pmod 2$, $h \geq 3$, denoted IKSP(v, h), is a PPBD $(V, \mathcal{B})$ with a *hole* $H \subset V$ of h elements, where $\mathcal{B}$ contains $(v-h)/2$ blocks of size four (quadruples) and $\lfloor (h-3)/2 \rfloor (v-h)/3 + (v-h)(v-4)/6$ blocks of size 3 (triples), so that no pair of elements from the hole appears in a block of $\mathcal{B}$. The blocks are partitioned into $\lfloor (h-3)/2 \rfloor$ parallel classes of $(v-h)/3$ triples each on $V \setminus H$, and $(v-h)/2$ parallel classes of one quadruple and $(v-4)/3$ triples on V. We call the parallel classes on $V \setminus H$ the *auxiliary parallel classes*.

When $h \in \{3, 4\}$ there are no auxiliary parallel classes in an IKSP, and hence an IKSP$(v, 3)$ is a KSP(v) when $v \equiv 1 \pmod 6$, and an IKSP$(v, 4)$ is a KSP(v) when $v \equiv 4 \pmod 6$. These are, however, special types of KSPs since the hole of size h need not appear in an arbitrary KSP. Table 19.1 gives a KSP(16) which is not an IKSP(16,4), using the first 16 letters of the alphabet as symbols. The set of uncovered pairs in this KSP(16) form the three 4-cycles (a,b,c,d), (e,f,g,h), and (i,j,k,l).

Parallel class					
#1	#2	#3	#4	#5	#6
ajmp	ebnp	fkop	ihmn	dgmo	clno
fdn	fcm	jgn	jbo	hlp	idp
eio	aho	elm	gcp	akn	bkm
khc	kig	aci	alf	jce	fhj
bgl	jld	bdh	edk	fib	ega

Table 19.1 *A KSP*(16)

An IKSP(v, h) covers all pairs outside the hole when v and h are odd, and all but $(v-h)/2$ of the pairs when v and h are even. In either case, however, the IKSP leaves pairs in the hole uncovered and hence these IKSPs are not simply a kind of pairwise balanced design.

Colbourn and Ling (1999) produce numerous small examples, and then employ the Fundamental Frame Construction. When a 3-frame of type $T = g_1 \cdots g_s$ exists, groups can be filled in using IKSPs:

Theorem 19.35 *$1 \leq i < s$, and a KSP$(g_s + x)$ exists, then a KSP$(x + \sum_{i=1}^{s} g_i)$ exists. If a 3-frame of type $g_1 \cdots g_s$ exists, IKSP$(g_i + x, x)$s exist for $1 \leq i < s$, and a KSP$(g_s + x)$ exists, then a KSP$(x + \sum_{i=1}^{s} g_i)$ exists.*

Proof For $1 \leq i < s$, there are $g_i/2$ frame parallel classes missing the group of size g_i, and the same number of parallel classes in the IKSP$(g_i + x, x)$ which contain a block of size four; match these up arbitrarily, placing the g_i points of the IKSP on the ith group of the frame and the x points in its hole on x new points.

Next, each IKSP contains $\lfloor (x-3)/2 \rfloor$ parallel classes of triples. Form unions of these with $\lfloor (x-3)/2 \rfloor$ parallel classes of the KSP$(g_s + x)$, to form $\lfloor (x-3)/2 \rfloor$ additional parallel classes. There remain $g_s/2$ parallel classes of the KSP(g_s+x), which can be matched arbitrarily with the $g_s/2$ frame parallel classes of the sth group to complete the construction. □

Putting the pieces together, we obtain:

Theorem 19.36 *A KSP(v) exists whenever* $v \equiv 4 \pmod 6$ *and* $v \geq 16$.

Proof Colbourn and Ling (1999) produce a KSP$(6t+4)$ when $2 \leq t < 12$ and for $t \in \{13, 14, 17\}$ by direct methods.

When an idempotent TD(5,n) exists (that is, when $n \geq 5$ and $n \notin \{6, 10\}$), remove all but m points in one group where $m \notin \{2, 3\}$. Then treat the groups as blocks, and the blocks of the parallel class of the idempotent TD as groups, to obtain a $\{4, 5, m, n\}$-GDD of type $5^m 4^{n-m}$. Use Theorem 19.17 giving weight 6 and using ingredient frames from Theorem 19.19 to produce a 3-frame of type $30^m 24^{n-m}$. With IKSP(28,4) and IKSP(34,4), Theorem 19.35 gives a KSP$(24n+6m+4)$. Some simple arithmetic now establishes that when $t \geq 20$ and $t \notin \{22, 23, 26, 27, 30, 31, 46, 47\}$, a KSP$(6t+4)$ exists.

Since an IKSP(22,4) exists and 3-frames of type 18^s exist for $s \geq 4$, Theorem 19.35 gives KSP$(6t+4)$s when $t \geq 12$ and $t \equiv 0 \pmod 3$. By the same token, since an IKSP(28,4) exists we obtain KSP$(6t+4)$s when $t \geq 16$ and $t \equiv 0 \pmod 4$.

If an IKSP(40,10), a KSP(40), and a 3-frame of type 30^s exist, so also does a KSP$(30s+10)$ by Theorem 19.35; this treats cases when $t \geq 21$ and $t \equiv 1 \pmod 5$. In the same manner, a 3-frame of type 54^5, an IKSP(70,16), and a KSP(70) are used to produce KSP$(6 \cdot 47 + 4)$.

Truncating one group of a TD(5,5) to 2 or 3 points and giving weight 6 yields 3-frames of types $30^4 12^1$ and $30^4 18^1$ which, together with IKSP(34,4), and either KSP(16) or KSP(22), handles the cases with $t \in \{22, 23\}$. Similarly, removing one point from a TD(5,4) gives a $\{4,5\}$-GDD of type $4^4 3^1$, which yields a 3-frame of type $24^4 18^1$ and hence a solution for KSP$(6 \cdot 19 + 4)$. □

Colbourn and Ling (1999) also address the case when $v \equiv 1 \pmod 6$. See also Phillips, Wallis, and Rees (1998) for further results on KSPs and on the analogous covering problem.

19.3.6 *Semiframes*

Rees (1991a) introduced a common generalization of resolvable GDDs and Kirkman frames. A *semiframe* SF$(p, d; g^u)$ is a 3-GDD of type g^u whose triples can be partitioned into p parallel classes, and d frame parallel classes missing each group. Simple counting shows that $2d(u-1)+2p = g(u-1)$, and when both parallel and frame parallel classes are present, it is necessary that $g \equiv 0 \pmod 3$ and $u \geq 4$. Rees (1991a), and Kageyama and Miao (1996a), give direct constructions along with constructions that generalize RGDD constructions, but the determination of possible values of the parameters (g, u, p, d) remains far from complete.

19.3.7 *Resolvable coverings*

Nearly Kirkman triple systems can be thought of as resolvable maximum packings by triples. In the same vein, we can ask for resolvable minimum coverings by triples for orders which are multiples of 6. Assaf, Mendelsohn, and Stinson (1987) employed Kirkman frames to settle existence of such coverings with a few possible exceptions:

Theorem 19.37 *A minimum covering by triples on 6n elements which is resolvable exists for all $n \geq 3$ except possibly when*

$$n \in \{6, 7, 8, 10, 11, 13, 14, 17, 22\}.$$

Lamken and Mills (1993) settled the possible exceptions that remained affirmatively.

19.3.8 $\{2,3\}$-*PBDs*

A more substantial generalization of nearly Kirkman triple systems was studied by Rees (1987). He established that:

Lemma 19.38 *Let $r \geq 1$ and $0 \leq f \leq 2r$, $(r, f) \neq (1, 2)$ or $(3, 6)$. There exists a $\{2,3\}$-GDD of type $(2r)^3$ which is resolvable into f parallel classes of blocks of size 3 and $x = 4r - 2f$ parallel classes of blocks of size 2.*

Rees (1987) went on to determine when a $\{2,3\}$-PBD exists having a specified number of parallel classes of triples, with all blocks of size two partitioned into 1-factors:

Theorem 19.39 *A $\{2,3\}$-PBD of order v resolved into ℓ parallel classes of triples and $v - 1 - 2\ell$ parallel classes of pairs exists if and only if $v \equiv 0 \pmod 6$ and $0 \leq \ell \leq (v-2)/2$, except when $v \in \{6, 12\}$ and $\ell = (v-2)/2$.*

The two exceptional cases are, of course, the nonexistent NKTSs.

Rees's interest in resolvable $\{2,3\}$-PBDs arose from the problem of determining the smallest number of blocks needed to construct a PBD on v points in which the largest block has size k. When $2k < v < 3k + 1$, this minimum is achieved by a PBD with one block of size k and all other blocks of sizes three or four, each meeting the block of size k. Deleting then the block of size k gives a $\{2,3\}$-PBD on $v - k$ points, with each deleted point of the k-block indicating a parallel class of the $\{2,3\}$-PBD. In this problem, there is no need for the parallel classes to contain blocks of the same size.

Rees and Wallis (1986) addressed the existence of such resolved $\{2,3\}$-PBDs:

Theorem 19.40 *When* $p \equiv 0 \pmod 6$, *there exists a* $\{2,3\}$*-PBD of order* p *with a resolution into* k *parallel classes, for each* $p/2 \leq k \leq p-1$ *except when* $p = 6$ *and* $k = 3$, *or* $p = 12$ *and* $k = 6$.

When parallel classes can contain both triangles and pairs, there is no requirement that the order be a multiple of six. Hence, Rees (1990) settled existence for orders congruent to 2 or 4 modulo 6. Rees (1990a) examines the cases for odd orders, and Rees (1991) finishes up the possible exceptions left, to obtain:

Theorem 19.41 *There exists a* $\{2,3\}$*-PBD of order* p *with a resolution into* k *parallel classes if and only if*

1. $p \equiv 0 \pmod 6$, $p/2 \leq k \leq p-1$, $(p,k) \notin \{(6,3),(12,6)\}$;
2. $p \equiv 1,5 \pmod 6$, $(p-1)/2 \leq k \leq p-4$, *and* $p(k-p+1) \equiv 0 \pmod 3$;
3. $p \equiv 2,4 \pmod 6$, $p/2 \leq k \leq p-1$, *and* $p(k-p+1) \equiv 0 \pmod 3$;
4. $p \equiv 3 \pmod 6$, $(p-1)/2 \leq k \leq p-4$, $(p,k) \notin \{(9,6),(3,1)\}$.

Wevrick and Vanstone (1996) examine the existence of resolvable $\{2,3\}$-PBDs in which each parallel class contains ρ pairs and τ triples; such PBDs are *class-uniformly resolvable designs*, or CURDs. They establish a number of existence results for CURDs.

19.4 Colouring triples and the chromatic index

Given a (partial) TS(v,λ) $\mathcal{D} = (V,\mathcal{B})$, a *colouring of triples* or *block-colouring* of $\mathcal{D}$ is a mapping $\psi : \mathcal{B} \mapsto C$ (the set of *colours*) so that if $\psi(B) = \psi(B')$ for $B, B' \in \mathcal{B}$, $B \neq B'$, then $B \cap B' = \varnothing$. If the set of colours has cardinality q, then ψ is a q-block-colouring. For each $c \in C$, $\psi^{-1}(c)$ is a *block colour class.* The *chromatic index* of $\mathcal{D}$, $\chi'(\mathcal{D})$, is the smallest q for which $\mathcal{D}$ admits a q-block-colouring.

Evidently, any block colour class is a PPC. Therefore $\chi'(\mathcal{D}) \geq |\mathcal{B}|/\lfloor v/3 \rfloor$ for any TS(v,λ) $\mathcal{D}$. When $v \equiv 3 \pmod 6$ and $\lambda = 1$, this inequality simply states that the chromatic index is at least $(v-1)/2$. Theorem 19.9 on KTSs can then be restated as a result on chromatic indices:

Theorem 19.42 *For every* $v \equiv 3 \pmod 6$, *there exists an STS*(v) *with minimum chromatic index* $(v-1)/2$.

Similarly, Theorem 19.10 can be restated:

Theorem 19.43 *For every* $v \equiv 0 \pmod 3$, $v \neq 6$, *there exists a TS*$(v,2)$ *with minimum chromatic index* $v-1$.

In the unique TS$(6,2)$, every two triples intersect, and hence this system has chromatic index 10.

19.4.1 *Hanani triple systems*

A *Hanani triple system* (HATS) is a Steiner triple system with a partition of its blocks into $(v-1)/2$ maximum parallel classes, and a single partial parallel class with $(v-1)/6$ blocks. The existence question for Hanani triple systems was settled by Vanstone, Stinson, Schellenberg, Rosa, Rees, Colbourn, Carter, and Carter (1993):

Theorem 19.44 *A Hanani triple system of order v exists if and only if $v \equiv 1 \pmod 6$ and $v \notin \{7, 13\}$.*

Some direct constructions are obtained when $v \equiv 7 \pmod{12}$ by finding a set of $(v-1)/3$ disjoint starter blocks on $\{\infty\} \cup (\mathbb{Z}_{(v-1)/2} \times \{0,1\})$ which miss the pure difference $((v-1)/6)_0$. The development of this base maximum parallel class modulo $((v-1)/2, -)$ gives $(v-1)/2$ maximum parallel classes, while the remaining pure difference gives the required partial parallel class. When $v \equiv 1 \pmod{12}$, a somewhat more complicated direct construction is also given.

For odd orders, Theorem 19.45, due to Hanani (1974), is a straightforward corollary of Theorem 19.44. Taking the 'short' PPCs in two Hanani triple systems of order v to be on disjoint sets of points, and forming the union of the two Hanani triple systems yields an almost resolvable twofold triple system of order v.

Theorem 19.45 *An almost resolvable twofold triple system of order v exists if and only if $v \equiv 1 \pmod 3$.*

19.4.2 *Orthogonal double covers*

When $v \equiv 1, 4 \pmod{12}$, there is an easy way to obtain an almost resolvable TTS(v) from an $S(2,4,v)$. If $(V, \mathcal{B})$ is an $S(2,4,v)$ then $R_x = \{\{a,b,c\} : \{x,a,b,c\} \in \mathcal{B}\}$ forms an almost parallel class. Then $(V, \bigcup_{x\in V} R_x)$ is an almost resolvable TTS(v), with $\{R_x : x \in V\}$ its corresponding resolution.

The system produced in this manner has the additional property that exactly one triple is shared by R_x and R_y whenever $x \neq y$ and $x, y \in V$. They represent, in effect, a special case of orthogonal double covers of the complete graph. To be precise, an *orthogonal double cover* of K_n by a graph G is a collection $\mathcal{G} = \{G_1, \ldots, G_n\}$ of spanning subgraphs of K_n, all of which are isomorphic. Moreover, every edge of K_n belongs to exactly two members of $\mathcal{G}$, and every two subgraphs in $\mathcal{G}$ share exactly one edge. Bennett and Wu (1990) and Ganter and Gronau (1991) show that an orthogonal double cover of K_n by G, where G has one isolated vertex and the remaining vertices are partitioned into triangles, also exists whenever the order is congruent to 7 or 10 modulo 12, with the exception of order 10 for which no such cover exists.

Demetrovics, Füredi, and Katona (1985) had earlier examined such systems in the study of representation of operations in a relational database system.

19.4.3 *Rosa triple systems*

Ling and Colbourn (1997) examine triple systems with minimum chromatic index; we discuss their results in this section. A Rosa triple system is a triple system of order congruent to 2 modulo 3 whose chromatic index is minimum, having the largest possible number of maximum parallel classes in such a block colouring. The existence of Rosa triple systems is settled completely. Together with known results on Kirkman, Hanani, and almost resolvable twofold triple systems, the existence of Rosa triple systems is used to settle completely the existence of triple systems with minimum chromatic index.

If the b blocks of a TS(v, λ) can be partitioned into $\lfloor b/\lfloor v/3 \rfloor \rfloor$ maximum parallel classes, and (when $b \not\equiv 0 \pmod{\lfloor v/3 \rfloor}$) one partial parallel class, the triple system is called *minci.*

A *Rosa triple system* RTS(v, λ) is a minci TS(v, λ) when $v \equiv 2 \pmod 3$. Rosa posed the question of the existence of such systems in general, because of the role they play in block colouring problems.

Following Ling and Colbourn (1997), the existence of Rosa triple systems is first settled when $v \equiv 5 \pmod 6$. The number of blocks in a TS$(6x+5, 3)$ is $(9x+9)(2x+1)+1$. So, if an RTS$(6x+5, 3)$ exists, the number of blocks not in a maximum parallel class is precisely one.

Let $V = \{0, 1, 2, \ldots, 6x+4\}$ and $H = \{6x, 6x+1, \ldots, 6x+4\}$. An RTS$(6x+5)-RTS(5)$ is defined to be a collection of blocks of size three so that every pair of points in V, but not both in H, appears in exactly three blocks. Each pair of points in H appears in no block. In addition, the blocks admit a partition so that there are nine sets of $2x$ blocks, and each set of $2x$ blocks partitions $V \setminus H$. The other blocks partition into maximum parallel classes on V.

Lemma 19.46 *There exists an RTS*$(5, 3)$.

Proof All 3-subsets of $\{1, 2, 3, 4, 5\}$ give a TS$(5, 3)$, which is an RTS as each maximum parallel class only contains one block. □

Lemma 19.47 *If there exists an RTS*$(6x+5)-$*RTS*(5)*, then there exists an RTS*$(6x+5, 3)$.

Proof In the RTS$(6x+5)-$RTS(5), there are exactly nine partial parallel classes missing the hole. Associate a different block of RTS$(5, 3)$ defined on H with each partial parallel class. This gives nine more maximum parallel classes; only one block is left over. □

Ling and Colbourn (1997) provide a 'small' example:

Lemma 19.48 *There exists an RTS*$(17)-$*RTS*(5).

Lemma 19.49 *There exists an RTS*$(v)-$*RTS*(5) *for* $v = 23$, 35, *and* 47.

Proof Write $v = 6x+5$. Let the points be $\mathbb{Z}_{6x} \cup \{X, \overline{X}, Y, \overline{Y}, \infty\}$. Define two permutations on this point set:

$$\begin{aligned} \alpha &= (0, 1, \ldots, 6x-1)(X, \overline{X})(Y, \overline{Y})(\infty) \\ \beta &= (0, 1, \ldots, 6x-1)(X)(\overline{X})(Y)(\overline{Y})(\infty) \end{aligned} \tag{19.1}$$

Now define three sets of blocks as follows. First, there are three *special* blocks, each consisting of three elements of $\mathbb{Z}_{6x}$ that are distinct modulo 3. Next there is a *half base class* of $x+1$ triples. This is a set of blocks containing $\{\infty, 0, 3x\}$, and a set of other triples all of whose elements are distinct and nonzero modulo

$3x$; of these, one triple contains X and one contains Y, while none contains $\overline{X}$ or $\overline{Y}$. Finally define a *whole base class* of $2x+1$ triples, of which $X, \overline{X}, Y, \overline{Y}, \infty$ are each contained in one and all other elements are distinct modulo $6x$. The primary requirement is that, calculating differences between every pair of elements from $\mathbb{Z}_{6x}$ among the special triples, and the triples of the half and the whole base classes, every nonzero element of $\mathbb{Z}_{6x}$ is found as a difference precisely three times.

Forming the orbits of the special triples and whole base class under β, and the triples of the half base class under α, yields a partial triple system in which every pair is covered three times, except those on $\{X, \overline{X}, Y, \overline{Y}, \infty\}$, which are uncovered. To show that the system is an RTS(v)$-$RTS(5), proceed as follows.

For each special triple B, and for $j \in \{0, 1, 2\}$, the blocks

$$\{\beta^{3i+j}(B) : 0 \leq i \leq 2x\}$$

form a parallel class (B has all three elements distinct modulo 3), and hence each special triple yields three frame parallel classes, each missing the hole $\{X, \overline{X}, Y, \overline{Y}, \infty\}$. Now for the half parallel class $\mathcal{H}$, since x is odd, form the set of blocks $\mathcal{H} \cup (\alpha^{3x}(\mathcal{H}))$, suppressing the duplication of $\{\infty, 0, 3x\}$. Since all elements in the half parallel class are distinct modulo $3x$, the result is a whole parallel class $\mathcal{P}$. Now $3x$ parallel classes are obtained as $\{\alpha^i(\mathcal{P}) : 0 \leq i < 3x\}$. Finally for the whole parallel class $\mathcal{F}$, form the $6x$ parallel classes $\{\beta^i(\mathcal{F}) : 0 \leq i < 6x\}$.

The special triples, half base classes, and whole base classes for the required cases are given in Table 19.2. □

Lemma 19.50 *There is an RTS*(11, 3).

Proof Use the ingredients in Table 19.3 as follows. Over $\mathbb{Z}_{10}$, the special block $B = \{0, 1, 2\}$ produces three maximum parallel classes $\{B+j, B+3+j, B+6+j\}$, for $j = 0, 1, 2$, with one block $(B+9)$ remaining. For the half class, add 5 to each block and form a maximum parallel class by suppressing the duplication of the block containing ∞. Develop modulo 10 to get five maximum parallel classes. Finally develop the whole class modulo 10 to get ten further maximum parallel classes. □

Lemma 19.51 *There exists an RTS*(29, 3).

Proof There is a cyclic TS(29,3) with starter blocks {0,1,2}, {0,4,8}, {0,5,10}, {0,6,12}, {0,10,20}, {0,1,22}, {2,9,17}, {3,5,18}, {4,24,27}, {7,16,19}, {8,11,25}, {10,14,21}, {12,23,28}, {13,15,26}. The last nine starter blocks form a base parallel class; develop modulo 29 to produce 29 parallel classes. Each of the five remaining base blocks has the form $\{0, a, 2a\}$. It is easy to see that, whenever *any two* blocks are removed from the 29 blocks in the orbit of {0,1,2}, the remaining 27 blocks partition into three parallel classes. Multiplying such a partition by a establishes the same statement for the orbit of $\{0, a, 2a\}$ (since a and 29 are relatively prime). It remains only to choose two blocks from the orbit of each of

Order	Part	
23	Special	{0,1,2}, {0,1,8}, {0,4,8}
	Half	{1,7,13}, {X,2,5}, {Y,3,8}, {∞,0,9}
	Whole	{3,12,14}, {8,11,13}, {X,1,4}, {$\overline{X}$,5,9}, {Y,10,15}, {$\overline{Y}$,2,9}, {∞,17,7}
35	Special	{0,1,2}, {0,1,8}, {0,4,8}
	Half	{6,12,20}, {9,18,28}, {11,23,25}, {X,1,4}, {Y,2,7}, {∞,0,15}
	Whole	{1,10,20}, {2,11,17}, {7,19,21}, {15,18,28}, {12,24,29}, {8,14,25}, {X,0,3}, {$\overline{X}$,5,9}, {Y,22,27}, {$\overline{Y}$,16,23}, {∞,6,13}
47	Special	{0,1,2}, {0,1,8}, {0,4,8}
	Half	{3,14,31}, {5,17,30}, {6,19,34}, {20,29,39}, {12,15,32} {X,1,4}, {Y,2,7}, {∞,0,21}
	Whole	{0,2,26}, {35,3,21}, {7,17,37}, {8,10,32}, {9,14,20}, {33,39,6}, {41,5,28}, {15,24,36}, {12,31,38}, {29,40,13} {X,1,4}, {$\overline{X}$,30,34}, {Y,22,27}, {$\overline{Y}$,16,23}, {∞,11,19}

Table 19.2 *RTS*(v)–*RTS*(5) *for small orders* v

Order	Part	
11	Special	{0,1,2}
	Half	{∞,0,5}, {1,3,7}
	Whole	{∞,0,1}, {2,5,8}, {7,9,4}

Table 19.3 *RTS*$(11, 3)$

the first five starter blocks, and from these ten blocks to form a parallel class leaving one block over. This extra parallel class is {19,20,21}, {26,27,28}, {0,4,25}, {9,13,17}, {5,10,15}, {6,11,16}, {1,18,24}, {2,8,14}, {3,12,22}; the extra block left over is {0,10,20}. □

Lemma 19.52 *There exists an* *RTS*$(41, 3)$.

Proof (Sketch) The proof is similar to that of Lemma 19.51. See Ling and Colbourn (1997). □

Recursive constructions are now introduced to settle the spectrum of Rosa triple systems when $v \equiv 5 \pmod 6$.

Lemma 19.53 *Suppose there exists a* $(3,3)$*-frame of type* $g_1 g_2 \cdots g_n$ *where the size of each group is a multiple of six. If, for each* $i = 1, 2, \ldots, n-1$, *there exists an* *RTS*$(g_i + 5)$–*RTS*(5) *and there exists an* *RTS*$(g_n + 5, 3)$, *then there is an* *RTS*$(g_1 + g_2 + \cdots + g_n + 5, 3)$.

Proof The number of frame parallel classes missing a group of size g_i is precisely $3g_i/2$, which is also the number of maximum parallel classes in an RTS$(g_i +$

5)–RTS(5). Associate each frame parallel class with a maximum parallel class to obtain a maximum parallel class in the TS$(v,3)$ where $v = g_1+g_2+\cdots+g_n+5$, for all groups where $i = 1, 2, \ldots, n-1$. Finally, there are nine frame parallel classes on each group; together with the frame parallel classes missing the last group, there is a set of $3g_n/2+9$ parallel classes missing the set of g_n+5 points. Simple counting shows that the number of maximum parallel classes in an RTS$(g_n+5,3)$ is precisely $3g_n/2+9$. Append each maximum parallel class in RTS$(g_n+5,3)$ to a frame parallel class. All but one block in the RTS$(g_n+5,3)$ are exhausted in this way; hence it is an RTS$(v,3)$. □

Lemma 19.54 *If $v \equiv 5 \pmod{12}$, then there is an RTS$(v,3)$.*

Proof The cases when $v = 5, 17, 29, 41$ are dealt with in Lemmas 19.46, 19.48, 19.51, and 19.52. There exists a $(3,3)$-frame of type 12^r for all $r \geq 4$ by Theorem 19.20. Apply Lemma 19.53 to obtain the result; the RTS(17)–RTS(5) exists by Lemma 19.48. □

Lemma 19.55 *Suppose there exists a 4-GDD of type $g_1g_2\cdots g_n$. Then there exists a $(3,1)$-frame of type $(2g_1)(2g_2)\cdots(2g_n)$ and hence a $(3,3)$-frame of type $(2g_1)(2g_2)\cdots(2g_n)$.*

Proof (Sketch) Use the Fundamental Frame construction, giving weight 2 to each point, using a $(3,1)$-frame of type 2^4. □

Lemma 19.56 *There exist 4-GDDs of type 6^43^1, 6^49^1, 6^63^1, 9^5, 9^56^1, $6^512^115^1$, 9^518^1, and 6^821^1.*

Proof A 4-GDD of type 9^5 is contained in Brouwer, Schrijver, and Hanani (1977). In addition, 4-GDDs of type 9^518^1 and 6^821^1 are obtained by completing a resolvable 3-GDD of type 9^5 and 6^8. For the remaining designs, see Rees and Stinson (1989). □

Lemma 19.57 *If $v \equiv 11 \pmod{12}$, then there exists an RTS$(v,3)$.*

Proof If $v = 11, 23, 35, 47$, apply Lemma 19.50 and Lemma 19.49. For $59 \leq v \leq 143$, use the 4-GDDs from Lemma 19.56 in Lemma 19.55 to obtain a $(3,3)$-frame, and apply Lemma 19.53 to yield KTS$(v,3)$s for all v in the interval. Finally, take a TD$(6,n)$ and truncate a group to obtain a $\{5,6,n\}$-GDD of type 5^nx^1 where $0 \leq x \leq n-1$. We can also truncate a group of size five to obtain a $\{4,5,6,n-1,n\}$-GDD of type $5^{n-1}x^1$. In these GDDs, give weight six to each point and apply the fundamental construction for frames (Theorem 19.17) to obtain a $(3,3)$-frame of type $(30)^n(6x)^1$ or $(30)^{n-1}(6x)^1$. Use this frame together with Lemma 19.53 and Lemma 19.49 to obtain an RTS$(30n+6x+5,3)$ or RTS$(30(n-1)+6x+5,3)$. This gives an RTS$(v,3)$ for all $v \geq 143$ and $v \equiv 11 \pmod{12}$. □

Again following Ling and Colbourn (1997), the existence of Rosa triple systems of order v when $v \equiv 2 \pmod 6$ is settled next. An RTS$(6x+2,6)$ has

$(18x+9)(2x)+2$ blocks; hence the number of blocks not contained in a maximum parallel class is precisely two.

The following holey structure is used. Let $V = \mathbb{Z}_6 \cup \{\infty_1, \infty_2\}$. Let $B = \{0,1,2\}$; then $B+i, B+3+i$ for $i = 0,1,2$ gives three parallel classes missing the points ∞_1 and ∞_2. Also, developing $\{\infty_1, 0, 2\}$ and $\{\infty_2, 1, 4\}$ over $\mathbb{Z}_6$ gives six maximum parallel classes on V. The three starter blocks give a TS(8, 2)−TS(2, 2) (a triple system on eight points with a hole of size two and index 2). Take three copies to obtain a set of blocks on eight points with 27 maximum parallel classes, nine of which miss the hole. Denote this configuration by RTS(8)−RTS(2).

Lemma 19.58 *There is an RTS$(v, 6)$ for $v = 8$, 14, and 20.*

Proof Employ the ingredients in Table 19.4 as follows. In each case, the specified triples are developed under the action of $\mathbb{Z}_{v-1}$. For each special triple B, three maximum parallel classes are formed as $\{\{B + 3i + j : 0 \le i < (v-2)/3\} : j = 0,1,2\}$. For each of the first, second, and third base classes, apply the action of $\mathbb{Z}_{v-1}$ to produce $v-1$ maximum parallel classes from each. □

Order	Part	
8	Special	{0,1,2}, {0,1,2}
	First	{∞,0,2}, {4,5,1}
	Second	{∞,0,3}, {1,2,6}
	Third	{∞,0,3}, {2,4,6}
14	Special	{0,1,2}, {0,1,2}
	First	{∞,0,2}, {3,7,8}, {9,10,5}, {4,6,11}
	Second	{∞,0,3}, {2,4,11}, {6,8,1}, {9,12,5}
	Third	{∞,0,3}, {6,9,1}, {8,12,2}, {11,4,7}
20	Special	{0,1,2}, {0,1,2}
	First	{∞,0,2}, {3,4,15}, {5,6,12}, {7,9,14}, {8,10,17}, {16,1,11}
	Second	{∞,0,3}, {2,6,8}, {17,1,12}, {7,10,15}, {5,11,14}, {9,13,18}
	Third	{∞,0,3}, {4,10,1}, {8,12,2}, {5,9,16}, {7,14,18}, {17,6,11}

Table 19.4 *RTS$(v, 6)$s for small orders*

Lemma 19.59 *If $v \equiv 2 \pmod 6$, there exists an RTS$(v, 6)$.*

Proof There exists a (3, 6)-frame of type 6^r for all $r \ge 4$ by Theorem 19.20. An RTS$(6r+2, 6)$ is constructed on the $6r$ points in the frame and two extra points. The blocks of an RTS(8)−RTS(2) are placed on each group with the two extra points aligned on ∞_1 and ∞_2. The number of frame parallel classes is 18 which is the same as the number of maximum parallel classes in the RTS(8)−RTS(2). Associate each frame parallel class with a maximum parallel class to obtain a maximum parallel class on the $6r+2$ points, for $r-1$ groups. In each group, nine parallel classes remain on the six points in the group. Together with the frame parallel classes, this yields 27 frame parallel classes missing the remaining eight points. Associate each frame parallel class with a maximum parallel class in an RTS$(8,6)$, to obtain 27 further maximum parallel classes on $6r+2$ points leaving two blocks unused. This is an RTS$(6r+2,6)$. This construction, together with Lemma 19.58, yields the result. □

19.4.4 *Minimum chromatic index*

Now the existence of triple systems with minimum chromatic index can be established; for the existence results used in the proof, recall Theorem 19.9 for Kirkman triple systems, Theorem 19.45 for almost resolvable triple systems, and Theorem 19.44 for Hanani triple systems.

Theorem 19.60 *There is a triple system TS(v,λ) that admits a block-colouring in*

$$\left\lceil \frac{\lambda v(v-1)/6}{\lfloor v/3 \rfloor} \right\rceil$$

colours, in which at most one colour class is not a maximum parallel class, whenever $\lambda \equiv 0 \pmod{\gcd(v-2,6)}$ and $v \geq 3$, except when $v=6$ and $\lambda \equiv 2 \pmod 4$, $(v,\lambda)=(13,1)$, or $v=7$ and $\lambda \in \{1,3,5\}$.

Proof When $v \equiv 3 \pmod 6$, simply take λ copies of a Kirkman triple system. When $v \equiv 0 \pmod 6$ and $v \neq 6$, take $\lambda/2$ copies of a resolvable TS$(v,2)$. There is a resolvable TS$(6,4)$. However, a simple counting argument shows that if a TS$(6,\lambda)$ is resolvable, then $\lambda \equiv 0 \pmod 4$. When $v \equiv 4 \pmod 6$, take $\lambda/2$ copies of an almost resolvable TS$(v,2)$. When $v \equiv 1 \pmod 6$ and $v \notin \{7,13\}$, take λ copies of a Hanani triple system of order v; in alternate copies, place the short parallel class of $(v-1)/6$ triples on two disjoint sets of $(v-1)/2$ points; then in pairs of copies, adjoin the two short parallel classes to form one maximum parallel class. The lack of Hanani triple systems of orders 7 and 13 poses a problem for higher indices. There are almost resolvable TS$(v,2)$s for $v \in \{7,13\}$, so take $\lambda/2$ copies whenever λ is even. There is a TS$(13,3)$ which has chromatic index 20, and has all but one colour class being a maximum parallel class; see Table 19.5. If $\lambda \geq 3$ is odd, the union of a minci TS$(13,3)$ and a minci TS$(13,\lambda-3)$ provides the required system. For TS$(7,\lambda)$, there are 1, 10, and 109 nonisomorphic systems for $\lambda = 1,3,5$, respectively. A complete backtrack verified that none is minci. However, a minci TS$(7,7)$ does exist, on $\{0,1,2,3,a,b,c\}$ with parallel classes as follows:

$$\begin{array}{llll}
\{0,2,a\},\{1,3,b\} & \{0,2,a\},\{1,3,b\} & \{0,2,b\},\{1,3,a\} & \{0,2,b\},\{1,3,c\} \\
\{0,2,c\},\{1,3,a\} & \{0,1,a\},\{2,3,c\} & \{0,1,c\},\{2,3,a\} & \{0,3,b\},\{1,2,c\} \\
\{0,3,c\},\{1,2,b\} & \{0,1,3\},\{a,b,c\} & \{0,2,3\},\{a,b,c\} & \{1,2,3\},\{a,b,c\}
\end{array}$$

$$\{\{1,2,a\},\{3,b,c\}\} \text{ under the permutation } (0\ 1\ 2\ 3)(a\ b\ c)$$

with short parallel class $\{\{0,1,2\}\}$. By unions, this gives a minci TS$(7,\lambda)$ for all odd $\lambda \geq 7$.

Now if $v \equiv 5 \pmod 6$, form an RTS$(v,3)$ in which the short parallel class contains just the triple $\{0,1,2\}$. Form $\lambda/3$ copies by adding $3i$ to each block of the RTS for $0 \leq i < \lambda/3$. Then the extra blocks in up to $(v-2)/3$ consecutive copies form a partial parallel class, so these are adjoined in runs of $(v-2)/3$, with the remaining extra blocks (if any) forming a short parallel class. When $v \equiv 2 \pmod 6$, proceed similarly, assuming the short parallel class is $\{\{0,1,2\},\{3,4,5\}\}$, and replace 3 by 6 in the argument given. □

{0,3,7} {1,10,11} {2,5,6} {4,9,12}	{0,2,8} {1,9,12} {4,6,11} {5,7,10}
{0,3,6} {1,2,11} {4,7,9} {8,10,12}	{0,4,11} {1,5,12} {2,6,7} {3,8,10}
{0,7,9} {1,10,12} {2,3,11} {4,6,8}	{1,5,11} {2,10,12} {3,4,9} {6,7,8}
{0,9,10} {1,2,7} {4,5,12} {6,8,11}	{0,8,12} {1,4,7} {2,3,6} {5,9,11}
{0,1,5} {3,7,11} {4,6,12} {8,9,10}	{0,3,5} {1,4,8} {2,9,12} {7,10,11}
{0,1,6} {2,8,9} {3,5,12} {4,10,11}	{0,4,8} {2,9,11} {3,7,12} {5,6,10}
{0,6,12} {1,3,9} {2,5,8} {4,7,10}	{0,9,10} {1,3,4} {5,8,11} {6,7,12}
{0,5,10} {1,3,8} {2,7,12} {6,9,11}	{0,11,12} {1,7,8} {2,3,10} {5,6,9}
{0,7,11} {1,6,10} {2,4,5} {3,8,9}	{0,2,4} {3,6,10} {5,7,9} {8,11,12}
{1,6,9} {2,4,10} {3,11,12} {5,7,8}	{0,1,2} {3,4,5}

Table 19.5 *A minci* $TS(13,3)$

19.4.5 *Upper bound*

We have examined designs having chromatic index as small as possible. How large can the chromatic index of a design be?

An elementary result is due to Colbourn and Colbourn (1982a).

Theorem 19.61 *If* $\mathcal{D}$ *is an* $S(2,k,v)$*, then* $\chi'(\mathcal{D}) \leq kv/(k-1)$.

Proof Construct the block-intersection graph of $\mathcal{D}$; its vertices are the blocks of $\mathcal{D}$, and two vertices are adjacent if the corresponding blocks have nonempty intersection. The chromatic index of $\mathcal{D}$ is the chromatic number of its block-intersection graph. The latter has maximum degree less than $kv/(k-1)$, and invoking the theorem of Brooks (1941) completes the proof. □

The bound of Theorem 19.61 is likely quite weak. Indeed, a well-known conjecture of Erdös, Faber, and Lovász given in Erdös (1981) can be formulated as the statement that $\chi'(\mathcal{D}) \leq v$ for $\mathcal{D}$ any 2-$(v,K,1)$-design (i.e. pairwise balanced design of index 1 with block sizes from the set K).

This inequality has been shown to hold for cyclic $S(2,k,v)$s by Colbourn and Colbourn (1982a), and for certain projective spaces by Beutelspacher, Jungnickel, and Vanstone (1989).

Pippenger and Spencer (1989) used a probabilistic method to show that the chromatic index of STS(v)s is asymptotic to $v/2$. This is much stronger than the general inequality, but no constructive proof of that inequality for STSs is known. The only 'numerical evidence' is provided by results for very small orders. So, for instance, of the 80 STS(15)s, four are resolvable, and thus have chromatic index 7. Colbourn (1982) calculated that 13 have chromatic index 8, and the remaining 63 have chromatic index 9.

If S is an STS(v) with $v \equiv 3 \pmod 6$, without a single PC (see Conjecture 19.4) then $\chi'(S) \geq (v+3)/2$, i.e. its chromatic index exceeds the minimum chromatic index by at least two. Similarly, the existence of STS(v) for $v \equiv 1 \pmod 6$ without a single APC (see Conjecture 19.5) would also imply the existence of STS(v)s with chromatic index exceeding the minimum chromatic index by at least two. On the other hand, no example seems to be known of any STS(v) with $v > 7$, whose chromatic index exceeds the minimum chromatic index by *more* than two.

19.5 Computational complexity

19.5.1 *The complexity of block-colouring*

The complexity of computing the chromatic index of designs is unknown. In practice, it appears quite hard to compute the chromatic index even for STSs of comparatively small orders.

A depth-first branch-and-bound method for computing the chromatic index of STSs was proposed by Colbourn (1982). Coupled with two heuristics, this method made it possible to determine the chromatic index of each of the 80 nonisomorphic STS(15)s. In the absence of a polynomial time algorithm for computing the chromatic index, we may search for algorithms guaranteed to run in polynomial time but possibly giving only approximate answers. Two fairly simple greedy methods were investigated by Colbourn and Colbourn (1983b). The block-by-block greedy method uses at most $(3/2)v$ colours for a block-colouring of an STS(v) but may in fact require $3\lfloor (v-1)/2 \rfloor - 2$ colours. Obviously, the performance of such a method depends heavily on the initial ordering of triples. The colour-by-colour greedy algorithm, having selected i colour classes, selects the $(i+1)$st colour class by taking a maximal colour class among the remaining uncoloured blocks; a colour class here is maximal if every uncoloured block intersects it, and there is no block whose deletion from the colour class enables the simultaneous addition of two uncoloured blocks. In this method, the first colour class is a maximal PPC, and so must contain at least $(v-1)/6$ blocks.

Both greedy algorithms can be improved by introducing certain heuristics (see Colbourn and Colbourn, 1983b). Applying these techniques to the 80 STS(15)s seems to indicate that these are not only efficient but also provide reasonably good colourings.

19.5.2 *Generating Kirkman triple systems quickly*

Hofmeister and Lefmann (1996) asked if it is possible to generate a Kirkman triple system of order $n \equiv 3 \pmod 6$ in time which is linear in the size of the output. Questions of this type have not been widely studied in design theory. However, when a design is used in an algorithmic setting (as Hofmeister and Lefmann (1996) use designs in the maximum cut in graphs problem), it is naturally important to generate the designs as quickly as possible. Here we describe a (previously unpublished) fast generation technique for Kirkman triple systems.

We assume that an integer $n \equiv 3 \pmod 6$ is provided as input, and we are to write out the $(n-1)/2$ parallel classes each containing $n/3$ blocks for a Kirkman triple system of order n. In total, $\binom{n}{2}$ integers from the set $\mathbb{Z}_n$ are to be written, each occurring $(n-1)/2$ times among the output. It follows that $\Theta(n^2 \log n)$ bits of output are to be written, and that $\Theta(n^2)$ integers from $\mathbb{Z}_n$ are to be written. It is convenient to drop the log factors, and suppose that we are employing a unit-cost random-access machine that can manipulate integers of value up to n in constant time.

Then our goal is to produce a $\Theta(n^2)$ time algorithm for generating a Kirkman triple system of order n. We use a standard construction technique, but employ only certain ingredients that we can generate easily. First we establish:

Lemma 19.62 *If* $m \equiv 1,5 \pmod 6$, *three mutually orthogonal latin squares (MOLS) can be generated in time* $\Theta(m^2)$.

Proof Define three squares L_1, L_2, L_3 by setting $L_i(j,k) = (j+ik) \bmod m$. Three squares so defined are mutually orthogonal when $m \equiv 1,5 \pmod 6$, and it is easy to see that they can be generated in time $\Theta(m^2)$. □

Next we show how to form a $\{4,5\}$-GDD of type m^4t^1 whenever $m \equiv 1,5 \pmod 6$ and $0 \le t \le m$. The basic strategy is to use Lemma 19.62 to produce three MOLS of side m, say L_1, L_2, L_3. Now on $\mathbb{Z}_m \times \{r,c,1,2,3\}$, define a collection of sets of size five by:

$$\{\{(j,r),(k,c),(L_1(j,k),1),(L_2(j,k),2),(L_3(j,k),3)\} : 0 \le j,k < n\}.$$

The result is a transversal design TD$(5,m)$ (a $\{5\}$-GDD of type m^5). Now remove $m-t$ of the points whose second component is 3. The result is a $\{4,5\}$-GDD of type m^4t^1. We have established:

Lemma 19.63 *If* $m \equiv 1,5 \pmod 6$ *and* $0 \le t \le m$, *a* $\{4,5\}$*-GDD of type* m^4t^1 *can be generated in time* $\Theta(m^2)$.

We next employ a standard inflation to turn the $\{4,5\}$-GDDs constructed into $\{4\}$-GDDs. There exist $\{4\}$-GDDs of type 3^4 and 3^5. Hence we have

Lemma 19.64 *If* $m \equiv 1,5 \pmod 6$ *and* $0 \le t \le m$, *a* $\{4\}$*-GDD of type* $(3m)^4(3t)^1$ *can be generated in time* $\Theta(m^2)$.

Proof Use Lemma 19.63 to make a $\{4,5\}$-GDD of type m^4t^1 on point set X. Now, on point set $X \times \{1,2,3\}$, whenever $B \subset X$ is a block of the $\{4,5\}$-GDD, form a $\{4\}$-GDD of type $3^{|B|}$ on $B \times \{1,2,3\}$, with the groups on $\{i\} \times \{1,2,3\}$ for $i \in B$. This can be done 'online', since as each block of the $\{4,5\}$-GDD is treated, the 9 or 15 blocks of the $\{4\}$-GDD can immediately be written. □

Lemma 19.64 can now be used repeatedly to form a pairwise balanced design with block sizes 4 and 7. Suppose the goal is to produce a $\{4,7\}$-PBD on p points, for some number $p \equiv 1 \pmod 3$. Such a PBD exists except when $p \in \{10,19\}$.

First write $p = 3s + 1$. Now write $s = 4m + t$, where $m \equiv 1, 5 \pmod 6$ and $0 \le t \le m$, and $t \notin \{3, 6\}$. To do this, first compute $\widehat{m} = \lfloor s/4 \rfloor$ and $\widehat{t} = s - 4\widehat{m}$. If $\widehat{m} \not\equiv 1, 5 \pmod 6$ or $\widehat{t} \in \{3, 6\}$, subtract 1 from $\widehat{m}$ and add 4 to $\widehat{t}$ repeatedly until the conditions are met or until $\widehat{t} > \widehat{m}$. When $s > 42$, this procedure always 'works': we can set $m = \widehat{m}$, and $t = \widehat{t}$, and the conditions of Lemma 19.64 are met. In fact, for all $s > 42$, the chosen value of t never exceeds 22.

Lemma 19.65 *For all $s \neq 3, 6$, a $\{4,7\}$-PBD on $3s + 1$ points can be generated in time $\Theta(s^2)$.*

Proof Precompute a table of $\{4, 7\}$-PBDs on $3a + 1$ points for $1 \le a \le 42$, $a \neq 3, 6$. This takes constant time independent of s. Now write $s = 4m + t$ as above, and use Lemma 19.64 to generate a $\{4\}$-GDD on a set X of $3s$ points with groups X_1, X_2, X_3, X_4 of size $3m$, and group Y of size $3t$. Add a new point ∞, and on $Y \cup \{\infty\}$, place a $\{4, 7\}$-PBD on $3t + 1$ points; this is found by table lookup in the precomputed solutions, since $t \le 22$ and $t \neq 3, 6$, so writing these blocks to output takes constant time.

Proceed recursively to handle the set X_1 in place of X, with the proviso that for each block written for X_1, a copy is written for each of X_2, X_3, and X_4.

The only issue is the timing. Suppose that Lemma 19.64 uses at most cm^2 time. Then the total time $T(s)$ used satisfies the inequality

$$T(s) \le c\lfloor s/4 \rfloor^2 + c' + 4T(\lfloor s/4 \rfloor),$$

which is $O(s^2)$. □

Now we are nearly done. One last inflation completes the task.

Lemma 19.66 *If $n \equiv 3 \pmod 6$, a Kirkman triple system on n points can be generated in $O(n^2)$ time.*

Proof Precompute solutions for $n = 21, 39$, so that these can be generated in constant time. Otherwise, write $n = 2s + 1$, whence $s \equiv 1 \pmod 3$, and apply Lemma 19.65 to generate a {4,7}-PBD on a set V of size s in $O(s^2)$ time; let $\mathcal{B}$ be the block set of the PBD. We form the KTS on $(V \times \{L, R\}) \cup \{\sigma\}$. We use a Kirkman triple system on 9 points on $(\{1, 2, 3, 4\} \times \{L, R\}) \cup \{\sigma\}$, and one on 15 points on $(\{1, 2, 3, 4, 5, 6, 7\} \times \{L, R\}) \cup \{\sigma\}$. For each block $B \in \mathcal{B}$, we form a copy of one of the two small KTS on $(B \times \{L, R\}) \cup \{\sigma\}$, by rewriting $\{1, 2, \ldots, |B|\}$ in the ingredient KTS as the $|B|$ elements of B. In so doing, we omit all blocks containing σ.

Once all blocks $B \in \mathcal{B}$ are handled in this way, write blocks $\{(i, L), (i, R), \sigma\}$ for $i \in V$. The output design is a TS$(n, 3)$, but its resolvability must be addressed.

It might be desirable that the system be written to output in a way that makes the resolution into parallel classes clear. Each of the s parallel classes contains exactly one of the blocks of the form $\{(i, L), (i, R), \sigma\}$ for $i \in V$, so we index the parallel classes by the elements of V in the obvious way. The remaining blocks are partitioned as follows. When block $B \in \mathcal{B}$ is treated in the proof of

Lemma 19.66, the KTS placed on $(B \times \{L, R\}) \cup \{\sigma\}$ has a partition of all of its blocks into parallel classes, and each block lies in a parallel class with exactly one of the blocks of the form $\{(i, L), (i, R), \sigma\}$. Its parallel class in the output design is then determined to be with the same reference block as in the ingredient KTS in which it appears. □

Thus, on output, each block could be 'tagged' to indicate its parallel class. Alternatively, by treating the blocks through each point of the {4,7}-PBD in turn, Lemma 19.66 could be modified to write all blocks in the same parallel class consecutively, with only a constant factor increase in its running time.

19.6 Enumeration of Kirkman triple systems

Two Kirkman triple systems $(V_1, \mathcal{B}_1, \mathcal{R}_1)$ and $(V_2, \mathcal{B}_2, \mathcal{R}_2)$ are *isomorphic* if there is a bijection $\pi : V_1 \mapsto V_2$ for which $\pi(\mathcal{B}_1) = \mathcal{B}_2$ and $\pi(\mathcal{R}_1) = \mathcal{R}_2$ (we abuse notation somewhat by using π to denote also the induced mappings on blocks and parallel classes).

Let $NK(v)$ be the number of nonisomorphic KTS(v). There is a unique STS(9), and it admits a unique resolution; thus $NK(9) = 1$. Mulder (1917) and Cole (1922) establish that four of the 80 STS(15) are resolvable, one underlying a unique KTS(15) and three each underlying two nonisomorphic KTS(15); thus $NK(15) = 7$. Ahrens (1901), Rosa (1963), and Mathon, Phelps, and Rosa (1983) list the seven systems explicitly. We repeat them in Table 19.6, listing blocks vertically and using the numbering of the underlying STSs.

Results on KTS(21) appear in Mathon, Phelps, and Rosa (1981), Mathon and Rosa (1981), and Petrenjuk and Petrenjuk (1983, 1988). Dejter, Franek, and Rosa (1996) prove that $NK(21) \geq 192$.

Current best lower bounds on $NK(v)$ for $v \leq 81$ are given in Mathon and Rosa (1996):

v	$NK(v)$	v	$NK(v)$	v	$NK(v)$	v	$NK(v)$
21	192	27	909	33	28	39	88
45	84	51	9419	57	1	63	82160
69	1	75	1	81	10^7		

The only general lower bound on $NK(v)$ is due to Stinson and Vanstone (1984).

Theorem 19.67 $NK(v) \geq 2^{(v-1)(v-3)/16}/v$ *for all* $v \equiv 3 \pmod 6$.

19.7 Kirkman triple systems with subsystems

A Kirkman triple system of order v can contain a Kirkman triple system of order $w < v$. It is required that the parallel classes of the larger Kirkman triple system extend the parallel classes of the smaller one. By considering the $(w-1)/2$ parallel classes of the smaller, we find that $(v-w)(w-1)/6$ triples lie in these parallel classes but do not meet the subsystem. It follows that $\binom{v-w}{2} - (v-w)(w-1)/2 \geq w(v-w)/2$, and so $v - w - 1 - (w - 1) \geq w$. Thus $v \geq 3w$. Rees and Stinson (1988, 1988a, 1989) showed that this necessary condition is sufficient:

#	Mon	Tues	Wed	Thurs	Fri	Sat	Sun
1a	adefg	abcdf	abcdg	abcfg	abcde	abcef	abceg
	bjhik	hemkj	jmehi	dlikh	fhlik	lidjh	ndhij
	cnmol	ignol	kofln	enjmo	gjomn	mkgon	ofklm
1b	adefg	abcdf	abcdg	abcfg	abcde	abceg	abcef
	bjhik	hemkj	jmehi	dilhj	flhij	ldikh	nhdik
	cnmol	ignol	kofln	ekonm	gnkmo	mfjno	ojglm
7a	adefg	abceg	abcdf	abcfg	abcde	abcef	abcdg
	bjikh	hdlkj	jemhi	dlihk	fmhij	lidhj	nheki
	comln	ifonm	kgnlo	enjmo	goknl	mkgon	ofjml
7b	adefg	abceg	abcdf	abcfg	abcde	abcdg	abcef
	bjikh	hdlkj	jemhi	dmhji	flikh	lheik	nidjn
	comln	ifonm	kgnlo	eoknl	gnjmo	mjfno	okglm
15a	abcfg	abcde	abceg	abcdf	adefg	abcdg	abcef
	dihmj	fhikl	ldkih	nejih	bhkji	hmejk	jldhi
	eklno	gjnmo	mfojn	ogmlk	conlm	iofnl	kngmo
15b	abcfg	abcde	abceg	abcdf	adefg	abcef	abcdg
	dihmj	fhikl	ldkih	nejih	bjhik	hmdkj	jlehi
	eklno	gjnmo	mfojn	ogmlk	cnmol	iognl	knfom
61	adefg	abcfg	abcde	abcdf	abcdg	abcef	abceg
	bijlh	dijhk	fhimk	heljk	jmehi	lkdhi	ndhij
	cknom	elmno	gjonl	ignom	kofln	mngoj	ofkml

Table 19.6 *The seven solutions of the Kirkman schoolgirl problem*

Theorem 19.68 *A Kirkman triple system of order v having a sub-KTS of order w exists if and only if $v, w \equiv 3 \pmod 6$ and $v \geq 3w$.*

Kirkman triple systems having subsystems that need not be resolvable have been less studied. Theorem 19.10 gives a KTS$(2q+1)$ containing an STS(q) whenever $q \equiv 1 \pmod 6$ is a prime power.

Using Lemma 19.11, Mullin, Stinson, and Vanstone (1982) observe that whenever $v \equiv 1 \pmod 6$ and there is a PBD of order v whose block sizes are prime powers congruent to 1 modulo 6, we obtain a KTS$(2v+1)$ containing an STS(v). Known results on such PBDs from Bennett, Gronau, Ling, and Mullin (1996) apply:

Theorem 19.69 *A PBD of order v with block sizes congruent to 1 modulo 6 and prime powers exists when $v \equiv 1 \pmod 6$, except possibly when $v \in \{$55, 115, 145, 205, 235, 265, 319, 355, 391, 415, 445, 451, 493, 649, 667, 685, 697, 745, 781, 799, 805, 1315$\}$.*

Mullin, Stinson, and Vanstone (1982) settle some of the remaining cases to obtain

Theorem 19.70 *A KTS$(2v+1)$ containing an STS(v) exists when $v \equiv 1 \pmod 6$, except possibly when $v \in \{115, 145, 205, 265, 355, 415, 649, 697\}$.*

The existence of KTSs with smaller sub-STSs appears not to have been studied.

19.8 Work points

1. Prove or disprove: every STS(v) contains a partial parallel class of size at least $(v-4)/3$.
2. Prove or disprove: for all $v \equiv 3 \pmod 6$, $v \geq 15$, there exists an STS(v) with no parallel class.
3. Prove or disprove: for all $v \equiv 1 \pmod 6$, $v \neq 13$, there exists an STS(v) with no almost parallel class.
4. Settle the remaining possible exceptions in Theorem 19.70.
5. Generalize Theorem 19.70 to determine when a KTS(v) exists having a sub-STS(w).
6. For which orders does an automorphism-free Kirkman triple system exist?
7. Prove that for every abstract group Γ, there is a Kirkman triple system whose automorphism group is isomorphic to Γ.
8. Given integers v, m and s, when does a PTS(v) exist whose blocks can be partitioned into s partial parallel classes each of size s?
   The solution when $m = \lfloor v/3 \rfloor$ and m is as large as possible was completed by Colbourn and Zhao (1999). Cases when v is small, and the application to optical and radio communication systems, have been examined by Colbourn, Yates, and Zhao (1999). See Zhao, Yates, and Yasukawa (1996) and Zhao (1998) for related work, and Atkin and Corrales (1989), Atkin, Fares, and Corrales (1989), and Fares, Abul-Shohoud, Raslan, and Nassef (1992) for similar codes based on triple systems.

20

ORTHOGONAL RESOLUTIONS

20.1 Introduction

Let $(V, \mathcal{B})$ be a TS(v, λ). Two resolutions $\mathcal{R}, \mathcal{S}$ of $(V, \mathcal{B})$ are *orthogonal* if $|R_i \cap S_j| \leq 1$ for any $R_i \in \mathcal{R}$ and any $S_j \in \mathcal{S}$. A TS(v, λ) admitting two orthogonal resolutions is *orthogonally resolvable* or *doubly resolvable*. More generally, a set $\mathcal{R} = \{\mathcal{R}_1, \ldots, \mathcal{R}_s\}$ of s resolutions of a TS(v, λ) is a *d-orthogonal set of resolutions* if for any d-subset $\{i_1, \ldots, i_d\}$ of $\{1, \ldots, s\}$ and any d parallel classes $R_{i_1,j_1}, R_{i_2,j_2}, \ldots, R_{i_d,j_d}$ where $R_{i_k,j_k} \in \mathcal{R}_{i_k}$, we have $|R_{i_1,j_1} \cap R_{i_2,j_2} \cap \cdots \cap R_{i_d,j_d}| \leq 1$. A d-orthogonal set $\mathcal{R}$ of resolutions is *regular of strength t* $(2 \leq t \leq d)$ if $\mathcal{R}$ is a t-orthogonal set but contains no $(t-1)$-orthogonal set. Thus 2-orthogonal is the same as orthogonal (a 2-orthogonal set is necessarily regular of strength 2).

The motivation for studying orthogonal resolutions of TS(v, λ)s (or, more generally, of designs) comes from their connection to the existence of various kinds of square and (multidimensional) cubical arrays.

A d-orthogonal set of resolutions of a TS(v, λ) gives rise to an interesting multidimensional array. A *multidimensional Kirkman design* of dimension d, index λ, and order v (briefly MKD(d, λ, v)) is a d-dimensional array such that

1. every cell of the array is either empty or contains a 3-subset of a v-set V,
2. every element of V is contained in exactly one cell of any $(d-1)$-dimensional subarray, and
3. the collection of 3-subsets in the nonempty cells is the collection of triples of a TS(v, λ) (the *underlying triple system* of the MKD).

An MKD(d, λ, v) is *regular of strength t* $(2 \leq t \leq d)$ if its projection on any t dimensions is an MKD(t, λ, v) but its projection on any $t-1$ dimensions is never an MKD. The following theorem is elementary.

Theorem 20.1 *A regular MKD(d, λ, v) of strength t with underlying triple system $(V, \mathcal{B})$ exists if and only if there exists a regular d-orthogonal set of d resolutions of strength t of $(V, \mathcal{B})$.*

Proof If $\{\mathcal{R}_1, \ldots, \mathcal{R}_d\}$ is a d-orthogonal set of resolutions of $(V, \mathcal{B})$ where $\mathcal{R}_i = \{R_{i,1}, \ldots, R_{i,s}\}$, place in the cell $(j_1, j_2, \ldots, j_d)$ of a d-dimensional array K the set $R_{1,j_1} \cap R_{2,j_2} \cap \cdots R_{d,j_d}$ which is either empty or is a triple of $\mathcal{B}$. Conversely, given an MKD(d, λ, v) K with an underlying triple system $(V, \mathcal{B})$, the triples of K in any $(d-1)$-dimensional hypercube obtained by fixing the coordinate i_c in the

ith coordinate direction ($i \in \{1, \ldots, d\}$) yield a resolution $\mathcal{R}_i$, and $\{\mathcal{R}_1, \ldots, \mathcal{R}_d\}$ is a d-orthogonal set. □

Several special cases of MKDs are of interest; we consider these in more detail.

20.2 Generalized Room squares of degree 3

A regular MKD with $d = 2$, $\lambda = v - 2$ (and, of course, $t = 2$) whose underlying triple system is the simple TS$(v, v-2)$ (that is, whose triples are all 3-subsets of a v-set, or, which is the same, the complete 3-uniform v-hypergraph), can be thought of as a straightforward generalization of a Room square, and is therefore a *generalized Room square* of degree 3 (and order v, briefly GRS(v)). It is a square array of side $\binom{v-1}{2}$ by $\binom{v-1}{2}$); each row (column) of a GRS(v) contains $v/3$ nonempty cells, and the triples in the nonempty cells of each row (column) form a parallel class. By Theorem 20.1, a GRS(v) exists if and only if there exists a pair of orthogonal 1-factorizations of the complete 3-uniform v-hypergraph.

A first example of a GRS was given by Baker (1973) for v=9. This is the smallest possible nontrivial GRS, as clearly a GRS(6) cannot exist. We present a different example of a GRS(9) (from Rosa, 1978).

Example 20.2 Two orthogonal 1-factorizations $\mathcal{F}$, $\mathcal{G}$ of the complete 3-uniform hypergraph ${K_9}^3$ on $V = \mathbb{Z}_8 \cup \{\infty\}$ are as follows. The base resolutions of $\mathcal{F}$ are:

$$\begin{aligned} F_1 &= \{\{0,1,\infty\}, \{2,3,6\}, \{4,5,7\}\}, \\ F_2 &= \{\{0,2,\infty\}, \{1,3,5\}, \{4,6,7\}\}, \\ F_3 &= \{\{0,3,\infty\}, \{2,5,6\}, \{1,4,7\}\}, \\ F_4 &= \{\{0,4,\infty\}, \{1,2,3\}, \{5,6,7\}\}, \end{aligned}$$

and the base resolutions of $\mathcal{G}$ are

$$\begin{aligned} G_1 &= \{\{0,1,\infty\}, \{3,4,7\}, \{2,5,6\}\}, \\ G_2 &= \{\{0,2,\infty\}, \{3,4,6\}, \{1,5,7\}\}, \\ G_3 &= \{\{0,3,\infty\}, \{4,5,6\}, \{1,2,7\}\}, \\ G_4 &= \{\{0,4,\infty\}, \{1,3,6\}, \{2,5,7\}\}. \end{aligned}$$

The corresponding GRS is of side 28.

Only finitely many GRSs have been constructed by a direct method similar to that of Example 20.2. The orders of these include v=12, 18, 24, 30, 42 (see Stiffler and Blake, 1975), v=15 (see Rosa, 1978), and $v = 21$ (see Colbourn and Rosa, 1995). However, the recursive constructions to follow yield GRSs for infinitely many orders.

Two Steiner systems S$(3, 5, v)$ $(V, \mathcal{B}_1), (V, \mathcal{B}_2)$ are *orthogonal* if

(i) $\mathcal{B}_1 \cap \mathcal{B}_2 = \varnothing$, and

(ii) whenever Q, Q' are two blocks of $\mathcal{B}_1$ such that $Q \cap Q' = P$, $|P|$=2, and R, R' are such that $(Q \setminus P) \cup R \in \mathcal{B}_2$, $(Q' \setminus P) \cup R' \in \mathcal{B}_2$ then $R \neq R'$.

The existence of orthogonal STSs implies the existence of (ordinary) Room squares (§14.4). For GRSs, we have the following analogue:

Theorem 20.3 *If there exist two orthogonal Steiner systems $S(3,5,v)$ then there exists a $GRS(v+1)$.*

Proof Label the rows and columns of a $\binom{v}{2}$ by $\binom{v}{2}$ array with 2-subsets of a v-set V. If $(V,\mathcal{B}_1),(V,\mathcal{B}_2)$ are two orthogonal Steiner systems $S(3,5,v)$, place the triple $\{a,b,c\}$ in the cell in row labelled $\{x,y\}$ and column labelled $\{u,v\}$ if $\{a,b,c,x,y\} \in \mathcal{B}_1$ and $\{a,b,c,u,v\} \in \mathcal{B}_2$. Place the triple $\{a,b,\infty\}$ (where ∞ is the additional element) in the 'diagonal' cell in row labelled $\{a,b\}$ and column labelled $\{a,b\}$. □

Unfortunately, the only order v for which a pair of orthogonal $S(3,5,v)$s is presently known to exist is v=17 (Zhu, 1984), but the GRS(18) is known already.

But the next theorem of Ganter (1976), which is a generalization of a result due to Stiffler and Blake (1975), yields several infinite classes of GRSs.

Theorem 20.4 *If there exists a 3-wise balanced design $S(3,K,v)$ and for each $k \in K$ there exists a $GRS(k+1)$ then there exists a $GRS(v+1)$.*

Corollary 20.5 *(Stiffler and Blake, 1975) If there exists a $GRS(k+1)$ and a Steiner system $S(3,k,v)$ then there exists a $GRS(v+1)$.*

In particular, there exists a GRS(v) for $v = q^a + 2$, $q \in \{7,13,16\}$, since an $S(3,q+1,q^a+1)$ is known to exist for every prime power q.

We also have the following multiplicative recursive construction from Rosa (1978).

Theorem 20.6 *If a $GRS(v)$ exists and v is even then there exists a $GRS(3v)$.*

However, the spectrum for GRSs remains to be determined. The smallest order v for which the existence of a GRS(v) is undecided is $v = 27$.

20.3 Kirkman squares

A *Kirkman square* is a (regular) MKD$(2,1,v)$ (briefly, a KS(v)). Its underlying design is an STS(v), and its existence is equivalent to the existence of a pair of orthogonal resolutions of this underlying STS(v) (see Theorem 20.1); the latter is said to be *doubly resolvable* or *orthogonally resolvable.*

The first example of a doubly resolvable STS was obtained by Mathon and Vanstone (1980, 1980a) who used PG(3,3) to construct a KS(27). Trivially, there exists no KS(9), and an exhaustive check of the four nonisomorphic resolvable STS(15)s and their resolutions reveals that there exists no KS(15). The existence of a KS(21) is in doubt; thus the example of Mathon and Vanstone yields the smallest order for which a KS is known to exist.

The main tool in direct constructions of Kirkman squares is the starter-adder method; see Janko and van Trung (1981), Vanstone and Rosa (1982), Rosa and Vanstone (1985), and Stinson and Vanstone (1985). Kirkman squares constructed by this method have the property that the automorphism group of the two orthogonal resolutions of the underlying STS is transitive on the parallel classes. Thus the order v of such a square must satisfy $v \equiv 3 \pmod{12}$. If $\mathcal{R} = \{R_1, \ldots, R_r\}$ and $\mathcal{S} = \{S_1, \ldots, S_r\}$ are two orthogonal resolutions of $(V, \mathcal{B})$ where $V = \mathbb{Z}_{(v-1)/2} \times \{0,1\} \cup \{\infty\}$, and $\alpha = (0_0 1_0 \cdots ((v-1)/2)_0)\ ((0_1 1_1 \cdots (v-1)/2)_1)(\infty)$, $\alpha R_i = R_{i+1}, \alpha S_i = S_{i+1}$, and if, say, $R_1 = \{B_1, \ldots, B_r\}$, $S_1 = \{C_1, \ldots, C_r\}$, and $C_i = B_i + a_i$ for $i = 1, \ldots, r$ then $(B_1, \ldots, B_r)$ is the *starter* and $(a_1, \ldots, a_r)$ is the *adder*, and the corresponding KS is a *starter-adder* Kirkman square. The orthogonality of $\mathcal{R}$ and $\mathcal{S}$ ensures that $a_i \neq a_j$ for $i \neq j$.

While enumerating all KTS(27)s with a cyclic automorphism of order 13 (thus transitive on parallel classes), Janko and van Trung (1981) found that there are exactly three nonisomorphic starter-adder KS(27)s. One of these is displayed in Example 20.7.

Example 20.7 A Kirkman square of order 27.

B_1	B_2	B_3	B_4	B_5
$\infty 0_0 0_1$	$2_0 9_0 12_1$	$3_0 7_0 11_1$	$5_0 10_0 6_1$	$4_0 3_1 9_1$
$a_1 = 0$	$a_2 = 1$	$a_3 = 5$	$a_4 = 3$	$a_5 = 6$
B_6	B_7	B_8	B_9	
$6_0 4_1 8_1$	$8_0 1_1 2_1$	$1_0 11_0 12_0$	$5_1 7_1 10_1$	
$a_6 = 10$	$a_7 = 7$	$a_8 = 8$	$a_9 = 12$	

Rosa and Vanstone (1985) and Stinson and Vanstone (1985) conducted a search for a restricted class of starter-adder KS(v)s for which the underlying STS(v)s contain a maximum order subsystem, i.e. a subsystem of order $(v-1)/2$. In this way, several KS(39)s and KS(51)s were found.

Baker (1984) gave a construction for an infinite class of starter-adder KSs.

Theorem 20.8 *For all* $m \equiv 1 \pmod 3$ *there exists a* KS$(2^{2m+2} - 1)$.

In particular, this gives a KS(63).

Fuji-Hara and Vanstone (1982) have shown that the affine space AG$(2^{i+1}, 3)$, $i \geq 1$, admits a skew resolution, and hence is doubly resolvable (as the affine resolution and a skew resolution are orthogonal). In particular, this gives a KS(81).

Another construction giving infinite classes of starter-adder KSs is obtained from the following theorem.

Theorem 20.9 *If there exists a starter-adder* KS$(2q+1)$ *on the set of elements* $\mathbb{F}_q \times \{0,1\} \cup \{\infty\}$ *then for each* $n \geq 1$ *there exists a* KS$(2q^n + 1)$ *on the set of elements* $\mathbb{F}_{q^n} \times \{0,1\} \cup \{\infty\}$.

We also have a PBD-closure theorem for Kirkman squares:

Theorem 20.10 *If there exists a* PBD$(v, K, 1)$ *and if for each* $k \in K$ *there exists a* KS$(2k+1)$ *then there exists a* KS$(2v+1)$.

Proof Let $(V, \mathcal{D})$ be a PBD$(v, K, 1)$. Put $W = V \times \{0,1\} \cup \{\infty\}$ where ∞ is a new element. Label a square S of side v with the elements of V. For a block $B \in \mathcal{D}$, $|B| = k$, let S_B be the subsquare of S whose rows and columns are labelled by the elements of B. Place a copy of a KS$(2k+1)$ (which exists by the assumption) whose elements are $B \times \{0,1\} \cup \{\infty\}$ on S_B in such a way that the diagonal cell (i,i) contains the triple $\{\infty, i_1, i_2\}$ for each $i \in B$.

Performing the above for each block $B \in D$ but placing for each $i \in V$ only one copy of the triple $\{\infty, i_1, i_2\}$ in the diagonal cell (i,i) yields a KS$(2v+1)$ whose set of elements is W. □

Theorem 20.10 together with some small known KSs suffices to prove the following asymptotic existence result.

Theorem 20.11 *For all $v \equiv 3 \pmod 6$, v sufficiently large, there exists a KS(v).*

Proof KS(v)s for v=39 and 81 are known to exist, so choose $K = \{19, 40\}$. By Wilson's theorem (Theorem 3.1), there exists v_0 such that for all $v \equiv 1 \pmod 3$ and $v > v_0$, there exists a PBD$(v, K, 1)$. □

Two further recursive constructions for Kirkman squares are due to Vanstone (1980) and to Colbourn, Curran, and Vanstone (1987). We state these as the next two theorems.

Theorem 20.12 *(The singular direct product for Kirkman squares). If there exists a KS$(2v_1+1)$, and a KS$(2v_2+1)$ with a sub-KS$(2v_3+1)$, and there exist three MOLS$(v_2 - v_3)$ then there exists a KS$(2[(2v_1+1)(v_2-v_3)+v_3]+1)$.*

Theorem 20.13 *(The Kirkman frame construction for Kirkman squares). If there exists a (t,u) Kirkman frame, a KS$(2(us+l)+1)$ with a sub-KS$(2l+1)$, and there exist three MOLS(s) then there exists a KS$(2t(us+l)+1)$.*

It has been conjectured that a KS(v) exists for all $v \equiv 3 \pmod 6$, $v \geq 27$. However, at present there are only six orders $v \leq 100$ for which KS(v) is known to exist: v=27, 39, 51, 63, 81, 87.

20.4 Kirkman cubes

A regular MKD$(3,1,v)$ is a *Kirkman cube*. It is *strong* (sKC) if it is of strength 2, and *weak* (wKC) if it is of strength 3.

The starter-adder method of direct construction is applicable to Kirkman cubes as well. The first example of a strong Kirkman cube, an sKC(255), was constructed by Vanstone (1983). The only other sKC obtained by a direct construction (of starter-adder type) is an sKC(39); see Rosa and Vanstone (1985). Stinson and Vanstone (1986) have constructed a set of six pairwise orthogonal packings in PG(5,2) which implies the existence of an sKC(63) (actually, this yields a 6-dimensional Kirkman cube of order 63 and strength 2). Nazarok (1995) gives a construction of four pairwise orthogonal resolutions of a KTS$(2q+1)$ for certain prime powers q; in particular, there exists a 4-dimensional Kirkman cube of order 87. Since the PBD-closure theorem of §20.3 holds for Kirkman cubes as well, Stinson and Vanstone (1986) obtain the following asymptotic result:

Theorem 20.14 *For every $v \equiv 3 \pmod{12}$, v sufficiently large, there exists an sKC(v).*

Lamken has shown that Theorem 20.14 holds for all sufficiently large orders v satisfying $v \equiv 3 \pmod 6$.

Weak Kirkman cubes are somewhat easier to come by. While there is no wKC(9), Rosa and Vanstone (1980) give a wKC(15). In fact, a wKC(v) is known to exist for all $v \equiv 3 \pmod 6$, $15 \leq v \leq 63$ except for v=33 and 57. Also, Theorem 20.11 holds for wKCs as well.

Theorem 20.15 *For every $v \equiv 3 \pmod 6$, $v \geq v_0$, there exists a wKC(v).*

Rosa and Vanstone (1983) have conjectured that v_0=15. Proving this, or at least obtaining a good bound on v_0, is contingent on the existence of a wKC(33) which is unknown at present.

20.5 Doubly resolvable twofold triple systems

All MKD(d, λ, v)s considered so far had λ=1. Next we consider the case where the underlying triple system is a twofold triple system TTS(v). The corresponding MKD($2, 2, v$) is a *twofold Kirkman square*; its side is $v-1$. Colbourn and Vanstone (1984) used the starter-adder method to construct twofold Kirkman squares of orders v=15, 18, 21, 24, 27, and 30. They chose the underlying TTS(v) to be 1-rotational; this is a convenient choice since the 1-rotational automorphism partitions the triples of a TTS(v) into $v/3$ orbits, and $v/3$ is also the number of triples in a parallel class. One of the resolutions is relatively easy to obtain, by choosing a starter containing one block from each orbit. In spite of the relative ease with which acceptable adders (those producing an orthogonal resolution) can be found for small orders, no direct construction for an infinite class of starter-adder obtained twofold Kirkman squares appears to be known.

An analogue of Theorems 20.12 and 20.15 (the asymptotic existence of twofold Kirkman squares for all sufficiently large orders $v \equiv 0 \pmod 3$) is also easily obtained. In fact, Lamken (1995) proves the following result:

Theorem 20.16 *A twofold Kirkman square of order v exists if and only if $v \equiv 0 \pmod 3$ except for $v \in \{6, 9\}$ and possibly for $v \in \{72, 78, 90, 114, 117, 126\}$.*

20.6 Room rectangles and parallelepipeds

In the definition of a d-orthogonal set of resolutions, the resolutions consist of parallel classes (which may be viewed as 1-designs on the underlying set V). Relaxing this condition by allowing the resolutions to consist of other designs led Kramer and Mesner (1974) to the following definition (actually, their definition is more general; we confine ourselves here to the case when the underlying design is still a triple system).

A *Room rectangle* RR($m, n; (v, v_1, v_2), (\lambda, \lambda_1, \lambda_2)$) is an $m \times n$ array R such that

1. each cell is either empty or contains a triple of an underlying $TS(v, \lambda)$ $(V, \mathcal{B})$,
2. each triple of $\mathcal{B}$ is contained in exactly one cell of the array, and
3. the triples in the nonempty cells of any row [column] are the triples of a $TS(v_1, \lambda_1)$ [$TS(v_2, \lambda_2)$].

Thus in a Room rectangle, a resolution does not necessarily consist of parallel classes but is rather a partition of the underlying design into subdesigns each of which is a $TS(v_i, \lambda_i)$. Here we permit λ=0, in which case the $TS(v_i, 0)$ must be a 1-design.

A *Room parallelepiped* $RP((n_1, \ldots, n_d); (v, v_1, \ldots, v_d); (\lambda, \lambda_1, \ldots, \lambda_d))$ of dimension d is an $n_1 \times \cdots \times n_d$ array R satisfying (1) and (2) above, and

4. the triples in the nonempty cells with fixed coordinate i_c in the ith coordinate direction are the triples of a $TS(v_i, \lambda_i)$.

We can define a d-orthogonal set of resolutions of this more general kind in a direct analogy to the definition given in §20.1. The following extension of Theorem 20.1 is then immediate.

Theorem 20.17 *A Room parallelepiped of dimension d exists if and only if there exists a d-orthogonal set of d resolutions of the underlying TS*(v, λ).

The definition of a Room rectangle allows us to extend the considerations of §20.4 to twofold triple systems $TS(v, 2)$ when $v \equiv 1 \pmod 3$. The underlying design in this case is an *almost resolvable* $TS(v, 2)$, and the orthogonal resolutions consist of almost parallel classes. The Room rectangle corresponding to such a doubly almost resolvable $TTS(v)$ is an $RR(v, v; (v, v-1, v-1); (2, 0, 0))$. An example for v=10 is taken from Colbourn and Vanstone (1984):

<table>
<tr><td>ABF</td><td></td><td></td><td></td><td>CHI</td><td></td><td></td><td>DEG</td><td></td><td></td></tr>
<tr><td>DIJ</td><td>BCG</td><td></td><td></td><td></td><td></td><td></td><td></td><td>AEH</td><td></td></tr>
<tr><td></td><td>EFJ</td><td>CDH</td><td></td><td></td><td></td><td></td><td></td><td></td><td>ABI</td></tr>
<tr><td></td><td></td><td>AFG</td><td>DEI</td><td></td><td>BCJ</td><td></td><td></td><td></td><td></td></tr>
<tr><td></td><td></td><td></td><td>BGH</td><td>AEJ</td><td></td><td>CDF</td><td></td><td></td><td></td></tr>
<tr><td></td><td></td><td></td><td>ACJ</td><td></td><td></td><td>BEH</td><td></td><td>FGI</td><td></td></tr>
<tr><td></td><td></td><td></td><td></td><td>BDF</td><td></td><td></td><td>ACI</td><td></td><td>GHJ</td></tr>
<tr><td>CEG</td><td></td><td></td><td></td><td></td><td>FHI</td><td></td><td></td><td>BDJ</td><td></td></tr>
<tr><td></td><td>ADH</td><td></td><td></td><td></td><td></td><td>GIJ</td><td></td><td></td><td>CEF</td></tr>
<tr><td></td><td></td><td>BEI</td><td></td><td></td><td>ADG</td><td></td><td>FHJ</td><td></td><td></td></tr>
</table>

Colbourn and Vanstone use the starter-adder method to find doubly almost resolvable $TTS(v)$s for v=13, 16, 19, 25, and 28. Lamken (1994) has the strongest result in this area:

Theorem 20.18 *A doubly near resolvable TTS*(v) *exists for* $v \equiv 1 \pmod 3$, $v \geq 10$, *except possibly when* $v \in \{34, 70, 85, 88, 115, 124, 133, 142\}$.

One of the earliest (and nicest) examples of Room parallelepipeds are the Steiner tableaus of Kramer and Mesner (1974a). Each of the three nonisomorphic

Steiner tableaus T(3,7) constructed by Kramer and Mesner is an RP((7,7,7); (9,9,9,9); (7,1,1,1)) which is regular of strength 3 (where the meaning of 'regular of strength t' is the same as that given at the beginning of §20.1). In other words, it is a 3-dimensional cube whose underlying design consists of all triples of a 9-set, and the collection of triples in the nonempty cells of any plane in any of the three directions is an STS(9). It follows that the seven planes in any of the three directions constitute a large set of STS(9)s. We refer the reader for further details to Kramer and Mesner (1974a).

Another interesting example is a (square) RR((8,8);(8,7,7);(5,1,1)) due to Mullin and Rosa (1977) from a pair of orthogonal S(3,4,8)s. The triples in the nonempty cells of any row (column) are the triples of an STS(7), and the collection of the eight rows (eight columns) constitutes an overlarge set of eight STS(7)s. This Room rectangle is:

	235	346	457	156	267	137	124
467		037	056	236	024	257	345
157	456		014	067	347	035	136
126	247	567		025	017	145	046
237	057	135	167		036	012	256
134	367	016	246	127		047	023
245	034	147	027	357	123		015
356	026	045	125	013	146	234	

Other interesting examples of Room rectangles and Room parallelepipeds (many of them 'proper', i.e. non-square and non-cube) are given by Kramer and Magliveras (1980), Kramer and Mesner (1980), Kramer, Kreher, Magliveras, and Mesner (1981), and Kramer, Kreher, and Mesner (1982).

20.7 Work points

1. What is the spectrum for GRS(v)s? In particular, does a GRS(27) exist?
2. Does a Kirkman square KS(v) exist for every $v \equiv 3 \pmod 6$, $v \geq 27$?
3. Does there exist a starter-adder KS(v) for every $v \equiv 3 \pmod 6$, $v \geq 27$?
4. Does there exist an sKC(v) for every sufficiently large $v \equiv 3 \pmod 6$?
5. Does there exist a wKC(v) for every $v \equiv 3 \pmod 6$, $v \geq 15$?
6. Settle the possible exceptions in Theorem 20.16.
7. Settle the possible exceptions in Theorem 20.18.
8. Is there a starter-adder generated twofold Kirkman square for every $v \equiv 0 \pmod 3$, $v \geq 15$?
9. A *Wingo square* is a $\lfloor (v-1)/2 \rfloor \times \lfloor (v-1)/2 \rfloor$ square in which each cell is either empty or contains one triple. The triples in the square form an MPT(v). The partition of the triples induced by the rows, and also by the columns, is a resolution of the MPT. Hence when $v \equiv 0 \pmod 6$, the MPT is a doubly resolvable NKTS. Smith (1977) gives a Wingo square with $v = 24$. Rosa (1980a) describes consequences of this. When do Wingo squares exist?

21

STSS WITH TWO SUBSYSTEMS

Colbourn, Oravas, and Rees (1999) address the existence of Steiner triple systems of order v that contain two subsystems, one of order u and the other of order w, which are either disjoint, or share precisely one element. This problem can be cast in quite general terms as follows.

An *incomplete Steiner triple system*, or *ISTS*, is a v-set V of elements, together with a set $\mathcal{H}$ of subsets of V called *holes*, and a set $\mathcal{B}$ of triples; the defining property is that every pair of elements in V appears either in at least one hole of $\mathcal{H}$ or in exactly one triple of V, but not both. Following Rees and Stinson (1989a), a *$\diamond$-incomplete Steiner triple system*, or *$\diamond$-$(v;w,u;z)$-ISTS*, $(V;W,U;\mathcal{B})$, is a v-set V; two subsets $W,U \subseteq V$ with $|W| = w$ and $|U| = u$ for which $|U \cap W| = z$; and a set $\mathcal{B}$ of 3-element subsets (*triples*) of V in which every pair of elements from V not contained within U or W appears in exactly one triple in $\mathcal{B}$, while pairs within U or W appear in no triple.

A Steiner triple system of order v containing disjoint subsystems of orders w and u is a $\diamond$-$(v;w,u;0)$-ISTS, while a Steiner triple system of order v containing subsystems of orders w and u sharing a single element is a $\diamond$-$(v;w,u;1)$-ISTS. The notion of $\diamond$-ISTSs permits the size of the intersection to exceed one, and also permits v, w, and u to have integral values that are not themselves orders of Steiner triple systems. We focus on the cases when $z = 0$ and $z = 1$ here, but permit v, w, and u to take on values that need not be orders of STSs.

Theorem 1.15 can be applied to establish a useful auxiliary result on latin squares:

Theorem 21.1 *A latin square of order n having a subsquare A of order a, and a subsquare B of order b, so that $a \geq b$ and A and B share no rows, columns, or symbols, exists if and only if $n \geq 2a + b$.*

Proof Consider one row of B, and examine the a cells in the columns of A; these cells contain a distinct symbols, none of which appear in A or B. Hence $n \geq 2a + b$.

Now suppose that $n = 2a + b + \varepsilon$. Form a partial latin square P with rows $\rho_1, \ldots, \rho_{a+b}$, columns $\gamma_1, \ldots, \gamma_{a+b}$, and symbols $\sigma_1, \ldots, \sigma_{a+b}$, with A on rows $\rho_1, \ldots, \rho_a$, columns $\gamma_1, \ldots, \gamma_a$, and symbols $\sigma_1, \ldots, \sigma_a$, and B on rows $\rho_{a+1}, \ldots, \rho_{a+b}$, columns $\gamma_{a+1}, \ldots, \gamma_{a+b}$, and symbols $\sigma_{a+1}, \ldots, \sigma_{a+b}$. Form a graph with $(\{\rho_1, \ldots, \rho_{a+b}\}, \{\gamma_1, \ldots, \gamma_{a+b}\})$ as a bipartition of its vertex set, and place an edge between ρ_i and γ_j exactly when $i \leq a < j$ or $j \leq a < i$ to form a graph G. G consists of two vertex-disjoint copies of $K_{a,b}$, and hence has a proper a-edge-colouring. Use Theorem 1.15 to form a proper $(a + \varepsilon)$-edge-colouring in which every colour is used at least $\lfloor 2ab/(a+\varepsilon)\rfloor$ times, and use the colours $\sigma_{a+b+1},\ldots,\sigma_n$. When edge $\{\rho_i,\gamma_j\}$ is assigned colour σ_k, place σ_k in cell (i,j) of P. Now P is an $(a+b)\times(a+b)$ array with all cells filled, containing symbols $\sigma_1,\ldots,\sigma_n$, in which each symbol appears a, b, $\lfloor 2ab/(a+\varepsilon)\rfloor$, or $\lceil 2ab/(a+\varepsilon)\rceil$ times. Ryser's theorem (Theorem 1.6) ensures that P can be completed to an $n\times n$ latin square if every symbol occurs at least $2a+2b-n=b-\varepsilon$ times. So it suffices to check that $\lfloor 2ab/(a+\varepsilon)\rfloor \geq b-\varepsilon$. When $\varepsilon<a$, the left-hand side is at least b; and when $\varepsilon\geq a\geq b$, the right-hand side is nonpositive, so the requirement is met. □

We employ a holey type of GDD first defined by Rees and Stinson (1989a), which generalizes the IGDDs of §6.5.4. An *incomplete group divisible design* (*IGDD*) of type $(g_1,h_1)^{t_1}\cdots(g_s,h_s)^{t_s}$ is a set V of $\sum_{i=1}^{s} t_i g_i$ points; a partition $\mathcal{G}$ of V into $\sum_{i=1}^{s} t_i$ *groups* with t_i groups having size g_i, for $1\leq i\leq s$; a *hole* $H\subset V$ which intersects the t_i groups of size g_i in h_i points each, for $1\leq i\leq s$; and a set $\mathcal{B}$ of triples for which no triple contains a pair inside a group or in the hole, but every other pair is contained in exactly one triple. By permitting a second hole H' disjoint from H, we obtain an IGDD of type $(g_1;h_1,\ell_1)^{t_1}\cdots(g_s;h_s,\ell_s)^{t_s}$ when H' intersects the t_i groups of size g_i in ℓ_i points each, for $1\leq i\leq s$.

21.1 Necessary conditions for ◇-ISTSs

Suppose that a ◇-$(v;w,u;z)$-ISTS $(V;W,U;\mathcal{B})$ exists. We suppose without loss of generality from this point on that $w\geq u>z$, and introduce the notation $(v;\{w,u\};z)$ for the type when no assumption about the order of the hole sizes is made. Let $W'=W\setminus U=\{q_1,\ldots,q_{w-z}\}$. Let $p\in U\setminus W$, and consider the triples of the form $\{p,q_i,r_i\}$ in $\mathcal{B}$ for $1\leq i\leq w-z$. Now $r_i\notin W\cup U$, and $r_i\neq r_j$ unless $i=j$. It follows that $V\setminus(W\cup U)$ contains at least $w-z$ elements, and hence that $v\geq 2w+u-2z$. This is the basic necessary condition on the size of the containing system.

Counting pairs that are to appear in triples of $\mathcal{B}$, we require that

$$\binom{v}{2}-\binom{w}{2}-\binom{u}{2}+\binom{z}{2}\equiv 0 \pmod 3,$$

since each triple contains three pairs. In a similar manner, by counting triples involving a specified element, we require that the number of pairs containing that element be even. Specifically, since an element in $V\setminus(W\cup U)$ appears in $v-1$ pairs, v must be odd. Similarly, since an element in $W\setminus U$ appears in $v-w$ pairs, w must be odd; and symmetrically, u must be odd. Finally, if there is an element in $U\cap W$, then z must be odd; however, $z=0$ is permitted since then no element at all is present in the intersection of U and W.

When the ◇-$(v;w,u;z)$-ISTS is to lead to an STS with subsystems of orders w and u intersecting in z elements, the z elements must themselves induce a subsystem and in this case $u\geq 2z+1$ is necessary. Consider the partition of V

into sets Z of size z, $W \setminus Z$ of size $w-z$, $U \setminus Z$ of size $u-z$, and $X = V\setminus(W\cup U)$ of size $x = v - w - u + z$. Triples with elements in Z contain $xz/2$ pairs from X, leaving $x(x-1-z)/2$ pairs to appear with elements from $W \setminus Z$ and $U \setminus Z$. It follows that

$$\frac{1}{2}x(x-1-z) \geq \frac{1}{2}(u-z)(x-w+z) + \frac{1}{2}(w-z)(x-u+z),$$

and hence that

$$(x-(w-z))(x-(u+1)) \geq (2z+1-u)(w-z). \tag{21.1}$$

When $x = w - z$ or equivalently $v = 2w + u - 2z$, the left-hand side of (21.1) is zero, so that $u \geq 2z + 1$. As v increases but w, u, and z remain constant, this relation between u and z is relaxed. When $u = 2z+1$, the right-hand side of (21.1) is zero, so that when $w - z < x < u + 1$, no $\diamond$-$(v; w, 2z+1; z)$-ISTS can exist. For example, while a $\diamond$-$(31; 15, 15; 7)$-ISTS exists, no $\diamond$-$(v; 15, 15; 7)$-ISTS exists when $v \in \{33, 37\}$, but they are not excluded by (21.1) when $v \geq 39$. Apparently, the necessary conditions are somewhat complicated by (21.1); fortunately, when $z \in \{0, 1\}$, (21.1) is always satisfied. Hence we do not examine (21.1) in more detail here.

Since the conditions are somewhat unwieldy, we translate the modulo 3 conditions and those modulo 2 in Table 21.1 to indicate, for admissible selections of u, w, and z, all possible congruence classes modulo 6 in which $v - (2w + u - 2z)$ can lie.

$z = 0$ or $z \equiv 3 \pmod 6$				$z \equiv 1 \pmod 6$				$z \equiv 5 \pmod 6$			
	u				u				u		
w	1	3	5	w	1	3	5	w	1	3	5
1	0,4	2,4	4	1	0,2	0,4	0	1	–	–	0,4
3	0,2	0,4	0	3	2,4	0,2	2	3	–	–	0,2
5	0	4	–	5	2	0	–	5	0,2	0,4	0

Table 21.1 *Possible excess (modulo 6) of order over the minimum*

Theorem 6.15 establishes sufficiency when $u \in \{1, 3\}$, so we may assume that $u \geq 5$. Theorem 6.16 establishes sufficiency when $w = u$ and $z = 0$.

21.2 Meeting the bound

When $v \geq w \geq u > z$, a $\diamond$-$(v; w, u; z)$-ISTS can exist only if $v \geq 2w + u - 2z$. In this section, we treat the case of equality. Rosa (1970) considered a restricted version of this case in the context of producing a $v \mapsto 2v + 7$ construction for STSs, which is given in §3.2.

Theorem 21.2 *A $\diamondsuit$-$(2w+u-2z; w, u; z)$-ISTS with $w \geq u > z \geq 0$ exists whenever v, w, and u are all odd, $u \geq 2z+1$, and $(w-z)(u-2z-1) \equiv 0 \pmod 3$.*

Proof We consider three cases for z.

Case 1 z is odd. By Theorem 6.15, there is a 3-GDD of type $1^{w-z}(w-u+z)^1$. Let V be its element set, $\mathcal{B}$ its block set, and G the group of size $w-u+z$. Let $G = \{\ell_1, \ldots, \ell_{w-u+z}\}$. Let $\mathcal{B}_1 = \{B \in \mathcal{B} : B \cap G = \varnothing\}$. For $1 \leq i \leq w-u+z$, form a 1-factor $F_i = \{\{p,q\} : \{\ell_i, p, q\} \in \mathcal{B}\}$. Now define $V' = V \backslash G$, a set of $w-z$ points on which the triples in $\mathcal{B}_1$ and the 1-factors $F_1, \ldots, F_{w-u+z}$ are defined. Define sets $W' = \{a_1, \ldots, a_{w-z}\}$, $U' = \{b_1, \ldots, b_{u-z}\}$, and $Z = \{c_1, \ldots, c_z\}$.

Define $\mathcal{B}_2 = \{\{c_i, x, y\} : \{x,y\} \in F_i, 1 \leq i \leq z\}$. Now form a graph $H = (V', E')$ where $E' = \bigcup_{i=z+1}^{w-u+z} F_i$. H is a $(w-u)$-regular graph having a partition into 1-factors. Since $w-z \geq w-u$, by Theorem 1.15 H has a partition $H_1, \ldots, H_{w-z}$ into matchings each of size $(w-u)/2$. Then define $\mathcal{B}_3 = \{\{a_i, x, y\} : \{x,y\} \in H_i, 1 \leq i \leq w-z\}$.

Next form a bipartite graph G' on $V' \cup W'$ whose edge set is $E(G') = \{\{x,y\} : x \in V', y \in W', \nexists B \in \mathcal{B}_3 \text{ s.t. } \{x,y\} \subset B\}$. Then G' is a regular graph of degree $u-z$; since it is bipartite, it has a partition $J_1, \ldots, J_{u-z}$ into 1-factors. So define $\mathcal{B}_4 = \{\{b_i, x, y\} : \{x,y\} \in J_i, 1 \leq i \leq u-z\}$. Then the required $\diamondsuit$-ISTS is

$$(V' \cup W' \cup U' \cup Z; W' \cup Z, U' \cup Z; \mathcal{B}_1 \cup \mathcal{B}_2 \cup \mathcal{B}_3 \cup \mathcal{B}_4).$$

Case 2 $z = 0$ and $w \equiv 3 \pmod 6$. There is a Kirkman triple system of order w, $(V', \mathcal{B})$, with parallel classes $\mathcal{P}_1, \ldots, \mathcal{P}_{(w-1)/2}$. Let $V' = \{a_1, \ldots, a_w\}$, $W = \{b_1, \ldots, b_w\}$, and $U = \{c_1, \ldots, c_u\}$. Define

$$\begin{aligned}
\mathcal{B}_1 &= \{\{a_i, a_j, a_k\} : \{a_i, a_j, a_k\} \in \mathcal{P}_\ell, 1 \leq \ell \leq (u-1)/2\} \\
\mathcal{B}_2 &= \{\{a_i, a_j, b_k\}, \{a_i, b_j, a_k\}, \{b_i, a_j, a_k\} : \\
&\quad\ \{a_i, a_j, a_k\} \in \mathcal{P}_\ell, (u+1)/2 \leq \ell \leq (w-1)/2\} \\
\mathcal{B}_3 &= \{\{c_u, a_i, b_i\} : 1 \leq i \leq w\} \\
\mathcal{B}_4 &= \{\{c_{2\ell}, a_i, b_j\}, \{c_{2\ell}, a_j, b_k\}, \{c_{2\ell}, a_k, b_i\}, \\
&\quad\ \{c_{2\ell-1}, a_i, b_k\}, \{c_{2\ell-1}, a_j, b_i\}, \{c_{2\ell-1}, a_k, b_j\} : \\
&\quad\ \{a_i, a_j, a_k\} \in \mathcal{P}_\ell, 1 \leq \ell \leq (u-1)/2\}.
\end{aligned}$$

Then $(V' \cup W \cup U; W, U; \mathcal{B}_1 \cup \mathcal{B}_2 \cup \mathcal{B}_3 \cup \mathcal{B}_4)$ is the required $\diamondsuit$-ISTS.

Case 3 $z = 0$ and $u \equiv 1 \pmod 6$. Let $n = (u-1)/6$. Form a (pure or hooked) Skolem n-sequence $\{(a_i, b_i) : 1 \leq i \leq n\}$. Let $V' = \{d_0, \ldots, d_{w-1}\}$, $W = \{e_0, \ldots, e_{w-1}\}$, and $U = \{c_1, \ldots, c_u\}$. Define $\mathcal{B}_1 = \{\{d_j, d_{j+i}, d_{j+n+b_i}\} : 1 \leq i \leq n\}$, reducing subscripts modulo w as needed. Now form the collection of pairs $\mathcal{P} = \{\{j, w-j\} : 2j \bmod w \notin \{i, n+a_i, n+b_i, w-i, w-n-a_i, w-n-b_i\} \text{ for } 1 \leq i \leq n\}$. $\mathcal{P}$ contains $(w-u)/2$ pairs, which are disjoint. Then define $\mathcal{B}_2 = \{\{e_j, d_{x+j}, d_{y+j}\} : \{x,y\} \in \mathcal{P}, 0 \leq j < w\}$, reducing subscripts modulo w as needed. Now form a graph G whose edge set contains all pairs $\{x,y\}$ with $x \in V'$ and $y \in W$ that do not appear in a triple of $\mathcal{B}_2$. G is u-regular and bipartite, and hence the remainder of the proof parallels Case 1 closely. □

21.3 Near the minimum

Whenever the numerical necessary conditions permit the bound on v, namely $v \geq 2w + u - 2z$, to be met, Theorem 21.2 produces the required $\diamond$-ISTS. However, from Table 21.1, we can see that for certain admissible choices of w, u, and z, the necessary conditions prohibit an ISTS with $v = 2w + u - 2z$, but permit one with $v = 2w + u - 2z + 2$ or $v = 2w + u - 2z + 4$. In this section, we examine these cases 'near' the minimum when $z \in \{0, 1\}$.

We begin by establishing the existence of a class of IGDDs.

Lemma 21.3 *Let $s \geq 4$ and $k \in \{0, 2, 4, 6\}$. Then for each $1 \leq t \leq s$ there exists an IGDD of type $(12 + k; 6, k)^1(18; 6, 6)^{t-1}(12; 6, 0)^{s-t}$.*

Proof The elements are as follows:

1. sets $H_1, \ldots, H_s$, each containing 6 elements;
2. sets $J_1, \ldots, J_s$, each containing 6 elements; and
3. sets $K_1, \ldots, K_t$, with $K_2, \ldots, K_t$ each containing 6 elements, and K_1 containing k elements.

We let ϕ be a bijection from elements in H_i to elements in J_i for $1 \leq i \leq s$. Groups are $H_i \cup J_i \cup K_i$ for $1 \leq i \leq t$, and $H_i \cup J_i$ for $t < i \leq s$. Holes are $\bigcup_{i=1}^{s} H_i$ and $\bigcup_{i=1}^{t} K_i$.

Now $s \geq 4$, and so a 1-Kirkman frame of type 6^s exists by Theorem 19.19. Form one with groups $H_1, \ldots, H_s$, and for $1 \leq i \leq s$, let $\mathcal{P}_{i1}$, $\mathcal{P}_{i2}$, and $\mathcal{P}_{i3}$ be the frame parallel classes missing group H_i. Now we form the triples. For $t + 1 \leq i \leq s$, and $1 \leq j \leq 3$, whenever $\{x, y, z\} \in \mathcal{P}_{ij}$, we add the triples $\{x, \phi(y), \phi(z)\}$, $\{\phi(x), y, \phi(z)\}$, and $\{\phi(x), \phi(y), z\}$. We do the same for $i = 1$ and $1 \leq j \leq k/2$. For $2 \leq i \leq t$ and $1 \leq j \leq 3$, whenever $\{x, y, z\} \in \mathcal{P}_{ij}$, we add the triple $\{\phi(x), \phi(y), \phi(z)\}$ and place the pairs $\{x, \phi(y)\}$, $\{x, \phi(z)\}$, $\{y, \phi(x)\}$, $\{y, \phi(z)\}$, $\{z, \phi(x)\}$, and $\{z, \phi(y)\}$ in a graph G_i. We do the same for $i = 1$ and $k/2 < j \leq 3$. Now for $1 \leq i \leq t$, G_i is a bipartite graph which is regular of degree $|K_i|$. Partition G_i into 1-factors, and add to all pairs of each of the $|K_i|$ 1-factors a different element of K_i to form triples. □

In addition to these IGDDs, we employ a type of ISTS with three holes. A *lattice ISTS* $\mathcal{L}_n(c; a, b; z)$ is an ISTS on $3n + c$ points with three holes. The first hole has size $n + a$, the second has size $n + b$, and the third has size c. The first and third holes share a points, the second and third share b points, and the first and second share z points; when $z > 0$, the latter are also contained in the third hole. The case when $n = 6$ is employed primarily, and then the subscript is omitted, to write $\mathcal{L}(c; a, b; z)$.

Theorem 21.4 *Let $z \in \{0, 1\}$ and let $a, b, c > 0$ with $c \geq a + b - z$. If there exist*

1. *an IGDD of type $(12 + k; 6, k)^1(18; 6, 6)^{t-1}(12; 6, 0)^{s-t}$,*
2. *a $\diamond$-$(12 + k + c; 6 + a, k + b; z)$-ISTS,*
3. *a $\diamond$-$(12 + c; 6 + a, c; a)$-ISTS, and*
4. *a lattice ISTS $\mathcal{L}(c; a, b; z)$,*

then there exists a $\diamond$-$(12s+6t-6+k+c; 6s+a, 6t-6+k+b; z)$-*ISTS.*

Proof To an IGDD of type $(12+k;6,k)^1(18;6,6)^{t-1}(12;6,0)^{s-t}$, adjoin c infinite points $\{\infty_1,\ldots,\infty_c\}$. For $t<i\leq s$, place on $H_i\cup J_i\cup\{\infty_1,\ldots,\infty_c\}$ the triples of a $\diamond$-$(12+c;6+a,c;a)$-ISTS, aligning the hole of size c on the infinite points, the hole of size $6+a$ on $H_i\cup\{\infty_1,\ldots,\infty_a\}$, and hence the common points on $\{\infty_1,\ldots,\infty_a\}$. For $2\leq i\leq t$, place the triples of the lattice ISTS $\mathcal{L}(c;a,b;z)$ on $H_i\cup J_i\cup K_i\cup\{\infty_1,\ldots,\infty_c\}$ aligning the third hole on the infinite points, and the first hole on $H_i\cup\{\infty_1,\ldots,\infty_a\}$. Align the second hole on $K_i\cup\{\infty_{a+1},\ldots,K_{a+b}\}$ if $z=0$, and on $K_i\cup\{\infty_a,\ldots,K_{a+b-1}\}$ if $z=1$. Finally, place on $H_1\cup J_1\cup K_1\cup\{\infty_1,\ldots,\infty_c\}$ the $\diamond$-$(12+k+c;6+a,k+b;z)$-ISTS aligning the hole of size $6+a$ on $H_1\cup\{\infty_1,\ldots,\infty_a\}$ and the hole of size k_b on $K_1\cup\{\infty_{a+1},\ldots,\infty_{a+b}\}$ if $z=0$, and on $K_1\cup\{\infty_a,\ldots,\infty_{a+b-1}\}$ if $z=1$. In the latter case, $\{\infty_a\}$ forms the intersection of the two holes. □

21.3.1 *Lattice ISTSs*

In applying Theorem 21.4, we require the IGDDs from Lemma 21.3, some $\diamond$-ISTSs, and the more complicated lattice ISTSs. In this section, we develop a uniform construction of Colbourn, Oravas, and Rees (1999) for the latter. An auxiliary result on PTSs, stated here without proof, is their first step:

Lemma 21.5 *Let n be an even integer. There is a partial STS on n points whose leave is a cyclic r-regular graph if and only if either $n\equiv 0 \pmod 6$ and r is odd; or $n\equiv 2 \pmod 6$ and $r\equiv 1 \pmod 6$; or $n\equiv 4 \pmod 6$ and $r\equiv 3 \pmod 6$.*

Theorem 21.6 *Let a,b be positive integers with $a\geq b$. Let $z\in\{0,1\}$. Let $m'=\max(2\lceil(a-z)/2\rceil, 2(a-b))$. Suppose that n is even, $n\geq 2a-b+1$, and that $2a-b\equiv 1 \pmod 6$ if $n\equiv 2 \pmod 6$ while $2a-b\equiv 3 \pmod 6$ if $n\equiv 4 \pmod 6$ (no condition on $2a-b$ is enforced when $n\equiv 0 \pmod 6$).*

Then for each even m with $0\leq m\leq 2n-2m'$, there is an $\mathcal{L}_n(2a+b-2z+m;a,b;z)$. If in addition a $\diamond$-$(2a+b-2z+m;a,b;z)$-ISTS exists, then so does a $\diamond$-$(3n+2a+b-2z+m;n+a,n+b;z)$-ISTS.

Proof Suppose first that $m>0$. Apply Lemma 21.5 to construct a partial STS $\mathcal{T}$ on $\mathbb{Z}_n$ whose leave G is a cyclic $(2a-b)$-regular graph. Let $\mathcal{R}=\{r_1,\ldots,r_{a-b}\}\subseteq\mathbb{Z}_n\setminus\{(n/2)-1,n-1\}$; $\mathcal{R}$ is a set of $a-b$ distinct residues, and we further require that the differences $\{\pm(r_2-r_1),\ldots,\pm(r_{a-b}-r_{a-b-1})\}$ are all distinct and not equal to $(n/2)$, and that all appear as differences among the edges of G. When $a=b$, $\mathcal{R}=\varnothing$.

Since $m>0$, we find $n>m'$ and hence that a set S of m' residues can be chosen from $\mathbb{Z}_n\setminus\{(n/2)-1,n-1\}$ so that $\mathcal{R}\subset S$ and $s\in S$ implies that $s+(n/2)\in S$. Then $\overline{S}=\mathbb{Z}_n\setminus S$ contains $\{(n/2)-1,n-1\}$, and $s\in\overline{S}$ implies that $s+(n/2)\in\overline{S}$. Since $\overline{S}$ has order $n-m'$, it follows from the proof of Theorem 2.1 in Rees (1987) that there is a 3-GDD $\mathcal{G}$ of type n^3m^1 on the point set $(\mathbb{Z}_n\times\{1,2,3\})\cup\{1,2,\ldots,m\}$ in which the triples that miss the group of size m partition into $n-(m/2)\geq m'$ holey parallel classes $\{T_i : i\in\mathcal{I}\subseteq\mathbb{Z}_n\}$ where

$$T_i = \begin{cases} \{\{j_1, (j+i)_2, (j+2i)_3\} : j \in \mathbb{Z}_n\} & \text{if } 0 \le i \le (n/2)-1, \\ \{\{j_1, (j+i)_2, (j+2i+1)_3\} : j \in \mathbb{Z}_n\} & \text{if } (n/2) \le i \le n-2. \end{cases}$$

Note that $S \subseteq \mathcal{I}$.

The lattice ISTS is constructed by adjoining $2a+b-2z$ new points to $\mathcal{G}$: Adjoin $\{w_1, \dots, w_{a-z}\}$ to $\mathbb{Z}_n \times \{1\}$, $\{\infty_1, \dots, \infty_{a-z}\}$ to $\mathbb{Z}_n \times \{2\}$, and $\{u_1, \dots, u_{b-z}\}$ to $\mathbb{Z}_n \times \{3\}$. When $z = 1$, adjoin a further point $w_a = u_b = \infty_a$ common to all three groups. Select a subset $S_{a-z} \subseteq S$ of size $a - z$ that contains $\mathcal{R}$ (recall that $|\mathcal{R}| = a - b \le a - z$). Break each triple in T_i, for $i \in S_{a-z}$, into its constituent pairs. Complete each of the 1-factors on $\mathbb{Z}_n \times \{2,3\}$ so generated using $w_1, \dots, w_{a-z}$, each of the 1-factors on $\mathbb{Z}_n \times \{1,3\}$ using $\infty_1, \dots, \infty_{a-z}$, and each of the $a - z - |\mathcal{R}| = b - z$ 1-factors on $\mathbb{Z}_n \times \{1,2\}$ arising from triples in $\{T_i : i \in S_{a-z} \setminus \mathcal{R}\}$ using $u_1, \dots, u_{b-z}$. Then develop each of the starter triples $\{0_1, (r_1)_2, (r_2)_2\}, \dots, \{0_1, (r_{a-b-1})_2, (r_{a-b})_2\}$ modulo n. Now construct a copy of the partial STS $\mathcal{T}$ on $\mathbb{Z}_n \times \{2\}$; then the edges in its leave G that remain are those *not* in the set $\{\pm(r_2 - r_1), \dots, \pm(r_{a-b} - r_{a-b-1})\}$, and those edges between them form a cyclic a-regular graph having a difference (namely $(n/2)$) of even order. So construct a 1-factorization of this a-regular graph and complete each of the 1-factors with an element from $\{\infty_1, \dots, \infty_a\}$. The required ISTS is now produced, with holes on $(\mathbb{Z}_n \times \{1\}) \cup \{w_1, \dots, w_a\}$, $(\mathbb{Z}_n \times \{3\}) \cup \{u_1, \dots, u_b\}$, and $\{w_1, \dots, w_{a-z}\} \cup \{\infty_1, \dots, \infty_a\} \cup \{u_1, \dots, u_{b-z}\} \cup \{1, 2, \dots, m\}$.

When $m = 0$ and $m' > 0$, we proceed similarly except that the set S of m' residues is required to *contain* $(n/2)-1$ and $n-1$. Retain all triples in the classes T_i for $i \in \overline{S}$, and manipulate the classes T_i for $i \in S_{a-z}$ to produce the ISTS, ensuring that $n - 1 \in S_{a-z}$.

When $m = m' = 0$, we find that $a = b = z = 1$ and the required ISTS is obtained by adjoining a common point to the groups of a 3-GDD of type n^3, and filling one group with an STS$(n+1)$.

The final statement of the theorem is obtained by filling the hole of size $c = 2a + b - 2z + m$ with the $\diamond$-$(c; a, b; z)$-ISTS. □

Corollary 21.7 *An $\mathcal{L}(c; a, b; z)$ exists for the sets of parameters in Table 21.2. Cases marked by $\star$ are those for which a $\diamond$-$(18 + c; 6 + a, 6 + b; z)$-ISTS can be constructed by filling the hole of size c in the lattice ISTS by a $\diamond$-$(c; a, b; z)$-ISTS. In addition, there exist $\diamond$-$(v; w, u; z)$-ISTSs for the parameter sets in Table 21.3.*

21.3.2 *Intersecting cases*

Lemma 21.8 *The following statements hold.*

1. *If $w \ge u > 1$ and either $w \equiv u \equiv 1 \pmod 6$, or $w \equiv 3 \pmod 6$ and u is odd, then a $\diamond$-$(2w + u; w, u; 1)$-ISTS exists.*
2. *If $w > u > 1$ and either $w \equiv 1 \pmod 6$ and $u \equiv 3 \pmod 6$, or $w \equiv 3 \pmod 6$ and $u \equiv 1 \pmod 6$, then a $\diamond$-$(2w + u + 2; w, u; 1)$-ISTS exists.*

Proof If $w = u$, form a 3-GDD of type $(w-1)^3 2^1$ by Theorem 6.15. Fill one group of size $w - 1$ with a 3-GDD of type $2^{(w-1)/2}$ to produce a 3-GDD of type

a	b	z	Range of m	Range of c
5	5	0	0	15
5	5	1	0,2,4	13,15,17
3	3	0	0,2,4	9*,11,13*
3	3	1	0,2,4,6,8	7*,9*,11,13*,15*
3	1	0	0,2,4	7*,9*,11
3	1	1	0,2,4	5,7*,9*
1	1	0	0,2,4,6,8	3*,5,7*,9*,11
1	1	1	0,2,4,6,8,10,12	1*,3*,5,7*,9*,11,13*

Table 21.2 *Parameters for Corollary* 21.7, *first statement*

v	$w =$ $n+a$	$u =$ $n+b$	n	a	b	z	m	Ingredient $\diamond$-$(c;a,b;z)$
63	21	19	18	3	1	1	4	(9;3,1;1)
63	21	19	18	3	1	0	2	(9;3,1;0)
65	23	19	18	5	1	1	2	(11;5,1;1)
71	23	21	18	5	3	0	4	(17;5,3;0)
71	23	21	18	5	3	0	4	(17;5,3;1)
101	35	27	24	11	3	1	6	(29;11,3;1)
107	35	33	30	5	3	0	4	(17;5,3;0)
107	35	33	30	5	3	1	6	(17;5,3;1)
113	41	31	30	11	1	1	2	(23;11,1;1)
119	41	33	30	11	3	1	6	(29;11,3,1)
125	41	39	36	5	3	1	6	(17;5,3;1)

Table 21.3 *Parameters for Corollary* 21.7, *second statement*

$(w-1)^2 2^{(w+1)/2}$. Add an infinite point to each group, turning groups of size two into triples, and retaining the two holes of size w intersecting in the infinite point. Henceforth $u < w$.

If $w \le 21$, form a 3-GDD of type $(u-1)^1(w-1)^1 2^{(w+1)/2}$ if $v = 2w+u$, or type $(u-1)^1(w-1)^1 2^{(w+3)/2}$ if $v = 2w+u+2$. The existence of these GDDs follows from Colbourn (1993), except when $v = 63$, $w = 21$, and $u = 19$ which is treated by Corollary 21.7. Add an infinite point and proceed as above.

Now $w \ge 25$. Write $s = \lfloor w/6 \rfloor$ and $t = \lceil (u-1)/6 \rceil$. Then $1 \le t \le s$ and $s \ge 4$, and so we apply Theorem 21.4 with $z = 1$ according to the selection of parameters in Table 21.4 for the first statement. For the second statement, employ the parameters in Table 21.5. When the source is listed as a 3-GDD, it is from Colbourn (1993). □

To complete all cases with $2w + u - 2 \le v < 2w + u + 4$, it remains only to consider the case when $w \equiv 5 \pmod 6$ and $u \equiv 1 \pmod 6$, when $v-(2w+u-2) = 2$. We defer this case to §21.3.4.

$w \bmod 6$	$u \bmod 6$	k	c	a	b	Input designs
1	1	6	3	1	1	◇-(21;7,7;1)
						◇-(15;7,3;1)
						$\mathcal{L}(3;1,1;1)$
3	1	6	7	3	1	◇-(25;9,7;1)
						◇-(19;9,7;3)
						$\mathcal{L}(7;3,1;1)$
3	3	2	7	3	1	◇-(21;9,3;1)
						◇-(19;9,7;3)
						$\mathcal{L}(7;3,1;1)$
3	5	4	7	3	1	◇-(23;9,5;1)
						◇-(19;9,7;3)
						$\mathcal{L}(7;3,1;1)$

Table 21.4 *Parameters for Lemma* 21.8, *first statement*

$w \bmod 6$	$u \bmod 6$	k	c	a	b	Input designs
1	3	0	7	1	3	◇-(19;7,3;1)
						◇-(19;7,7;1)
						$\mathcal{L}(7;1,3;1)$
3	1	6	9	3	1	◇-(27;9,7;1)
						◇-(21;9,9;3)
						$\mathcal{L}(9;3,1;1)$

Table 21.5 *Parameters for Lemma* 21.8, *second statement*

21.3.3 *Disjoint cases*

Lemma 21.9 *The following statements hold.*

1. *If* $w > u > 1$ *and either* $w \equiv 1 \pmod 6$ *and* $u \equiv 3 \pmod 6$, *or* $w \equiv 3 \pmod 6$ *and* $u \equiv 1 \pmod 6$, *then a* ◇-$(2w+u+2; w, u; 0)$-*ISTS exists.*
2. *If* $w \geq u > 1$, $w \equiv 1 \pmod 6$, *and* u *is odd, then a* ◇-$(2w+u+4; w, u; 0)$-*ISTS exists.*

Proof If $w = u$, Theorem 6.16 produces the required design. Henceforth $u < w$. If $w \leq 21$, form a 3-GDD of type $u^1 w^1 1^{w+2}$ if $v = 2w+u+2$, or type $u^1 w^1 2^{w+4}$ if $v = 2w+u+4$. The existence of these GDDs follows from Colbourn (1993), except when $v = 63$, $w = 21$, and $u = 19$ which is treated by Corollary 21.7.

Now $w \geq 25$. Write $s = \lfloor w/6$ and $t = \lceil (u-1)/6 \rceil$. Then $1 \leq t \leq s$ and $s \geq 4$, and so we apply Theorem 21.4 with $z = 0$ according to the selection of parameters in Table 21.6 for the first statement. For the second statement, employ the parameters in Table 21.7. □

It remains to treat two cases to handle all possibilities when $2w + u \leq v < 2w + u + 6$. The first has $w \equiv u \equiv 3 \pmod 6$ and $v - (2w+u) = 4$. The second

w mod 6	u mod 6	k	c	a	b	Input designs
1	3	0	7	1	3	◇-(19;7,3;0)
						◇-(19;7,7;1)
						$\mathcal{L}(7;1,3;0)$
3	1	6	9	3	1	◇-(27;9,7;0)
						◇-(21;9,9;3)
						$\mathcal{L}(9;3,1;0)$

Table 21.6 *Parameters for Lemma* 21.9, *first statement*

w mod 6	u mod 6	k	c	a	b	Input designs
1	1	6	7	1	1	◇-(25;7,7;1)
						◇-(19;7,7;1)
						$\mathcal{L}(7;1,1;0)$
1	3	2	7	1	1	◇-(21;7,3;0)
						◇-(19;7,7;1)
						$\mathcal{L}(7;1,1;0)$
1	5	4	7	1	1	◇-(23;7,5;0)
						◇-(19;7,7;1)
						$\mathcal{L}(7;1,1;0)$

Table 21.7 *Parameters for Lemma* 21.9, *second statement*

has $w \equiv 5 \pmod 6$, $u \equiv 3 \pmod 6$, and $v - (2w + u) = 4$. To apply the strategy of Lemma 21.8 in the first case would employ 13 infinite points, but then an STS(25) with a sub-STS(13) would be needed, and none can exist. Hence we require more sophisticated variations on the theme of Lemma 21.8.

21.3.4 *Eframes*

Colbourn, Oravas, and Rees (1999) define, find, and apply a new class of frames. An (n, w)-*eframe* is a 3-GDD of type $6^n w^1$ whose blocks are partitioned into n 3-frame parallel classes, each associated with one of the groups of size 6, and $w/2$ 1-frame parallel classes associated with the group of size w.

Theorem 21.10 *Let* $n \equiv 0, 1 \pmod 3$, *and suppose that there exists a TD(5,n). Let* $0 \le w < n$ *when* $n \equiv 1 \pmod 3$ *and* $1 \le w \le n$ *when* $n \equiv 0 \pmod 3$. *Then there exists an* $(n, 2w)$-*eframe.*

Proof Without loss of generality the TD(5,n) is on $\{0, \dots, n-1\} \times \{1, 2, 3, 4, 5\}$, with groups on $\{0, \dots, n-1\} \times \{i\}$ for $i \in \{1, 2, 3, 4, 5\}$. Let $\mathcal{B}$ be the block set of this TD, and assume that $\{j_1, j_2, j_3, j_4, 0_5\} \in \mathcal{B}$ for $0 \le j < n$; we use the convention that j_i denotes $(j, i) \in \{0, \dots, n-1\} \times \{1, 2, 3, 4, 5\}$. Now on $\{0, \dots, n-1\} \times \{1, 2, 3\}$, define n frame parallel classes

$$\mathcal{H}_k = \{\{a_1, b_2, c_3\} : \{a_1, b_2, c_3, k_4, \ell_5\} \in \mathcal{B}, 1 \le \ell < n\},$$

for $0 \le k < n$. Then $\mathcal{H}_k$ is a frame parallel class with respect to the 'group' $\{k_1, k_2, k_3\}$. Further define a set of (genuine) parallel classes

$$\mathcal{P}_\ell = \{\{a_1, b_2, c_3\} : \{a_1, b_2, c_3, k_4, \ell_5\} \in \mathcal{B}, 0 \le k < n\}.$$

for $1 \le \ell < n$. We can verify that $\mathcal{H}_k \cap \mathcal{P}_\ell$ contains exactly one triple for $0 \le k < n$ and $1 \le \ell < n$.

Now write $\beta = w$ if $n \equiv 1 \pmod 3$ and $\beta = w - 1$ if $n \equiv 0 \pmod 3$. We construct the required eframe on

$$((\{0, \ldots, n-1\} \times \{1,2,3\}) \cup \{\infty_1, \ldots, \infty_w\}) \times \{0,1\},$$

where $\{a\} \times \{1,2,3\} \times \{0,1\}$ forms a group of size 6 for $0 \le a < n$ (and we denote its 3-frame parallel class by $\mathcal{C}_a$), and $\{\infty_1, \ldots, \infty_w\} \times \{0,1\}$ forms the group of size $2w$ (and we denote its w 1-frame parallel classes by $\mathcal{D}_1, \ldots, \mathcal{D}_w$).

For $0 \le k < n$ and $1 \le \ell < n$, consider the triple $\{a_1, b_2, c_3\} \in \mathcal{H}_k \cap \mathcal{P}_\ell$. If $\ell > \beta$, form a 3-GDD of type 2^3 on $\{a_1, b_2, c_3\} \times \{0,1\}$ aligning groups on $\{x\} \times \{0,1\}$ for $x \in \{a_1, b_2, c_3\}$; place the four triples so defined in $\mathcal{C}_k$. If instead $\ell \le \beta$, place the two triples $\{a_1, b_2, c_3\} \times \{i\}$ for $i = 0,1$ in $\mathcal{D}_\ell$; then place the triples $\{x_i, y_{i-1}, (\infty_\ell, i)\}$ for $i = 0,1$ and $(x,y) \in \{(a_1,b_2),(b_2,c_3),(c_3,a_1)\}$ in $\mathcal{C}_k$.

Now if $n \equiv 1 \pmod 3$, for $i \in \{1,2,3\}$, form a Kirkman frame of type 2^n on $\{0, \ldots, n-1\} \times \{i\} \times \{0,1\}$; for $0 \le j < n$, place the triples associated with the frame parallel class of the group $\{j\} \times \{i\} \times \{0,1\}$ in the class $\mathcal{C}_j$.

If instead $n \equiv 0 \pmod 3$, for $i \in \{1,2,3\}$, form a Kirkman frame of type 2^n on $((\{0, \ldots, n-1\} \times \{i\}) \cup \{\infty_w\}) \times \{0,1\}$; for $0 \le j < n$, place the triples associated with the frame parallel class of the group $\{j\} \times \{i\} \times \{0,1\}$ in the class $\mathcal{C}_j$, and finally place the associated with the frame parallel class of the group $\{\infty_w\} \times \{0,1\}$ in the class $\mathcal{D}_w$. The result is the required eframe. □

We apply this to the case of interest, when the extra group has four elements.

Corollary 21.11 *Let $n \ge 4$ and $n \notin \{5,6,8,10,11,14,17\}$. Then an $(n,2)$-eframe and an $(n,4)$-eframe exist.*

Proof If $n \equiv 0,1 \pmod 3$, apply Theorem 21.10 with $w \in \{1,2\}$. Otherwise, if $n \ge 20$, apply Theorem 21.10 to form an $(n-4,26)$- or an $(n-4,28)$-eframe, and fill the hole of size 26 or 28 with a $(4,2)$- or a $(4,4)$-eframe. □

Lemma 21.12 *The following statements hold.*

1. *If $w > u > 1$, $w \equiv 5 \pmod 6$, and $u \equiv 1 \pmod 6$, then a $\diamond$-$(2w + u; w, u; 1)$-ISTS exists.*
2. *If $w > u > 1$, $w \equiv 5 \pmod 6$, and $u \equiv 3 \pmod 6$, then a $\diamond$-$(2w + u + 4; w, u; 0)$-ISTS exists.*
3. *If $w > u > 1$, $w \equiv 5 \pmod 6$, and $u \equiv 3 \pmod 6$, then a $\diamond$-$(2w + u + 4; w, u; 1)$-ISTS exists.*

Proof Write $s = (w+1)/6$ and $t = \lceil \frac{u}{6} \rceil$. We treat various cases for values of s.

Case 1 $s \leq 4$.
Form a 3-GDD of type $(u-1)^1(w-1)^1 2^{3s}$, $u^1 w^1 1^{6s+3}$, or $(u-1)^1(w-1)^1 2^{3s+2}$ for the three statements in the theorem, respectively. These GDDs are from Colbourn (1993), with the following exceptions: (v, w, u) = (65,23,19) for the first statement; (71,23,21) and (65,23,15) for the second and third statements. For (65,23,19), and (71,23,21) for intersecting and disjoint, Corollary 21.7 provides the designs. For (65,23,15) in the intersecting case, adjoin an infinite point to the groups of a 3-GDD of type $14^3 22^1$ from Theorem 6.15. For (65,23,15) in the disjoint case, form a latin square of order 21 having disjoint subsquares of orders 7 and 5 by Theorem 21.1, and so produce an IGDD of type $(21; 7, 5)^3$. Adjoin two infinite points and fill groups using a $\diamond$-(23;9,5;0)-ISTS from Theorem 21.2.

Case 2 $s \geq 5$ and $s - 1 \notin \{6, 7, 9, 11, 12, 15, 18\}$.
Using Corollary 21.11, form an $(s-1, 4)$-eframe, and use it to construct an IGDD of type $(8+k; 4, k)^1(18; 6, 6)^{t-1}(12; 6, 0)^{s-t}$ where $k \in \{0, 2, 4\}$. Fill in the holes of this IGDD with ingredients from Table 21.8.

w mod 6	u mod 6	k	c	a	b	Input designs
5	1	0	3	1	1	$\diamond$-(11;5,1;1)
						$\diamond$-(15;7,3;1)
						$\mathcal{L}(3; 1, 1; 1)$
5	3	2	7	1	1	$\diamond$-(17;5,3;0)
						$\diamond$-(19;7,7;1)
						$\mathcal{L}(7; 1, 1; 0)$
5	3	2	7	1	1	$\diamond$-(17;5,3;1)
						$\diamond$-(19;7,7;1)
						$\mathcal{L}(7; 1, 1; 1)$

Table 21.8 *Ingredients for Lemma* 21.12, *Case* 2.

Case 3 $s \in \{9, 11, 15, 18\}$.
Start with an $(s-2, 10)$-eframe, which exists by Theorem 21.10. We suppose that $t \geq 2$ since when $u = 3$, Theorem 6.15 gives a 3-GDD of type $1^{w+7} w^1$ which (by simple counting) contains a triple disjoint from the hole, and hence a $\diamond$-$(2w+7; w, 3; 0)$-ISTS exists. Construct from the eframe an IGDD of type $(20+k; 10, k)^1(18; 6, 6)^{t-2}(12; 6, 0)^{s-2}$ and take $k = 6$ when $u \equiv 1 \pmod 6$ and $k = 8$ when $u \equiv 3 \pmod 6$. Now proceed as above, taking $(c, a, b) = (3, 1, 1)$ when $u \equiv 1 \pmod 6$ and $(c, a, b) = (7, 1, 1)$ when $u \equiv 3 \pmod 6$. In Table 21.8, the first input design is replaced as follows, using designs from Colbourn (1993): for the first statement, by $\diamond$-(29;11,7;1) from a 3-GDD of type $6^1 10^1 2^6$; for the second statement, by $\diamond$-(35;11,9;0) from a 3-GDD of type $9^1 11^1 1^{15}$; and for the third statement, by a $\diamond$-(35;11,9;1) from a 3-GDD of type $8^1 10^1 2^8$. The remaining input designs remain the same as in Case 2.

Case 4 $s = 12$.
If $t \in \{1, 2\}$, the designs are $\diamond$-$(v; w, u; z)$-ISTSs for $(v, w, u, z) =$ (149,71,7,1), (149,71,3,0), (155,71,9,0), (149,71,3,1), (155,71,9,1). For (149,71,7,1), use a 3-GDD of type $6^{13}70^1$ from Theorem 6.15. For (149,71,3,0), employ a 3-GDD of type $1^{78}71^1$ from Theorem 6.15. The solution for (155,71,9,0) is in Table 21.9. For (149,71,3,1), employ a 3-GDD of type $1^{78}71^1$, which has a triple disjoint from the hole. For (155,71,9,1), see Lemma 21.17. When $t \geq 3$, produce an IGDD of type $(32+k; 16, k)^1(18; 6, 6)^{t-3}(12; 6, 0)^{s-t}$ using a (9,16)-eframe from Theorem 21.10. Take $k = 12$ and $(c, a, b) = (3, 1, 1)$ when $u \equiv 1 \pmod 6$, and $k = 14$ and $(c, a, b) = (7, 1, 1)$ when $u \equiv 3 \pmod 6$. The relevant first input designs, all from Colbourn (1993), are $\diamond$-(47;17,13;1) from a 3-GDD of type $12^1 16^1 2^9$; $\diamond$-(53;17,15;0) from a 3-GDD of type $15^1 17^1 1^{21}$; and $\diamond$-(53;17,15;1) from a 3-GDD of type $14^1 16^1 2^{11}$.
Case 5 $s \in \{6, 7\}$.
For the first statement, when $w = 35$ and $u = 7$ add an infinite point to a 3-GDD of type $6^7 34^1$ from Theorem 6.15. When $w = 41$ and $u = 13$, see Lemma 21.17. When $w = 41$ and $u = 31$, see Corollary 21.7. For the remaining cases, see Table 21.9.

For the second statement, when $u = 3$ form a 3-GDD of type $w^1 1^{w+7}$ by Theorem 6.15 and observe that there is a triple disjoint from the hole. When $w = 35$ and $u = 15$, see Lemma 21.17. When $w = 35$ and $u = 21$, form a latin square of order 31 with disjoint subsquares of orders 11 and 7 by Theorem 21.1, and so produce an IGDD of type $(31; 11, 7)^3$; adjoin two infinite points and fill groups using a $\diamond$-(33;13,7;0)-ISTS. When $w = 35$ and $u = 33$, see Corollary 21.7. For the remaining cases, see Table 21.9.

For the third statement, $w = 35$, when $u \in \{3, 9, 21\}$ form 3-GDDs of types $2^{21}34^1$, $8^6 34^1$, or $20^3 34^1$ from Theorem 6.15; when $u \in \{27, 33\}$, see Corollary 21.7; and for $u = 15$ see Lemma 21.17. Finally for $w = 41$ and $u = 6t + 0$, $0 \lesssim t \lesssim 3$, use a Kirkman frame of type 10^4 to yield an IGDD of type $(28; 10, 8)^t(22; 10, 2)^{4-t}$, and adjoin seven infinite points, using $\diamond$-(29;11,7;1) and an ISTS on 35 points having concurrent holes of sizes 11, 9, and 7 from a 3-GDD of type $2^5 6^1 8^1 10^1$ from Colbourn (1993). For $w = 41$ and $u \in \{33, 39\}$, see Corollary 21.7. □

Lemma 21.13 *Let* $w \equiv 3 \pmod 6$, $u \equiv 3 \pmod 6$, *and* $w > u > 1$. *Then there exists a* $\diamond$-$(2w + u + 4; w, u; 0)$-*ISTS.*

Proof If $w = u$, apply Theorem 6.16. So suppose that $u < w$. Write $s = (w+3)/6$ and $t = (u+3)/6$. Proceed exactly as in the Lemma 21.12, employing instead $(s-1, 2)$-, $(s-2, 8)$-, and $(s-3, 14)$-eframes to yield IGDDs of types

1. $(6; 2, 2)^1(18; 6, 6)^{t-1}(12; 6, 0)^{s-t}$,
2. $(24; 8, 8)^1(18; 6, 6)^{t-2}(12; 6, 0)^{s-t}$, and
3. when $s = 12$, either
   - $(36; 14, 8)^1(12; 6, 0)^9$ $(t = 2)$ or
   - $(42; 14, 14)^1(18; 6, 6)^{t-3}(12; 6, 0)^{12-t}$ $(t \geq 3)$.

Set $(c, a, b) = (7, 1, 1)$. The relevant new input designs are, respectively, $\diamond$-(13;3,3;0), $\diamond$-(31;9,9;0), $\diamond$-(43;15,9;0), and $\diamond$-(49;15,15;0). Theorem 6.16 provides all but the third, which arises from a 3-GDD of type $9^1 15^1 1^{19}$ from Colbourn (1993). The remaining input designs, $\diamond$-(19;7,7;1) and $\mathcal{L}(7; 1, 1; 0)$ are as in Lemma 21.12.

Now let $w = 33$. If $u = 3$, a 3-GDD of type $33^1 1^{40}$ from Theorem 6.15 contains a triple disjoint from the hole. When $u \geq 9$, we employ the frame method using a Kirkman frame of type 8^4 to yield an IGDD of type $(24; 8, 8)^t(18; 8, 2)^{4-t}$. Adjoin seven infinite points, one of which is to be incorporated in the hole of size $u = 6t + 9$, for $0 \leq t \leq 3$, and a second to be incorporated into the hole of size $w = 33 = 8 \cdot 4 + 1$. We require as input designs $\diamond$-(25;9,3;0) from a 3-GDD of type $1^{16}9^1$ from Theorem 6.15; $\diamond$-(25;9,7;1) from a 3-GDD of type $6^1 8^1 2^5$ from Colbourn (1993) (which contains a triple meeting the 6-hole but missing the 8-hole, as required); and an ISTS on 31 points having two disjoint holes of size 9 and a third hole of size 7 meeting both 9-holes in a single point from Theorem 21.6 with $n = 8$, $a = b = 1$, $z = 0$, and $m = 4$.

Now let $w = 39$. By Theorem 21.1, form a latin square of order $26 + 2t$ with two disjoint subsquares of orders 13 and $2t$, and so produce an IGDD of type $(26 + 2t; 13, 2t)^3$. Adjoin seven infinite points and fill in the groups. We require an ISTS on $33 + 2t$ points having holes of sizes $2t + 3$ and 7 intersecting in three (infinite) points, and having a third hole of size 13 which is disjoint from the first two holes. These designs are constructed by hillclimbing as in Colbourn (1993).

Finally, for $w \leq 21$ form a 3-GDD of type $u^1 w^1 1^{w+4}$ from Colbourn (1993), except when $v = 61$, $w = 21$, and $u = 15$. For this last case, form a latin square of order 18 having disjoint subsquares of orders 4 and 7 using Theorem 21.1, to produce an IGDD of type $(18; 7, 4)^3$. Adjoin seven infinite points, and fill in groups, using an ISTS on 25 points with two holes of size 7 meeting in 3 points, and and third hole of size 7 disjoint from the first two. This design is produced by hillclimbing. □

21.3.5 *Incomplete group divisible designs*

Colbourn, Oravas, and Rees (1999) construct some classes of IGDDs of type $(g, h)^3(f, 0)^1$, abbreviated to $(g, h)^3 f^1$. Simple parity and congruence conditions require that g must be even unless $f = 0$, and that f must be even. In addition, $f \leq 2(g - h)$.

First they prove an existence result:

Theorem 21.14 *If a TD(5,n) exists, then so does*

1. *an IGDD of type* $(2n + 2m; 2m)^3(2x)^1$ *when* $0 \leq m \leq n$ *and* $0 \leq x \leq 2n$;
2. *an IGDD of type* $(3n + n + 2m, n + 2m)^3(2x)^1$ *when* $0 \leq m \leq n$ *and* $0 \leq x \leq 3n$.

Proof For the first statement, use two groups of the TD(5,n) to form two partitions into parallel classes $\mathcal{P}_1, \ldots, \mathcal{P}_n$ and $\mathcal{Q}_1, \ldots, \mathcal{Q}_n$ on the remaining three groups forming a TD(3,n). Observe that $\mathcal{Q}_i \cap \mathcal{P}_j$ contains a single triple T_{ij}. Now select $\beta_1, \ldots, \beta_n$ so that $\beta_j \in \{0,2\}$ and $2m = \sum_{j=1}^{n} \beta_j$. Then select $\alpha_1, \ldots, \alpha_n$ so that $\alpha_i \in \{0,2,4\}$ and $2x = \sum_{i=1}^{n} \alpha_i$.

There exists an IGDD of type $(2+\beta_j;\beta_j)^3\alpha_i^1$ whenever $\beta_j \in \{0,2\}$ and $\alpha_i \in \{0,2,4\}$; these are easily produced. Each point of the TD(3,n) is given weight two. For each parallel class $\mathcal{P}_j$, add β_j elements to each of the three groups. For each parallel class $\mathcal{Q}_i$, add α_i elements in the fourth group (which is added). Then the triple T_{ij} is replaced by an IGDD of type $(2+\beta_j;\beta_j)^3\alpha_i^1$ on the relevant elements.

For the second statement, proceed in the same manner using weight 3 instead of 2, using IGDDs of type $(3+y;y)^3z^1$ for $y \in \{1,3\}$ and $z \in \{0,2,4,6\}$. □

Theorem 21.15 *Let n,m,x be odd and $0 \le m,x \le n$. Then there exists an IGDD of type $(n+m;m)^3(2n)^1$.*

Proof Form on $\{0,1,\ldots,n-1\} \times \{1,2,3\}$ the set of triples obtained from $\{\{0_1,i_2,(2i)_3\} : 0 \le i < n\}$ by developing modulo $(n,-)$. Let $|M| = m$ and $M \subseteq \{0,\ldots,n-1\}$, so that $0 \in M$. Retain the triples in orbits from $\{\{0_1,i_2,(2i)_3\} : i \in M\}$ and remove the rest. Add m elements $\{x_1,\ldots,x_m\} \times \{1,2,3\}$. Now on $(\{0,\ldots,n-1\} \cup \{x_1,\ldots,x_m\}) \times \{1,2,3\}$ define a graph G which contains all edges of the form $\{i_k,(x_j,\ell)\}$ for $0 \le i < n$, $1 \le j \le m$, $k,\ell \in \{1,2,3\}$ and $k \ne \ell$, along with all edges in the orbits of $\{0_1,i_2\}$, $\{0_2,i_3\}$, and $\{0_1,(2i)_3\}$ for $0 \le i < n$ and $i \notin M$. G is a $(2n)$-regular graph. If we show that G has a partition into $2n$ 1-factors, each can be used to form triples with an infinite point, and the infinite points collected to form a group, to yield the required IGDD. So it remains to partition G into $2n$ 1-factors. We rewrite the orbit representatives of the pairs to form two 1-factors which can each be developed modulo n to form the needed $2n$ 1-factors. Elements among the x_is remain fixed in this development. For $1 \le i < n$, form the pairs $\{\{i_2,(2i)_3\},\{(n-i)_1,i_3\},\{(n-2i)_1,(n-i)_2\}\}$; take their union to form a graph H. Now H is a collection of $(n-1)/2$ vertex-disjoint hexagons of the form $((2i)_1,i_2,(2i)_3,(-2i)_1,(-i)_2,(-2i)_3)$ for $1 \le i \le (n-1)/2$; moreover, every orbit of pairs with nonzero difference is represented exactly once in H. Considering only those edges of H which arise from a triple defined by those $i \notin M$ gives a subgraph H' on $3(n-m)$ edges. Since H was a vertex-disjoint union of hexagons, H' is bipartite. Now partition H' into two subgraphs H_1 and H_2, both having $(n-m)/2$ edges between elements subscripted a and b for $(a,b) \in \{(1,2),(1,3),(2,3)\}$. For $k=1,2$, extend H_k to a 1-factor J_k by adding edges as follows: For $i=1,2,3$, let $y_{1,i},\ldots,y_{m,i}$ be the vertices among $\{0_i,\ldots,(n-1)_i\}$ which do not appear in an edge of H_k; then add edges $\{\{y_{j,i},(x_j,k+i)\} : 1 \le i \le m\}$, reducing the second component modulo 3 into the range $\{1,2,3\}$. Develop J_1 and J_2 modulo n to obtain the required $2n$ 1-factors. □

Lemma 21.16 *Let $z \in \{0,1\}$. If a $\diamond$-$(n+m+\gamma;m+z,\gamma;z)$-ISTS exists, and an IGDD of type $(n+m,m)^3x^1$ exists, then a $\diamond$-$(3n+3m+x+\gamma;3m+z,x+\gamma;z)$-ISTS exists.*

Proof Add γ new elements to the IGDD, placing a $\diamond$-$(n+m+\gamma; m+z, \gamma; z)$-ISTS on each group of size $n+m$ together with the γ new points, aligning the hole of size γ on the new points and that of size $m+z$ on the hole of the IGDD together with z of the new points. The result has a hole of size $x+\gamma$ on the points of the fourth group of the IGDD together with the new points, and a hole of size $3m+z$ on the points in the hole of the IGDD together with z of the new points. □

Table 21.9 gives constructions for $\diamond$-$(v; w, u; 1)$-ISTSs by Lemma 21.16.

v	z	$w = x+\gamma$	$u = 3m+z$	n	m	x	γ	Input $\diamond$-design
89	1	41	7	14	2	26	15	(31;15,3;1)
101	1	41	19	14	6	28	13	(33;13,7;1)
107	1	41	25	14	8	26	15	(37;15,9;1)
119	1	41	37	14	12	28	13	(39;13,13;1)
$71+6t$ $(2 \le t \le 5)$	1	35	$6t+1$	12	$2t$	22	13	$(25+2t; 13, 2t+1; 1)$
83	0	35	9	13	3	26	9	(25;9,3;0)
101	0	35	27	13	9	26	9	(31;9,9;0)
95	0	41	9	15	3	30	11	(29;11,3;0)
$89+6t$ $(2 \le t \le 6)$	0	41	$6t+3$	15	$2t+1$	26	15	$(31+2t; 15, 2t+1; 0)$
155	0	71	9	25	3	50	21	(49;21,3;0)

Table 21.9 *Cases via IGDDs*

In a few cases, the general techniques of Colbourn, Oravas, and Rees (1999) do not appear to provide the $\diamond$-ISTSs needed. Hence they resort to an heuristic computational method, which is a simple adaptation of the hillclimbing method used in Colbourn (1993):

Lemma 21.17 *$\diamond$-$(v; w, u; z)$-ISTSs exist for*

$$(v, w, u, z) \in \{(89, 35, 15, 0), (95, 41, 13, 0), (89, 35, 15, 1), (155, 71, 9, 1)\}.$$

21.4 Consequences

In previous sections, we examined (in substantial detail) the existence of $\diamond$-ISTSs when the order is at or near the minimum. Colbourn, Oravas, and Rees (1999) also examine some of the consequences of this determination. A main tool is an easy one:

Lemma 21.18 *If a $\diamond$-$(v; \{w, b\}; c)$-ISTS exists and a $\diamond$-$(b; u, c; z)$-ISTS exists, then a $\diamond$-$(v; \{w, u\}, z)$-ISTS exists.*

Proof Place the $\diamond$-$(b; u, c; z)$-ISTS on the hole of size b in the $\diamond$-$(v; \{w, b\}; c)$-ISTS, aligning the holes of size c. $\square$

This simple lemma, together with earlier results, has surprising consequences:

Theorem 21.19 *Let $z \in \{0,1\}$. Let $v \geq 2(\max(w, 2u+1)) + w - 2$ and the necessary congruence conditions on v, w, u, z be met for a $\diamond$-$(v; w, u; z)$-ISTS. Then the $\diamond$-ISTS exists.*

Proof Choose a, g, and x so that $2a + x - 2g = v - w$, $g \in \{1,3\}$, and $g > 1$ only if $x = 0$, $x \in \{0,2,4,6\}$, $a \geq \max(w, 2u+1)$, $a \equiv 1 \pmod 2$, and $\binom{a}{2} \equiv \binom{u}{2} \pmod 3$. We choose $x = 6$ only when $w \equiv 3 \pmod 6$, $u \equiv 5 \pmod 6$, and $v - w \equiv 2 \pmod 6$. Form a $\diamond$-$(2a + x + w - 2g; a, w; g)$-ISTS and a $\diamond$-$(a; u, g; z)$-ISTS and apply Lemma 21.18. That the parameters can be chosen, except when $w \equiv 3 \pmod 6$ and $u \equiv 5 \pmod 6$, to meet the conditions shown to be sufficient for the existence of the two $\diamond$-ISTSs is routine. When $u \equiv w \equiv 1 \pmod 6$, for example, the choices for $(a \bmod 6, x, g)$ to produce the four required congruence classes for $v - w$ modulo 12 are: (1,0,1) for 0 mod 12; (1,2,1) for 2 mod 12; (3,2,1) for 6 mod 12; and (3,4,1) for 8 mod 12. When $w \equiv 1 \pmod 6$ and $u \equiv 5 \pmod 6$, the two required congruence classes for $v - w$ modulo 12 are obtained by setting $(a \bmod 6, x, g)$ to be (5,2,1) for 10 mod 12, and (5,0,3) for 4 mod 12. Other cases are similar. When $w \equiv 3 \pmod 6$ and $u \equiv 5 \pmod 6$, choosing $(a \bmod 6, x, g)$ to be (5,0,1) handles the case when $v - w \equiv 8 \pmod{12}$, and the case when $v - w \equiv 2 \pmod{12}$ is handled by $(a \bmod 6, x, g) = (5, 6, 1)$, but no smaller value of x 'works' here. $\square$

When $w \geq 2u + 1$, Theorem 21.19 treats all cases when $v \geq 3w$. Colbourn, Oravas, and Rees (1999) improve this result somewhat.

Theorem 21.20 *Let $z \in \{0,1\}$. Let $w \geq 2u + 1$, and $v \geq 2w + 2u + 1$. If the necessary congruence conditions on v, w, u, z are met, then a $\diamond$-$(v; w, u; z)$-ISTS exists.*

Proof If $v \geq 3w - 2$, apply Theorem 21.19. Otherwise, choose b and x so that $x \in \{0,2,4\}$, $b + x = v - 2w + 2$, $b \geq 2u + 1$, $b \equiv 1 \pmod 2$, and $\binom{b}{2} \equiv \binom{u}{2} \pmod 3$. Form a $\diamond$-$(2w + x + b - 2; w, b; 1)$-ISTS and a $\diamond$-$(b; u, z; z)$-ISTS and apply Lemma 21.18. That the parameters can be chosen to meet the conditions shown to be sufficient for the existence of the two $\diamond$-ISTSs is routine. $\square$

When $w \geq 3u$, Colbourn, Oravas, and Rees (1999) complete the determination:

Theorem 21.21 *Let $z \in \{0,1\}$. Let $w \geq 3u$, and $v \geq 2w + u$. If the necessary congruence conditions on v, w, u, z are met, then a $\diamond$-$(v; w, u; z)$-ISTS exists.*

Proof If $v \geq 2w + 2u + 1$, apply Theorem 21.20. Otherwise, choose d and x so that $x \in \{0,2,4\}$, $d - x = 2w + 2u - v$, $d \equiv 1 \pmod 2$. Form a $\diamond$-$(2w + 2u + x - d; w, 2u + x + d; d)$-ISTS by Theorem 21.2, and a $\diamond$-$(2u + x + d; u, d; z)$-ISTS and apply Lemma 21.18. Again, the selection of appropriate values for d and x can easily be verified. $\square$

It appears plausible that the conditions developed earlier suffice when $z \in \{0, 1\}$ for the existence of $\diamond$-$(v; w, u; z)$-ISTSs, and in particular that the condition that $w \geq 3u$ is not necessary in the statement of Theorem 21.21. However, it also appears that even when $z \in \{0, 1\}$, the solution of cases at and near the minimum, together with known embedding results, does not suffice to furnish a proof. In addition, the technique used at and near the minimum is at present restricted to such values by the need to have a 'small' number of infinite elements. For this reason the complete solution of the existence of $\diamond$-ISTSs appears to require new techniques.

21.5 Work points

1. Extend Theorem 21.21 to replace the condition that $w \geq 3u$ by the condition that $w \geq u$.
2. Determine the necessary conditions for an STS to exist having two subsystems of sizes w and u that intersect in z elements.

22

NESTED AND DERIVED TRIPLE SYSTEMS

In this chapter, we examine extensions of triple systems to designs with larger block size. We have already seen a number of problems involving the placement of the blocks of a triple system (or partial triple system) into a 'larger design'. For example, embedding permits us to increase the number of elements, immersing increases the index, and enclosing increases both. Here we consider operations that increase the block size.

22.1 Nested triple systems

Preece (1967) and Federer (1972) introduced two different ways in which the blocks of a design can be 'nested' within the blocks of another design. Kageyama and Miao (1997) unified these definitions. Suppose that $(V, \mathcal{B})$ is a design of order v, having block size k, index λ, and b blocks. Let $\mathcal{B} = \{B_1, \ldots, B_b\}$. Choose integers s and η_m, α_m, and β_m for $1 \leq m \leq s$, so that $\sum_{m=1}^{s} \eta_m \alpha_m \beta_m \leq k$. Partition each block B_ℓ into $1+\sum_{m=1}^{s} \alpha_m \beta_m$ classes, D^ℓ_{mij} of size η_m for $1 \leq m \leq s$, $1 \leq i \leq \alpha_m$, and $1 \leq j \leq \beta_m$, and one final class $\overline{D}_\ell$ of size $k - \sum_{m=1}^{s} \alpha_m \beta_m$. Next define $\mathcal{B}_{mj} = \{D^\ell_{mij} : 1 \leq \ell \leq b, 1 \leq i \leq \alpha_m\}$. If, for every choice of m satisfying $1 \leq m \leq s$ and j satisfying $1 \leq j \leq \beta_m$, we find that $\mathcal{B}_{mj}$ forms the blocks of a design on v points with block size η_m, then $(V, \mathcal{B})$ is a *nested design* of form $(\eta_1^{\alpha_1})^{\beta_1} \cdots (\eta_s^{\alpha_s})^{\beta_s}$.

This general definition is somewhat unwieldy, but specializes to the notion of Preece (1967) when $m = 1$, $\beta_1 = 1$, and $\eta_1 \alpha_1 = k$; in this case, blocks of size k are partitioned into subblocks, and the union of *all* of the subblocks forms a design. See Ghosh (1996) for numerous statistical applications. The general definition specializes to that of Federer (1972) when $m = 1$, $\alpha_1 = \beta_1 = 1$, and $2 \leq \eta_1 < k$, in which case only one subblock is chosen from each block.

We consider one of the simpler cases first, which is a nesting along the lines of Federer (1972). Let $(V, \mathcal{B})$ be a TS(v, λ), with $\mathcal{B} = \{B_1, \ldots, B_b\}$. A (v, k, μ)-design $(V, \mathcal{D})$ with $\mathcal{D} = \{D_1, \ldots, D_{b'}\}$ is a $(k, \mu; 3, \lambda)$-*nesting* of $(V, \mathcal{B})$ provided that $b = b'$ and $B_i \subseteq D_i$ for $1 \leq i \leq b$. Using the general definition, this design is a nested design of form $(3^1)^1$. For such a nesting to exist, the condition $b = b'$ implies that

$$\frac{\mu}{k(k-1)} = \frac{\lambda}{6} \quad \text{or} \quad \mu = \frac{\lambda k(k-1)}{6}. \tag{22.1}$$

Moreover, since the designs $(V, \mathcal{B})$ and $(V, \mathcal{D})$ must meet the necessary conditions for existence, we have:

$$6|\lambda v(v-1), 2|\lambda(v-1), k(k-1)|\mu v(v-1) \quad \text{and} \quad k-1|\mu(v-1). \tag{22.2}$$

The simplest nontrivial case arises when $k = 4$. Then a $(4, \mu; 3, \lambda)$-nesting can exist only when $\mu = 2\lambda$ by (22.1). In addition, by (22.2), as well as the existence of a $TS(v, \lambda)$ we require that $\lambda(v-1) \equiv 0 \pmod 3$. Since nesting into designs with block size four has been the most extensively studied, we often call a triple system $TS(v, \lambda)$ *nested* when it has a $(4, 2\lambda; 3, \lambda)$-nesting. Evidently a nested $TS(v, \lambda)$ can exist only when

$$\begin{array}{ll} v \equiv 1 \pmod 6 & \text{if } \lambda \equiv 1, 5 \pmod 6 \\ v \equiv 1 \pmod 3 & \text{if } \lambda \equiv 2, 4 \pmod 6 \\ v \equiv 1 \pmod 2, v \neq 3 & \text{if } \lambda \equiv 3 \pmod 6 \\ v \notin \{2, 3\} & \text{if } \lambda \equiv 0 \pmod 6 \end{array}$$

Let us consider a simple example. Consider an STS(7). The necessary conditions are met for this design to be nested in a $(7, 4, 2)$-design. Here is such a nesting:

0	1	3	6
1	2	4	0
2	3	5	1
3	4	6	2
4	5	0	3
5	6	1	4
6	0	2	5

This example illustrates a further interesting property. When the $TS(v, \lambda)$ and its $(4, 2\lambda; 3, \lambda)$-nesting share a v-cycle as an automorphism, the (cyclic) $TS(v, \lambda)$ is *cyclically nested.* In this case, we can present starter blocks. A cyclically nested STS(13) is given by starter blocks $\{\{7, 8, 11|0\}, \{1, 3, 9|0\}\}$.

Before considering sufficiency, we remark on an interesting connection with block colouring. For each block of a $(4, 2\lambda; 3, \lambda)$-nesting, let the *nesting element* be the element of the block that does not appear in the block of the $TS(v, \lambda)$. Now consider all blocks of the $TS(v, \lambda)$ having the same nesting element. In this collection of triples, every element occurs at most λ times. Hence when $\lambda = 1$, colouring blocks the same when they have the same nesting element, we obtain a (very special) proper block-colouring of the triple system using v colours. Each colour class contains $(v-1)/6$ disjoint triples, and when element j appears in a triple of colour i, element i does *not* occur in a triple of colour j.

22.1.1 *Index one*

Longyear (1986) introduced nesting in a combinatorial setting. She notes that the finite field construction (Theorem 2.11) yields a nested $STS(p^n)$ for p a prime and $n \geq 1$:

Theorem 22.1 *Let p be a prime, $n \geq 1$, $p^n \equiv 1 \pmod 6$, and ω be a primitive root of $\mathbb{F}_{p^n}$. Writing $6t = p^n - 1$, the starter blocks $\{\{0, \omega^i, \omega^{2t+i}, \omega^{4t+i}\} : 0 \leq i < t\}$ form a nested $STS(p^n)$.*

Proof Theorem 2.11 shows that these starter blocks with the 0 omitted form an STS(p^n). Thus we need only verify that

$$\{\{\pm\omega^i, \pm\omega^{2t+i}, \pm\omega^{4t+i}\} : 0 \leq i < t\} = \mathbb{F}_{p^n} \setminus \{0\};$$

this is immediate. □

Theorem 22.1 yields cyclically nested triple systems whenever the order is prime. Remarkably, Novák (1975) conjectures a very much stronger statement:

Conjecture 22.2 *Every cyclic STS(v), $v \equiv 1 \pmod 6$, can be cyclically nested.*

This conjecture, if true, would settle the existence of nested STS(v)s in a very satisfying manner. However, while no general results are available, we report a previously unpublished result:

Theorem 22.3 *Every cyclic STS(v) with $v \equiv 1 \pmod 6$ and $v \leq 61$ can be cyclically nested.*

Table 22.1 reports the numbers of cyclic nestings of the cyclic STS(v)s, indicating by the entry $a(b)$ that exactly b of the nonisomorphic cyclic STS(v)s admit exactly a cyclic nestings. The table provides convincing evidence for Conjecture 22.2, in view of the increasing numbers of cyclic nestings for each system. However, this computational evidence has not led to even partial general results. A striking pattern emerges in the numbers of cyclic nestings. The number of systems having a specific number of cyclic nestings appears frequently to be a power of 2. To account for this, we establish the following:

Lemma 22.4 *Let $\mathcal{B}$ be a set of starter blocks for a (partial) cyclic STS(v), Let $\widehat{\mathcal{B}}$ be obtained from $\mathcal{B}$ by multiplying any number of starter blocks by -1 (equivalently, replacing certain starter blocks $\{a, b, c\}$ by $\{v-a, v-b, v-c\}$). Then $\mathcal{B}$ has a cyclic nesting if and only if $\widehat{\mathcal{B}}$ has a cyclic nesting.*

Proof If d is the nesting element for starter block $\{a, b, c\}$, choose $v - d$ as the nesting element for $\{v-a, v-b, v-c\}$. □

The lemma establishes that we need only consider solutions to Heffter's first difference problem, rather than considering all different cyclic STS(v)s arising from each solution.

To treat nested STSs when $v \equiv 1 \pmod 6$, Stinson (1985) examined a different route. A $\{3\}$-GDD is *nested* if a fourth point is adjoined to each triple so that two points from different groups appear in precisely two of the resulting blocks of size four. Stinson (1985) establishes the existence of nested $\{3\}$-GDDs of type 2^4 and 2^7.

Theorem 22.5 *For all $v \equiv 1 \pmod 6$, there is a nested STS(v).*

7	1(1)
13	1(1)
19	1(2), 5(2)
25	14(8), 20(4)
31	22(4), 53(16), 55(16), 59(16), 74(16), 77(12)
37	219(16), 237(32), 247(32), 248(32), 249(32), 251(32), 253(38), 255(32), 260(32), 262(32), 264(32), 266(32), 268(16), 271(32), 273(16), 274(32), 275(32), 277(16), 278(32), 279(16), 285(32), 287(32), 288(64), 291(32), 292(32), 307(12), 317(12), 340(16), 351(16), 363(6)

Table 22.1 *Numbers of cyclic nestings*

Proof If $v \in \{7, 13, 19, 37\}$, apply Theorem 22.1. Otherwise write $v = 6t + 1$. Form a $\{4, 7\}$-GDD of type 3^t, by deleting a point from a $(3t + 1, 4, 1)$-design if $t \equiv 0, 1 \pmod 4$, or by deleting a point not in a 7-block from a PBD with block sizes 4 and 7 if $t \equiv 2, 3 \pmod 4$ (from Brouwer, 1979). Give weight two to each point of this $\{4, 7\}$-GDD and apply the Fundamental Construction (Theorem 3.3) using the nested GDDs of type 2^4 and 2^7 to obtain a nested $\{3\}$-GDD of type 6^t. Now append a new point ∞, and on the six points of each group together with ∞ place a nested STS(7). □

Indeed, at the present time, no example of an STS(v), $v \equiv 1 \pmod 6$, is known that *cannot* be nested.

The notion of cyclic nesting can be generalized to cyclic 3-GDDs of type 3^n as well. Here the nesting element cannot be chosen so as to use differences 0 or $\pm n$. An analogue of Conjecture 22.2 cannot hold here, however:

Lemma 22.6 *There is a cyclic* $\{3\}$*-GDD of type* 3^n *that cannot be cyclically nested whenever* $n \equiv 13 \pmod{18}$.

Proof Write $n = 18t + 13$. Choose a cyclic STS($54t + 39$) having a cyclic sub-STS($18t + 13$) (which exists by Lemma 7.16). Remove the short orbit, and consider the remaining starter blocks $\mathcal{B}$. Each starter block handles either three differences that are multiples of 3 (and their negatives), or three that are 1 modulo 3 and three that are 2 modulo 3 (including the negatives). For the latter type, no matter what nesting element is chosen, the differences accounted for fall equally among the residue classes modulo 3. Thus, in treating the first type, there remain $6t + 4$, $6t + 5$, and $6t + 5$ differences in the 0, 1, 2 modulo 3 residue classes, respectively. But then there is no way to assign nesting elements to the $3t + 2$ starter blocks of the first type, since each choice of a nesting element accounts either for six differences that are multiples of 3, or three each of those congruent to ± 1 modulo 3. □

22.1.2 *Higher indices*

It suffices to treat the cases $\lambda = 2$, 3, and 6. Colbourn and Colbourn (1983c) establish sufficiency in these cases.

Theorem 22.7 *A nested TS$(v, 2)$ exists if and only if $v \equiv 1 \pmod 3$.*

Proof Theorem 19.45 establishes that whenever $v \equiv 1 \pmod 3$, there exists an almost resolvable TS$(v, 2)$. Let $P_1, \ldots, P_v$ be the almost parallel classes of such a resolution, with the element i missing from the almost parallel class P_i for $1 \leq i \leq v$. Then form a $(v, 4, 4)$-design by adding element i to each triple $\{x, y, z\} \in P_i$, for $1 \leq i \leq v$. □

Almost resolvable systems give a convenient way to produce nestings for $\lambda = 2$, but not all nestings arise in this way.

When $\lambda \in \{3, 6\}$, the ingredients are even simpler; we give the construction for $\lambda = 6$ here.

Theorem 22.8 *A nested TS$(v, 3)$ exists if and only if $v \equiv 1 \pmod 2$ and $v \neq 3$.*

Theorem 22.9 *A nested TS$(v, 6)$ exists if and only if $v \geq 4$.*

Proof Brayton, Coppersmith, and Hoffman (1974) establish that if $v \geq 4$ and $v \neq 6$, there is a self-orthogonal latin square of order v. Let L be such a square with rows, columns, and symbols indexed by a v-set V. Then

$$\{\{i, j, L(i, j), L(j, i)\} : 1 \leq i, j \leq v, i \neq j\}$$

is the set of blocks of a $(v, 4, 12)$-design; nested in it is the TS$(v, 6)$ $\{\{i, j, L(i, j)\} : 1 \leq i, j \leq v, i \neq j\}$. The remaining case, $v = 6$, is settled affirmatively in Colbourn and Colbourn (1983c). □

22.1.3 *Nesting GDDs and partial triple systems*

When $v \equiv 3 \pmod 6$, no STS(v) can be nested, as a consequence of (22.2). Lindner and Rodger (1987) (see also Lindner, 1987a) consider the question of how close we can come to such a nesting. Their examination leads to a nesting of $\{3\}$-GDDs of type 3^t.

The construction employs skew Room squares (see §1.3.6).

Theorem 22.10 *For all $t \geq 7$, t odd, there exists a nested $\{3\}$-GDD of type 3^t.*

Proof Let R be a skew Room square of order t based on $\{0, \ldots, t\}$. Form a set of triples on $\{1, \ldots, t\} \times \mathbb{Z}_3$ by including

$$\{\{x_i, y_i, r_{i+1}\} : i \in \mathbb{Z}_3, \{x, y\} \text{ in row } r \text{ of } R\}.$$

This is a $\{3\}$-GDD with groups on $\{x\} \times \mathbb{Z}_3$ for $1 \leq x \leq t$. Now nest each block $\{x_i, y_i, r_{i+1}\}$ using the element c_{i+1}, where c is the column of R in which $\{x, y\}$ appears. □

Lindner and Rodger (1987) describe this as a nesting of a partial triple system whose leave is a set of $v/3$ disjoint triangles. This suggests a generalization to nesting partial triple systems of order 3 (mod 6) having a prescribed quadratic leave. See Phelps and Rodger (1993).

22.1.4 *Nesting into larger blocks*

Numerous more general nestings have been examined; see, for example, Ghosh (1996). We focus on cases involving triple systems here. Kageyama and Miao (1999b) settle the existence of nested designs of form $(2^1 3^1)^1$ when $k = 5$. In this case, every block has five elements partitioned into a class of size three and a class of size two. They establish:

Theorem 22.11 *A nested design of form* $(2^1 3^1)^1$ *of order* v*, block size five, and index* λ *exists if and only if* $v \geq 5$*, and*

$$\lambda \equiv \begin{cases} 0 \pmod{10} & \text{if } v \equiv 1 \pmod 2 \\ 0 \pmod{20} & \text{if } v \equiv 0 \pmod 2 \end{cases}$$

The same conditions apply when the form desired is simply $(3^1)^1$, and the block size is five.

Hishida, Ishikawa, Jimbo, Kageyama, and Kuriki (1999), Kageyama and Miao (1996, 1996b, 1998, 1999a), and Miao, Kageyama, and Duan (1996) report on related studies of nested designs.

22.2 Compatible Steiner triple systems

Consider nesting a Steiner triple system STS(v) into a design with block size $k > 4$. For a $(k, \mu; 3, 1)$-nesting to exist, by (22.1) whenever $k \equiv 0, 1 \pmod 3$, μ is an integer, $\mu = k(k-1)/6$. Provided that an STS(v) exists, when $k \equiv 0, 1 \pmod 3$, the only condition in (22.2) that applies is that $k - 1 | k(k-1)(v-1)/6$. Thus when $k \equiv 0 \pmod 3$, the necessary conditions admit the possibility of a $(k, k(k-1)/6; 3, 1)$-nesting; when $k \equiv 1 \pmod 3$, the necessary conditions require only that $v \equiv 1 \pmod 6$ for a $(k, k(k-1)/6; 3, 1)$-nesting to exist. These simple computations suggest an intriguing possibility, which we examine next.

Let $(V, \mathcal{B})$ and $(V, \mathcal{D})$ be two STS(v)s with $\mathcal{B} = \{B_1, \ldots, B_b\}$ and $\mathcal{D} = \{D_1, \ldots, D_b\}$. Form a collection of blocks $\mathcal{S} = \{B_i \cup D_i : 1 \leq i \leq b\}$. If $\mathcal{S}$ is the collection of blocks of a $(v, 6, 5)$-design, then the Steiner triple systems $\mathcal{B}$ and $\mathcal{D}$ are *compatible*. Naturally, if an STS(v) has a compatible STS(v), it has a $(6, 5; 3, 1)$-nesting.

If for two such triple systems $\mathcal{B}$ and $\mathcal{D}$, we can find elements $n_1, \ldots, n_b$ for which $\{B_i \cup \{n_i\} : 1 \leq i \leq b\}$ and $\{D_i \cup \{n_i\} : 1 \leq i \leq b\}$, then $\mathcal{B}$ and $\mathcal{D}$ can be *compatibly nested.* Evidently, if two STS(v)s can be compatibly nested, each has a $(7, 7; 3, 1)$-nesting.

These notions can be extended. A set of ℓ pairwise compatible STS(v)s yields a $(3\ell, \ell(3\ell - 1)/2; 3, 1)$-nesting of each; and when ℓ pairwise compatible STS(v)s admit a common nesting, each has a $(3\ell + 1, (3\ell + 1)\ell/2; 3, 1)$-nesting. Carrying this to its extreme, when $v \equiv 3 \pmod 6$, a *compatible minimal partition* is a collection of $b = v(v-1)/6$ ordered partitions $P_1, \ldots, P_b$ of $\{1, \ldots, v\}$, where

$$P_j = (C_{j,1}, \ldots, C_{j,v/3})$$

and for every $1 \leq c < d \leq v/3$,

1. $\{C_{c,i} : 1 \leq i \leq b\}$ is an STS(v);
2. $\{C_{d,i} : 1 \leq i \leq b\}$ is an STS(v); and
3. these two STS(v)s are compatible.

Table 22.2 gives a compatible minimal partition for $v = 9$.

0	1	2	3	4	5	6	7	8
3	4	5	6	7	8	0	1	2
6	7	8	0	1	2	3	4	5
0	3	6	1	4	7	2	5	8
1	4	7	2	5	8	0	3	6
2	5	8	0	3	6	1	4	7
0	4	8	1	5	6	2	3	7
1	5	6	2	3	7	0	4	8
2	3	7	0	4	8	1	5	6
0	5	7	1	3	8	2	4	6
1	3	8	2	4	6	0	5	7
2	4	6	0	5	7	1	3	8

Table 22.2 *A compatible minimal partition for* $v = 9$

Similarly, when $v \equiv 1 \pmod 6$, a *compatibly nested minimal partition* is a set of $(v-1)/3$ mutually compatible STS(v)s sharing a common nesting. Table 22.3 gives an example for $v = 7$.

2	4	5	0	1	3	6
3	5	6	1	2	4	0
0	4	6	2	3	5	1
0	1	5	3	4	6	2
1	2	6	4	5	0	3
0	2	3	5	6	1	4
1	3	4	6	0	2	5

Table 22.3 *A compatibly nested minimal partition for* $n = 7$

This topic is largely uninvestigated. However, for compatibly nested STS(v)s, we can obtain a finite field construction:

Theorem 22.12 *Let* ω *be a primitive root of* $\mathbb{F}_{p^n}$, $p^n = 6t + 1$. *Let* $\mathcal{S} = \{\omega^0, \omega^{2t}, \omega^{4t}\}$, *and* $\mathcal{S}_i = \omega^i \mathcal{S}$. *Form an ordered partition* $\mathcal{P}$ *of* $\mathbb{F}_{p^n}$,

$$(x_0 = 0, \mathcal{S}_0, \mathcal{S}_1, \ldots, \mathcal{S}_{2t-1}).$$

Now form t partitions $\mathcal{P}_j = \omega^j \mathcal{P}$ *for* $0 \le j < t$. *Then*

$$\{\mathcal{P}_j + x : x \in \mathbb{F}_{p^n}, 0 \le j < t\}$$

is a compatibly nested minimal partition of order p^n.

Proof Each ordered partition $\mathcal{P}_j + x$ is indeed a partition with singleton (nesting) element x. For each remaining class, the triples form a TS$(p^n, 1)$ that is correctly nested by the singleton class (see Theorem 22.1 for comparison). Consider the triples generated by two triples, say $\mathcal{S}_c$ and $\mathcal{S}_d$, of $\mathcal{P}$. The differences arising between their elements are

$$\pm\omega^c\omega^{2it}(\omega^{2jt+d-c} - 1) \text{ for } 0 \le i, j < 3$$

none of which is zero provided that $c \ne d$. Moreover, the differences arising in partitions $\mathcal{P}_j$, $0 \le j < t$, then represent every nonzero difference in $\mathbb{F}_{p^n}$ three times each. Thus applying the additive structure of $\mathbb{F}_{p^n}$ to form partitions $\mathcal{P}_j + x$, we find every pair between the two STS(v)s three times each as required. □

Let $P_{1,6}$ be the set of all prime powers congruent to 1 (mod 6). Recall Theorem 19.69, which gives the orders for which a PBD with all block sizes in $P_{1,6}$ exists. That theorem, together with Theorem 22.12, yields:

Theorem 22.13 *Two compatibly nested STS(v)s exist for all* $v \equiv 1 \pmod 6$, *with the possible exceptions as in Theorem 19.69.*

Proof Suppose that a PBD of order v exists with block sizes in $P_{1,6}$. Then for each block B of this PBD, replace the block by a pair of compatibly nested STS$(|B|)$s from Theorem 22.12. □

For two compatible STS(v)s, PBD-closure asserts that there exists a constant v_0 such that for all $v \ge v_0$, $v \equiv 1, 3 \pmod 6$, there exists a PBD with block sizes in $\{9\} \cup P_{1,6}$. Thus for v sufficiently large, the necessary condition $v \equiv 1, 3$ (mod 6) for compatible STS(v)s is sufficient. No estimate of v_0 seems to be available for this set of block sizes.

22.3 Perpendicular arrays of triple systems

Mendelsohn (1980) introduced a combinatorial structure that exhibits more structure than a nested STS(v). A *perpendicular array of triple systems* PATS(v) is a $v(v-1)/6 \times 4$ array; treating the rows as unordered, we obtain the blocks of a $(v, 4, 2)$-design, and striking out any column yields an array in which the rows are the blocks of a STS(v). Of course, any PATS(v) yields a nested STS(v) by striking out any one column.

Mendelsohn (1980) establishes the nonexistence of a PATS(7) or a PATS(13), but he settled a number of cases ($v \in \{19, 25, 31, 49, 61, 67\}$) affirmatively. The set $\{v : \exists PATS(v)\}$ is PBD-closed. Thus Mendelsohn's constructions underlie an asymptotic existence result for PATS.

Stinson (1986) found solutions for a number of further small orders. More importantly, he shows that there is an 8×4 array on eight points whose rows are the blocks of a $\{4\}$-GDD of type 2^4, while striking out any column yields the rows of a $\{3\}$-GDD of type 2^4. Using this ingredient, he observes the following:

Lemma 22.14 *If there is a* $\{4\}$*-GDD of order* v *and type* $g_1^{t_1} \cdots g_\ell^{t_\ell}$*, and for each* $1 \leq i \leq \ell$ *there is a PATS*$(2g_i + 1)$*, then there exists a PATS*$(2v + 1)$.

By truncating transversal designs TD$(5, m)$ to form $\{4,5\}$-GDDs of type $m^4 s^1$, and applying the Fundamental Construction with weight 3 using $\{4\}$-GDDs of type 3^4 and 3^5, Stinson constructs a $\{4\}$-GDD of type $(3m)^4(3s)^1$. Provided that a PATS$(6m+1)$ and a PATS$(6s+1)$ exist, Lemma 22.14 gives a PATS$(24m + 6s + 1)$. This is the main device used in proving the best result at this point concerning PATS, in Stinson (1986):

Theorem 22.15 *A PATS*(v) *exists whenever* $v \equiv 1 \pmod 6$, $v \geq 19$, *except possibly when* $v \in \{43, 55, 85, 133\}$.

22.4 Derived triple systems

We have considered numerous ways of extending triple systems to designs with larger block size, larger index, and more elements. Here we examine one last extension: to t-designs with $t > 2$.

Recall that a *quadruple system* QS(v, λ) is a pair $(V, \mathcal{B})$ in which $\mathcal{B}$ is a collection of 4-element subsets of V that contain each 3-element subset of V λ times. A *Steiner quadruple system*, or SQS(v), is a QS$(v, 1)$. Now let $\mathcal{B}_x = \{B \setminus \{x\} : x \in B \in \mathcal{B}\}$. The 3-wise balance of $\mathcal{B}$ ensures that $(V \setminus \{x\}, \mathcal{B}_x)$ is a TS$(v-1, \lambda)$. This is a *derived* triple system of the QS(v, λ).

Hanani (1960) proved that:

Theorem 22.16 *A QS*(v, λ) *exists if and only if* $4|\lambda\binom{v}{3}$, $3|\lambda\binom{v-1}{2}$, $2|\lambda(v-2)$, *and* $v \notin \{2, 3\}$.

A trivial corollary is that there exists a derived TS(v, λ) if and only if a TS(v, λ) exists and $\lambda(v+1)v(v-1) \equiv 0 \pmod{24}$; the latter condition requires that if $v \equiv 4 \pmod 8$, $\lambda \equiv 0 \pmod 2$, and that if $v \equiv 2 \pmod 4$, $\lambda \equiv 0 \pmod 4$. We summarize:

Lemma 22.17 *A derived TS*(v, λ) *exists whenever a TS*(v, λ) *exists except when*

1. $v \equiv 6, 10 \pmod{12}$ *and* $\lambda \equiv 2 \pmod 4$*; and*
2. $v \equiv 2 \pmod{12}$ *and* $\lambda \equiv 6 \pmod{12}$.

Since the necessary conditions for a derived STS(v) are the same as for an STS(v), it is natural to ask for a characterization of derived STS(v)s. The unique STS(7) and STS(9) are necessarily derived. Mendelsohn and Hung (1974) established that both STS(13)s are derived. After much preliminary effort by Phelps

(1975, 1976, 1976a), Gibbons (1976), and Grannell and Griggs (1985), Diener, Schmitt, and de Vries (1985) established that all 80 STS(15)s are derived. It remains open at the present time whether every STS(v) is derived.

Phelps (1976) describes a large general class of derived STS(v)s:

Theorem 22.18 *An STS*$(2v+1)$ *having a sub-STS*(v) *that is derived is itself derived.*

Proof Phelps's proof employs perhaps the most widely used recursive construction for quadruple systems, the doubling construction. We describe this first. Let V and W be two disjoint sets of size v, and let $(V, \mathcal{B})$ and $(W, \mathcal{D})$ be QS$(v, 1)$. Let $\mathcal{F} = \{F_1, \ldots, F_{v-1}\}$ be a 1-factorization on V, and $\mathcal{G} = \{G_1, \ldots, G_{v-1}\}$ be a 1-factorization on W. Let π be a permutation of $\{1, \ldots, n-1\}$. Then

$$\mathcal{B} \cup \mathcal{D} \cup \{\{w, x, y, z\} : \{w, x\} \in F_i, \{y, z\} \in G_{\pi(i)}, 1 \leq i < n\}$$

forms the blocks of an SQS$(2v)$ on $V \cup X$.

Now to employ this construction, let $(S, \mathcal{T})$ be an STS$(2v+1)$ having a sub-STS(v) $(P, \mathcal{Q})$. Now let $W = S \setminus P$ and $V = P \cup \{\infty\}$. By our hypothesis that $(P, \mathcal{Q})$ is derived, there is an SQS$(v+1)$ $(V, \mathcal{B}$ whose derived triple system for the point ∞ is $(P, \mathcal{Q})$.

Next observe that deleting the points of P from the triple system $(S, \mathcal{T})$ leaves a 1-factorization $\mathcal{G} = \{G_1, \ldots, G_{v-1}\}$ on W. Now choose an SQS$(v+1)$ $(W, \mathcal{D})$ arbitrarily on W, and choose an arbitrary 1-factorization $\mathcal{F} = \{F_1, \ldots, F_{v-1}\}$ on V, labelling the 1-factors so that F_i contains the edge $\{\infty, i\}$ for $1 \leq i < v$. Applying the doubling construction with π as the identity then yields an SQS$(2v+2)$ whose derived triple system for the point ∞ is $(S, \mathcal{T})$ as required. □

Phelps (1976a) extends this theorem to handle some cases when no maximum subsystem is present, when an STS$(2v+1)$ has some blocks carrying $v(v-1)/6-1$ triples. His argument proceeds in a similar way using doubling, using a quadrilateral exchange to recover the STS$(2v+1)$ after applying the doubling construction.

Another large class of derived triple systems is obtained by using direct and indirect products. Hanani (1960) showed that if an STS(v) is derived, so also is the direct product of this triple system with a single block. Aliev (1967), Phelps (1976a), and Rokowska (1977) all generalize this result to prove:

Theorem 22.19 *The direct product of derived triple systems is derived.*

Lindner (1975) and Mendelsohn (1978) extend this further to singular direct products of triple systems.

22.4.1 *Heterogeneity and homogeneity*

In general, two derived TS(v, λ)s from a QS$(v+1, \lambda)$ need not be isomorphic. If all are isomorphic, the quadruple system is *homogeneous*, while if all are pairwise nonisomorphic, the quadruple system is *heterogeneous*. Lindner and Rosa

(1976) found the first heterogeneous quadruple system, an SQS(20), and prove that a heterogeneous SQS(v) can be doubled to form a heterogeneous SQS($2v$). Subsequently, Gibbons (1976) found a heterogeneous SQS(16). Fu (1984) produced heterogeneous SQS(v)s for all $v = 7 \cdot 2^k$, $k \geq 2$. Little is known about the spectrum of heterogeneous Steiner quadruple systems; however, the following conjecture of Hartman and Phelps (1992) appears to be quite plausible:

Conjecture 22.20 *Almost all Steiner quadruple systems are heterogeneous and therefore also automorphism-free.*

Turning to homogeneous SQS(v)s, de Vries (1987) shows that of the 80 STS(15)s, at least 69 of them are derived systems of homogeneous SQS(16)s. Not all STS(v)s are so derived, however, as the projective space PG(3, 2) is not derived from a homogeneous SQS(16). Naturally, any SQS(v) admitting a transitive automorphism group is homogeneous; this provides essentially all known general results on homogeneous systems. It seems reasonable to expect that a homogeneous Steiner quadruple system exists for every order $v \equiv 2, 4 \pmod 6$, since for $v \notin \{8, 14, 16\}$ a cyclic system is conjectured to exist (Hartman and Phelps, 1992).

22.5 Work points

1. Prove or disprove Conjecture 22.2.
2. A weaker version of Conjecture 22.2 is that every cyclic STS can be made to have disjoint starter blocks. Prove or disprove this weaker conjecture.
3. For which orders do compatible minimal partitions exist?
4. For which orders do compatibly nested minimal partitions exist?
5. Is every STS(v) derived?
6. Prove or disprove Conjecture 22.20.

23

DECOMPOSABILITY

If the triples of a TS(v, λ) $(V, \mathcal{B})$ can be partitioned into two classes $\mathcal{B}_1$ and $\mathcal{B}_2$, so that for $i \in \{1, 2\}$, $(V, \mathcal{B}_i)$ is a TS(v, λ_i) (so that $\lambda_1 + \lambda_2 = \lambda$), then $(V, \mathcal{B})$ is *decomposable* (or *reducible*). If no such partition exists, it is *indecomposable* (or *irreducible*). For example, each of the four nonisomorphic TS$(7, 2)$s is decomposable while Kirkman (1853) already established that the only TS$(7, 3)$ without repeated triples is indecomposable. Any TS(v, λ) with $\lambda = \lambda_{\min}$ is trivially indecomposable.

23.1 Existence

Kramer (1974) was apparently the first to consider indecomposable triple systems. For which values of v and λ does an indecomposable TS(v, λ) exist? This problem is not completely settled.

Kramer (1974) determined the spectrum for simple indecomposable TS$(v, 2)$s and indecomposable TS$(v, 3)$s. He employed a simple observation. If a TS(v, λ) has a sub-TS(w, λ), any decomposition of the TS(v, λ) into systems of indices λ_1 and λ_2 induces a decomposition with the same indices of the sub-TS(w, λ). Thus if an indecomposable TS(w, λ) is found, the existence of indecomposable TS(v, λ)s for all λ-admissible orders $v \geq 2w + 1$ is immediate by Theorem 6.9. Kramer (1974) developed specialized embedding constructions to avoid the introduction of repeated triples; however, now we can use Shen's theorem (Theorem 6.22) in place of Theorem 6.9. Provided that the TS(w, λ) is simple and indecomposable, the embedding TS(v, λ) is also.

Subsequently, Colbourn and Rosa (1985a) determined the spectrum for simple indecomposable TS$(v, 4)$s; Archdeacon and Dinitz (1993) and Zhang (1994) independently settled index five with few exceptions; and Dinitz (1990) and Milici (1990) determined the solution for simple indecomposable TS$(v, 6)$s, again except possibly for a few small values of v. The original proofs can now be greatly simplified thanks to Theorem 6.22. Typically, we have to produce a simple indecomposable TS(v, λ) for the smallest value of v for which it exists, as well as for other admissible values in the interval $[v, 2v]$.

Theorem 23.1 *(Indecomposable TSs with small index)*

1. *A simple indecomposable TTS(v) exists if and only if*
$$v \equiv 0, 1 \pmod 3, \; v \geq 4, \; v \neq 7.$$
2. *A simple indecomposable TS$(v, 3)$ exists if and only if*
$$v \equiv 1 \pmod 2, \; v \geq 5.$$

3. *A simple indecomposable TS$(v, 4)$ exists if and only if*

$$v \equiv 0, 1 \pmod 3, \quad v \geq 10.$$

4. *A simple indecomposable TS$(v, 5)$ exists if*

$$v \equiv 1, 3 \pmod 6, \quad v \geq 13,$$

except possibly for $v \in \{13, 15\}$.

5. *A simple indecomposable TS$(v, 6)$ exists if and only if*

$$v \geq 8, \quad v \neq 9, 10,$$

except possibly when $v \in \{11, 12, 13, 15, 16\}$.

Proof We treat the statements separately.

1. The unique TS$(4, 2)$ is trivially indecomposable; Theorem 6.22 produces simple indecomposable TS$(v, 2)$s for all $v \equiv 0, 1 \pmod 3$, $v \geq 9$. The unique TS$(6, 2)$ is also (trivially) indecomposable while none of the four nonisomorphic TS$(7, 2)$s is indecomposable.
2. The unique TS$(5, 3)$ is trivially indecomposable; Theorem 6.22 gives simple indecomposable TS$(v, 3)$s for all $v \equiv 1 \pmod 2$, $v \geq 11$. The unique TS$(7, 3)$ without repeated triples in Example 7.1 is indecomposable. A simple indecomposable TS$(9, 3)$ is given in Example 23.3.
3. No simple indecomposable TS$(v, 4)$ exists for $v = 6, 7$, or 9. Indeed, a simple TS$(6, 4)$ contains all possible triples and so necessarily decomposes into two TS$(6, 2)$s. No simple indecomposable TS$(7, 4)$ exists (for example, as a consequence of Theorem 14.1), and Harnau (1980) showed that no simple indecomposable TS$(9, 4)$ exists (see also Mathon and Lomas, 1992). A simple indecomposable TS$(10, 4)$ is given in Example 23.4. Theorem 6.22 then gives simple indecomposable TS$(v, 4)$s for all $v \equiv 0, 1 \pmod 3$, $v \geq 21$. To complete the proof, we still need to produce examples of simple indecomposable TS$(v, 4)$s for $v = 12, 13, 15, 16, 18$, and 19. For $v = 16$ and 18, examples are obtained by embedding a TS$(7, 4)$ into TS$(16, 4)$, and TS$(18, 4)$, respectively, again using Theorem 6.22. Although the TS$(7, 4)$ is decomposable, it can decompose only into a TS$(7, 3)$ and an STS(7) while any TS$(16, 4)$ or TS$(18, 4)$ can decompose only into two TS$(v, 2)$s. Thus the resulting triple systems are necessarily indecomposable. For orders $v = 12, 13, 15$, and 19, examples of simple indecomposable TS$(v, 4)$s appear in Colbourn and Rosa (1985a).
4. Zhang (1994) exhibits simple indecomposable TS$(v, 5)$s for

$$v \in \{19, 21, 25, 27, 31, 33, 37\};$$

Theorem 6.22 handles the remaining cases.

5. Any TS$(v,6)$ with $v < 8$ must contain repeated triples. Any TS$(v,6)$ with $v \equiv 2 \pmod 6$ is necessarily indecomposable. Applying Theorem 6.22 to the simple indecomposable TS$(8,6)$ yields simple indecomposable TS$(v,6)$s for all $v \geq 17$. Since every STS(9) has a disjoint mate (by Theorem 14.1), a simple indecomposable TS$(9,6)$ cannot exist. An examination of the complement of each of the 384 nonisomorphic TS$(10,2)$s without repeated triples (from Colbourn, Colbourn, Harms, and Rosa, 1983) reveals that each is decomposable; thus no simple indecomposable TS$(10,6)$ exists. Thus the existence of simple indecomposable TS$(v,6)$s remains in doubt only for $v = 11, 12, 13, 15$, and 16. □

Corollary 23.2 *The complete design TS$(v, v-2)$ is indecomposable if and only if $v = 3, 4, 5,$ or* 8.

Harms, Colbourn, and Ivanov (1987) show that of the 332 nonisomorphic simple TS$(9,3)$s, 172 are indecomposable.

Example 23.3 (Kramer, 1974) A simple indecomposable TS$(9,3)$.

The triples are 127, 128, 129, 136, 138, 139, 146, 147, 149, 156, 157, 158, 234, 235, 237, 245, 247, 256, 268, 269, 289, 345, 348, 356, 369, 378, 379, 459, 467, 468, 489, 578, 579, 589, 678, 679.

Colbourn and Rosa (1985a) found the following:

Example 23.4 A simple indecomposable TS$(10,4)$.

016, 017, 018, 019, 026, 027, 028, 029, 035, 037, 038, 039, 045, 046, 048, 049, 056, 057, 126, 127, 128, 129, 135, 136, 138, 139, 145, 147, 148, 149, 156, 157, 234, 235, 236, 237, 245, 246, 247, 258, 259, 289, 345, 348, 349, 367, 369, 378, 467, 468, 479, 568, 569, 589, 678, 679, 689, 789.

Colbourn and Colbourn (1985) were the first to show that indecomposable TS(v,λ)s exist for every odd λ. This observation is quite straightforward, using two powerful results. First, an r-regular graph having no regular factor exists whenever $r \geq 3$ is odd (Lemma 1.24). Let G be a λ-regular graph having no regular factor, and suppose that G has n vertices where $n+1$ is λ-admissible. It then follows that G is a neighbourhood of an element in a TS$(n+1,\lambda)$ by Theorem 12.8. Any decomposition of this TS$(n+1,\lambda)$ necessarily partitions G into two regular subgraphs, and hence no such decomposition can exist.

Archdeacon and Dinitz (1993) established the existence of simple indecomposable TS(v,λ)s for arbitrary λ, and, moreover, provided a specific bound on $v_0(\lambda)$ such that if v is admissible and $v > v_0(\lambda)$ then there exists a simple indecomposable TS(v,λ). We present their proof.

To handle the case of arbitrary λ, we extend the concept of decomposability to partial triple systems. If an indecomposable PTS(v,λ) is embedded in a TS(v,λ), the resulting system is also indecomposable. Since various powerful embedding theorems are available (see Chapter 11), the problem thus reduces to one of obtaining an indecomposable PTS(v, λ) of a 'reasonable' size. The following construction of Archdeacon and Dinitz provides such a PTS.

Lemma 23.5 *For every λ there exists a simple indecomposable PTS(v, λ) of order $v = \lambda(\lambda + 2)$.*

Proof Let $V = \{1, 2, \ldots, \lambda\} \cup \{a, b\}$, and let $\mathcal{B}_0 = \{\{i, a, b\} : 1 \le i \le \lambda\} \cup \{\{i, i+1, a\} : 1 \le i \le \lambda-1, i \text{ even}\} \cup \{\{i, i+1, b\} : 1 \le i \le \lambda-1, i \text{ odd}\}$. For the sake of simplicity, call the triples containing both a and b *white*, and the remaining triples *black*. Enlarge now $(V, \mathcal{B}_0)$ as follows. Let the new set of elements be $W = V \times I_\lambda$ where $I_\lambda = \{1, 2, \ldots, \lambda\}$, and let us say that a point $(v, i) \in V \times I_\lambda$ lies above the point $v \in V$. Let further $\mathcal{P} = \{\{(i, k_1), (i+1, k_2), (x, k_3)\} : \{i, i+1, x\} \in \mathcal{B}_0, x \in \{a, b\}, (k_1, k_2, k_3) \ne (1, 1, 1)\} \cup \{\{(i, 1), (a, 1), (b, 1)\} : 1 \le i \le \lambda\}$, and let us again say that the triples of $\mathcal{P}$ lie above the corresponding triples of $\mathcal{B}_0$ whose colour (black or white) they inherit. Clearly, $(W, \mathcal{P})$ is a PTS(v, λ) without repeated triples. It remains to be shown only that $(W, \mathcal{P})$ is indecomposable.

Consider now the white triples $t_i = \{(i, 1), (a, 1), (b, 1)\}$. Let the edges above $(b, 1)$ be coloured red, and those above $(b, 2)$ blue. Then every black triple above $\{1, 2, b\}$ contains one red and one blue edge, t_1 is the only triple with a red but not a blue edge, and t_2 is the only triple with a blue but not a red edge.

Assume now that $(W, \mathcal{P})$ is decomposable into $(W, \mathcal{P}_1)$ and $(W, \mathcal{P}_2)$, i.e. $\mathcal{P} = \mathcal{P}_1 \cup \mathcal{P}_2$, $\lambda = \lambda_1 + \lambda_2$. Each red edge and each blue edge appears in exactly λ triples in $\mathcal{P}$, hence each appears in exactly λ_i triples in $\mathcal{P}_i$. It follows that the total number of red edges appearing in triples of $\mathcal{P}_i$ is the same as the total number of blue edges in triples of $\mathcal{P}_i$. Each black triple of $\mathcal{P}$ either adds 1 to both sums or adds 0 to both sums. Without loss of generality, assume $t_1 \in \mathcal{P}_1$. Then there is one more red edge than blue edge in $\mathcal{P}_1$, and for the sums to be equal, there must be a triple with a blue but not a red edge. But the only such triple is t_2, and thus in our decomposition the PTS containing the triple t_1 must also contain the triple t_2. A similar argument shows that the PTS containing the triple t_2 must contain the triple t_3. Continuing in this fashion, we see that one PTS in our decomposition contains t_i for each i, $1 \le i \le \lambda$. But these white triples contain the edge $\{(a, 1), (b, 1)\}$ a total of λ times, thus $\lambda_1 = \lambda$ which means that $(W, \mathcal{P})$ is indecomposable. □

The existence of an indecomposable PTS(v, λ) and Theorem 11.14 lead to the following $O(\lambda^4)$ bound.

Theorem 23.6 *For any $\lambda \ge 1$, a simple indecomposable TS(v, λ) exists for all admissible $v \ge v_0(\lambda)$ where $v_0(\lambda) \le 6(2\lfloor \lambda/2 \rfloor + 1)((3\lambda + 2)\lambda(\lambda + 2) + 1) + 1$.*

Proof The indecomposable simple PTS(v, λ) constructed in Theorem 23.5 has $v = \lambda(\lambda+2)$ elements. By Rodger's theorem (Theorem 11.14) it can be embedded in a TS(w, λ) without introducing any repeated triples, for some $w \le 3(2\lfloor \lambda/2 \rfloor + 1)((3\lambda + 2)v + 1)$. This TS$(w, \lambda)$ can in turn be embedded by Theorem 6.22 into a TS(u, λ) for each admissible $u \ge 2w + 1$ which has no repeated triples and is indecomposable. □

Zhang (1996a) gives a substantial improvement in this bound, to establish the existence of an indecomposable simple TS(v, λ) whenever $v \geq 24\lambda - 5$.

23.2 The complexity of decomposition

Given a TS(v, λ), how difficult is it to determine whether it is decomposable? One necessary condition for decomposability is provided by the neighbourhood graph (see Chapter 12). If a TS(v, λ), say (V, B), is to be decomposable into a TS(v, λ_1) and a TS(v, λ_2), the neighbourhood graph of each element in (V, B) must be factorable into two regular factors of degree λ_1 and λ_2, respectively. But while this condition is clearly necessary for a TS(v, λ) to be decomposable, it is not sufficient even for $\lambda = 2$. The neighbourhood graph of each element of the TS$(9, 2)$ in Example 23.7, given in Morgan (1977) and Street (1985), consists of even cycles only, and thus is factorable into two 1-factors.

Example 23.7 An indecomposable TS$(9, 2)$.

128, 129, 134, 139, 145, 156, 167, 178, 236, 237, 248, 249, 256, 257, 347, 358, 359, 368, 458, 467, 469, 579, 689, 789.

Nevertheless, this TS$(9, 2)$ is indecomposable. Other examples are provided by any indecomposable triple system with even $\lambda > 2$: the neighbourhood graph of any element is λ-regular, and thus Petersen's theorem (Theorem 1.22) guarantees that it can be factorized into 2-factors.

Nevertheless, deciding whether a TS$(v, 2)$ is decomposable can be done efficiently, by considering its block intersection graph; this was observed, in effect, already in Kramer (1974).

Theorem 23.8 *A TS$(v, 2)$ is decomposable if and only if its block intersection graph is bipartite.*

Proof If a TTS(v) $(V, \mathcal{B})$ is decomposable into two STS(v)s, say $(V, \mathcal{B}_1)$, $(V, \mathcal{B}_2)$, then in its block intersection graph the set of vertices corresponding to $\mathcal{B}_i, i = 1, 2$, is independent. Thus $\mathcal{B} = \mathcal{B}_1 \cup \mathcal{B}_2$ is a bipartition of the block intersection graph. Conversely, if the block intersection graph of $(V, \mathcal{B})$ is bipartite and $\mathcal{B} = \mathcal{B}_1 \cup \mathcal{B}_2$ is its bipartition, then $(V, \mathcal{B}_i), i = 1, 2$, is an STS$(v)$, and $(V, \mathcal{B})$ is therefore decomposable. □

Corollary 23.9 *Deciding whether a TS$(v, 2)$ is decomposable can be done in polynomial time.*

Proof Deciding whether a graph is bipartite can be done in linear time. □

While determining decomposability of TS(v, λ) when $\lambda = 2$ can be carried out efficiently, this is apparently no longer the case when $\lambda = 3$. This was proved by Colbourn and Colbourn (1983, 1985):

Theorem 23.10 *Deciding whether a (simple) TS(v, λ) contains an STS(v) is NP-complete, for every fixed $\lambda \geq 3$.*

Proof We follow the proof of Colbourn and Colbourn (1983). Membership in NP is immediate, since a nondeterministically chosen subdesign STS(v) can be verified in polynomial time. To show completeness, we reduce the known NP-complete problem of λ-edge-colourability of λ-regular graphs (see Garey and Johnson, 1979) to our problem. Given an arbitrary n-vertex λ-regular graph G, we first determine a size for a latin background for G (as defined in §9.2). When $2n-1 \equiv 1$ or $3 \pmod 6$, let $v = 2n-1$, otherwise let $v = 2n+1$. Using Lemma 9.4, next construct a latin background $LB[G; v+1, v+1]$, L, in polynomial time. Produce λ disjoint latin backgrounds $L_1, \ldots, L_\lambda$ by repeatedly applying the permutation $(1\ 2 \cdots \lambda)(\lambda+1 \cdots v)(v+1)$ to the elements of L. Using these latin backgrounds, construct a simple TS$(2v+1, \lambda)$ with elements $\{x_1, \ldots, x_v, y_1, \ldots, y_{v+1}\}$, and with triples taken as follows:

1. On the elements $\{x_1, \ldots, x_v\}$, place λ disjoint STS(v)s; these exist for all $v \geq \max(9, \lambda+2)$, $v \equiv 1, 3 \pmod 6$ by Theorem 15.15.
2. For $1 \leq i \leq j \leq v+1$, if the (i,j) entry of one of the latin backgrounds is k, include the triple $\{x_k, y_i, y_j\}$.
3. For $1 \leq i \leq j \leq v+1$, if the (i,j) cell of the latin background L is empty, include the triples $\{\{x_k, y_i, y_j\} : 1 \leq k \leq \lambda\}$.

That the set of triples so defined forms a simple TS$(2v+1, \lambda)$ is easily verified, and this design can be constructed in polynomial time. To establish NP-completeness it remains to be shown only that the constructed design contains an STS$(2v+1)$ if and only if the graph G is λ-edge-colourable; further, this depends only on the triples of type (3) above.

Suppose we have an λ-edge-colouring of G. Then the triples $\{\{x_k, y_i, y_j\} : \{y_i, y_j\}$ has colour $k\}$, together with the set of triples of one of the disjoint STS(v)s, and the triples arising from one of the disjoint latin backgrounds is the set of triples of an STS$(2v+1)$.

Conversely, suppose that our TS$(2v+1, \lambda)$ contains an STS$(2v+1)$. In the latter, the pairs appearing in triples with x_i $(i = 1, \ldots, \lambda)$ form a 1-factor of G. Moreover, these 1-factors are all disjoint, and hence cover all edges of G. Since there are λ disjoint 1-factors, and $\lambda v/2$ edges in a λ-regular graph, the 1-factors comprise a λ-edge-colouring of G. □

Corollary 23.11 *Deciding whether a simple TS$(v,3)$ is decomposable is NP-complete.*

Proof A decomposable TS$(v,3)$ must contain an STS(v). □

Theorem 23.10 by itself does not imply that deciding whether a TS$(v,4)$ is decomposable is NP-complete. For example, a TS$(v,4)$ may decompose into two TS$(v,2)$s. In fact, any TS$(v,4)$ obtained by a construction similar to that of Theorem 23.10 is decomposable in this way. Nevertheless, using a somewhat different line of attack, Colbourn and Colbourn (1983) prove the following:

Theorem 23.12 *Deciding whether a simple TS$(v,4)$ is decomposable is NP-complete.*

Their proof employs essentially the observation that the simple TS$(7,4)$, while decomposable, has no decomposition into two TS$(7,2)$s. Thus a TS$(w,4)$ containing a TS$(v,4)$ and also the simple TS$(7,4)$, if decomposable, must contain an STS(w); and hence the TS$(v,4)$ must contain an STS(v).

23.3 A finite basis theorem

When repeated triples are permitted, there is no *a priori* bound on the index λ of an indecomposable TS(v,λ). Engel (1980) established that there are only finitely many indecomposable TS(v,λ)s for a fixed order v. We follow the proof in Street (1985) here.

Consider the set

$$\binom{V}{3} = \{T_1, \ldots, T_{\binom{v}{3}}\}$$

of all 3-subsets of V. A TS(v,λ) $(V,\mathcal{B})$ can be represented by its *characteristic vector*

$$\mathbf{x} = (x_1, \ldots, x_{\binom{v}{3}}),$$

where x_i is the number of occurrences of T_i in $\mathcal{B}$. Form a partial order on characteristic vectors of triple systems on point set V, by defining $\mathbf{x} \preceq \mathbf{y}$ if and only if $x_i \leq y_i$ for all $1 \leq i \leq \binom{v}{3}$. Evidently, the minimal vectors under $\preceq$ are precisely the indecomposable systems.

Theorem 23.13 *For v fixed, the number of indecomposable TS(v,λ)s is finite.*

Proof Let $\mathcal{X}$ be the infinite set of all characteristic vectors of TS(v,λ)s with $\lambda > 0$. Then $(\mathcal{X},\preceq)$ is a partial order. It suffices to prove that this poset has no antichains (set of incomparable elements) of infinite cardinality. Suppose to the contrary that $\mathcal{A}$ is an infinite antichain. Choose an infinite sequence of elements from $\mathcal{A}$ whose first coordinate is nondecreasing. Then choose an infinite subsequence whose second coordinate is also nondecreasing, and repeat this for each coordinate in turn. The sequence so produced is an infinite chain (set of mutually comparable elements) in the poset, and thus $\mathcal{A}$ cannot be an infinite antichain as supposed. □

While establishing finiteness, this technique appears to be of little use in determining the largest possible index for an indecomposable TS(v,λ).

23.4 Partitions into indecomposable systems

The solution to the large set problem for triple systems has an interesting corollary. Whenever $v-2 = \lambda_1 + \lambda_2 + \ldots + \lambda_s$, and for $1 \leq i \leq s$, $\lambda_i > 0$ and $\lambda_i \equiv 0 \pmod{\gcd(v-2,6)}$, the complete design on $\binom{v}{3}$ triples can be partitioned into s triple systems, with indices $\lambda_1, \ldots, \lambda_s$. Call such a partition a $\{\lambda_1, \ldots, \lambda_s\}$-*partition.*

Mesner apparently first asked for which choices of $\{\lambda_1, \ldots, \lambda_s\}$ there exists a $\{\lambda_1, \ldots, \lambda_s\}$-partition in which each of the triple systems is indecomposable.

Such a partition is an *indecomposable partition* of type $\{\lambda_1, \ldots, \lambda_s\}$. Evidently, large sets of triple systems provide one class of indecomposable partitions. However, when larger parts are permitted, very little is known. In the absence of a good bound on the largest index of an indecomposable simple triple system on v elements, results have concentrated in two areas: the solution for small orders, and the case when all part sizes are 'small'. For small orders, Gronau and Rentner (1985) and Colbourn and Harms (1987) establish that all types realized by indecomposable partitions of order at most 10 are:

Order	Type
6	$\{2,2,2\}$
7	$\{1,1,3\}$
9	$\{1,1,1,1,1,1,1\}$, $\{1,1,1,1,1,2\}$, $\{1,1,1,1,3\}$
	$\{1,1,1,2,2\}$, $\{1,1,2,3\}$, $\{1,2,2,2\}$, $\{1,3,3\}$, $\{2,2,3\}$
10	$\{2,2,2,2\}$, $\{2,2,4\}$, $\{4,4\}$

From this, it is evident that a $\mathrm{TS}(v, \lambda)$ may have indecomposable partitions into numerous different types. When $\lambda = 2$, of course this is not the case, as the system is either decomposable (type $\{1,1\}$) or it is not ($\{2\}$). Already for $\lambda = 3$, however, a decomposable $\mathrm{TS}(v, 3)$ may have an indecomposable partition of type $\{1,2\}$ or $\{1,1,1\}$; Colbourn and Harms (1987) establish that exactly one of the 332 simple $\mathrm{TS}(9,3)$s admits *both* types of indecomposable partitions.

In the same way that indecomposable partitions generalize large sets, we can generalize the notion of overlarge sets. To do this, define a *pointed triple system* $\mathrm{POTS}(v, \lambda)$ to be a $\mathrm{PTS}(v, \lambda)$ $(V, \mathcal{B})$ in which there is a subset L of V, $|L| = \lambda$, so that

1. if $x, y \in V \setminus L$, $\{x, y\}$ appears in λ triples of $\mathcal{B}$;
2. if $x \in V \setminus L$ and $y \in L$, $\{x, y\}$ appears in $\lambda - 1$ triples of $\mathcal{B}$; and
3. if $x, y \in L$, $\{x, y\}$ appears in $\lambda - 2$ triples of $\mathcal{B}$.

When $\lambda = 1$, a $\mathrm{POTS}(v, \lambda)$ is just an $\mathrm{STS}(v-1)$ with an additional element that is in no triples. A $\mathrm{POTS}(v, \lambda)$ is *indecomposable* exactly when it is indecomposable as a $\mathrm{PTS}(v, \lambda)$. An *indecomposable near-partition* of type $\Lambda = \{\lambda_1, \ldots, \lambda_s\}$ is a partition of all $\binom{v}{3}$ triples into indecomposable $\mathrm{POTS}(v, \lambda_i)$s, $1 \le i \le s$. Evidently, $\lambda_1 + \cdots + \lambda_s = v$ in an indecomposable near-partition. Overlarge sets are precisely near-partitions in which each system is a $\mathrm{POTS}(v, 1)$, and hence give examples of near-partitions which are trivially indecomposable. Colbourn and Harms (1987) produce indecomposable near-partitions of order 10 having all types consisting of i $\mathrm{POTS}(10, 2)$s and $10 - 2i$ $\mathrm{POTS}(10, 1)$s, $0 \le i \le 5$, $i \ne 1$. They use a simple doubling construction to prove:

Lemma 23.14 *If there exist two indecomposable near-partitions of order v, having types Λ_1 and Λ_2, then there exists an indecomposable near-partition of order $2v$ having type $\Lambda_1 \cup \Lambda_2$.*

They then adapt Theorem 15.8 to establish:

Theorem 23.15 *If there exists an indecomposable partition of odd order v and type Λ_1, and an indecomposable near-partition of order $v+1$ and type Λ_2, then there exists an indecomposable partition of order $2v+1$ and type $\Lambda_1 \cup \Lambda_2$.*

Together with the information on orders 9 and 10, these doubling constructions establish:

Theorem 23.16 *Let $v = 5 \cdot 2^i - 1$, $i \geq 1$. Let Λ be a partition of $v-2$ into parts all equal to 1 or 2. Then there is an indecomposable partition of order v and type Λ.*

Essentially, all that is known about the indecomposable partition problem in general follows from the constructions here, and from those constructions for large sets that preserve 'subsystems'.

23.5 Separations of triple systems

Consider an α-parallel class in a triple system, i.e. a set of triples in which each element occurs exactly α times. An α-parallel class is *separable* if there exist positive integers α_1, α_2 with $\alpha_1 + \alpha_2 = \alpha$ so that the α-parallel class can be partitioned into an α_1-parallel class and an α_2-parallel class. When $v \not\equiv 0 \pmod 3$, every α-parallel class has $\alpha \equiv 0 \pmod 3$, so some such classes cannot be separated for purely numerical reasons.

A triple system TS(v, λ) is itself a $\lambda(v-1)/2$-parallel class, so it is natural to ask about separations of triple systems. An $(\alpha_1, \ldots, \alpha_\ell)$*-separation* of a triple system $(V, \mathcal{B})$ is a partition of $\mathcal{B}$ into classes $\mathcal{B}_1, \ldots, \mathcal{B}_\ell$ in which $\mathcal{B}_i$ is an α_i-parallel class for $1 \leq i \leq \ell$. Evidently, a $(1, \ldots, 1)$-separation is a resolution, and hence by combining parallel classes we can produce any partition of $\lambda(v-1)/2$ for a resolvable triple system. For Steiner triple systems with $v \equiv 1 \pmod 6$, a $(3, \ldots, 3)$-separation can be formed by taking the orbits of a cyclic STS(v). In this way, all partitions into multiples of three can be realized.

The problems become more interesting if one requires that the classes in the partition themselves have no further separation; this is an *atomic separation.* For Steiner triple systems, the natural question is to determine for which partitions $\alpha_1 + \cdots + \alpha_\ell = \frac{v-1}{2}$ there exists a Steiner triple system of order v having an $(\alpha_1, \ldots, \alpha_\ell)$-atomic separation. Colbourn (1992) examined this question, reporting computational results for $v = 15$, shown in Tables 23.1 and 23.2.

For $v \in \{1, 3, 7\}$, an STS(v) is trivially nonseparable. For $v = 9$, the STS(v) is resolvable, and admits only a (1,1,1,1)-atomic separation. For $v = 13$, both STS(v)s have a 3-parallel class and hence admit only (3,3)-atomic separations. The first interesting case is therefore $v = 15$. For each partition of 7 (written vertically and in exponential notation in the tables), and for each STS(15) numbered as in Tables 5.8—5.12, a star ('$\star$') indicates that the system admits an atomic separation of the corresponding type. A blank indicates that no such atomic separation exists.

Design #19 realizes nine different types of atomic separations, while design #16 admits only one. Indeed, design #16 is the only one having no 2-parallel class, while 10 designs (numbers 14, 16, 24, 26, 29, 35, 40, 52, 59, and 62) have no 1-parallel class. Every STS(15) has a 3-parallel class, and indeed each has a nonseparable 3-parallel class. Most surprisingly, there are nonseparable 4- and 5-parallel classes, and every partition of 7 with maximum part at most five except for 1+1+1+1+1+2 is realized by at least one system. There are no nonseparable 6-parallel or 7-parallel classes.

No general theory is available here. In particular, no example is known of an STS(v) for $v \geq 7$ that has no 3-parallel class (perhaps separable). No example of an STS(v) for $v \geq 9$ is known which is itself nonseparable.

23.6 Work points

1. Does an indecomposable TS(13,5) and an indecomposable TS(15,5) exist?
2. Does there exist an indecomposable TS(v, 6) for $v = 11$, 12, 13, 15, 16?
3. What is the complexity of deciding whether a TS(v, λ) is decomposable for arbitrary $\lambda > 4$?
4. Determine the largest index $\mu(v)$ for which an indecomposable *simple* TS($v, \mu(v)$) exists. For $v \notin \{3, 4, 5, 7, 8\}$, can $\mu(v)$ be greater than $(v-2)/2$?
5. What is the largest index $\nu(v)$ for which an indecomposable TS($v, \nu(v)$) exists?
6. Prove or disprove: for $v \equiv 1, 3 \pmod 6$, $v \geq 9$, every partition of $v - 2$ in which each part is 1 or 2 is the type of an indecomposable partition.
7. Is there a nonseparable Steiner triple system of order greater than 7? If not, what is the largest α for which a nonseparable α-parallel class can appear in an STS(v)?
8. Is there a Steiner triple system of order greater than 3 having (a) no 3-parallel class? (b) no 1-, 2-, or 3-parallel class?

#	3^1 4^1	2^1 5^1	2^2 3^1	1^1 3^2	1^1 2^1 4^1	1^2 5^1	1^1 2^3	1^2 2^1 3^1	1^3 4^1	1^3 2^2	1^4 3^1	1^7
1				$\star$			$\star$	$\star$				$\star$
2			$\star$	$\star$			$\star$	$\star$		$\star$		
3	$\star$			$\star$			$\star$					
4	$\star$		$\star$	$\star$			$\star$	$\star$				
5			$\star$	$\star$			$\star$	$\star$		$\star$		
6	$\star$		$\star$	$\star$			$\star$	$\star$		$\star$		
7				$\star$			$\star$	$\star$			$\star$	$\star$
8	$\star$		$\star$	$\star$	$\star$		$\star$					
9	$\star$		$\star$	$\star$	$\star$		$\star$					
10	$\star$		$\star$	$\star$	$\star$		$\star$	$\star$				
11	$\star$		$\star$	$\star$	$\star$		$\star$	$\star$		$\star$		
12	$\star$		$\star$	$\star$	$\star$		$\star$					
13	$\star$		$\star$	$\star$	$\star$		$\star$					
14	$\star$		$\star$									
15	$\star$		$\star$	$\star$	$\star$		$\star$	$\star$		$\star$		
16	$\star$											
17	$\star$		$\star$	$\star$	$\star$		$\star$	$\star$		$\star$		
18	$\star$		$\star$	$\star$	$\star$		$\star$	$\star$				
19	$\star$		$\star$	$\star$			$\star$	$\star$	$\star$	$\star$	$\star$	$\star$
20	$\star$		$\star$	$\star$	$\star$		$\star$					
21	$\star$		$\star$	$\star$	$\star$		$\star$					
22	$\star$		$\star$	$\star$	$\star$		$\star$	$\star$		$\star$	$\star$	
23	$\star$		$\star$	$\star$	$\star$		$\star$					
24	$\star$		$\star$									
25	$\star$	$\star$	$\star$	$\star$	$\star$		$\star$					
26	$\star$		$\star$									
27	$\star$		$\star$	$\star$	$\star$		$\star$	$\star$				
28	$\star$	$\star$	$\star$	$\star$	$\star$		$\star$					
29	$\star$	$\star$	$\star$									
30	$\star$		$\star$	$\star$	$\star$		$\star$	$\star$		$\star$		
31	$\star$		$\star$	$\star$	$\star$		$\star$	$\star$				
32	$\star$	$\star$	$\star$	$\star$	$\star$		$\star$	$\star$				
33	$\star$	$\star$	$\star$	$\star$	$\star$		$\star$					
34	$\star$	$\star$	$\star$	$\star$	$\star$							
35	$\star$	$\star$	$\star$									
36	$\star$		$\star$	$\star$	$\star$		$\star$					
37	$\star$		$\star$		$\star$		$\star$					
38	$\star$	$\star$	$\star$	$\star$	$\star$		$\star$	$\star$				
39	$\star$	$\star$	$\star$	$\star$	$\star$							
40	$\star$	$\star$	$\star$									

Table 23.1 *Separations of STS*(15)*s I*

#	$3^1 4^1$	$2^1 5^1$	$2^2 3^1$	$1^1 3^2$	$1^1 2^1 4^1$	$1^2 5^1$	$1^1 2^3$	$1^2 2^1 3^1$	$1^3 4^1$	$1^3 2^2$	$1^4 3^1$	1^7
41	⋆	⋆	⋆	⋆	⋆							
42	⋆		⋆	⋆	⋆		⋆					
43	⋆		⋆	⋆	⋆		⋆	⋆		⋆		
44	⋆		⋆	⋆	⋆		⋆					
45	⋆		⋆	⋆	⋆	⋆	⋆					
46	⋆	⋆	⋆	⋆	⋆		⋆					
47	⋆	⋆	⋆	⋆	⋆							
48	⋆	⋆	⋆	⋆	⋆		⋆					
49	⋆		⋆	⋆	⋆		⋆					
50	⋆	⋆	⋆	⋆	⋆		⋆	⋆	⋆			
51	⋆	⋆	⋆	⋆	⋆		⋆	⋆				
52	⋆	⋆	⋆									
53	⋆	⋆	⋆	⋆	⋆		⋆					
54	⋆	⋆	⋆	⋆	⋆			⋆				
55	⋆	⋆	⋆	⋆	⋆		⋆					
56	⋆	⋆	⋆	⋆	⋆							
57	⋆		⋆	⋆	⋆		⋆	⋆	⋆			
58	⋆	⋆	⋆	⋆	⋆		⋆	⋆	⋆			
59	⋆		⋆									
60	⋆		⋆	⋆	⋆		⋆	⋆				
61	⋆		⋆	⋆			⋆	⋆				⋆
62	⋆	⋆	⋆									
63	⋆		⋆	⋆	⋆		⋆	⋆	⋆	⋆		
64	⋆		⋆	⋆	⋆					⋆		
65	⋆		⋆	⋆	⋆		⋆					
66	⋆	⋆	⋆	⋆	⋆		⋆	⋆				
67	⋆	⋆	⋆	⋆	⋆		⋆	⋆	⋆			
68	⋆		⋆	⋆	⋆		⋆					
69	⋆	⋆	⋆	⋆	⋆		⋆	⋆	⋆			
70	⋆		⋆	⋆	⋆	⋆	⋆					
71	⋆		⋆	⋆	⋆		⋆	⋆				
72	⋆	⋆	⋆	⋆	⋆		⋆	⋆				
73	⋆	⋆	⋆	⋆	⋆		⋆	⋆				
74	⋆		⋆	⋆	⋆	⋆	⋆					
75	⋆		⋆	⋆	⋆		⋆	⋆	⋆			
76	⋆		⋆	⋆	⋆		⋆					
77	⋆	⋆	⋆	⋆								
78	⋆		⋆	⋆	⋆		⋆	⋆	⋆	⋆		
79	⋆	⋆		⋆			⋆	⋆			⋆	
80				⋆			⋆	⋆				

Table 23.2 *Separations of STS*(15)*s II*

24

DIRECTED TRIPLE SYSTEMS

In this and the next chapter, we examine extensions of designs (particularly triple systems), in which blocks are taken to contain *ordered* rather than unordered pairs. We primarily consider two different types of ordered blocks. The *transitive triple* (a, b, c) is said to contain the ordered pairs (a, b), (a, c) and (b, c). The *cyclic triple* $\langle a, b, c\rangle$ contains the ordered pairs (a, b), (b, c) and (c, a). We examine directed triple systems in which the blocks are transitive triples, Mendelsohn triple systems in which the blocks are cyclic triples, and hybrid triple systems in which the blocks are permitted to be either.

Hung and Mendelsohn (1973) first introduced a generalization of Steiner triple systems in which blocks are transitive triples. A *directed triple system* of *order* v and *index* λ, or DTS(v, λ), is a pair $(V, \mathcal{D})$. V is a v-set of elements, and $\mathcal{D}$ is a collection of transitive triples on V, with the property that every *ordered* pair (x, y) of elements of V appears in precisely λ of the transitive triples. Directed triple systems have been studied extensively, often under the name 'transitive triple systems'. When the parameter λ is omitted, $\lambda = 1$ by convention. In this chapter, we concentrate on the relationships between triple systems and DTSs.

We first give some small examples.

DTS$(4, 1)$	develop $(i, i+2, i+1)$ modulo 4
DTS$(6, 1)$	develop $(0, \infty, 4)$ and $(0, 1, 3)$ modulo 5
DTS$(7, 1)$	develop (0,1,3) and (0,6,4) modulo 7
DTS$(5, 3)$	develop (0,1,2), (0,2,4), (0,3,1), and (0,4,3) modulo 5
DTS$(8, 3)$	develop (0,4,1), (0,4,2), (0,4,3), (0,1,7), (0,2,1), (0,5,3), and (0,2,5) modulo 8

Every DTS(v) $(V, \mathcal{B})$ has a *converse*, $(V, \mathcal{B}^R)$, where

$$\mathcal{B}^R = \{(a, b, c) : (c, b, a) \in \mathcal{B}\}.$$

The converse of a DTS is also a DTS; a DTS that is isomorphic to its converse is *self-converse*.

24.1 Existence I: basic constructions

The necessary condition for a DTS(v, λ) to exist is simply that the number of ordered pairs $\lambda v(v-1)$ to occur in blocks be divisible by three. Hence we require $v \equiv 0, 1 \pmod 3$ for $\lambda \equiv 1, 2 \pmod 3$, and we require only $v \neq 2$ for $\lambda \equiv 0 \pmod 3$.

Hung and Mendelsohn (1973) settled existence for $\lambda = 1$, and Seberry and Skillicorn (1980) settled existence for all λ. We present a number of different existence proofs for directed triple systems, in part because each affords some flexibility that the others do not. First we study simple recursive constructions.

Lemma 24.1 *If there exists a DTS(v, λ), there exists a DTS$(2v + 1, \lambda)$.*

Proof Let $(V, \mathcal{B})$ be a DTS(v, λ). Let L be a $(v + 1) \times (v + 1)$ latin square with constant diagonal, whose rows and columns are indexed by a $(v + 1)$-set X disjoint from V and whose offdiagonal symbols are from V. Form a set $\mathcal{C}$ of transitive triples as follows: For each $i \in X$ and each $j \in X$, $j \neq i$, add $(i, L(i, j), j)$ to $\mathcal{C}$ λ times. Then $(V \cup X, \mathcal{B} \cup \mathcal{C})$ is a DTS$(2v + 1, \lambda)$. □

Lemma 24.2 *If there exists a DTS(v, λ), there exists a DTS$(2v + 4, \lambda)$.*

Proof Let $(V, \mathcal{B})$ be a DTS(v, λ), and let $V = \{z_1, \ldots, z_v\}$. Let $X = \mathbb{Z}_{v+4}$ be disjoint from V. Form v disjoint sets $S_1, \ldots, S_v$ of ordered pairs on X, by taking $S_i = \{(a, b) : b - a \equiv i \pmod{v + 4}\}$. Now form a collection $\mathcal{C}$ of blocks by including

1. for each $1 \leq i \leq v$, and each $(a, b) \in S_i$, the triple (a, z_i, b) λ times; and
2. for each $i \in \mathbb{Z}_{v+4}$, the triple $(i, v + 2 + i, v + 1 + i)$ λ times. □

These suffice to settle existence:

Theorem 24.3 *A DTS(v, λ) exists if and only if either $\lambda \equiv 1, 2 \pmod 3$ and $v \equiv 0, 1 \pmod 3$, or $\lambda \equiv 0 \pmod 3$ and $v \neq 2$.*

Proof Since the union of a DTS(v, λ_1) and a DTS(v, λ_2) on the same element set yields a DTS$(v, \lambda_1 + \lambda_2)$, we need only treat the cases when $\lambda = 1$ and $v \equiv 0, 1 \pmod 3$, or $\lambda = 3$ and $v \equiv 2 \pmod 3$. We proceed inductively on the order.

We have already seen a DTS$(5, 3)$ and a DTS$(8, 3)$. Suppose $v \notin \{5, 8\}$, and λ is minimum. If v is odd, write $v = 2t + 1$. Now $t \neq 2$ and hence a DTS(t, λ) exists; applying the $2v + 1$ construction gives a DTS(v, λ). Similarly, if v is even, write $v = 2t + 4$ and apply the $2v + 4$ construction. □

Existence of DTS can also be established directly using tripling constructions. We explore this method next:

Lemma 24.4 *There exist a DTS$(3m)$, a DTS$(3m + 1)$, and a DTS$(3m + 2, 3)$ for all $m \geq 1$.*

Proof We have already seen a DTS(6), a DTS(7), and a DTS$(8, 3)$, so we may assume that $m \neq 2$. Let L_0, L_1, L_2 be three (not necessarily distinct) $m \times m$ idempotent latin squares, each with symbols from X, and rows and columns indexed by X. On $X \times \mathbb{Z}_3$, we form a collection $\mathcal{D}$ of transitive triples. For $x \in X$ and $i \in \mathbb{Z}_3$, we write (x, i) as x_i. Let $a, b \in X$, $a \neq b$, and $c^{(i)} = L_i(a, b)$ for $i \in \mathbb{Z}_3$. Add to $\mathcal{D}$ the triples $(a_i, c^{(i)}_{i+1}, b_i)$ for each $i \in \mathbb{Z}_3$.

$\mathcal{D}$ now contains all pairs of the form (a_i, b_i), (a_i, b_{i+1}), and (a_i, b_{i+2}) for $a \neq b$; it contains no pairs of the form (a_i, a_j). To complete $\mathcal{D}$ to a DTS$(3m)$, on each $\{a\} \times \mathbb{Z}_3$ for $a \in X$ place the triples of a DTS(3).

To obtain instead a DTS($3m+1$), add one further element ∞. Then for $a \in X$, on $(\{a\} \times \mathbb{Z}_3) \cup \{\infty\}$, place the triples of a DTS(4).

The construction of DTS($3m+2, 3$) is slightly more complicated. Repeat each triple of $\mathcal{D}$ three times. Add two further elements, α and β. For a single element $x \in X$, place the triples of a DTS(5, 3) on $(\{x\} \times \mathbb{Z}_3) \cup \{\alpha, \beta\}$. For all elements $a \in X$, $a \neq x$, include the blocks (α, a_0, a_1), (a_0, α, a_2), (a_1, a_2, α), (β, a_2, a_1), (a_1, β, a_0), and (a_2, a_0, β) each three times. The result is a DTS($3m+2, 3$). □

In this tripling construction, subsquares of L lead to sub-DTS of the resulting DTS.

Directed triple systems can also be easily constructed using PBDs. Let $(V, \mathcal{B})$ be a (v, K)-PBD, and suppose that for each $k \in K$ there is a DTS(k, λ). For each block $B \in \mathcal{B}$, place the triples of a DTS ($|B|, \lambda$) on the elements of B. The result is a DTS(v, λ). There exists a $(v, \{3, 4\})$-PBD for all $v \equiv 0, 1 \pmod 3$, $v \neq 6$ (Hanani, 1975); hence we obtain a further proof of the existence of DTS(v)s. Moreover, $(v, \{3, 4, 5\})$-PBDs exist for all $v \geq 0$ except for $v \in \{2, 6, 8\}$ (Hanani, 1975), giving a further proof of the existence of DTS($v, 3$)s.

Kang, Chang, and Yang (1994) establish a stronger existence result:

Theorem 24.5 *A self-converse DTS(v) exists if and only if $v \equiv 0, 1 \pmod 3$ and $v \neq 6$.*

This has been generalized to all indices; see Kang (1995).

24.2 Existence II: directing with conflict resolution

Suppose that we are given a directed triple system $(V, \mathcal{B})$ of order v and index λ, and replace each triple $(a, b, c) \in \mathcal{B}$ by the triple $\{a, b, c\}$. Since for every ordered pair $a, b \in V$, (a, b) and (b, a) each appear in λ transitive triples, the resulting collection of triples contains every unordered pair from V precisely 2λ times. This produces a triple system TS($v, 2\lambda$), the *underlying triple system* of the DTS(v, λ). A triple system TS($v, 2\lambda$) is *directable* if it is the underlying triple system of some DTS(v, λ).

This reveals that the study of directed triple systems is closely connected with that of triple systems; in fact, the connection is closer than might be expected at first. Colbourn and Colbourn (1983a) prove that every twofold triple system TS($v, 2$) is directable; Colbourn and Harms (1983) extend the technique to prove that every triple system with even index is directable.

In this section, we describe a simple method that, given a triple system $(V, \mathcal{B})$ with even index, produces a directed triple system having $(V, \mathcal{B})$ as its underlying triple system. The theorem that every triple system with even index is directable can then be deduced from the correctness of the algorithm.

Let $(V, \mathcal{B})$ be a triple system of order v and index 2λ. Replace each block $\{a, b, c\} \in \mathcal{B}$ by a transitive triple (a, b, c); there are six choices for the transitive triple, and one is chosen arbirarily. Let $\mathcal{D}$ be the collection of transitive triples chosen in this way. In general, an *arbitrarily directed triple system* ADTS(v, λ)

is a pair $(V, \mathcal{D})$ with $\mathcal{D}$ a collection of transitive triples, and whose underlying triple system is a $\mathrm{TS}(v, 2\lambda)$.

In general, $(V, \mathcal{D})$ fails to be a $\mathrm{DTS}(v, \lambda)$. The reason is quite simple. An ordered pair may appear in any number of transitive triples of $\mathcal{D}$ from 0 to 2λ, and does not in general appear in precisely λ as required. A pair (x, y) appearing in $s > \lambda$ transitive triples of $\mathcal{D}$ is a *conflict*, and the *severity* of the conflict is $s - \lambda$. The *deficiency* of $\mathcal{D}$ is the sum over all pairs of the severity of the conflict (if any) for that pair.

Our strategy is to transform $(V, \mathcal{D})$ into a collection of transitive triples having deficiency zero, that is, a $\mathrm{DTS}(v, \lambda)$. To do this, it suffices to show that we can always reduce the deficiency if it is nonzero.

Lemma 24.6 *Let $(V, \mathcal{D})$ be an $ADTS(v, \lambda)$ whose deficiency is nonzero. Then there is an $ADTS(v, \lambda)$ $(V, \mathcal{D}')$ having the same underlying triple system and having smaller deficiency.*

Proof $(V, \mathcal{D})$ must have a conflict. Let (c, x_0) be a pair that occurs $s > \lambda$ times. Now consider some triple of $\mathcal{D}$ containing the pair (c, x_0), and a third element x_1. This triple must be one of (c, x_0, x_1), (x_1, c, x_0), or (c, x_1, x_0). In the first case, replacing the triple by (x_0, c, x_1) reduces the severity of this conflict, and does not affect the existence or severity of any other conflict; hence the deficiency is reduced by one. In the second case, replacing the triple by (x_1, x_0, c) has the same effect.

In the final case, we cannot guarantee to reduce deficiency by interchanging c and x_0 in this triple. We therefore examine a sequence of triples all containing c. The first triple in the sequence is (c, x_1, x_0); at a general step, the last triple in the sequence has the form (c, x_i, x_{i-1}) or (x_{i-1}, x_i, c). Let us consider the first situation; the second is symmetric. We attempt to find a triple containing (x_i, c) that is not already in the sequence. If found, this triple contains a third element x_{i+1}. If this triple is of the form (x_i, c, x_{i+1}), we reorder the triple as (c, x_i, x_{i+1}); and if the triple is of the form (x_{i+1}, c, x_i), we reorder it as (x_{i+1}, x_i, c). We then reorder all previous triples in the sequence, replacing (c, x_i, x_{i-1}) by (x_i, x_{i-1}, c), and replacing (x_{i-1}, x_i, c) by (c, x_{i-1}, x_i). In either case, we terminate. A third case remains, in which the triple found is of the form (x_i, x_{i+1}, c); in that event, we continue extending the sequence of triples.

It we cannot find an unused triple containing the pair (x_i, c), we instead choose an unused triple containing (c, x_i) and proceed with constructing the sequence. Since (c, x_i) and (x_i, c) appear an even number of times in total in $\mathcal{D}$, we can only fail to be able to extend the sequence when $x_i = x_0$, and *all* triples containing (c, x_0) or (x_0, c) appear in the sequence.

We must show two things. First, if the sequence construction terminates 'early', and all triples of the sequence reordered as described, the deficiency is reduced. Second, if the sequence construction terminates because all triples involving c and x_0 are used, we must describe a suitable reordering to reduce the deficiency.

Suppose that the sequence terminates 'early'; carry out the reordering described. Consider an element x_i that appears (possibly many times) in the sequence. Every occurrence of (c, x_i) is replaced by an occurrence of (x_i, c) and vice versa; moreover, one can occur more frequently than the other in the sequence only when all occurrences in $\mathcal{D}$ of the more infrequent one appear in the sequence; hence no conflict can be introduced, and any existing conflict may be reduced or left unchanged. The same argument holds for x_0, with the added observation that when the sequence terminates early, (c, x_0) must appear more frequently in the sequence than does (x_0, c) and hence the conflict is reduced.

If instead the sequence terminates because all triples involving c and x_0 are exhausted, consider the final triple. It is either (c, x_0, x_s) or (x_s, x_0, c). The latter could only occur if there was no conflict on (c, x_0) to begin with, which cannot be. On the other hand, if (c, x_0, x_s) is the final triple, we simply replace it with (x_0, c, x_s), leaving all other triples unchanged, for a reduction of one in the deficiency. □

Repeated application of Lemma 24.6 yields:

Theorem 24.7 *(Directability) Every triple system with even index is directable.*

24.3 Existence III: the eulerian algorithm

In this section, we examine a second proof of the Directability Theorem (24.7), which provides an optimal algorithm for directing a triple system with even index. Our first proof is constructive, and so leads to a directing algorithm. In fact, the algorithm examines $O(\lambda v)$ triples to guarantee a reduction in deficiency. Since the initial deficiency is $O(\lambda v^2)$, the resulting algorithm can be implemented in $O(\lambda^2 v^3)$ time.

Can a triple system with even index be directed more efficiently? Harms and Colbourn (1983, 1985) study this problem, and develop an $O(\lambda v^2)$ time algorithm for directing. Their algorithm is the subject of this section.

The essential idea is to decide which of the six alternative transitive triples to use in such a way that it is never necessary to alter the choice for a triple already processed, and so that no conflict is ever introduced.

We begin with a TS$(v, 2\lambda)$ $(V, \mathcal{B})$ in which V is a set of elements with the (arbitrary) total order $\prec$. A *segment* of $\mathcal{B}$ for $v \in V$, denoted $S(v)$, is the collection of blocks $\{\{v, x, y\} : v \prec x \prec y\}$. The *segment graph* $G(v)$ is the multigraph with edges $\{\{x, y\} : \{v, x, y\} \in S(v)\}$. A *subsegment* is a collection of blocks corresponding to a connected component of a segment graph. Every block appears in precisely one segment, and hence is represented by one edge in exactly one of the segment graphs.

The method directs blocks of B a segment at a time, always directing those in the segment for $S(w)$ before those in $S(v)$ if $v \prec w$. To process a segment $S(v)$, we first partition it into subsegments $S_1(v), \ldots, S_k(v)$. Now for each subsegment $S_\ell(v)$ in turn, form the subsegment graph $G_\ell(v)$. This subsegment graph may have vertices of odd degree; if it does, introduce a 1-factor of *virtual* edges on the vertices of odd degree to form a connected multigraph $A_\ell(v)$ that has all vertex degrees even. Let t be the number of edges in $A_\ell(v)$.

$A_\ell(v)$ has an *eulerian circuit*, a sequence $(e_1, v_1, \ldots, e_t, v_t)$ where e_i and e_j are distinct edges for $i \neq j$, v_i is a vertex of $A_i(v)$, $e_i = \{v_{i-1}, v_i\}$ for $2 \leq i \leq t$, and $e_1 = \{v_1, v_t\}$. We employ the edge ordering induced by a fixed eulerian circuit in $A_\ell(v)$.

We employ a function **next** mapping undirected pairs to directed pairs that operates as follows. If the ordered pair (x, y), $x \prec y$, has already appeared in λ transitive triples produced, $\texttt{next}(\{x, y\}) = (y, x)$; otherwise, $\texttt{next}(\{x, y\}) = (x, y)$. (The outcome of evaluating **next** depends on the decisions made thus far in processing previous segments and subsegments, and also on earlier choices in processing the current subsegment.)

First we treat the case when t is even. We produce transitive triples for all triples in $S_\ell(v)$ as follows for $1 \leq i \leq t/2$:

1. for $e_{2i} = \{x, y\}$, if $\texttt{next}(e_{2i}) = (x, y)$, form the triple (v, x, y); otherwise form the triple (v, y, x).
2. for $e_{2i-1} = \{x, y\}$, if $\texttt{next}(e_{2i-1}) = (x, y)$, form the triple (x, y, v); otherwise form the triple (y, x, v).

When a transitive triple is formed, it is added to the directed triple system being produced whenever the edge is not virtual.

When t is odd, we cannot simply alternate placing v at the beginning and at the end of the transitive triples produced. Hence we treat one triple, that for e_1, differently. If $\texttt{next}(e_1) = (v_1, v_t)$, we form the triple (v_1, v, v_t); otherwise we employ instead the eulerian circuit $(e_1, v_t, e_t, \ldots, e_2, v_1)$. Then place v at the beginning or end of the remaining triples, using (1) and (2) above.

The alternation ensures that just after processing $S(v)$, the number of occurrences of an ordered pair containing v cannot exceed λ; all occurrences of it in segments yet to be processed are dealt with using the **next** function, which ensures that the pair can never be included more than λ times.

The time required to direct a triple system in this way is a linear function of the number of blocks, since the identification of segments and subsegments, the formation of $A_\ell(G)$, and finding eulerian circuits are all solvable by linear time algorithms.

We also remark on an important consequence of this algorithm:

Theorem 24.8 *Every partial triple system of index 2λ underlies a partial directed triple system of index λ.*

The flexibility in directing a triple system can best be seen by observing that it is sufficient for the **next** function to choose either (x, y) *or* (y, x) if neither has yet appeared λ times.

In fact, the large number of choices in the algorithm enables us to prove:

Theorem 24.9 *Every simple $TS(v, 2\lambda)$ underlies at least four $DTS(v, \lambda)$s that are mutually disjoint.*

Proof Given a simple TS$(v, 2\lambda)$, we modify the eulerian algorithm to produce two disjoint DTS(v, λ)s D_1 and D_2, whose converses are also disjoint from D_1 and D_2. We produce D_1 by applying the eulerian algorithm above, ensuring that every odd length eulerian circuit $(e_1, v_1, \ldots, e_t, v_t)$ that is used has $\mathtt{next}(e_1) = (v_1, v_t)$ and $\mathtt{next}(e_2) = (v_2, v_1)$ (this is always possible since the length is odd). Then produce D_2 in the same way, but for *every* eulerian circuit $(e_1, v_1, \ldots, e_t, v_t)$ used in producing D_1, use instead the eulerian circuit $(e_2, v_2, \ldots, e_t, v_t, e_1, v_1)$ in producing D_2.

Now consider the triples of D_1 and D_2 arising from processing $A_\ell(v)$. When $A_\ell(v)$ has an even length eulerian circuit with edge $\{x, y\}$ having $\mathtt{next}$ equal to (x, y), one of (v, x, y) and (x, y, v) appears in D_1, and the other in D_2. The same holds for edges $e_3, \ldots, e_t$ of an odd length eulerian circuit. In an odd length circuit, e_1 produces (v_1, v, v_t) in D_1 but produces (v_1, v_t, v) in D_2; e_2 produces (v, v_2, v_1) in D_1, but produces (v_2, v, v_1) in D_2. Hence D_1 and D_2 are disjoint.

Moreover, the converses are also disjoint from D_1 and D_2, provided the converse of a triple in D_1 does not appear in D_2, or vice versa. Since the TS$(v, 2\lambda)$ contains no repeated blocks, no such converse triple appears in D_1 or D_2. □

It may happen that every simple TS$(v, 2\lambda)$ supports six disjoint DTS(v, λ)s, but while it is easy to produce particular TS(v, λ)s with this property (for example, any TS(v, λ) that can be partitioned into STS(v)s), nothing seems to be known in general. Harms and Colbourn (1983a) state the following:

Conjecture 24.10 *Every simple TS$(v, 2)$ supports six disjoint DTS(v)s.*

They establish that every simple TS$(v, 2)$ underlies two disjoint DTS$(v, 3)$s, and also underlies three disjoint DTS$(v, 2)$s. Moreover, they establish validity of Conjecture 24.10 for all $v \leq 10$.

Rodger (1997) gives another proof of directability, which we introduce next. Let $(V, \mathcal{B})$ be a partial TS$(v, 2\lambda)$. Arbitrarily impose a fixed, total ordering $(\prec)$ on the elements of $V = \{x_1, \ldots, x_v\}$, so that $x_i \prec x_{i+1}$ for $1 \leq i < v$. We form a bipartite graph G whose vertices W are the $\binom{v}{2}$ unordered pairs of elements in V, together with a vertex t_B for each $B \in \mathcal{B}$. Edges are determined as follows. For every triple $B = \{a, b, c\} \in \mathcal{B}$, suppose without loss of generality that $a \prec b \prec c$. Then place edges between vertex t_B and the vertices $\{a, b\}$ and $\{b, c\}$ of G. Using the colouring algorithm of de Werra (1971), assign each edge of G one of the colours white or black, so that at every vertex of G the number of incident white edges differs from the number of incident black edges by at most one. Now if pair $\{x, y\}$ with $x \prec y$ appears in triple B, it is directed as (x, y) if the edge between $\{x, y\}$ and t_B is white, and as (y, x) if black. Rodger's crucial observation is that if we have edge $\{\{a, b\}, \{b, c\}\}$ in G, if we choose to order the two corresponding pairs either as $\{(a, b), (c, b)\}$ or as $\{(b, a), (b, c)\}$, then we remain free to order the last pair in either direction and still we obtain a directed triple. Therefore for edge $\{\{a, b\}, \{b, c\}\}$ in G, if it is white we select $\{(a, b), (c, b)\}$ and if black, $\{(b, a), (b, c)\}$. It remains only to choose the ordering of the third pair in each triple, but any selection results in a collection of directed triples, so we simply process them one by one assigning an ordering so as to keep occurrences of each pair as balanced as possible with the occurrences of the same pair reversed.

This technique has a useful corollary, which also follows directly from the eulerian algorithm:

Lemma 24.11 *A partial TS$(v, 2\lambda)$ can be directed to form a partial DTS(v, λ) in which, for every pair $\{x, y\}$ of elements, the number of occurrences of (x, y) in the DTS differs from the number of occurrences of (y, x) in the DTS by at most one.*

24.4 Consequences of directability

It is natural to extend the large body of results on triple systems to directed triple systems. The Directability Theorem 24.7 and Theorem 24.8 provide a suitable vehicle for extending many results immediately.

The chromatic number and chromatic index of a DTS(v, λ) are the same as those for the underlying TS$(v, 2\lambda)$. Hence results concerning colouring and resolvability from Chapters 18 and 19 carry over to DTSs. In a similar way, every sub-DTS(w, λ) of a DTS(v, λ) corresponds to a sub-TS$(w, 2\lambda)$ in the underlying TS$(v, 2\lambda)$; as a consequence of Stern's theorem (Stern, 1979a; Theorem 6.9), we obtain:

Theorem 24.12 *Let $v \geq 2w + 1$, $v, w \neq 2$. Then if* $\gcd(v - 2, 3) \mid \lambda$ *and* $\gcd(w - 2, 3) \mid \lambda$, *a DTS$(v, \lambda)$ exists having a sub-DTS(w, λ).*

Embedding results for triple systems with higher index also carry over to DTSs. In fact, the correspondence between triple systems and DTSs can be used to translate results from DTSs to triple systems as well. Hamm (1983, 1983a) developed an embedding of a partial DTS(w) in a DTS(v) where $v = O(w^2)$; her method exploits the ordering of the pairs to effect the embedding. In view of Theorem 24.8, any such result on embedding partial DTSs can be applied to embedding partial TS$(v, 2)$s. Hamm, Lindner, and Rodger (1983) improved these results to obtain a linear embedding; subsequently, Colbourn, Hamm, and Rodger (1984) used Theorem 24.8 to extend to partial DTS with higher index. Johansson's result, Theorem 11.12, together with Lemma 24.11, yields the best possible embedding result:

Theorem 24.13 *A partial DTS(v, λ) can be embedded in a DTS(w, λ) provided $w \geq 2v + 1$ and* $\lambda w(w - 1) \equiv 0 \pmod 3$.

The existence of simple TS(v, λ)s has a further consequence. A DTS(v, λ) is *simple* if it has no repeated triples, and *pure* if its underlying TS$(v, 2\lambda)$ has no repeated triples (Di Paola and Nemeth, 1976). From the Directability Theorem 24.7 and the existence of simple TS(v, λ)s, we obtain:

Theorem 24.14 *Let $v \geq 3$. Let* $\lambda \equiv 0 \pmod{\gcd(v - 2, 3)}$ *and* $\lambda \leq (v - 2)/2$. *Then a pure DTS(v, λ) exists.*

24.5 Automorphisms

The fact that every triple system underlies a directed triple system provides a powerful vehicle for proving results about directed triple systems. However, the directing process need not preserve important properties of the triple system, such as its automorphism group. Harms (1985) describes a conflict resolution algorithm for directing a cyclic triple system with index 2, producing a *cyclic* directed triple system with index 1. Colbourn (1991) describes a general method for preserving automorphism groups, that yields yet another directing algorithm. We examine this method first in the context where the trivial automorphism is the only group assumed.

Let $(V, \mathcal{B})$ be a $\mathrm{TS}(v, 2\lambda)$. For each triple $t = \{x, y, z\} \in \mathcal{B}$, *arbitrarily* choose a *standard* cyclic triple $\langle x, y, z\rangle$ for t. For each unordered pair $\{x, y\} \subset V$, arbitrarily choose a *default* order (x, y) for the pair. Form a graph $\mathcal{G}(\mathcal{B})$ as follows:

- every triple $t_i \in \mathcal{B}$ is represented by a vertex (we blur the distinction between the triple and the vertex representing it, calling both t_i);
- every unordered pair p_j is represented by a vertex;
- if the default order of pair p_j appears in the standard triple for t_i, add an edge between t_i and p_j; and
- if the reverse of the default order of pair p_j appears in triple t_i, add a new vertex m_{ij} and make m_{ij} adjacent to t_i and to p_j.

To direct the system, we must choose an ordering for each occurrence of an unordered pair in a triple of $\mathcal{B}$. For the choice to give a $\mathrm{DTS}(v, \lambda)$, it must choose each unordered pair λ times in the default order (and hence λ times in the reverse order as well). It must also choose edge orderings so that not all edges are ordered so as to produce the standard triple, since it is cyclic and not transitive. Moreover, we must not produce the converse of the standard triple, since it is also cyclic. However, the other six choices for a triple are all transitive. We need only avoid the standard triples and their converses.

To produce such a choice of orderings for the pairs, suppose that there is an edge-colouring of $\mathcal{G}(\mathcal{B})$ in two colours, white and black, so that at each vertex the number of edges of each colour is as nearly equal as possible. We interpret such an edge-colouring as a direction for the triple system as follows. For each triple t_i, and each pair p_j contained in it, either $\{t_i, p_j\}$ is an edge, or there is a vertex m_{ij} and edges $\{t_i, m_{ij}\}$ and $\{m_{ij}, p_j\}$. In the first case, if the edge $\{t_i, p_j\}$ is white, choose the default order for p_j in t_i; if black, choose the reverse of the default. In the second case, examine instead the edge $\{t_i, m_{ij}\}$; if it is white, choose the reverse of the default, and if black choose the default.

Since every vertex m_{ij} has degree two, it is incident with one white edge and one black edge; then for a vertex t_i to be balanced, we cannot choose all three edges at it of the same colour, ensuring that we never choose the standard triple or its converse. Finally, each white edge at vertex p_j corresponds to a use of the default order of p_j, and each black edge to a use of the reverse. Since p_j is incident with λ edges of each colour, the pairs are used the correct number of times in each order.

All that remains is to produce the edge-colouring. A simple greedy algorithm suffices to produce such a colouring. Starting at any vertex t_i, find a maximal path of uncoloured edges. Such a path must terminate at another vertex t_k representing a triple. Alternate white and black along this path starting at t_i with a colour that is as yet unused at t_i if possible. The balance in the edge-colouring is immediate for vertices p_j and m_{ij}. For a vertex t_k, it may not appear as the end of three paths since the paths chosen are maximal. It therefore either appears as an interior vertex of some path, in which case it is balanced, or it appears as the starting vertex of three paths. Then our choice to start with an as yet unused colour ensures that it is balanced.

This gives a second linear time algorithm for directing triple systems and partial triple systems that appears to be quite different from the other methods. Most importantly, it provides a vehicle for incorporating information about the automorphism group.

Let us suppose that $(V, \mathcal{B})$ has an automorphism group Γ. The action of the group determines orbits of unordered pairs, and orbits of triples in $\mathcal{B}$. In order to direct the system so as to preserve the group Γ, it is necessary that we choose edge orderings in such a way that if $\pi \in \Gamma$ maps $\{x, y, z\} \in \mathcal{B}$ to $\{a, b, c\}$, we choose a transitive triple (x, y, z) and a triple (a, b, c) so that $\pi : (x, y, z) \mapsto (a, b, c)$. To account for this, we construct a graph $\mathcal{G}(\mathcal{B}, \Gamma)$ whose vertices are the *orbits* of unordered pairs, and the *orbits* of triples of $\mathcal{B}$, under Γ. The situation is complicated by the fact that a triple may contain two or three pairs from the same orbit.

Colbourn (1991) determines conditions that are necessary and sufficient for a TS$(v, 2)$ with group Γ to be directable as a DTS(v) with group Γ. The basic condition is that for no triple $\{x, y, z\} \in \mathcal{B}$ does an automorphism $\pi \in \Gamma$ fix two elements x, y and map the third element z to a different element w; for if this happens, the pair $\{x, y\}$ appears in the same order in both triples, which cannot be. The second condition is that if for a triple $\{x, y, z\}$ we have $\pi(\{x, y, z\}) = \{x, y, z\}$, π fixes at least one element of $\{x, y, z\}$. If it were to permute the three cyclically, the automorphism π induces a cyclic triple on $\{x, y, z\}$.

The existence of directed triple systems with cyclic and rotational automorphisms has been settled. For cyclic systems, Cho, Han, and Kang (1986) prove:

Theorem 24.15 *A DTS(v, λ) having a cyclic automorphism exists if and only if*

1. $\lambda \equiv 1, 5 \pmod 6$ *and* $v \equiv 1, 4, 7 \pmod{12}$;
2. $\lambda \equiv 2, 4 \pmod 6$ *and* $v \equiv 1 \pmod 3$;
3. $\lambda \equiv 3 \pmod 6$ *and* $v \equiv 0, 1, 3 \pmod 4$; *or*
4. $\lambda \equiv 0 \pmod 6$ *and* $v \neq 2$.

A similar theorem by Cho, Chae, and Hwang (1987) settles existence for k-rotational DTS(v)s.

24.6 The number of directed triple systems

How many nonisomorphic DTS(v)s are there? We have seen that every TTS is directable as a DTS(v), and further that every automorphism of the DTS is an automorphism of the underlying TTS. Hence we have a result due to Phelps and Lindner (1984):

Theorem 24.16 *For v sufficiently large, $v \equiv 0, 1 \pmod 3$, the number of nonisomorphic DTS(v) is* $\exp((v^2 \ln v/3)(1 + o(1)))$.

Proof To see that this is an upper bound, consider the following algorithm that constructs $b = v(v-1)/3$ blocks. Assume that the v elements are totally ordered. Start with an empty collection of blocks $\mathcal{B}$. Form the next block by finding the first ordered pair (a, b) that has not yet appeared in a block. Now choose a third element c, and choose one of the blocks (a, b, c), (a, c, b), (c, a, b) to add to $\mathcal{B}$. Repeat this process until $\mathcal{B}$ has b triples. At least one representative of each isomorphism class of DTS(v)s is constructed in this way, along with some other sets of triples that need not be DTSs. The number of ways to make choices in this algorithm, $(3(v-2))^{(v(v-1)/3)}$, is an upper bound on the number of DTS(v)s. Hence the number of DTS(v)s is bounded above by $\exp((v^2 \ln v/3) + (v^2 \ln 3/3))$.

For the lower bound, Phelps and Lindner (1984) implicitly show that the number of nonisomorphic TTS(v)s is asymptotic to $\exp((v^2 \ln v/3)(1 + o(1)))$. Since directing nonisomorphic TTSs yields nonisomorphic DTSs, the theorem follows by applying the Directability Theorem 24.7. □

This result appears to indicate that the numbers of TTS(v)s and DTS(v)s are essentially the same; however, we must be careful! It is possible that *each* TTS(v) underlies a number of nonisomorphic TTS(v)s that is exponential in v, without violating the bounds of the theorem. This raises an interesting open question. How many (how few) nonisomorphic DTS(v) can a TTS(v) underlie? A naive lower bound is 1. A naive upper bound is $2^{v(v-1)/2}$ since for each unordered pair, there are two consistent orderings in the two blocks containing the pair.

For certain TTS(v)s, the number of corresponding DTS(v)s can be exponentially large. To illustrate this, take an automorphism-free STS(v), and replace each block independently by one of the three distinct DTS(3)s. Each selection results in a nonisomorphic DTS(v), and hence the TTS(v)s obtained by duplicating each block of the STS underlies precisely $3^{v(v-1)/6}$ nonisomorphic DTS(v)s.

It might happen that *every* TTS(v) underlies exponentially many nonisomorphic DTS(v)s, but nothing in this direction is known at the present time. An indication of the explosive growth in the number of DTS(v)s is obtained by considering the exact number of DTS(v)s for small orders v. Colbourn and Colbourn (1981a) catalogue the DTS(v)s for $v \leq 7$, establishing the results in Table 24.1.

Recall that the corresponding numbers for TTS(v)s are 1, 1, 1, and 4; the increase by directing is dramatic! Only two of the 32 DTS(6)s have nontrivial automorphism group, as do only 221 of the 2368 DTS(7)s. Moreover, only 10 of the 221 have more than two automorphisms.

Order	Number
3	1
4	3
6	32
7	2368

Table 24.1 *Nonisomorphic DTS(v)s*

The relative lack of automorphisms is not surprising; the block ordering alone ensures that there are at least three orbits of ordered pairs under the action of the automorphism group: if (a,b,c) is a triple of the system, (a,b), (a,c), and (b,c) are surely in different orbits. This *a priori* information that the pairs are distinguished into three classes leads to a practical heuristic algorithm for testing isomorphism of DTSs. Colbourn and Colbourn (1982) describe such an algorithm; we outline their approach here.

Given a DTS(v) $(V,\mathcal{B})$, the method attempts to order the elements using invariants that are preserved by automorphisms; if found, the total ordering can be taken as a canonical ordering of the vertices. Renaming the ith element in the canonical ordering to i then produces a unique *canonical form* for the DTS; two DTSs have the same canonical form if and only if they are isomorphic.

Initially, each block $B_i \in \mathcal{B}$ is assigned label 1, and each element is assigned the label 1. We *refine* the labelling as follows:

1. For each element x, let ℓ_x be the current label of x. Form three multisets S_x^1, S_x^2, S_x^3 where S_x^i consists of the current labels of all blocks containing x in position i. Then $pos(x) = (\ell_x, S_x^1, S_x^2, S_x^3)$ is the current *position vector* of x.

2. Sort the position vectors for all elements as follows. The position vector for x precedes that for y if $\ell_x < \ell_y$. When $\ell_x = \ell_y$, let i be minimum subject to $S_x^i \neq S_y^i$; if no such i exists, the position vectors are equal. Otherwise, if the sets differ in size, we choose the smaller to precede the larger. If the sets are equal in size, consider their symmetric difference; if the smallest element in their symmetric difference is in S_x^i, x precedes y and otherwise y precedes x.

3. Assign new labels to each vertex equal to the rank of their position vector on the sorted list (equal position vectors having the same rank).

4. Now we refine the block labelling. Each block (x,y,z) is assigned the invariant (ℓ_x,ℓ_y,ℓ_z). Sort these invariants, and assign the new label of the block to be the rank of the invariant on the sorted list (equal invariants having the same rank).

After this refinement, the labelling produced is invariant under automorphisms of the DTS. Repeat the refinement until the labels remain unchanged by the refinement process. Remarkably, the refinement often terminates with all elements having distinct labels, that is, a canonical labelling. Of the 2368 DTS(7)s, this refinement produced a canonical labelling for 1960 of them.

Refinement fails completely whenever every element appears the same number of times in the first, second, and third positions of blocks; we call such DTSs *regular*. Thus it fails completely on cyclic systems, which are regular. However, the failure of the algorithm need not be a consequence of automorphisms; Colbourn and Colbourn (1982) establish that a regular DTS(v) exists if and only if $v \equiv 1 \pmod 3$.

24.7 Intersections and large sets

In this section, we study the intersection problem and the large set problem for DTS(v)s. The *intersection* problem for directed triple systems is to determine the possible numbers of triples in common between two DTS(v)s. Two DTS(v)s $(V, \mathcal{B})$ and $(V, \mathcal{D})$ *intersect* in r triples if $|\mathcal{B} \cap \mathcal{D}| = r$.

Directing a TTS(v) does not preserve intersections. However, Lindner and Wallis (1982), and independently Fu (1983), proved:

Theorem 24.17 *For $v \equiv 0,1 \pmod 3$, there exist two DTS(v)s on the same element set intersecting in s triples if and only if $s \in \{0, 1, \ldots, v(v-1)/3\}$, $s \neq v(v-1)/3 - 1$.*

Proof Necessity follows from the observation that if two DTS(v)s agree in all but one triple, the last triple is forced to agree. For sufficiency, we first settle a small case. Consider the four DTS(4)s:

I	II	III	IV
(0,1,2)	(0,1,2)	(0,1,2)	(2,1,0)
(1,0,3)	(1,0,3)	(1,3,0)	(3,0,1)
(3,2,0)	(2,3,0)	(2,0,3)	(0,2,3)
(2,3,1)	(3,2,1)	(3,2,1)	(1,3,2)

I intersects with I, II, III, and IV in 4, 2, 1, and 0 triples, respectively.

For $v \geq 10$, $v \equiv 0,1 \pmod 3$, there exists a $(v, \{3,4\})$-PBD having at least one block of size 4 (see, for example, Colbourn, Rosa and Stinson, 1990). Form a DTS(v) by replacing each triple of the PBD by a DTS(3), and each quadruple by a DTS(4); form a second DTS(v) by replacing using DTS(3) having 0 or 2 triples in common with that in the first system, and DTS(4) having 0, 1, 2, or 4.

It remains to handle the cases when $v \in \{6, 7, 9\}$; see Lindner and Wallis (1982). □

Now we turn to the large set problem. Let $\mathcal{T}_v$ be the set of all transitive triples on a set of v elements. Can $\mathcal{T}_v$ be partitioned into $3(v-2)$ pairwise disjoint DTS(v) for all $v \equiv 0,1 \pmod 3$? When possible, such a set is a *large set* of DTS(v)s. Failing that, what is the largest number, $D_d(v)$, of disjoint DTS(v)s that exist?

We first exhibit large sets for some small orders; we write the DTS(v)s on elements $\{0, \ldots, v-1\}$, using ijk to represent the transitive triple (i, j, k). For

$v = 3$, {{102,201}, {012,210}, {021,120}} is a large set. For $v = 4$, form the DTS(v) $\mathcal{B} = \{012,103,230,321\}$, and let $\pi = (0)(123)$; then $\mathcal{B}$, $\pi\mathcal{B}$, $\pi^2\mathcal{B}$ and their converses form a large set.

For $v = 6$, consider the following six disjoint DTS(v)s:

S_1	S_2	S_3	S_4	S_5	S_6
102	213	324	435	540	051
203	314	425	530	041	152
304	415	520	031	142	253
405	510	021	132	243	354
501	012	123	234	345	450
214	420	430	042	052	204
325	532	541	154	103	310
431	043	053	205	215	421
542	254	104	410	320	032
153	305	315	521	531	143

$S_1, \ldots, S_6$ and their converses form a large set of DTS(6)s.

For $v = 7$, consider the DTS(7):

$$\{013, 026, 045, 356, 234, 146, 125, 320, 510, 640, 421, 543, 631, 652\}.$$

Applying the cyclic permutation (0123456), this DTS(7) yields seven disjoint DTS(7)s. Now consider:

$$\{325, 201, 506, 246, 136, 403, 154, 452, 102, 605, 531, 641, 304, 623\}.$$

Applying the same cyclic permutation yields seven further disjoint DTS(7)s. Finally, developing starter blocks $\{031, 302\}$ modulo 7 yields one final DTS(7) that is disjoint from the first 14. For odd orders, we have a simple result:

Lemma 24.18 *For* $v \equiv 1, 3 \pmod 6$, *there exists a large set of DTS(v)s.*

Proof We have already seen an example of a large set of DTS(7)s. For all $v \equiv 1, 3 \pmod 6$, $v \neq 7$, there is a large set of Steiner triple systems by Theorem 15.15. Each triple can be replaced independently by one of the three distinct DTS(3)s, and hence for all such orders, we obtain $D_d(v) = 3(v-2)$, i.e. a large set of DTS(v)s. □

The large set problem would be routine if Conjecture 24.10 were known to be true, for then one could form a large set of $(v-2)/2$ TS($v, 2$)s using Theorem 15.4 and then form six disjoint DTSs from each. Even in the absence of a proof for the conjecture, a simple lower bound on $D_d(v)$ for even v can be found. For $v \equiv 0, 4 \pmod 6$, the set of all triples can be partitioned into $(v-2)/2$ disjoint TTS(v)s (Teirlinck, 1975b). Applying Theorem 24.9 to each (simple) TTS in the partition, we have $2(v-2)$ disjoint DTS(v)s. Thus $D_d(v) \geq 2(v-2)$. However, a much stronger result can be obtained, leading to large sets.

Lindner and Street (1983) give recursive constructions for disjoint DTS(v)s, which we examine next.

A set $\mathcal{D}$ of transitive triples on a v-set has a *separation* D_1, D_2, D_3 if

- D_i is a partial DTS(v), $i \in \{1,2,3\}$; and
- $D_1 \cup D_2 \cup D_3 = \mathcal{D}$ and $D_i \cap D_j = \varnothing$ for $i \neq j$.

We first prove an easy preliminary result:

Lemma 24.19 *For $t > 1$, the set of $3t$ transitive triples*

$$\mathcal{D} = \{(\infty, x_i, x_{i+1}), (x_i, \infty, x_{i+1}), (x_i, x_{i+1}, \infty) : i \in \mathbb{Z}_t\}$$

has a separation D_1, D_2, D_3.

Proof Let

$$\begin{aligned} B_1(i) &= \{(\infty, x_i, x_{i+1})\}, \\ B_2(i) &= \{x_i, \infty, x_{i+1})\}, \text{ and} \\ B_3(i) &= \{x_i, x_{i+1}, \infty)\}. \end{aligned}$$

For $i, j \in \mathbb{Z}_t$, $i \neq j$, we can place $B_r(i)$ and $B_s(j)$ in the same D_k without violating the first two conditions *except* when $j = i \pm 1$ and $(r,s) = (1,1)$ or (3,3). Thus we need only construct for every $t \geq 2$ a $t \times 3$ array $A = (a_{ij})$ with rows indexed by $\mathbb{Z}_t$ and columns by $\mathbb{Z}_3$, and with the properties that

- every row of A contains 1, 2, and 3; and
- if $a_{ij} = 1$, $a_{i+1,j} \neq 1$ and if $a_{ij} = 3$, $a_{i+1,j} \neq 3$ (subscripts mod t).

Given such a matrix A, we partition the set $\mathcal{D}$ into three classes; the jth class contains the transitive triples $B_{a_{ij}}(i, i+1)$ for $i \in \mathbb{Z}_t$.

Finally, it is an easy matter to construct the required matrix A. For $\ell = 0, \ldots, \lfloor t/3 \rfloor - 1$, set $a_{3i,0} = a_{3i+1,2} = a_{3i+2,1} = 1$, $a_{3i,1} = a_{3i+1,0} = a_{3i+2,2} = 2$, and $a_{3i,2} = a_{3i+1,1} = a_{3i+2,0} = 3$. If $t \equiv 1 \pmod 3$, set $a_{t-1,0} = 2$, $a_{t-1,1} = 3$, and $a_{t-1,2} = 1$. If $t \equiv 2 \pmod 3$, set $a_{t-2,0} = a_{t-1,2} = 1$, $a_{t-2,1} = a_{t-1,1} = 2$, and $a_{t-2,2} = a_{t-1,0} = 3$. The verification of the conditions is now just a tedious exercise. □

Let L be an idempotent $n \times n$ latin square. Let $i, j \in \{0,1,2\}$, $i \neq j$ and let x, y be two distinct symbols from L. Let $t_{ij}(x,y)$ contain the three transitive triples on $\{(x,i), (y,i), (L(x,y), j)\}$ containing the ordered pair $((x,i),(y,i))$. Let T_{ij} be the union of $t_{ij}(x,y)$ over all x, y distinct symbols of L.

L is a *directed latin square* if T_{ij} has a separation. When available, a separation $T_{ij}^1, T_{ij}^2, T_{ij}^3$ enables us to modify the tripling construction in Lemma 24.4 to obtain three disjoint DTS($3v$)s, taking $T_{01}^k \cup T_{12}^k \cup T_{20}^k$ for $k = 1,2,3$, and placing three disjoint DTS(3)s on the elements $\{x\} \times \mathbb{Z}_3$ in the tripling construction. We are therefore concerned with finding directed latin squares; perhaps surprisingly, Lindner and Street (1984) prove:

Theorem 24.20 *Every idempotent latin square is directed.*

Proof Let L be an $n \times n$ idempotent latin square. Without loss of generality, we consider the set T_{01}. For each symbol x in L, form a directed graph $\mathcal{D}(x)$ whose vertices are the symbols of L, and whose arcs are the pairs (y, z) for which $L(y, z) = x$. In $\mathcal{D}(x)$, every vertex has indegree and outdegree exactly one; moreover, since L is idempotent, $\mathcal{D}(x)$ has exactly one loop, and it appears on the vertex x. Omitting this loop, $\mathcal{D}(x)$ consists of directed cycles $C_1, \ldots, C_r$, each of length at least two.

Consider a particular cycle C_m of $\mathcal{D}(x)$. Let $C_m = (y_0, \ldots, y_{t-1})$; that is, C_m consists of arcs (y_i, y_{i+1}) for $0 \leq i < t$, subscripts mod t. The ordered pairs of C_m account for all transitive triples of T_{01} of the forms:

$$((x,1),(y_i,0),(y_{i+1},0)),$$
$$((y_i,0),(x,1),(y_{i+1},0)), \text{ and}$$
$$((y_i,0),(y_{i+1},0),(x,1)).$$

Apply Lemma 24.19 (with $\infty = (x, 1)$) to separate this set; the separation of each cycle of each $\mathcal{D}(x)$ can be carried out independently to separate T_{01}. □

Now we apply the tripling construction to prove:

Theorem 24.21 *$D_d(3v) \geq 6v + D_d(v)$ for all $v \geq 3$, $v \neq 6$.*

Proof Let L be an idempotent latin square with rows, columns, and symbols indexed by $\mathbb{Z}_v$, and let $\mathcal{B}$ be a resolvable TD$(3, v)$ with elements from $\mathbb{Z}_v$. Let $P_0, \ldots, P_v - 1$ be the parallel classes of the transversal design. For parallel class $P_\ell = \{(i, c_i, s_i) : i \in \mathbb{Z}_v\}$, define three permutations $\alpha_\ell : i \mapsto c_i$, $\beta_\ell : c_i \mapsto s_i$, and $\gamma_\ell : s_i \mapsto i$.

The application of a permutation π to a set T_{ij} of transitive triples, denoted $T_{ij}\pi$, is accomplished by *fixing* all elements (x, i), and mapping elements of the form (x, j) to $(\pi(x), j)$.

Now Theorem 24.20 guarantees that the latin square L is directed, and hence for each i, j, T_{ij} can be partitioned into three sets A_{ij}, B_{ij}, C_{ij}. For $\ell \in \mathbb{Z}_v$, form three disjoint DTS$(3v)$s D_ℓ^1, D_ℓ^2, and D_ℓ^3 as follows:

1. For each $(i, c_i, s_i) \in P_\ell$, form three disjoint DTS(3)s X_1, X_2, X_3 on the points $\{(i, 0), (c_i, 1), (s_i, 2)\}$, and include the triples of X_i in D_ℓ^i for $i = 1, 2, 3$.
2. Then include:

$$A_{01}\alpha_\ell \cup A_{12}\beta_\ell \cup A_{20}\gamma_\ell \subseteq D_\ell^1;$$

$$B_{01}\alpha_\ell \cup B_{12}\beta_\ell \cup B_{20}\gamma_\ell \subseteq D_\ell^2;$$

$$C_{01}\alpha_\ell \cup C_{12}\beta_\ell \cup C_{20}\gamma_\ell \subseteq D_\ell^3.$$

It is routine to verify that the $3v$ DTS(v)s $\{D_\ell^i\}$, $\ell \in \mathbb{Z}_v$, $i = 1, 2, 3$, are pairwise disjoint.

To obtain a second (disjoint) collection of $3v$ DTS$(3v)$s, we apply the automorphism $\sigma : i \mapsto i + 1 \pmod{v}$ to each symbol of L, and to the elements of the third group of the resolvable transversal design. Then we form $3v$ DTS(v)s E_ℓ^i for $\ell \in \mathbb{Z}_v$ and $i = 1, 2, 3$ as follows:

1'. Form three disjoint DTS(3)s Y_1, Y_2, Y_3 on $\{(i,0),(c_i,1),(s_i+1,2)\}$ for each $(i,c_i,s_i) \in P_\ell$; include the triples of Y_i in E_ℓ^i for $i = 1,2,3$.

2'. Then include:

$$A_{02}(\sigma \cdot \gamma_\ell^{-1}) \cup A^21(\beta_\ell^{-1} \cdot \sigma^{-1}) \cup A_{10}(\alpha_\ell^{-1}) \subseteq E_\ell^1;$$

$$B_{02}(\sigma \cdot \gamma_\ell^{-1}) \cup B^21(\beta_\ell^{-1} \cdot \sigma^{-1}) \cup B_{10}(\alpha_\ell^{-1}) \subseteq E_\ell^2;$$

$$C_{02}(\sigma \cdot \gamma_\ell^{-1}) \cup C^21(\beta_\ell^{-1} \cdot \sigma^{-1}) \cup C_{10}(\alpha_\ell^{-1}) \subseteq E_\ell^3.$$

The $3v$ DTS($3v$)s E_ℓ^i for $\ell \in \mathbb{Z}_v$, $i = 1,2,3$, that result are disjoint, and disjoint from the D_ℓ^is constructed earlier.

Now for each $k \in \{2,3,\ldots,v-1\}$, form a set of unordered triples R_k, containing $\{(i,0),(j,1),(\sigma^k(L(i,j)),2)\}$ for $i,j \in \mathbb{Z}_v$. Replacing each unordered triple by three disjoint DTS(3)s causes R_k to split into three disjoint *partial* DTS($3v$)s that are disjoint from the $6v$ already produced. Finally, we complete $D_d(v)$ of these $3(v-2)$ partial DTSs using $D_d(v)$ disjoint DTS(v)s on each $\mathbb{Z}_v \times \{i\}$ for $i \in \mathbb{Z}_3$. □

A very similar argument establishes an analogous result when $v \equiv 1 \pmod 3$:

Theorem 24.22 $D_d(3v+1) \geq 6v + D_d(v+1)$ *for* $v \geq 3$, $v \neq 6$.

Proof We sketch the proof, using notation from the proof of Theorem 24.21. We form disjoint DTS($3v+1$)s on $(\mathbb{Z}_v \times \mathbb{Z}_3) \cup \{\infty\}$. Form permutations α_ℓ, β_ℓ, and γ_ℓ as before. For each $1 \leq \ell \leq v$, form six disjoint DTS(v)s D_ℓ^i, $1 \leq i \leq 6$, as follows:

1. For each $(i,c_i,s_i) \in P_\ell$, form six disjoint DTS(4)s $Z_1,\ldots,Z_6$ on the points $\{(i,0),(c_i,1),(s_i,2),\infty\}$; include Z_i in D_ℓ^i for $i = 1,\ldots,6$.
2. Then include:

$$A_{01}\alpha \cup A_{12}\beta \cup A_{20}\gamma \subseteq D_\ell^1;$$

$$B_{01}\alpha \cup B_{12}\beta \cup B_{20}\gamma \subseteq D_\ell^2;$$

$$C_{01}\alpha \cup C_{12}\beta \cup C_{20}\gamma \subseteq D_\ell^3;$$

$$A_{02}\gamma^{-1} \cup A_{21}\beta^{-1} \cup A_{10}\alpha^{-1} \subseteq D_\ell^4;$$

$$B_{02}\gamma^{-1} \cup B_{21}\beta^{-1} \cup B_{10}\alpha^{-1} \subseteq D_\ell^5;$$

$$C_{02}\gamma^{-1} \cup C_{21}\beta^{-1} \cup C_{10}\alpha^{-1} \subseteq D_\ell^6.$$

The $6v$ DTS($3v+1$)s so constructed are disjoint. We construct $3(v-1)$ *partial* DTS($3v+1$)s on $\mathbb{Z}_v \times \mathbb{Z}_3$ by constructing three disjoint systems for each σ^k, $k = 1,\ldots,v-1$. $D_d(v+1)$ of these are then completed to disjoint DTS($3v+1$)s by placing, on each $(\mathbb{Z}_v \times \{i\}) \cup \{\infty\}$, for $i \in \mathbb{Z}_3$, one of the $D_d(v+1)$ disjoint DTS($v+1$)s. □

Theorems 24.21 and 24.22 provide a suitable multiplication by three that preserves large sets, and opens the question of multiplication by other orders. Lindner (1987) develops general recursive constructions; we study these next.

Theorem 24.23 *If there exists a large set of DTS$(v+2)$s, $v \neq 2$, and $n \equiv 1,5 \pmod 6$, there exists a large set of DTS$(nv+2)$s.*

Proof We construct $3nv$ disjoint DTS$(nv+2)$s on $(\mathbb{Z}_v \times \mathbb{Z}_n) \cup \{\infty_1, \infty_2\}$, Let L be an idempotent latin square with rows, columns, and symbols indexed by $\mathbb{Z}_v$, and let $\sigma : i \mapsto i+1 \pmod v$. We construct $3nv$ disjoint systems $S(k,d,j)$ for $k \in \mathbb{Z}_v$, $d \in \mathbb{Z}_n$, and $j \in \{1,2,3\}$, using the following ingredients:

1. If $a+a+a \equiv d \pmod n$ for $a \in \mathbb{Z}_n$, form a large set $\{X(k,a,j) : k \in \mathbb{Z}_v; j \in \{1,2,3\}\}$ on $(\mathbb{Z}_v \times \{a\}) \cup \{\infty_1, \infty_2\}$.
2. For $a,b,c \in \mathbb{Z}_n$, $a < b < c$, satisfying $a+b+c \equiv d \pmod n$, and $x,y \in \mathbb{Z}_v$, let $\{Z_{xy}^{abc}(k,d,j) : j \in \{1,2,3\}\}$ be three disjoint DTS(3)s on $\{(x,a),(y,b),(L(x,y)\sigma^k, c)\}$. Then let
$$Z(k,d,j) = \bigcup_{x,y \in \mathbb{Z}_n} \bigcup_{\substack{a+b+c \equiv d \pmod n \\ a<b<c}} Z_{xy}^{abc}(k,d,j).$$
3. For each $a,b \in \mathbb{Z}_n$, $a \neq b$, satisfying $a+a+b \equiv d \pmod n$, form the set T_{ab} of transitive triples based on the latin square L; use Theorem 24.20 to partition the triples into three sets T_{ab}^j, $j \in \{1,2,3\}$; then let
$$Y(k,d,j) = \bigcup_{\substack{a+a+b \equiv d \pmod n \\ a \neq b}} T_{ab}^j \sigma^k,$$
where the application of permutation σ^k only maps (x,b) to $(x\sigma^k, b)$.
4. It remains to treat all triples containing precisely one of ∞_1 or ∞_2. We adopt the strategy that in each DTS produced, such triples containing ∞_2 are the converses of those containing ∞_1 (with ∞_2 substituted for ∞_1 in the converse). Let $R_i(k,d)$ be the set
$$\{(\infty_1,(i,a),(i\sigma^k,b)),((i,a),\infty_1,(i\sigma^k,b)),((i,a),(i\sigma^k,b),\infty_1) :$$
$$2a+b \equiv d \pmod n)\}.$$
Lemma 24.19 gives a separation $R_i(k,d,j)$, $j \in \{1,2,3\}$ for this set. Set
$$R(k,d,j) = \bigcup_{i \in \mathbb{Z}_v} R_i(k,d,j).$$
Then $R'(k,d,j)$ contains the converse of each triple in $R(k,d,j)$, replacing ∞_1 by ∞_2 throughout.

Finally we have all of the needed ingredients. Form the $3nv$ DTS$(nv+2)$s by taking
$$S(k,d,j) = X(k,d,j) \cup Y(k,d,j) \cup Z(k,d,j) \cup R(k,d,j) \cup R'(k,d,j)$$
for all $k \in \mathbb{Z}_v$, $d \in \mathbb{Z}_n$, and $j \in \{1,2,3\}$.

We leave as an exercise the verifications that each $S(k,d,j)$ is a DTS$(nv+2)$, and that two distinct ones are disjoint. □

Since there is a (trivial) idempotent latin square of order 1, and there is a large set of DTS(3)s, Theorem 24.23 establishes that there is a large set of DTS$(n+2)$s for all $n \equiv 1,5 \pmod 6$. This gives a very direct proof of Lemma 24.18, without recourse to the solution of the apparently more difficult solution for large sets of Steiner triple systems.

The proof of Theorem 24.23 does not apply when $v = 2$, simply because there is no idempotent latin square of order 2. Despite this, Lindner (1987) adapts the method to handle this exceptional situation:

Theorem 24.24 *For $n \equiv 1,5 \pmod 6$, there exists a large set of DTS$(2n+2)$s.*

Proof We proceed as in the proof of Theorem 24.23, using a large set of DTS(4)s to define $X(k,d,j)$, and defining $Z(k,d,j)$ as before. At this point, what remains is precisely those triples, for each $d \in \mathbb{Z}_n$, on $\{\infty_i,(x,a),(y,b)\}$ for $i \in \{1,2\}$, $x,y \in \mathbb{Z}_2$, and those based on $\{(0,a),(1,a),(y,b)\}$ for $y \in \mathbb{Z}_2$, for all a,b satisfying $a+a+b \equiv d \pmod n$.

$A_1(a)$	$A_2(a)$	$A_3(a)$
$(\infty_1,(0,a),(0,b))$	$((0,b),\infty_1,(0,a))$	$((0,a),(0,b),\infty_1)$
$(\infty_1,(1,a),(1,b))$	$((1,b),\infty_1,(1,a))$	$((1,a),(1,b),\infty_1)$
$(\infty_2,(1,a),(0,b))$	$((0,b),\infty_2,(1,a))$	$((0,b),(1,a),\infty_2)$
$(\infty_2,(0,a),(1,b))$	$((1,b),\infty_2,(0,a))$	$((1,b),(0,a),\infty_2)$
$((0,b),(0,a),(1,a))$	$((0,a),(1,a),(0,b))$	$((1,a),(0,b),(0,a))$
$((1,b),(1,a),(0,a))$	$((1,a),(0,a),(1,b))$	$((0,a),(1,b),(1,a))$

Table 24.2 *Disjoint sets for Theorem* 24.24

Consider the (disjoint) sets in Table 24.2. Further define $B_j(a)$ to be the set of triples obtained by replacing each occurrence of (y,b) in $A_j(a)$ with $(1-y,b)$. The six sets of triples $A_j(a)$, $B_j(a)$, $j \in \{1,2,3\}$ are then pairwise disjoint.

$A_r(a)$ and $A_s(b)$ $(a+a+b \equiv d \pmod n)$ can be included in the same partial DTS if and only if (r,s) is neither (1,1) nor (3,3). Hence using the method of the proof of Lemma 24.19, we can separate all triples appearing in the $A_j(a)$ and hence also all triples in the $B_j(a)$, to obtain, for every $d \in \mathbb{Z}_n$, six disjoint sets $\{Y(k,d,j)\}$ for $k \in \mathbb{Z}_2$ and $j \in \{1,2,3\}$.

Finally, we set $S(k,d,j) = X(k,d,j) \cup Y(k,d,j) \cup Z(k,d,j)$ for all $k \in \mathbb{Z}_2$, $d \in \mathbb{Z}_n$, and $j \in \{1,2,3\}$ to form a large set of $6n$ DTS$(2n+2)$s. □

Corollary 24.25 *If for every $\alpha \geq 0$ there exists a large set of DTS$(2^\alpha+2)$s, then for every $v \equiv 0,1 \pmod 3$ there exists a large set of DTS(v)s.*

Proof Write $v-2 = 2^\alpha\beta$ for β odd. Since $v-2 \equiv 1,2 \pmod 3$, $\beta \equiv 1,5 \pmod 6$. □

The large set problem is thus reduced to the problem for relatively few orders: those with $v = 2^\alpha + 2$. For $\alpha = 0$, 1, and 2, these are large sets of DTS(3)s, DTS(4)s, and DTS(6)s. For $\alpha \equiv 0, 4 \pmod 6$, $\alpha > 4$, a large set can be produced with the $3v$ construction (Theorem 24.21). For $\alpha \equiv 3, 5 \pmod 6$, a large set can be produced with the $3v + 1$ construction (Theorem 24.22). For $\alpha = 4$, Lindner and Street (1984) produce a large set for order 18. The complete solution thus rests on the solution for orders $2^\alpha + 2$, $\alpha \equiv 1, 2 \pmod 6$, $\alpha \geq 7$. Kang and Chang (1992) complete the solution by developing a recursive construction of a large set of DTS($16v + 2$)s from a large set of DTS($v + 2$)s. Their method parallels the proof of Theorem 24.23 closely. In summary:

Theorem 24.26 *For all* $v \equiv 0, 1 \pmod 3$, *there exists a large set of disjoint* $DTS(v)$*s.*

The solution of the large set problem for larger λ requires only a complete solution for $\lambda = 3$ and $v \equiv 2 \pmod 3$. When $v \equiv 5 \pmod 6$, we can replace each block of a large set of TS($v, 3$)s from Theorem 15.5 using three disjoint DTS(3)s to obtain a large set of DTS($v, 3$)s. Chang (1996) completes the solution when $v \equiv 2 \pmod 6$; see Kang (1995).

24.8 Work points

1. How many (and how few) nonisomorphic DTS(v, λ)s can a given TS($v, 2\lambda$) underlie?
2. Prove or disprove Conjecture 24.10 which states that every TS($v, 2\lambda$) supports six disjoint DTS(v, λ)s.

25

MENDELSOHN TRIPLE SYSTEMS

In this chapter, we consider a second type of ordering: designs in which the blocks are cyclic triples. A *Mendelsohn triple system* of *order* v and *index* λ is a pair $(V, \mathcal{B})$. V is a v-set of *elements* and $\mathcal{B}$ is a collection of cyclic triples on V; every ordered pair of distinct elements from V appears in precisely λ of the cyclic triples of $\mathcal{B}$. The abbreviation MTS(v, λ) is typically employed, and the parameter λ is often omitted when $\lambda = 1$. Mendelsohn (1971) first studied such designs under the name 'cyclic triple systems', which unfortunately can be confused with triple systems having a cyclic automorphism.

We begin again with some small examples.

Example 25.1 Small MTS(v, λ)s.

MTS$(3,1)$	$\{\langle 0,1,2\rangle, \langle 2,1,0\rangle\}$
MTS$(4,1)$	develop $\langle 0,1,2\rangle$ modulo 4
MTS$(5,3)$	$\{\langle i, i+j, i+2j\rangle : i \in \mathbb{Z}_5, j \in \mathbb{Z}_5 \setminus \{0\}\}$
MTS$(6,1)$	does not exist
MTS$(7,1)$	develop $\langle 0,1,3\rangle$ and $\langle 0,3,1\rangle$ modulo 7
MTS$(8,3)$	to the blocks of any MTS$(7,2)$ on $\{0,1,\ldots,6\}$, add the blocks obtained by developing the starter blocks $\langle 0,1,\infty\rangle$, $\langle 0,2,\infty\rangle$, $\langle 0,4,\infty\rangle$, $\langle 0,3,1\rangle$ modulo 7

Every MTS(v, λ) has an *underlying* TS$(v, 2\lambda)$, obtained by replacing each cyclic triple $\langle a,b,c\rangle$ by $\{a,b,c\}$. Moreover, every MTS(v, λ) has a *converse* obtained by replacing each triple $\langle a,b,c\rangle$ by $\langle a,c,b\rangle$ $(= \langle c,b,a\rangle)$. The converse of an MTS(v, λ) is again an MTS(v, λ); if isomorphic, the MTS is *self-converse.*

A Mendelsohn triple system is *simple* if all of its triples are distinct as cyclic triples; it is *pure* if its underlying TS$(v, 2\lambda)$ is simple.

25.1 Existence: recursions and PBD constructions

Mendelsohn (1971) first settled existence for MTS(v)s. We introduce simple existence proofs here.

Lemma 25.2 *If there exists an MTS(v, λ), there exists an MTS$(2v+1, \lambda)$.*

Proof Let $(V, \mathcal{B})$ be an MTS(v, λ). Let L be a $(v+1) \times (v+1)$ latin square with constant diagonal, whose rows and columns are indexed by a $(v+1)$-set X that is disjoint from V, and whose offdiagonal systems are in V. Then form a set of cyclic triples $\mathcal{C}$ by including $\mathcal{B}$ and $\langle i, L(i,j), j\rangle$ λ times for all $i \neq j$, $i, j \in X$. $\mathcal{C}$ forms the triples of an MTS$(2v+1, \lambda)$ on $V \cup X$. □

This $2v+1$ construction is identical to that for directed triple systems, *except* that the triples are taken to be cyclic. We can also mimic the $2v+4$ construction that we employed for DTSs:

Lemma 25.3 *If there exists an MTS*(v,λ), *there exists an MTS*$(2v+4,\lambda)$.

Proof We choose an MTS(v,λ) on element set V, with $V \cap \mathbb{Z}_{v+4} = \varnothing$. Now choose *any* three distinct nonzero integers $i,j,k \in \mathbb{Z}_{v+4}$ satisfying $i+j+k \equiv 0 \pmod{v+4}$; take $v+4$ cyclic triples λ times each, obtained by developing $\langle 0,i,i+j\rangle \bmod v+4$. Each integer $d \in \mathbb{Z}_{v+4} \setminus \{0,i,j,k\}$ then defines a directed 1-factor of ordered pairs $\{(i,i+d) : i \in \mathbb{Z}_{v+4}\}$. This gives v directed 1-factors that are pairwise disjoint, and do not contain any pairs from the chosen cyclic triples. Associate the v directed 1-factors $F_1,\ldots,F_v$ with the v elements $x_1,\ldots,x_v$ of V. Form triples by including (λ times each) the triples $\langle a,x_i,b\rangle$ for $(a,b) \in F_i$, $1 \le i \le v$. □

The tripling constructions for DTSs (§24.1) also carry over to MTSs directly, as do the PBD constructions. Hence we have the main existence theorem for Mendelsohn triple systems:

Theorem 25.4 *Let* $v \ge 0$, $v \ne 2$. *Let* $\lambda \equiv 0 \pmod{\gcd(v-2,3)}$. *Then an MTS*$(v,\lambda)$ *exists except when* $v = 6$ *and* $\lambda = 1$.

Proof For $v \equiv 0,1 \pmod 3$, $v \ne 6$, there is a $\{3,4\}$-PBD of order v (Hanani, 1975). Replacing each block b of the PBD by an MTS$(|b|)$ on the same points yields an MTS(v); taking each triple λ times gives an MTS(v,λ). For $v = 6$, replacing each block of a TS$(6,2)$ by the two triples of an MTS(3) gives an MTS$(6,2)$. An MTS$(6,3)$ is obtained by taking

$$\{\langle 0,1,2\rangle, \langle 0,2,4\rangle, \langle 0,3,1\rangle, \langle \infty,0,2\rangle, \langle \infty,0,4\rangle, \langle \infty,0,4\rangle\}$$

as starter blocks on $\mathbb{Z}_5 \cup \{\infty\}$. Unions of MTS$(6,2)$s and MTS$(6,3)$s yield MTS$(6,\lambda)$s for all $\lambda \ge 2$.

When $v \equiv 2 \pmod 3$, there is a $\{3,4,5\}$-PBD except when $v \in \{2,8\}$ (see Hanani, 1975). Replace each block b in such a PBD by the triples of an MTS$(|b|,3)$ to obtain an MTS$(v,3)$. Repeating each triple $\lambda/3$ times then yields the desired result. Example 25.1 gives an MTS$(8,3)$. □

Bennett and Mendelsohn (1978) show a stronger result when $\lambda = 1$:

Theorem 25.5 *For* $v \equiv 0,1 \pmod 3$, $v \notin \{3,6\}$, *there exists a pure MTS*(v).

Chang, Yang, and Kang (1996) settle the existence of self-converse MTSs:

Theorem 25.6 *A self-converse MTS*(v) *exists if and only if* $v \equiv 0,1 \pmod 3$ *and* $v \ne 6$.

This has been generalized to all indices; see Kang (1995).

25.2 Orienting a triple system

Mendelsohn (1971) observes that not every TS$(v,2)$ is the underlying system of an MTS(v). We call a TS$(v,2\lambda)$ *orientable* if it underlies an MTS(v,λ), *nonorientable* otherwise. The production of an MTS(v,λ) from its underlying triple system is *orienting*. There are two simple observations about orienting that permit us to test whether a triple system is orientable. The first is that once the direction of a single pair $\{x,y\}$ in a triple $\{x,y,z\}$ is selected, the cyclic ordering of the triple is uniquely determined. The second is that, in a TS$(v,2)$, once the cyclic ordering of one triple containing the pair is selected, the cyclic ordering of the other triple containing $\{x,y\}$ is uniquely determined.

Intuitively, to check orientability of a TS$(v,2)$, select the cyclic ordering of one triple, and repeatedly select the ordering for those triples whose ordering is forced by those already ordered. There are three possible outcomes: we may succeed in ordering all triples and find an MTS(v), or we may order some of the triples and find that no others are forced, or we may force a single triple to be ordered in two different ways. In the latter case, the TS$(v,2)$ cannot be oriented at all. Only one choice is made for the ordering of the first triple, and it can be made without loss of generality since if there is an orientation containing $\langle x,y,z\rangle$, the converse is an orientation containing $\langle x,z,y\rangle$.

In the case that no contradiction arises but not all triples are ordered, we observe that the ordering of the triples already fixed in no way impacts our freedom to select the ordering of the remaining triples; hence we can simply remove the triples already ordered, and apply the same procedure to the remaining triples. In this way, eventually a contradiction establishing nonorientability is reached, or all triples are ordered consistently to form an MTS.

This procedure can be implemented very efficiently to test a given TS$(v,2)$ for orientability. We pursue this next, following the presentation by Colbourn (1985). We form a boolean formula that evaluates to `true` if and only if the TS$(v,2)$ is orientable. Let $t_1,\dots,t_b$ be the triples of the TS$(v,2)$, and associate with each triple t_i a boolean (`true` or `false`) variable x_i. For each triple $t_i=\{x,y,z\}$, choose *arbitrarily* a *standard* cyclic triple $\hat{t}_i=\langle x,y,z\rangle$. For each $1\le i<j\le b$, we form a clause X_{ij} that represents the consistency of the ordering for t_i and t_j. If t_i and t_j intersect in zero or one element, the triples can be chosen independently, and so X_{ij} is null. If t_i and t_j have two elements in common, say $\{a,b\}$, form the clause $X_{ij}=(x_i\vee x_j)\wedge(\bar{x}_i\vee\bar{x}_j)$ if (a,b) is contained in both $\hat{t}_i$ and $\hat{t}_j$ or in neither; otherwise set $X_{ij}=(x_i\vee\bar{x}_j)\wedge(\bar{x}_i\vee x_j)$. Each variable x_i has the interpretation that when `true`, t_i is to be taken in the standard order, and when `false` in the converse. For X_{ij} to be `true`, the pair $\{a,b\}$ must be taken in different orders in t_i and in t_j. If two triples have three elements in common, we proceed as for two common elements.

Now consider the formula

$$\bigwedge_{1\le i<j\le b} X_{ij}.$$

Any orientation of the TS$(v,2)$ corresponds to a satisfying assignment of this formula; and any satisfying assignment ensures that every pair is chosen once in each order, hence inducing an orientation.

It may appear that an easy problem, orientation, has been transformed into a much harder problem, satisfiability of boolean formulas. However, the particular formula produced is the conjunction of clauses, each of which is the disjunction of two literals; the formula is in *2-conjunctive normal form.* Even, Itai, and Shamir (1976) give a linear time algorithm for determining satisfiability of boolean formulas in 2-conjunctive normal form. Hence we have:

Theorem 25.7 *There is an algorithm for determining whether a TS*$(v, 2)$ *can be oriented whose running time is linear in the number of blocks of the triple system.*

Bennett and Mendelsohn (1978b) show that for every order $v \equiv 0, 1 \pmod 3$, $v \geq 9$, there is a nonorientable TS$(v, 2)$, and hence the method for determining orientability is essential. In fact, the study of Mendelsohn triple systems derives much of its interest from the observation that orientability is apparently a subtle property.

In general, we cannot hope to obtain the analogue of the statement that all triple systems of even index are directable. Nevertheless, we might hope to expect a computationally tractable characterization of the orientable triple systems. However, Colbourn (1986) has shown that determining whether a TS$(v, 4)$ underlies an MTS$(v, 2)$ is NP-complete. The essential difficulty in orienting for higher index is that, while choosing the order of one pair still forces the ordering on a triple, fixing the ordering on a triple does not *by itself* force the ordering on any other triples.

25.3 The number of Mendelsohn triple systems

Perhaps not surprisingly, the number of nonisomorphic Mendelsohn triple systems is an exponential function of the order. The rate of growth in the number $M(v)$ of MTS(v)s has been determined by Phelps and Lindner (1984). They employed the tripling constructions to produce MTS$(3m)$s and MTS$(3m{+}1)$s. Using the tripling constructions, it is immediate that the number of distinct MTS$(3m)$s is at least the cube of the number $I(m)$ of idempotent $m \times m$ latin squares. Hence $M(3m) \geq (I(m))^3/(3m)!$; similarly, $M(3m+1) \geq (I(m))^3/(3m+1)!$.

Phelps and Lindner (1984) establish that $I(m) = \exp((m^2 \ln m)(1 + o(1)))$. As a result, they obtain the asymptotic number of MTS(v)s:

Theorem 25.8 *For v sufficiently large, the number $M(v)$ of MTS(v)s for* $v \equiv 0, 1 \pmod 3$ *is* $\exp(((v^2 \ln v)/3)(1 + o(1)))$.

Proof Write $v = 3m$ or $v = 3m + 1$. We must show that

$$\lim_{v \to \infty}(\ln M(v)/(v^2 \ln v)) = 1/3.$$

We treat the case $v = 3m$, the case for $3m + 1$ being similar. Since $M(v) \geq (I(m))^3/v!$, we have

$$\lim_{m\to\infty} M(3m) \geq \lim_{m\to\infty} [3 \ln I(m) - 3m \ln(3m)].$$

Then

$$\lim_{m\to\infty} \frac{M(3m)}{(3m)^2 \ln(3m)} \geq \lim_{m\to\infty} \left(\frac{3 \ln I(m)}{9m^2(\ln m + \ln 3)} - \frac{1}{3m} \right) \geq \frac{1}{3}.$$

To show that the limit is at *most* $1/3$, we observe that constructing a collection of $v(v-1)/3$ cyclic triples in a greedy manner by choosing any of the $2v-4$ cyclic triples containing the smallest ordered pair that is as yet uncovered produces all distinct Mendelsohn triple systems. The number of distinct collections produced by this greedy algorithm is bounded by $2v^{v(v-1)/3}$, and hence the upper bound follows. □

The asymptotic growth rates of twofold, directed, and Mendelsohn triple systems are all the same. However, the asymptotics do not reflect the differences between directed and Mendelsohn triple systems. Many twofold triple systems underlie no MTS, and many have only two orientations (i.e. an orientation and its converse). Nevertheless, Phelps and Lindner (1984) establish a stronger result. They establish that the number $I_5(m)$ of idempotent latin squares having a common 5×5 subsquare is also $\exp((m^2 \ln m)(1 + o(1)))$. In conjunction with the tripling constructions, this establishes that the number of MTS(v)s *having a nontrivial subdesign* is also $\exp(((v^2 \ln v)/3)(1 + o(1)))$. The underlying twofold triple system of such an MTS necessarily has at least four orientations.

It is unknown at the present time whether or not the ratio of the number of orientable TS(v, 2)s to the number of TS(v, 2)s approaches 0 or 1 (or something in between).

Exact enumeration results for MTS(v)s were found by Mendelsohn (1971) for $v \leq 7$, Mathon and Rosa (1977) for $v = 9$, and by Ganter, Mathon, and Rosa (1978, 1978a) for $v = 10$. Denny (1998) corrected the enumeration for $v = 10$, and completed the enumeration for $v = 12$. We tabulate the numbers of nonisomorphic MTS(v)s here:

Nonisomorphic MTS(v)s	
Order	Number
3	1
4	1
6	0
7	3
9	18
10	143
12	4905693

Denny (1998) reports the statistics in Table 25.1 about the MTS(12) designs, where for each size $|\Gamma|$ of the automorphism group of the underlying TS(12,2), the number of orientable systems with this group order is given as Nd, and the number of nonisomorphic MTS(12)s that result as Nm.

$\|\Gamma\|$	Nd	Nm	$\|\Gamma\|$	Nd	Nm
1536	1	3	576	1	2
432	1	7	192	3	4
144	1	4	128	2	5
72	2	6	64	2	6
54	1	1	48	6	12
36	1	2	32	9	15
24	7	15	18	2	3
16	24	40	12	15	38
9	1	1	8	77	173
6	106	196	4	482	867
3	531	647	2	11011	15003
1	4679953	4888643	total	4692239	4905693

Table 25.1 *Automorphisms of MTS(12)s*

25.4 Cyclic and rotational MTSs

Hoffman and Lindner (1981) prove a strong partial result concerning the existence of MTS(v)s having a cyclic automorphism:

Theorem 25.9 *When* $v \not\equiv 0 \pmod 3$, *there is a partial cyclic MTS(v) having* $2\lfloor v/6 \rfloor$ *starter blocks.*

Proof A difference triple for a cyclic MTS(v) is a cyclic triple $\langle a, b, c\rangle$ with $a+b+c \equiv 0 \pmod v$; a corresponding starter block is then $\langle 0, a, a+b\rangle$. We must produce $2\lfloor v/6 \rfloor$ disjoint difference triples in $\mathbb{Z}_v \setminus \{0\}$.

Write $v = 6x+3y+z$ with $x \geq 0$, $y \in \{0,1\}$, and $z \in \{1,2\}$. Form x difference triples by taking $\langle t, x+y+t, 5x+2y+z-2t\rangle$ for $1 \leq t \leq x$, and form x disjoint difference triples by taking $\langle 2x+y+t, 5x+3y+z+t-1, 5x+2y+z-2t+1\rangle$ for $1 \leq t \leq x$. This gives $2x = 2\lfloor v/6 \rfloor$ difference triples as required. □

Corollary 25.10 *A cyclic MTS(v) exists for all* $v \equiv 1 \pmod 6$.

Shen (1999) completely settles the existence of cyclic MTS(v, λ)s:

Theorem 25.11 *A cyclic MTS(v, λ) exists only when*

$$\begin{array}{lll} v \equiv 1,3 \pmod 6 & \text{if} & \lambda \equiv 1,5 \pmod 6, \\ v \equiv 0,1 \pmod 3 & \text{if} & \lambda \equiv 2,4 \pmod 6, \\ v \equiv 1 \pmod 2 & \text{if} & \lambda \equiv 3 \pmod 6, \\ v \geq 3 & \text{if} & \lambda \equiv 0 \pmod 6. \end{array}$$

Moreover, these conditions are sufficient with precisely three exceptions, when $(v, \lambda) \in \{(6,2), (9,1), (9,2)\}$

Cho (1986) studies rotational Mendelsohn triple systems:

Lemma 25.12 *A 1-rotational MTS(v) exists if and only if* $v \equiv 1,3,4 \pmod 6$ *and* $v \neq 6$.

Assume π is a 1-rotational automorphism of an MTS(v). Then π^k is also an automorphism of the MTS(v) for any integer k. Since π^k is a k-rotational permutation provided $v \equiv 1 \pmod{k}$, any 1-rotational MTS(v) is also k-rotational if $v \equiv 1 \pmod{k}$. Since a 3-rotational MTS(10) appears in Cho (1986), Lemma 25.12 yields:

Corollary 25.13 *A k-rotational MTS(v) exists whenever $v \equiv 1, 3, 4 \pmod{6}$ and $v \equiv 1 \pmod{k}$, except when $k = 1$ and $v = 10$.*

Jiang and Colbourn (1998) complete the solution, showing:

Theorem 25.14 *There exists a k-rotational MTS(v) if and only if $v \equiv 0, 1 \pmod{3}$ and $v \equiv 1 \pmod{k}$, except when $k = 1$ and $v \equiv 0 \pmod{6}$ or $v = 10$.*

Brand and Huffman (1987a, b) study Mendelsohn systems admitting the affine group.

25.5 Subsystems, intersection, and large sets

Hoffman and Lindner (1981) determine precisely when an MTS(v) can contain an MTS(w) as a subdesign. We prove the result more generally for arbitrary index:

Theorem 25.15 *For $\lambda \geq 1$, $v, w \geq 0$, $v, w \neq 2$, $v \geq 2w + 1$, and $v, w \equiv 0, 1 \pmod{3}$ if $\lambda \not\equiv 0 \pmod{3}$, there exists an MTS($v, \lambda$) having an MTS($w, \lambda$) as a subdesign provided $v, w \neq 6$ if $\lambda = 1$.*

Proof If $w < 3$, the statement is immediate from the existence of an MTS(v, λ). We leave the easy case when $w = 3$ to the reader, so assume that $w \geq 4$. We first treat the case when $v - w \not\equiv 0 \pmod{3}$. Using Theorem 25.9, find $m = 2\lfloor (v - w)/6 \rfloor$ difference triples for a partial cyclic MTS($v - w$) on $\mathbb{Z}_{v-w}$. Write $\hat{w} = w - (v - 1 - 3m)$. Since $v - 1 - 3m \leq 4$, $\hat{w} \geq 0$. Moreover, $\lambda\hat{w} \equiv 0 \pmod{3}$, and $\hat{w} \leq 3m$ since $w \leq v - 1$.

Form a collection of λm difference triples on $\mathbb{Z}_{v-w}$ by repeating each difference triple λ times. Select any set of $\lambda m - \hat{w}\lambda/3$ of these to form starter blocks on $\mathbb{Z}_{v-w}$. There remain $\lambda(v-1-3m)+\lambda\hat{w} = \lambda w$ differences on $\mathbb{Z}_{v-w}$ unaccounted for by these starter blocks. By (arbitrarily) grouping these differences in w groups of λ differences each, the ordered pairs not yet covered define w λ-regular directed graphs $D_1, \ldots, D_w$ on $\mathbb{Z}_{v-w}$. Add w new elements $X = \{x_1, \ldots, x_w\}$. Then form cyclic triples $\langle x_i, a, b \rangle$ for each $1 \leq i \leq w$ and $(a, b) \in D_i$.

Finally, place the triples of an MTS(w, λ) on X.

When $v - w \equiv 0 \pmod{3}$, we instead employ the tripling construction. If we employ an idempotent latin square of order m with an idempotent subsquare of order s, we obtain MTS($3m$)s, MTS($3m + 1$)s, and MTS($3m + 2, 3$)s having subdesigns of order $3s$, $3s + 1$, and $3s + 2$, respectively. Hilton (1973) constructs such idempotent quasigroups for all $m \geq 2s + 1$, provided $s \neq 2$ (no idempotent 2×2 latin square exists). Then to treat the cases when $w \in \{6, 7, 8\}$, instead form an $m \times m$ square having a 2×2 subsquare on two elements $\{a, b\}$ with all other elements idempotent; this is easily done for $m \geq 5$. Applying $3m$, $3m+1$, or $3m+2$ constructions then constructs an MTS$(3m, \lambda)$, MTS$(3m+1, \lambda)$, or MTS$(3m+2, \lambda)$ omitting an MTS$(6, \lambda)$, MTS$(7, \lambda)$ or MTS$(8, \lambda)$, respectively. This hole is filled with the appropriate subdesign, which exists provided $\lambda > 1$ if $w = 6$. □

The analogous result for incomplete MTS(v)s was proved by Bennett and Chen (1991). An *incomplete Mendelsohn triple system* MTS$(v, w; \lambda)$ is a partition of all ordered pairs on a v-set, *except* those on a subset of size w, into cyclic triples. Bennett and Chen (1991) establish:

Theorem 25.16 *An MTS$(v, w; 1)$ exists for all $v, w \geq 1$ provided $v \geq 2w+1$ and $(v-w)(v-2w-1) \equiv 0 \pmod 3$, except when $v = 6$ and $w = 1$.*

Bennett and Zhu (1990) establish a related result. A *holey Mendelsohn triple system* HMTS(v, λ) of type m^h is a partition of λ copies of all ordered pairs on a v-set, $v = mh$, excepting those on h disjoint subsets each of size m, into cyclic triples. They prove:

Theorem 25.17 *An HMTS(mh, λ) of type m^h exists if and only if $\lambda h(h-1)m^2 \equiv 0 \pmod 3$, except when $\lambda = 1$, $m = 1$, and $h = 6$.*

The *intersection problem* for Mendelsohn triple systems is to determine the spectrum of numbers of blocks in common between two MTS(v)s. Hoffman and Lindner (1982) give a complete solution for this problem, which we examine next.

Two MTS(v)s $(V, \mathcal{B})$ and $(V, \mathcal{D})$ *intersect* in r cyclic triples if $|\mathcal{B} \cap \mathcal{D}| = r$. It is immediate that $r \geq 0$, and $r \leq t_v = v(v-1)/3$. Write $r = t_v - x$. Consider the two sets $\mathcal{B}' = \mathcal{B} \setminus \mathcal{D}$ and $\mathcal{D}' = \mathcal{D} \setminus \mathcal{B}$. $\mathcal{B}'$ and $\mathcal{D}'$ contain x cyclic triples each; moreover, the sets of cyclic triples cover precisely the same ordered pairs of elements. The sets are *mutually balanced.* In addition, $\mathcal{B}'$ and $\mathcal{D}'$ are disjoint. If $x \in \{1, 2, 3, 5\}$, no such disjoint mutually balanced sets exist; hence we define

$$I_m(v) = \{0, 1, \ldots, t_v\} \setminus \{t_v - 5, t_v - 3, t_v - 2, t_v - 1\}.$$

$I_m(v)$ is the set of potential intersection sizes for MTS(v)s.

Hoffman and Lindner (1982) prove:

Theorem 25.18 *Let $v \equiv 0, 1 \pmod 3$, $v \neq 6$. There exist two MTS(v)s intersecting in r triples if and only if $r \in I_m(v)$.*

Proof The proof is recursive, and hence requires the solution for a number of small orders. We refer the reader to Hoffman and Lindner (1982) for the solution for all $v \leq 18$. So suppose that $v > 18$. Write $v = 2w + s$ for $s \in \{1, 4\}$ and $w \equiv 0, 1 \pmod 3$. We have $w \geq 9$. Using the $2v+1$ construction (Lemma 25.2), we observe that *any* MTS(w) can be chosen; in addition, we can permute the offdiagonal symbols of the latin square used. We can fix any number ℓ of the offdiagonal symbols in this permutation for $0 \leq \ell \leq v$ *except* $\ell = v - 1$. Then we obtain that if $r' \in I_m(w)$, there are two MTS(v)s intersecting in $r' + \ell(v+1)$ triples.

Using the $2v+4$ construction (Lemma 25.3), we observe that the w directed 1-factors can be attached to the w elements of V in two different ways, having $\ell(w+4)$ triples in common for $0 \le \ell \le w$, $\ell \ne w-1$. The triples on $\mathbb{Z}_{w+4}$ can be chosen the same way in each MTS, or chosen once as $\langle 0, i, i+j \rangle$ and once as $\langle 0, i, i+k \rangle$. Hence for $r' \in I_m(w)$, $0 \le k \le 1$, $0 \le \ell \le w$, $\ell \ne w-1$, we have $r' + \ell(w+4) + k(w+4)$ as the intersection size of two MTS$(2w+4)$s.

In either case, it is routine to check that if $r \in I_m(v)$, we can choose r', ℓ, and k to obtain the intersection size r via the doubling constructions. □

Now we turn to a more difficult problem. On a v-set V, there are $v(v-1)(v-2)/3$ distinct cyclic triples. An MTS(v) has $v(v-1)/3$ cyclic triples. Hence it is possible in principle for the set of all cyclic triples on a v-set to be partitioned into $v-2$ MTS(v)s; such a partition is a *large set* of MTSs. $D_m(v)$ denotes the maximum number of disjoint MTS(v)s on the same v-set; the existence of a large set is equivalent to $D_m(v) = v-2$.

A large set of MTS(3)s contains a single MTS. A large set of MTS(4)s contains two MTSs, namely the unique MTS(4) and its converse (which is isomorphic). Teirlinck (1991) completed the solution for the analogous large set problem for Steiner triple systems (Theorem 15.15). Replacing each triple by an MTS(3) yields:

Lemma 25.19 *Let $v \equiv 1,3 \pmod 6$ and $v \ne 7$. Then there exists a large set of MTS(v)s.*

There is no large set of STS(7)s, but Colbourn and Colbourn (1981) establish that there *is* a large set of MTS(7)s.

Hence the large set problem for MTS(v)s remains for $v \equiv 0,4 \pmod 6$, $v \ge 10$. Lindner (1981) made the first substantial inroads into this problem. The results parallel those for directed triple systems quite closely, and so we simply state the main results:

Theorem 25.20 *For $v \equiv 0,1 \pmod 3$,*

$$D_m(3v) \ge 2v + D_m(v) \quad \text{and} \quad D_m(3v+1) \ge 2v + D_m(v+1).$$

Teirlinck and Lindner (1988) and Kang and Chang (1993) establish:

Theorem 25.21 *Let $v \equiv 1,5 \pmod 6$, and suppose that a large set of MTS$(n+2)$s exist. Then a large set of MTS$(nv+2)$s exist.*

Wu (1987) develops a construction for the case when $n = 2$ in this theorem, using a different approach. Kang (1991) establishes that there is a large set of MTS(2^n+2)s for all $n \ge 3$, and Kang and Chang (1993) establish that for $k \equiv 1,2 \pmod 3$, $k > 1$, a large set of MTS$(72k+6)$s exists.

Kang and Lei (1993) completed the determination:

Theorem 25.22 *A large set of MTS(v)s exists if and only if $v \equiv 0,1 \pmod 3$ and $v \ne 6$.*

When $v \equiv 2 \pmod 3$, a *large set* is a partition of all cyclic triples into $(v-2)/3$ MTS$(v,3)$s; for $v \equiv 5 \pmod 6$, such large sets are obtained from large sets of TS$(v,3)$s (Theorem 15.5). For $v \equiv 2 \pmod 6$, the problem is addressed in Kang and Lei (1993):

Theorem 25.23 *A large set of MTS$(v,3)$s exists if and only if $v \equiv 2 \pmod 3$ and $v \geq 5$.*

25.6 Embedding partial MTSs

Lindner (1971) showed that every partial MTS(v) can be embedded finitely; in particular, he showed that it can be embedded in an MTS(w) for some $w \leq 2^{2v}$. Subsequently, Lindner and Cruse (1976) dramatically improved this result; we outline their construction here. We require a theorem of Andersen, Hilton, and Rodger (1982):

Theorem 25.24 *A partial idempotent latin square of order m can be embedded in an idempotent latin square of order n provided $n \geq 2m+1$.*

An MTS(v) $(V, \mathcal{B})$ has an associated *semisymmetric latin square M* defined by $M(x,x) = x$ and $M(x,y) = z$ if $\langle x,y,z \rangle \in \mathcal{B}$. Evidently every semisymmetric latin square is idempotent. Semisymmetric latin squares have been studied by Mendelsohn (1970), Lindner and Mendelsohn (1973), Steedley (1974), Radó (1974), and Bennett (1982a).

Now consider a *partial* MTS(v); it defines a *partial* semisymmetric latin square M of order v. We can employ the embedding for partial idempotent squares to obtain:

Theorem 25.25 *Let $(V, \mathcal{B})$ be a partial MTS(v). Then for all $w \geq 6v+3$, $w \equiv 0, 1 \pmod 3$, there exists an MTS(w) that embeds $(V, \mathcal{B})$.*

Proof Apply Theorem 25.24 to embed the partial semisymmetric latin square M in a $u \times u$ *idempotent* latin square L for $u = \lfloor w/3 \rfloor$, whence $u \geq 2v+1$. We use a modified tripling construction to produce an MTS$(3u)$ or MTS$(3u+1)$. Suppose that L has rows, columns, and symbols indexed by a u-set U. We form an MTS$(3u)$ on $U \times \mathbb{Z}_3$ that contains all cyclic triples in $\mathcal{B}$. If $M(x,y)$ is defined, let $M(x,y) = z$. Include the cyclic triples $\langle x_i, y_j, z_k \rangle$ for $i+j+k \equiv 0 \pmod 3$, $i,j,k \in \mathbb{Z}_3$. If $M(x,y)$ is not defined, let $L(x,y) = z$; take the cyclic triples $\langle x_i, y_i, z_{i+1} \rangle$ for $i \in \mathbb{Z}_3$. The set of cyclic triples contains all triples of $\mathcal{B}$ on each of $U \times \{0\}$, $U \times \{1\}$, and $U \times \{2\}$. Adding MTS(3)s on $\{x\} \times \mathbb{Z}_3$ for $x \in U$ completes the system. The construction of an MTS$(3u+1)$ is instead completed by adding a further point ∞, and placing MTS(4)s on $\{\infty\} \cup \{\{x\} \times \mathbb{Z}_3\}$ for all $x \in U$. □

The best possible result of this type would be to obtain an embedding of a partial MTS(v) in an MTS(w) for all $w \geq 2v+1$, $w \equiv 0, 1 \pmod 3$. To see this, form a partial MTS(v), $v \geq 9$, by taking an MTS(v) having a sub-MTS(4) on $\{w,x,y,z\}$, removing the subsystem, and adding the triples $\langle w,x,y \rangle$ and

$\langle w, y, x \rangle$. To embed the resulting partial system, some new element α is required to appear with the pair (x, z); but then each element of the partial MTS(v) must appear with α.

Rodger (1986a, 1987) improved upon the Lindner–Cruse embedding to obtain:

Theorem 25.26 *A partial MTS(v) can be embedded in an MTS(w) for all $w \geq 4v$, $w \equiv 0, 1 \pmod 3$.*

Hamm (1984) examines the problem of *intersection-preserving* embedding for partial MTSs: given two partial MTS(v)s, embed both in MTSs of the same size so that the triples in common between the two containing MTSs are identical to those in common between the two partial MTSs. Using a variant of the Lindner–Cruse construction, she establishes:

Theorem 25.27 *Two partial MTS(v)s have intersection-preserving embeddings into two MTS(w)s for all $w \geq 6v + 3$, $w \equiv 0, 1 \pmod 3$.*

Rodger (1986) considers a related question, showing that:

Theorem 25.28 *A partial pure MTS(v, λ) has an embedding in an MTS(w, λ) which is also pure, for some $w \leq 3(2\lfloor \lambda/2 \rfloor + 1)(12\lambda + 4)v + 1$.*

When the containing MTS need not be pure, Rodger (1986) shows:

Theorem 25.29 *A partial MTS(v, λ) having α repeated cyclic triples can be embedded in an MTS(w, λ) having α repeated triples for some*

$$w \leq 3(2\lfloor \lambda/2 \rfloor + 1)((3\lambda + 2)v + 1).$$

25.7 Resolvable and almost resolvable MTSs

An *almost parallel class* in an MTS(v, λ) with $v \equiv 1 \pmod 3$ is a set of $(v-1)/3$ disjoint cyclic triples in the system. A partition of the set of all triples of the MTS into almost parallel classes is a *resolution*, and the MTS is *almost resolvable.* The existence of almost resolvable MTS(v)s has been settled by Bennett, Mendelsohn, and Mendelsohn (1980) and Bennett and Sotteau (1981). We describe the construction of almost resolvable MTSs here.

Lemma 25.30 *Let $v = p^\alpha$, p a prime, and $v \equiv 1 \pmod 3$. Then there exists an almost resolvable MTS(v).*

Proof Let ω be a cube root of unity modulo p^α. For every $y \in \mathbb{F}_{p^\alpha}$ define the *set* P_y of triples

$$\{\langle x, x\omega + y(1-\omega), (x\omega + y(1-\omega))\omega + y(1-\omega)\rangle \mid x \in \mathbb{F}_{p^\alpha}, y \neq x\}.$$

Then P_y is an almost parallel class missing the element y, and the union of the v parallel classes is an MTS(v). □

Next we construct almost resolvable MTS(v)s of other orders:

Lemma 25.31 *The set of orders for which an almost resolvable MTS(v) exists is PBD-closed.*

Proof Let $(V, \mathcal{B})$ be a (v, K)-PBD, and define on each block $B \in \mathcal{B}$ an almost resolvable MTS $M(B)$. For each element $x \in B$, there is a unique almost parallel class of $M(B)$ that misses x; we denote the class by $M(B, x)$. Now define an almost parallel class P_x on V, that misses x, by taking the union of $M(B, x)$ over all blocks $B \in \mathcal{B}$ that contain x. The set of almost parallel classes $\{P_x \mid x \in V\}$ is a resolution of an MTS(v). □

Ganter, Mathon, and Rosa (1978) give some almost resolvable MTS(10)s, and the prime power construction provides almost resolvable MTS(v)s for $v \in \{4, 7, 19\}$. Together with the existence of $\{4, 7\}$-PBDs for all $v \equiv 1 \pmod 3$, $v \neq 10, 19$, from Brouwer (1979), we have:

Theorem 25.32 *An almost resolvable MTS(v) exists for all* $v \equiv 1 \pmod 3$

When $v \equiv 0 \pmod 3$, a *parallel class* is a set of $v/3$ disjoint triples, and a *resolution* is a partition of the triples of the MTS into parallel classes; an MTS having such a resolution is *resolvable*. When $v \equiv 3 \pmod 6$, Ray-Chaudhuri and Wilson (1971) establish that there is a Kirkman triple system. By replacing each triple by an MTS(3), we obtain resolvable MTS(v)s for all $v \equiv 3 \pmod 6$. Bermond, Germa, and Sotteau (1979) (see also Bermond, 1974) complete the determination of existence, establishing:

Theorem 25.33 *There exists a resolvable MTS(v) if and only if*

$$v \equiv 0 \pmod 3 \text{ and } v \neq 6.$$

Their proof relies on a multiplication by four; we give a more general construction here:

Lemma 25.34 *If a resolvable MTS(v) exists, and* $w \equiv 1 \pmod 3$, *a resolvable MTS(vw) exists.*

Proof Let $(V, \mathcal{B})$ be a resolvable MTS(v), and $P_1, \ldots, P_{v-1}$ its parallel classes. We form a resolvable MTS(vw) on $V \times W$, $|W| = w$. Using P_i, $i \geq 2$, we form w parallel classes as follows. If $\langle x, y, z \rangle \in P_i$, form a resolvable TD$(3, w)$ on $\{x, y, z\} \times W$ having w parallel classes. Orient each triple $\{(x, a), (y, b), (z, c)\}$ as $\langle (x, a), (y, b), (z, c) \rangle$. This gives $w(v-2)$ of the parallel classes, so $2w - 1$ more are needed. For P_1, for each $\langle x, y, z \rangle \in P_1$, we form a resolvable TD$(3, w)$ on $\{x, y, z\} \times W$; here we assume that $\{\{(x, w), (y, w), (z, w)\} \mid w \in W\}$ forms a parallel class P of the resolvable transversal design. For the $w - 1$ parallel classes of the transversal design other than P, we proceed as above to form parallel classes.

At this point, w more parallel classes are required. We place on each $\{x\} \times W$ an almost resolvable MTS(w). Then for each $w \in W$, we form a parallel class by taking $\langle (x, w), (y, w), (z, w) \rangle$ for each $\langle x, y, z \rangle \in P_1$, *and* the almost parallel class in $\{x\} \times W$ *missing* (x, w) for each $x \in V$. □

Bermond, Germa, and Sotteau (1979) apply this product construction using $w = 4$. When $v \in \{18, 24, 42\}$, they construct resolvable MTS(v)s. They also employ a direct construction for orders $v \equiv 6 \pmod{12}$, $v \geq 54$.

25.8 Hybrid triple systems

We have seen that not every triple system is orientable; hence given a TS($v, 2\lambda$), we can in general only orient *some* of its blocks as cyclic triples, and then take the remainder as directed triples. Colbourn, Pulleyblank, and Rosa (1989) generalized DTSs and MTSs along these lines. A *hybrid triple system* of *order* v and index λ, or HYTS(v, λ), is a v-set V together with a collection of triples $\mathcal{B}$, some of which are cyclic and some transitive; every ordered pair occurs in precisely λ of the triples.

Two main questions arise. First, in constructing an HYTS, what flexibility is there in choosing the number of cyclic triples in the system? Second, given a fixed TS($v, 2\lambda$), how much of it can we guarantee to take as cyclic triples before we are forced to take the remainder as transitive triples? We examine what is known about these two questions here.

A hybrid triple system HYTS(v, λ) is c-hybrid if it has precisely c cyclic triples (and necessarily $\lambda v(v-1)/3 - c$ transitive triples). We denote by $C(v, \lambda)$ the set of integers $\{c \mid \exists\ c\text{-HYTS}(v, \lambda)\}$.

Colbourn, Pulleyblank, and Rosa (1989) settled existence for c-HYTS(v):

Theorem 25.35 *For all $v \equiv 0, 1 \pmod 3$, $v \neq 6$, there exists a c-hybrid triple system of order v and index 1 if and only if $0 \leq c \leq v(v-1)/3$ and $v \neq v(v-1)/3 - 1$.*

Proof It is easy to verify that $C(3, 1) = \{0, 2\}$ and $C(4, 1) = \{0, 1, 2, 4\}$. For $v \equiv 0, 1 \pmod 3$, $v \geq 10$, there is a $\{3, 4\}$-PBD of order v having at least one block of size 4. Hence we obtain the desired result for $v \geq 10$. Colbourn, Pulleyblank, and Rosa (1989) give the required solutions for $v \in \{7, 9\}$. □

Heinrich (1991) proves a similar result for HYTS(v, λ)s:

Theorem 25.36 *Let $\lambda \mid gcd(v-2, 3)$. Then if $0 \leq c \leq \lambda v(v-1)/3$ and $c \neq \lambda v(v-1)/3 - 1$, there exists a c-HYTS(v, λ) except for $v = 6$, $\lambda = 1$, and $c \in \{8, 10\}$.*

Every triple system TS($v, 2\lambda$) is directable, but not every such system is orientable. Here we address the question of how much of an arbitrary TS($v, 2$) can be oriented. In other words, for an arbitrary TS($v, 2$), what is the *largest* c for which the triple system underlies a c-HYTS(v)? Let $mc(v)$ denote the minimum, over all TS($v, 2$), of the maximum c for which the system underlies a c-HYTS(v).

There is a unique TS($6, 2$), and it is not orientable. In fact, this triple system underlies a 7-HYTS(6), and $c = 7$ is the maximum possible. Hence $mc(6) = 7$. Colbourn, Pulleyblank, and Rosa (1989) use this to prove:

Theorem 25.37 *Let $\epsilon > 0$ and v be sufficiently large. Then*

$$mc(v) \leq \left(\frac{7}{10} + \epsilon\right) \frac{v(v-1)}{3}.$$

Proof Colbourn, Pulleyblank, and Rosa (1989) establish that for $\delta > 0$, there is a $\{6,7\}$-PBD in which the number b_7 of blocks of size 7 satisfies $b_7 \leq \delta b_6$, where b_6 is the number of blocks of size 6, provided v is sufficiently large; see also Colbourn and Rödl (1989). Replace each 6-block by a TS(6, 2), and each 7-block by a TS(7, 2). The resulting TS(v, 2) has at most $7b_6 + 14b_7$ cyclic triples among $10b_6 + 14b_7$ triples, so that

$$mc(v) \leq \frac{7 + 14\delta}{10 + 14\delta} \frac{v(v-1)}{3}.$$

Choosing δ sufficiently small yields $mc(v) \leq ((7/10) + \epsilon)\, v(v-1)/3$. □

The amalgamation of many copies of the TS(6, 2) in this way ensures that a large fraction of the triple system cannot be chosen as cyclic triples. In general, consider the *block intersection graph G* obtained by representing each triple of a TS(v, 2) by a vertex, and placing an edge between two vertices if and only if the corresponding triples share at least two elements. M. Colbourn and Johnstone (1984) prove that G consists of connected components each of which are 3-connected. In the $(7/10)+\epsilon$ upper bound, the triple system constructed has block intersection graph consisting of components on 10 and 14 vertices. It appears plausible that *if G* is restricted to have few components, the upper bound of Theorem 25.37 can be raised.

Colbourn, Pulleyblank, and Rosa (1989) prove that any partial triple system with $\lambda = 2$ whose block intersection graph is acyclic is orientable as a partial MTS(v). This is a simple greedy algorithm. They further prove that every 3-connected cubic graph with more than eight vertices has an acyclic subgraph containing at least two-thirds of the vertices. These results together establish that:

Theorem 25.38 *Every TTS(v, 2) underlies a c-HYTS(v) with $c \geq 2v(v-1)/9$.*

It is reasonable to expect that the fraction in Theorem 25.38 should in fact be 7/10. Moreover, if the TTS is restricted to have block intersection graph with few components, the maximum orientability seems likely to increase further.

Most problems on hybrid triple systems can be settled by considering only cyclic or transitive triples. However, the large set problem for hybrid triple systems remains interesting since the number of systems in a large set is $4(v-2)$ (as expected, the sum of the number $v-2$ of systems in a large set of Mendelsohn triple systems and the number $3(v-2)$ of systems in a large set of directed triple systems). Kang and Lei (1996) solve this problem when the hybrid triple systems are required to be neither Mendelsohn nor directed.

25.9 Work points

1. Asymptotically, what is the ratio of the number of orientable TTSs to the total number of TTSs?
2. Prove or disprove that at least 7/10 of the triples of a TS$(v, 2)$ can always be oriented.
3. Asymptotically, what is the ratio of numbers of nonisomorphic MTS(v)s to nonisomorphic TS$(v, 2)$s?

BIBLIOGRAPHY

This bibliography lists both references cited in the main text, and other publications concerning triple systems. Since some publications listed in this bibliography do not concern triple systems directly, and are included as a result of citation in the text, we have indicated those supplementary references by placing the symbol † after the year.

For each paper concerning triple systems, review information for *Mathematical Reviews* of the American Mathematical Society is given when available. Older reviews are in the format MR nn#mmmm, where nn is the volume number and mmmm is the (sequentially assigned) review number; newer reviews are in the format MR yyv:ccmmm, where yy is the year, v is a letter designating the issue, cc is the classification number, and mmm is a (sequentially assigned) review number. Review information is also given for *Zentralblatt für Mathematik und ihre Grenzgebiete.* Older reviews are given in the format Z nn, pp, where nn is the volume number and pp the page number. Newer reviews are in the format Z nnn.ccmmm, where nnn is the volume number, cc is the classification number, and mmm is a (sequentially assigned) review number.

This bibliography is intended to include all published material whose primary focus is on triple systems, and consequently not all references are cited in the text. Preprints are included, but the year of publication is included as 1999 to indicate that the publication data is incomplete at the time of this book's publication.

ABBOTT, H.L. / HARE, D.R.

1989 On the construction of color critical linear hypergraphs, *Ars Combinat.* 28, 71–82. MR 90k:05111; Z 736.05057. Addendum: *Ars Combinat.* 31 (1991) 125–126. MR 92c:05112.

ABEL, R.J.R.

1996 Difference families, in: *CRC Handbook of Combinatorial Designs* (Ed. C.J. Colbourn and J.H. Dinitz), CRC Press, Boca Raton, FL, pp. 270–287. Z 851.05026.

ABEL, R.J.R. / BROUWER, A.E. / COLBOURN, C.J. / DINITZ, J.H.

1996† Mutually Orthogonal Latin Squares (MOLS), in: *CRC Handbook of Combinatorial Designs* (Ed. C.J. Colbourn and J.H. Dinitz), CRC Press, Boca Raton, FL, pp. 111–141.

ABEL, R.J.R. / COLBOURN, C.J. / YIN, J.X. / ZHANG, H.

1997† Existence of incomplete transversal designs with block size five and any index λ, *Des. Codes Crypt.* 10, 275–307.

ABEL, R.J.R. / FURINO, S.C.

1996 Resolvable and near-resolvable designs, in: *CRC Handbook of Combinatorial Designs* (Ed. C.J. Colbourn and J.H. Dinitz), CRC Press, Boca Raton, FL, pp. 87–94. Z 855.05014.

ABEL, R.J.R. / GREIG, M.

1996 BIBDs with small block size, in: *CRC Handbook of Combinatorial Designs* (Ed. C.J. Colbourn and J.H. Dinitz), CRC Press, Boca Raton, FL, pp. 41–46. Z 855. 05015.

ABRHAM, J.

1986† Exponential lower bounds for the number of Skolem and extremal Langford sequences, *Ars Combinat.* 22, 187–198.

ABRHAM, J. / KOTZIG, A.

1981† Skolem sequences and additive permutations, *Discrete Math.* 37, 143–146.

ADAMS, P.

1993 Lambda-fold 2-perfect bowties, *Utilitas Math.* 44, 243–253. MR 94j:05096; Z 792.05026.

ADAMS, P. / BILLINGTON, E.J.

1996 Lambda-fold 3-perfect 9-cycle systems, *Ars Combinat.* 44, 241–262. MR 97m: 05055; Z 889.05062.

ADAMS, P. / BILLINGTON, E.J. / LINDNER, C.C.

1992 The spectrum of 3-perfect 9-cycle systems, *Australas. J. Combin.* 5, 103–108. MR 93a:05076; Z 774.05075. Corrigendum: *Australas. J. Combin.* 8, 277. MR 94f: 05088; Z 791.05076.

1994 k-perfect $3k$-cycle systems, *J. Comb. Math. Comb. Comput.* 15, 141–154. MR 94m:05050; Z 806.05045.

ADAMS, P. / BILLINGTON, E.J. / RODGER, C.A.

1994 Pasch decompositions of lambda-fold triple systems, *J. Comb. Math. Comb. Comput.* 15, 53–63. MR 94m:05034; Z 806.05016.

ADAMS, P. / BRYANT, D.E. / KHODKAR, A.

1998 3,5-cycle decompositions, *J. Combin. Des.* 6, 91–110. MR 98m:05111.

AHLSWEDE, R. / ALON, N. / ERDÖS, P.L. / RUSZINKÓ, M. / SZÉKELY, L.

1997 Intersecting systems, *Combin. Probab. Comput.* 6, 127–137. MR 98c:05154.

AHRENS, W.

1901 *Mathematische Unterhaltungen und Spiele*, B.G. Teubner, Leipzig. (2. Aufl., B.G. Teubner, Leipzig, 1918) 2. Band, Kap. XIV: Anordnungsprobleme.

AIGNER, M. / TRIESCH, E. / TUZA, Z.

1992 Irregular assignments and vertex-distinguishing edge-coloring of graphs, *Ann. Discrete Math.* 52, 1–9. MR 93h:05066; Z 769.05035.

AJOODANI-NAMINI, S.

1989 *On the Spectrum of Support Sizes of Triple Systems*, M.S. Thesis, University of Tehran.

AJOODANI-NAMINI, S. / KHOSROVSHAHI, G.B.

1995 On a conjecture of A. Hartman, in: *Combinatorics Advances* (Ed. C.J. Colbourn and E. Mahmoodian), Kluwer Academic, Dordrecht, pp. 1–12. MR 96j:05021; Z 837. 05020.

AJOODANI-NAMINI, S. / KHOSROVSHAHI, G.B. / SHOKOUFANDEH, A.

1996 Intersections of triple systems: small orders, *J. Comb. Math. Comb. Comput.* 20, 33–52. MR 96m:05028; Z 852.05024.

AKBARI, S. / KHOSROVSHAHI, G.B. / MAYSOORI, C. / SHARIARI, S.

1997 On maximum size anti-Pasch sets of triples, in: *Geometry, Combinatorial Designs and Related Structures* (Ed. J.W.P. Hirschfeld, S.S. Magliveras, M.J. de Resmini), Cambridge University Press, pp. 7–14. Z 891.05011.

AL DHAHIR, M.W. / ABDUL ELAH, M.S. / JUNGNICKEL, D.

1987 8_3 in PG(2,q), *Arch. Math.* 49, 141–150. MR 88j:51014; Z 595.51003.

ALEKSEJEV, V.E.

1967 O skolemovskom metode postrojenija cikličeskich sistem trojek Štejnera, *Mat. Zametki* 2, 145–156. MR 35#5341; Z 179, 28. (English translation: Skolem method of constructing cyclic Steiner triple systems, *Math. Notes* 2, 571–576.)

1974 O čisle sistem trojek Štejnera, *Mat. Zametki* 15, 767–774. MR 50#12754; Z 291. 05006. (English translation: On the number of Steiner triple systems, *Math. Notes* 15, 461–464.)

ALIEV, I.Š.O.

1967 Simmetričeskije algebry i sistemy Štejnera, *Dokl. Akad. Nauk SSSR* 174, 511–513. MR 35#2756; Z 213, 291. (English translation: Symmetric algebras and Steiner systems, *Soviet Math. Dokl.* 8, 651–653.)

1972 Kombinatornyje schemy i algebry, *Sib. Math. Ž.* 13, 499–509. MR 46#5140; Z 285.05025. (English translation: Combinatorial designs and algebras, *Sib. Math. J.* 13, 341–348.)

ALIEV, I.Š.O. / SEIDEN, E.

1969 Steiner triple systems and strongly regular graphs, *J. Combinat. Theory* 6, 33–39. MR 38#4362; Z 175, 503.

ALLENBY, R.J.T. / REDFERN, E.J.

1989† *Introduction to Number Theory and Computing*, Computer Science Press, Potomac MD.

ALON, N. / KIM, J.-H. / SPENCER, J.

1997 Nearly perfect matchings in regular simple hypergraphs, *Israel J. Math.* 100, 171–187. MR 98k:05112; Z 882.05107.

ALPERT, S.L.

1975 Two-fold triple systems and graph imbeddings, *J. Combinat. Theory (A)* 18, 101–107. MR 51#185; Z 299.05015.

ALSPACH, B. / HARE, D.R.

1991 Edge-pancyclic block-intersection graphs, *Discrete Math.* 97, 17–24. MR 93a: 05081; Z 758.05010.

ALSPACH, B. / HEINRICH, K. / MOHAR, B.

1990† A note on hamilton cycles in block intersection graphs, *Finite Geometries and Designs*, Amer. Math. Soc., pp. 1–4.

ALSPACH, B. / VARMA, B.N.

1979 On almost Steiner triple systems, *Congressus Numer.* 24, 937.

ALTSHULER, A.

1997 Construction and representation of neighborly manifolds, *J. Combinat. Theory (A)* 77, 246–267. MR 97j:57025; Z 869.52004.

ALTSHULER, A. / BOKOWSKI, J. / SCHUCHERT, J.

1994 Spatial polyhedra without diagonals, *Israel J. Math.* 86, 373–396. MR 95b: 52025; Z 799.52011.

1994a Sphere systems and neighborly spatial polyhedra with 10 vertices, *Suppl. Rend. Circ. Mat. Palermo II* 35, 15–28. MR 95h:52028; Z 808.52009.

1996 Neighborly 2-manifolds with 12 vertices, *J. Combinat. Theory (A)* 75, 148–162. MR 97g:52036; Z 855.52002.

ALTSHULER, A. / BREHM, U.

1992 Neighborly maps with few vertices, *Discrete Comput. Geom.* 8, 93–104. MR 93b: 52016; Z 756.52014.

ANDERSEN, L.D.

1979 *Latin Squares and their Generalizations*, Ph.D. Thesis, University of Reading. Part II: Embedding partial Steiner triple systems.

1982 Embedding latin squares with prescribed diagonal, *Ann. Discrete Math.* 15, 9–26. MR 86a:05021; Z 503.05009.

ANDERSEN, L.D. / HILTON, A.J.W.

1997 Triangulations of 3-way regular tripartite graphs of degree 4, with applications to orthogonal latin squares, *Discrete Math.* 167/168, 17–34. MR 98g:05025; Z 876.05015.

ANDERSEN, L.D. / HILTON, A.J.W. / MENDELSOHN, E.

1980 Embedding partial Steiner triple systems, *Proc. London Math. Soc.* (3) 41, 557–576. MR 82a:05010; Z 461.05010.

ANDERSEN, L.D. / HILTON, A.J.W. / RODGER, C.A.

1982† A solution to the embedding problem for partial idempotent latin squares, *J. London Math. Soc. (2)* 26, 21–27.

ANDERSEN, L.D. / MENDELSOHN, E.

1984 Embedding latin squares and multi-dimensional Howell designs into Steiner triple systems, *Ann. Discrete Math.* 20, 47–54. MR 86i:05031; Z 563.05015.

ANDERSON, I.

1974 *A First Course in Combinatorial Mathematics*, Clarendon Press, Oxford. Chapter 7: Steiner systems and sphere packings, pp. 97–116. MR 49#2402; Z 268.05001. 2nd edition, 1989. MR 90i:05001.

1979 Infinite families of bi-embedding numbers, *J. Graph Theory* 3, 263–268. MR 81b:05034; Z 417.05024.

1982 On the toroidal thickness of graphs, *J. Graph Theory* 6, 177–184. MR 83e:05044; Z 482.05035; Z 487.05024.

1982a Triples, current graphs and biembeddings, *Aequat. Math.* 24, 230–242. MR 84m:05026; Z 519.05025.

1983 Ladder graphs, *Ars Combinat.* 16A, 27–32. MR 85d:05101; Z 544.05010.

1990 *Combinatorial Designs: Construction Methods*, Ellis Horwood, Chichester and Halsted Press, New York. Z 691.05002.

1991 Kirkman and GK_{2n}, *Bull. Inst. Combin. Appl.* 3, 111–112. Z 828.05047.

1995 Cyclic designs in the 1850s: The work of the Rev. R.R. Anstice, *Bull. Inst. Combin. Appl.* 15, 41–46. MR 97a:01028; Z 854.05001.

1996 Some 2-rotational and cyclic designs, *J. Combin. Des.* 4, 247–254. MR 97m:05025.

ANDERSON, I. / WHITE, A.T.

1978 Current graphs and bi-embeddings, *J. Graph Theory* 2, 131–140. MR 58#0546; Z 418.05020.

ANSTICE, R.R.

1852 On a problem in combinations, *Cambr. and Dublin Math. J.* 7, 279–292.

1853 On a problem in combinations, *Cambr. and Dublin Math. J.* 8, 149–154.

ARCHDEACON, D. / DINITZ, J.

1993 Indecomposable triple systems exist for all lambda, *Discrete Math.* 113, 1–6. MR 94c:05013; Z 777.05019.

ARMANIOUS, M.H.

1989 Die Varietät der auflösbaren Ternare der Ordung 2, *Beiträge Algebra Geom.* 28, 33–37. MR 90h:05017; Z 699.51006.

1992 The associated geometric category of Steiner triple systems, *Period. Math. Hungar.* 25, 257–261. MR 94e:51009; Z 779.18008.

1997 Construction of nilpotent sloops of class n, *Discrete Math.* 171, 17–25. Z 884. 05016.

ARONHOLD, S.

1850 Zur Theorie der homogenen Functionen dritter Grades von drei Variabeln, *J. Reine Angew. Math.* 39, 140–159.

ASSAF, A.M.

1990 Modified group divisible designs, *Ars Combinat.* 29, 13–20. MR 91c:05048; Z 702.05014.

ASSAF, A.M. / HARTMAN, A.

1989 Resolvable group divisible designs with block size 3, *Discrete Math.* 77, 5–20. MR 90m:05018; Z 714.05007.

ASSAF, A.M. / MENDELSOHN, E. / STINSON, D.R.

1987 On resolvable coverings of pairs by triples, *Utilitas Math.* 32, 67–74. MR 88k: 05043; Z 649.05024.

ASSMUS, E.F. JR.

1995 On 2-ranks of Steiner triple systems, *Electronic J. Combinat.* 2, #R9, 35 pp. MR 96f:05032; Z 815.05016.

1998 Linearly derived Steiner triple systems, *Des. Codes Crypt.* 13, 31–49.

ASSMUS, E.F. JR. / KEY, J.D.

1992 *Designs and Their Codes*, Cambridge University Press, Cambridge. Chapter 8: Steiner systems. MR 93j:51003; Z 762.05001.

ASSMUS, E.F. JR. / MATTSON, H.F., JR.

1966 On the number of inequivalent Steiner triple systems, *J. Combinat. Theory* 1, 301–305. MR 34#4168; Z 158, 15. Errata: *J. Combinat. Theory* 2 (1967) 394. MR 35 #4116.

ATKIN, G.E. / CORRALES, H.P.

1989 An efficient modulation/coding scheme for MFSK systems on bandwidth constrained channels, *IEEE Trans. Sel. Areas Commun.* SAC-7, 1396–1401.

ATKIN, G.E. / FARES, D.A. / CORRALES, H.P.

1989 Coded multipulse position modulation in a noisy optical channel, *Microwave Optical Technol. Lett.* 2, 336–340.

AUMANN, G.

1987 Steiner-Tripel-Systeme, *Praxis Math.* 29, 337–340, 365–366. MR 88i:05033.

AVIS, D.

1980 A note on some computationally difficult set covering problems, *Math. Program.* 18, 138–145. MR 83a:90110; Z 429.90047.

BAARTMANS, A. / LANDJEV, I. / TONCHEV, V.D.

1996 On the binary codes of Steiner triple systems, *Des. Codes Crypt.* 8, 29–43. MR 97m:05026.

BAARTMANS, A. / WALLIS, W.D. / YUCAS, J.

1992 A geometric construction of the Steiner system S(4,7,23), *Discrete Math.* 102, 177–186. MR 93i:05014; Z 762.51004.

BAARTMANS, A. / YUCAS, J.

1992 On the existence of a double extension of $PG(3,2)$, *Ars Combinat.* 33, 145–156. MR 93i:05012; Z 761.51002.

1993 Kirkman years in PG(3,2), *Australas. J. Combin.* 7, 129–132. MR 93k:05036; Z 776.05015.

BABAI, L.

1974† On the minimum order of graphs with given group, *Canad. Math. Bull.* 17, 467–470.

1979† Endomorphisms of sub- and factorsemigroups, in: *Algebraic theory of semigroups*, North-Holland, Amsterdam, 1979, pp. 43–50.

1980 Almost all Steiner triple systems are asymmetric, in: *Topics on Steiner Systems* (Ed. C.C. Lindner and A. Rosa), *Ann. Discrete Math.* 7, 37–39. Z 462.05013.

BABAI, L. / LUKS, E.M.

1983† Canonical labeling of graphs, in: *Proc. Fifteenth ACM Symp. Theory of Computing*, pp. 171–183.

BAGCHI, S. / BAGCHI, B.

1992 Designs from pairs of finite fields. I: A cyclic unital U(6) and other regular Steiner 2-designs, *J. Combinat. Theory (A)* 52, 51–61. MR 93k:05024; Z 679.05008.

BAKER, C.A.

1995 Extended Skolem sequences, *J. Combin. Des.* 3, 363–379. MR 96i:05002; Z 885. 05040.

BAKER, M.J.C.

1973 A new type of magic square, *Math. Gaz.* 57, 132–133. Z 263.05022.

BAKER, R.D.

1976 Partitioning the planes of $AG_{2m}(2)$ into 2-designs, *Discrete Math.* 15, 205–211. MR 54#4999; Z 326.05013.

1978 Quasigroups and tactical systems, *Aequat. Math.* 18, 296–303. MR 80d:05018; Z 393.05017.

BAKER, R.D. / WILSON, R.M.

1977 Nearly Kirkman triple systems, *Utilitas Math.* 11, 289–296. MR 56#133; Z 362. 05030.

BALL, W.W.R.

1939 *Mathematical Recreations and Problems*, 1st ed., London 1892 (French translation by J. Fitz-Patrick, Hermann, Paris 1898: Récréations et problèmes mathématiques.) 4th ed. *Mathematical Recreations and Essays*, London 1905. Z 22, 1.

BALL, W.W.R. / COXETER, H.S.M.

1974 *Mathematical Recreations and Essays*, 12th ed., University of Toronto Press, Toronto. Chapter 10: Combinatorial designs (by J.J. Seidel). MR 50#4229.

BANERJEE, S. / KAGEYAMA, S.

1996 Non-isomorphic solutions of group divisible designs with $k = 3$ and $\lambda_1 = 0$, *J. Stat. Plann. Infer.* 56, 23–32. MR 97m:05019; Z 873.05013.

BANERJEE, S. / MALHOTRA, R. / KAGEYAMA, S.

1993 Methods of constructing group divisible designs with $k = 3$ and 4, *Commun. Stat. Theory Methods* 22, 2831–2842. Z 785.62079.

BARANYAI, Z.

1975 On the factorization of the complete uniform hypergraph, *Infinite and Finite Sets, Colloq. Math. Soc. J. Bolyai* 10, Keszthely 1973, North-Holland, Amsterdam, 91–108. MR 54#5047; Z 306.05137.

BARRAU, J.A.

1908 Over drietalstelsels, in het bijzonder die van dertien elementen, *Kon. Akad. Wetensch. Amst. Verslag Wis-en Natuurk. Afd.* 17, 274–279. (On triple systems, particularly those of thirteen elements, *Kon. Akad. Wetensch. Amst. Proc. Sect. Sci.* 11, 290–295.)

1908a Over de combinatorische opgave van Steiner, *Kon. Akad. Wetensch. Amst. Verslag Wis-en Natuurk. Afd.* 17, 318–326. (On the combinatory problem of Steiner, *Kon. Akad. Wetensch. Amst. Proc. Sect. Sci.* 11, 352–360.)

1909 Tripelstelsels van vijftien elementen, *Handel. Nederl. Natuur-en Geneesk. Congres* 12, 185–189.

BATTEN, L.M.

1989 Translation triple systems, *Geom. Ded.* 32, 119–123. MR 90i:51008; Z 686. 51003.

1989a Minimally projectively embeddable Steiner systems, *Discrete Math.* 77, 21–27. MR 91b:51007; Z 701.05008.

BAYS, S.

1917 Une question de Cayley relative au problème des triades de Steiner, *Enseignement Math.* 19, 57–67.

1917a Sur les systèmes de triples de 13 éléments, *Enseignement Math.* 19, 332–333.

1917b Sur les systèmes cycliques de triples de Steiner, *C.R. Acad. Sci. Paris* Ser. A. 165, 543–545.

1920 Sur les systèmes cycliques de triples de Steiner, *C.R. Acad. Sci. Paris* Ser. A. 171, 1363–1365.

1922 Sur la généralisation du problème de triples de Steiner, *Enseignement Math.* 22, 66–67.

1922a Sur les systèmes cycliques de triples de Steiner, *C.R. Acad. Sci. Paris* Ser. A. 175, 936–939.

1923 Recherche des systèmes cycliques de triples de Steiner différents pour N premier (ou puissance de nombre premier) de la forme $6n+1$, *J. Math. Pures Appl.* (9) 2, 73–98.

1923a Sur les systèmes cycliques de triples de Steiner, *Ann. Sci. Ecole Norm. Sup.* (3) 40, 55–96.

1925 Sur les systèmes cycliques de triples de Steiner différents pour N premier (ou puissance de nombre premier) de la forme $6n+1$, *Ann. Fac. Sci. Univ. Toulouse* (3) 17, 23–61.

1931 Sur les systèmes cycliques de triples de Steiner différents pour N premier (ou puissance de nombre premier) de la forme $6n+1$, I. *Comment. Math. Helv.* 2, 294–305. II–III. *Comment Math. Helv.* 3, 22–41; Z 1, 264; IV–V. *Comment Math. Helv.* 3, 122–147. Z 2, 182; VI. *Comment Math. Helv.* 3, 307–325. Z 3, 100.

1932 Sur les systèmes cycliques de triples de Steiner différents pour N premier de la forme $6n+1$, *Comment. Math. Helv.* 4, 183 194. Z 5, 49.

1935 Sur le nombre de systèmes cycliques de triples différents pour chaque classe, *Actes Soc. Helvét. Sci. Nat.* 116, 275–276. Z 13, 99.

1935a Sur les systèmes de caractéristiques appartenant à $d = 3$, *Actes Soc. Helvét. Sci. Nat.* 116, 276–277. Z 13, 99.

BAYS, S. / BELHÔTE, G.

1933 Sur les systèmes cycliques de triples de Steiner différents pour N premier de la forme $6n+1$, *Comment. Math. Helv.* 6, 28–46. Z 7, 195.

BEENKER, G.F.M. / GERARDS, A.M.H. / PENNING, P.

1978 A construction of disjoint Steiner triple systems, TH Report 78-WSK-01, Technological Univ. Eindhoven, 4 pp. Z 376.05006.

BEHZAD, M. / MAHMOODIAN, E.S.

1991 Graphs versus designs — A quasisurvey, in: *Graph Theory, Combinatorics and Applications*, Vol. 1, Wiley, New York, pp. 125–142. Z 840.05096.

BEINEKE, L.W.

1996 Graph decompositions, *Congressus Numer.* 115, 213–226. MR 97i:05101.

BELHÔTE, G.

1933 *Contribution a l'étude des systèmes cycliques de triples de Steiner différents pour N premier de la forme* $6n + 1$, Thèse de doctorat, Université de Fribourg.

BÉNÉTEAU, L.

1975 Boucles de Moufang commutatives d'exposant 3 et quasi-groupes de Steiner distributifs, *C.R. Acad. Sci. Paris* Ser. A 281, 75–76. MR 58#16335; Z 319.50026.

1980 Topics about 3-Moufang loops and Hall triple systems, *Simon Stevin* 54, 107–124. MR 81j:05035; Z 434.05014.

1980a Une classe particuliére de matroides parfaits, *Ann. Discrete Math.* 8, 229–232. MR 82d:05046; Z 445.05037.

1981 Les systèmes triples de Hall de dimension 4, *Europ. J. Combinat.* 2, 205–212. MR 84e:05030; Z 472.51002.

1983 Hall triple systems and related topics, *Ann. Discrete Math.* 18, 55–60. MR 84f: 05021; Z 503.05008.

1983a L'irreductibilité centrale dans les boucles de Moufang commutatives et dans les systèmes triples de Hall, *Discrete Math.* 45, 31–44. MR 84i:20074; Z 509.20054.

1984 The Hall triple systems of small class, *Europ. J. Combinat.* 5, 1–5. MR 86b: 05013; Z 539.05016.

1986 Symplectic geometry, quasigroups, and Steiner systems, *Ann. Discrete Math.* 30, 9–14. MR 87j:05036; Z 595.51008.

1996 Hall triple systems, in: *CRC Handbook of Combinatorial Designs* (Ed. C.J. Colbourn and J.H. Dinitz), CRC Press, Boca Raton, FL, pp. 377–381. Z 851.05022.

BÉNÉTEAU, L. / LACAZE, J.

1985 The groups related to the smallest non-affine Hall triple system, *J. Geometry* 24, 1–5. MR 86c:20007; Z 568.51014.

1986 Designs arising from symplectic geometry, in: *Applied Algebra, Algorithmics and Error-Correcting Codes*, Proc. 2nd Int. Conf. Toulouse 1984, Lect. Notes Comp. Sci. 228, Springer, Berlin, 198–205. MR 88f:51008; Z 603.20057.

1988 Symplectic trilinear forms and related designs and quasigroups, *Commun. Algebra* 16, 1035–1051. MR 89f:20048; Z 643.20049.

BENNETT, F.E.

1976 *Construction of Extended Triple Systems*, Ph.D. Thesis, University of Manitoba.

1978 Extended cyclic triple systems, *Discrete Math.* 24, 139–146. MR 80a:05025; Z 386.05012.

1982 Direct constructions for perfect 3-cyclic designs, *Ann. Discrete Math.* 15, 63–68. MR 85m:05029; Z 499.05011.

1982a Self-orthogonal semisymmetric quasigroups, *J. Combinat. Theory (A)* 33, 117–119. MR 84h:20069; Z 486.05016.

BENNETT, F.E. / CHEN, M.

1991 Incomplete perfect Mendelsohn designs, *Ars Combinat.* 31, 211–216. MR 92d: 05014; Z 748.05015.

BENNETT, F.E. / GRONAU, H.-D.O.F. / LING, A.C.H. / MULLIN, R.C.

1996 PBD-closure, in: *CRC Handbook of Combinatorial Designs* (Ed. C.J. Colbourn and J.H. Dinitz), CRC Press, Boca Raton, FL, pp. 203–213. Z 845.05011.

BENNETT, F.E. / MAHMOODI, A.

1996 Directed designs, in: *CRC Handbook of Combinatorial Designs* (Ed. C.J. Colbourn and J.H. Dinitz), CRC Press, Boca Raton, FL, pp. 317–321. Z 848.05016.

BENNETT, F.E. / MENDELSOHN, E. / MENDELSOHN, N.S.

1980 Resolvable perfect cyclic designs, *J. Combinat. Theory (A)* 29, 142–150. MR 82b:05030; Z 444.05021.

BENNETT, F.E. / MENDELSOHN, N.S.

1978 On pure cyclic triple systems and semisymmetric quasigroups, *Ars Combinat.* 5, 13–22. MR 58#28241; Z 415.05005.

1978a On the existence of extended triple systems, *Utilitas Math.* 14, 249–267. MR 81g:05038; Z 392.05013.

1978b Some remarks on 2-designs $S_2(2,3,v)$, *Congressus Numer.* 21, 119–127. MR 80c:05024; Z 408.05006.

1979 Constructions of totally symmetric quasigroups containing a specified number of idempotents. *Utilitas Math.* 15, 33–50. MR 80h:20104; Z 407.20059.

BENNETT, F.E. / SHEN, H.

1991 On indecomposable pure Mendelsohn triple systems, *Discrete Math.* 97, 47–57. MR 93e:05010; Z 758.05021.

BENNETT, F.E. / SOTTEAU, D.

1981 Almost resolvable decomposition of K_n^*, *J. Combinat. Theory (B)* 30, 228–232. MR 82g:05033; Z 509.05042.

BENNETT, F.E. / WEI, R.

1993 Embeddings of resolvable Mendelsohn triple systems, *J. Combin. Des.* 1, 281–299. MR 95j:05039; Z 886.05029.

BENNETT, F.E. / WEI, R. / ZHU, L.

1997 Resolvable Mendelsohn triple systems with equal-sized holes, *J. Combin. Des.* 5, 329–340. MR 98h:05029.

BENNETT, F.E. / WU, L.

1990 On minimum matrix representation of closure operations, *Discrete Appl. Math.* 26, 25–40. MR 91h:05032; Z 724.05011.

BENNETT, F.E. / ZHANG, H. / ZHU, L.

1996 Self-orthogonal Mendelsohn triple systems, *J. Combinat. Theory (A)* 73, 207–218. MR 97b:05020; Z 842.05008.

BENNETT, F.E. / ZHU, L.

1990 Perfect Mendelsohn designs with equal sized holes, *J. Comb. Math. Comb. Comput.* 8, 181–186. MR 91h:05003; Z 734.05015.

BERGE, C.

1958 *Théorie des graphes et ses applications*, Dunod, Paris. MR 21#1608; Z 88, 154. 2nd ed., Dunod, Paris, 1963. MR 27#5246; Z 121, 401. (English translation: *The Theory of Graphs and its Applications*, Wiley, New York, 1962. MR 24A#2381; Z 124, 362. Russian translation: *Teorija grafov i jejo primenenija*, Izd. inostr. lit., Moskva, 1962.)

1964 Graph theory, *Amer. Math. Monthly* 71, 471–481. MR 30#3458; Z 121, 401.

1973† *Graphs and Hypergraphs*, North-Holland, Amsterdam.

BERMOND, J.-C.

1974 An application of the solution of Kirkman's schoolgirl problem: the decomposition of the symmetric oriented complete graph into 3-circuits, *Discrete Math.* 8, 301–304. MR 49#4850; Z 281.05105.

BERMOND, J.-C. / BROUWER, A.E. / GERMA, A.

1978 Systèmes de triplets et différences associées, in: *Combinatoires et Théorie des Graphes* (Ed. J.-C. Bermond *et al.*) Gauthier–Villars, Paris, pp. 35–38. MR 80j:05020; Z 412.05012.

BERMOND, J.-C. / GERMA, A. / SOTTEAU, D.

1979 Resolvable decomposition of K_n^*, *J. Combinat. Theory (A)* 26, 179–185. MR 80k:05083; Z 411.05027.

BERTANI, L.

1975 Construzione di spazi di Steiner regolari, *Bollet. U.M.I.* (4) 11, 370–374. MR 51 #10120; Z 359.05015.

BETH, T.

1974 Algebraische Auflösungsalgorithmen für einige unendliche Familien von 3-Designs, *Le Matematiche* 29, 105–135. Z 303.05016.

BETH, T. / JUNGNICKEL, D.

1982 Variations on seven points: an introduction to the scope and methods of coding theory and finite geometry, *Aequat. Math.* 25, 153–176. MR 85b:05044; Z 546.05001.

BETH, T. / JUNGNICKEL, D. / LENZ, H.

1985 *Design Theory*, Bibliographisches Institut Mannheim; also Cambridge University Press, Cambridge, 1986. MR 86j:05026; Z 569.05002.

BEUTELSPACHER, A. / GANTER B.

1986 Zur Vervollständigung von Dreiecksräumen, *Arch. Math.* 47, 182–184. MR 87k:05030; Z 598.05016.

BEUTELSPACHER, A. / JUNGNICKEL, D. / VANSTONE, S.A.

1989† On the chromatic index of a finite projective space, *Geom. Ded.* 32, 313–318.

BHAT-NAYAK, V.N. / KANE, V.D.

1984 Construction and irreducibility of quasi-multiple designs, *Ars Combinat.* 18, 113–129. MR 87e:05023; Z 556.05007.

1984a A technique for obtaining nonisomorphic balanced incomplete block designs, *Utilitas Math.* 26, 235–247. MR 86g:05011; Z 547.05017.

1988 Two families of simple and irreducible BIBDs, *Discrete Math.* 68, 311–313. MR 89h:05013; Z 644.05007.

1993 Some families of BIBDs, *J. Indian Math. Soc.* 59, 23–29. MR 94j:05017; Z 859.05017.

BHAT-NAYAK, V.N. / SARVATE, D.G.

1996 Simple and irreducible group divisible designs with block size three, *Congressus Numer.* 114, 179–192. MR 97g:05018.

BHATTACHARYA, K.N.

1943 A note on two-fold triple systems, *Sankhya* 6, 313–314. MR 5, 29; Z 60, 313.

BIGELOW, D.C.

1990 *Enclosings of Latin Squares and Triple Systems*, Ph.D. Thesis, Pure Mathematics, University of Waterloo, Canada.

BIGELOW, D.C. / COLBOURN, C.J.

1991 Faithful enclosings of triple systems: Doubling the index, *Acta Math. Univ. Comen.* 60, 133–151. MR 93a:05039; Z 767.05021.

1992 Faithful enclosings of triple systems: A generalization of Stern's theorem, in: *Graphs, Matrices, and Designs* (Ed. R. Rees) Dekker, pp. 31–42. MR 94f:05017; Z 792.05027.

BIGGS, N.L.

1979 The schoolgirls problem, *Math. Spectrum* 12, 7–11.

1981 T.P. Kirkman, mathematician, *Bull. London Math. Soc.* 13, 97–120. MR 84h:01044; Z 454.01015.

BIGGS, N.L. / LLOYD, E.K. / WILSON, R.J.

1995 The history of combinatorics, in: *Handbook of Combinatorics* (Ed. R.L. Graham, M. Grötschel, and L. Lovász), Elsevier, Amsterdam, pp. 2163–2198. MR 96j:05002; Z 847.05001.

BILANIUK, S. / MENDELSOHN, E.

1984 A survey of colouring of Steiner systems, *Congressus Numer.* 43, 127–140. MR 86j:05027; Z 558.51007.

BILLINGTON, E.J.

1982 On lambda coverings of pairs by triples, repeated elements allowed in triples, *Utilitas Math.* 21C, 187–203. MR 84b:05021; Z 503.05019.

1982a Construction of some irreducible designs, in: *Combinatorial Math. IX*, Springer, pp. 182–196. MR 83m:05023; Z 488.05010.

1982b Further constructions of irreducible designs, *Congressus Numer.* 35, 77–89. MR 84m:05016; Z 513.05010.

1982c More balanced ternary designs with block size three, *Discrete Math.* 39, 9–21. MR 84b:05020; Z 481.05011.

1984 Balanced n-ary designs: A combinatorial survey and some new results, *Ars Combinat.* 17A, 37–72. MR 85i:05030; Z 537.05004.

1985 Balanced ternary designs with block size three, any Λ and any R, *Aequat. Math.* 29, 244–289. MR 87a:05026; Z 629.05014.

1985a On Mendelsohn and directed triple systems with repeated elements allowed in triples, *J. Stat. Plann. Inf.* 12, 75–79. MR 86k:05027; Z 572.05020.

1987 Cyclic balanced ternary designs with block size three and index two, *Ars Combinat.* 23B, 215–232. MR 88e:05014; Z 621.05005.

1988 Cyclic balanced ternary designs with block size three, *Ars Combinat.* 26A, 35–56. MR 90f:05030; Z 676.05012.

1993 The intersection problem for combinatorial designs, *Congressus Numer.* 92, 33–54. MR 95b:05020.

BILLINGTON, E.J. / FU, C.M. / FU, H.L.

1997 2-colouring $\{C_3, C_4\}$-designs, *Bull. Inst. Combin. Appl.* 20, 62–64. MR 98d:05024; Z 874.05008.

BILLINGTON, E.J. / HOFFMAN, D.G.

1989 The number of repeated blocks in balanced ternary designs with block size three, *J. Comb. Math. Comb. Comput.* 6, 163–172. MR 90i:05027; Z 693.05010.

1991 The number of repeated blocks in balanced ternary designs with block size three II, *Discrete Math.* 92, 25–38. MR 92m:05048; Z 751.05019.

1992 Pairs of simple balanced ternary designs with prescribed numbers of triples in common, *Australas. J. Combin.* 5, 59–71. MR 93a:05015; Z 776.05007.

BILLINGTON, E.J. / LINDNER, C.C.

1992 The spectrum for lambda-fold 2-perfect 6-cycle systems, *Europ. J. Combinat.* 13, 5–14. MR 93b:05128; Z 754.05046.

1994 The spectrum for 2-perfect bowtie systems, *Discrete Math.* 135, 61–68. MR 95m:05034; Z 815.05018.

1996 Maximum packings of uniform group divisible triple systems, *J. Combin. Des.* 4, 397–404. MR 97g:05132.

1998 Maximum packings of bowtie designs, *J. Comb. Math. Comb. Comput.* 27, 227–249.

BILLINGTON, E.J. / LO FARO, G.

1998 Repeated blocks in indecomposable twofold extended triple systems, *Australas. J. Combin.* 18, 209–218.

BILLINGTON, E.J. / MAHMOODIAN, E.S.

1993 Multi-set designs and numbers of common triples, *Graphs Combin.* 9, 105–115. MR 94e:05067; Z 778.05018.

1993a Numbers of common blocks in Mendelsohn and directed triple systems with repeated elements, *Bull. Iranian Math. Soc.* 19, 29–44. MR 95c:05032; Z 806.05017.

BILLINGTON, E.J. / PIKE, D.

1998 Decomposing block-intersection graphs of Steiner triple systems into triangles, *Australas. J. Combin.* 18, 51–64.

BILLINGTON, E.J. / STANTON, R.G.

1982 Bipackings and lambda packings of pairs into triples, repeated elements allowed in triples, *J. Combinat. Inf. System Sci.* 7, 83–90. MR 84f:05034; Z 494.0516.

BILLS, S.

1867 Solution of a problem proposed by W. Lea, *Educational Times Reprints* 8, 32–33.

BLAKE-WILSON, S. / PHELPS, K.T.

1999 Constant-weight codes and group-divisible designs, *Des. Codes Crypt.*

BLISS, G.A.

1933 Eliakim Hastings Moore, *Bull. Amer. Math. Soc.* 39, 831–838.

1934 The scientific work of Eliakim Hastings Moore, *Bull. Amer. Math. Soc.* 40, 501–514.

BLOCK, R.E.

1967† On the orbits of collineation groups, *Math. Zeit.* 96, 353–399.

BOKOWSKI, J. / EGGERT, A.

1991 Toutes les réalisations du tore de Möbius avec sept sommets, *Structural Topology* 17, 59–78. MR 93b:52017; Z 748.51016.

BOL'BOT, A.D.

1967 Equationally complete varieties of totally symmetric quasigroups, *Algebra i Logika* 6, 13–19. MR 35#6536.

BOLING, P.

1997 *The Bowtie Algorithm for Steiner Triple Systems*, Ph.D. Thesis, Auburn University.

BONDY, J.A. / MURTY, U.S.R.

1976† *Graph Theory with Applications*, Macmillan, New York.

BOOTH, T.R.

1976 *Construction of Disjoint Steiner Triple Systems from Steiner Quadruple Systems*, M.S. Thesis, Auburn University.

BOSÁK, J.

1986 *Rozklady grafov*, Veda, Bratislava. (English translation by J. Širáň and M. Škoviera: *Decompositions of Graphs*, Kluwer, Dordrecht, 1990. MR 91i:05085; Z 701. 05042.)

BOSE, R.C.

1939 On the construction of balanced incomplete block designs, *Ann. Eugenics* 9, 353–399. MR 1, 199; Z 23, 1.

1960 On a method of constructing Steiner's triple systems, *Contributions to Probability and Statistics*, Stanford University Press, Stanford, CA, pp. 133–141. MR 22 #11449; Z 94, 329.

1963 On the application of finite projective geometry for deriving a certain series of balanced Kirkman arrangements, in: *The Golden Jubilee Commemoration Volume* (1958–1959), Part II, Calcutta Math. Soc., pp. 341–354. MR 27#4769; Z 116, 112.

1979 Combinatorial problems of experimental design I. Incomplete block designs, *Proc. Symp. Applied Math.* (Amer. Math. Soc.) 34, 47–68. MR 80g:62054; Z 426. 05002.

BRAND, N.

1985 Isomorphic designs that are not multiplier equivalent, *Discrete Math.* 57, 159–165. MR 87d:05039; Z 587.05014.

1985a Some combinatorial isomorphism theorems, *Congressus Numer.* 50, 155–163. MR 87g:05112; Z 595.05013.

1988 Design isomorphisms and group isomorphisms, *Geom. Ded.* 27, 281–294. MR 89i:05045; Z 671.05009.

1989 On the Bays–Lambossy theorem, *Discrete Math.* 78, 217–222. MR 90j:05029; Z 694.05011.

BRAND, N. / HUFFMAN, W.C.

1984 Topological invariants of 2-designs arising from difference families, *J. Combinat. Theory (A)* 36, 253–278. MR 85g:05028; Z 538.05009.

1987 Constructions and topological invariants of 2-$(v, 3, \lambda)$ designs with group actions, *Europ. J. Combinat.* 8, 5–26. MR 88g:05022; Z 629.05009.

1987a Invariants and constructions of Mendelsohn designs, *Geom. Ded.* 22, 173–196. MR 88a:05021; Z 649.05014.

1987b Mendelsohn designs admitting the affine group, *Graphs Combin.* 3, 313–324. MR 88k:05024; Z 659.05022.

BRAY, A.

1882 The fifteen schoolgirls, *Knowledge* 2, 80–81.

1883 Twenty-one school-girl puzzle, *Knowledge* 2, 268.

BRAYTON, R.K. / COPPERSMITH, D. / HOFFMAN, A.J.

1974† Self-orthogonal Latin squares of all orders $n \neq 2, 3, 6$, *Bull. Amer. Math. Soc.* 80, 116–118.

BREACH, D.R. / STREET, A.P.

1985 Irreducible designs from supplementary difference sets, *Bull. Austral. Math. Soc.* 31, 105–115. MR 86b:05014; Z 548.05011.

BRICKELL, E.F. / WEI, V.K.

1987 Optical orthogonal codes and cyclic block designs, *Congressus Numer.* 58, 175–192. MR 89d:05024; Z 668.94012.

BROOKS, R.L.

1941† On colouring the nodes of a network, *Proc. Cambridge Phil. Soc.* 37, 194–197.

BROUWER, A.E.

1976 A note on the covering of all triples on 7 points with Steiner triple systems, Mathematisch Centrum Amsterdam, ZN 63/76, 1–8.

1977 Steiner triple systems without forbidden subconfigurations, Mathematisch Centrum Amsterdam, ZW 104/77, 8. Z 367.05011.

1978 Two new nearly Kirkman triple systems, *Utilitas Math.* 13, 311–314. MR 58 #10527; Z 379.05008.

1979† Optimal packings of K_4's into a K_n, *J. Combinatorial Theory (A)* 26, 278–297.

1980 A series of separable designs with application to pairwise orthogonal Latin squares, *Europ. J. Combinat.* 1, 39–41. MR 81i:05035; Z 472.05013.

1981 The linear spaces on 15 points, *Ars Combinat.* 12, 3–35. MR 83b:05031; Z 481.05019.

1981a On the size of a maximum transversal in a Steiner triple system, *Can. J. Math.* 33, 1202–1204. MR 83e:05016; Z 481.05016.

1995 Block designs, in: *Handbook of Combinatorics*, Elsevier, Amsterdam, 693–745. MR 97b:05014; Z 848.05008.

BROUWER, A.E. / SCHRIJVER, A. / HANANI, H.

1977† Group divisible designs with block-size four, *Discrete Math.* 20 (1977/78), 1–10.

BROUWER, A.E. / SHEARER, J.B. / SLOANE, N.J.A. / SMITH, W.D.

1990 A new table of constant weight codes, *IEEE Trans. Inform. Theory* 36, 1334–1380. MR 91h:94028; Z 713.94017.

BROWN, R.B.

1985 Quadratic algebras from Steiner triple systems, *Commun. Alg.* 13, 781–793. MR 86f:17006; Z 587.17002.

BROWN, T.C. / BUHLER, J.P.

1982 A density version of a geometric Ramsey theorem, *J. Combinat. Theory (A)* 32, 20–34. MR 83d:05010; Z 476.51008.

BROWN, W.G. / ERDÖS, P. / SÓS, V.T.

1971 Some extremal problems on r-graphs, in: *New Directions in the Theory of Graphs*, Academic Press, New York, pp. 53–63. MR 50#4376; Z 258.05132.

BRUCK, R.H.

1963 What is a loop? in: *Studies in Modern Algebra* (Ed. A.A. Albert), Prentice Hall, Englewood Cliffs, NJ, pp. 59–99. Z 199, 52.

BRUEN, A.A. / HADDAD, L. / WEHLAU, D.

1998 Caps and colouring Steiner triple systems, *Designs Codes Crypt.* 13, 51–55. MR 98k:05022; Z 892.05010.

BRUNEL, G.

1894 Remarques sur les systèmes de triades, *Mém. Soc. Sci. Phys. Nat. Bordeaux* (4) 5, 47.

1895 Sur les systèmes de triades formées avec $6n+1$ éléments, *C.R. Assoc. Fr. Av. Sci.* 24.

1896 Sur les triades formées avec $6n-1$ et $6n-2$ éléments, *Proces-verbaux Soc. Sci. Phys. Nat. Bordeaux*, 40–43.

1897 Sur un problème combinatoire de Steiner, *Proces-verbaux Soc. Sci. Phys. Nat. Bordeaux*, 37–41.

1902 Sur les deux systèmes de triades de treize elements, *J. Math. Soc. Sci. Phys. Nat. Bordeaux* (6) 2, 1–23.

BRYANT, D.E.

1996 A special class of nested Steiner triple systems, *Discrete Math.* 152, 315–320. MR 97a:05032; Z 851.05023.

1997 On the volume of trades in triple systems, *Australas. J. Combin.* 15, 161–176. MR 97m:05027; Z 880.05017.

BRYANT, D.E. / KHODKAR, A.

1997 A census of $(9;1;3,2)$ balanced ternary designs, *J. Comb. Math. Comb. Comput.* 23, 153–160. MR 97h:05035.

BRYANT, D.E. / OATES-WILLIAMS, S.

1994 Constructing identities for finite quasigroups, *Comm. Algebra* 22, 1783–1795. MR 94k:20120; Z 803.20050.

1995 Mendelsohn designs associated with a class of idempotent quasigroups, *Discrete Math.* 138, 161–167. MR 95m:05059; Z 818.05026.

BUEKENHOUT, F. / DELANDTSHEER, A. / DOYEN, J. / KLEIDMAN, P.B. / LIEBECK, M.W. / SAXL, J.

1990 Linear spaces with flag-transitive automorphism groups, *Geom. Dedicata* 36, 89–94. MR 91j:20009; Z 707.51017.

BUGGENHAUT, J. VAN

1974 On the existence of 2-designs $S_2(2,3,v)$ without repeated blocks, *Discrete Math.* 8, 105–109. MR 48#8259; Z 276.05021.

1974a Existence and constructions of 2-designs $S_3(2,3,v)$ without repeated blocks, *J. Geometry* 4, 1–10. MR 48#8260; Z 271.05006.

BURATTI, M.

1999 1-rotational Kirkman triple systems generate dicyclic Steiner 2-designs with block size 4.

1999a 1-rotational Steiner triple systems over arbitrary groups.

BURATTI, M. / ZUANNI, F.

1999 The 1-rotational Kirkman triple systems of order 33, *J. Stat. Plann. Infer.*

BURNSIDE, W.

1894 On an application of the theory of groups to Kirkman's problem, *Messenger Math.* (2) 23, 137–143.

BUROSCH, G.

1979 Über die Anzahl elementarer BUB der vollständigen (6,3)-Familie, *Rostock. Math. Kolloq.* 11, 5–11. MR 81c:05013; Z 431.62049.

BUROSCH, G. / GRONAU, H.-D. / RASCH, D.

1986 A short proof of a theorem of Hedayat and Khosrovshahi on 2-(6,3,λ) designs, *Rostock Math. Kolloq.* 30, 63–66. MR 88k:62136; Z 609.05010.

BUSSEMAKER, F.C. / SEIDEL, J.J.

1970 Symmetric Hadamard matrices of order 36, *Ann. N.Y. Acad. Sci.* 175, 66–79. MR 44#3894; Z 239.05010.

1970a Symmetric Hadamard matrices of order 36, Technological University Eindhoven, Report 70 WSK–02, 68 pp. MR 43#1863.

BUSSEY, W.H.

1909 On the tactical problem of Steiner, *Bull. Amer. Math. Soc.* 16, 19–22.

1914 The tactical problem of Steiner, *Amer. Math. Monthly* 21, 3–12.

BUTLER, R.A.R. / HOFFMAN, D.G.

1992 Intersections of group divisible triple systems, *Ars Combinat.* 34, 268–288. MR 93i:05022; Z 770.05014.

CALAHAN, R.S.

1990 *Automorphisms of Steiner Triple Systems with Holes*, Ph.D. Thesis, Auburn University.

1993 Steiner triple systems with a given automorphism, *Ars Combinat.* 36, 107–117. MR 94f:05018; Z 793.05018.

CALAHAN, R.S. / GARDNER, R.B.

1991 3-transrotational Steiner triple systems. *Congressus Numer.* 81, 65–68. MR 92j: 05024; Z 765.05013.

1992 A special case of bicyclic Steiner triple systems, *Congressus Numer.* 88, 77–80. MR 93j:05016; Z 795.05028.

CALAHAN, R.S. / HOFFMAN, D.G.

1990 Automorphisms of Steiner triple systems, *Congressus Numer.* 70, 181–182. MR 91b:05035; Z 696.05010.

CALAHAN-ZIJLSTRA, R.S. / GARDNER, R.B.

1994 Bicyclic Steiner triple systems, *Discrete Math.* 128, 35–44. MR 95c:05019; Z 797.05015.

1996 Reverse directed triple systems, *J. Comb. Math. Comb. Comput.* 21, 179–186. MR 97a:05033; Z 864.05014.

CAMERON, P.J.

1975 Minimal edge-colourings of complete graphs, *J. London Math. Soc.* (2) 11, 337–346. MR 54#10080; Z 312.05107.

1976 *Parallelisms of Complete Designs*, London Math. Soc. Lecture Note Series No. 23, Cambridge University Press, Cambridge. MR 54#7269; Z 333.05007.

1980 Extremal results and configuration theorems for Steiner systems, in: *Topics on Steiner Systems* (Ed. C.C. Lindner and A. Rosa), *Ann. Discrete Math.* 7, 43–63. Z 441.51006.

1985 Four lectures on projective geometry, in: *Finite Geometries, Lecture Notes Pure Appl. Math.*, Marcel Dekker, New York, 27–63. MR 87e:51014; Z 591.51012.

1994 An orbit theorem for Steiner triple systems, *Discrete Math.* 125, 97–100. MR 94k:05031; Z 809.05092.

1994a Infinite linear spaces, *Discrete Math.* 129, 29–41. MR 95e:51006; Z 803.51001.

CAMINA, A.R.

1994† A survey of the automorphism groups of block designs, *J. Combin. Des.* 2, 79–100.

CAMINA, A.R. / GAGEN, T.M.

1984† Block transitive automorphism groups of designs, *J. Algebra* 86, 549–554.

CAMINA, A.R. / MILLER, A.A.

1996 Line closed sets of Steiner triple systems and of classical linear spaces, *J. Stat. Plann. Infer.* 56, 65–77. MR 97m:05028; Z 873.05016.

CAMINA, A.R. / MISCHKE, S.

1996† Line-transitive automorphism groups of linear spaces, *Electron. J. Combinat.* 3, #R3.

CARMICHAEL, R.D.

1932 Note on triple systems, *Bull. Amer. Math. Soc.* 38, 695–696. Z 5, 340.

1937 *Introduction to the Theory of Groups of Finite Order*, Ginn, Boston. (Reprinted by Dover, New York, 1956). Z 19, 197.

CARNES, N.P.

1994 Cyclic automorphisms of Mendelsohn triple systems, *Discrete Math.* 126, 29–45. MR 94m:05035; Z 791.05011.

CARNES, N.P. / DYE, A. / REED, J.F.

1996 Cyclic antiautomorphisms of directed triple systems, *J. Combin. Des.* 4, 105–115. MR 96m:05029.

CARPENTER, L.L. / KEY, J.D.

1995 On Hadamard matrices from resolvable Steiner designs, *Congressus Numer.* 108, 53–63. MR 96h:05034.

CARPMAEL, E.

1881 Some solutions of Kirkman's 15-school-girl problem, *Proc. London Math. Soc.* 12, 148–156.

CARTWRIGHT, M.

1986 Crackly phones and the schoolgirl problem, *New Scientist* 111, 36–40.

CAYLEY, A.

1849 On the triple tangent planes of surfaces of the third order, *Cambridge and Dublin Math. J.* 4, 118–132.

1850 On the triadic arrangements of seven and fifteen things, *London Edinburgh and Dublin Philos. Mag. and J. Sci.* (3) 37, 50–53 (*Collected Mathematical Papers* I, 481–484).

1863 On a tactical theorem relating to the triads of fifteen things, *London, Edinburgh and Dublin Philos. Mag. and J. Sci.* (4) 25, 59–61 (*Collected Mathematical Papers* V, 95–97).

ČERNÝ, A. / HORÁK, P. / WALLIS, W.D.

1997 Kirkman's school projects, *Discrete Math.* 167/168, 189–196. MR 97m:05061; Z 873.05015.

CHAFFER, R. / EGGEN M.

1975 Strong finite embeddability for classes of quasigroups, *Algebra Univ.* 5, 257–262. MR 52#10939; Z 323.08007.

CHANG, Y.-X.

1992 On the spectrum for golf designs (in Chinese), *J. Hebei Normal College (Nat. Sci.)*, 8–15.

1993 A recursive construction for LD*-designs (in Chinese), *Acta Math. Appl. Sinica* 16, 201–207. MR 96d:05013.

1994 The existence spectrum for golf designs (in Chinese), *Acta Math. Sinica* 37, 728–734. MR 96j:05027; Z 813.05014.

1995 A completion of the spectrum for large sets of disjoint ordered triple systems, *Ars Combinat.* 39, 276–280. MR 95k:05024; Z 831.05013.

1996 On large sets of directed triple systems, *J. Stat. Plann. Infer.* 51, 137–142. MR 97b:05021; Z 849.05008.

CHANG, Y.-X. / KANG, Q.

1991 On the large sets of pairwise disjoint transitive triple systems, *Chinese Sci. Bull.* 36, 1278.

CHANG, Y.-X. / KANG, Q. / LEI, J.

1991 The existence of LD-designs and LD*-designs, *Acta Math. Appl. Sinica* 14, 565–567. Z 768.05010.

CHANG, Y.-X. / LO FARO, G.

1999 Intersection numbers of Kirkman triple systems.

CHANG, Y.-X. / YANG, G.-H. / KANG, Q.

1996 The spectrum of self-converse MTS, *Ars Combinat.* 44, 273–281. MR 97g:05028; Z 889.05019.

CHEE, Y.M.

1991 Graphical t-designs with block sizes three and four, *Discrete Math.* 91, 201–206. MR 93c:05020; Z 748.05035.

1996 Graphical designs, in: *CRC Handbook of Combinatorial Designs* (Ed. C.J. Colbourn and J.H. Dinitz), CRC Press, Boca Raton, FL, pp. 366–369. Z 856.05020.

CHEE, Y.M. / COLBOURN, C.J. / GALLANT, R.P. / LING, A.C.H.

1997 On a problem of Hartman and Heinrich concerning pairwise balanced designs with holes, *J. Comb. Math. Comb. Comput.* 23, 121–128. MR 97h:05020; Z 872.05005.

CHEE, Y.M. / COLBOURN, C.J. / KREHER, D.L.

1990 Simple t-designs with $v \leq 30$, *Ars Combinat.* 29, 193–258. MR 91b:05028; Z 719.05011.

CHEE, Y.M. / COLBOURN, C.J. / LING, A.C.H.

1997 Weakly union-free twofold triple systems, *Annals Combinat.* 1, 215–225.

1999 Asymptotically optimal erasure-resilient codes for large disk arrays.

CHEE, Y.M. / LIM, A.

1989 Quadrilateral-free Steiner triple systems of orders 31 and 37, *J. Comb. Inf. Syst. Sci.* 14, 244–248. MR 92c:05026; Z 855.05024.

CHEE, Y.M. / ROYLE, G.F.

1991 The 2-rotational Steiner triple systems of order 25, *Discrete Math.* 97, 93–100. MR 92i:05041; Z 756.05026.

1992 Enumeration of small nonisomorphic 1-rotational twofold triple systems, *Math. Comput.* 59, 609–612. MR 93a:05025; Z 771.05018.

CHEN, D. / LINDNER, C.C. / STINSON, D.R.

1992 Further results on large sets of disjoint group-divisible designs, *Discrete Math.* 110, 35–42. MR 94b:05038; Z 778.05011.

CHEN, D. / STANTON, R.G. / STINSON, D.R.

1991 Disjoint packings on 6k+5 points, *Utilitas Math.* 40, 129–138. MR 93b:05013; Z 754.05056.

CHEN, D. / STINSON, D.R.

1990 Recent results on combinatorial constructions for threshold schemes, *Australas. J. Combin.* 1, 29–48. MR 92g:05026; Z 758.05033.

1993 On the construction of large sets of disjoint group-divisible designs, *Ars Combinat.* 35, 103–115. MR 93m:05020; Z 796.05006.

CHEN, K. / GE, G. / ZHU, L.

1999 Generalized Steiner triple systems with group size five.

1999a Starters and related codes, *J. Stat. Plann. Infer.*

CHO, C.J.

1982 Rotational triple systems, *Ars Combinat.* 13, 203–209. MR 83i:05014; Z 504. 05013.

1982a Rotational Steiner triple systems, *Discrete Math.* 42, 153–159. MR 84b:05025; Z 505.05012.

1983 *Combinatorial Designs with Prescribed Automorphism Types*, Ph.D. Thesis, McMaster University, vii+241 pp.

1986 Rotational Mendelsohn triple systems, *Kyungpook Math. J.* 26, 5–9. MR 87j: 05027; Z 626.05008.

1991 Rotational extended triple systems, *Kyungpook Math. J.* 31, 219–234. MR 93a: 05026; Z 778.05019.

1992 Regular extended triple systems, *Kyungpook Math. J.* 32, 115–124. MR 93i: 05035; Z 778.05020.

CHO, C.J. / CHAE, Y. / HWANG, S.G.

1987 Rotational directed triple systems, *J. Korean Math. Soc.* 24, 133–142. MR 88j: 05012; Z 779.05004.

CHO, C.J. / HAN, Y.-H. / KANG, S.-H.

1986 Cyclic directed triple systems, *J. Korean Math. Soc.* 23, 117–125. MR 88d: 05022; Z 617.05014.

CHU, W.S.

1993 Homogeneous embeddings of disjoint Steiner triple systems (in Chinese), *J. Shanghai Jiao Tong Univ.* 27, 57–68. MR 94k:05032.

1996 Homogeneous embeddings of disjoint Mendelsohn triple systems, *J. Stat. Plann. Infer.* 51, 143–155. MR 97b:05022; Z 852.05023.

1997 The cardinality of the intersection of simple directed triple systems with $\lambda = 2$ (in Chinese), *J. Shanghai Jiaotong Univ.* 31, 92–96. MR 98j:05026.

1998 Embeddings of simple directed triple systems, *Acta Math. Sinica* 14, 135–138.

1999 Intersections of simple twofold directed triple systems.

1999a Existence of threefold directed triple systems without 3-times repeated triples.

CHVÁTAL, V. / ERDÖS, P.

1972† A note on hamiltonian circuits, *Discrete Math.* 2, 111–113.

CLAPHAM, P.C.

1976 Steiner triple systems with block-transitive automorphism groups, *Discrete Math.* 14, 121–131. MR 53#12971; Z 323.05012.

CLAUSEN, T.

1853 Über eine combinatorische Aufgabe, *Archiv Math. Phys.* (1) 21, 93–96.

CLAYTON, R.

1987 On twofold triple systems and their automorphism groups, *J. Combinat. Theory (A)* 46, 306–313. MR 89a:05026; Z 654.05008.

CLEBSCH, A.

1876 *Vorlesungen über Geometrie, I. Geometrie der Ebene*, Teubner, Leipzig. 5. Die Curven dritter Ordnung und dritter Klasse.

1895 Zum Gedächtnis an Julius Plücker, in: *Julius Plückers Gesammelte Wissenschaftliche Abhandlungen I* (Ed. A. Schoenflies and A. Clebsch), Teubner, Leipzig, ix–xxxv.

COCKAYNE, E.J. / MILLER, G.G. / PRINS, G.

1972† An interpolation theorem for partitions which are complete with respect to hereditary properties, *J. Combinat. Theory* (B) 13, 290–297.

COKER, G.D. / GARDNER, R.B.

1995 Some rotational automorphisms of Mendelsohn triple and quadruple systems, *Utilitas Math.* 47, 117–127. MR 95m:05035; Z 831.05016.

COLBOURN, C.J.

1981 Disjoint cyclic Steiner triple systems, *Congressus Numer.* 32, 205–212. MR 84d:045040; Z 494.05008.

1982 Computing the chromatic index of Steiner triple systems, *Computer J.* 25, 338–339. MR 84a:05008.

1982a Distinct cyclic Steiner triple systems, *Utilitas Math.* 22, 103–126. MR 84d:05041; Z 511.65010.

1983 Embedding partial Steiner triple systems is NP-complete, *J. Combinat. Theory (A)* 35, 100–105. MR 84k:68033; Z 529.68020.

1986 Orienting triple systems is NP-complete, *Ars Combinat.* 22, 155–163. MR 88a:05036; Z 614.05013.

1987 Directing and orienting triple systems, in: *Designs and Finite Geometries* (Kyoto 1986), Surikaisekikenkyusho Kokyuroku No. 607, pp. 33–38. MR 88k:05045.

1987a Leaves, excesses, and neighbourhoods, *Acta Univ. Carol. Math. Phys.* 28, 41–47. MR 89h:05020; Z 639.05006.

1987b Realizing small leaves of partial triple systems, *Ars Combinat.* 23A, 91–94. MR 88d:05027; Z 646.05008.

1989 Simple neighbourhoods in triple systems, *J. Combinat. Theory (A)* 52, 10–19. MR 90i:05014; Z 713.05011.

1991 Automorphisms of directed triple systems, *Bull. Austral. Math. Soc.* (A) 43, 257–264. MR 92d:05020; Z 703.05017.

1992 Separations of Steiner triple systems: some questions. *Bull Inst. Comb. Appl.* 6, 53–56. MR 93i:05023; Z 828.05015.

1993 Small group divisible designs with block size three, *J. Comb. Math. Comb. Comput.* 14, 153–172. MR 94c:05037; Z 793.05012.

1996 Winning the lottery, in: *CRC Handbook of Combinatorial Designs* (Ed. C.J. Colbourn and J.H. Dinitz), CRC Press, Boca Raton, FL, pp. 578–584. Z 864.05008.

1999 Minimum weights of point codes of Steiner triple systems, *J. Stat. Plann. Infer.*

1999a Weakly union-free maximum packings, *Annals Combinat.*

COLBOURN, C.J. / COLBOURN, M.J.

1981 Disjoint cyclic Mendelsohn triple systems, *Ars Combinat.* 11, 3–8. MR 83c:05031; Z 468.61002.

1982 A recursive constructions for 1-rotational Steiner 2-designs, *Congressus Numer.* 35, 163–167. MR 85b:05029; Z 514.05016.

1982a The chromatic index of cyclic Steiner 2-designs, *Internat. J. Math. Math. Sci.* 5, 823–825. MR 84b:05022; Z 501.05011.

1983 Decomposition of block designs: computational issues, *Combinatorial Math. X*, Lecture Notes Math. 1036, Springer, 141–146. MR 85h:05017; Z 523.05011.

1983a Every twofold triple system can be directed, *J. Combinat. Theory (A)* 34, 375–378. MR 84h:05016; Z 508.05019.

1983b Greedy colourings of Steiner triple systems, *Ann. Discr. Math.* 18, 201–207. MR 84f:05030; Z 546.05009.

1983c Nested triple systems, *Ars Combinat.* 16, 27–34. MR 85d:05035; Z 538.05007.

1985 The computational complexity of decomposing block designs, *Ann. Discrete Math.* 27, 345–350. MR 87f:05018; Z 597.05011.

COLBOURN, C.J. / COLBOURN, M.J. / HARMS, J.J. / ROSA, A.

1983 A complete census of (10,3,2) block designs and of Mendelsohn triple systems of order ten. III. (10,3,2) block designs without repeated blocks, *Congressus Numer.* 37, 211–234. MR 84j:05018; Z 549.05008.

COLBOURN, C.J. / COLBOURN, M.J. / PHELPS, K.T. / RÖDL, V.

1982 Colouring block designs is NP-complete, *SIAM J. Alg. Discr. Meth.* 3, 305–307. MR 84b:68038; Z 497.05022.

COLBOURN, C.J. / COLBOURN, M.J. / ROSA, A.

1983 Completing small partial triple systems, *Discrete Math.* 45, 165–179. MR 84g:05028; Z 517.05011.

COLBOURN, C.J. / COLBOURN, M.J. / STINSON, D.R.

1985 The computational complexity of finding subdesigns in combinatorial designs, *Ann. Discrete Math.* 26, 59–66. MR 87g:68020; Z 574.68037.

COLBOURN, C.J. / CURRAN, D. / VANSTONE, S.A.

1987 Recursive constructions for Kirkman squares with block size three, *Utilitas Math.* 32, 169–174. MR 89d:05045; Z 637.05006.

COLBOURN, C.J. / CUSACK, C.A. / KREHER, D.L.

1995 Partial Steiner triple systems with equal-sized holes, *J. Combinat. Theory (A)* 70, 56–65. MR 96b:05030; Z 818.05013.

COLBOURN, C.J. / DINITZ, J.H.

1997 Complete arcs in Steiner triple systems, *J. Combinat. Theory (A)* 80, 320–333. MR 98j.05027a; Z 884.05017. Erratum: *J. Combinat. Theory (A)* 81 (1998) 262. MR 98j.05027b.

1999 Generating sets in Steiner triple systems, *Math. Slovaca.*

COLBOURN, C.J. / DINITZ, J.H. / STINSON, D.R.

1991 Spanning sets and scattering sets in Steiner triple systems, *J. Combinat. Theory (A)* 57, 46–59. MR 92b:05011; Z 765.05016.

1999 Quorum systems constructed from combinatorial designs.

COLBOURN, C.J. / GIBBONS, P.B.

1998 Uniform orthogonal group divisible designs with block size three, *New Zealand J. Math.* 27, 15–33.

COLBOURN, C.J. / GIBBONS, P.B. / MATHON, R. / MULLIN, R.C. / ROSA, A.

1994 The spectrum of orthogonal Steiner triple systems, *Canad. J. Math.* 46, 239–252. MR 95e:05019; Z 809.05018.

COLBOURN, C.J. / HADDAD, L. / LINEK, V.

1996 Equitable embeddings of Steiner triple systems, *J. Combinat. Theory (A)* 73, 229–247. MR 96m:05030; Z 840.05010.

1997 Balanced Steiner triple systems, *J. Combinat. Theory (A)* 78, 292–302. MR 98m:05020; Z 883.05021.

COLBOURN, C.J. / HAMM, R.C.

1986 Embedding and enclosing partial triple systems and $\lambda = 3$, *Ars Combinat.* 21, 111–117. MR 87g:05026; Z 625.05011.

COLBOURN, C.J. / HAMM, R.C. / LINDNER, C.C. / RODGER, C.A.

1986 Embedding partial graph designs, block designs, and triple systems with $\lambda > 1$, *Canad. Math. Bull.* 29, 385–391. MR 88d:05023; Z 556.05008.

COLBOURN, C.J. / HAMM, R.C. / RODGER, C.A.

1984 Small embeddings of partial directed triple systems and partial triple systems with even λ, *J. Combinat. Theory (A)* 37, 363–369. MR 86g:05012; Z 559.05010.

COLBOURN, C.J. / HAMM, R.C. / ROSA, A.

1985 Embedding, immersing, and enclosing, *Congressus Numer.* 47, 229–236. MR 87i:05044; Z 648.05012.

COLBOURN, C.J. / HARMS, J.J.

1983 Directing triple systems, *Ars Combinat.* 15, 261–266. MR 85b:05036; Z 516. 05008.

1987 Partitions into indecomposable triple systems, *Ann. Discrete Math.* 34, 107–118. MR 89g:05020; Z 646.05007.

COLBOURN, C.J. / HOFFMAN, D.G. / REES, R.

1992 A new class of group divisible designs with block size three, *J. Combinat. Theory (A)* 59, 73–89. MR 93f:05015; Z 759.05012.

COLBOURN, C.J. / JIANG, Z.

1996 The spectrum for rotational Steiner triple systems, *J. Combin. Des.* 4, 205–217. MR 97b:05023.

COLBOURN, C.J. / JUNGNICKEL, D. / ROSA, A.

1988 The strong chromatic number of partial triple systems, *Discrete Appl. Math.* 20, 31–38. MR 89c:05021; Z 642.05022.

COLBOURN, C.J. / KOCAY, W.L. / STINSON, D.R.

1986 Some NP-complete problems for hypergraph degree sequences, *Discrete Appl. Math.* 14, 239–254. MR 87j:05127; Z 594.05051.

COLBOURN, C.J. / LINDNER, C.C.

1992 Support sizes of triple systems, *J. Combinat. Theory (A)* 61, 193–210. MR 94b: 05025; Z 772.05012.

COLBOURN, C.J. / LING, A.C.H.

1998 A class of partial triple systems with applications in survey sampling, *Commun. Stat. Theory Meth.* 27, 1009–1018. MR 98m:62020.

1998a Point code minimum Steiner triple systems, *Des. Codes Crypt.* 14, 141–146.

1999 Kirkman school project designs, *Discrete Math.*

COLBOURN, C.J. / MAGLIVERAS, S.S. / MATHON, R.A.

1992 Transitive Steiner and Kirkman triple systems of order 27, *Math. Comput.* 58, 441–449, S23–S27. MR 92e:05016; Z 754.05012,05013.

COLBOURN, C.J. / MAGLIVERAS, S.S. / STINSON, D.R.

1992 Steiner triple systems of order 19 with nontrivial automorphism group. *Math. Comput.* 59, 283–295, S25–S27. MR 92k:05022; Z 767.05025.

COLBOURN, C.J. / MAHMOODIAN, E.S.

1988 Support sizes of $(12, 3, \lambda)$-designs, *Technical Report* 1–1367, Sharif University of Technology, Iran.

1990 The spectrum of support sizes for threefold triple systems, *Discrete Math.* 83, 9–19. MR 91f:05016; Z 705.05009.

1993 Support sizes of sixfold triple systems, *Discrete Math.* 115, 103–131. MR 94i:05014; Z 779.05005.

COLBOURN, C.J. / MANSON, K.E. / WALLIS, W.D.

1984 Frames for twofold triple systems, *Ars Combinat.* 17, 69–78. MR 85g:05047; Z 553.05014.

COLBOURN, C.J. / MATHON, R.A.

1987 Leave graphs of small maximal partial triple systems, *J. Comb. Math. Comb. Comput.* 2, 13–28. MR 88m:05013; Z 639.05007.

1996 Steiner systems, in: *CRC Handbook of Combinatorial Designs* (Ed. C.J. Colbourn and J.H. Dinitz), CRC Press, Boca Raton, FL, pp. 66–75. Z 851.05020.

COLBOURN, C.J. / MATHON, R.A. / ROSA, A. / SHALABY, N.

1991 The fine structure of threefold triple systems: $v \equiv 1, 3 \pmod 6$, *Discrete Math.* 92, 49–64. MR 92k:05021; Z 765.05015.

COLBOURN, C.J. / MATHON, R.A. / SHALABY, N.

1991 The fine structure of threefold triple systems: $v \equiv 5 \pmod 6$, *Australasian J. Combin* 3, 75–92. MR 92i:05042; Z 768.05011.

COLBOURN, C.J. / MCKAY, B.D.

1987 Cubic neighbourhoods in triple systems, *Ann. Discrete Math.* 34, 119–136. MR 89g:05021; Z 644.05010.

COLBOURN, C.J. / MENDELSOHN, E. / PRAEGER, C.E. / TONCHEV, V.D.

1992 Concerning multiplier automorphisms of cyclic Steiner triple systems. *Des. Codes Crypt.* 2, 237–251. MR 93g:05020; Z 765.05014.

COLBOURN, C.J. / MENDELSOHN, E. / ROSA, A.

1988 Extending the concept of decomposability for triple systems, *Ann. Discrete Math.* 37, 107–116. MR 88m:05013; Z 661.05016.

COLBOURN, C.J. / MENDELSOHN, E. / ROSA, A. / ŠIRÁŇ, J.

1994 Anti-mitre Steiner triple systems, *Graphs and Combinatorics* 10, 215–224. MR 95j:05040; Z 815.05017.

COLBOURN, C.J. / MILICI, S.

1989 Support sizes of triple systems with small index, *J. Comb. Math. Comb. Comput.* 6, 155–161. MR 90j:05030; Z 708.05010.

COLBOURN, C.J. / OORSCHOT, P.C. VAN

1989 Applications of combinatorial designs in computer science, *ACM Computing Surveys* 21, 223–250.

COLBOURN, C.J. / ORAVAS, M.A. / REES, R.S.

1999 Steiner triple systems with disjoint or intersecting subsystems.

COLBOURN, C.J. / PHELPS, K.T. / DE RESMINI, M.J. / ROSA, A.

1991 Partitioning Steiner triple systems into complete arcs, *Discrete Math.* 89, 149–160. MR 92f:05022; Z 765.05011.

COLBOURN, C.J. / PULLEYBLANK, W.R. / ROSA, A.

1989 Hybrid triple systems and cubic feedback sets, *Graphs Combinat.* 5, 15–28. MR 89m:05015; Z 676.05015.

COLBOURN, C.J. / RÖDL, V.

1989† Percentages in pairwise balanced designs, *Discrete Math.* 77, 57–63.

COLBOURN, C.J. / ROSA, A.

1985 Maximal partial Steiner triple systems of order $v \leq 11$, *Ars Combinat.* 20, 5–28. MR 87f:005032; Z 607.05014.

1985a Indecomposable triple systems with $\lambda = 4$, *Studia Sci. Math. Hungar.* 20, 139–144. MR 88i:05022; Z 527.05011; Z 608.05014.

1986 Quadratic leaves of maximal partial triple systems, *Graphs Combin.* 2, 317–337. MR 89i:05054; Z 609.05009.

1987 Quadratic excesses of coverings by triples, *Ars Combinat.* 24, 23–30. MR 89c: 05027; Z 637.05008.

1987a Element neighbourhoods in twofold triple systems, *J. Geometry* 30, 36–41. MR 88i:05023; Z 624.05011.

1991 Leaves, excesses and neighbourhoods in triple systems, *Austral. J. Comb.* 4, 143–178. MR 92i:05043; Z 767.05023.

1991a† Support sizes of λ-factorizations, in: *Combinatorics88, Proc. Int. Conf. on Incidence Geometries and Combinatorial Structures*, Vol. I, pp. 249–265.

1992 Directed and Mendelsohn triple systems, in: *Contemporary Design Theory: A Collection of Surveys* (Ed. J.H. Dinitz and D.R. Stinson), Wiley, New York, pp. 97–136. Z 767.05026.

1994 Problem 192. Perfect Steiner triple systems, *Discrete Math.* 125, 413.

1995 Orthogonal resolutions of triple systems, *Australas. J. Combin.* 12, 259–269. MR 96h:05022; Z 851.05021.

COLBOURN, C.J. / ROSA, A. / STINSON, D.R.

1991 Pairwise balanced designs with block sizes three and four, *Canad. J. Math.* 43, 673–704. MR 92g:05028; Z 759.05010.

COLBOURN, C.J. / ROSA, A. / ZNÁM, Š.

1993 The spectrum of maximal partial Steiner triple systems, *Des. Codes Crypt.* 3, 209–219. MR 94f:05019; Z 783.05016.

COLBOURN, C.J. / STINSON, D.R. / TEIRLINCK, L.

1992 A parallelization of Miller's $n^{\log n}$ isomorphism technique, *Inf. Proc. Letters* 42, 223–228. Z 769.05085.

COLBOURN, C.J. / VANSTONE, S.A.

1982 Doubly resolvable twofold triple systems, *Congressus Numer.* 34, 219–223. MR 83m:05024; Z 554.05009.

COLBOURN, C.J. / YATES, K.W. / ZHAO, S.

1999 Constructions for an MT-MFSK multiplexing technique for radio communications with bursty traffic.

COLBOURN, C.J. / ZHAO, S.

1999 Maximum Kirkman signal sets for synchronous uni-polar multi-user communication systems.

COLBOURN, M.J.

1979 An analysis technique for Steiner triple systems, *Congressus Numer.* 24, 289–303. MR 83c:05019; Z 426.05010.

1980 *Cyclic Block Designs: Computational Aspects of their Construction and Analysis,* Ph.D. Thesis, Department of Computer Science, University of Toronto.

1985 Algorithmic aspects of combinatorial designs: a survey, *Ann. Discrete Math.* 26, 67–136. MR 87f:05019; Z 597.05010.

COLBOURN, M.J. / COLBOURN, C.J.

1980 Cyclic Steiner systems having multiplier automorphisms, *Utilitas Math.* 17, 127–149. MR 81h:05022; Z 455.05020.

1981a Concerning the complexity of deciding isomorphism of block designs, *Discrete Appl. Math.* 3, 155–162. MR 82f:05031; Z 475.05075.

1981b Cyclic block designs with block size 3, *Europ. J. Combinat.* 2, 21–26. MR 82g: 05018; Z 464.05010.

1981c Some small directed triple systems, *Congressus Numer.* 30, 247–255. MR 83b: 05041; Z 509.05029.

1982b The analysis of directed triple systems by refinement, *Ann. Discrete Math.* 15, 97–103. MR 85m:05011; Z 495.065014.

1984 Recursive constructions for cyclic block designs, *J. Stat. Plann. Infer.* 10, 97–103. MR 86a:62115; Z 553.05018.

COLBOURN, M.J. / COLBOURN, C.J. / ROSENBAUM, W.L.

1982 Trains: An invariant for Steiner triple systems, *Ars Combinat.* 13, 149–162. MR 83j:05013; Z 506.05012.

COLBOURN, M.J. / JOHNSTONE, J.K.

1984 Twofold triple systems with a minimal change property, *Ars Combinat.* 18, 151–160. MR 87e:05025; Z 554.05008.

COLBOURN, M.J. / MATHON, R.A.

1980 On cyclic Steiner 2-designs, in: *Topics on Steiner Systems* (Ed. C.C. Lindner and A. Rosa), *Ann. Discrete Mathematics* 7, 215–253. Z 438.05012.

COLE, F.N.

1913 The triad systems of thirteen letters, *Trans. Amer. Math. Soc.* 14, 1–5.

1922 Kirkman parades, *Bull. Amer. Math. Soc.* 28, 435–437.

COLE, F.N. / CUMMINGS, L.D. / WHITE, H.S.

1917 The complete enumeration of triad systems in 15 elements, *Proc. Nat. Acad. Sci. U.S.A.* 3, 197–199.

CONWELL, H.H.

1910 The three space PG(3,2) and its groups, *Ann. Math.* 11, 60–76.

COOPER, D.S.

1982 Maximal collections of Steiner triple systems of order 9. (Unpublished note.)

CORRADI, K. / HAJNAL, A.

1963† On the maximal number of independent circuits in a graph, *Acta Math. Acad. Sci. Hungar.* 14, 423–439.

COUPLAND, J.

1972 *Steiner Triple Systems*, M.Sc. Thesis, University College of Wales, Aberystwyth.

1975 *On the Construction of Certain Steiner Systems*, Ph.D. Thesis, University College of Wales, Aberystwyth.

COXETER, H.S.M.

1943 The map-coloring of nonorientable surfaces, *Duke Math. J.* 10, 293–304.

CROWE, D.W.

1972 Steiner triple systems, Heawood's torus coloring, Csaszar's polyhedron, Room designs, and bridge tournaments, *Delta* 3, 27–32. MR 47#6505; Z 242.05140.

CRUSE, A.B.

1974† On embedding incomplete symmetric latin squares, *J. Combinat. Theory (A)* 16, 18–22.

CUMMINGS, L.D.

1913 Note on the groups for triple systems, *Bull. Amer. Math. Soc.* 12, 355–356.

1914 *On a Method of Comparison for Triple-Systems*, Ph.D. Thesis, Bryn Mawr College.

1914a On a method of comparison for triple-systems, *Trans. Amer. Math. Soc.* 15, 311–327.

1918 An undervalued Kirkman paper, *Bull. Amer. Math. Soc.* 24, 336–339.

1919 The trains for the 36 groupless triad systems on 15 elements, *Bull. Amer. Math. Soc.* 25, 321–324.

CUMMINGS, L.D. / WHITE, H.S.

1916 Groupless triad systems on fifteen elements, *Bull. Amer. Math. Soc.* 22, 12–16.

CURRAN, D.J. / VANSTONE, S.A.

1989 Doubly resolvable designs from generalized Bhaskar Rao designs, *Discrete Math.* 73, 49–63. MR 89k:05016; Z 677.05011.

DANZIGER, P.

1997 Uniform restricted resolvable designs with $r = 3$, *Ars Combinat.* 46, 161–176. MR 98d:05024.

DANZIGER, P. / GRANNELL, M.J. / GRIGGS, T.S. / ROSA, A.

1998 On the 2-parallel chromatic index of Steiner triple systems, *Australas. J. Combin.* 17, 109–131.

DANZIGER, P. / MENDELSOHN, E. / GRANNELL, M.J. / GRIGGS, T.S.

1996 Five-line configurations in Steiner triple systems, *Utilitas Math.* 49, 153–159. MR 97h:05025; Z 858.05019.

DAS, A. / GUPTA, S.

1997 Optimal block designs for triallel cross experiments, *Commun. Stat. Theory Meth.* 26, 1767–1777. MR 98b:62136.

DAS, P.K.

1989 *Almost Selfcomplementary Graphs and Extensions*, Ph.D. Thesis, McMaster University.

DAS, P.K. / ROSA, A.

1992 Halving Steiner triple systems, *Discrete Math.* 109, 59–67. MR 93h:05019; Z 783.05017.

DAVIES, D.H.

1987 *Automorphisms of Designs*, Ph.D. Thesis, University of East Anglia.

DAVIS, E.W.

1897 A geometric picture of the fifteen school-girl problem, *Ann. Math.* 11, 156–157.

DE BRANDES, M. / PHELPS, K.T. / RÖDL, V.

1982 Coloring Steiner triple systems, *SIAM J. Alg. Discr. Meth.* 3, 241–249. MR 83e:05027; Z 501.05036.

DE BRANDES, M. / RÖDL, V.

1984 Steiner triple systems with small maximal independent sets, *Ars Combinat.* 17, 15–19. MR 85i:05045; Z 548.51010.

DEHERDER, R.

1976 *Recouvrements*, Thèse de doctorat, Université Libre de Bruxelles, 212 pp.

DEHON, M.

1977 Designs et hyperplans, *J. Combinat. Theory (A)* 23, 264–274. MR 57#5781; Z 403.05020.

1979 Planar Steiner triple systems, *J. Geometry* 12, 1–9. MR 80b:05015; Z 369.05006; Z 386.05011.

1980 *Designs et Sous-Designs*, Thése présenteé en vue de l'obtention du grade de Docteur en Sciences, Faculté des Sciences, Université Libre de Bruxelles. vi+187 pp.

1980a Ranks of incidence matrices of t-designs $S_\lambda(t, t+1, v)$, *European J. Comb.* 1, 97–100. MR 81m:05042; Z 447.05017.

1980b Ranks of incidence matrices of $(2,3)$-coverings, *Ars Combinat.* 9, 131–147. MR 82g:05026; Z 447.05016.

1983 On the existence of 2-designs $S_\lambda(2,3,v)$ without repeated blocks, *Discrete Math.* 43, 155–171. MR 84d:05030; Z 502.05008.

DEHON, M. / TEIRLINCK, L.

1987 The geometry of subspaces of an $S(\lambda; 2, 3, v)$, *Ann. Discrete Math.* 34, 137–144. MR 88k:05026; Z 635.05013.

DEJTER, I.J. / FRANEK, F. / ROSA, A.

1996 A completion conjecture for Kirkman triple systems, *Utilitas Math.* 50, 97–102. MR 97i:05013; Z 866.05012.

DEKKER, J.C.E.

1983 Isols and balanced block designs with $\lambda = 1$, *Compositio Math.* 49, 79–93. MR 85e:03107; Z 524.05012.

DELANDTSHEER, A.

1985 Finite 2-homogeneous linear spaces, in: *Finite Geometries* (Ed. C.A. Baker, L.M. Batten), Marcel Dekker, New York, pp. 91–93. MR 87a:51002; Z 583.51010.

DELANDTSHEER, A. / DOYEN, J.

1989† Most block-transitive t-designs are point-primitive, *Geom. Dedicata* 29, 307–310.

DELANDTSHEER, A. / DOYEN, J. / SIEMONS, J. / TAMBURINI, C.

1986 Doubly homogeneous 2-$(v, k, 1)$ designs, *J. Combinat. Theory (A)* 43, 140–145. MR 88b:05023; Z 647.05011.

DE LAUNEY, W.

1992 Signing balanced incomplete block designs over the group of order two, *J. Comb. Math. Comb. Comput.* 11, 73–82. MR 93a:05021; Z 764.05011.

DE LAUNEY, W. / SARVATE, D.G. / SEBERRY, J.

1985 Generalised Bhaskar Rao designs with block size 3 over Z_4, *Ars Combinat.* 19A, 273–285. MR 86j:05030; Z 565.05013.

DEMBOWSKI, P.

1968 *Finite Geometries*, Springer, Berlin. MR 38#1597; Z 137, 401.

DEMETROVICS, J. / FÜREDI, Z. / KATONA, G.O.H.

1985 Minimum matrix representation of closure operations, *Discrete Appl. Math.* 11, 115–128. MR 86k:05002; Z 577.68084.

DÉNES, J. / KEEDWELL, A.D.

1976 *Latin Squares and their Applications*, Academic Press, New York. MR 50 #4338; Z 283.05014.

1991 *Latin Squares. New Developments in the Theory and Applications, Annals Discrete Math.* 46, North-Holland, Amsterdam. MR 92h:05022; Z 754.05017.

DÉNES, J. / MULLEN, G.L. / SUCHOWER, S.J.

1994 A note on power sets of latin squares, *J. Comb. Math. Comb. Comput.* 16, 27–31. MR 95j:05045; Z 819.05017.

DENNISTON, R.H.F.

1974 Double resolvability of some complete 3-designs, *Manuscripta Math.* 12, 105–112. MR 50#1910; Z 276.05022.

1974a Some packings with Steiner triple systems, *Discrete Math.* 9, 213–227. MR 51 #5350; Z 306.05007.

1974b Sylvester's problem of the 15 school-girls, *Discrete Math.* 9, 229–233. MR 51 #5322; Z 285.05002.

1979 Further cases of double resolvability, *J. Combinat. Theory (A)* 26, 298–303. MR 80f:05010; Z 409.05019.

1979a Four doubly resolvable complete 3-designs, *Ars Combinat.* 7, 265–272. MR 81m: 05022; Z 455.05018.

1980 Non-isomorphic reverse Steiner triple systems of order 19, in: *Topics on Steiner systems* (Ed. C.C. Lindner and A. Rosa), *Ann. Discrete Math.* 7, 255–264. Z 457. 05008.

DENNY, P.C.

1998 *Search and Enumeration Techniques for Incidence Structures*, M.Sc. Thesis, Computer Science, University of Auckland.

DENNY, P.C. / GIBBONS, P.B.

1999 Case studies and new results in combinatorial enumeration.

DE PASQUALE, V.

1899 Sui sistemi ternari di 13 elementi, *Rend. R. Ist. Lombardo Sci. Lett.* (2) 32, 213–221.

DE RESMINI, M.J.

1989 On complete arcs in Steiner systems $S(2,3,v)$ and $S(2,4,v)$, *Discrete Math.* 77, 65–73. MR 90j:05045; Z 722.05020.

1991 Some naive constructions of $S(2,3,v)$ and $S(2,4,v)$, *Discrete Math.* 97, 149–153. MR 92m:05017; Z 766.05012.

1992 The fractal-like Steiner triple system, *Ann. Discrete Math.* 52, 119–126. MR 93h:05020; Z 766.51012.

1994 Some partitions of S(2,3,v^2) and S(2,4,v^2), *Australas. J. Combin.* 10, 289–291. MR 95k:05021; Z 819.05013.

DEVILLERS, A. / DOYEN, J.

1998 Homogeneous and ultrahomogeneous linear spaces, *J. Combinat. Theory (A)* 84, 236–241.

DE VRIES, H.L.

1984 Historical notes on Steiner systems, *Discrete Math.* 52, 293–297. MR 86j:05033; Z 551.01005.

1987 Some Steiner quadruple systems S(3,4,16) such that all 16 derived triple systems S(2,3,15) are isomorphic, *Ars Combinat.* 24A, 107–129. MR 89b:05041; Z 664.05010.

DE VRIES, J.

1894 Zur Theorie der Tripelsysteme, *Rend. Circ. Mat. Palermo* 8, 222–226.

DE WERRA, D.

1971† Equitable colorations of graphs, *Rev. Fran. Inf. Rech. Oper.* 5, 3–8.

1975† On a particular conference scheduling problem, *INFOR* 13, 308–315.

DEZA, M.

1980 Finite commmutative Moufang loops, related matroids and association schemes, *Congressus Numer.* 26, 3–15. MR 81m:05043; Z 456.05022.

DEZA, M. / GRISHUKHIN, V.

1998 Once more about 80 Steiner triple systems on 15 points, *J. Stat. Plann. Infer.* 72, 165–184

DEZA, M. / HUANG, T.

1996 ℓ_1-Embeddability of some block graphs and cycloids, *Bull. Inst. Math. Acad. Sinica* 24, 87–102. MR 97c:05039; Z 858.05015.

DEZA, M. / SABIDUSSI, G.

1990 Combinatorial structures arising from commutative Moufang loops, in: *Quasigroups and Loops: Theory and Applications*, Heldermann, Berlin, pp. 151–160. Z 723.20042.

DEZA, M. / SINGHI, N.M.

1980 Some properties of perfect matroid designs, *Ann. Discrete Math.* 6, 57–76. MR 82d:05047; Z 457.05018.

DICKSON, L.E.

1916 Applications of Finite Groups, in: *Theory and Applications of Finite Groups*, Part III (Ed. G.A. Miller, H.F. Blichfeldt, and L.E. Dickson) Wiley, New York.

DI DOMENICO, M.C.

1996 On atypical Steiner systems, *J. Combin. Inf. Syst. Sci.* 21, 71–80.

DIENER, I. / SCHMITT, E. / DE VRIES, H.L.

1985 All 80 Steiner triple systems on 15 elements are derived, *Discrete Math.* 55, 13–19. MR 86h:05028; Z 564.05010.

DINITZ, J.H.

1989 Indecomposable triple systems with $\lambda = 6$, *J. Comb. Math. Comb. Comput.* 5, 139–142. MR 90c:05022; Z 679.05010.

DINITZ, J.H. / RODNEY, P.

1997 Disjoint difference families with block size 3, *Utilitas Math.* 52, 153–160. Z 892.05012.

DINITZ, J.H. / STINSON, D.R.

1992 Room squares and related designs, in: *Contemporary Design Theory: A Collection of Surveys* (Ed. J.H. Dinitz and D.R. Stinson), Wiley, pp. 137–204. Z 768.05015.

DIPAOLA, J.W.

1969 When is a totally symmetric loop a group? *Amer. Math. Monthly* 76, 249–252. MR 39#1583; Z 196, 40

1970 Steiner triples and totally non-symmetric loops, in: *Combinatorial Structures and their Applications*, Gordon and Breach, New York, pp. 59–61. MR 41#8260; Z 263.05018.

1994 New quasigroups from partial triple systems, *Congressus Numer.* 99, 105–109. Z 810.05016.

1996 Large sets of mutually almost disjoint Steiner triple systems, *Congressus Numer.* 114, 173–177.

1996a Steiner triples and difference triples, *Congressus Numer.* 121, 123–128. MR 97m:05029; Z 896.05010.

DIPAOLA, J.W. / GROPP, H.

1989 Hyperbolic graphs from hyperbolic planes, *Congressus Numer.* 68, 23–44. MR 90c:51013; Z 699.05052.

DIPAOLA, J.W. / NEMETH, E.

1976 Generalized triple systems and medial quasigroups, *Congressus Numer.* 17, 289–306. MR 55#2599; Z 359.20072.

1979 Applications of parallelisms, *Ann. N.Y. Acad. Sci.* 319, 153–163. MR 81g:05091; Z 481.05015.

DIRAC, G.A.

1952† Some theorems on abstract graphs, *Proc. London Math. Soc.* 2, 69–81.

DIXON, A.C.

1894 Note on Kirkman's problem, *Messenger Math.* 23, 88–89.

DOEHLERT, D.H.

1965 Balanced sets of incomplete block designs of block size three, *Technometrics* 7, 561–577.

DOMORJAD, A.P.

1961 *Matematatičeskije igry i razvlečenija*, Fizmatgiz, Moskva. Chapter 20: Sostavlenije raspisanij, pp. 109–112. (English translation: *Mathematical Games and Pastimes*, Pergamon, Oxford, 1964. Z 116, 1.). MR 23A#761.

DONOVAN, D.

1984 The quiddity of the isomorphism classes of some balanced ternary designs: including some anomalies, *Ars Combinat.* 17A, 133–144. MR 85f:05016; Z 542.05014.

1990 Single laws for two subvarieties of squags, *Bull. Austral. Math. Soc.* 42, 157–165. MR 91f:20077; Z 697.05017.

1990a The centre of a sloop, *Australas. J. Combin.* 1, 83–90. MR 92h:05019; Z 756.05027.

DONOVAN, D. / OATES-WILLIAMS, S.

1991 Single laws for sloops and squags, *Discrete Math.* 92, 79–83. MR 93g:20131; Z 745.20060.

DONOVAN, D. / RAHILLY, A.

1992 The central spectrum of the order of a Steiner loop, *Southeast Asian Bull. Math.* 16, 115–121. MR 93j:20140; Z 789.20078.

DORFMAN, R.

1943† The detection of defective members of large populations, *Ann. Math. Stat.* 14, 436–440.

DOYEN, J.

1969 Sur la structure de certains systèmes triples de Steiner, *Math. Z.* 111, 289–300. MR 40#53; Z 182, 27.

1970 On the number of non-isomorphic Steiner systems $S(2, in : m, n)$, *Combinatorial Structures and their Applications*, Gordon and Breach, New York, pp. 63–64. MR 43#4695; Z 247.05010.

1970a Problem 30, in: *Combinatorial Structures and their Applications*, Gordon and Breach, New York, p. 504.

1970b Sur la croissance du nombre de systèmes triples de Steiner non isomorphes, *J. Combinat. Theory* 8, 424–441. MR 41#1555; Z 192, 333.

1970c Systèmes triples de Steiner non engendrés par tous leurs triangles, *Math. Z.* 118, 197–206. MR 43#70; Z 201, 533.

1972 A note on reverse Steiner triple systems, *Discrete Math.* 1, 315–319. MR 45 #8542; Z 272.05013.

1972a Constructions of disjoint Steiner triple systems, *Proc. Amer. Math. Soc.* 32, 409–416. MR 45#4990; Z 215, 335.

1973 Recent results on Steiner triple systems, *Finite Geometric Structures and their Applications* (C.I.M.E. II Ciclo Bressanone 1972), Edizioni Cremonese, Roma, 201–210. MR 49#10567; Z 264.05008.

1976 Recent developments in the theory of Steiner systems, *Teorie Combinatorie I*, Colloq. Internaz., Atti dei Convegni Lincei 17, Accademia Nazionale dei Lincei, Roma, pp. 277–285. MR 55#10286; Z 348.05001.

1981 Linear spaces and Steiner systems, in: *Geometries and Groups*, Proc. Colloq. F.U. Berlin, May 1981 (Ed. M. Aigner and D. Jungnickel) Lecture Notes in Math. 893, Springer, New York, pp. 30–42. MR 83f:51014; Z 478.05011.

1989 Designs and automorphism groups, in: *Surveys in Combinatorics 1989* (Ed. J. Siemons), Cambridge University Press, pp. 75–83. MR 91e:05015; Z 722.05015.

DOYEN, J. / HUBAUT, X. / VANDENSAVEL, M.

1978 Ranks of incidence matrices of Steiner triple systems, *Math. Z.* 163, 251–259. MR #80h:05012; Z 373.05011; Z 383.05007.

DOYEN, J. / ROSA, A.

1973 A bibliography and survey of Steiner systems, *Bollet. U.M.I.* (4) 7, 392–419. MR 48#118; Z 273.05001.

1978 An extended bibliography and survey of Steiner systems, *Congressus Numer.* 20, 297–361. MR 80g:51009; Z 473.05001.

1980 An updated bibliography of Steiner systems, in: *Topics on Steiner Systems* (Ed. C.C. Lindner and A. Rosa), *Ann. Discrete Math.* 7, 317–349. Z 442.51006.

DOYEN, J. / VALETTE, G.

1971 On the number of nonisomorphic Steiner triple systems, *Math. Z.* 120, 178–192. MR 43#4695; Z 203, 308; Z 208, 24.

DOYEN, J. / WILSON, R.M.

1973 Embeddings of Steiner triple systems, *Discrete Math.* 5, 229–239. MR 48#5881; Z 263.05017.

DRAKE, D.A. / HO, C.Y.

1988 Projective extensions of Kirkman systems as substructures of projective planes, *J. Combinat. Theory (A)* 48, 197–208. MR 89i:51010; Z 646.51017.

DRAKE, D.A. / LARSON, J.A.

1992 The nonexistence of certain projective planes of order 15, *Experiment. Math.* 1, 65–69. MR 93h:51013; Z 764.51007.

DRIESSEN, L.M.H.E. / FREDERIX, G.H.M. / VAN LINT, J.H.

1976 Linear codes supported by Steiner triple systems, *Ars Combinat.* 1, 33–42. MR 54#7106; Z 337.94006.

DU, D.-Z. / HWANG, F.K.

1993 *Combinatorial Group Testing and its Applications*, World Scientific. MR 97m: 90038; Z 867.90060.

DUCROCQ, P.M. / STERBOUL, F.

1978 On *G*-triple systems, *Publications du Laboratoire de Calcul de l'Université des Sciences et Techniques de Lille*, No. 103, 18 pp.

1980 Les *G*-systèmes triples, *Ann. Discrete Math.* 9, 141–145. MR 83g:05018; Z 459. 05034.

DUDEK, J.

1987 On Steiner quasigroups, *Math. Slovaca* 37, 71–83. MR 88h:20097; Z 612.20042.

DUDEK, J. / GALUSZKA, J.

1995 A characterization of distributive Steiner quasigroups and semilattices, *Discuss. Math. Alg. Stochastic Methods* 15, 101–119. MR 97b:20098; Z 835.08004.

DUDENEY, H.E.

1910 On Kirkman's school-girl problem, *Educational Times Reprints* (2) 14 (1908) 97–99; 15 (1909) 17–19; 17, 35–38, 53.

DUHEM, P.

1902 Notice sur la vie et les travaux de Georges Brunel (1856–1900), *Mém. Soc. Sci. Phys. Nat. Bordeaux* (6) 2, I–XCVII.

DUKES, P. / MENDELSOHN, E.

1999 Skew-orthogonal Steiner triple systems.

ECKENSTEIN, O.

1910 Note on Kirkman's school-girl problem, *Educational Times Reprints* (2) 16 (1909) 76–77, 17, 38–39, 49–53.

1912 Bibliography of Kirkman's school-girl problem, *Messenger Math.* (2) 41, 33–36.

EGORYCHEV, G.P.

1981† The solution of van der Waerden's problem for permanents, *Adv. Math.* 42, 299–305.

EHRMANN, R.C.

1973 Kirkman's school-girls as points in a finite geometry, *Notices Amer. Math. Soc.* 20, A–530.

1975 Projective space walk for Kirkman's school-girls, *Math. Teacher* 68, 64–69.

EMCH, A.

1929 Triple and multiple systems, their geometric configurations and groups, *Trans. Amer. Math. Soc.* 31, 25–42.

1930 Finite groups and their geometric representation, *J. Reine Angew. Math.* 162, 238–255.

1930a Über eine besondere Klasse von algebraischen Flächen, *Comment. Math. Helv.* 2, 99–115.

ENGEL, K.

1980 Über die Anzahl elementarer, teilweise balancierter unvollständiger Blockpläne, *Rostock. Math. Kolloq.* 13, 19–41. MR 82e:05027; Z 464.62072.

1981 Optimalitätsaussagen über Tripelsysteme, *Rostock. Math. Kolloq.* 17, 17–26. MR 84h:05017; Z 482.05028.

1989 On the number of t-(v,k,λ) designs, *Ars Combinat.* 28, 273–277. MR 91b: 05029; Z 703.05011.

ENGEL, K. / GRONAU, H.-D.O.F.

1988 On 2-$(6,3,\lambda)$ designs, *Rostock. Math. Kolloq.* 34, 37–48. MR 90b:05021; Z 656. 05013.

ERDÖS, P.

1964 On extremal problems of graphs and generalized graphs, *Israel J. Math.* 2, 183–190. MR 32#1134; Z 129.39905.

1976 Problems and results in combinatorial analysis, *Creation in Mathematics* 9, 25. MR 57#5764; Z 361.05037.

1981† On the combinatorial problems I would most like to see solved, *Combinatorica* 1, 25–42.

ERDÖS, P. / FRANKL, P. / FÜREDI, Z.

1985† Families of finite sets in which no set is contained in the union of r others, *Israel J. Math.* 51, 70–89.

ERDÖS, P. / HAJNAL, A.

1966 On chromatic number of graphs and set systems, *Acta Math. Acad. Sci. Hung.* 17, 61–99. MR 33#1247.

ESLAMI, Z. / KHOSROVSHAHI, G.B. / TAYFEH-REZAIE, B.

1999 The enumeration of ways of halving the 2-(10,3,8) design.

ETHERINGTON, I.M.H.

1965 Quasigroups and cubic curves, *Proc. Edinburgh Math. Soc.* (2) 14, 273–291. MR 33#4170; Z 133, 275.

ETZION, T.

1992 Optimal partitions for triples, *J. Combinat. Theory (A)* 59, 161–176. MR 93b: 05018; Z 757.05011.

1992a Partitions of triples into optimal packings. *J. Combinat. Theory (A)* 59, 269–284. MR 93h:05048; Z 757.05012.

1994 Large sets of coverings, *J. Combin. Des.* 2, 359–374. MR 95e:05030; Z 833. 05021.

1997 Optimal constant weight codes over Z_k and generalized designs, *Discrete Math.* 169, 55–82. MR 98g:05032.

ETZION, T. / VARDY, A.

1998 On perfect codes and tilings: problems and solutions, *SIAM J. Discrete Math.* 11, 205–223.

EUGENI, F. / GIONFRIDDO, M.

1989 On the minimum number of blocks of a maximal partial spread in $STS(v)$ and $SQS(v)$, *J. Geometry* 36, 37–48. MR 90i:05015; Z 698.05017.

EVANS, T.

1982 Universal-algebraic aspects of combinatorics, in: *Universal Algebra* North-Holland, Amsterdam, pp. 241–266. MR 84d:08004; Z 484.05014.

EVEN, S. / ITAI, A. / SHAMIR, A.

1976† On the complexity of timetable and multicommodity flow problems, *SIAM J. Computing* 5, 691–703.

EXOO, G.

1991 Three color Ramsey number of $K_4 - e$, *Discrete Math.* 89, 301–305. MR 92c: 05110; Z 728.05045.

FALIKMAN, D.I.

1982† Proof of the van der Waerden conjecture on the permanent of a doubly stochastic matrix (in Russian), *Mat. Zametki* 29, 931–938 and 957.

FANO, G.

1892 Sui postulati fondamentali della geometria proiettiva, *Giornale di Matimatiche* 30, 114–124.

FARES, D.A. / ABUL-SHOHOUD, W.H. / RASLAN, N.A. / NASSEF, M.A.

1992 $\delta_{\max}$ detection of multipulse signaling in noisy optical channels, *Microwave Optical Tech. Lett.* 5, 269–273.

FEDERER, W.T.

1972† Construction of classes of experimental designs using transversals in Latin squares and Hedayat's sum composition method, in: *Statistical Papers in Honor of G.W. Snedecor* (Ed. T.A. Bancroft), pp. 91–114.

FEO, T.A. / RESENDE, M.G.C.

1989 A probabilistic heuristic for a computationally difficult set covering problem, *Oper. Res. Lett.* 8, 67–71. MR 90c:90168; Z 675.90073.

FERRERO, G.

1972 Gruppi di Steiner e sistemi fini, *Matematiche* 27, 167–190. MR 48#1944; Z 264. 05011.

1972a Sui gruppi che amettono funzioni di Steiner, Rend. *Ist. Mat. Univ. Trieste* 4, 156–170. MR 48#5883; Z 62.20045.

1972b Deformazioni, raffinamenti e composizioni di funzioni di Steiner (I), *Riv. Mat. Univ. Parma* (3) 1, 125–142. MR 51#2937; Z 344.05010.

1975 Su un problema relativo ai sistemi di Steiner disgiunti, *Rend. Ist. Mat. Univ. Trieste* 7, 58–64. MR 52#13424; Z 319.05009.

1977 Deformazioni, raffinamenti e composizioni di Steiner (II), *Riv. Mat. Univ. Parma* (4) 3, 151–157. MR 58#5268; Z 395.51009.

1978 Automorfismi di neocorpi di Steiner, *Sistemi Binari e loro Applicazioni*, Conv. Taormina/Messina, 35–45. Z 572.05016.

FERRERO, G. / GALLINA, G.

1978 Funzioni di Steiner con parecchi moltiplicatori, *Riv. Mat. Univ. Parma* (4) 4, 475–484. MR 82k:20065; Z 405.05011.

FERRERO, G. / SUPPA, A.

1971 Sistemi, annelloidi e funzioni di Steiner, *Atti Sem. Mat. Fiz. Univ. Modena* 20, 272–280. Z 249.05001.

FISCHER, B.

1967 Eine Kennzeichnung der symmetrischen Gruppen vom Grade 6 und 7, *Math. Z.* 95, 288–298. MR 34#5909; Z 139, 251.

FISHER, R.A.

1925† *Statistical Methods for Research Workers*, Oliver and Boyd, Edinburgh.

1940 An examination of the different possible solutions of a problem in incomplete blocks, *Ann. Eugenics* 10, 52–75. MR 1, 348.

FISHER, R.A. / YATES, F.

1938 *Statistical Tables for Biological, Agricultural and Medical Research*, Oliver and Boyd, Edinburgh, 5th ed. 1957, 25–29, 90–93. Z 79, 352. 6th ed., Hafner, New York, 1963.

FISKE, T.S.

1927 Frank Nelson Cole, *Bull. Amer. Math. Soc.* 33, 773–777.

FITTING, F.

1911 Die Herstellung von Tripelsystemen für jede Anzahl von Elementen, *Nieuw Arch. Wisk.* (2) 9, 359–369.

1913 Aufstellung aller voneinander unabhängigen Lösungen des Kirkman'schen 15-Pensionatsdamen-Problems, *Nieuw Arch. Wisk.* (2) 10, 244–251.

FOLKMAN, J. / FULKERSON, D.R.

1969† Edge colorings in bipartite graphs, in: *Combinatorial Mathematics and its Applications*, UNC Press, Chapel Hill, NC, pp. 561–577.

FOODY, W.M.

1980 *Properties, Construction and Applications of BIB Designs with Repeated Blocks*, Ph.D. Thesis, University of Illinois.

FOODY, W. / HEDAYAT, A.

1977 On theory and applications of BIB designs with repeated blocks, *Ann. Statist.* 5, 932–945. MR 57#1788. Corrigendum: *Ann. Statist.* 7 (1979) 925. MR 80k:62091.

1988 A graphical proof of the nonexistence of BIB$(7, b, r, 3, \lambda|16)$ designs, *J. Stat. Plann. Infer.* 20, 77–90. MR 89k:62097; Z 662.05011.

FORT, M.K. / HEDLUND, G.A.

1958 Minimal coverings of pairs by triples, *Pacific J. Math.* 8, 709–719. MR 21 #2595; Z 84, 14.

FRANCEL, M. / JOHN, D.

1998 Counting balanced ternary designs, *Ars Combinat.* 49, 161–184.

FRANCEL, M.A. / SARVATE, D.G.

1996 Ternary designs with the maximum or near maximum number of replication numbers, *Australas. J. Combin.* 13, 283-294. Z 852.05021.

1998 One and two-block configurations in balanced ternary designs, *Ars Combinat.* 49, 217–224.

FRANEK, F.

1987 Isomorphisms of infinite Steiner triple systems II, Techn. Report, Dept. of Comp. Sci. and Systems, McMaster University, 19 pp.

1994 Isomorphisms of infinite Steiner triple systems, *Ars Combinat.* 38, 7–25. MR 95m:05036; Z 816.05014.

1999 On cliques in spanning graphs of projective Steiner triple systems.

FRANEK, F. / MATHON, R. / ROSA, A.

1989 Maximal partial triple systems with hexagonal leave, *J. Comb. Math. Comb. Comput.* 5, 105–118. MR 90c:05023; Z 749.05017.

1991 On a class of linear spaces with 16 points, *Ars Combinat.* 31, 97–104. MR 92d: 05016; Z 767.05017.

1992 Maximal partial triple systems of order 13 with a quadratic leave: the case of 9 edges, *J. Comb. Math. Comb. Comput.* 11, 55–60. MR 93b:05019; Z 767.05027.

1995 Maximal sets of triangle-factors in K_{15}, *J. Comb. Math. Comb. Comput.* 17, 111–124. MR 95j:05059; Z 831.05050.

FRANEK, F. / ROSA, A. / GRIGGS, T.S.

1997 Large sets of mutually almost disjoint Steiner triple systems not from Steiner quadruple systems, *Des. Codes Crypt.* 12, 59–67. MR 98h:05030; Z 891.05010.

FRANKL, P. / FÜREDI, Z.

1984 A new extremal property of Steiner triple systems, *Discrete Math.* 48, 205–212. MR 85d:05050; Z 553.05019.

FRASCELLA, W.J.

1965 A generalization of Sierpinski's theorem on Steiner triples and the axiom of choice, *Notre Dame J. Formal Logic* 6, 163–179. MR 33#2552; Z 268.04007. Corrigendum and addendum: *Notre Dame J. Formal Logic* 6, 323–324. MR 35#2750; Z 268.04008.

1966 The construction of a Steiner triple system on sets of the power of the continuum without the axiom of choice, *Notre Dame J. Formal Logic* 7, 196–202. MR 35 #2751; Z 147, 262.

1973 A stronger theorem concerning the non-existence of combinatorial designs on infinite sets, *Notre Dame J. Formal Logic* 14, 554–558. MR 49#47; Z 265.04004.

FRENZ, T.C.

1989 *Computing Techniques for the Enumeration of Cyclic Steiner Systems*, M.Sc. Thesis, Rochester Institute of Technology, 72 pp.

FRENZ, T.C. / KREHER, D.L.

1992 An algorithm for enumerating distinct cyclic Steiner systems, *J. Comb. Math. Comb. Comput.* 11, 13–22. MR 93a:05040; Z 755.05009.

FRIEDLAND, S.

1979† A lower bound for the permanent of a doubly stochastic matrix, *Ann. Math.* 110, 167–176.

FROBENIUS, G.

1917† Über zerlegbare Determinanten, *Sitzungsber. König. Preuss. Akad. Wiss.* 27, 274–277.

FROST, A.H.

1871 General solution and extension of the problem of the fifteen school-girls, *Quart. J. Pure. Appl. Math.* 11, 26–37.

FU, H.L.

1979 *A Heterogeneous Quadruple System of Order 28*, M.Sc. Thesis, Auburn University, 24 pp.

1980† *On the Construction of Certain Types of Latin Squares having Prescribed Intersections*, Ph.D. Thesis, Auburn University.

1983 Directed triple systems having a prescribed number of triples in common, *Tamkang J. Math.* 14, 85–90. MR 85g:05048; Z 515.05020.

1984 A heterogeneous quadruple system of order $14 \cdot 2^k$, *Tamkang J. Math.* 15, 103–114. MR 87g:05060; Z 556.005011.

FU, H.L. / LINDNER, C.C. / RODGER, C.A.

1997 The Doyen–Wilson theorem for minimum coverings with triples, *J. Combin. Des.* 5, 341–352. MR 99a:05013.

1998 Two Doyen–Wilson theorems for maximum packings with triples, *Discrete Math.* 178, 63–71.

FU, H.L. / RODGER, C.A.

1998 Group divisible designs with two associate classes: $n = 2$ or $m = 2$, *J. Combinat. Theory (A)* 83, 94–117.

FU, H.L. / RODGER, C.A. / SARVATE, D.G.

1999 The existence of group divisible designs with first and second associates having block size 3, *Ars Combinat.*

FUGÈRE, J. / HADDAD, L. / WEHLAU, D.

1994 5-chromatic Steiner triple systems, *J. Combin. Des.* 2, 287–299. MR 95d:05022; Z 829.05011.

FUJI-HARA, R.

1984 Baer subplanes in finite projective plane and Kirkman cubes, *Enumeration and Design* (Ed. D.M. Jackson and S.A. Vanstone), Academic Press, 141–156. MR 86g: 51024; Z 558.51010.

FUJI-HARA, R. / VANSTONE, S.A.

1980 On the spectrum of doubly resolvable Kirkman systems, *Congressus Numer.* 28, 399–407. MR 82j:05039; Z 456.05020.

1982 On a line partitioning problem for PG$(2k, q)$, *Rend. Sem. Mat. Brescia* 7, 337–341. MR 85h:51019; Z 545.51009.

FULKERSON, D.R. / NEMHAUSER, G.L. / TROTTER, L.E. JR.

1974 Two computationally difficult set covering problems that arise in computing the 1-width of incidence matrices of Steiner triple systems, *Math. Prog. Study* 2, 72–81. Z 353.90060.

FÜREDI, Z.

1991 Maximal independent subsets in Steiner systems and in planar sets, *SIAM J. Discrete Math.* 4, 196–199. MR 92c:52006; Z 741.51006.

FURINO, S.C.

1990 *α-Resolvable Structures*, Ph.D. Thesis, University of Waterloo.

FURINO, S.C. / MIAO, Y. / YIN, J.X.

1996 *Frames and Resolvable Designs*, CRC Press, Boca Raton, FL. MR 97d:05030; Z 855.62061.

FURINO, S.C. / MULLIN, R.C.

1993† Block designs with large holes and α resolvable BIBDs, *J. Combin. Des.* 1, 101–112.

GALLANT, R.P. / JIANG, Z. / LING, A.C.H.

1999 The spectrum of cyclic group divisible designs with block size three, *J. Combin. Des.*

GALLINA, G.

1977 Su certi sistemi di Steiner non disgiunti, *Riv. Mat. Univ. Parma* (4) 3, 199–202. MR 58#277; Z 391.05007.

1978 Gruppi ammettenti funzioni di Steiner, *Bollet. U.M.I.* (5) 15, 225–230. MR 58 #5269; Z 387.05001.

1978a Sull'esistenza di certe funzioni di Steiner, *Riv. Mat. Univ. Parma* (4) 4, 215–219. MR 82k:20066; Z 406.51008.

GANTER, B.

1976† *Combinatorial Designs and Algebras*, TH Darmstadt, Preprint No.270.

GANTER, B. / GRONAU, H.-D.O.F.

1991 On two conjectures of Demetrovics, Füredi and Katona concerning partitions, *Discrete Math.* 88, 149–155. MR 92f:05013; Z 739.05004.

GANTER, B. / GÜLZOW, A. / MATHON, R. / ROSA, A.

1978 A complete census of (10,3,2) designs and of Mendelsohn triple systems of order ten. IV. (10,3,2) designs with repeated blocks, *Kasseler Math. Schriften* 5/78, 41 pp.

GANTER, B. / MATHON, R. / ROSA, A.

1978 A complete census of (10,3,2)-block designs and of Mendelsohn triple systems of order ten. I. Mendelsohn triple systems without repeated blocks, *Congressus Numer.* 20, 383–398. MR 81b:05020a; Z 483.05016.

1978a A complete census of (10,3,2) designs and of Mendelsohn triple systems of order ten. II. Mendelsohn triple systems with repeated blocks, *Congressus Numer.* 22, 181–204. MR 81b:05020b; Z 434.05010.

GANTER, B. / METZ, R.

1977 Kombinatorische Algebra: Koordinatisierung von Blockplänen, in: *Beiträge zur geometrischen Algebra* Lehrbücher u. Monographien Exakt. Wiss. Math. Reihe, Bd. 22, Birkhäuser, Basel, pp. 111–124. MR 58#21674; Z 398.05018.

GANTER, B. / WERNER, H.

1975 Equational classes of Steiner systems, *Algebra Univ.* 5, 125–140. MR 53#7908a; Z 312.08002.

1980 Co-ordinating Steiner systems, in: *Topics on Steiner Systems* (Ed. C.C. Lindner and A. Rosa), *Ann. Discrete Math.* 7, 3–24. Z 437.51007.

GARDNER, M.

1980 Mathematical games. What unifies dinner guests, strolling school-girls and handcuffed prisoners? *Scientific American* 242, 16–22.

GARDNER, R.B.

1987 *Automorphisms of Steiner Triple Systems*, M.S. Thesis, Auburn University.

1992 Steiner triple systems with near-rotational automorphisms, *J. Combinat. Theory (A)* 61, 322–327. MR 93g:05021; Z 778.05021.

1992a Transrotational directed triple systems, *Congressus Numer.* 90, 97–102. MR 93m:05030; Z 795.05029.

1993 A note on a class of Steiner triple systems, *Ars Combinat.* 36, 157–160. MR 94f: 05020; Z 793.05019.

1994 Steiner triple systems with transrotational automorphisms, *Discrete Math.* 131, 99–104. MR 95f:05017; Z 805.05009.

1994a Near-rotational directed triple systems, *Ars Combinat.* 38, 137–143. MR 95m: 05037; Z 816.05015.

1998 Bicyclic directed triple systems, *Ars Combinat.* 49, 249–257.

GARDNER, R.B. / MICALE, B. / PENNISI, M. / ZIJLSTRA, R.

1997 Cyclic and rotational hybrid triple systems, *Discrete Math.* 171, 121–139. MR 98f:05027; repeated in *Discrete Math.* 175, 143–161. MR 98j.05028; Z 879.05010; Z 887.05009.

GAREY, M.R. / JOHNSON, D.S.

1979† *Computers and Intractability*, Freeman and Sons.

GE, G.

1999 Generalized Steiner triple systems with group size eleven.

GE, G. / WU, D.

1999 Generalized Steiner triple systems with group size ten.

GEISER, C.F.

1869 Über die Doppeltangenten einer ebenen Curve vierten Grades, *Math. Ann.* 1, 129–138.

GENMA, M. / MISHIMA, M. / JIMBO, M.

1997 Cyclic resolvability of cyclic Steiner 2-designs, *J. Combin. Des.* 5, 177–187.

GIBBONS, P.B.

1976 *Computing Techniques for the Construction and Analysis of Block Designs*, Ph.D. Thesis, University of Toronto. (Department of Computer Science, University of Toronto, Technical Report No. 92).

1985 A census of orthogonal Steiner triple systems of order 15, *Ann. Discrete Math.* 26, 165–182. MR 87g:05041; Z 582.05011.

1996 Computational methods in design theory, in: *CRC Handbook of Combinatorial Designs* (Ed. C.J. Colbourn and J.H. Dinitz), CRC Press, Boca Raton, FL, pp. 718–740.

GIBBONS, P.B. / MATHON, R.

1987 Construction methods for Bhaskar Rao and related designs, *J. Austral. Math. Soc.* (A) 42, 5–30. MR 87j:05028; Z 615.05013.

1993 The use of hill-climbing to construct orthogonal Steiner triple systems, *J. Combin. Des.* 1, 27–50. MR 95i:05019; Z 817.05014.

GIBBONS, P.B. / MATHON, R. / CORNEIL, D.G.

1977 Computing techniques for the construction and analysis of block designs, *Utilitas Math.* 11, 161–192. MR 58#10491; Z 352.05009.

GIBBONS, P.B. / MENDELSOHN, E.

1996 The construction of antipodal triple systems by simulated annealing, *Discrete Math.* 155, 59–76. MR 97d:05035; Z 864.05013.

GIBBONS, P.B. / MENDELSOHN, E. / SHEN, H.

1994 Antipodal triple systems, *Australas. J. Combin.* 9, 137–151. MR 95b:05031; Z 807.05010.

GIONFRIDDO, M.

1986 Sui sistemi di Steiner, *Quad. Sem. Geom. Comb. Fac. Ing. L'Aquila* No. 9, 17 pp.

1987 On the number of pairwise disjoint blocks in a Steiner system, *Ann. Discrete Math.* 34, 189–195. MR 89b:05031; Z 644.05011.

1989 Partial parallel classes in Steiner systems, *Colloq. Math.* 57, 221–227. MR 90j:05035; Z 702.05026.

GIONFRIDDO, M. / LO FARO, G.

1989 Resolvable S_2(2,3,12)s and families of S_Λ (3,6,v)s, *J. Combin. Inform. System Sci.* 14, 25–32. MR 91j:05015; Z 705.05007.

GIONFRIDDO, M. / MILICI, S.

1996 A result concerning two conjectures of Berge and Chvatal, *Discrete Math.* 155, 77–79. MR 97d:05036; Z 858.05082.

GIONFRIDDO, M. / MILICI, S. / VACIRCA, V.

1984 On disjoint partial triple systems, *Rend. Circ. Mat. Palermo* 33, 170–184. MR 86d:05019; Z 562.05016.

GIONFRIDDO, M. / TUZA, ZS.

1994 On conjectures of Berge and Chvátal, *Discrete Math.* 124, 79–86. MR 94m:05080; Z 789.05073.

GIZATULLIN, M.K.

1996 Generalizations of Cole's systems, *Serdica Math. J.* 22, 45–56. MR 97b:05024; Z 858.05017.

GOUDARZI, M.G.

1967 Sur le problème de Kirkman, *Acad. Roy. Belg. Bull. Cl. Sci.* 53, 183–199. MR 35 #6019; Z 146, 417.

1967a Sur le problème de Kirkman généralisé, *Acad. Roy. Belg. Bull. Cl. Sci.* 53, 1379–1384. MR 38#6444.

GOULDEN, I.P. / PIETSCH, CH.

1999 A generating function for the number of 2-(6,3,2n) block designs.

GOWER, R.A.H.

1991 Nonisomorphic 2-perfect 6-cycle systems of order 13. *Bull. Austral. Math. Soc.* 44, 381–385. MR 92k:05035; Z 721.05047.

1993 Minimal defining sets in a family of Steiner triple systems, *Australas. J. Combin.* 8, 55–73. MR 94j:05025; Z 790.05014.

1995 *Critical Sets in Latin Squares and their Applications to Minimal Defining Sets of Designs*, Ph.D. Thesis, University of Queensland. (Abstract in *Bull. Austral. Math. Soc.* 53 (1996), 527–528. Z 879.05014.)

1997 Defining sets for the Steiner triple systems from affine spaces, *J. Combin. Des.* 5, 155–175. MR 98h:51007.

GRABLE, D.A.

1997 On random greedy triangle packing, *Electron. J. Combinat.* 4, #R11. MR 98a:05031; Z 885.05051.

1999 More-than-nearly-perfect packings and partial designs.

GRABLE, D.A. / PHELPS, K.T.

1996 Random methods in design theory: a survey, *J. Combin. Des.* 4, 255–273. MR 97d:05031.

GRABLE, D.A. / PHELPS, K.T. / RÖDL, V.

1995 The minimum independence number for designs, *Combinatorica* 15, 175–185. MR 96f:05026; Z 824.05005.

GRAHAM, R.L. / LI, S.-Y.R. / LI, W.-C.W.

1980† On the structure of t-designs, *SIAM J. Alg. Disc. Meth.* 1, 8–14.

GRANNELL, M.J. / GRIGGS, T.S.

1985 Derived Steiner triple systems of order 15, *Ann. Discrete Math.* 26, 183–208. MR 87j:05037; Z 582:05012.

1986 Product constructions for cyclic block designs, II. Steiner 2-designs, *J. Combinat. Theory (A)* 42, 179–183. MR 87j:05029; Z 592.05009.

1988 Some applications of computers in design theory, in: *Computers in Math. Research*, Oxford University Press, New York, 135–148. MR 89m:05016; Z 652.05014.

1999 Configurations in Steiner triple systems.

GRANNELL, M.J. / GRIGGS, T.S. / MENDELSOHN, E.

1994 A small basis for four-line configurations in Steiner triple systems, *J. Combin. Des.* 3, 51–59. MR 95j:05041; Z 820.05010.

GRANNELL, M.J. / GRIGGS, T.S. / MURPHY, J.P.

1992 Equivalence classes of Steiner triple systems, *Congressus Numer.* 86, 19–25. MR 93e:05011; Z 773.05023.

1997 Twin Steiner triple systems, *Discrete Math.* 167/168, 341–352. MR 97m:05030; Z 869.05013.

1999 Some new perfect Steiner triple systems.

1999a Switching cycles in Steiner triple systems, *Utilitas Math.*

GRANNELL, M.J. / GRIGGS, T.S. / PHELAN, J.S.

1987 Countably infinite Steiner triple systems, *Ars Combinat.* 24B, 189–216. MR 89c:05018; Z 724.05009.

1988 A new look at an old construction for Steiner triple systems, *Ars Combinat.* 25A, 55–60. MR 89d:05036; Z 656.05017.

1995 Continuity of Mendelsohn and Steiner triple systems, *Discrete Math.* 138, 247–253. MR 95m:05038; Z 820.05011.

GRANNELL, M.J. / GRIGGS, T.S. / ŠIRÁŇ, J.

1998 Surface embeddings of Steiner triple systems, *J. Combin. Des.* 6, 325–336.

GRANNELL, M.J. / GRIGGS, T.S. / WALLACE, J.

1999 The smallest defining set of a Steiner triple system, *Utilitas Math.*

GRAVER, J.E. / JURKAT, W.B.

1973† The modular structure of integral designs, *J. Combinat. Theory (A)* 15, 75–90.

GRAVES, L.M.

1962 A finite Bolyai–Lobachevsky plane, *Amer. Math. Monthly* 69, 130–132. MR 26 #4252; Z 106, 143.

GRAY, B.D.

1997 Smallest defining sets of designs associated with PG(d,2), *Australas. J. Combin.* 16, 87–98. MR 98f:05022; Z 883.05018.

GRAY, B.D. / RAMSAY, C.

1998 On the number of Pasch configurations in a Steiner triple system, *Bull. Inst. Combin. Appl.* 24, 105–112.

GRAY, K.

1988 On the resolvability of B[3,3;9] block designs containing no repeated blocks, *J. Comb. Math. Comb. Comput.* 3, 105–120. MR 89f:05030; Z 658.05012.

1990 On the minimum number of blocks defining a design, *Bull. Austral. Math. Soc.* (A) 41, 97–112. MR 91e:05016; Z 679.05006.

GREENHILL, C.

1993 An algorithm for finding the smallest defining sets for t-designs, *J. Comb. Math. Comb. Comput.* 14, 39–60. MR 94e:05039; Z 814.05013.

GREIG, M.

1999† Designs from projective planes and PBD bases, *J. Combin. Des.*

GRIDGEMAN, N.T.

1973 Kirkman, Thomas Penyngton, *Dictionary of Scientific Biography VII*, Charles Scribner, New York, 383–384.

GRIGGS, T.S. / DE RESMINI, M.J. / ROSA, A.

1992 Decomposing Steiner triple systems into four-line configurations, *Ann. Discrete Math.* 52, 215–226. MR 93m:05031; Z 769.05021.

GRIGGS, T.S. / MENDELSOHN, E. / ROSA, A.

1994 Simultaneous decompositions of Steiner triple systems, *Ars Combinat.* 37, 157–173. MR 95d:05023; Z 807.05009.

GRIGGS, T.S. / MURPHY, J.P.

1993 101 Anti-Pasch Steiner triple systems of order 19, *J. Comb. Math. Comb. Comput.* 13, 129–142. MR 94a:05021; Z 778.05022.

GRIGGS, T.S. / MURPHY, J.P. / PHELAN, J.S.

1990 Anti-Pasch Steiner triple systems. *J. Comb. Inf. Syst. Sci.* 15, 79–84. MR 92g: 05040; Z 741.05009.

GRIGGS, T.S. / ROSA, A.

1994 Avoidance in triple systems, *Acta Math. Univ. Comen.* 63, 117–131. MR 96f: 05033; Z 830.05010.

1995 Large sets of Steiner triple systems, in: *Surveys in Combinatorics 1995* (Ed. P. Rowlinson), London Math. Soc. Lecture Note Ser. 218, Cambridge University Press, pp. 25–39. MR 96j:05026; Z 831.05014.

1996 An invariant for one-factorizations of the complete graph, *Ars Combinat.* 42, 177–188. MR 97h:05135; Z 856.05078.

GRINBERG, E.J. / IKAUNIECE, B.A.

1975 O postrojenii polnykh sistem trojek Štejnera, *Voprosy Kibernetiki* 16, 21–29.

GRISHUKHIN, V.

1997 Regular two-graphs from the even unimodular lattice $E_8 \oplus E_8$, *Europ. J. Combinat.* 18, 391–407. MR 98f:05152; Z 879.05019.

GRONAU, H.-D.O.F.

1985 A survey of results on the number of t-(v,k,λ)-designs, *Ann. Discrete Math.* 26, 209–220. MR 87g:05031; Z 585.05001.

GRONAU, H.-D.O.F. / MULLIN, R.C. / PIETSCH, CH.

1995 The closure of all subsets of $\{3,4,\cdots,10\}$ which include 3, *Ars Combinat.* 41, 129–161. MR 96m:05023; Z 844.05018.

1996 Pairwise balanced designs as linear spaces, in: *CRC Handbook of Combinatorial Designs* (Ed. C.J. Colbourn and J.H. Dinitz), CRC Press, Boca Raton, FL, pp. 213–220. Z 845.05010.

GRONAU, H.-D.O.F. / PRESTIN, J.

1982 Some results on designs with repeated blocks, *Rostock Math. Kolloq.* 21, 15–38. MR 85h:05018; Z 521.05021.

GRONAU, H.-D.O.F. / RENTNER, I.

1985 On the decomposition of the set of all k-element subsets of a v-element set into indecomposable t-(v,k,λ) designs, *Rostock. Math. Kolloq.* 28, 49–54. MR 87g:05032; Z 599.05009.

GROPP, H.

1987 'Il Metodo di Martinetti' or Configurations and Steiner systems S(2,4,25), *Ars Combinat.* 24B, 179–188. MR 89c:05022; Z 652.05008.

1990 Configurations and the Tutte conjecture, *Ars Combinat.* 29A, 171–177. Z 739. 05016.

1991 The history of Steiner systems S(2,3,13), *Mitt. Math. Ges. Hamburg* 12, 849–861. MR 93c:05002; Z 767.05028.

1991a Blocking sets in configurations n_3, *Mitt. Math. Sem. Giessen* 201, 59–72. MR 92h:51015; Z 742.51009.

1992 Non-symmetric configurations with deficiencies 1 and 2, *Ann. Discrete Math.* 52, 227–239. MR 93k:05037; Z 767.05034.

1994 Nonsymmetric configurations with natural index, *Discrete Math.* 124, 87–98. MR95f:05021; Z 791.05012.

1996 Configurations, in: *CRC Handbook of Combinatorial Designs* (Ed. C.J. Colbourn and J.H. Dinitz), CRC Press, Boca Raton, FL, pp. 253–256. Z 864.05024.

GROSS, K.B.

1975 On the maximal number of orthogonal Steiner triple systems, *Congressus Numer.* 13, 233–240. MR 52#7923; Z 326.05011.

1975a On the maximal number of pairwise orthogonal Steiner triple systems, *J. Combinat. Theory (A)* 19, 256–263. MR 52#2917; Z 334.05016.

GRÜTTMÜLLER, M.

1996 On the number of indecomposable block designs, *Australas. J. Combin.* 14, 181–186. MR 97g:05022; Z 863.05015.

1999 On the PBD-closure of sets containing 3.

1999a On the existence of mandatory representation designs with block sizes 3 and k.

GUELZOW, A.

1990 *Some Classes of E-Minimal Algebras of Affine Type: Nilpotent Squags, p-Groups and Nilpotent SQS-Skeins*, Ph.D. Thesis, University of Manitoba.

1996 Representation of finite nilpotent squags, *Discrete Math.* 154, 63–76. MR 97d: 05037; Z 853.20048.

GUO, H.T.

1999 Intersections of simple triple systems.

GUSTAVSSON, T.

1991 *Decompositions of Large Graphs and Digraphs with High Minimum Degree*, Doctoral Thesis, University of Stockholm, 100 pp.

GUY, R.K.

1969 A many-faceted problem of Zarankiewicz, *The Many Facets of Graph Theory*, Springer, Berlin, pp. 129–148. MR 41#91; Z 186, 18.

1976 Packing [1,n] with solutions of ax+by=cz: the unity of combinatorics, *Teorie Combinatorie II*, Colloq. Internaz., Atti dei Conveggni Lincei, Roma 1973, Accad. Naz. Lincei, pp. 173–179. MR 57#9565; Z 361.05012.

1995 The unity of combinatorics, in: *Combinatorics Advances* (Ed. C.J. Colbourn and E. Mahmoodian), Kluwer Academic, Dordrecht, pp. 129–159. MR 96k:05001; Z 836.05001.

GYÁRFÁS, A. / SCHELP, R.H.

1998 A communication problem and directed triple systems, *Discrete Appl. Math.* 85, 139–147.

HADDAD, L.

1999 On the chromatic numbers of Steiner triple systems, *J. Combin. Des.* 7, 1–10.

1999a Colouring affine and projective geometries.

HADDAD, L. / RÖDL, V.

1994 Unbalanced Steiner triple systems, *J. Combinat. Theory (A)* 66, 1–16. MR 95b:05032; Z 806.05015.

HADDAD, L. / ZHOU, H.

1995 Star chromatic numbers of hypergraphs and partial Steiner triple systems, *Discrete Math.* 146, 45–58. MR 97b:05056; Z 845.05016.

HAGGARD, G.

1972 On the function $N(3,2,v)$, in: *Proc. 3rd Southeastern Conf. Combinatorics. Graph Theory and Computing*, Boca Raton, pp. 243–250. MR 50#121; Z 265.05005.

HALL, J.I.

1972 *Steiner Triple Systems and 2-Transitive Groups*, M.Sc. Thesis, Oxford University.

1974 Steiner triple systems with geometric minimally generated subsystems, *Quart. J. Math. Oxford Ser.* (2) 25, 41–50. MR 49#10569; Z 279.05017.

1980 On identifying $PG(3,2)$ and the complete 3-design on seven points, in: *Topics on Steiner Systems* (Ed. C.C. Lindner and A. Rosa), *Ann. Discrete Math.* 7, 131–141. Z 443.05022.

1980a On the order of Hall triple systems, *J. Combinat. Theory (A)* 29, 261–262. MR 81h:05024; Z 445.05023.

HALL, J.I. / UDDING, J.T.

1976 On pairs of Steiner triple systems intersecting in subsystems, Technological University Eindhoven Report 76-WSK-04, 22 pp. Z 351.05010.

1977 On intersection of pairs of Steiner triple systems, *Indag. Math.* 39. (*Proc. Koninkl. Nederl. Akad. Wetensch. Ser.* (A) 80, 87–100.) MR 56#5317; Z 354.05027.

HALL, M. JR.

1958 A survey of combinatorial analysis, *Some Aspects of Probability*, Wiley, New York, pp. 35–104. MR 22#2556; Z 86, 12. (Russian translation: *Kombinatornyj analiz*, Izdat. inostr. lit., Moskva, 1963. Z 192, 89.)

1960 Automorphisms of Steiner triple systems, *IBM J. Res. Develop.* 4, 460–472. MR 23A#1282; Z 100, 18.

1960a Current studies on combinatorial designs, in: *Proc. Symp. Appl. Math.* 10, Amer. Math. Soc., Providence, RI, pp. 1–14. MR 25#4020; Z 96, 122.

1962 Automorphisms of Steiner triple systems, in: *Proc. Symp. Pure Math.* 6, Amer. Math. Soc., Providence, RI, pp. 47–66. MR 24A#2890; Z 114, 12.

1964 Block designs, in: *Applied Combinatorial Mathematics* (Ed. E.F. Beckenbach), Wiley, New York, pp. 369–405. Z 158, 16.

1967 *Combinatorial Theory*, Blaisdell, Waltham MA. MR 37#80; Z 196, 24. 2nd ed., Wiley, New York, 1986. MR 87j:05001; Z 323.3011. (Russian translation: *Kombinatorika*, Mir. Moskva, 1970. MR 42#4411.)

1967a Group theory and block designs, in: *Proc. Internat. Conf. Theory of Groups* (Canberra 1965), Gordon and Breach, New York, 115–144. MR 36#2514; Z 323. 2011. (Russian translation: Teorija grupp i blok-schemy, *Kombinatornyj analiz* 1 (1971) 10–34. Z 285.20006.)

1976 Construction of combinatorial designs, in: *Teorie Combinatorie I*, Colloq. Internaz., Atti dei Convegni Lincei 17, Accad. Naz. dei Lincei, Roma, pp. 13–41. MR 55 #12529; Z 385.05002.

1980 Group problems arising from combinatorics, *Finite Groups*, Proc. Sympos. Pure Math. 37, 445–456. MR 82a:20005; Z 449.20005.

1985 Steiner triple systems with a doubly transitive automorphism group, *J. Combinat. Theory (A)* 38, 192–202. MR 86h:05029; Z 579.05017.

HALL, M. JR. / SWIFT, J.D.

1955 Determination of Steiner triple systems of order 15, *Math. Tables Aids Comput.* 9, 146–152. MR 18, 192; Z 68.24902..

HALL, P.

1935† On representatives of subsets, *J. London Math. Soc.* 10, 26–30.

HAMADA, N.

1981 The geometric structure and the *p*-rank of an affine triple system derived from a nonassociative Moufang loop with the maximum associative center, *J. Combinat. Theory (A)* 30, 285–297. MR 82k:05021; Z 504.05017.

HAMM, R.C.

1983 Embedding partial transitive triple systems, *Congressus Numer.* 39, 447–453. MR 85m:05019; Z 535.05017.

1983a *Embedding Theorems for Triple Systems*, Ph.D. Thesis, Auburn University.

1984 Intersection preserving embeddings of partial Mendelsohn triple systems, *Congressus Numer.* 44, 117–125. MR 86f:05023; Z 575.05010.

HAMM, R.C. / LINDNER, C.C. / RODGER, C.A.

1983 Linear embeddings of partial directed triple systems with $\lambda = 1$ and partial triple systems with $\lambda = 2$, *Ars Combinat.* 16, 11–16. MR 85d:05039; Z 537.05016.

HAMMER, J. / SARVATE, D.G.

1992 Generalizations of results of Stanton on BIBDs, *Discrete Math.* 102, 149–154. MR 93d:05020; Z 785.05015.

HAN, J.X.

1990 Complete solutions to a class of simplified schoolgirls problems of Kirkman (in Chinese), *Math. Practice Theory* 3, 70–77. MR 91j:05022.

HANANI, H.

1960 A note on Steiner triple systems, *Math. Scand.* 154–156. MR 23A#2330; Z 100, 18.

1961 The existence and construction of balanced incomplete block designs, *Ann. Math. Statist.* 32, 361–386. MR 29#4161; Z 107, 361.

1964 On covering of balanced incomplete block designs, *Canad. J. Math.* 16, 615–625. MR 29#4706; Z 192, 92.

1974 On resolvable balanced incomplete block designs, *J. Combinat. Theory* 17, 275–289. MR 50#6883; Z 305.05010.

1975 Balanced incomplete block designs and related designs, *Discrete Math.* 11, 255–369. MR 52#2918; Z 361.62067.

1976 Resolvable designs, in: *Teorie Combinatorie II*, Colloq. Internaz., Atti dei Convegni Lincei, Roma 1973, Accad. Naz. dei Lincei, Rome, pp. 249–252. MR 58#21676; Z 399.05014.

1984 On the original Steiner systems, *Discrete Math.* 51, 309–310. MR 85i:05034; Z 555.05020.

HANANI, H. / RAY-CHAUDHURI, D.K. / WILSON, R.M.

1972 On resolvable designs, *Discrete Math.* 3, 343–357. MR 46#8857; Z 263.05016.

HARE, D.R.

1995 Cycles in the block-intersection graph of pairwise balanced designs, *Discrete Math.* 137, 211–221. MR 95m:05027; Z 813.05041.

HARMS, J.J.

1985 Directing cyclic triple systems, *Ann. Discrete Math.* 26, 221–225. MR 87f:05020; Z 577.05014.

HARMS, J.J. / COLBOURN, C.J.

1983 An efficient algorithm for directing triple systems, *Congressus Numer.* 39, 455–461. MR 85m:0500; Z 532.05007.

1983a Partitions into directed triple systems, *Ars Combinat.* 16, 21–25. MR 85d:05040; Z 542.05017.

1985 An optimal algorithm for directing triple systems using Eulerian circuits, *Ann. Discrete Math.* 27, 433–438. MR 87g:05152; Z 582.05013.

HARMS, J.J. / COLBOURN, C.J. / IVANOV, A.V.

1987 A census of (9,3,3) block designs without repeated blocks, *Congressus Numer.* 57, 147–170. MR 88c:05024; Z 633.05010.

HARNAU, W.

1979 Die Anzahl paarweise nichtisomorpher, elementarer, wiederholungsfreier 2-(9, 3, λ)-Blockpläne, I, *Rostock. Math. Kolloq.* 11, 75–83. MR 81e:05036; Z 464.62073.

1980 Die Anzahl paarweise nichtisomorpher, elementarer, wiederholungsfreier 2-(9, 3, λ)-Blockpläne, II, *Rostock. Math. Kolloq.* 13, 43–47. MR 82d:05025; Z 475.62058.

HARTMAN, A.

1978 *Resolvable Designs*, M.Sc. Thesis, Technion, Haifa.

1979 A survey of existence problems for resolvable designs, *Southeast Asian Math. Bull.*, special issue, 268–277. MR 81f:05025; Z 444.05019.

1986 Partial triple systems and edge colourings, *Discrete Math.* 62, 183–196. MR 88e:05017; Z 656.05016.

1987 Halving the complete design, *Ann. Discrete Math.* 34, 207–224. MR 89b:05033; Z 643.05013.

HARTMAN, A. / HEINRICH, K.

1993 Pairwise balanced designs with holes, in: *Graphs, Matrices and Designs* (Ed. R. Rees), Dekker, pp. 171–204. MR 94e:05042; Z 789.05004.

HARTMAN, A. / HOFFMAN, D.G.

1987 Steiner triple systems with an involution, *Europ. J. Combinat.* 8, 371–378. MR 89c:05019; Z 675.05010.

HARTMAN, A. / MENDELSOHN, E.

1986 The last of the triple systems, *Ars Combinat.* 22, 25–41. MR 88e:05024; Z 615.05032.

HARTMAN, A. / MENDELSOHN, E. / ROSA, A.

1984 On the strong Lindner conjecture, *Ars Combinat.* 18, 139–150. MR 87e:05029; Z 559.05012.

HARTMAN, A. / PHELPS, K.T.

1992 Steiner quadruple systems, in: *Contemporary Design Theory* (Ed. J.H. Dinitz and D.R. Stinson) Wiley, New York, pp. 205–240. Z 765.05017.

HARTMAN, A. / YEHUDAI, Z.

1990 Greedesigns, *Ars Combinat.* 29A, 69–76. MR 97f:05038; Z 723.05017.

HARTSFIELD, N. / JACKSON, B. / RINGEL, G.

1995 The splitting number of the complete graph, *Graphs Combinat.* 1, 311–329.

HASSE, H.

1931 Gruppentheoretischer Beweis eines Satzes über gewisse Tripelsysteme, *Norsk Mat. Tidsskr.* 13, 105–107. Z 3, 385.

HEATHCOTE, G.

1993 Linear spaces on 16 points, *J. Combin. Des.* 1, 359–378. MR 95j:51018; Z 817. 05021.

HEDAYAT, A.

1971 A set of three orthogonal Latin squares of order 15 associated with a Kirkman–Steiner triple system of order 15, *Ann. Math. Statist.* 42, 1803.

1972 An algebraic property of the totally symmetric loops associated with Kirkman–Steiner triple systems, *Pacific J. Math.* 40, 305–310. MR 46#3340; Z 241.05014.

1990 The theory of trade-off for t-designs, in: *Coding Theory and Design Theory, Part II: Design Theory* (Ed. D.K. Ray-Chaudhuri) Springer, New York, pp. 101–126. MR 91d:05016; Z 721.05008.

HEDAYAT, A. / HWANG, H.L.

1983† An algorithm for generating a basis for trades on t-designs, *Commun. Stat. (Simul. Comp.)* 12, 109–125.

1984 BIB$(8,56,21,3,6)$ and BIB$(10,30,9,3,2)$ designs with repeated blocks, *J. Combinat. Theory (A)* 36, 73–91. MR 85h:05020; Z 523.05009.

HEDAYAT, A. / KHOSROVSHAHI, G.B.

1981 An algebraic study of BIB designs: a complete solution for $v = 6$ and $k = 3$, *J. Combinat. Theory (A)* 30, 43–52. MR 82e05029; Z 469.05014.

HEDAYAT, A. / LANDGEV, I.N. / TONCHEV, V.D.

1989 Results on the support of BIB designs, *J. Stat. Plann. Infer.* 22, 295–306. MR 90d:05037; Z 674.05006.

HEDAYAT, A. / LI, S.-Y.R.

1979 The trade-off method in the construction of BIB designs with variable support sizes, *Ann. Statist.* 7, 1277–1287. MR 81b:62092; Z 424.62052.

1980 Combinatorial topology and the trade off method in BIB designs, *Ann. Discrete Math.* 6, 189–200. MR 82a:05014; Z 455.05016.

HEDAYAT, A. / PESOTAN, H.

1985 A study of BIB designs through support matrices, *J. Stat. Plann. Infer.* 11, 363–372. MR 87f:05021; Z 564.05007.

HEDAYAT, A. / RAKTOE, B.L.

1970 On the equivalence of resolvable Kirkman–Steiner triple systems and sets of mutually orthogonal Latin squares, *Ann. Math. Statist.* 41, 1140.

HEDAYAT, A. / RAO, C.R. / STUFKEN, J.

1988 Sampling plans excluding contiguous units, *J. Stat. Plann. Infer.* 19, 159–170. MR 90f:62031; Z 642.62003.

1988a Designs in survey sampling avoiding contiguous units, in: *Handbook in Statistics* 6 (Sampling) (Ed. P.B. Krishnaiah and C.R. Rao) Elsevier, Amsterdam, pp. 575–583.

HEFFTER, L.

1891 Über das Problem der Nachbargebiete, *Math. Annalen* 38, 479–508.

1896 Über Nachbarconfigurationen,Tripelsysteme und metacyklische Gruppen,*Deutsche Mathem. Vereinig. Jahresber.* 5, 67–69.

1897 Über Tripelsysteme, *Math. Ann.* 49, 101–112.

HEIDEMA, C.E.

1978 Generalized pseudosurface embeddings of graphs and associated block designs, *Lecture Notes in Mathematics* 642, 255–270.

HEINRICH, K.

1982 Disjoint subquasigroups, *Proc. London Math. Soc.* 45, 547–563. MR 84b:05029; Z 506.05014.

1991 Simple direct constructions for hybrid triple designs, *Discrete Math.* 97, 223–227. MR 93a:05027; Z 742.05015.

HELLERSTEIN, L. / GIBSON, G.A. / KARP, R.M. / KATZ, R.H. / PATTERSON, D.A.

1994 Coding techniques for handling failures in large disk arrays, *Algorithmica* 12, 182–208.

HENDERSON, A.

1911 *The Twenty-Seven Lines upon the Cubic Surface*, Hafner, New York.

HENDERSON, M.

1964 Generalized finite planes, *Colloq. Math.* 13, 29–36. MR 30#5210; Z 125, 100.

HERGERT, F.

1984 *Algebraische Methoden für nichtlineare Codes*, Dissertation, TH Darmstadt, 108 pp.

HESSE, O.

1844 Über die Wendepuncte der Curve dritter Ordnung, *J. Reine Angew. Math.* 28, 97–107.

1847 Algebraische Auflösung derjenigen Gleichungen 9ten Grades, deren Wurzel die Eigneschaft haben, dafs eine gegebene rationale und symmetrische Function $\theta(x_\lambda, x_\mu)$ je zweier Wurzeln x_λ, x_μ eine dritte Wurzel x_κ giebt, so dafs gleichzeitig: $x_\kappa = \theta(x_\lambda, x_\mu)$, $x_\lambda = \theta(x_\mu, x_\kappa)$, $x_\mu = \theta(x_\kappa, x_\lambda)$, *J. Reine Angew. Math.* 34, 193–208.

1848 Über Curven dritter Ordnung und die Kegelschnitte, welche diesen Curven in die verschiedenen Puncten berühen, *J. Reine Angew. Math.* 36, 143–176.

HEXEL, E.

1984 On residual systems of Steiner systems, *Indian J. Math.* 26, 29–41. MR 87a: 05029; Z 632.05014.

HIGMAN, D.G. / MCLAUGHLIN, J.E.

1961† Geometric ABA-groups, *Illinois J. Math.* 5, 382–397.

HILTON, A.J.W.

1969 On Steiner and similar triple systems, *Math. Scand.* 24, 208–216. MR 41#3296; Z 188, 40.

1972 A simplification of Moore's proof of the existence of Steiner triple systems, *J. Combinat. Theory (A)* 13, 422–425. MR 46#1616; Z 244.05010.

1973† Embedding an incomplete diagonal latin square in a complete diagonal latin square, *J. Combinat. Theory (A)* 15, 121–128.

1977 On the Szamkolowicz–Doyen classification of Steiner triple systems, *Proc. London Math. Soc.* (3) 24, 102–116. MR 55#128; Z 358.05010.

1978 Dimension in linear spaces, in: *Problèmes Combinatoireset Théorie des Graphes*, Colloq. Internat. C.N.R.S. No. 260, Editions du CNRS, Paris, pp. 233–236. MR 80g: 05012; Z 412.05018.

1987 Embedding partial triple systems, *J. Comb. Math. Comb. Comput.* 2, 77–95. MR 89g:05029; Z 638.05008.

HILTON, A.J.W. / RODGER, C.A.

1990 Edge-colouring graphs and embedding partial triple systems of even index, *Cycles and Rays*, NATO ASI, Kluwer, Dordrecht, pp. 101–112. MR 92c:05027; Z 715. 05021.

1991 The embedding of partial triple systems when 4 divides λ, *J. Combinat. Theory (A)* 56, 109–137. MR 91i:05029; Z 758.05022.

HILTON, A.J.W. / RODGER, C.A. / WOJCIECHOWSKI, J.

1992 Prospects for good embeddings of pairs of partial orthogonal latin squares and of partial Kirkman triple systems, *J. Comb. Math. Comb. Comput.* 11, 83–91. MR 92m:05038; Z 766.05013

HILTON, A.J.W. / TEIRLINCK, L.

1980 Dimension in triple systems, in: *Topics on Steiner Systems* (Ed. C.C. Lindner and A. Rosa), *Ann. Discrete Math.* 7, 73–87. Z 441.05011.

HIRSCHFELD, J.W.P.

1992 *Finite Projective Spaces of Three Dimensions*, Clarendon Press, Oxford. Chapter 17: Spans, spreads and packings. MR 87j:51013; Z 574.51001.

HISHIDA, T. / ISHIKAWA, K. / JIMBO, M. / KAGEYAMA, S. / KURIKI, S.

1999 Non-existence of a nested BIB design NB(10,15,2,3).

HOFFMAN, A.J. / RICHARDSON, M.

1961 Block design games, *Canad. J. Math.* 13, 110–128. MR 23B#2067; Z 101, 372.

HOFFMAN, D.G.

1999 An extension of the Doyen–Wilson theorem.

HOFFMAN, D.G. / LINDNER, C.C.

1981 Embeddings of Mendelsohn triple systems, *Ars Combinat.* 11, 265–269. MR 83c:05034; Z 488.05020.

1982 Mendelsohn triple systems having a prescribed number of triples in common, *Europ. J. Combinat.* 3, 51–61. MR 83h:05026; Z 486.05018.

1987 The flower intersection problem for Steiner triple systems, *Ann. Discrete Math.* 34, 243–248. MR 88k:05034; Z 675.05009.

HOFFMAN, D.G. / RODGER, C.A.

1984 Embedding partial perpendicular arrays of strength 3, *Congressus Numer.* 44, 55–60. MR 89h:05031; Z 558.05011.

1987 Embedding totally symmetric quasigroups, *Ann. Discrete Math.* 34, 249–257. MR 88m:20131; Z 645.20049.

1996† On the number of edge-disjoint one factors and the existence of k-factors in complete multipartite graphs, *Discrete Math.* 160, 177–187.

HOFFMAN, D.G. / RODGER, C.A. / ROSA, A.

1993† Maximal sets of 2-factors and Hamiltonian cycles, *J. Combinat. Theory (B)* 57, 69–76.

HOFMEISTER, T. / LEFMANN, H.

1996 A combinatorial design approach to MAXCUT, *Random Structures and Algorithms* 9, 163–175. Z 863.05037. Preliminary version: *Lecture Notes in Computer Science* 1046, 441–452. MR 98g:05141.

HOLYER, I.

1981† The NP-completeness of edge-coloring, *SIAM J. Computing* 10, 718–720.

1981a† The NP-completeness of some edge-partition problems, *SIAM J. Computing* 10, 713–717.

HORÁK, P. / PHILLIPS, N. / WALLIS, W.D. / YUCAS, J.

1997 Counting frequencies of configurations in Steiner triple systems, *Ars Combinatoria* 46, 65–75. MR 98h:05031.

HORÁK, P. / PIKE, D. / RAINES, M.

1999 Hamilton cycles in block intersection graphs of triple systems, *J. Combin. Des.*

HORÁK, P. / ROSA, A.

1988 Decomposing Steiner triple systems into small configurations, *Ars Combinat.* 26, 91–105. MR 90a:05029; Z 780.05008.

HORNER, J.

1868 On triads of once-paired elements, *Quart J. Pure Appl. Math.* 9, 15–18.

HORTON, J.D.

1970 Variations on a theme by Moore, in: *Proc. Louisiana Conf. Combinatorics, Graph Theory and Computing*, Louisiana State University, Baton Rouge, pp. 146–166. MR 46#8864; Z 248.05018.

HORVÁTH, T.

1984 Die Zahl der Ovale in der Bolyai–Lobachevsky Ebene $< 3, 3 >$, *Potsdamer Forsch.* B 41, 95–98. Z 604.51006.

1986 Die Zahl der Ovale in der Bolyai–Lobachevsky Ebene S$< 3, 3 >$, *Ann. Univ. Sci. Budapest Rol. Eőtvős Sect. Math.* 28, 233–239. MR 87k:51017; Z 599.51004.

HSU, D.F.

1980 *Cyclic Neofields and Combinatorial Designs*, Springer, New York. MR 84g:05001; Z 443.16022.

HSU, D.F. / SHEN, H.

1994 On the existence of perfect Mendelsohn designs without repeated blocks, *Ars Combinat.* 37, 129–139. MR 95b:05026; Z 810.05014.

HUANG, C. / MENDELSOHN, E. / ROSA, A.

1982 On partially resolvable t-partitions, *Ann. Discrete Math.* 12, 169–183. MR 87e:05036; Z 495.05013.

HUANG, W.-C. / KAO, C.-Y. / HORNG, J.-T.

1994 A genetic algorithm approach for set covering problems, in: *Proc. First IEEE Conf. Evolutionary Computation*, Vol. 2, IEEE, pp. 569–574.

HUFFMAN, W.C.

1986 On automorphism groups of certain 2-$(v, 3, 2)$ designs induced by group actions, *Discrete Math.* 59, 283–296. MR 87f:05022; Z 586.05005.

1996 The equivalence of two cyclic objects on *pq* elements, *Discrete Math.* 154, 103–127. MR 97d:05286; Z 859.05014.

HUGHES, D.R. / PIPER, F.C.

1985 *Design Theory*, Cambridge University Press, Cambridge. MR 87e:05022; Z 561.05009.

HUNG, S.H.Y. / MENDELSOHN, N.S.

1973 Directed triple systems, *J. Combinat. Theory (A)* 14, 310–318. MR 47#3190; Z 263.05020.

HWANG, F.K. / LIN, S.

1974 A direct method to construct triple systems, *J. Combinat. Theory (A)* 17, 84–94. MR 52#127; Z 282.05010.

HWANG, F.K. / SÓS, V.

1987† Nonadaptive hypergeometric group testing, *Stud. Sci. Math. Hung.* 22, 257–263.

INNAMORATI, S. / LO FARO, G.

1992 Constructing Steiner triple systems partially embedded in a projective plane, *J. Inform. Optim. Sci.* 13, 267–277. MR 93g:05023; Z 767.51002.

IRVING, R.W.

1974 Generalized Ramsey numbers for small graphs, *Discrete Math.* 9, 251–264. MR 51#12591; Z 287.05104.

IVANOV, A.V.

1983 Konstruktivnoye perečislenie sistem incidentnosti, III, *Rostock. Math. Kolloq.* 24, 43–62. MR 86j:05029b; Z 529.05009.

1985 Constructive enumeration of incidence systems, *Ann. Discrete Math.* 26, 227–246. MR 87g:05033; Z 578.05007.

IVANOV, A.V. / FARADŽEV, I.A.

1983 Konstruktivnoye perečislenie sistem incidentnosti, I, *Rostock. Math. Kolloq.* 24, 4–22. MR 86j:05029a; Z 529.05007.

JANKO, Z. / TRAN VAN TRUNG

1980 On projective planes of order 12 which have a subplane of order 3, I, *J. Combinat. Theory (A)* 29, 254–256. MR 82f:51005; Z 446.51006.

1981 On projective planes of order 12 with an automorphism of order 3, II, *Geom. Dedicata* 11, 257–284. MR 83d:05022a; Z 467.51010.

JEURISSEN, R.H.

1988 A proof by graphs that PSL(2,7) ≅ PSL(3,2), *Discrete Math.* 70, 315–317. MR 89f:05094; Z 654.20012.

1995 Special sets of lines in PG(3,2), *Lin. Alg. Appl.* 226–228, 617–638. MR 96f:51013; Z 839.51008.

JEURISSEN, R.H. / VAN OS, C.H. / STEENBRINK, J.H.M.

1994 The configuration of bitangents of the Klein curve, *Discrete Math.* 132, 83–96. MR 96d:14050; Z 808.05059.

JIANG, Z.

1995 *Rotational Steiner Triple Systems*, Ph.D. Thesis, University of Waterloo.

1995a $(f,2)$-Rotational Steiner triple systems, in: *Combinatorics Advances* (Ed. C.J. Colbourn and E.S. Mahmoodian), Kluwer Academic, pp. 181–206. MR 96h:05023; Z 838.05014.

1996 Concerning cyclic group divisible designs with block size three, *Australas. J. Combin.* 13, 227–245. MR 96k:05021; Z 847.05013.

1998 On rotational Steiner triple systems: five, seven and eleven, *Ars Combinat.* 47, 201–221. MR 98k:05023.

JIANG, Z. / COLBOURN, C.J.

1998 Completing the spectrum of rotational Mendelsohn triple systems, *Discrete Math.* 183, 153–160. MR 99a:05014.

JIANG, Z. / MCLEISH, M.

1996 Directed triple systems with a class of automorphisms, *Ars Combinat.* 44, 229–239. MR 97g:05029.

1999 Concerning extended triple systems with rotational automorphisms, *Utilitas Math.*

JIMBO, M.

1983 A recursive construction of 1-rotational Steiner 2-designs, *Aequat. Math.*, 26, 184–190. MR 85g:05030; Z 572.005015.

1993 Recursive constructions for cyclic BIB designs and their generalizations, *Discrete Math.* 116, 79–95. MR 94e:05043; Z 778.05013.

JIMBO, M. / KURIKI, S.

1983 On a composition of cyclic 2-designs, *Discrete Math.* 43, 249–255. MR 85a:05014; Z 526.05011.

JIMBO, M. / VANSTONE, S.A.

1984 Recursive constructions for resolvable and doubly resolvable 1-rotational Steiner 2-designs, *Utilitas Math.* 26, 45–61. MR 86a:05017; Z 558.05009.

JOHANSSON, A.

1999 A note on extending partial triple systems.

JOHNSEN, E.C. / STORER, T.

1974 Combinatorial structures in loops, II. Commutative inverse-property cyclic neofields of prime-power order, *Pacific J. Math.* 52, 115–127. MR 53#12982; Z 256.05008; Z 286.05007.

1974a Combinatorial structures in loops, IV. Steiner triple systems in neofields, *Math. Z.* 138, 1–14. MR 56–5319; Z 276.05019; Z 285.05022.

JOHNSON, D.M. / MENDELSOHN, N.S.

1972 Extended triple systems, *Aequat. Math.* 8, 291–298. MR 47#3211; Z 256.20096.

JOHNSON, D.S.

1982 The NP-completeness column: an ongoing guide, *J. Algorithms* 3, 182–195. MR 83i:68072a; Z 494.68049.

JUNGNICKEL, D.

1978 Composition theorems for difference families and regular planes, *Discrete Math.* 23, 151–158. MR 82j:05027; Z 389.05017.

1989 Design Theory: an update, *Ars Combinat.* 28, 129–199. MR 91b:05054; Z 704.05005.

JUNGNICKEL, D. / MULLIN, R.C. / VANSTONE, S.A.

1991 The spectrum of α-resolvable block designs with block-size 3, *Discrete Math.* 97, 269–277. MR 92k:05016; Z 753.05008.

JUNGNICKEL, D. / PICKERT, G.

1996 A life's work in geometry: an homage to Hanfried Lenz, *Des. Codes Crypt.* 8, 9–22. MR 97k:01036; Z 867.01032.

JUNGNICKEL, D. / VANSTONE, S.A.

1991 Triple systems in $PG(2,q)$, *Discrete Math.* 92, 131–135. MR 92k:05023; Z 764.05012.

KAGEYAMA, S. / MIAO, Y.

1996 The spectrum of nested designs with block size three and four, *Congressus Numer.* 114, 73–80. MR 97g:05024.

1996a Some constructions of semiframes, *Ars Combinat.* 43, 17–31. MR 98f:05036; Z 881.05015.

1996b Nested directed BIB designs with block size three or four, *J. Ind. Soc. Agric. Stat.* 49, 99–110. MR 98j:62059.

1997 Nested designs with block size five and subblock size two, *J. Stat. Plann. Infer.* 64, 125–139. Z 885.05021.

1999 On nested group divisible designs with block size three or four, *Proceedings of the Eighth Quadrennial International Conference on Graph Theory, Combinatorics, Algorithms amd Applications.*

1999a How to nest a perfect Mendelsohn design with block size three or four, *J. Combin. Info. Syst. Sci.*

1999b Existence of nested designs with block size five.

KALBFLEISCH, J.G. / STANTON, R.G.

1968 Maximal and minimal coverings of $(k-1)$-tuples by k-tuples, *Pacific J. Math.* 28, 131–140. MR 38#2032; Z 182, 28.

KANG, Q.

1987 A recurrence construction of LD-designs, *Acta Math. Appl. Sinica* 10, 481–490. MR 89c:05014.

1989 *Large Sets of Block Designs*, Doctoral Thesis, Eindhoven University, 85 pp. MR 90g:05037; Z 668.05008.

1990 A generalization of Mendelsohn triple systems, *Ars Combinat.* 29C, 207–215. MR 97e:05041.

1991 The construction of large sets of disjoint Mendelsohn triple systems of order 2^n+2, *Discrete Math.* 90, 199–205. MR 92j:05025; also in: *Science Bulletin (China)* 35 (1990) 801–805. MR 91g:05018; Z 734.05022.

1995 *Large Sets of Triple Systems and Related Designs*, Science Press, New York. MR 96g:05023; Z 823.05014.

1996 Large sets of resolvable MTS and DTS of orders p^n+2, *J. Combin. Des.* 4, 301–321. MR 97g:05030.

1997 Large sets of three types of oriented triple systems, *J. Stat. Plann. Infer.* 58, 151–176. MR 98e:05013.

KANG, Q. / CHANG, Y.-X.

1990 Symmetric Mendelsohn triple systems and large sets of disjoint Mendelsohn triple systems, in: *Combinatorial Designs and Applications* (Ed. W.D. Wallis, H. Shen, W. Wei, and L. Zhu), Marcel Dekker, New York, pp. 69–78. MR 92a:05024; Z 736.05018; abstract in: *Science Bulletin (China)* 34, 1226–1228. MR 90m:05035; Z 682.05016.

1992 A completion of the spectrum for large sets of disjoint transitive triple systems, *J. Combinat. Theory (A)* 60, 287–294. MR 93i:05024; Z 783.05019.

1992a Further results about large sets of disjoint Mendelsohn triple systems II, *Chinese Sci. Bull.* 37, 90–91.

1993 Further results about large sets of disjoint Mendelsohn triple systems, *Discrete Math.* 118, 263–268. MR 94c:05014; Z 782.05012. (also *Science Bull. (China)* 37, 90–91. II. *Science Bull. (China)* 38, 436–437. Z 779.05006.)

KANG, Q. / CHANG, Y.-X. / YANG, G.-H.

1994 The spectrum of self-converse DTS, *J. Combin. Des.* 2, 415–425. MR 95g:05029; Z 829.05013.

KANG, Q. / LEI, J.-G.

1993 A completion of the spectrum for large sets of disjoint Mendelsohn triple systems, *Bull. Inst. Combin. Appl.* 9, 14–26. MR 94g:05017.

1996 On large sets of resolvable and almost resolvable oriented triple systems, *J. Combin. Des.* 4, 95–104. MR 97d:05038.

1996a On large sets of hybrid triple systems, *J. Stat. Plann. Infer.* 51, 181–188. MR 97b:05025; Z 855.05022.

KANG, Q. / LEI, J.-G. / CHANG, Y.-X.

1994 The spectrum of large sets of disjoint Mendelsohn triple systems with any index, *J. Combin. Des.* 2, 351–358. MR 95e:05014; Z 834.05010.

KANTOR, W.M.

1975 2-transitive designs, in: *Combinatorics*, Reidel, Dordrecht, pp. 365–418. MR 51 #12558; Z 362.05031.

1985 Homogeneous designs and geometric lattices, *J. Combinat. Theory (A)* 38, 66–74. MR 87c:51007; Z 559.05015.

KARMARKAR, N. / RAMAKRISHNAN, K. / RESENDE, M.G.C.

1991 An interior point algorithm to solve computationally difficult set covering problems, *Math. Programming (B)* 52, 597–618. MR 93b:90050; Z 753.90046.

KÁRTESZI, F. / HORVÁTH, T.

1986 Einige Bemerkungen bezüglich der Struktur von endlichen Bolyai–Lobachevsky Ebenen, *Ann. Univ. Sci. Budapest Rol. Eőtvős Sect. Math.* 28, 263–270. MR 87m: 51014; Z 599.51003.

KAUFMANN, P.B.

1926 *Studien über zyklische Dreiersysteme der Form $N = 6n + 3$*, Inaugural-Dissertation der Math.-Natur. Fakultät der Universität Freiburg in der Schweiz, Sarnen.

KEEDWELL, A.D.

1984 Circuit designs and Latin squares, *Ars Combinat.* 17, 79–90. MR 85i:05036; Z 549.05013.

KEPKA, T.

1970 F-quasigroups isotopic to Moufang loops, *Czechosl. Math. J.* 29 (104), 62–83. MR 80b:20095; Z 444.20067.

1978 Distributive Steiner quasigroups of order 3^5, *CMUC* 19, 389–401. MR 58#6032; Z 373.20062.

1980 Commutative Moufang loops and distributive Steiner quasigroups nilpotent of class 3, *Commentat. Math. Univ. Carol.* 21, 355–370. MR 81i:20103; Z 439.20048.

KEY, J.D.

1986 Extensions of Steiner systems, University of Nebraska–Lincoln, Dept. of Computer Science, Report Series #27, 8 pp.

1992 Extendable Steiner designs, *Geom. Dedicata* 41, 201–205. MR 93h:05036; Z 778. 05014.

1993 Designs, codes and rigidity theorems, *Congressus Numer.* 95, 105–115. MR 94m:05024; Z 812.51007.

1994 Ternary codes of Steiner triple systems, *J. Combin. Des.* 2, 25–30. MR 94k: 05033; Z 819.05012.

KEY, J.D. / SHOBE, F.D.

1997 Some transitive Steiner triple systems of Bagchi and Bagchi, *J. Stat. Plann. Infer.* 58, 79–86. MR 98i:05020; Z 873.05017.

KEY, J.D. / SHULT, E.E.

1984 Steiner triple systems with doubly transitive automorphism groups: a corollary to the classification theorem for finite simple groups, *J. Combinat. Theory (A)* 36, 105–110. MR 85f:05020; Z 541.05014.

KEY, J.D. / SULLIVAN, F.E.

1993 Codes of Steiner triple and quadruple systems, *Des. Codes Crypt.* 3, 117–125. MR 94k:05050; Z 779.51004. Correction: *Des. Codes Crypt.* 7 (1996), 175. Z 841. 51006.

1995 Steiner triple systems with many affine hyperplanes, *Congressus Numer.* 107, 105–112.

KEY, J.D. / WAGNER, A.

1986 On an infinite class of Steiner systems constructed from affine spaces, *Arch. Math.* 47, 376–378. Z 584.05012; Z 595.05012.

KHODKAR, A.

1993 The fine structure of balanced ternary designs with block size three, *Utilitas Math.* 44, 197–230. MR 94f:05014; Z 791.05009.

1995 Smallest defining sets for the 36 nonisomorphic twofold triple systems of order 9, *J. Comb. Math. Comb. Comput.* 17, 209–215. MR 95k:05025; Z 821.05004.

1995a Twofold triple systems having a prescribed number of triples in common, *Utilitas Math.* 48, 203–214. MR 96k:05022; Z 845.05015.

1995b Numbers of common triples in simple balanced ternary designs, *Ars Combinat.* 40, 235–246. MR 96d:05030; Z 841.05010.

1996 The fine structure of $(v,3)$ directed triple systems: $v \equiv 0$ or 1 (mod 3), *Ars Combinat.* 43, 213–224. MR 97e:05042; Z 881.05023.

KHOSROVSHAHI, G.B.

1990 On trades and designs, *Comput. Statist. Data Analysis* 10, 163–167. Z 825. 62657.

KHOSROVSHAHI, G.B. / AJOODANI-NAMINI, S.

1990 A new basis for trades, *SIAM J. Discrete Math.* 3, 363–372. MR 91m:05023; Z 714.05010.

1991 Combining t-designs, *J. Combinat Theory (A)* 58, 26–34. MR 92m:05023; Z 735. 05021.

KHOSROVSHAHI, G.B. / MAHMOODIAN, E.S.

1987 A table of BIB(10,30t,6t,3,2t) and BIB(11,55t,15t,3,3t) designs, Technical Report 2–1365, Sharif University of Technology, Iran.

1987a A linear algebraic algorithm for reducing the support size of t-designs and to generate a basis for trades, *Commun. Stat. (Simul. Comp.)* 16, 1015–1038. Z 695. 62203.

1988 On BIB designs with various support sizes for $v = 9$ and $k = 3$, *Commun. Stat. (Simul. Comp.)* 17, 765–770. Z 695.62195.

KHOSROVSHAHI, G.B. / YOUSEFI-AZARI, H.

1996 Construction of some cyclic anti-Pasch Steiner triple systems of small orders, *J. Combin. Inf. Syst. Sci.* 21, 1–4.

KILLGROVE, R. / ROBERTS, C. / STERNFELD, R. / TAMEZ, R. / DERBY, R. / KIEL, D.

1996 Latin squares and other configurations, *Congressus Numer.* 117, 161–174. MR 98f:05030; Z 897.05017.

KILLGROVE, R. / TAMEZ, R. / STERNFELD, R. / GORDON, C. / GRAHAM, G. / PARKER, E.T. / KAUFFMAN, T.

1992 Local isotopy and local motions in Steiner systems, *Congressus Numer.* 89, 129–151. MR 93m:05032; Z 797.05014.

KING, J.D.

1969 A characterization of some double transitive groups, *Math. Z.* 107 (1968), 43–48. Correction: 112, 393–394. MR 41#312; Z 164, 25.

KIRKMAN, T.P.

1847 On a problem in combinations, *Cambridge and Dublin Math. J.* 2, 191–204.

1850 Query VI, *Lady's and Gentleman's Diary*, 48.

1850a Note on a unanswered prize question, *Cambridge and Dublin Math. J.* 5, 255–262.

1850b On the triads made with fifteen things, *London, Edinburgh and Dublin Philos. Mag. and J. Sci.* (3) 37, 169–171.

1851 Solution to Query VI, *Lady's and Gentleman's Diary*, 48.

1853 Theorems on combinations, *Cambridge and Dublin Math. J.* 8, 38–45.

1857 On the perfect r-partitions of $r^2 - r + 1$, *Transactions of the Historic Society of Lancashire and Cheshire*, 127–142.

1862 On the puzzle of the fifteen young ladies, *London, Edinburgh and Dublin Philos. Mag. and J. Sci.* (4) 23, 198–204.

1869 Solutions of three problems proposed by W. Lea, *Educational Time Reprints* 11, 97–99.

1874 Solutions of two questions (3143 and 4005), *Educational Times Reprints* 19 (1874), 71–72; 20, 36.

1887 Two questions on polyhedra, *Math. Quest. Educational Times* 47, 125–127.

KIRKMAN, T.P. / WOODALL, H.J.

1891 Solution of a problem, *Educational Times Reprints* 55, 60.

KLEIN, F.

1879 Über die Transformation siebenter Ordnung der elliptischen Funktionen, *Math. Annalen* 14, 428–471.

KLOSSEK, S.

1975 Kommutative Spiegelungsräume, *Mitt. Math. Sem. Giessen*, Heft 117, Giessen, 72 pp. MR#10966; Z 317.20043.

KNUTH, D.E.

1973 *The Art of Computer Programming, Vol. 3: Sorting and Searching,* Addison-Wesley, Reading, MA, Chapter 6.5, pp. 565–570. MR 56#4281; Z 302.68010.

KOCAY, W.L.

1992 Reconstructing graphs as subsumed graphs of hypergraphs and some self-complementary triple systems, *Graphs Combinat.* 8, 259–276. MR 93h:05117; Z 759. 05064.

KOCAY, W.L. / STINSON, D.R. / VANSTONE, S.A.

1985 On strong starters in cyclic groups, *Discrete Math.* 56, 45–60. MR 87a:05045; Z 578.20014.

KOLMOGOROV, A.N. / YUSHKEVICH, A.P.

1981† *Mathematika XIX Veka: Geometriya, Teoriya Analitcheskikh Funksii*, Izsalet'stvo Nauka, Moscow. (English translation: *Mathematics of the 19th Century*, Birkhäuser, Basel, 1996.)

KONG, Q.

1994 On the girl student problem of Kirkman — a manual solution of the seven weeks arrangement (in Chinese), *Math. Pract. Theory*, No. 3, 79–80. Z 816.05016.

KÖNIG, D.

1916† Über Graphen und ihre Anwendung auf Determinantentheorie und Mengenlehre, *Math. Ann.* 77, 453–465.

KORCHMÁROS, G.

1982 Moufang loops of odd prime exponent p and a class of linear spaces of order p, *J. Combinat. Theory (A)* 32, 336–349. MR 83k:20081; Z 484.51010.

KORDECKI, W.

1983 On the groups defined by the configurations $C(n, v)$, *Discuss. Math.* 6, 7–9. Z 546.05016.

KOROVINA, N.P.

1982 Cikličeskie sistemy troek Štejnera vtorogo tipa, *Kombinatornoalgebr. metody v prikl. mat.*, Gor'kij, 76–88. MR 85j:05009.

KOSZAREK, CH.

1983 *Paare von isomorphen Steinerschen Tripelsystemen mit vorgeschriebener Anzahl gemeinsamer Blöcke*, Diplomarbeit, Institut für Numerischehe und Angewandte Mathematik, Georg-August-Universität Göttingen, 53 pp.

KOTZIG, A. / LINDNER, C.C. / ROSA, A.

1975 Latin squares with no subsquares of order two and disjoint Steiner triple systems, *Utilitas Math.* 7, 287–294. MR 53#5331; Z 311.05013.

KOTZIG, A. / ROSA, A.

1974 Nearly Kirkman systems, *Congressus Numer.* 10, 607–614. MR 51#190; Z 311. 05023.

KOTZIG, A. / TURGEON, J.

1975† On certain constructions for Latin squares with no Latin subsquares of order two, *Discrete Math.* 16, 263–270.

KOWALEWSKI, G.

1930 *Alte und neue mathematische Spiele. Eine Einführung in die Unterhaltungsmathematik*, Teubner, Leipzig, Kap. IV. 13. Das Problem der fünfzehn Pensionatsdamen.

KRAÏTCHIK, M.

1930 *La mathématique des jeux ou récréations mathématiques*, Stevens, Bruxelles. Chapter 18: Jeux de répartition, pp. 540–545.

1942 *Mathematical Recreations*, Norton, New York. Reprinted by Dover, New York, 1953.

KRAMER, E.S.

1974 Indecomposable triple systems, *Discrete Math.*, 8, 173–180. MR 48#10863; Z 276.05020.

1977 Some triple system partitions for prime powers, *Utilitas Math.* 12, 113–116. MR 57#16099; Z 267.05010.

KRAMER, E.S. / KREHER, D.L. / MAGLIVERAS, S.S. / MESNER, D.M.

1981 An assortment of Room-type designs, *Ars Combinat.* 11, 9–29. MR 83i:05022; Z 473.05017.

KRAMER, E.S. / KREHER, D.L. / MESNER, D.M.

1982 Some crowded Room rectangles, *Ars Combinat.* 13, 71–85. MR 84b:05036, Z 497.05015.

KRAMER, E.S. / MAGLIVERAS, S.S.

1980 A $57 \times 57 \times 57$ Room-type design, *Ars Combinat.* 9, 163–166. MR 81e:05045; Z 439.05016.

KRAMER, E.S. / MESNER, D.M.

1974 Intersections among Steiner systems, *J. Combinat. Theory (A)* 16, 273–285. MR 49#78; Z 282.05011.

1974a On Steiner tableaus, *Discrete Math.* 10, 123–131. MR 50#1933; Z 305.05014.

1975 The possible (impossible) systems of 11 disjoint $S(2,3,13)$s ($S(3,4,14)$s) with automorphism of order 11, *Utilitas Math.* 7, 55–58. MR 51#7898; Z 309.05014.

1980 A special 52×52 Room-type design, *Ars Combinat.* 9, 39–42. MR 81e:05044; Z 439.05014.

KREHER, D.L. / STINSON, D.R. / ZHU, L.

1997 On the maximum number of fixed points in automorphisms of prime order of 2-$(v,k,1)$ designs, *Annals Combinat.* 1, 227–243.

KÜHNEL, W.

1998 Topological aspects of twofold triple systems, *Expositiones Math.* 16, 289–332.

KUNKLE, T. / SARVATE, D.G.

1995 On ternary designs with a specified number of blocks with repeated elements, *Ars Combinat.* 40, 129–142. MR 96i:05025; Z 841.05011.

1996 Balanced (part) ternary designs, in: *CRC Handbook of Combinatorial Designs* (Ed. C.J. Colbourn and J.H. Dinitz), CRC Press, Boca Raton, FL, pp. 233–238. Z 868.05014.

KURIKI, S. / JIMBO, M.

1983 On 1-rotational $S_\lambda(2,3,r)$ designs, *Discrete Math.* 46, 33–40. MR 85b:05033; Z 519.05007.

LACAZE, J.

1984 Automorphism groups of the Hall triple systems of class 2, *Rend. Sem. Mat. Brescia* 7, 445–451. MR 86c:05037; Z 543.51007.

LACAZE, J. / BÉNÉTEAU, L.

1983 The automorphism group of the smallest non-affine Hall triple system, *Ann. Discrete Math.* 17, 387–391. MR 87d:05002; Z 521.05016.

LAM, C. / MIAO, Y.

1999 Cyclically resolvable cyclic Steiner triple systems of order 21 and 39.

1999a $(C_k \oplus G, k, \lambda)$ difference families.

LAM, C. / SEBERRY, J.

1984 Generalized Bhaskar Rao designs, *J. Stat. Plann. Infer.* 10, 83–95. MR 86h: 05026; Z 552.05012.

LAMKEN, E.R.

1987 On classes of doubly resolvable (v,3,2)-BIBDs from balanced tournament designs, *Ars Combinat.* 24, 85–91. MR 88k:05028; Z 647.05014.

1991 3-complementary frames and doubly near resolvable $(v,3,2)$-BIBDs, *Discrete Math.* 88, 59–78. MR 92e:05023; Z 722.05018.

1991a On near generalized balanced tournament designs, *Discrete Math.* 97, 279–294. Z 741.05017.

1993 Existence results for generalized balanced tournament designs with block size 3, *Des. Codes Crypt.* 3, 33–61. MR 93m:05022; Z 791.05010.

1994 The existence of doubly near resolvable (v,3,2)-BIBDs, *J. Combin. Des.* 2, 427–440. MR 95m:05030; Z 822.05003.

1995 The existence of doubly resolvable $(v,3,2)$-BIBDs, *J. Combinat. Theory (A)* 72, 50–76. MR 96m:05024; Z 836.05008.

1996 Balanced tournament designs, in: *CRC Handbook of Combinatorial Designs* (Ed. C.J. Colbourn and J.H. Dinitz), CRC Press, Boca Raton, FL, pp. 238–241. Z 849.05018.

1997 The existence of partitioned generalized balanced tournament designs with block size 3, *Des. Codes Crypt.* 11, 37–71. MR 98a:05042; Z 869.05012.

1998 The existence of $\mathrm{KS}_3(v;2,4)$, *Discrete Math.* 186, 195–216.

LAMKEN, E.R. / MILLS, W.H.

1993 Resolvable coverings, *Congressus Numer.* 96, 21–26. MR 94k:05052; Z 801. 05020.

LAMKEN, E.R. / REES, R. / VANSTONE, S.A.

1991 Class-uniformly resolvable pairwise balanced designs with block sizes two and three, *Discrete Math.* 92, 197–209. MR 93a:05020; Z 749.05011.

LAMKEN, E.R. / VANSTONE, S.A.

1988 The existence of a class of Kirkman squares of index 2, *J. Austral. Math. Soc.* (A) 44, 33–41. MR 89e:05049; Z 668.05016.

1990 Balanced tournament designs and resolvable (v,3,2)-BIBDs, *Discrete Math.* 83, 37–47. MR 91e:05020; Z 705.05005.

1992 Balanced tournament designs with almost orthogonal resolutions, *J. Austral. Math. Soc. (A)* 49, 175–195. MR 91d:05032; Z 739.05018.

1993 Existence results for doubly near resolvable (v,3,2)-BIBDs, *Discrete Math.* 120, 135–148. MR 94e:05045; Z 787.05009.

LAMKEN, E.R. / WILSON, R.M.

1999 Decompositions of edge-colored complete graphs.

LANDGEV, I.N.

1987 Balanced incomplete 2-(11,3,3) designs with repeating blocks (in Bulgarian), in: *Mathematics and Mathematical Education*, Sunny Beach, Bulg. Acad. Sci., Sofia, pp. 487–492. MR 89d:05030.

1989 Nondecomposable block designs with v=7 and λ=3 (in Bulgarian), in: *Math. and Math. Education* (Albena, 1989), Bulg. Akad. Nauk, Sofia, pp. 386–389. MR 90f: 05018.

LANG, K. / ZEITLER, H.

1985 Zur Klassifikation von Steinerzahlen, *Resultate Math.* 8, 47–53. MR 87a:05034; Z 579.51006.

LAWLESS, J.F.

1971 Pairwise balanced designs and the construction of certain combinatorial systems, in: *Proc. 2nd Louisiana Conf. Combinatorics, Graph Theory and Computing*, Baton Rouge, pp. 353–366. MR 47#6507; Z 291.05009.

LEE, S.M.

1982 Simple algebraic structures associated with Steiner triple systems, *Math. Sem. Notes Kobe Univ.* 10, 519–527. MR 85e:05024; Z 527.08002.

LEFMANN, H. / PHELPS, K.T. / RÖDL, V.

1993 Extremal problems for triple systems, *J. Combin. Des.* 1, 379–394. MR 95h: 05027; Z 817.05015.

LEFMANN, H. / PUDLÁK, P. / SACICKÝ, P.

1997 On sparse parity check matrices, *Des. Codes Crypt.* 12, 107–130.

LEI, J.-G.

1995 The existence spectrum of large sets of 3-GDDs (in Chinese), *Kexue Tongbao* 40, 2029–2030.

1996 Further results on transitive Kirkman triple systems, *J. Stat. Plann. Infer.* 51, 189–194. MR 97b:05026; Z 855.05023.

1997 Completing the spectrum for LGDD(m^v), *J. Combin. Des.* 5, 1–11. MR 97h: 05021.

1999 On the large sets of disjoint Kirkman triple systems.

LEI, J.-G. / KANG, Q.

1993 On large sets of disjoint 3-GDD, *J. Hebei Normal College* 2, 14–21.

LENZ, H.

1979 Blockpläne und verwandte Inzidenzstrukturen, *Jahrbuch Überblicke Math.*, 79–105. MR 81c:05014; Z 408.05005.

1982 A note on arcs in Steiner triple systems, *J. Combin. Inform. System Sci.* 7, 317–319. MR 85e:05025; Z 531.05011.

1991 Half a century of design theory, *Mitt. Math. Ges. Hamburg* 12, 579–593. MR 92m:05003; Z 756.05017.

LENZ, H. / ZEITLER, H.

1982 Arcs and ovals in Steiner triple systems, *Combinatorial Theory* (Ed. D. Jungnickel, K. Vedder), Lecture Notes Math. 969, Springer, 229–250. MR 84g:05029; Z 498.05012.

LENZI, D.

1990 Steiner's triple systems and a particular type of alternative or flexible algebras, *Bollet. Un. Mat. Ital.* (7) 4A, 377–384. MR 92b:17002; Z 725.17001.

1993 On some geometrical properties of alternative Cayley–Steiner algebras, *Pure Math. Appl. (A)* 3, 225–232. MR 94k:17001; Z 830.17002.

LEVAN, M. / PHELPS, K.T.

1999 Switching equivalence classes of perfect codes.

LEVENSHTEIN, V.I.

1990 Perfect codes in the metric of deletions and insertions (in Russian), *Diskret. Mat.* 3, 3–20. MR 92h:94033.

LEVI, F.W.

1942 *Finite Geometrical Systems*, University of Calcutta, India. Z 060.32304.

LIEBLER, R.A. / MAGLIVERAS, S.S. / TSARANOV, S.V.

1997 Block transitive resolutions of t-designs and Room rectangles, *J. Stat. Plann. Infer.* 58, 119–133. MR 98k:05036; Z 883.05014.

LIMBOS, M.

1980 Projective embeddings of small 'Steiner triple systems', in *Topics on Steiner Systems* (Ed. C.C. Lindner and A. Rosa), *Ann. Discrete Math.* 7, 151–173. Z 444. 05018.

LINDNER, C.C.

1971 Construction of quasigroups using the singular direct product, *Proc. Amer. Math. Soc.* 29, 263–266. MR 43#6354; Z 217, 83.

1971a Embedding partial idempotent Latin squares, *J. Combinat. Theory (A)* 10, 240–245. MR 43#1862.

1971b Finite partial cyclic triple systems can be finitely embedded, *Algebra Univ.* 1, 93–96. MR 46#1617; Z 219.05011.

1971c Identities preserved by the singular direct product, *Algebra Univ.* 1, 86–89. MR 44#5401; Z 221.20099.

1972 Identities preserved by the singular direct product, II, *Algebra Univ.* 2, 113–117. MR 46#5519; Z 249.20038.

1972a Finite embedding theorems for partial Latin squares, quasigroups and loops, *J. Combinat. Theory (A)* 13, 339–345. MR 47#3200; Z 246.05012.

1974 Construction of nonisomorphic reverse Steiner quasigroups, *Discrete Math.* 7, 281–288. MR 48#11379; Z 275.20118.

1974a A simple construction of disjoint and almost disjoint Steiner triple systems, *J. Combinat. Theory (A)* 17, 204–209. MR 55#10288; Z 287.05021.

1974b Construction of Steiner triple systems having exactly one triple in common, *Canad. J. Math.* 26, 225–232. MR 50#1914; Z 287.05020.

1974c Disjoint finite partial triple systems can be embedded in disjoint finite Steiner triple systems, *J. Combinat. Theory (A)* 18, 126–129. MR 50#9615; Z 298.05017.

1975 Some remarks on the Steiner triple systems associated with Steiner quadruple systems, *Colloq. Math.* 32, 301–306. MR 53#5329; Z 274.05013; Z 306.05006.

1975a On the structure of the Steiner triple systems derived from Steiner quadruple systems, *Colloq. Math.* 34, 137–142. MR 54#2499; Z 321.05016.

1975b Construction of Steiner quadruple systems having large numbers of nonisomorphic associated Steiner triple systems, *Proc. Amer. Math. Soc.* 49, 256–260. MR 52 #10448; Z 304.05006.

1975c A partial Steiner triple system of order n can be embedded in a Steiner triple system of order $6n + 3$, *J. Combinat. Theory (A)* 18, 349–351. MR 52#129; Z 304. 05005.

1975d A brief up-to-date survey of finite embedding theorems for partial quasigroups, *Congressus Numer.* 13, 53–78. MR 52#7929; Z 323.20074.

1976 Embedding block designs into resolvable block designs, *Ars Combinat.* 1, 215–219. MR 54#2495; Z 333.05006.

1978 How to embed a partial Steiner system, in: *Conference on Binary Systems and their Applications Taormina*, Tecnoprint, Bologna, pp. 79–86. Z 566.05011.

1980 A survey of embedding theorems for Steiner systems, *Topics on Steiner System* (Ed. C.C. Lindner and A. Rosa), *Ann. Discrete Math.* 7, 175–202. Z 457.05007.

1981 On the number of disjoint Mendelsohn triple systems, *J. Combinat. Theory (A)* 30, 326–330. MR 82k:05019; Z 483.05015.

1983 Quasigroup identities and orthogonal arrays, in: *Surveys in Combinatorics* (Southampton 1983), London Math. Soc. Lect. Note Ser. 82, Cambridge University Press, pp. 77–105. MR 86f:05033; Z 522.05014.

1987 Construction of large sets of pairwise disjoint transitive triple systems II, *Discrete Math.* 65, 65–74. MR 88g:05037; Z 631.05012.

1987a How to nest a Steiner triple system, *Ars Combinat.* 24B, 77–92. MR 89d:05037; Z 673.05012.

1990 Graph decompositions and quasigroup identities, *Le Matematiche* 45, 83–118. MR 93g:05029; Z 735.05065.

1992 A brief history of embedding of partial odd cycle systems, *J. Comb. Math. Comb. Comput.* 12, 201–214. MR 93i:05040; Z 769.05022.

1996 Embedding partial quasigroups, in: *CRC Handbook of Combinatorial Designs* (Ed. C.J. Colbourn and J.H. Dinitz), CRC Press, Boca Raton, FL, pp. 323–326. Z 847.05031.

1997 A survey of small embeddings for partial cycle systems, in: *Geometry, Combinatorial Designs and Related Structures* (Ed. J.W.P. Hirschfeld, S.S. Magliveras, M.J. de Resmini), Cambridge University Press, pp. 123–147. Z 892.05025.

LINDNER, C.C. / CRUSE, A.B.

1976 Small embeddings for partial semisymmetric and totally symmetric quasigroups, *J. London Math. Soc.* 12, 479–484. MR 55#5779; Z 338.20099.

LINDNER, C.C. / EVANS, T.

1977 *Finite Embedding Theorems for Partial Designs and Algebras*, Seminaire de Mathématiques Superieures, Université de Montréal, Presses de l'Université de Montréal, 196 pp. MR 57#208; Z 363.05017.

LINDNER, C.C. / MENDELSOHN, E. / RODGER, C.A.

1991 The intersection problem for minimum coverings of K_n by triples, *Australas. J. Combin.* 4, 179–197. MR 92k:05039; Z 767.05029.

LINDNER, C.C. / MENDELSOHN, E. / ROSA, A.

1976 On the number of 1-factorizations of the complete graph, *J. Combinat. Theory (B)* 20, 265–282. MR 58#10617; Z 293.05156; Z 321.05131.

LINDNER, C.C. / MENDELSOHN, N.S.

1973 Construction of perpendicular Steiner quasigroups, *Aequat. Math.* 9, 150–156. MR 48#6305; Z 267.20067.

LINDNER, C.C. / PHELPS, K.T.

1978 A note on partial parallel classes in Steiner systems, *Discrete Math.* 24, 109–112. MR 80c:05054; Z 401.05020.

LINDNER, C.C. / PHELPS, K.T. / RODGER, C.A.

1991 The spectrum for 2-perfect 6-cycle systems, *J. Combinat. Theory (A)* 57, 76–85. MR 92m:05025; Z 758.05015.

LINDNER, C.C. / RODGER, C.A.

1987 The spectrum of nested Steiner triple systems of order 3 (mod 6), *Ars Combinat.* 23, 75–79. MR 88h:05026; Z 621.05006.

1989 The construction of 2-fold triangulations, *J. Comb. Math. Comb. Comput.* 5, 3–13. MR 90g:05039; Z 748.05026.

1992 Decompositions into cycles II: cycle systems, in: *Contemporary Design Theory: A Collection of Surveys* (Ed. J.H. Dinitz and D.R. Stinson), Wiley, pp. 325–369. Z 774.05078.

1994 The complete solution of the problem of equationally defining closed trail systems of K_n, *Bull. Inst. Combin. Appl.* 11, 84–95. MR 95k:20111; Z 809.05017.

1997 *Design Theory*, CRC Press, Boca Raton, FL.

1999 The intersection problem for maximum packings of triples.

LINDNER, C.C. / ROSA, A.

1975 Construction of large sets of almost disjoint Steiner triple systems, *Canad. J. Math.* 27, 256–260. MR 51#5334; Z 321.05017.

1975a On the existence of automorphism free Steiner triple systems, *J. Algebra* 34, 430–443. MR 53#2707; Z 308.05014.

1975b Finite embedding theorems for partial Steiner triple systems, *Discrete Math.* 13, 31–39. MR 52#128; Z 309.05013.

1975c Steiner triple systems having a prescribed number of triples in common, *Canad. J. Math.* 27, 1166–1175. MR 56#11819; Z 336.05008. Corrigendum: 30 (1978), 896. MR 58#278.

1976 Finite embedding theorems for partial triple systems with $\lambda > 1$, *Ars Combinat.* 1, 159–166. MR 53#12973; Z 334.05021.

1976a Steiner quadruple systems all of whose derived Steiner triple systems are nonisomorphic, *J. Combinat. Theory (A)* 21, 35–43. MR 53#12974; Z 332.05013.

1977 A partial Room square can be embedded in a Room square, *J. Combinat. Theory (A)* 22, 97–102. MR 55#5467; Z 349.05014.

LINDNER, C.C. / STRALEY, T.H.

1971 Construction of quasigroups containing a specified number of subquasigroups of a given order, *Algebra Univ.* 1, 238–247. MR 45#3618; Z 229.20074.

1976 A note on nonisomorphic reverse Steiner quasigroups, *Publicationes Math.* 23, 85–87. MR 54#443; Z 359.20071.

LINDNER, C.C. / STREET, A.P.

1983 Construction of large sets of pairwise disjoint transitive triple systems, *Europ. J. Combinat.* 4, 335–346. MR 85g:05050; Z 533.05010.

1984 Ordered triple systems and transitive quasigroups, *Ars Combinat.* 17A, 297–306. MR 85e:05026; Z 542.05019.

1991 The Stern–Lenz theorem: background and applications, *Congressus Numer.* 80, 5–21. MR 92m:05030; Z 754.05059.

LINDNER, C.C. / WALLIS, W.D.

1982 Embeddings and prescribed intersections of transitive triple systems, *Ann. Discrete Math.* 15, 265–272. MR 86j:05031; Z 498.05017.

LINDSTRÖM, B.

1988 Any STS(13) contains the non-Pappus configuration, *Utilitas Math.* 33, 193–194. MR 89d:05038; Z 647.51005.

LING, A.C.H.

1997 *Pairwise Balanced Designs and Related Codes*, Ph.D. Thesis, University of Waterloo.

1997a A direct product construction for 5-sparse triple systems, *J. Combin. Des.* 5, 443–447. MR 98k:05024.

LING, A.C.H. / COLBOURN, C.J.

1997 Rosa triple systems, in: *Geometry, Combinatorial Designs and Related Structures* (Ed. J.W.P. Hirschfeld, S.S. Magliveras, M.J. de Resmini), Cambridge University Press, pp. 149–159. Z 882.05025.

LING, A.C.H. / COLBOURN, C.J. / GRANNELL, M.J. / GRIGGS, T.S.
1999 Construction techniques for anti–Pasch Steiner triple systems.
LIU, K.
1984 *Large Sets of Disjoint Steiner Triple Systems* (in Chinese), M.Sc. Thesis, Beijing University.
LIZZIO, A.
1990 On the parameter $D(v, t_v - 15)$ for STS(v), *J. Comb. Inf. Syst. Sci.* 15, 29–33. MR 92g:05041; Z 757.05024.
LIZZIO, A. / MILICI, S.
1983 Constructions of disjoint and mutually balanced partial Steiner triple systems, *Boll. Un. Matem. Ital.* A(6) 2, 183–191. MR 85f:05022; Z 516.05011.
1985 On some pairs of partial triple systems, *Rend. Istit. Mat. Univ. Trieste* 17, 21–29. MR 87m:05032; Z 616.05012.
LIZZIO, A. / MILICI, S. / QUATTROCCHI, G.
1990 On a decomposition of $2K_n$, *J. Combinat. Inf. Syst. Sci.* 15, 204–210. MR 92f: 05077; Z 760.05072.
LIZZIO, A. / QUATTROCCHI, G.
1995 On complete arcs in $H(v, 3, 1)$, *Boll. Un. Mat. Ital. A* 9, 549–656. MR 97a: 05027; Z 848.05010.
LLOYD, E.K.
1995 The reaction graph of the Fano plane, in: *Combinatorics and Graph Theory 95, Hefei, China, 1995* (Ed. Ku Tung Shin) World Scientific, Singapore, pp. 260–274. MR 98k:05129.
LO FARO, G.
1983 On the size of partial parallel classes in Steiner systems STS(19) and STS(27), *Discrete Math.* 45, 307–312. MR 84m:05019; Z 522.05012.
1985 Partial parallel classes in Steiner system $S(2, 3, 19)$, *J. Inform. Optim. Sci.* 6, 133–136. MR 86m:05021; Z 606.05011.
1987 Kirkman triple systems having a prescribed number of triples in common, *Ars Combinat.* 24B, 9–21. MR 89e:05028; Z 679.05009.
1995 A class of extended triple systems and numbers of common triples, *Australas. J. Combin.* 12, 209–224. MR 96e:05023; Z 838.05013.
1997 Constructing 3-chromatic Steiner triple systems, *Discrete Math.* 174, 187–190. MR 98h:05032.
LONGYEAR, J.Q.
1986 Nested group divisible designs and small nested designs, *J. Stat. Plann. Infer.* 13, 81–87. MR 87f:62143; Z 591.05007.
LORENA, L. / DE SOUZA LOPES, L.
1997 Genetic algorithms applied to computationally difficult set covering problems, *J. Oper. Res. Soc.* 48, 440–445. Z 887.90119.
LORIMER, P.
1974 A class of block designs having the same parameters as the designs of points and lines in a projective 3-space, in: *Combinatorial Mathematics*, Lecture Notes in Math. 403, Springer, Berlin, pp. 73–78. MR 52#130; Z 319.05010.
LOVÁSZ, L.
1968 On chromatic number of finite set-systems, *Acta Math. Acad., Sci. Hung.* 19, 59–67. MR 36#3673; Z 157.55203.
1968a Graphs and set systems, in: *Beiträge zur Graphentheorie*, Teubner, Leipzig, pp. 99–106. MR 39#5417; Z 175.21101.
LOVÁSZ, L. / PLUMMER, M.D.

1986† *Matching Theory*, North-Holland, Amsterdam.

Lu, J.X.

1965 Construction methods for balanced incomplete block designs and resolvable balanced incomplete block designs (in Chinese; unpublished); in: *Collected Works of Lu Jiaxi on Combinatorial Designs* (Ed. L. Wu, L. Zhu, and Q. Kang) Inner Mongolia People's Press, Hunhot, Inner Mongolia, 1990, pp. 1–24.

1983 On large sets of disjoint Steiner triple systems, I, *J. Combinat. Theory (A)* 34, 140–146. MR 84e:05031a; Z 514.05018.

1983a On large sets of disjoint Steiner triple systems, II, *J. Combinat. Theory (A)* 34, 147–155. MR 84e:05031b; Z 514.05019.

1983b On large sets of disjoint Steiner triple systems, III, *J. Combinat. Theory (A)* 34, 156–182. MR 84e:05031c; Z 514.05020.

1984 On large sets of disjoint Steiner triple systems, IV, *J. Combinat. Theory (A)* 37, 136–163. MR 86a:05019; Z 545.05014.

1984a On large sets of disjoint Steiner triple systems, V, *J. Combinat. Theory (A)* 37, 164–188. MR 86a:05019; Z 545.05015.

1984b On large sets of disjoint Steiner triple systems, VI, *J. Combinat. Theory (A)* 37, 189–192. MR 86a:05019; Z 557.05012.

1984c On large sets of disjoint Steiner triple systems, VII. (Unpublished manuscript.)

Lu, Q.

1995 Concerning special pure Mendelsohn triple systems, *J. Comb. Math. Comb. Comput.* 17, 125–128. MR 96b:05043; Z 828.05014.

Lucas, E.

1883 *Récréations mathématiques*, Vol. 2, Gauthier–Villars, Paris. (Sixième récréation: Les jeux de demoiselles, 161–197.)

Luks, E.M.

1982† Isomorphism of graphs of bounded valence can be tested in polynomial time, *J. Comp. Syst. Sci.* 25, 42–65.

Lüneburg, H.

1963 Steinersche Tripelsysteme mit fahnentransitiver Kollineationsgruppe, *Math. Ann.* 149, 261–270. MR 26#2933; Z 105, 131.

Luo, J.J.

1986 A historical review of research on some topics in Steiner systems, with a survey of the background of the scientific work of Jia Xi Lu (in Chinese), *Adv. Math. (Beijing)* 15, 175–184. MR 88f:05023.

Lyu, L.-T. / Lin, M.-C.

1994 On noncatastrophic binary (d, k) trellis codes with large free distances, *Proc. 1994 IEEE Int. Symp. Information Theory*, IEEE, 345.

Macavei, M.

1972 Algoritme grafice pentru construirea sistemelor de triplete Steiner, *Studii Cercet. Mat.* 24, 351–357. MR 51#2940; Z 238.05006.

Macdonald, S.O.

1976 A characterization of affine triple systems, *Ars Combinat.* 2, 23–24. MR 55 #2600; Z 355.05005.

Macfarlane, A.

1916 *Lectures on Ten British Mathematicians of the Nineteenth Century*, Wiley, New York. 5: Cayley, 8: Sylvester, 9: Kirkman

MacNeish, H.F.

1922† Euler squares, *Ann. Math.* 23, 221–227.

Mammana, M.F.

1996 On the upper edge-colourings for Steiner systems, *J. Combin. Inf. Syst. Sci.* 21, 81–86.

MANNINO, C. / SASSANO, A.

1994 An exact algorithm for the maximum cardinality stable set problem, *Comput. Optim. Appl.* 3, 243–258. MR 95f:05105; Z 821.90131.

1995 Solving hard set covering problems, *Oper. Res. Lett.* 18, 1–5. Z 857.90093.

MARCZAK, A.W.

1997 On nondistributive Steiner quasigroups, *Colloq. Math.* 74, 135–145. MR 98e: 20071; Z 891.05012.

MARINO, M.C.

1984 On Mendelsohn triple systems (in Italian), *Atti Sem. Mat. Fis. Univ. Modena* 33, 11–19. MR 87b:05041; Z 582.05008.

MARKOV, A.A.

1965 Ob odnoj kombinatornoj zadače, *Problemy Kibernet.* 15, 263–266. MR 35 #1497.

MARSDEN, E.

1883 The school-girls' problem, *Knowledge* 3, 183.

MATHON, R.A.

1991 Computational methods in design theory, in: *Surveys in Combinatorics 1991* (Ed. A.D. Keedwell), Cambridge University Press, Cambridge, pp. 101–117. MR 93c: 05024; Z 736.05014.

1996 Computational methods in design theory, in: *Computational and Constructive Design Theory* (Ed. W.D. Wallis) Kluwer, pp. 29–48. Z 856.05010.

1997 Searching for spreads and packings, in: *Geometry, Combinatorial Designs and Related Structures* (Ed. J.W.P. Hirschfeld, S.S. Magliveras, M.J. de Resmini), Cambridge University Press, pp. 161–176. Z 887.51009.

MATHON, R.A. / LOMAS, D.

1992 A census of 2-(9,3,3) designs, *Australas. J. Combin.* 5, 145–158. MR 93a:05023; Z 770.05009.

MATHON, R.A. / PHELPS, K.T. / ROSA, A.

1981 A class of Steiner triple systems of order 21 and associated Kirkman systems, *Math. Comput.* 37, 209–222. MR 83k:05022; Z 473.05009. Addendum: *Math. Comput.* 64 (1995), 1355–1356. Z 824.05008.

1983 Small Steiner triple systems and their properties, *Ars Combinat.* 15, 3–110. MR 85e:05027; Z 516.05010.

MATHON, R.A. / PIETSCH, CH.

1999 On the family of 2-(7,3,λ) designs.

MATHON, R.A. / ROSA, A.

1977 A census of Mendelsohn triple systems of order nine, *Ars Combinat.* 4, 309–315. MR 57#2933; Z 442.05006.

1983 A census of 1-factorizations of K_9^3: Solutions with group of order > 4, *Ars Combinat.* 16, 129–147. MR 85b:05137; Z 539.05019.

1984 The 4-rotational Steiner and Kirkman triple systems of order 21, *Ars Combinat.* 17A, 241–250. MR 85e:05028; Z 537.05010.

1985 Tables of parameters of BIBDs with $r \leq 41$ including existence, enumeration, and resolvability results, *Ann. Discrete Math.* 26, 275–308. MR 87a:05003; Z 579. 05016.

1996 2-(v, k, λ) designs of small order, in: *CRC Handbook of Combinatorial Designs* (Ed. C.J. Colbourn and J.H. Dinitz), CRC Press, Boca Raton, FL, pp. 3–41. Z 845. 05008.

MATHON, R.A. / STREET, A.P.

1995 Partitions of sets of two-fold triple systems, and their relation to some strongly regular graphs, *Graphs Combin.* 11, 347–366. MR 97d:05039; Z 837.05021.

1997 Partitions of sets of designs on seven, eight, and nine points, *J. Stat. Plann. Infer.* 58, 135–150. MR 98e:05014.

1997a Overlarge sets and partial geometries, *J. Geometry* 60, 85–104. MR 98k:51013; Z 897.51003.

MATHON, R.A. / VANSTONE, S.A.

1980 On the existence of doubly resolvable Kirkman systems and equidistant permutation arrays, *Discrete Math.* 30, 157–172. MR 81h:05025; Z 447.05021.

1980a Doubly resolvable Kirkman systems, *Congressus Numer.* 29, 611–625. MR 82e:05030; Z 459.05021.

MCDIARMID, C.J.H.

1972† The solution of a time-tabling problem, *J. Inst. Math. Appl.* 9, 23–34.

MCLEISH, M.

1975† On the existence of Latin squares with no subsquares of order two, *Utilitas Math.* 8 41–53.

MENDELSOHN, E.

1975 On the groups of automorphisms of Steiner triple and quadruple systems, *Congressus Numer.* 13, 255–264. MR 52#5438; Z 322.05012.

1978 The smallest non-derived Steiner triple system is simple as a loop, *Algebra Univ.* 8, 256–259. MR 57#2934; Z 375.20059.

1978a On the groups of automorphisms of Steiner triple and quadruple systems, *J. Combinat. Theory (A)* 25, 97–104. MR 80d:05010; Z 399.05010.

1978b Every (finite) group is the group of automorphism of a (finite) strongly regular graph, *Ars Combinat.* 6, 75–86. MR 81a:05065; Z 455.05035.

1980 Perpendicular arrays of triple systems, in: *Topics on Steiner Systems* (Ed. C.C. Lindner and A. Rosa), *Ann. Discrete Mathematics* 7, 89–94. Z 431.05010.

1985 The work of Lu Jia Xi, *J. Math. Res. Expos.* 5, 143–144.

1996 Mendelsohn designs, in: *CRC Handbook of Combinatorial Designs* (Ed. C.J. Colbourn and J.H. Dinitz), CRC Press, Boca Raton, FL, pp. 388–393. Z 845.05021.

MENDELSOHN, E. / ASSAF, A.

1987 On the spectrum of imbrical designs, *Ann. Discrete Math.* 34, 363–369. MR 88k:05058; Z 636.05007.

MENDELSOHN, E. / REES, R.

1988 Mandatory representation designs, *J. Combinat. Theory (A)* 49, 349–362. MR 89i:05050; Z 661.05014.

MENDELSOHN, E. / ROSA, A.

1983 Embedding maximal packings of triples, *Congressus Numer.* 40, 235–247. MR 85i:05065; Z 534.05014.

1985† One-factorizations of the complete graph — a survey, *J. Graph Theory* 9 (1985), 43–65.

1997 Ubiquitous configurations in Steiner triple systems, *J. Combin. Des.* 5, 13–31. MR 97h:05026.

MENDELSOHN, E. / SHALABY, N.

1990 Decomposing the complement of 3-nets (latin squares) into triples (A natural generalization of GDDs), *Ars Combinat.* 30, 209–224. MR 91m:05044; Z 738.05020.

MENDELSOHN, E. / SHALABY, N. / SHEN, H.

1991 Nuclear designs, *Ars Combinat.* 32, 225–238. MR 92j:05042; Z 751.05020.

1992 All leaves and excesses are realizable for $k = 3$ and all λ, in: *Proc. Fourth Czechoslovak Symp. Combin., Graph Theory and Complexity* (Ed. J. Nešestril and M. Fiedler), *Ann. Discrete Math.* 51, pp. 223–228. MR 93j:05033; Z 771.05019.

MENDELSOHN, E. / SHEN, H.

1987 A construction of resolvable group divisible designs with block size 3, *Ars Combinat.* 24, 39–43. MR 88k:05048; Z 638.05012.

MENDELSOHN, N.S.

1969 Combinatorial designs as models of universal algebras, in: *Recent Progress in Combinatorics*, Proc. 3rd Waterloo Conf. 1968, pp. 123–132. MR 41#85; Z 192, 333.

1970 Orthogonal Steiner systems, *Aequat. Math.* 5, 268–272. MR 44#1587; Z 212, 35.

1971 A natural generalization of Steiner triple systems, in: *Computers in Number Theory*, Academic Press, New York, pp. 323–338. MR 48#122; Z 216, 301.

1971a A single groupoid identity for Steiner loops, *Aequat. Math.* 6, 228–230. MR 45 #6969; Z 244.20087.

MERTELSMANN, A.F.H.

1898 Das Problem der 15 Pensionatsdamen, *Zeitschr. Math. Phys.* 43, 329–334.

METZ, R. / WERNER, H.

1976 On automorphism groups of finite Steiner triple systems, in: *Teorie Combinatorie* II, Colloq. Internat. Roma 1973, Atti dei Convegni Lincei 17, Roma, pp. 373–376. MR 57#12253; Z 441.05010.

MIAO, Y.

1997 *Construction of Block Designs with Resolvability or Nested Structure*, D.Sc. Thesis, Hiroshima University, Japan.

MIAO, Y. / KAGEYAMA, S. / DUAN, X.

1996 Further constructions of nested group divisible designs, *J. Japan Stat. Soc.* 26, 231–239. MR 98f:05025; Z 884.62083.

MIAO, Y. / ZHU, L.

1989 Existence of incomplete group divisible designs, *J. Comb. Math. Comb. Comput.* 6, 33–49. MR 90i:05017; Z 688.05009.

MICALE, B. / PENNISI, M.

1993 Cyclic and rotational oriented triple systems, *Ars Combinat.* 35, 65–68. MR 94a:05022; Z 783.05020.

1997 On the directed triple systems with a given automorphism, *Australas. J. Combin.* 15, 233–240. MR 97m:05031; Z 880.05018.

1998 The spectrum of d-cyclic oriented triple systems, *Ars Combinat.* 48, 219–223. MR 98m:05019.

MIGLIORATO, R.

1988 Codici corretori e sistemi di terne di Steiner, *Rend. Mat.* (7) 8, 265–275. MR 90f: 94022; Z 724.94015.

MIGLIORI, G.

1987 Sets of type $(0, 2, q/2)$ in a projective plane and Steiner triple system (in Italian), *Rend. Math.* (7) 7, 77–82. MR 90h:51011; Z 682.51005.

MILAZZO, L.

1997 The monochromatic block number, *Discrete Math.* 165/166, 487–496.

MILAZZO, L. / TUZA, ZS.

1997 Upper chromatic number of Steiner triple and quadruple systems, *Discrete Math.* 174, 247–259. MR 98k:05025.

1998 Strict colorings for classes of Steiner triple systems, *Discrete Math.* 182, 233–243. MR 98i:05021.

MILICI, S.

1986 On the parameter $D(v, t_v - 13)$ for Steiner triple systems, *Ann. Discrete Math.* 30, 311–330. MR 88c:05030; Z 594.05018.

1989 On the existence of MPT-free $S_3(2,3,v)$'s. *Utilitas Math.* 35, 169–175. MR 90c: 05027; Z 668.05009.

1991 Indecomposable $S_6(2,3,v)$'s, *J. Combinat. Theory (A)* 56, 16–26. MR 91k: 05020; Z 729.05008.

1991a On the existence of $S_3(2,3,v)$'s without two-times repeated blocks, in: *Combinatorics 88*, Mediterranean Press, 1991, pp. 253–272. MR 94a:05018.

MILICI, S. / QUATTROCCHI, G.

1983 Alcune condizioni necessarie per l'esistenza di tre DMB PTS con elementi di grado 2, *Mathematiche (Catania)* 38, 113–132. MR 89b:05046; Z 624.05012.

1985 $D(v, t_v - 14)$ per sistemi di terne di Steiner, *Matematiche (Catania)* 40, 93–105. MR 90a:05030; Z 655.51006.

1986 Some results on the maximum number of STSs such that any two of them intersect in the same block-set, *J. Inform. Optim. Sci.* 7, 291–302. MR 87k:05033; Z 607.05013.

1987 On the intersection problem for three Steiner triple systems, *Ars Combinat.* 24A, 175–194. MR 89f:05038; Z 673.05013.

1988 The spectrum for three-times repeated blocks in a $S_3(2,3,v)$, *J. Combinat. Theory (A)* 49, 117–128. MR 89f:05039; Z 657.05012.

1990 Ovals and hyperovals in maximum partial triple systems, *J. Combinat. Inf. System Sci.* 15, 218–232. MR 92i:05044; Z 744.51010.

1997 The fine structure of threefold directed triple systems, *Australas. J. Combin.* 16, 125–154. MR 98h:05033.

MILICI, S. / QUATTROCCHI, G. / SHEN, H.

1995 Embeddings of simple maximum packings of triples with λ even, *Discrete Math.* 145, 191–200. MR 97a:05064; Z 843.05015.

1996 Support sizes of directed triple systems, *J. Combinat. Theory (A)* 75, 198–216. MR 97e:05043; Z 858.05018.

MILICI, S. / TUZA, ZS.

1994 The spectrum of λ-times repeated blocks for TS(v,λ), *Discrete Math.* 129, 159–166. MR 95b:05033; Z 807.05011.

1995 Coverable graphs, *Studia Sci. Math. Hungar.* 30, 329–344. MR 96f:05149; Z 844. 05077.

1996 Cycle systems without 2-colorings, *J. Combin. Des.* 4, 135–142. MR 97a:05045.

MILLER, G.L.

1978 On the $n^{\log n}$ isomorphism technique, in: *Proc. Tenth Annual ACM Sympos. Theory of Computing*, San Diego, CA, pp. 51–58.

MILLS, S.

1980 Thomas Kirkman — the mathematical cleric of Croft, *Memoirs of the Manchester Literary and Philosophical Society* 120, 100–109.

MILLS, W.H.

1973 Covering problems, *Congressus Numer.* 8, 23–52. MR 50#12771; Z 318.05003.

MISHIMA, M. / MIAO, Y. / KAGEYAMA, S. / JIMBO, M.

1998 Constructions of nested directed BIB designs, *Australas. J. Combin* 18, 157–172.

MITSCHKE, A. / WERNER, H.

1973 On groupoids representable by vector spaces over finite fields, *Arch. Math.* 24, 14–20. MR 47#5148; Z 255.20049.

MÖBIUS, A.
1886 Zur Theorie der Polyëder und der Elementarverwandtschaft, in: *Gesammelte Werke II* (Ed. F. Klein), Hirzel, Leipzig, pp. 519-559.
MOORE, E.H.
1893 Concerning triple systems, *Math. Ann.* 43, 271–285.
1895 The group of holoedric transformation into itself of a given group, *Bull. Amer. Math. Soc.* 1, 61–66.
1895a Concerning triple systems, *Rend. Circ. Math. Palermo* 9, 86.
1896 Tactical memoranda I–III, *Amer. J. Math.* 18, 264–303.
1897 Concerning regular triple systems, *Bull. Amer. Math. Soc.* 4, 11–16.
1898 Concerning abelian-regular transitive triple systems, *Math. Ann.* 50, 225–240.
1899 Concerning the general equations of the seventh and eighth degrees, *Math. Ann.* 51, 417–444.
MORAN, A.T.
1997 *Block Designs and their Defining Sets*, Ph.D. Thesis, University of Queensland, Australia.
MORGAN, E.J.
1977 Some small quasi-multiple designs, *Ars Combinat.* 3, 233–250. MR 56#15456; Z 394.05005.
1980 Balanced ternary designs with block size three, in: *Combinatorial Mathematics VII*, Lecture Notes in Math. 748, Springer, Berlin, pp. 79–91. MR 85a:05029; Z 454. 05017.
MORGAN, J.P.
1996 Nested designs, in: *Handbook of Statistics 13: Design and Analysis of Experiments* (Ed. S. Ghosh and C.R. Rao), Elsevier, pp. 939–976.
MOUETTE, L.
1956 Groupements en triades. Problème de Kirkman généralisé, *Mathesis* 65, 46–52. MR 17, 1173; Z 71, 13.
1957 Recherches sur la théorie des triades, *Mathesis* 66, 283–287. MR 20#5743; Z 79, 10.
MOURA, L.
1999 *Polyhedral Aspects of Combinatorial Designs*, Ph.D. Thesis, Computer Science, University of Toronto.
MUKERJEE, R. / JIMBO, M. / KAGEYAMA, S.
1987 On cyclic semiregular group divisible designs, *Osaka J. Math.* 24, 395–407. MR 89h:05018; Z 644.62078.
MULDER, P.
1917 *Kirkman-Systemen*, Academisch Proefschrift ter verkrijging van den graad van doctor in de Wis-en Natuurkunde aan de Rijksuniversiteit te Groningen, Leiden.
MÜLLER, P.
1982 *Durchschnittseigenschaften von Steiner-Tripel-Systemen*, Dissertation, Universität Bayreuth, 78 pp. Z 532.51001.
MULLIN, R.C. / GRONAU, H.D.-O.F.
1996† PBDs, frames, and resolvability, in: *CRC Handbook of Combinatorial Designs* (C.J. Colbourn, J.H. Dinitz; eds.), CRC Press, Boca Raton, FL, pp. 224–226.
MULLIN, R.C. / NEMETH, E.
1969 On furnishing Room squares, *J. Combinat. Theory* 7, 266–272. MR 41#5228; Z 186, 19.

1970 On the nonexistence of orthogonal Steiner systems of order 9, *Canad. Math. Bull.* 13, 131–134. MR 41#3297; Z 195, 31.

MULLIN, R.C. / POPLOVE, A.L. / ZHU, L.

1987 Decomposition of Steiner triple systems into triangles, *J. Comb. Math. Comb. Comput.* 1, 149–174. MR 88e:05018; Z 651.05017.

MULLIN, R.C. / ROSA, A.

1977 Orthogonal Steiner systems and generalized Room squares, *Congressus Numer.* 18, 315–323. MR 58#27560; Z 482.05021.

MULLIN, R.C. / STINSON, D.R.

1987 Pairwise balanced designs with block sizes $6t+1$, *Graphs Combinat.* 3, 365–377. MR 88j:05008; Z 659.05015.

MULLIN, R.C. / STINSON, D.R. / VANSTONE, S.A.

1982 Kirkman triple systems containing maximum subdesigns, *Utilitas Math.* 21C, 283–300. MR 83j:05012; Z 498.05013.

MULLIN, R.C. / VANSTONE, S.A.

1980 Steiner systems and Room squares, in: *Topics on Steiner Systems* (Ed. C.C. Lindner and A. Rosa), *Ann. Discrete Math.* 7, 95–104. Z 461.05012.

MULLIN, R.C. / WALLIS, W.D.

1975† The existence of Room squares, *Aequat. Math.* 13, 1–7.

NAGELL, T.

1963 Thoralf Skolem in memoriam, *Acta Math.* 110, i–xi and 303. Z 112.00316.

NANDI, H.K.

1946 Enumeration of nonisomorphic solutions of balanced incomplete block designs, *Sankhya* 7, 305–312. MR 8, 127d; Z 060.31411.

NASH-WILLIAMS, C.ST.J.A.

1970 Problem, in: *Combinatorial Theory and its Applications III* (Ed. P. Erdös, A. Rényi, and V.T. Sós), North-Holland, Amsterdam, 1970, pp. 1179–1182.

1972 Simple constructions for balanced incomplete block designs with block size three, *J. Combinat. Theory (A)* 13, 1–6, MR 46#3337; Z 258.05012.

NAZAROK, A.V.

1993 On the existence of multidimensional Kirkman hypercubes of a certain class (in Ukranian), *Dopovidi Akad. Nauk Ukrainy* 7, 34–35. MR 94k:05040.

1995 Orthogonal starters and multidimensional Kirkman hypercubes (in Russian), *Diskret. Analyz.* 7, 136–139. (English: *Discrete Math. Appl.* 5 (1995), 541–545.) MR 97a:05034; Z 847.05014.

NEPOROŽNEV, I.P. / PETRENJUK, A.JA.

1972 Konstruktivnoje perečislenije sistem grupp par i oglavlennyje sistemy trojek Štejnera, *Kombinatornyj Analiz* 2, 17–37. MR 58#21682a.

1974 Konstruktivnoje perečislenije sistem grupp par i oglavlennyje sistemy trojek Štejnera (prodolzhenije), *Kombinatornyj Analiz* 3, Izdatel'stvo MGU, Moskva, 28–42. MR 58#21682b.

1980 Konstruktivnoje perečislenije sistem grupp par i oglavlennyje sistemy trojek Štejnera, *Kombinatornyj Analiz* 5, Izdatel'stvo MGU, Moskva, 99–102. MR 85g: 05016.

NEŠETŘIL, J. / PHELPS, K.T. / RÖDL, V.

1983 On the achromatic number of simple hypergraphs, *Ars Combinat.* 16, 95–102. MR 85e:05133; Z 539.05047.

NEŠETŘIL, J. / RÖDL, V.

1987 Strong Ramsey theorems for Steiner systems, *Trans. Amer. Math. Soc.* 303, 183–192. MR 89b:05127; Z 632.05040.

NEŠETŘILOVÁ, H. / NEŠETŘIL, J.

1972 Poznámka ke Kirkmanovu problému školních dívek, *Dějiny Vědy Techn.* 5, 171–173.

NETTO, E.

1881 *Die Substitutionentheorie und ihre Anwendungen in der Algebra*, Teubner, Leipzig. (English translation by F.N. Cole: *The Theory of Substitutions and its Applications to Algebra*, A.A. Register Publ. Co., 1892. Reprinted by Chelsea, New York, 1964.) Chapter XII: Equations with rational relations between three roots.

1893 Zur Theorie der Tripelsysteme, *Math. Ann.* 42, 143–152.

1898 Kombinatorik, *Encyklopädie der mathematischen Wissenschaften I*, Heft 1, 28–46. Teubner, Leipzig.

1901 *Lehrbuch der Combinatorik*, Teubner, Leipzig. 2nd ed. (with notes by V. Brun and Th. Skolem), 1927. Reprinted by Chelsea, New York, 1968. MR 20#1632.

NEUMAIER, A.

1979 Strongly regular graphs with smallest eigenvalue $-m$, *Arch. Math.* 33, 392–400. MR 81e:05118; Z 415.05028.

NOETHER, M.

1879 Über die Gleichungen achten Grades und ihr Auftreten in der Theorie der Curven vierter Ordnung, *Math. Ann.* 15, 89–110.

1895 Note über die Siebensysteme von Kegelschnitten, welche durch die Berührungspunkte der Doppeltangenten einer ebenen Curve vierter Ordnung gehen, *Math. Ann.* 46, 545–556.

NOVÁK, J.

1963 Příspěvek k teorii kombinací, *Časopis Pěst. Mat.* 88, 129–141. MR 28#2977; Z 138, 244.

1970 Maximální systémy trojic z 12 prvku, *Mathematics (Geometry and Graph Theory)*, Univ. Karlova, Praha, 105–110. MR 43#3141; Z 216, 302.

1970a Über gewisse Tripel-Systeme, in: *Combinatorial Theory and its Applications* III, Colloq. Math. Soc. J. Bolyai 4, North-Holland, Amsterdam, pp. 821–828. MR 45 #8543; Z 217, 18.

1974 Edge bases of complete uniform hypergraphs, *Mat. Čas.* 24, 43–57. MR 59 #9710; Z 273.05132.

1975 A note on disjoint cyclic Steiner triple systems, in: *Recent Advances in Graph Theory*, Academia, Praha, pp. 439–440. MR 52#2920; Z 365.05012.

NOVÁK, J. / VILD, J.

1966 Některé maximální systémy trojic, *Sborník vědeckých prací vys. školy strojní a textilní v Liberci*, pp. 29–36.

NOWAKOWSKI, R.J.

1975† *Generalizations of the Langford–Skolem Problem*, M.Sc. Thesis, University of Calgary.

ODIJK, M.A. / VAN MAAREN, M.

1998 Improved solutions to the Steiner triple covering problem, *Inform. Proc. Lett.* 65, 67–69.

O'KEEFE, E.S.

1961 Verification of a conjecture of Th. Skolem, *Math. Scand.* 9, 80–82. MR 23A #2331; Z 105, 250.

OPENCOMB, W.E.

1984 On the intricacy of combinatorial construction problems, *Discrete Math.* 50, 71–97. MR 86a:05002; Z 548.05013.

ORAVAS, M.A.

1993 *On the Existence of Almost Uniform Linear Spaces*, Ph.D. Thesis, McMaster University.

O'SHAUGHNESSY, C.D.

1968 A Room design of order 14, *Canad. Math. Bull.* 11, 191–194. MR 36#3940.

1972 On Room squares of order $6m+2$, *J. Combinat. Theory (A)* 13, 306–314. MR 46#1619; Z 246.05018.

OSTROM, T.G. / SHERK, F.A.

1964 Finite projective planes with affine planes, *Canad. Math. Bull.* 7, 549–559. MR 30#2390; Z 125, 384.

PÁLFY, P.P.

1987 Isomorphism problem for relational structures with a cyclic automorphism, *Europ. J. Combinat.* 8, 35–43. MR 88i:05097; Z 614.05049.

PARKER, E.T.

1975 On sets of pairwise disjoint blocks in Steiner triple systems, *J. Combinat. Theory (A)* 19, 113–114. MR 51#12559; Z 302.05012.

PEIRCE, B.

1860 Cyclic solutions of the school-girl puzzle. *Astronomical Journal* (U.S.A.) 6, 169–174.

PELIKÁN, J.

1970 Properties of balanced incomplete block designs, in: *Combinatorial Theory and its Applications* III, Colloq. Math. Soc. J. Bolyai 4, North-Holland, Amsterdam, pp. 869–889. MR 46#8859; Z 217, 20.

PELTESOHN, R.

1936 *Das Turnierproblem für Spiele zu je dreien*, Dissertation Berlin, August Pries, Leipzig. Z 13, 338.

1939 Eine Lösung der beiden Heffterschen Differenzenprobleme, *Compositio Math.* 6, 251–257. Z 20, 39.

PENNISI, M. / ROMANO, R.

1991 Parallel classes in $S_\lambda(2,3,v)$, *Pure Math. Appl.* (A) 2, 135–139. MR 92i:05045; Z 758.05024.

PERFECT, H.

1995 The Revd. Thomas Penyngton Kirkman FRS (1806–1895): schoolgirl parades — but much more! *Math. Spectrum* 28, 1–6.

PETER-ORTH, C.

1987 On solving computationally difficult set covering problems, *Ars Combinat.* 24A, 71–96. MR 89d:05039; Z 676.05018.

PETERSEN, J.

1891† Die Theorie der regulären Graphen, *Acta Math.* 15, 193–220.

PETRENJUK, A.JA.

1970 Nekotorye ocenki čisel Mano, *Vestnik Moskov. Univ. Mat. Mekh.* 25, 31–35. (English translation: Some bounds for Mano numbers, *Vestn. Mosk. Univ.* 25 (1970), 87–90.) MR 44#2634.

1971 O neizomorfnych oglavlennych sistemach trojek Štejnera i sistemach grupp par, *Kibernetika (Kiev)*, No. 4, 93–102. Z 226.05014.

1972 Priznaki neizomorfnosti sistem trojek Štejnera, *Ukr. Mat. Ž.* 24, 772–780, 861. (English translation: Tests for nonisomorphic Steiner triple systems, *Ukr. Mat. J.* 24, 620–626.) MR 48#123; Z 256.05007.

PETRENJUK, A.JA. / PETRENJUK, V.I.

1988 Some more nonisomorphic KRs STS(21). (Unpublished manuscript.)

PETRENJUK, L.P. / PETRENJUK, A.JA.

1976 Postrojenie nekotorych klassov kubičeskich grafov i neizomorfnost' Kirkmanovych sistem trojek, *Kombinatornyj Analiz* 4, Izdat. Mosk. Univ., Moskva, 73–77. MR 56#15499; Z 436.05059.

1980 An enumeration method for non-isomorphic combinatorial designs, in: *Topics on Steiner Systems* (Ed. C.C. Lindner and A. Rosa), *Ann. Discrete Math.* 7, 265–276. Z 457.05016.

1983 Metod H-preobrazovanii v primenenii k Kirkmanovym rozrešeniam Štejnerovykh sistem, *Kirovogr. Inst. S.-Kh. Mašinostr.*, Kirovograd, 26 pp.

1983a O semeistve neizomorfnykh kirkmanovykh rozrešenii odnoi Šteinerovoi sistemy trojek, *Kombinatorno-algebr. metody v prikladnoi matematike*, Gor'kii, 176–192. Z 568.05009.

1983b O neizomorfnykh B(25,4,1) i KST(21), Kirovogr. Inst. S.-Kh. Mašinostr., Kirovograd, 9 pp.

PETRENJUK, V.I. / PETRENJUK, A.JA.

1995 Pro novi kirkmanovi systemy trijok poriadku 21, in: *Naučnyje Trudy Akademii*, Vyp. I, Gos. Letn. Akad. Ukr., Kirovograd, pp. 134–145.

PHELAN, J.S.

1991 *Anti-Pasch and Pasch Steiner Triple Systems*, M.Phil. Thesis, Lancashire Polytechnic.

PHELPS, K.T.

1975 *Derived Triple Systems of Order 15*, M.Sc. Thesis, Auburn University, Alabama, 56 pp.

1976 Some sufficient conditions for a Steiner triple system to be a derived triple system, *J. Combinat. Theory (A)* 20, 393–397. MR 53#12975; Z 329.05010.

1976a Some derived Steiner triple systems, *Discrete Math.* 16, 343–352. MR54#10041; Z 351.05011.

1979 Latin square graphs and their automorphism groups, *Ars Combinat.* 7, 273–299. MR 81e:05082; Z 421.05036.

1980 A survey of derived triple systems, in: *Topics on Steiner Systems* (Ed. C.C. Lindner and A. Rosa), *Ann. Discrete Math.* 7, 105–114. Z 456.05017.

1980a On the number of commutative latin squares, *Ars Combinat.* 10, 311–322. MR 82c:05026; Z 462.05014.

1980b Automorphism free latin square graphs, *Discrete Math.* 31, 193–200. MR 81m:05037; Z 442.05061.

1983 A combinatorial construction of perfect codes, *SIAM J. Alg. Disc. Meth.* 4, 398–403. MR 85e:94024; Z 549.94029.

1986 Isomorphism of cyclic twofold triple systems, *Ars Combinat.* 22, 207–210. MR 87m:05030; Z 634.05013.

1987 A construction of cyclic Steiner triple systems of order p^n, *Discrete Math.* 67, 107–110. MR 89b:05047; Z 636.05010.

1987a Isomorphism problems for cyclic block designs, *Ann. Discrete Math.* 34, 385–391. MR 88k:05030; Z 647.05008.

1987b Isomorphism of circulant combinatorial structures, *Ars Combinat.* 24A, 195–210. MR 89f:05041; Z 721.05010.

1990 Cyclic large sets of Steiner triple systems of order 15, *Math. Comp.* 55, 821–824 and S1–S6. MR 91b:05036; Z 708.05011.

PHELPS, K.T. / LINDNER, C.C.

1984 On the number of Mendelsohn and transitive triple systems, *Europ. J. Combinat.* 5, 239–242. MR 86a:05020; Z 559.05011.

PHELPS, K.T. / DE RESMINI, M.J.

1992 Partitioning twofold triple systems into complete arcs, *Discrete Math.* 104, 273–280. MR 93d:05023; Z 783.05021.

PHELPS, K.T. / RODGER, C.A.

1993 Nesting partial Steiner triple systems with 2-regular leave graphs, *Discrete Math.* 112, 165–172. MR 94d:05014; Z 785.05017.

PHELPS, K.T. / RÖDL, V.

1984 On the algorithmic complexity of coloring simple hypergraphs and Steiner triple systems, *Combinatorica* 4, 79–88. MR 85d:05052; Z 535.68030.

1986 Steiner triple systems with minimum independence number, *Ars Combinat.* 21, 167–172. MR 87j:05038; Z 625.05012.

PHELPS, K.T. / ROSA, A.

1981 Steiner triple systems with rotational automorphisms, *Discrete Math.* 33, 57–66. MR 82a:05018; Z 463.05014.

PHELPS, K.T. / ROSA, A. / MENDELSOHN, E.

1989 Cyclic Steiner triple systems with cyclic subsystems, *Europ. J. Combinat.* 10, 363–367. MR 90d:05042; Z 726.05018.

PHELPS, K.T. / YIN, C.

1997 Generalized Steiner systems with block size three and group size $g \equiv 3 \pmod 6$, *J. Combin. Des.* 5, 417–432. MR 98k:05021.

1999 Generalized Steiner systems with block three and group size four, *Ars Combinat.*

PHILLIPS, N.C.K. / WALLIS, W.D. / REES, R.S.

1998 Kirkman packing and covering designs, *J. Combin. Math. Combin. Comput.* 28, 299–325.

PICKERT, G.

1982 From triangular graphs to Steiner triple systems and projective spaces, in: *Proc. First Internat. Conf. on Mathematics* (King Saud University Riyadh), pp. 65–80. Z 571.51004.

1995 Konstruktion zweier (15,3,1)-Blockpläne, *Abh. Math. Sem. Univ. Hamburg* 65, 165–173. MR 96j:51018; Z 859.51002.

1996 Extension of gravity centers configuration to Steiner triple systems, *Des. Codes Crypt.* 8, 203–214. MR 97b:05027; Z 867.05005.

PIERCE, W.A.

1953 The impossibility of Fano's configuration in a projective plane with eight points per line, *Proc. Amer. Math. Soc.* 4, 908–912. MR 15, 460; Z 52, 165.

PIETSCH, CH.

1993 *Über die Enumeration von Inzidenzstrukturen,* Dissertation, Univ. Rostock.

1994 On the enumeration of 2-(7,3,λ) block designs, *J. Comb. Math. Comb. Comput.* 16, 103–114. MR 95i:05023; Z 815.05013.

PIGOZZI, D. / SICHLER, J.

1985 Homomorphisms of partial and of complete Steiner triple systems and quasigroups, in: *Universal Algebra and Lattice Theory,* Lecture Notes Math. 1149, Springer, Berlin, 224–237. MR 87d:05037; Z 572.20057.

PIPPENGER, N. / SPENCER, J.

1989 Asymptotic behavior of the chromatic index for hypergraphs, *J. Combinat. Theory (A)* 51, 24–42. MR 90h:05091; Z 729.05038.

PLÜCKER, J.

1835 *System der analytischen Geometrie, auf neue Betrachtungsweisen gegründet, und insbesondere eine ausführliche Theorie der Curven dritter Ordnung enthaltend*, Duncker and Humblot, Berlin.

1839 *Theorie der algebraischen Curven, gegründet auf eine neue Behandlungsweise der analytischen Geometrie*, Marcus, Bonn.

1847 Über Curven dritter Ordnung und analytische Beweisführung, *J. Reine Angew. Math.* 34, 329–336.

PONNUSWAMY, K.N. / SRINIVASAN, M.R.

1991 Construction of partial triallel crosses (PTC) using a class of balanced incomplete block designs (BIBD), *Commun. Stat. Theory Methods* 20, 3315–3323. MR 93b: 62108.

PONNUSWAMY, K.N. / SUBBARAYAN, A.

1990 On the application of Steiner and related triple systems to plant breeding experiments, *Proc. R.C. Bose Symposium on Probability, Statistics and Design of Experiments*, Wiley Eastern, New Delhi, pp. 581–594.

POWER, J.

1867 On the problem of the fifteen school girls, *Quart. J. Pure Appl. Math.* 8, 236–251. MR 85d:20084.

PRAEGER, C.E. / STREET, A.P.

1981 Designs from finite groups, *Congressus Numer.* 31, 165–185. MR 83e:05024; Z 514.05015.

PREECE, D.A.

1967† Nested balanced incomplete block designs, *Biometrika* 54, 479–486.

1992 Enumeration of some 7 × 15 Youden squares and construction of some 7 × 15 double Youden rectangles, *Utilitas Math.* 41, 51–62. MR 92m:05041; Z 753.05006.

PREECE, D.A. / VOWDEN, B.J. / HUGHES JONES, R. / RODGER, C.A. / VOWDEN, C.J.

1995 Choreographing designs, *Math. Sci.* 20, 15–32. MR 96g:05036; Z 830.05015.

PRIMROSE, E.J.F.

1976 Kirkman's schoolgirls' in modern dress, *Math. Gaz.* 60, 292–293. MR 58#23991; Z 343.05004.

PRINCE, A.R.

1983 Steiner triple systems of order 19 constructed from the Steiner triple system of order 9, *Proc. Roy. Soc. Edinburgh* (A) 94, 89–92. MR 89f:05022; Z 514.05017.

1986 Steiner triple systems of order 19 associated with a certain type of projective plane of order 10, *Periodica Math. Hungar.* 17, 177–184. MR 87j:05039; Z 614.51006.

PUKANOW-KURPIŃSKA, K.

1986 Systemy trójek, *Prace Naukowe Acad. Ekonom. we Wroclawú* 357, 43–66.

PUKHAREV, N.K.

1965 Ob A_n^k-algebrach i regul'arnych konečnych ploskostjach, *Sib. Mat. Ž.*, 6, 892–899. MR 32#8249; Z 148, 408.

QUACKENBUSH, R.W.

1974 Near-Boolean algebras I: Combinatorial aspects, *Discrete Math.* 10, 301–308. MR 51#314; Z 326.05015.

1976 Varieties of Steiner loops and Steiner quasigroups, *Canad. J. Math.* 28, 1187–1198. MR 54#12946; Z 359.20070.

1980 Algebraic speculations about Steiner systems, in: *Topics on Steiner Systems* (Ed. C.C. Lindner and A. Rosa), *Ann. Discrete Math.* 7, 25–35. Z 437.51008.

1984 Universal algebra and combinatorics, in: *Universal Algebra and its Links with Logic, Algebra and Computer Science* (Darmstadt), R & E Res. Ex. Math. 4, Heldermann, Berlin, pp. 150–167. MR 87g:05021; Z 554.08005.

1999 Nilpotent block designs I: Basic concepts for Steiner triple and quadruple systems.

QUATTROCCHI, G.

1986 Sul massimo numero di DMB PTS aventi 12 blocchi e immergibili in un STS, *Riv. Mat. Univ. Parma* (4), 12, 41–51. MR 89a:05032; Z 624.05016.

1987 Sul parametro D(13,14) per sistemi di terne di Steiner, *Matematiche (Catania)*, 39, 61–80. MR 89b:05048; Z 634.05015.

1989 Intersections among maximum partial triple systems, *J. Combinat. Inf. System Sci.* 14, 192–201. MR 92e:05030; Z 855.05025.

1990 On the intersection of two S(3,4,v) having a same derived triple system, *Discrete Math.* 82, 279–286. MR 91f:05029; Z 699.05010.

1991 Threefold triple systems with blocks at most two-times repeated, in: *Combinatorics 88*, Mediterranean Press, 1991, pp. 331–348. MR 94a:05023.

QUATTROCCHI, G. / MILICI, S.

1993 Repeated blocks in maximum packing of triples with index 2, *Discrete Math.* 115, 287–291. MR 94b:05049; Z 778.05023.

QUATTROCCHI, G. / ZEITLER, H.

1993 Hyperovals in Steiner triple systems, *J. Geometry* 47, 125–130. MR 94e:05051; Z 782.05013.

QUATTROCCHI, P.

1966 Un metodo per la construzione di certe strutture che generalizzano la nozione di piano grafico, *Le Matematiche* 21, 377–386. MR 34#2482; Z 147, 199.

QUIRING, D.

1979 A construction of disjoint Steiner triple systems, *J. Combinat. Theory (A)*, 27, 407–408. MR 81j:05026; Z 419.05005.

RADO, F.

1974 On semi-symmetric quasigroups, *Aequat. Math.*, 11, 250–255. MR 53#13459; Z 293.20059.

RAGHAVACHARI, M.

1977 Some properties of the width of the incidence matrices of balanced incomplete block designs and Steiner triple systems, *Sankhya* (B), 39, 181–188. MR 81a:05025; Z 411.62051.

RAHILLY, A.

1992 Constructing designs using the union method, *Australas. J. Combin.* 6, 7–21. MR 93j:05014; Z 773.05016.

RAHMAN, M.

1983 Asymptotically double error-correcting codes based on triple systems, *Utilitas Math.*, 24, 145–154. MR 84m:94024; Z 522.94016.

RAHN, J.M.

1985 The genus of a block design, *Congressus Numer.* 50, 255–268. MR 87e:05065; Z 595.05032.

RAINES, M.E.

1996 More on embedding partial totally symmetric quasigroups, *Australas. J. Comb.* 14, 297–306. MR 98b:20116; Z 866.05013.

RAINES, M.E. / RODGER, C.A.

1997 Matchings in the leave of equitable partial Steiner triple systems, *J. Comb. Math. Comb. Comput.* 24, 115–118. MR 97m:05032; Z 883.05019.

1997a Embedding partial extended triple systems and totally symmetric quasigroups, *Discrete Math.* 176, 211–222. MR 99a:05015.

RAMSAY, C.

1997 On a family of STS(v) whose smallest defining sets contain at least b/4 blocks, *Bull. Inst. Combin. Appl.* 20, 91–94. MR 97m:05033; Z 876.05007.

1997a An algorithm for completing partials, with an application to the smallest defining sets of the STS(15)s, *Util. Math.* 52, 205–221. MR 98i:05017.

RAO, C.R.

1961 A study of BIB designs with replications 11 to 15, *Sankhya* Ser. A, 23, 117–127. MR 25#732; Z 99, 354.

RAUSCHE, A.

1988 On the existence of special block designs, *Rostock. Math. Kolloq.* 35, 13–20. MR 90c:05031; Z 688.05012.

RAY-CHAUDHURI, D.K.

1981 Affine triple systems, in: *Combinatorics and Graph Theory* (Ed. S.B. Rao), Lecture Notes in Math. 885. Springer, New York, pp. 60–69. MR 83g:05019; Z 474. 05010.

RAY-CHAUDHURI, D.K. / WILSON, R.M.

1970 On the existence of resolvable balanced incomplete block designs, in: *Combinatorial Structures and their Applications*, Gordon and Breach, New York, pp. 331–341. MR 42#1678; Z 265.05004.

1971 Solution of Kirkman's school-girl problem, *Proc. Symp. Pure Math.*, 19, Amer. Math. Soc., Providence, RI, 187–203. MR 47#3195; Z 248.05009.

1973 The existence of resolvable block designs, in: *A Survey of Combinatorial Theory*, North-Holland, Amsterdam, Chapter 30, pp. 361–375. MR 50#12761; Z 274. 05010.

READ, R.C.

1977† Every one a winner or how to avoid isomorphism search when cataloguing combinatorial configurations, *Ann. Discrete Math.* 2, 107–120.

REES, R.

1987 Uniformly resolvable pairwise balanced designs with blocksizes two and three, *J. Combinat. Theory (A)*, 45, 207–225. MR 88g:05027; Z 659.05016.

1990 The existence of restricted resolvable designs I. (1,2)-factorizations of K_{2n}, *Discrete Math.* 81, 49–80. MR 91f:05019a; Z 702.05011.

1990a The existence of restricted resolvable designs II. (1,2)-factorizations of K_{2n+1}, *Discrete Math.* 81, 263–301. MR 91f:05019b; Z 702.05012.

1991 The spectrum of restricted resolvable designs with $r = 2$, *Discrete Math.* 92, 305–320. MR 93e:05007; Z 748.05037.

1991a Semiframes and nearframes, in: *Combinatorics 88*, Mediterranean Press, 1991, pp. 359–367. MR 94a:05036.

1993 Two new direct product-type constructions for resolvable group-divisible designs, *J. Combin. Des.* 1, 15–26. MR 95h:05024; Z 817.05008.

1998 Maximal sets of triangle-factors on $v = 6m$ vertices, *J. Combin. Des.* 6, 235–244.

1998a Maximal sets of triangle-factors on $v = 6m + 3$ vertices, *J. Combin. Des.* 6, 309–323.

1999 Mandatory representation designs MR$(3, k; v)$ with k even and v odd, *J. Stat. Plann. Infer.*

REES, R. / ROSA, A. / WALLIS, W.D.

1994 Maximal sets of triangle-factors, *Australas. J. Combin.* 9, 67–108. MR 95c: 05101; Z 806.05053.

REES, R. / SHALABY, N. / SHARARY, A.

1999 Indecomposable Skolem sequences.

REES, R. / STINSON, D.R.

1987 On resolvable group-divisible designs with block size 3, *Ars Combinat.*, 23, 107–120. MR 88d:05025; Z 621.05004.

1988 Kirkman triple systems with maximum subsystems, *Ars Combinat.*, 25, 125–132. MR 89e:05039; Z 679.05011.

1988a On the existence of Kirkman triple systems containing Kirkman subsystems, *Ars Combinat.* 26, 3–16. MR 90a:05031; Z 702.05013.

1989 On combinatorial designs with subdesigns, *Discrete Math.* 77, 259–280. MR 90k:05032; Z 694.05010.

1989a† On the existence of incomplete designs of block size four having one hole, *Utilitas Math.* 35, 119–152.

REES, R. / WALLIS, W.D.

1986 A class of resolvable pairwise balanced designs, *Congressus Numer.* 55, 211–220. MR 88c:05028; Z 611.05008.

REICHARD, S.

1997 An algorithm for the construction of partial geometries with given point graphs, TU Dresden, Report MATH-AL-12-1997, 12 pp.

REISS, M.

1859 Ueber eine Steinersche combinatorische Aufgabe, welche im 45sten Bande dieses Journals, Seite 181, gestellt worden ist, *J. Reine Angew. Math.* 56, 326–344.

RINGEL, G.

1961 Über das Problem der Nachbargebiete auf orientierbaren Flächen, *Abh. Math. Sem. Univ. Hamburg* 25, 105–127. MR 23A#2876; Z 182, 266.

1965 Die toroidale Dicke des vollständigen Graphen, *Math. Z.* 87, 19–26. MR 30 #2489; Z 132, 213.

1975† *Map Color Theorem,* Springer.

ROBINSON, R.M.

1971 Triple systems with prescribed subsystems, *Notices Amer. Math. Soc.* 18, 637.

1975 The structure of certain triple systems, *Math. Comput.* 29, 223–241. MR 52 #5440; Z 293.05015.

RODGER, C.A.

1986 Triple systems with a fixed number of repeated triples, *J. Austral. Math. Soc.* (A) 41, 180–187. MR 88e:05016; Z 615.05015.

1986a Graph theory and embeddings, *Ars Combinat.* 21A, 247–256. MR 87f:05075; Z 592.05010.

1987 Embedding partial Mendelsohn triple systems, *Discrete Math.* 65, 187–196. MR 89e:05035; Z 663.05011

1995 Linear spaces with many small lines, *Discrete Math.* 129, 167–180. MR 95h: 05035; Z 805.51008.

1997 A solution for the embedding problem for partial transitive triple systems, *Bull. London Math. Soc.* 29, 299–302. MR 98a:05024; Z 874.05010.

1999 Balanced transitive orientations, *J. Comb. Math. Comb. Comput.*

RODGER, C.A. / STUBBS, S.J.

1987 Embedding partial triple systems, *J. Combinat. Theory (A)* 44, 241–252. MR 88a:05039; Z 659.05027. Erratum: *J. Combinat. Theory (A)* 66 (1994), 182–183. MR 95a:05016; Z 795.05030.

RÖDL, V.

1985† On a packing and covering problem, *Europ. J. Combinat.* 5, 69–78.

RÖDL, V. / THOMA, L.

1996 Asymptotic packing and the random greedy algorithm, *Random Struc. Algor.* 8, 161–177. Z 856.05074.

RODNEY, P.

1996 Resolvable $\times\omega$-bounded designs, *Utilitas Math.* 50, 215–231. MR 97g:05027; Z 864.05010.

ROGERS, D.G.

1985 Triangulating the remnants of complete graphs, *Southeast Asian Bull. Math.* 9, 86–88. MR 88f:05040; Z 612.05058.

ROKOWSKA, B.

1971 Some remarks on the number of different triple systems of Steiner, *Colloq. Math.* 22, 317–324. MR 44#3890; Z 218.05008.

1972 On the number of different triple systems of Steiner, *Prace Naukowe Instytutu Matematyki i Fizyki Teoretycznej Politechniki Wroclawskiej* Nr. 6, 41–57. MR 50 #6885; Z 254.05005.

1973 On the number of non-isomorphic Steiner triples, *Colloq. Math.* 27, 149–160. MR 48#3759; Z 254.05006.

1977 Non-isomorphic Steiner triples with subsystems, *Colloq. Math.* 38, 153–164. MR 57#142; Z 377.05004.

ROSA, A.

1963 Ispol'zovanie grafov dlja rešenija zadači Kirkmana, *Mat.-Fyz. Čas.* 13, 105–113. MR 28#1615; Z 145, 441.

1963a Zametka k odnoj zadače po teorii grafov, *Mat-Fyz. Čas.* 13, 238–239. MR 32 #2349; Z 145, 441.

1966 Poznámka o cyklických Steinerových systémoch trojíc, *Mat.-Fyz. Čas.* 16, 285–290. MR 35#2759; Z 147, 267.

1970 On the chromatic number of Steiner triple systems, in: *Combinatorial Structures and their Applications*, Gordon and Breach, New York, pp. 369–371. MR 43 #6110; Z 264.05012.

1970a Steiner triple systems and their chromatic number, *Acta Fac. Rerum Natur. Univ. Comen. Math.* 24, 159–174. MR 45#87; Z 213, 262.

1972 On reverse Steiner triple systems, *Discrete Math.* 2, 61–71. MR 45#4996; Z 242. 05016.

1974 On the falsity of a conjecture on orthogonal Steiner triple systems, *J. Combinat. Theory (A)* 16, 126–128. MR 48#10845; Z 277.05008.

1975 A theorem on the maximum number of disjoint Steiner triple systems, *J. Combinat. Theory (A)* 18, 305–312. MR 51#191; Z 306.05008.

1975a Algebraic properties of designs and recursive constructions, *Congressus Numer.* 13, 183–200. MR 56#8391; Z 328.05012.

1978 On generalized Room squares, in: *Problèmes Combinatoire et Théorie des Graphes*, Colloq. Internat. CNRS 260, Paris, pp. 353–358. MR 81h:05028; Z 412. 05015.

1979 Generalized Howell designs, *Ann. N.Y. Acad. Sci.* 319, 484–489. MR 81j:05033; Z 483.05021.

1980 Intersection properties of Steiner systems, in: *Topics on Steiner Systems* (Ed. C.C. Lindner and A. Rosa). *Ann. Discrete Math.* 7, 115–128. Z 438.05013.

1980a Room squares generalized, *Ann. Discrete Math.* 8, 43–57. MR 82b:05036; Z 445. 05026.

1986 Some results on twofold designs with block size three and four, *Designs and Finite Geometries*, Kyoto, Surikaisekiken., Kokyuroku No. 607, 5–21. MR 88i:05029.

1987 Colouring problems in combinatorial designs, *Congressus Numer.* 56, 45–52. MR 88g:05028; Z 633.05011.

1987a Repeated blocks in indecomposable twofold triple systems, *Discrete Math.* 65, 261–276. MR 88h:05025; Z 655.05011.

1988 Cyclic Steiner triple systems and labellings of triangular cacti, *Scientia Ser. A: Math. Sci.* 1, 87–95. Z 732.05007.

1992 Configurations in triple systems: avoidance and decompositions, *Le Matematiche (Catania)* 47, 313–314. Z 795.05027.

ROSA, A. / COLBOURN, C.J.

1992 Colorings of block designs, in: *Contemporary Design Theory: A Collection of Surveys* (Ed. J.H. Dinitz and D.R. Stinson), Wiley, New York, pp. 402–430. Z 767. 05016.

ROSA, A. / HOFFMAN, D.

1986 The number of repeated blocks in twofold triple systems, *J. Combinat. Theory (A)* 41, 61–88. MR 87g:05062; Z 591.05006.

ROSA, A. / VANSTONE, S.A.

1983 Kirkman cubes, *Ann. Discrete Math.* 18, 699–712. MR 84f:05023; Z 545.05019.

1985 On the existence of strong Kirkman cubes of order 39 and blocks size 3, *Ann. Discrete Math.* 26, 306–319. MR 87g:05049; Z 582.05010.

ROSELLE, D.P.

1971 Distributions of integers into s-tuples with given differences, in: *Conf. Numerical Methods*, Winnipeg, pp. 31–42. MR 49#211; Z 267.05018.

ROTH, R.

1979 The catalogue of Hall triple systems of small orders, *Notices Amer. Math. Soc.* 26, Abstract 763–17–5, A54.

1980 *Hall Triple systems and Commutative Moufang Exponent 3 Loops*, Ph.D. Thesis, Ohio State University.

ROTH, R. / RAY-CHAUDHURI, D.K.

1984 Hall triple systems and commutative Moufang exponent 3 loops: The case of nilpotence class 2, *J. Combinat. Theory (A)* 36, 129–162. MR 85k:20208; Z 527. 05018.

ROY, B.K.

1982 Construction of (M,S) optimal design for block size 3, *J. Stat. Plann. Infer.* 7, 35–37. MR 84m:62102; Z 494.62071.

ROY, B.K. / SHAH, K.R.

1984 On the optimality of a class of minimal covering designs, *J. Stat. Plann. Infer.* 10, 189–194. Z 565.62047.

ROYAL SOCIETY OF LONDON

1870† *Catalogue of Scientific Papers I–VI.* (1800–1863), Cambridge University Press, Cambridge.

1879† *Catalogue of Scientific Papers VII–VIII.* (1864–1873), Cambridge University Press, Cambridge.

1894† *Catalogue of Scientific Papers IX–XI.*(1874–1883), Cambridge University Press, Cambridge.

1908† *Catalogue of Scientific Papers: Index*, Cambridge University Press, Cambridge.

1918† *Catalogue of Scientific Papers XII–XIX.* (1884–1900), Cambridge University Press, Cambridge.

ROYLE, G.

1989 1000000 2-(12,3,2) designs, *Abstracts Amer. Math. Soc.* 10, 482.
RUMOV, B.T.
1974 Some imbedding theorems for block designs balanced with respect to pairs (in Russian), *Mat. Zametki* 16, 173–184. MR 51#12560; Z 327.05017. (English translation: *Math. Notes* 16, 686–692. Z 336.05086.)
1975 O postrojenii odnoj beskonečnoj serii sistem trojek Kirkmana, *Voprosy Kibernetiki* 16, 77–78.
RUZSA, I.Z. / SZEMERÉDI, E.
1978 Triple systems with no six points carrying three triangles, in: *Colloq. Math. Soc. J. Bolyai 18, Combinatorics*, Keszthely 1976, pp. 939–945. MR 80c:05016; Z 393. 05031.
RYSER, H.J.
1951† A combinatorial theorem with an application to Latin rectangles, *Proc. Amer. Math. Soc.* 2, 550–552.
1963 *Combinatorial Mathematics* (Carus Math. Monograph 14), Wiley, New York. MR 27#51; Z 112, 248. (Russian translation: *Kombinatornaja matematika*, Mir, Moskva, 1966. Z 146, 11. French translation: *Mathématiques combinatoires*, Dunod, Paris, 1969. Z 165, 327.)
1969† Permanents and systems of distinct representatives, in: *Combinatorial Mathematics and its Applications*, UNC Press, Chapel Hill, pp. 55–68.
SADE, A.
1965 Quasigroups demi-symétriques, *Ann. Soc. Sci. Bruxelles* Ser. I, 79, 133–143. MR 34#2760; Z 132, 265.
SAINTE-LAGUË, A.
1943 *Avec des Nombres et des Lignes (Récréations Mathématiques)*, Vuibert, Paris, 2nd ed., Deuxième partie, Chapter 5: Les triades, pp. 197–209.
SALMON, G.
1849 On the triple tangent planes to a surface of the third order, *Cambridge and Dublin Math. J.* 4, 252–260.
1850 Lettre de Dublin à l'éditeur de ce journal, *J. Reine Angew. Math.* 39, 365–366.
1862 *A Treatise on the Higher Plane Curves*, Dublin.
SANDU, N.I.
1987 Infinite irreducible systems of identities of commutative Moufang loops and distributive Steiner quasigroups (in Russian), *Izv. Akad. Nauk SSSR Ser. Mat.* 51, 171–188. MR 88b:20110; Z 615.20055.
SARVATE, D.G.
1984 All directed GDDs with block size three, $\lambda_1 = 0$, exist, *Utilitas Math.* 26, 311–317. MR 86c:05034; Z 527.05017.
1985 A note on equi-neighboured block designs, *Utilitas Math.* 28, 91–98. MR 87d: 05036; Z 561.05010.
1985a Some results on directed and cyclic designs, *Ars Combinat.* 19A, 179–190. MR 86i:05028; Z 563.05011.
1986 Block designs without repeated blocks, *Ars Combinat.* 21, 71–87. MR 87f: 05025; Z 607.05010.
1986a All simple BIBDs with block size 3 exist, *Ars Combinat.* 21A, 257–270. MR 87f: 05026; Z 586.05003.
SARVATE, D.G. / HAMM, R.C.
1991 Row-column directed block designs, *Discrete Math.* 92, 321–328. MR 92k: 05019; Z 756.05022.
SAUER, N. / SCHÖNHEIM, J.

1969 Maximal subsets of a given set having no triple in common with a Steiner triple system on the set, *Canad. Math. Bull.* 12, 777–778. MR 40#5456; Z 194, 10.

SAUER, N. / STONE, M.G.

1982 No solution to Kirkman's schoolgirl problem is vertex transitive, *Abstracts Amer. Math. Soc.* 3, No. 5, 388 (*82T–05–399).

SAUER, N. / STONE, M.G. / WEEDMARK, R.H.

1983 Every finite algebra with congruence lattice M_7 has principal congruences, in: *Universal Algebra and Lattice Theory*, Lecture Notes in Math. 1004, Springer, Berlin, pp. 273–292. MR 85h:08005; Z 518.06004.

SAVUR, S.R.

1939 A note on the arrangement of incomplete blocks, when $k = 3$ and $\lambda = 1$, *Ann. Eugenics* 9, 45–49.

SAXL, J.

1981 On points and triples of Steiner triple systems, *Arch. Math.* 36, 558–564. MR 82j:20012; Z 479.05012.

SCAPELLATO, R.

1981 Sulla regolarita dei sistemi di Steiner distributivi, *Riv. Mat. Univ. Parma* (4), 7, 405–408. MR 83j:05014; Z 488.51004.

1981a Sull'esistenza di funzioni di Steiner, *Boll. Un. Mat. Ital.* (5), A18, 127–131. MR 82c:20048; Z 535.20007.

1982 On the existence of Steiner functions, in: *Proc. Conf. on Near-rings and Near-fields*, San Benedetto del Tronto, Italy, Univ. Parma, pp. 201–204.

1982a On a class of distributive Steiner quasigroups, *CMUC* 23, 313–317. MR 83j: 20071; Z 503.20034.

1982b Gruppi infiniti che amettono funzioni di Steiner, *Riv. Mat. Univ. Parma* (4), 8, 355–360. MR 85b:20080; Z 534.20036.

SCHEID, F.

1960 A tournament problem, *Amer. Math. Monthly* 67, 39–41. MR 23A#72; Z 93, 15.

SCHELLENBERG, P.J. / STINSON, D.R.

1989 Threshold schemes from combinatorial designs, *J. Comb. Math. Comb. Comput.* 5, 143–160. MR 90c:05032; Z 674.94008.

SCHMITT, E.

1990 On strongly regular graphs and configurations n_3, *J. Combin. Inf. Syst. Sci.* 15, 1–9. MR 92j:05185; Z 748.05096.

SCHÖNHEIM, J.

1964 On coverings, *Pacific J. Math.* 14, 1405–1411. MR 30#1934.

1966 On maximal systems of k-tuples, *Studia Sci. Math. Hung.* 1, 363–368. MR 34 #2485; Z 146, 14.

1969 On the number of mutually disjoint triples in Steiner systems and related maximal packing and minimal covering systems, in: *Recent Progress in Combinatorics*, Academic Press, New York, pp. 311–318. MR 40#5457; Z 196, 26.

SCHREIBER, S.

1973 Covering all triples on n marks by disjoint Steiner systems, *J. Combinat. Theory (A)* 15, 347–350. MR 48#10846; Z 271.05021.

1974 Some balanced complete block designs, *Israel J. Math.* 18, 31–37. MR 50#141; Z 292.05004.

1989 Cyclical Steiner triple systems orthogonal to their opposites, *Discrete Math.* 77, 281–284. MR 90m:05026; Z 726.05019.

1989a Symmetric quasigroups of odd order, *Discrete Math.* 77, 285–297. MR 90k: 20106; Z 726.05020.

SCHULTZ, D.J. / PARNES, M. / SRINIVASAN, R.

1993 Further applications of d-complete designs to group testing, *J. Combin. Inform. System Sci.* 18, 31–41. MR 95m:05032; Z 855.05018.

SCHWENK, J.

1993 Erweiterte Tripelsysteme, *Mitt. Math. Sem. Giessen* 214, 111 pp. MR 95c: 05020; Z 803.05007.

1995 A classification of abelian quasigroups, *Rend. Mat. (7)* 15, 161–172. MR 97c: 20101; Z 831.05015.

SEBERRY, J.

1982 Some families of partially balanced incomplete block designs, in: *Combinatorial Mathematics IX* (Brisbane, 1981), *Lecture Notes in Math.* 952, Springer, Berlin, pp. 378–386. MR 84d:05038; Z 488.05021.

1984 Regular group dvisible designs and Bhaskar Rao designs with block size three, *J. Stat. Plann. Infer.* 10, 69–82. MR 86b:05012; Z 552.05011.

1985 Generalized Bhaskar Rao designs of block size three, *J. Stat. Plann. Infer.* 11, 373–379. MR 87a:05032; Z 572.05021.

SEBERRY, J. / SKILLICORN, D.

1980 All directed BIBDs with $k = 3$ exist, *J. Combinat. Theory (A)* 29, 244–248. MR 81h:05026; Z 439.05011.

SEE, K. / SONG, S.Y. / STUFKEN, J.

1997 On a class of partially balanced incomplete block designs with applications in survey sampling, *Commun. Stat. Theory Meth.* 26, 1–13.

SEIDENBERG, A.

1968† *Elements of the Theory of Algebraic Curves*, Addison-Wesley, Reading MA. Chapters 9 and 10, Cubics.

SEMPLE, J.G. / KNEEBONE, G.T.

1959† *Algebraic Curves*, Clarendon, Oxford. Chapter 1, Plane algebraic curves.

SEMPLE, J.G. / ROTH, L.

1949† *Introduction to Algebraic Geometry*, Clarendon, Oxford.

SEVERN, E.A.

1984 *Maximal Partial Steiner Triple Systems*, Ph.D. Thesis, Department of Mathematics, University of Toronto. (Dept. of Computer Science, University of Toronto, Technical Report #172/84).

SHALABY, N.

1994 The existence of near-Skolem sequences and hooked near-Skolem sequences, *Discrete Math.* 135, 303–319. MR 96e:05027; Z 818.05008.

1996 Skolem sequences, in: *CRC Handbook of Combinatorial Designs* (Ed. C.J. Colbourn and J.H. Dinitz), CRC Press, Boca Raton, FL, pp. 457–462. Z 844.05031.

SHALABY, N. / YIN, J.

1999 Nested optimal λ-packings and λ-coverings of pairs with triples, *Des. Codes Crypt.*

SHARRY, M.J.

1992 *Large and Overlarge Sets of Block Designs*, Ph.D. Thesis, University of Queensland, Australia.

SHARRY, M.J. / STREET, A.P.

1988 Partitioning sets of triples into designs, *Ars Combinat.* 26B, 51–66. MR 90a: 05027; Z 685.05006.

1988a Partitioning sets of triples into designs II, *J. Comb. Math. Comb. Comput.* 4, 53–68. MR 90m:05023; Z 676.05014.

1991 One-factorizations and large sets of designs, *Bull. Inst. Combin. Appl.* 2, 71–78. MR 92c:05025; Z 828.05049.

1991a Partitioning sets of blocks into designs, *Australas. J. Combin.* 3, 111–140. MR 92c:05024; Z 763.05012.

1991b A doubling construction for overlarge sets of Steiner triple systems, *Ars Combinat.* 32, 143–151. MR 92j:05026; Z 756.05030.

1993 Cyclic and rotational overlarge sets of some small Steiner systems, *Utilitas Math.* 44, 115–130. MR 94m:05055; Z 792.05025.

1996 Another look at large sets of Steiner triple systems, in: *Computational and Constructive Design Theory* (Ed. W.D. Wallis) Kluwer Academic, pp. 255–335. Z 879.05009.

SHEN, H.

1990 Existence of indecomposable B(3,6;v) without repeated blocks (in Chinese), *J. Shanghai Jiao Tong Univ.* 24, 1–7. MR 91m:05027.

1990a Resolvable twofold triple systems without repeated triples, *Chinese Science Bull.* 35, 89–92. MR 92e:05017; Chinese version: *Kexue Tongbao* 33 (1988), 1855–1857. MR 90d:05039; Z 756.05030.

1992 Indecomposable triple systems without repeated blocks, *Ars Combinat.* 33, 305–310; abstract in *Advances Math. (China)* 18 (1989), 375–377. MR 93d:05024; Z 773.05024.

1992a Embeddings of simple triple systems, *Sci. China A* 35, 283–291. MR 93i:05025; Z 776.05017.

1992b Embeddings of pure Mendelsohn triple systems (in Chinese), *J. Shanghai Jiaotong Univ.* 26, 85–89. MR 93d:05025.

1993 On the embeddings of Mendelsohn triple systems, *Acta Math. Sinica New Ser.* 9, 246–251. MR 95e:05028; Z 795.05031.

1995 Embeddings of pure Mendelsohn triple systems and pure directed triple systems, *J. Combin. Des.* 3, 41–50. MR 95h:05028; Z 829.05012. Abstract: *Advances Math. (China)* 20 (1991), 250–252.

1997 The intersection problem for Steiner and Kirkman triple systems, *Atti Accad. Peloritana Pericolanti Cl. Sci. Fis. Mat. Nat.* 73, 57–60. Z 883.05020.

1999 Existence of cyclic Mendelsohn triple systems, *J. Comb. Infor. Syst. Sci.*

1999a Intersections of Kirkman triple systems, *J. Stat. Plann. Infer.*

SHEN, H. / CHU, W.S.

1994 Asymptotic existence of nearly Kirkman systems, *Sci. China (A)* 37, 555–562. MR 95h:05025; Z 804.05011.

SHEN, H. / WANG, Y.

1992 On the embeddings of resolvable triple systems; abstract in *Science Bull. (China)* 37, 475.

SHEN, H. / WEN, H.

1994 The spectrum of support sizes of triple systems (in Chinese), *J. Shanghai Jiaotong Univ.* 28, 90–95. MR 95a:05011.

SIERPÍNSKI, W.

1946 Sur un problème de triades, *C.R. Soc. Sci. Varsovie* 33–38, 13–16. MR 8, 570, Z 60, 130.

SIMMONS, G.J.

1981 Two-fold triple systems and embeddings of K_v in 2-manifolds, *Sankhya* (B) 43, 112–134. MR 83i:05015; Z 511.05011.

SINGH, J.J.

1982 Some Bhaskar Rao designs and applications for $k = 3$, $\lambda = 2$, *University of Indore Research J. Science* 7, 8–15.

SINHA, K.

1992 A construction of balanced ternary designs, *Ars Combinat.* 33, 276–277. Z 771.05015.

SKOLEM, TH.

1927 Om en del kombinatoriske problemer, *Norsk Mat. Tidsskr.* 9, 87–105.

1931 Über einige besondere Tripelsysteme mit Anwendung auf die Reproduktion gewisser Quadratsummen bei Multiplikation, *Norsk. Mat. Tidsskr.* 13, 41–51. Z 3, 145.

1957 On certain distributions of integers in pairs with given differences, *Math. Scand.* 5, 57–68. MR 19, 1159; Z 84, 43.

1958 Some remarks on the triple systems of Steiner, *Math. Scand.* 6, 273–280. MR 21 #5582; Z 105, 250.

SMITH, J.D.H.

1978 On the nilpotence class of commutative Moufang loops, *Math. Proc. Cambridge Phil. Soc.* 84, 387–404. MR 58#16932; Z 393.20054.

1985 Commutative Moufang loops: the first 50 years, *Algeb. Groups Geom.* 2, 209–234. MR 87h:20130; Z 586.20034.

SMITH, P.

1977 A doubly divisible nearly Kirkman system, *Discrete Math.* 18, 93–96. MR 57 #5783; Z 391.05014.

SOBCZYK, A.

1967 Multi-functions associated with Steiner systems, *Notices Amer. Math. Soc.* 14, 925–926.

SOBEL, M.

1968† Binomial and hypergeometric group testing, *Stud. Sci. Math. Hung.* 3, 19–42.

SOBEL, M. / GROLL, P.H.

1959† Group testing to estimate efficiently all defective items in a binomial sample, *Bell Syst. Tech. J.* 38, 1179–1252.

1966† Binomial group testing with an unknown proportion of defectives, *Technometrics* 8, 631–656.

SOBOCIŃSKI, B.

1964 A theorem of Sierpínski on triads and the axiom of choice, *Notre Dame J. Formal Logic* 5, 51–58. MR 31#3343; Z 134, 248.

SOLÉ, P.

1997 Codes, lattices, and Steiner systems, *Electronic J. Combinat.* 4, 5 pp. #R6. MR 98a:05025; Z 885.05028.

SÓS, V.T. / ERDÖS, P. / BROWN, W.G.

1973 On the existence of triangulated spheres and 3-graphs and related problems, *Period. Math. Hungar.* 3, 221–228. MR 48#2003; Z 258.05132.

SPENCE, E.

1992 A complete classification of symmetric (31,10,3) designs, *Des. Codes Crypt.* 2, 127–136. MR 93c:05025; Z 770.05012.

SPENCER, J.

1968 Maximal consistent families of triples, *J. Combinat. Theory* 5, 1–8. MR 37 #2616; Z 157, 32.

1995 Asymptotic packing via a branching process, *Random Struc. Algor.* 7, 167–172. MR 96m:05154; Z 847.60068.

SPIELMAN, D.A.

1996 Faster isomorphism testing of strongly regular graphs, in: *Proc. Symp. Theory of Computing (STOC) 1996*, ACM, pp. 576–584.

SPOTTISWOODE, W.

1852 On a problem in combinatorial analysis, *London, Edinburgh and Dublin Philos. Mag. and J. Sci.* (4) 3, 348–354.

STANTON, R.G. / GOULDEN, I.P.

1981 Graph factorization, general triple systems and cyclic triple systems, *Aequat. Math.* 22, 11–28. MR 83h:05016; Z 466.05011.

STANTON R.G. / ROGERS, M.J.

1982 Packing and coverings by triples, *Ars Combinat.* 13, 61–69. MR 84i:05042; Z 501.05023.

STANTON, R.G. / ROGERS, M.J. / QUINN, R.F. / COWAN, D.D.

1983 Bipackings of pairs into triples, and isomorphism classes of small bipackings, *J. Austral. Math. Soc.* (A) 34, 214–228. MR 84j:05047; Z 515.05026.

STEEDLEY, D.

1974 Separable quasigroups, *Aequat. Math.* 11, 189–195. MR 53#8310; Z 193.20058.

STEINER, J.

1846 Sätze über Curven zweiter und dritter Ordnung, *J. Reine Angew. Math.* 32, 300–304.

1846a Geometrische Lehrsätze, *J. Reine Angew. Math.* 32, 182–184.

1853 Combinatorische Aufgabe, *J. Reine Angew. Math.* 45, 181–182 (Gesammelte Werke II, 435–438).

STEINITZ, E.

1894† *Über die Construction der Configurationen n_3*, Dissertation, Breslau.

STERBOUL, F.

1978 A problem in constructive combinatorics and related questions, in: *Combinatorics* (Ed. A. Hajnal, V. Sós), Colloquia Math. Soc. J. Bolyai 18, North-Holland, Amsterdam, pp. 1049–1064. MR 81h:05037; Z 387.05019.

STERN, G.

1979 *Direkte Konstruktionen spezieller Tripelsysteme*, Dissertation, FU Berlin, 41 pp.

1979a Tripelsysteme mit Untersystemen, *Arch. Math.* 33, 204–208. MR 81f:05048; Z 435.05009.

STERN, G. / LENZ, H.

1980 Steiner triple systems with given subspaces; another proof of the Doyen–Wilson theorem, *Boll. Un. Mat. Ital.* A(5) 17, 109–114. MR 81f:05049; Z 501.05012.

STIFFLER, J.J. / BLAKE, I.F.

1975 An infinite class of generalized Room squares, *Discrete Math.* 12, 159–163. MR 51#5343; Z 317.05018.

STINSON, D.R.

1983 A comparison of two invariants for Steiner triple systems: fragments and trains, *Ars Combinat.* 16, 69–76. MR 85f:05023; Z 552.05013; Z 562.05015.

1985 Isomorphism testing of Steiner triple systems: canonical forms, *Ars Combinat.* 19, 213–218. MR 87b:05033; Z 583.05014.

1985a Hill-climbing algorithms for the construction of combinatorial designs, *Ann. Discrete Math.* 26, 321–334. MR 87f:05033; Z 591.05009.

1985b The spectrum of nested Steiner triple systems, *Graphs Combinat.* 1, 189–191. MR 89e:05040; Z 583.05015.

1986 Concerning the spectrum of perpendicular arrays of triple systems, *Discrete Math.* 61, 305–310. MR 87i:05042; Z 609.05013.

1987 Frames for Kirkman triple systems, *Discrete Math.* 65, 289–300. MR 88f:05024; Z 651.05015.

1989 A new proof of the Doyen–Wilson theorem, *J. Austral. Math. Soc.* (A) 47, 32–42. MR 90c:05035; Z 702.05020.

1991 A survey of Kirkman triple systems and related designs, *Discrete Math.* 92, 371–393. MR 92h:05020; Z 751.05012.

STINSON, D.R. / FERCH, H.

1985 2000000 Steiner triple systems of order 19, *Math. Comput.* 44, 533–535. MR 86e: 05014; Z 564.05011.

STINSON D.R. / SEAH, E.

1986 284457 Steiner triple systems of order 19 contain a subsystem of order 9, *Math. Comput.* 46, 717–729. MR 87c:05027; Z 594.05017.

STINSON, D.R. / VANSTONE, S.A.

1984 A note on nonisomorphic Kirkman triple systems, *J. Combin. Inf. System Sci.* 9, 113–116. MR 89g:05030; Z 631.05007.

1985 A Kirkman square of order 51 and block size 3, *Discrete Math.* 55, 107–111. MR 86i:05029; Z 563.05017.

1985a Some non-isomorphic Kirkman triple systems of order 39 and 51, *Utilitas Math.* 27, 199–205. MR 86k:05018; Z 582.05009.

1986 Orthogonal packings in PG(5,2), *Aequat. Math.* 31, 159–168. MR 87m:51027; Z 606.05010.

1988 A combinatorial approach to threshold schemes, *SIAM J. Discrete Math.* 1, 230–236. MR 89d:05035; Z 667.05008.

STINSON, D.R. / WALLIS, W.D.

1983 Snappy constructions for triple systems, *Austral. Math. Soc. Gaz.* 10, 84–88. MR 85e:05029; Z 531.05012.

1983a Two-fold triple systems without repeated blocks, *Discrete Math.* 47, 125–128. MR 85i:05041; Z 523.05010.

1987 Graphs which are not leaves of maximal partial triple systems, *Ann. Discrete Math.* 34, 449–460. MR 89a:05033; Z 655.05053.

STINSON, D.R. / WEI, Y.J.

1992 Some results on quadrilaterals in Steiner triple systems, *Discrete Math.* 105, 207–219. MR 93h:05021; Z 783.05022.

STINSON, D.R. / ZHU, L.

1991 Orthogonal Steiner triple systems of order $6m+3$, *Ars Combinat.* 31, 33–63. MR 92c:05028; Z 759.05016.

STOJAKOVIĆ, Z.

1994 Derived Mendelsohn triple systems, *Bull. Austral. Math. Soc.* 49, 413–419. MR 95d:05024; Z 805.05008.

1996 Single identities for Mendelsohn and Steiner quasigroups, *Bull. Austral. Math. Soc. (A)* 53, 419–424. MR 97d:20093.

STONG, R.A.

1985† On 1-factorizability of Cayley graphs, *J. Combinat. Theory (B)* 39, 298–307.

STRALEY, T.H.

1972 Construction of Steiner quasigroups containing a specified number of subquasigroups of a given order, *J. Combinat. Theory (A)* 13, 374–382. MR 46#3672; Z 248.20092.

STREET, A.P.

1980 Some designs with block size three, in: *Combinatorial Mathematics VIII* (Ed. R.W. Robinson, G.W. Southern, W.D. Wallis), Springer, Berlin, pp. 224–237. MR 82f:05014; Z 454.05014.

1985 A survey of irreducible incomplete block designs, *Ars Combinat.* 19A, 43–60. MR 86f:05026; Z 562.05011.

1992 Defining sets for t-designs and critical sets for Latin squares, *New Zealand J. Math.* 21, 133–144. MR 93c:05026; Z 758.05030.

1995 Defining sets for block designs: an update, in: *Combinatorics Advances* (Ed. C.J. Colbourn and E. Mahmoodian), Kluwer Academic, Dordrecht, pp. 307–320. MR 96h:05021; Z 837.05017.

1996 Trades and defining sets, in: *CRC Handbook of Combinatorial Designs* (Ed. C.J. Colbourn and J.H. Dinitz), CRC Press, Boca Raton, FL, pp. 474–478. Z 847.05011.

STREET, A.P. / STREET, D.J.

1987 *Combinatorics of Experimental Design*, Clarendon, Oxford. MR 88k:05022; Z 622.05001.

STREET, A.P. / WALLIS, W.D.

1977 *Combinatorial Theory: An Introduction*, CBRC, Winnipeg. MR 56#2831; Z 364.05001.

1982 *Combinatorics: A First Course*, CBRC, Winnipeg. Chapter 18: Systems of pairs and triples, and related designs.

SUBBARAYAN, A.

1988 *Steiner Triple System Applications to Plant Breeding Experiments*, Ph.D. Thesis, Univ. Madras.

SULLIVAN, F.E.

1993 Some comments on binary codes of Steiner triple systems, *Congressus Numer.* 93, 215–222. MR 95c:05021; Z 810.05017.

SUZUKI, H.

1990 2-designs over GF(2^m), *Graphs Combin.* 6, 293–296. MR 93m:51005; Z 719.05010.

SWIFT, J.D.

1960 Isomorph rejection in exhaustive search techniques, in: *Proc. Symp. Appl. Math.* 10, Amer. Math. Soc., Providence, RI, pp. 195–200. MR 22#5145; Z 96, 5.

1968 Quasi-Steiner systems, *Atti Accad. Naz. Lincei Rend. Cl. Fis. Mat. natur.* (8) 44, 40–44. MR 37#6196; Z 159, 303.

1969 A generalized Steiner problem, *Rend. Mat.* (6) 2, 563–569. MR 40#7125; Z 322.05014.

SYLVESTER, J.J.

1861 Note on the historical origin of the unsymmetrical six-valued function of six letters, *London, Edinburgh and Dublin Philos. Mag. and J. Sci.* 21, 369–377 (*Collected Mathematical Papers* II, 264–271).

1861a On a problem in tactic which serves to disclose the existence of a four-valued function of three sets of three letters each, *London, Edinburgh and Dublin Philos. Mag. and J. Sci.* (4) 21, 515–520 (*Collected Mathematical Papers* II, 277–285).

1861b Concluding paper on tactic, *London, Edinburgh and Dublin Philos. Mag. and J. Sci.* (4) 22, 45–54 (*Collected Mathematical Papers* II, 277–285).

1861c Remark on the tactic of nine elements, *London, Edinburgh and Dublin Philos. Mag. and J. Sci.* (4) 22, 144–147 (*Collected Mathematical Papers* II, 286–289).

1875 Proposed problem, *Educational Times*, 193.

1876 On the fifteen young ladies problem, *Proc. London Math. Soc.* 7, 235–236.

1880 Solution of a problem, *Educational Times Reprints* 33, 53.

1893 Note on a nine schoolgirl's problem, *Messenger Math.* (2) 22, 159–160. Correction: 192. (*Collected Mathematical Papers* IV, 732–733).

SZAMKOLOWICZ, L.

1962 On the problem of existence of finite regular planes, *Colloq. Math.* 9, 245–250. MR 25#5441; Z 106, 143.

1963 Remarks on finite regular planes, *Colloq. Math.* 10, 31–37. MR 29#1571; Z 118, 152.

1964 Sur une classification des triplets de Steiner, *Atti Accad. Naz. Lincei Rend. Cl. Sci. Fis. Mat. Natur.* (8) 36, 125–128. MR 31#660; Z 137, 147.

1965 Alcuni problemi della teoria dei sistemi di Steiner, *Rend. Mat. e Appl.* (5) 24, 348–359. MR 34#7387; Z 146, 14.

1969 On Steiner manifolds, *Colloq. Math.* 20, 45–51. MR 39#1584.

TAKEUCHI, K.

1962 A table of difference sets generating balanced incomplete block designs, *Rev. Inst. Internat. Statist.* 30, 361–366. MR 28#1712; Z 109, 122.

TANNENBAUM, P.

1975 Abelian inverse property neofields, *Congressus Numer.* 13, 279–281. MR 52 #5842; Z 318.12112.

1976 Abelian Steiner triple systems, *Canad. J. Math.* 28, 1251–1268. MR 54#10468; Z 353.05012.

1977 Regular Steiner triple systems, in: *Proc. 2nd Caribbean Conf. Combinatorics and Computing*, University of the West Indies, Cave Hill, Barbados (Ed. R.C. Read and C.C. Cadogan), pp. 215–218.

TARSKI, A.

1975† An interpolation theorem for irredundant bases of closure systems, *Discrete Math.* 12, 185–192.

TEIRLINCK, L.

1973 On the maximum number of disjoint Steiner triple systems, *Discrete Math.* 6, 299–300. MR 48#5885; Z 266.05006.

1973a The existence of reverse Steiner triple systems, *Discrete Math.* 6, 301–302. MR 48#8263; Z 266.05007.

1975 On linear spaces in which every plane is either projective or affine, *Geometriae Dedicata* 4, 39–44. MR 52#5441; Z 309.50014.

1975a A simplification of the proof of the existence of reverse Steiner triple systems of order congruent to 1 modulo 24, *Discrete Math.* 13, 297–298. MR 51#12562; Z 311. 05011.

1975b On the maximal number of disjoint triple systems, *J. Geometry* 6, 93–96. MR 51 #5333; Z 289.05016; Z 302.05011.

1976 *Combinatorial Structures*, Thesis, Vrije Universiteit Brussel, Departement voor Wiskunde, 106 pp.

1977 On making two Steiner triple systems disjoint, *J. Combinat. Theory (A)* 23, 349–350. MR 56#8394; Z 401.05018.

1977a Planes and hyperplanes of 2-coverings, *Bull. Soc. Math. Belg.* 29, 73–81. MR 58 #5285; Z 415.51003.

1979 On Steiner spaces, *J. Combinat. Theory (A)* 26, 103–114. MR 80j:51011; Z 411. 05015.

1980 On projective and affine hyperplanes, *J. Combinat. Theory (A)* 28, 290–306. MR 81j:05038; Z 444.05026.

1984 On large sets of disjoint quadruple systems, *Ars Combinat.* 17, 173–176. MR 85i:05042; Z 554.05007.

1990 On the use of pairwise balanced designs and closure spaces in the construction of structures of degree at least 3, *Le Matematiche* 65, 197–218. MR 93h:05042; Z 734. 05020.

1991 A completion of Lu's determination of the spectrum for large sets of disjoint Steiner triple systems, *J. Combinat. Theory (A)* 57, 302–305. MR 92c:05029; Z 742. 05016.

1992 Large sets of disjoint designs and related structures, in: *Contemporary Design Theory: A Collection of Surveys* (Ed. J.H. Dinitz and D.R. Stinson), Wiley, pp. 561–592. Z 805.05012.

1993 Large sets with holes, *J. Combin. Des.* 1, 69–94. MR 95i:05026; Z 818.05014.

TEIRLINCK, L. / LINDNER, C.C.

1988 The construction of large sets of idempotent quasigroups, *Europ. J. Combinat.* 9, 83–89. MR 89e:20123; Z 644.20041.

THOMAS, S.

1986 Triangle transitive Steiner triple systems, *Geom. Dedicata* 21, 19–20. MR 87j: 05040; Z 589.51014.

1987 Designs over finite fields, *Geom. Dedicata* 24, 237–242. MR 88g:51016; Z 627. 51013.

TIAN, Z.H.

1997 The existence spectrum of the self-converse directed triple system DTS(v, λ), *Acta Math. Appl. Sinica* 20, 531–534.

TONCHEV, V.D.

1987 Steiner triple systems of order 21 with automorphisms of order 7, *Ars Combinat.* 23, 93–96. MR 88d:05028; Z 611.05009. Correction: *Ars Combinat.* 39 (1995), 3. MR 96b:05031; Z 824.05009.

1987a Transitive Steiner triple systems of order 25, *Discrete Math.* 67, 211–214. MR 88i:05034; Z 619.05009.

1996 Codes, in: *CRC Handbook of Combinatorial Designs* (Ed. C.J. Colbourn and J.H. Dinitz), CRC Press, Boca Raton, FL, pp. 517–543. Z 854.94016.

TONCHEV, V.D. / VANSTONE, S.A.

1992 On Kirkman triple systems of order 33, *Discrete Math.* 106/107, 493–496. MR 93e:05012; Z 773.05025.

TONCHEV, V.D. / WEISHAAR, R.S.

1997 Steiner triple systems of order 15 and their codes, *J. Stat. Plann. Infer.* 58, 207–216. MR 97k:05032.

TREASH, C.A.

1970 A method of obtaining Steiner triple systems, *Ann. N.Y. Acad. Sci.* 175, 383–384. MR 42#97.

1971 The completion of finite incomplete Steiner triple systems with applications to loop theory, *J. Combinat. Theory (A)* 10, 259–265. MR 43#397; Z 217, 20.

TURAN, P.

1941† An extremal problem in graph theory (in Hungarian), *Mat. Fiz. Lapok* 48, 436–452.

TUTTE, W.T.

1947† The factorization of linear graphs, *J. London Math. Soc.* 22, 107–111.

1952† The factors of graphs, *Canad. J. Math.* 4, 314–328.

TUZA, ZS.

1995 Zero-sum block designs and graph labelings, *J. Combin. Des.* 3, 89–99. MR 95m:05206.

URLAND, E.

1997 Parallel algorithm for computing the fragment vector in Steiner triple systems, in: *Proc. European Conf. Parallel Processing EURO-PAR '96*, Springer, Berlin, pp. 809–812.

1997a Parallel computing of fragment vector in Steiner triple systems, in: *Proc. 2nd Int. Conf. Vector and Parallel Processing VECPAR '96*, Springer, Berlin, pp. 326–337.

1999 A linear basis for the 7-line configurations, *J. Comb. Math. Comb. Comput.*

VAKIL, F.

1990 *Non-adaptive Group Testing for Hypergeometric Problems*, Ph.D. Thesis, Temple University.

VAKIL, F. / PARNES, M.

1994 On the structure of a class of sets useful in non-adaptive group testing, *J. Stat. Plann. Infer.* 39, 57–69. MR 95f:62142; Z 797.62068.

VAKIL, F. / PARNES, M. / RAGHAVARAO, D.

1990† Group testing with at most two defectives when every item is included in exactly two groups, *Utilitas Math.* 38, 161–164.

VAN LINT, J.H.

1973 Block designs with repeated blocks and $(b, r, \lambda) = 1$, *J. Combinat. Theory (A)* 15, 288–309. MR 48#3757; Z 285.05014.

VAN LINT, J.H. / RYSER, H.J.

1972 Block designs with repeated blocks, *Discrete Math.* 3, 381–396. MR 46#8858; Z 251.05011.

VAN LINT, J.H. / TONCHEV, V.D.

1993 A class of non-embeddable designs, *J. Combinat. Theory (A)* 62, 252–260. MR 93m:05023; Z 772.05006.

VAN LINT, J.H. / WILSON, R.M.

1992 *A Course in Combinatorics*, Cambridge University Press, Cambridge. Chapter 19: Designs. MR 94g:05003; Z 769.05001.

VANSTONE, S.A.

1980 Doubly resolvable designs, *Discrete Math.* 29, 77–86. MR 81f:05030; Z 447.05010.

1982 On mutually orthogonal resolutions and near-resolutions, *Ann. Discrete Math.* 15, 357–369. MR 85m:05018; Z 512.05007.

1983 A note on the existence of strong Kirkman cubes, *Ann. Discrete Math.* 17, 629–632. MR 87i:05051; Z 521.05015.

VANSTONE, S.A. / ROSA, A.

1982 Starter-adder techniques for Kirkman squares and Kirkman cubes of small sides, *Ars Combinat.* 14, 199–212. MR 84g:05037; Z 506.05013.

VANSTONE, S.A. / STINSON, D.R. / SCHELLENBERG, P.J. / ROSA, A. / REES, R. / COLBOURN, C.J. / CARTER, M.J. / CARTER, J.E.

1993 Hanani triple systems, *Israel J. Math.* 83, 305–319. MR 94g:05018; Z 783.05023.

VEBLEN, O. / YOUNG, J.W.

1916† *Projective Geometry*, Ginn and Co., Boston.

VERRALL, H.

1994 Hamiltonian decompositions of complete 3-uniform hypergraphs,*Discrete Math.* 132, 333–348. MR 95j:05141; Z 805.05060.

VIZING, V.G.

1964† On an estimate of the chromatic class of a p-graph (in Russian), *Diskret Analiz.* 3, 25–30.

VOLOSHIN, V.I.

1995† On the upper chromatic number of a hypergraph, *Australas. J. Combin.* 11, 25–45.

VUČKOVIČ, V.

1965 Note on a theorem of W. Sierpinski, *Notre Dame J. Formal Logic* 6, 180–182. MR 33#3949; Z 201, 338.

VYAS, R.

1982 Some Bhaskar Rao designs and applications for $k = 3$, $\lambda = 4$, *University of Indore Research J. Science* 7, 16–25.

WALKER, R.J.

1950† *Algebraic Curves*, Princeton University Press, Princeton. Chapter 3, Plane algebraic curves.

WALLIS, W.D.

1981† The smallest regular graphs without one-factors, *Ars Combinat.* 11, 295–300.

1983 The problem of the hospitable golfers, *Ars Combinat.* 15, 149–152. MR 84m:05020; Z 531.05017.

1988 *Combinatorial Designs*, Dekker, New York. MR 89d:05023; Z 637.05004.

1992 One-factorizations of complete graphs, in: *Contemporary Design Theory: A Collection of Surveys* (Ed. J.H. Dinitz and D.R. Stinson), Wiley, pp. 593–631. Z 783.05084.

1994 Large sets and Room squares, *Bull. Inst. Combin. Appl.* 12, 93–94. MR 95m:05039; Z 839.05018.

1997 *One-Factorizations*, Kluwer Academic Press. Z 863.05064.

WALLIS, W.D. / ZHANG, G.-H.

1992 Small clique coloring numbers, *Bull. Inst. Combin. Appl.* 5, 107–111. MR 93e:05035; Z 828.05050.

1993 On the partition and coloring of a graph by cliques, *Discrete Math.* 120, 191–203. MR 94f:05059; Z 784.05027.

WALTER, S.

1983 *Geordnete Steinersche Triplesysteme*, Dissertation, Naturwiss. Fakult. Friedrich-Alexander-Univ. Erlangen-Nürnberg, 104 S. Z 614.51005.

WANG, G.X.

1991 Existence of simple cyclic triple systems (in Chinese), *J. Shanghai Jiaotong Univ.* 25, No. 5, 95–100. MR 92m:05032.

1996 Existence of pure Mendelsohn triple systems with $\lambda = 4$ or 5 (in Chinese), *J. Shanghai Jiaotong Univ.* 29, 116–121. MR 96m:05031.

WANG, J. / YIN, J.

1999 Existence of holey 3-GDDs of type $(u, g^t w^1)$.

WANG, J. / ZHU, L.

1991 Doubly nested triple systems and nested B[4,3λ;v], *J. Comb. Math. Comb. Comput.* 9, 129–140. MR 92d:05021; Z 749.05014.

WARNOCK, W.G.

1939 Triple systems as ruled quadrics, *Bull. Amer. Math. Soc.* 45, 476–480. Z 21, 151.

WEBB, B.S.

1995 Cycle structures of automorphisms of 2-$(v,k,1)$ designs, *J. Combin. Des.* 3, 341–348. MR 96f:05031.

WEBB, T.M.

1977 *Some Constructions of Sets of Mutually Almost Disjoint Steiner Triple Systems*, M.S. Thesis, Auburn University.

WEI, R.

1993 Group-divisible designs with equal-sized holes, *Ars Combinat.* 35, 315–324. MR 94b:05042; Z 786.05012.

WEICHSEL, P.M.

1994 Dominating sets in n-cubes, *J. Graph Theory* 18, 479–488. MR 95h:05125; Z 806.05061.

WEIDEMAN, C. / RAGHAVARAO, D.

1987 Some optimum non-adaptive hypergeometric group testing designs for identifying two defectives, *J. Stat. Plann. Infer.* 16, 55–61. MR 88i:62137; Z 613.62100.

1987a Non-adaptive hypergeometric group testing designs for identifying at most two defectives, *Commun. Stat.* 16A, 2991–3006. Z 634.62074.

WERNER, H.

1972 A unique factorization theorem for Steiner triple systems, *Technische Hochschule Darmstadt Preprint*, Nr. 19.

WEVRICK, D.

1990 *On the Existence of a Sub-Class of Restricted Resolvable Partitions*, M.Math. Thesis, University of Waterloo.

WEVRICK, D. / VANSTONE, S.A.

1996 Class-uniformly resolvable designs with block sizes 2 and 3, *J. Combin. Des.* 4, 295–300. MR 97a:05031.

WHITE, A.T.

1978 Block designs and graph imbeddings, *J. Combinat. Theory (B)* 25, 166–183. MR 58#21685; Z 382.05011.

WHITE, H.S.

1913 Triple-systems as transformations, and their paths among triads, *Trans. Amer. Math. Soc.* 14, 6–13.

1915 The synthesis of triad systems Δ_t in t elements, in particular for $t = 31$, *Proc. Nat. Acad. Sci. U.S.A.* 1, 4–6.

1915a The multitude of triad systems on 31 letters, *Trans. Amer. Math. Soc.* 16, 13–19.

1934 Construction of a groupless and headless triad system on 31 elements, *Bull. Amer. Math. Soc.* 40, 829–832. Z 10, 289.

WHITE, H.S. / COLE, F.N. / CUMMINGS, L.D.

1919 Complete classification of the triad systems on fifteen elements, *Memoirs Nat. Acad. Sci. U.S.A.* 14, 2nd memoir, 1–89.

WILSON, R.J. / LLOYD, E.K.

1994 Combinatorics, in: *Companion Encyclopedia of the History and Philosophy of the Mathematical Sciences* (Ed. I. Grattan-Guinness), Routledge, London, pp. 952–965.

WILSON, R.M.

1972 Cyclotomy and difference families in elementary abelian groups, *J. Number Theory* 4, 17–47. MR 46#8860; Z 259.05011.

1973† The necessary conditions for t-designs are sufficient for something, *Utilitas Math.* 4, 207–215.

1974 Nonisomorphic Steiner triple systems, *Math. Z.* 135, 303–313. MR 49#4803; Z 264.05009.

1974a Some partitions of all triples into Steiner triple systems, in: *Hypergraph Seminar*, Lecture Notes in Math. 411, Springer, Berlin, pp. 267–277. MR 51#5336; Z 311. 05010.

WITT, E.

1938 Über Steinersche systeme, *Abh. Math. Sem. Hansischen Univ.* 12, 267–275.

WITTMANN, P.

1997 Vertex-distinguishing edge-colourings of 2-regular graphs, *Discrete Appl. Math.* 79, 265–277. MR 99a:05128; Z 882.05064.

WOOLBRIGHT, D.E.

1980 On the size of partial parallel classes in Steiner systems, in: *Topics on Steiner Systems* (Ed. C.C. Lindner and A. Rosa), *Ann. Discrete Math.* 7, 203–211. Z 461. 05011.

WOOLHOUSE, W.S.B.

1844 Prize question 1733, *Lady's and Gentleman's Diary.*

1861 On the Rev. T.P. Kirkman's problem respecting certain triadic arrangements of fifteen symbols, *London, Edinburgh and Dublin Philos. Mag. and J. Sci.* (4) 22, 510–515.

1862 On triadic combinations of 15 symbols, *Lady's and Gentleman's Diary*, 84–88. Reprinted in *Assurance Magazine* 10, 275–281.

1863 On triadic combinations, *Lady's and Gentleman's Diary*, 79–90.

WOOLHOUSE, W.S.B. / BILLS, S.

1867 Solution of a problem (2261), *Educational Times Reprints* 8, 76–83.

WU, D. / ZHU, L. / GE, G.

1999 Generalized Steiner triple systems with group size g = 7, 8.

WU, L.

1987 On large sets of disjoint Mendelsohn triple systems, *J. Suzhou Univ.* 3, 111–114.

1990 Large sets of KTS(v), in: *Combinatorial Designs and Applications* (Ed. W.D. Wallis, H. Shen, W. Wei, and L. Zhu), Marcel Dekker, New York, pp. 175–178. MR 91m:05030; Z 736.05019.

WYNN, H.P.

1977† Convex sets of finite population plans, *Ann. Statist.* 5, 414–418.

XIA, W.-Q.

1987 On the existence of k-rotational $S(2,3,v)$ (in Chinese), *J. Tongji Univ.* 15, 521–524. MR 92e:05018; Z 719.05012.

1987a Some necessary and sufficient conditions for the existence of orthogonal Steiner triple systems (in Chinese), *J. Huazhong Univ. Sci. Tech.* 15, No. 6, Suppl., 87–90. MR 90a:05032.

YALAVIGI, C.C.

1970 Kirkman's school girls problem, *Math. Education* 4, 15–18. MR 42#7531.

1978 Direct construction of Kirkman's designs for $k = 3$, *Math. Education* Sect. A 12, 28–30. MR 80d:05013; Z 388.05001.

YAN, D.H.

1994 Embeddings of pure handcuffed triple systems (in Chinese), *J. Shanghai Jiaotong Univ.* 28, 110–116. MR 96a:05024.

YAN, D.H. / SHEN, H.

1996 The spectrum of support sizes for MTS(v, 2) and for MTS(v, 3) (in Chinese), *J. Shanghai Jiaotong Univ.* 30, 9–15. MR 98i:05022.

YATES, F.

1936 Incomplete randomized blocks, *Ann. Eugen.* 7, 121–140.

YIN, J. / WANG, J.

1997 (3, λ)-group divisible covering designs, *Australas. J. Combin.* 15, 61–70. MR 98a:05023; Z 880.05013.

YOSHIZAWA, M.

1980 On block-schematic Steiner systems $S(t, t+1, v)$, *Proc. Japan. Acad. Ser. A. Math. Sci.* 56, 348–350. MR 81k:05018; Z 461.05009.

1981 On block-schematic Steiner systems $S(t, t+1, v)$, *Canad. J. Math.* 33, 1432–1438. MR 83c:05021; Z 461.05009.

YOUNG, H.P.

1973 Affine triple systems, in: *Finite Geometric Structures and their Applications*, Edizioni Cremonese, Roma, pp. 265–282. MR 50#142a; Z 297.05031.

1973a Affine triple systems and matroid designs, *Math. Z.* 132, 343–359. MR 50 #142b; Z 263.05019.

YOUNG, H.P. / EDMONDS, J.

1973 Matroid designs, *J. Res. Nat. Bur. Standards* 77B, 15–44. MR 52#2929; Z 269. 05010.

ZAPPA, P.

1996 Triplets of Latin squares, *Boll. Un. Mat. Ital. A* 10. 63–69. MR 97b:05038; Z 859.05023.

ZEITLER, H.

1973 Moderne Mathematik am Tetraeder, *Didaktik der Mathematik* 1, 32–45.

1974 Konstruktion spezieller Steiner-Tripel-Systeme, *Math.-Phys. Semesterber.* 21, 206–233. MR 52#132; Z 359.50016.

1984 Ovale in STS (13), *Math. Semesterber.* 31, 195–214. MR 86e:51009; Z 551. 05023.

1985 On the dimension of Steiner triple systems, *Ars Combinat.* 20A, 51–57. MR 87d: 05038; Z 606.05012.

1986 Ovals in Steiner triple systems, *Ann. Discrete Math.* 30, 373–382. MR 87m: 05033; Z 589.05019.

1987 Reguläre Polygone und ebene Steiner-Tripelsysteme, *Resultate Math.* 11, 359–373. MR 88m:05016; Z 645.51008.

1987a Die Dimension bei der Konstruktion von Steiner-Tripelsystemen mit Zentralverfahren, *Bayreuther Math. Schr.* 25, 1–19. MR 89c:05020; Z 671.51005.

1988 Eine Klasse spezieller gemischter Ovale in STS(v), *J. Geometry* 33, 185–202. MR 89j:51006; Z 653.51007.

1988a Klassisches und Modernes über Steiner Tripelsysteme, *Elem. Math.* 43, 97–112. MR 89f:05040; Z 699.05011.

1988b Nochmals: Hyperovale in Steiner-Tripel-Systemen, *Math. Semesterber.* 35, 127–131. MR 90c:51013; Z 659.51011.

1988c Miszellen zum Thema Ovale in Steiner-Tripel-Systemen, *Ann. Univ. Sci. Budapest Rol. Eötvös Nom., Section Math.* 31, 3–17. MR 90e:51014; Z 676.51004.

1989 Modifications of the "central-method" to construct Steiner triple systems, *Discrete Math.* 77, 367–384. MR 90m:05027.

1990 Telecommunication and incidence structures, *Results Math.* 17, 169–177. MR 91b:90085; Z 694.51009.

1992 Anzahl maximaler Untersysteme in Steiner-Tripelsystemen, *Math. Semesterber.* 39, 163–183. MR 93i:05026; Z 797.05017.

1993 Die Catania-Aufgabe und ihre Entwicklung, *Mitt. Math. Ges. Hamburg* 13, 79–92. MR 94k:05034; Z 797.05016.

1995 Algebraische Kurven und Steiner-Tripelsysteme, *Praxis Math.* 37, 16–20.

1996 Was kann man mit Quasigruppen anfangen? *Mitt. Math. Sem. Hamburg* 15, 67–88.

ZEITLER, H. / LENZ, H.

1984 Regular ovals in Steiner triple systems, *J. Combinat. Inform. Syst. Sci.* 9, 155–162. MR 90c:05036; Z 645.51007.

1985 Hyperovale in Steiner-Tripel-Systemen, *Math. Semesterber.* 32, 19–49. MR 86j:51013; Z 567.51010.

ZHANG, J.

1997 The spectrum of the simple self-converse Mendelsohn triple system MTS(v, λ), *Acta Math. Appl. Sinica* 20, 487–497.

ZHANG, X.

1994 Indecomposable triple systems with $\lambda = 5$, *J. Comb. Math. Comb. Comput.* 16, 153–162. MR 95h:05026; Z 827.05011.

1996 Construction of orthogonal group divisible designs, *J. Comb. Math. Comb. Comput.* 20, 121–128. MR 96i:05030; Z 851.05017.

1996a Construction for indecomposable simple (v, k, λ)-BIBDs, *Discrete Math.* 156, 317–322. MR 96d:05034; Z 859.05018.

ZHAO, S.

1998 *Application of BIBDs in MT-MFSK Signal Set Design for Multiplexing Bursty Sources*, Ph.D. Thesis, University of Technology Sydney.

ZHAO, S. / YATES, K.W. / YASUKAWA, K.

1996 Application of Kirkman designs in joint detection multiple access schemes, in: *Proc. Int. Symp. Spread Spectrum Techniques and Applications*, Vol. 2, IEEE, pp. 857–861.

ZHENG, S.Q.

1995 Sparse hypernetworks based on Steiner triple systems, in: *Proc. 24th Conf. Parallel Processing* (Ed. P. Banerjee) CRC Press, Boca Raton, FL, pp. 218–223.

1998 An abstract model for optical interconnection networks, in: *Parallel Computing Using Optical Interconnections* (Ed. K. Li, Y. Pan, and S.Q. Zheng), Kluwer Academic, Norwell, MA, pp. 139–162.

ZHU, L.

1980 A construction for orthogonal Steiner triple systems, *Ars Combinat.* 9, 253–262. MR 81j:05027; Z 444.05017.

1984† A pair of orthogonal Steiner systems S(3,5,17), *Discrete Math.* 31, 111–113.

1989 Perfect Mendelsohn designs, *J. Comb. Math. Comb. Comput.* 5, 43–54. MR 90c:05033; Z 667.05009.

1991 Problem 145, *Discrete Math.* 97, 419–420.

ZIRAKZADEH, A.

1969 A model for the finite projective spaces with three points on every line, *Amer. Math. Monthly* 76, 774–778. MR 40#3419; Z 181, 233.

ZULAUF, K.

1897 *Über Tripelsysteme von 13 Elementen*, Dissertation Giessen, Wintersche Buchdruckerei, Darmstadt.

INDEX

The manufacturer's authorised representative in the EU for product safety is Oxford University Press España S.A. of el Parque Empresarial San Fernando de Henares, Avenida de Castilla, 2 – 28830 Madrid (www.oup.es/en or product.safety@oup.com). OUP España S.A. also acts as importer into Spain of products made by the manufacturer.

www.ingramcontent.com/pod-product-compliance
Lightning Source LLC
LaVergne TN
LVHW011856200926
848484LV00002B/462
* 9 7 8 0 1 9 8 5 3 5 7 6 8 *